Wave Propagation in Materials for Modern Applications

Edited by

Andrey Petrin

Wave Propagation in Materials for Modern Applications
Edited by Andrey Petrin

Republished by InTech
Janeza Trdine 9, 51000 Rijeka, Croatia

Additional hard copies can be obtained from orders@intechopen.com

Wave Propagation in Materials for Modern Applications, Edited by Andrey Petrin
p. cm.
ISBN 978-953-7619-65-7

For EU product safety concerns:
IN TECH d.o.o., Prolaz Marije Krucifikse Kozulić 3, 51000 Rijeka, Croatia,
info@intechopen.com or visit our website at intechopen.com.

Meet the editor

Andrey B. Petrin was born in 1961. He graduated from the Moscow Institute of Physics and Technology, Moscow, Russia, in 1983 and received the Ph.D. degree in physics and mathematics in 1988. In 1983, he joined the S. P. Korolev Space Corporation "Energy" Russia, where he has participated in several research projects on antennas and EM propagation. In 1990, he joined the Moscow State University of Forestry, Russia, as a Lecturer of General Physics. Since 1996, he has been with the Joint Institute for High Temperatures of Russian Academy of Sciences, Moscow, as a Leading Scientific Researcher of Theoretical Department. His scientific interests include propagation of electromagnetic waves, electrodynamics of plasma, gasdynamics, ECR microwave gas discharges, electrowetting phenomena, electric field singularities on geometric singularities such as microtips and microwedges, electric field induced instabilities of free fluid surface, nanofocusing of light energy, heat transfer in nanoscale. He is the author of more than 45 papers in scientific journals and 4 U.S. patents: www.petrin.nm.ru.

Preface

In the recent decades, there has been growing interest in micro- and nanotechnology. The advances in nanotechnology give rise new applications and new types of materials with unique electromagnetic and mechanical properties. This book is devoted to the modern methods in electrodynamics and acoustics which have been developed to describe wave propagation in these modern materials and nanodevices.

The book consists of original works of the leading scientists in the field of wave propagation which produced new theoretical and experimental methods in this field of research and obtained new and important results.

The first part of the book consists of chapters with general mathematical methods and approaches to the problem of wave propagation. A special attention is attracted to the advanced numerical methods fruitfully applied in the field of wave propagation.

The next chapters of the book are combined in several parts according to the materials in which the considering waves propagate.

So, the second part of the book is devoted to the problems of wave propagation in newly developed metamaterials, micro- and nanostructures and porous media. In this part the interested reader find important and fundamental results on electromagnetic wave propagation in media with negative refraction index and electromagnetic imaging in devices based on the materials.

The third part of the book is devoted to the problems of wave propagation in elastic and piezoelectric media. In the forth part the works on the problems of wave propagation in plasma are collected. The fifth, sixth and seventh parts are devoted to the problems of wave propagation in media with chemical reactions, in nonlinear and disperse media, respectively. And finally, in the eighth part of the book some experimental methods in wave propagations are considered.

It is necessary to emphasis that this book is not a textbook. It is important that the results combined in it are taken "from the desks of researchers". So, I am sure that in this book the interested and actively working readers (scientists, engineers and students) find many interesting results and new ideas.

Editor

Andrey Petrin

Joint Institute for High Temperatures of Russian Academy of Science, Russia

a_petrin@mail.ru

Contents

I. Wave Propagation.
Fundamental Approaches to the Problem.

1

A Volume Integral Equation Method for the Direct/Inverse Problem in Elastic Wave Scattering Phenomena

Terumi Touhei
Department of Civil Engineering, Tokyo University of Science, Japan

1. Introduction

The analysis of elastic wave propagation and scattering is an important issue in fields such as earthquake engineering, nondestructive testing, and exploration for energy resources. Since the 1980s, the boundary integral equation method has played an important role in the analysis of both forward and inverse scattering problems. For example, Colton and Kress (1998) presented a survey of a vast number of articles on forward and inverse scattering analyses. They also presented integral equation methods for acoustic and electromagnetic wave propagation, based on the theory of operators (1983 and 1998). Recently, Guzina, Fata and Bonnet (2003), Fata and Guzina (2004), and Guzina and Chikichev (2007) have dealt with inverse scattering problems in elastodynamics.

The type of volume integral equation known as the Lippmann–Schwinger equation (Colton & Kress, 1998) has been an efficient tool for theoretical investigation in the field of quantum mechanics (see, for example, Ikebe, 1960). Several applications of the volume integral equation to scattering analysis for classical mechanics have also appeared. For example, Hudson and Heritage (1981) used the Born approximation of the solution of the volume integral equation obtained from the background structure of the wave field for the seismic scattering problem. Recently, Zaeytijd, Bogaert, and Franchois (2008) presented the MLFMA-FFT method for analyzing electro-magnetic waves, and Yang, Abubaker, van den Berg et al. (2008) used a CG-FFT approach to solve elastic scattering problems. These methods were used to establish a fast algorithm to solve the volume integral equation via a Fast Fourier transform, which is used for efficient calculation of the convolution integral.

In this chapter, another method for the volume integral equation is presented for the direct forward and inverse elastic wave scattering problems for 3-D elastic full space. The starting point of the analysis is the volume integral equation in the wavenumber domain, which includes the operators of the Fourier integral and its inverse transforms. This starting point yields a different method from previous approaches (for example, Yang et al., 2008). By replacing these operators with discrete Fourier transforms, the volume integral equation in the wavenumber domain can be treated as a Fredholm equation of the 2nd kind with a non-Hermitian operator on a finite dimensional vector space, which is to be solved by the Krylov subspace iterative scheme (Touhei et al, 2009). As a result, the derivation of the coefficient matrix for the volume integral equation is not necessary. Furthermore, by means of the Fast

Fourier transform, a fast method for the volume integral equation can be established. The method itself can be extended to the scattering problem for a 3-D elastic half space (Touhei, 2009). This chapter also presents the possibility of the volume integral equation method for 3-D elastic half space by constructing a generalized Fourier transform for the half space.
An important property of the volume integral equation in the wavenumber domain is that it separates the scattered wave field from the fluctuation of the medium. This property yields the possibility of inverse scattering analysis. There are several methods for inverse scattering analysis that make use of the volume integral equation (for example, Kleinman and van den Berg (1992); Colton & Kress (1998)). These methods can be used to investigate the relationship between the far field patterns and the fluctuation of the medium in the volume integral equation in the space domain. Under these circumstances, for the inverse scattering analysis, the possibility of solving the volume integral equation in the wavenumber domain should also be investigated.
In this chapter, basic equations for elastic wave propagation are first presented in order to prepare the formulation. After clarifying the properties of the volume integral equation in the wavenumber domain, a method for solving the volume integral equation is developed.

2. Basic equations for elastic wave propagation

Figures 1(a) and (b) show the concept of the problem discussed in this chapter. Figure 1(a) shows a 3-D elastic full space, in which a plane incident wave is propagating along the x_3 axis towards an inhomogeneous region where material properties fluctuate with respect to their reference values. Figure 1(b) is a 3-D elastic half space. Here, waves from a point source propagate towards an inhomogeneous region. Scattered waves are generated by the interactions between the incident waves and the fluctuating areas. This chapter considers a volume integral equation method for solving the scattering problem for both a 3-D elastic full space and a half space. At this stage, basic equations are presented as the starting point of the formulation.

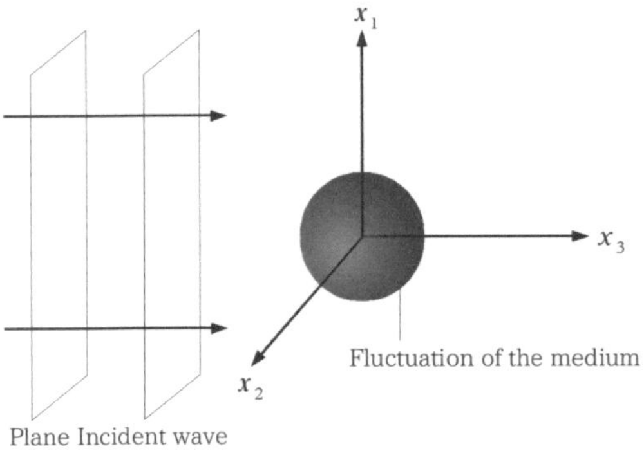

(a) Scattering problem in a 3-D elastic full space

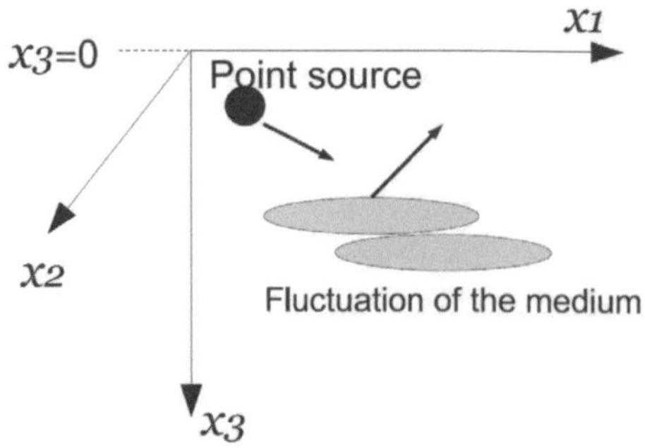

(b) Scattering problem in a 3-D elastic half space

Fig. 1. Concept of the analyzed model.

A Cartesian coordinate system is used for the wave field. A spatial point in the wave field is expressed as:

$$x = (x_1, x_2, x_3) \tag{1}$$

where the subscript index indicates the component of the vector. For the case in which an elastic half space is considered, x_3 denotes the vertical coordinate with the positive direction downward, where the free surface boundary is denoted by $x_3 = 0$.
The fluctuation of the medium is expressed by the Lamé constants so that:

$$\begin{aligned} \lambda(x) &= \lambda_0 + \tilde{\lambda}(x) \\ \mu(x) &= \mu_0 + \tilde{\mu}(x) \end{aligned} \tag{2}$$

where λ_0 and μ_0 are the background Lamé constants of the wave field, and $\tilde{\lambda}$ and $\tilde{\mu}$ are their fluctuations. The back ground Lamé constants are positive and bounded. The magnitudes of the fluctuations are assumed to satisfy

$$|\tilde{\lambda}(x)| < \lambda_0, \quad |\tilde{\mu}(x)| < \mu_0 \tag{3}$$

Let the time factor of the wave field be $\exp(-i\omega t)$, where ω is the circular frequency and t is the time. Then, the equilibrium equation of the wave field taking into account the effects of a point source is expressed as:

$$\partial_j \sigma_{ij} + \rho\omega^2 u_i = -q_i \delta(x - x_s) \tag{4}$$

where σ_{ij} is the stress tensor, ∂_j is the partial differential operator, ρ is the mass density, u_i is the total displacement field, q_i is the amplitude of the point source, x_s is the position at which the point source is applied, and $\delta(\cdot)$ is the Dirac delta function. The subscript indices i and j in Eq. (4) are the components of the Cartesian coordinate system to which the summation convention is applied. The constitutive equation showing the relationship between the stress and strain tensors is as follows:

$$\sigma_{ij} = \lambda \delta_{ij} \epsilon_{kk} + 2\mu \epsilon_{ij} \tag{5}$$

where δ_{ij} is the Kronecker delta, and ϵ_{ij} is the strain tensor given by

$$\epsilon_{ij} = (1/2)\left(\partial_i u_j + \partial_j u_i\right) \tag{6}$$

Substituting Eqs. (6) and (5) into Eq. (4) yields the following governing equation for the current problem:

$$\left(L_{ij}(\partial_1, \partial_2, \partial_3) + \delta_{ij}\rho\omega^2\right) u_j(x) = N_{ij}(\partial_1, \partial_2, \partial_3, x) u_j(x) - q_i \delta(x - x_s) \tag{7}$$

where $L_{ij}(\partial_1, \partial_2, \partial_3)$ and $N_{ij}(\partial_1, \partial_2, \partial_3, x)$ are the differential operators constructed by the background Lamé constants and their fluctuations, respectively. The explicit forms of the operators L_{ij} and N_{ij} are given by

$$L_{ij}(\partial_1, \partial_2, \partial_3) = (\lambda_0 + \mu_0)\partial_i \partial_j + \mu_0 \delta_{ij} \partial_k \partial_k \tag{8}$$

$$\begin{aligned} N_{ij}(\partial_1, \partial_2, \partial_3, x) &= -\left(\tilde{\lambda}(x) + \tilde{\mu}(x)\right)\partial_i\partial_j - \delta_{ij}\tilde{\mu}(x)\partial_k\partial_k \\ &\quad -\partial_i\tilde{\lambda}(x)\partial_j - \delta_{ij}\partial_k\tilde{\mu}(x)\partial_k - \partial_j\tilde{\mu}(x)\partial_i \end{aligned} \tag{9}$$

For the case in which an elastic half space is considered, the free boundary conditions are necessary and are expressed by

$$P_{ij}u_j(x) = 0, \quad (\text{at } x_3 = 0) \tag{10}$$

where P_{ij} is the operator describing the free boundary condition having the following components:

$$[P_{ij}] = \begin{bmatrix} \mu(x)\partial_3 & 0 & \mu(x)\partial_1 \\ 0 & \mu(x)\partial_3 & \mu(x)\partial_2 \\ \lambda(x)\partial_1 & \lambda(x)\partial_2 & (\lambda(x) + 2\mu(x))\partial_3 \end{bmatrix} \tag{11}$$

3. Method for forward and inverse scattering analysis in the elastic full space based on the volume integral equation

3.1 Definition of the forward and inverse scattering problem

Now, we deal with the concept of the problem shown in Fig. 1(a). The forward and inverse problem for a 3-D elastic full space will be discussed based on the volume integral equation. The forward and inverse scattering problems considered in this section can be described as follows:

Definition 1 *The forward scattering problem is to determine the scattered wave field from information about the regions of fluctuation, the background structure of the wave field, and the plane incident wave.*

Definition 2 *The inverse scattering problem involves reconstructing the fluctuating areas from information about the scattered waves, the background structure of the wave field, and the plane incident wave.*

To clarify the above problems mathematically, the volume integral equation is obtained from Eq. (7). Assume that the right-hand side of Eq. (7) is the inhomogeneous term. Since there is no point source in the wave field shown in Fig. 1(a), the solution of Eq. (7) is expressed by the following volume integral equation:

$$u_i(x) = F_i(x) - \int_{\mathbb{R}^3} G_{ij}(x,y)N_{jk}(\partial_1, \partial_2, \partial_3, y)u_k(y)dy \tag{12}$$

where F_i and G_{ij} are the plane incident wave and the Green's function, respectively, which satisfy the following equations:

$$\left(L_{ij}(\partial_1, \partial_2, \partial_3) + \delta_{ij}\rho\omega^2\right)F_j(x) = 0 \tag{13}$$

$$\left(L_{ij}(\partial_1, \partial_2, \partial_3) + \delta_{ij}\rho\omega^2\right)G_{jk}(x,y) = -\delta_{ik}\delta(x-y), \quad (x, y \in \mathbb{R}^3) \tag{14}$$

It is convenient to express the volume integral equation in terms of the scattered wave field

$$v_i(x) = u_i(x) - F_i(x) \tag{15}$$

which becomes:

$$\begin{aligned} v_i(x) &= -\int_{\mathbb{R}^3} G_{ij}(x,y) N_{jk}(\partial_1, \partial_2, \partial_3, y) F_k(y) dy \\ &\quad - \int_{\mathbb{R}^3} G_{ij}(x,y) N_{jk}(\partial_1, \partial_2, \partial_3, y) v_k(y) dy \end{aligned} \tag{16}$$

By means of Eq. (16), the forward and inverse scattering problems considered in this section can be stated mathematically. The forward scattering problem is to determine v_i after G_{ij}, N_{jk}, and F_k have been obtained. The inverse scattering problem determines $\tilde{\lambda}$ and $\tilde{\mu}$ in N_{jk} in Eq. (16) after G_{ij}, v_i, and F_k have been obtained. In the remainder of this section, a method for dealing with Eq. (16) is described.

3.2 The Fourier transform and its application to the volume integral equation

The following Fourier integral and its inverse transforms:

$$\begin{aligned} \left(\mathscr{F} u_i\right)(\xi) &= \frac{1}{\sqrt{2\pi}^3} \int_{\mathbb{R}^3} u_i(x) \exp(-ix \cdot \xi) dx \\ \left(\mathscr{F}^{-1} \hat{u}_i\right)(x) &= \frac{1}{\sqrt{2\pi}^3} \int_{\mathbb{R}^3} \hat{u}_i(\xi) \exp(ix \cdot \xi) d\xi \end{aligned} \tag{17}$$

play an important role in the formulation, where $\xi = (\xi_1, \xi_2, \xi_3) \in \mathbb{R}^3$ is a point in the wavenumber space, $x \cdot \xi$ is the scalar product defined by

$$x \cdot \xi = x_1\xi_1 + x_2\xi_2 + x_3\xi_3 \tag{18}$$

and $\mathscr{F}$ and $\mathscr{F}^{-1}$ are the operators for the Fourier transform and the inverse Fourier transform, respectively. In the following formulation, the symbol ˆ attached to a function is used to express the Fourier transform of the function. For example, $\hat{u}_i$ denotes the Fourier transform of u_i. The domain of the operators for $\mathscr{F}$ and $\mathscr{F}^{-1}$ defined in Eq. (17) is assumed to be $L^2(\mathbb{R}^3)$, so that the convergence of the integrals should be understood in the sense of the limit in the mean. In the following formulation, the domain of $\mathscr{F}$ and $\mathscr{F}^{-1}$ for the Green's function is assumed to be extended from $L^2(\mathbb{R}^3)$ to the distribution (Hörmander, 1983).
The Fourier transform of the equation for the Green's function defined by Eq. (14) becomes

$$L_{ij}(i\xi_1, i\xi_2, i\xi_3)\hat{G}_{jk}(\xi, y) = -\frac{1}{\sqrt{2\pi}^3}\delta_{ik} \exp(-i\xi \cdot y) \tag{19}$$

Equation (19) yields

$$\hat{G}_{ij}(\xi, y) = \frac{1}{\sqrt{2\pi}^3} \exp(-i\xi \cdot y)\hat{h}_{ij}(\xi) \tag{20}$$

where $\hat{h}_{ij}(\xi)$ is expressed by

$$
\begin{aligned}
\hat{h}_{ij}(\xi) &= \frac{\delta_{ij}}{\mu_0(|\xi|^2 - k_T^2 - i\epsilon)} \\
&\quad - \frac{\xi_i \xi_j}{2\mu_0(1-\nu_0)} \frac{1}{(|\xi|^2 - k_T^2 - i\epsilon)(|\xi|^2 - k_L^2 - i\epsilon)}
\end{aligned} \tag{21}
$$

In Eq. (21), ν_0 is the Poisson ratio obtained from the back ground Lamé constants λ_0 and μ_0, k_L, and k_T are the wavenumber of the P and S waves obtained from

$$
\begin{aligned}
k_L &= \frac{\omega}{c_L} \\
k_T &= \frac{\omega}{c_T}
\end{aligned} \tag{22}
$$

$|\xi|^2$ is given by

$$
|\xi|^2 = \xi_1^2 + \xi_2^2 + \xi_3^2 \tag{23}
$$

and ϵ is an infinitesimally small positive number. Note that c_T and c_L in Eq. (22) are the S and P wave velocities, respectively, for the background structure of the wave field defined by

$$
c_T = \sqrt{\frac{\mu_0}{\rho}} \tag{24}
$$

and

$$
c_L = \sqrt{\frac{\lambda_0 + 2\mu_0}{\rho}} \tag{25}
$$

respectively.

Next, let us investigate the Fourier transform of function w_i in the following form:

$$
w_i(x) = \int_{\mathbb{R}^3} G_{ij}(x,y) f_j(y) dy \tag{26}
$$

to obtain the Fourier transform of the volume integral equation. Note that $f_j(y)$ is in $\mathscr{S}(\mathbb{R}^3)$, i.e., the space of rapidly decreasing functions (Reed & Simon, 1975), then changing the order of integration yields

$$
\begin{aligned}
\hat{w}_i(\xi) &= \frac{1}{\sqrt{2\pi}^3} \int_{\mathbb{R}^3} \left[\int_{\mathbb{R}^3} G_{ij}(x,y) f_j(y) dy \right] \exp(-ix\cdot\xi) dx \\
&= \hat{h}_{ij}(\xi) \frac{1}{\sqrt{2\pi}^3} \int_{\mathbb{R}^3} f_j(y) \exp(-i\xi\cdot y) dy = \hat{h}_{ij}(\xi) \hat{f}_j(\xi)
\end{aligned} \tag{27}
$$

In particular, the Fourier transform of w_i can be separated into the product of $\hat{h}_{ij}$ and $\hat{f}_j$. As reported in a previous study (Hörmander, 1983), f_j can be extended to distributions with compact support. According to Eq. (27), the Fourier transform of the volume integral equation shown in Eq. (16) becomes:

$$\hat{v}_i(\xi) = -\hat{h}_{ij}(\xi)\Big(\mathscr{F}N_{jk}F_k\Big)(\xi) - \hat{h}_{ij}(\xi)\Big(\mathscr{F}N_{jk}\mathscr{F}^{-1}\hat{v}_k\Big)(\xi) \tag{28}$$

For the case in which an explicit form of the plane incident wave is obtained, $N_{jk}F_k$ on the right-hand side of Eq. (28) can be simplified. As an example, a plane incident pressure (P) wave propagating along the x_3 axis has the following form:

$$F_i(x) = a\partial_i \exp(ik_L x_3) \tag{29}$$

where a is the amplitude of the P wave potential. In this case, $N_{jk}F_k$ can be expressed as

$$N_{jk}F_k(x) = r_j(x)\exp(i\xi_p \cdot x) \tag{30}$$

where

$$\begin{aligned} r_1(x) &= ak_L^2\partial_1\tilde{\lambda}(x) \\ r_2(x) &= ak_L^2\partial_2\tilde{\lambda}(x) \\ r_3(x) &= ak_L^2(\partial_3 + ik_L)\Big(\tilde{\lambda}(x) + 2\tilde{\mu}(x)\Big) \end{aligned} \tag{31}$$

Note that ξ_p is the wavenumber vector of the plane incident wave having the following components:

$$\xi_p = (0, 0, k_L) \tag{32}$$

As a result, Eq. (28) can be rewritten as

$$\hat{v}_i(\xi) = -\hat{h}_{ij}(\xi)\hat{r}_j(\xi - \xi_p) - \hat{h}_{ij}(\xi)\Big(\mathscr{F}N_{jk}\mathscr{F}^{-1}\hat{v}_k\Big)(\xi) \tag{33}$$

A method for forward and inverse scattering analysis is developed in the following based on Eq. (33).

3.3 Method for forward scattering analysis

Let us rewrite Eq. (33) in the following form:

$$\hat{v}_i(\xi) = \hat{f}_i(\xi) - A_{ik}\hat{v}_k \tag{34}$$

where $\hat{f}_i$ is given by

$$\hat{f}_i(\xi) = -\hat{h}_{ij}(\xi)\hat{r}_j(\xi - \xi_p) \tag{35}$$

which can be treated as a given function and A_{ik} is the linear operator such that

$$A_{ik} = \hat{h}_{ij}(\xi)\mathscr{F}N_{jk}\mathscr{F}^{-1} \tag{36}$$

Equation (34) clearly shows a Fredholm integral equation of the second kind, in which the linear operator is constructed by the multiplication operator $\hat{h}_{ij}$, the Fourier transform and the inverse Fourier transform, and the differential operator N_{jk}. For the actual numerical calculations in this chapter, the Fourier transform and its inverse Fourier transform are dealt with by means of the discrete Fourier transform. Naturally, the discrete Fourier transform is evaluated by means of an FFT. Let us denote the operators for the discrete Fourier transforms as $\mathscr{F}_D$ and $\mathscr{F}_D^{-1}$. For the operators $\mathscr{F}_D$ and $\mathscr{F}_D^{-1}$, the subsets in $\mathbb{R}^3$ below are defined as follows:

$$D_x = \left\{ (n_1\Delta x_1, n_2\Delta x_2, n_3\Delta x_3) \,\middle|\, n_1 \in \mathbb{N}_1, n_2 \in \mathbb{N}_2, n_3 \in \mathbb{N}_3 \right\} \subset \mathbb{R}^3 \tag{37}$$

$$D_\xi = \left\{ (n_1\Delta \xi_1, n_2\Delta \xi_2, n_3\Delta \xi_3) \,\middle|\, n_1 \in \mathbb{N}_1, n_2 \in \mathbb{N}_2, n_3 \in \mathbb{N}_3 \right\} \subset \mathbb{R}^3 \tag{38}$$

These subsets define a finite number of grid points, where Δx_j , (j = 1, 2, 3) is the interval of the grid in the space domain, $\Delta\xi_j$, (j = 1, 2, 3) is the interval of the grid in the wavenumber space, and $\mathbb{N}_1$, $\mathbb{N}_2$, and $\mathbb{N}_3$ are sets of integers defined by

$$\begin{aligned} \mathbb{N}_1 &= \{n \mid -N_1/2 \le n < N_1/2\} \\ \mathbb{N}_2 &= \{n \mid -N_2/2 \le n < N_2/2\} \\ \mathbb{N}_3 &= \{n \mid -N_3/2 \le n < N_3/2\} \end{aligned} \tag{39}$$

where (N_1, N_2, N_3) defines the number of grid points in $\mathbb{R}^3$. For the discrete Fourier transform, note that there is a relationship between Δx_j and $\Delta\xi_j$ such that

$$\Delta x_j \, \Delta\xi_j = \frac{2\pi}{N_j}, \quad (j = 1, 2, 3) \tag{40}$$

The explicit form of the discrete Fourier transform and the inverse Fourier transform are expressed as

$$\begin{aligned} (\mathscr{F}_D u_{(D)})(\xi^{(l)}) &= \frac{\Delta x}{\sqrt{2\pi}^3} \sum_{k \in \mathbb{N}_1 \times \mathbb{N}_2 \times \mathbb{N}_3} u_{(D)}(x^{(k)}) \exp(-ix^{(k)} \cdot \xi^{(l)}) \\ (\mathscr{F}_D^{-1} \hat{u}_{(D)})(x^{(k)}) &= \frac{\Delta\xi}{\sqrt{2\pi}^3} \sum_{l \in \mathbb{N}_1 \times \mathbb{N}_2 \times \mathbb{N}_3} \hat{u}_{(D)}(\xi^{(l)}) \exp(ix^{(k)} \cdot \xi^{(l)}) \end{aligned} \tag{41}$$

where Δx and $\Delta\xi$ are denoted by

$$\Delta x = \Delta x_1 \, \Delta x_2 \, \Delta x_3, \quad \Delta\xi = \Delta\xi_1 \, \Delta\xi_2 \, \Delta\xi_3 \tag{42}$$

and $x^{(k)}$ and $\xi^{(l)}$ expressed by

$$x^{(k)} = (x_1^{(k)}, x_2^{(k)}, x_3^{(k)}), \quad \xi^{(l)} = (\xi_1^{(l)}, \xi_2^{(l)}, \xi_3^{(l)}) \tag{43}$$

are the points in D_x of the k-th grid and in D_ξ of the l-th grid, respectively. In addition, u_D and $\hat{u}_D$ are the discrete functions defined on the grids D_x and D_ξ.
Based on the discrete Fourier transform, the derivative of a function can be calculated. For example, $\partial_j f(x)$ is expressed by

$$\partial_j f(x) = \left(\mathscr{F}_D^{-1}(i\xi_j \mathscr{F}_D f) \right)(x), \quad x \in D_x, \ \xi \in D_\xi \tag{44}$$

Therefore, treatments for the operator N_{jk} are also made possible by the discrete Fourier transform. Let $N_{(D)jk}$ be the discretization of the operator for N_{jk} by means of the discrete Fourier transform. Then, the discretization for the operator $\mathscr{A}_{ij}$ is defined by

$$A_{(D)ik} = \hat{h}_{ij}(\xi)\mathscr{F}_D N_{(D)jk}\mathscr{F}_D^{-1} \tag{45}$$

As a result of the discretization, Eq. (34) becomes

$$\hat{v}_{(D)i}(\xi) = \hat{f}_{(D)i}(\xi) - A_{(D)ij}\hat{v}_{(D)j}(\xi), \qquad (\xi \in D_\xi) \tag{46}$$

The domain and range of the linear operator in Eq. (45) are in the set of functions defined on a finite number of grids in the wavenumber space D_ξ. Namely, the domain and range for the operator are finite dimensional vector spaces. Note that the operator $N_{(D)jk}$ included in $\mathscr{A}_{(D)ij}$ is bounded because the differential operators are approximated by the discrete Fourier transform. For the case in which the domain and range of the operator are finite dimensional vector spaces, the operator has matrix representations (Kato, 1980). Therefore, a technique for the linear algebraic equation, such as the Krylov subspace iteration method (Barrett et al., 1994), is applicable to Eq. (46). Krylov subspace iteration methods have been developed for systems of algebraic equations in matrix form:

$$A\boldsymbol{x} = \boldsymbol{b} \tag{47}$$

where A is the matrix, and $\boldsymbol{x}$ and $\boldsymbol{b}$ are unknown and given vectors, respectively. The Krylov subspace is defined by

$$K_m = \text{span}\,\{\boldsymbol{b}, A\boldsymbol{b}, A^2\boldsymbol{b}\ldots, A^m\boldsymbol{b}\} \tag{48}$$

where m is the number of iterations. The Krylov subspace iteration method determines the coefficients of the recurrence formula to approximate the solution from the orthonormal basis of K_m during the iterative procedure. Note that matrix A can be regarded as the linear transform on a finite dimensional vector space. In this way, the construction of the Krylov subspace is possible, even if the linear transform is obtained using discrete Fourier transforms. Namely, it is possible to solve Eq. (46) by the Krylov subspace iteration method, where the Krylov subspace is constructed by FFT. As a result, a fast method for the volume integral equation without the derivation of the matrix is expected to be established. The current method is also expected to use less computer memory for numerical analysis. Since the operator $A_{(D)ij}$ is non-Hermitian due to the presence of $N_{(D)jk}$, the Bi-CGSTAB method (Barrett et al., 1994) is selected for the solution of Eq. (46).

3.4 Method for inverse scattering analysis

According to Eq. (31) the explicit form of $\hat{r}_j(\xi - \xi_p)$ shown as the first term on the right-hand side of Eq. (33) becomes:

$$\begin{aligned}
\hat{r}_1(\xi - \xi_p) &= ak_L^2 i\xi_1\hat{\tilde{\lambda}}(\xi - \xi_p)\\
\hat{r}_2(\xi - \xi_p) &= ak_L^2 i\xi_2\hat{\tilde{\lambda}}(\xi - \xi_p)\\
\hat{r}_3(\xi - \xi_p) &= ak_L^2 i\xi_3\left(\hat{\tilde{\lambda}}(\xi - \xi_p) + 2\hat{\tilde{\mu}}(\xi - \xi_p)\right)
\end{aligned} \tag{49}$$

Based on Eq. (49), $\hat{r}_j$ is found to be the function describing the fluctuation of the medium. Therefore, the inverse scattering analysis becomes possible if $\hat{r}_j$ is obtained from Eq. (33) after the scattered wave field $\hat{v}_i$ and the background structure of the wave field represented by $\hat{h}_{ij}$ have been provided. We introduce the vector Q_i, such that

$$r_i(x) = ak_L^3\lambda_0 Q_i(x) \tag{50}$$

to obtain the equation for the inverse scattering analysis in dimensionless form. Let us multiply both sides of Eq. (33) by $\left(-\hat{h}_{ij}^{-1}/(a\lambda_0 k_L^3)\right)$, which yields

$$\hat{\gamma}_j(\xi) \quad = \quad \hat{Q}_j(\xi - \xi_p) + \frac{1}{ak_L^3\lambda_0}\left(\mathscr{F}N_{jk}\mathscr{F}^{-1}\hat{v}_k\right)(\xi) \tag{51}$$

where $\hat{\gamma}_j$ is defined by

$$\hat{\gamma}_j(\xi) = -\hat{h}_{ji}^{-1}(\xi)\hat{v}_i(\xi)/(a\lambda_0 k_L^3) \tag{52}$$

Next, let the second term of Eq. (51) be modified to obtain the following:

$$\frac{1}{ak_L^3\lambda_0}\mathscr{F}N_{jk}\mathscr{F}^{-1}\hat{v}_k(\xi) = \mathscr{F}M_{jk}\mathscr{F}^{-1}\hat{Q}_k(\xi) \tag{53}$$

where M_{jk} is the differential operator determined by the scattered wave field. The remainder of this section describes how to obtain an explicit form of M_{jk}, so that Eq. (51) can be used to obtain $\hat{Q}$, which makes the estimation of the fluctuation of the medium possible. In order to obtain the explicit form of M_{jk}, a_j, which is defined as being equal to $N_{jk}v_k$, can be expressed as follows:

$$\alpha_j \quad = \quad -(\tilde{\lambda} + \tilde{\mu})\partial_j\Delta_v - \tilde{\mu}\eta_j - (\partial_j\tilde{\lambda})\Delta_v - 2(\partial_k\tilde{\mu})\epsilon_{kj} \tag{54}$$

where Δ_v and η_j are defined by

$$\begin{aligned} \Delta_v &= \partial_l v_l \\ \eta_j &= (\partial_1^2 + \partial_2^2 + \partial_3^2)v_j \end{aligned} \tag{55}$$

and ϵ_{jk} is the strain tensor due to the scattered wave field defined by Eq. (6). Let the separation of the fluctuation of the medium and the scattered wave field for α_j be denoted by

$$\alpha_j = m_{jk}p_k \tag{56}$$

where p_k is the state vector for the fluctuation of the medium, the components of which are

$$p_1 = \partial_1\tilde{\lambda}(x)/k_L, \quad p_2 = \tilde{\lambda}(x), \quad p_3 = \tilde{\mu}(x) \tag{57}$$

and m_{jk} is the differential operator that includes the effects of the scattered wave field, so that

$$[m_{jk}] = \begin{bmatrix} k_L\Delta_v & \partial_1\Delta_v & \partial_1\Delta_v + \eta_1 + 2\epsilon_{1l}\partial_l \\ 0 & \Delta_v\partial_2 + \partial_2\Delta_v & \partial_2\Delta_v + \eta_2 + 2\epsilon_{2l}\partial_l \\ 0 & \Delta_v\partial_3 + \partial_3\Delta_v & \partial_3\Delta_v + \eta_3 + 2\epsilon_{3l}\partial_l \end{bmatrix} \tag{58}$$

Likewise, let the separation of the fluctuation of the medium and the scattered wave field for Q_j defined by Eq. (50) be denoted as follows:

$$Q_j = \kappa_{jk} p_k \tag{59}$$

where κ_{jk} is the operator that includes the effects of the scattered wave field, so that:

$$[\kappa_{jk}] = \frac{1}{k_L \lambda_0} \begin{bmatrix} k_L & 0 & 0 \\ 0 & \partial_2 & 0 \\ 0 & (\partial_3 + ik_L) & 2(\partial_3 + ik_L) \end{bmatrix} \tag{60}$$

According to Eqs. (59) and (60), the formal representation of the relationship between p_j and Q_j becomes

$$p_j = s_{jk} Q_k \tag{61}$$

where s_{jk} is the inverse of κ_{jk}, the components of which are

$$[s_{jk}] = \lambda_0 k_L \begin{bmatrix} k_L^{-1} & 0 & 0 \\ 0 & \partial_2^{-1} & 0 \\ 0 & -(1/2)\partial_2^{-1} & (1/2)(\partial_3 + ik_L)^{-1} \end{bmatrix} \tag{62}$$

Based on Eqs. (56) and (59), the following relationship can be derived:

$$\alpha_j = N_{jk} v_k = m_{jk} p_k = m_{jk} s_{kl} R_l \tag{63}$$

As a result, the operator M_{jk} defined by Eq. (53) can be constructed as follows:

$$M_{jl} = \frac{1}{a k_L^3 \lambda_0} m_{jk} s_{kl} \tag{64}$$

By means of the operator, Eq. (51) is modified to obtain

$$\hat{\gamma}_i(\xi) = \hat{Q}_i(\xi - \xi_p) + \left(\mathscr{F} M_{ij} \mathscr{F}^{-1} \right) \hat{Q}_j(\xi) \tag{65}$$

At this point, we have two tasks involving Eq. (65). One is to modify Eq. (65) to obtain a Fredholm equation of the second kind. The other task is to clarify the treatment of the operator s_{jk}, which includes ∂_2^{-1} and $(\partial_3 + ik_L)^{-1}$. To modify Eq. (65) to obtain a Fredholm equation of a second kind, the shift operator $S(\xi_p)$ defined by

$$S(\xi_p) \hat{Q}_i(\xi - \xi_p) = \hat{Q}_i(\xi) \tag{66}$$

is introduced. An explicit form of the shift operator can be obtained in terms of the Fourier transform, so that

$$S(\xi_p) = \mathscr{F} \exp(ix \cdot \xi_p) \mathscr{F}^{-1} \tag{67}$$

Application of the shift operator to both sides of Eq. (65) yields

$$S(\xi_p)\,\hat{\gamma}_i(\xi) = \hat{Q}_i(\xi) + \left(\mathscr{F}\exp(ix\cdot\xi_p)M_{ij}\mathscr{F}^{-1}\hat{Q}_j\right)(\xi) \tag{68}$$

which clearly has the form of a Fredholm equation of the second kind.
To clarify the treatments of ∂_2^{-1} in s_{jk}, consider the following equation:

$$\partial_2 u(x) = f(x), \quad f \in \mathscr{S}(\mathbb{R}^3) \tag{69}$$

Formally, it is possible to write the solution of the equation as

$$u(x) = \partial_2^{-1} f(x) = \int_{-\infty}^{\infty} H(x_2 - x_2')f(x_1, x_2', x_3)dx_2' \tag{70}$$

where H is the unit step function. The Fourier transform of H can be expressed as

$$\hat{H}(\xi_2) = -\frac{i}{\sqrt{2\pi}}\left(\text{p.v.}\frac{1}{\xi_2} + i\pi\delta(\xi_2)\right) \tag{71}$$

where p.v. denotes Cauchy's principal value.
Equation (71) can also be expressed as (Friedlander & Joshin, 1998),

$$\hat{H}(\xi_2) = \frac{1}{\sqrt{2\pi}}\frac{1}{i\xi_2 + \epsilon} \tag{72}$$

and, therefore, the Fourier transform for $u(x)$ in Eq. (70) becomes

$$\hat{u}(\xi) = \frac{1}{i\xi_2 + \epsilon}\hat{f}(\xi) \tag{73}$$

The treatment of ∂_2^{-1} is resolved by means of Eq. (73), which is represented by

$$\partial_2^{-1} f = \mathscr{F}^{-1}\frac{1}{i\xi_2 + \epsilon}\mathscr{F} f \tag{74}$$

Likewise, we obtain

$$(\partial_3 + ik_L)^{-1} f = \int_{-\infty}^{\infty} H(x_3 - x_3')\exp\left(ik_L(x_3 - x_3')\right)f(x_1, x_2, x_3')dx_3' \tag{75}$$

which yields

$$(\partial_3 + ik_L)^{-1} f = \mathscr{F}^{-1}\frac{1}{i\xi_3 + ik_L + \epsilon}\mathscr{F} f \tag{76}$$

As can be seen from Eqs. (74) and (76), ∂_2^{-1} and $(\partial_3 + ik_L)^{-1}$ in the operator s_{ij} can be dealt with and resolved in terms of the Fourier transform. As a result of the above procedure, the treatment of the differential operator M_{ij} defined by Eq. (53) can also be handled by the Fourier transform. After all, as in the formulation of the forward scattering problem, Eq. (68) can be discretized into the following form:

$$S_D(\xi_p)\,\hat{\gamma}_{(D)i}(\xi) = \hat{Q}_{(D)i}(\xi) + B_{(D)ij}\hat{Q}_{(D)j}, \quad \xi \in D_\xi \tag{77}$$

where $B_{(D)ij}$ is the operator expressed by

$$B_{(D)ij} = \mathscr{F}_D \exp(ix \cdot \xi_p) M_{(D)ij} \mathscr{F}_D^{-1} \tag{78}$$

The Krylov subspace iteration technique is also applied to Eq. (77) in the analysis. As a result of the above procedure, a fast method for the analysis of the inverse scattering is expected to be established.

3.5 Numerical example

A numerical example for a multiple scattering problem in a 3-D elastic full space is presented. The fluctuations in the x_1–x_2 and x_1–x_3 planes are shown in Figs. 2(a) and 2(b), respectively, where the maximum amplitudes of $\tilde{\lambda}$ and $\tilde{\mu}$ are 0.18 GPa. These fluctuations are smooth, so that they have continuous spatial derivatives. The background structure of the wave field for the Lamé constants is set such that λ_0 = 4 GPa and μ_0 = 2 GPa, and the mass density is set to ρ = 2 g/cm^3. The background velocity of the P and S waves are 2 and 1 km/s, respectively. The analyzed frequency is f = 1 Hz, and the amplitude of the potential for the incident P wave is $a = 1.0 \times 10^5$ cm^2. The intervals of the grids in the space domain for the discrete Fourier transform are set by $\Delta x_j = 0.25$(km), (j = 1, 2, 3), and the number of intervals of the grids in the space domain for the discrete Fourier transform are set by $N_j = 256$, (j = 1, 2, 3). As a result, the intervals of the grid in the wavenumber space become $\Delta\xi_j = 2\pi/(N_j \times \Delta x_j) \approx 0.098$ km^{-1}. In addition, ϵ for the Green's function in the wavenumber domain shown in Eq. (21) is set to 0.2.

Figures 3(a) and 3(b) show the amplitudes of the scattered waves in the $x_1 - x_2$ and $x_1 - x_3$ planes, respectively. According to Fig. 3(a), the scattered waves are prominent in the regions in which fluctuations of the medium are present. The regions for the high amplitudes of the scattered waves are found to be separated due to the locations of the fluctuations of the medium. Therefore, the effects of multiple scattering are not very significant here. The reflection of the waves due to the incident wave is found to be small because of the smooth fluctuations. According to Fig. 3(b), forward scattering is noticeable with the narrow directionality in the x_3 direction. Interference of the scattered waves can be observed in the far field range of regions of the fluctuation.

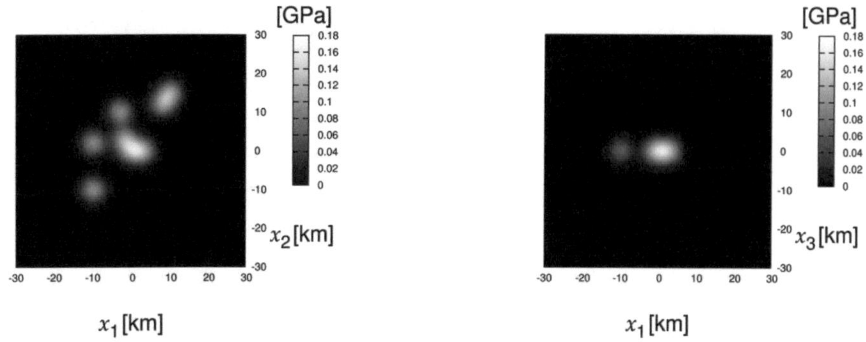

(a) Fluctuations of Lamé constants $\tilde{\lambda}$ and $\tilde{\mu}$ in the x_1 - x_2 plane.

(b) Fluctuations of Lamé constants $\tilde{\lambda}$ and $\tilde{\mu}$ in the x_1 - x_3 plane.

Fig. 2. Analyzed model of smooth fluctuations.

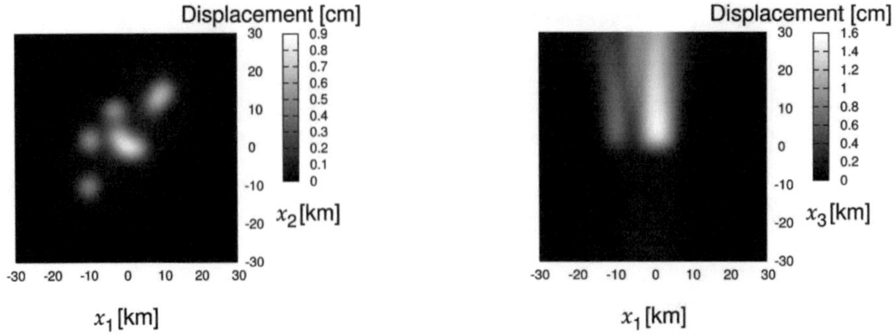

(a) Amplitudes of scattered waves in the x_1 - x_2 plane.
(b) Amplitudes of the scattered waves in the x_1 - x_3 plane.

Fig. 3. Results of the forward scattering analysis due to smooth fluctuations.

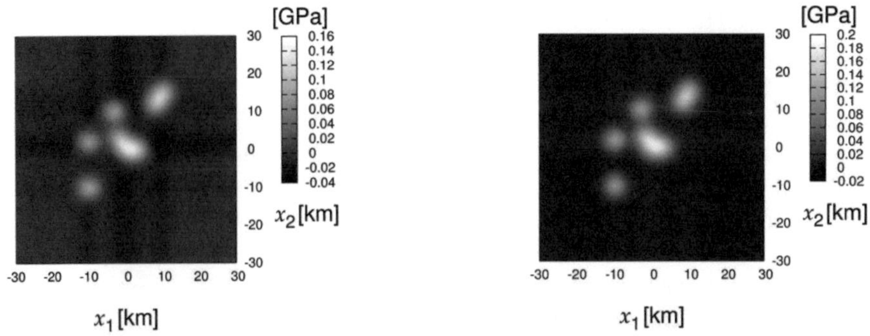

(a) Reconstruction of $\tilde{\lambda}$ in the x_1 - x_2 plane.
(b) Reconstruction of $\tilde{\mu}$ in the x_1 - x_2 plane.

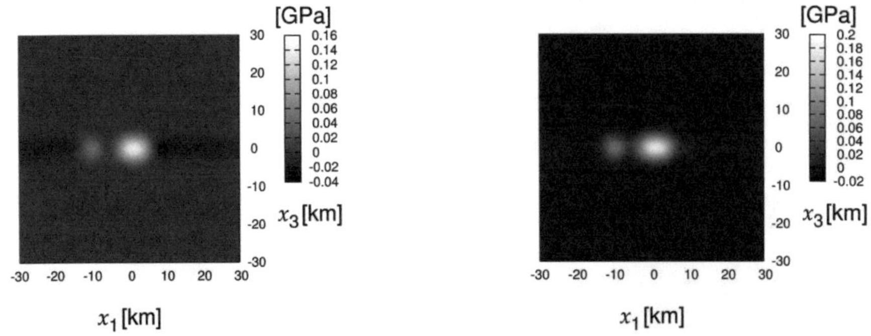

(c) Reconstruction of $\tilde{\lambda}$ in the x_1 - x_3 plane.
(d) Reconstruction of $\tilde{\mu}$ in the x_1 - x_3 plane.

Fig. 4. Results of the inverse scattering analysis due to smooth fluctuations.

The results of the inverse scattering analysis in the x_1−x_2 and x_1−x_3 planes are shown in Figs. 4(a) through 4(d). For the analysis, ϵ for expressing ∂_2^{-1} and $(\partial_3 + ik_L)^{-1}$ in the operator M_{jk} was set to 0.01. Figures 4(a) through 4(d) show that the amplitudes and locations for the fluctuations were successfully reconstructed from the scattered wave field. Namely, Eq. (77) is effective and available for the inverse scattering analysis for the case in which the entire scattered wave field is provided.

For an AMD Opteron 2.4 GHz processor, and using the ACML library for the FFT and Bi-CGSTAB method for the Krylov subspace iteration technique, the required CPU time was only two minutes, where two iterations of the Bi-CGSTAB method were needed to obtain the solutions.

4. Volume integral equation method for an elastic half space

In this section, we deal with the concept of the analyzed model shown in Fig. 1(b), which is the scattering problem in an elastic half space. As shown in Eq. (16), the volume integral equation for the problem in terms of the scattered wave field can be expressed by

$$v_i(x) = -\int_{\mathbb{R}^3_+} G_{ij}(x,y)N_{jk}(y)F_k(y)dy - \int_{\mathbb{R}^3_+} G_{ij}(x,y)N_{jk}(y)v_k(y)dy \tag{79}$$

where G is the Green's function in an elastic half space, and F_i is the wave from the point source, expressed as

$$F_i(x) = G_{ij}(x, x_s)q_j \tag{80}$$

Equation (79) can be solved by means of the Fourier transform constructed for elastic wave propagation for a half space. This section explains this transform for the integral equation for an elastic half space and its application to the volume integral equation.

4.1 Transforms for the elastic wave equation in a half space for horizontal components

First, in order to determine an appropriate transform for the elastic wave equation in a half space, the following equation:

$$\left[L_{ij}(\partial_1, \partial_2, \partial_3) + \rho\omega^2\delta_{ij}\right]u_j(x) = -f_i(x), \quad (x \in \mathbb{R}^2 \times \mathbb{R}_+ = \mathbb{R}^3_+) \tag{81}$$

together with the following boundary condition:

$$P^{(0)}_{ij}u_j(x) = 0, \quad (\text{at } x_3 = 0) \tag{82}$$

are investigated, where $P^{(0)}_{ij}$ is the operator describing the free boundary condition, the components of which are

$$[P^{0)}_{ij}] = \begin{bmatrix} \mu_0\partial_3 & 0 & \mu_0\partial_1 \\ 0 & \mu_0\partial_3 & \mu_0\partial_2 \\ \lambda_0\partial_1 & \lambda_0\partial_2 & (\lambda_0 + 2\mu_0)\partial_3 \end{bmatrix} \tag{83}$$

The force density f_i and the displacement field u_i are assumed to be in $L_2\,(\mathbb{R}^3_+)$. The scalar product of the function in $L_2(\mathbb{R}^3_+)$ is defined as

$$\left(u_i, v_i\right)_{L_2(\mathbb{R}^3_+)} = \int_{\mathbb{R}^3_+} [u_1^*(x)v_1(x) + u_2^*(x)v_2(x) + u_3^*(x)v_3(x)]dx \tag{84}$$

The following Fourier integral transform for the displacement field for the horizontal components is introduced for Eq. (81):

$$\begin{aligned} \left(\mathscr{F}^{(h)}u_i\right)(\xi_1,\xi_2,x_3) &= \frac{1}{2\pi}\int_{-\infty}^{\infty}\int_{-\infty}^{\infty} u_i(x_1,x_2,x_3)\exp\left[-i(x_1\xi_1+x_2\xi_2)\right]dx_1dx_2 \\ \left(\mathscr{F}^{(h)^{-1}}\hat{u}_i\right)(x_1,x_2,x_3) &= \frac{1}{2\pi}\int_{-\infty}^{\infty}\int_{-\infty}^{\infty} \hat{u}_i(\xi_1,\xi_2,x_3)\exp\left[i(x_1\xi_1+x_2\xi_2)\right]d\xi_1d\xi_2 \end{aligned} \quad (85)$$

where ξ_1 and ξ_2 are the horizontal coordinates of the wavenumber space. Note that the convergence of the integrals shown in Eq. (85) should be understood in the sense of the limit in the mean. According to the Fourier transform given by Eq. (85), Eq. (81) is transformed into the following:

$$L_{ij}(i\xi_1,i\xi_2,\partial_3)\hat{u}_j(\hat{x}) = -\hat{f}_i(\hat{x}) \quad (86)$$

where $\hat{u}_j$ and $\hat{f}_i$ in this section are define by

$$\hat{u}_j = \mathscr{F}^{(h)}u_j, \quad \hat{f}_i = \mathscr{F}^{(h)}f_i \quad (87)$$

respectively, and $\hat{x}$ is given by

$$\hat{x} = (\xi_1,\xi_2,x_3) \quad (88)$$

The Stokes-Helmholtz decomposition (Aki & Richards, 1980) is introduced in order to make the treatments for Eq. (86) more comprehensive. In general, the Stokes-Helmholtz decomposition of the displacement field u_i is expressed as:

$$u_i(x) = \partial_i\phi + \epsilon_{ijk}\epsilon_{kl3}\partial_j\partial_l\psi + \epsilon_{ij3}\partial_j\chi \quad (89)$$

where ϕ, ψ, and χ are the scalar potentials for the P, SV, and SH waves, respectively, and ϵ_{ijk} is the Eddington epsilon. The Fourier transform of Eq. (89) is as follows:

$$\begin{aligned} \hat{u}(\hat{x}) &= (i\xi_1\phi + i\xi_1\partial_3\psi + i\xi_2\chi)e_1 \\ &\quad +(i\xi_2\phi + i\xi_2\partial_3\psi - i\xi_1\chi)e_2 \\ &\quad +(\partial_3\phi + \xi_r^2\psi)e_3 \end{aligned} \quad (90)$$

where e_j, (j = 1, 2, 3) are the base vectors for the 3-D displacement field, $\hat{u} = \hat{u}_j e_j$, and $\xi_r = \sqrt{\xi_1^2+\xi_2^2}$. From Eq. (90), the wave field can be decomposed into the P-SV and SH waves by introducing the new base vectors e_i' defined by $e_i' = T_{ij}e_j$, where T_{ij} is expressed as

$$[T_{ij}] = \begin{bmatrix} 0 & 0 & 1 \\ ic & is & 0 \\ is & -ic & 0 \end{bmatrix} \quad (91)$$

where $c = \xi_1/\xi_r$ and $s = \xi_2/\xi_r$ for the case in which $\xi_r \neq 0$ and $c = 1$ and $s = 0$ for the case $\xi_r = 0$. Note that it is possible to impose arbitrary values on c and s when $\xi_r = 0$, because, based on Eq. (90), $\hat{u}_1 = \hat{u}_2 = 0$.

The linear transform T_{ij} defined by Eq. (91) is unitary and has the property whereby $T^*_{ik}T_{ij} = \delta_{kj}$. Equations (81) and (82) are transformed as follows by means of T_{ij} and the Fourier transform shown in Eq. (85):

$$\left(-\mathscr{A}_{ij} + \rho\omega^2\delta_{ij}\right)\hat{\hat{u}}_j(\hat{x}) = -T^*_{ik}\ \hat{f}_i(\hat{x}) \tag{92}$$

$$\mathscr{P}_{ij}\hat{\hat{u}}_j(\hat{x}) = 0, \quad \text{at } x_3 = 0 \tag{93}$$

where the operators $\mathscr{A}_{ij}$ and $\mathscr{P}_{ij}$ are obtained from:

$$\begin{aligned} \mathscr{A}_{ij} &= -T^*_{im}L_{mk}(i\xi_1, i\xi_2, \partial_3)T_{jk} \\ \mathscr{P}_{ij} &= T^*_{im}P_{mk}T_{jk} \end{aligned} \tag{94}$$

The relationship between $\hat{u}_i$ and $\hat{\hat{u}}_i$ is as follows:

$$\hat{\hat{u}}_i = T^*_{im}\hat{u}_m, \quad \hat{u}_m = T_{jm}\hat{\hat{u}}_j \tag{95}$$

The components of the operators $\mathscr{A}_{ij}$ and $\mathscr{P}_{ij}$ are

$$[\mathscr{A}_{ij}] = \begin{bmatrix} -(\lambda_0 + 2\mu_0)\partial_3^2 + \mu\xi_r^2 & (\lambda_0 + \mu_0)\xi_r\partial_3 & 0 \\ -(\lambda_0 + \mu_0)\xi_r\partial_3 & -\mu_0\partial_3^2 + (\lambda_0 + 2\mu_0)\xi_r^2 & 0 \\ 0 & 0 & -\mu_0\partial_3^2 + \mu_0\xi_r^2 \end{bmatrix} \tag{96}$$

$$[\mathscr{P}_{ij}] = \begin{bmatrix} (\lambda_0 + 2\mu_0)\partial_3 & -\lambda_0\xi_r & 0 \\ \mu_0\xi_r & \mu_0\partial_3 & 0 \\ 0 & 0 & \mu_0\partial_3 \end{bmatrix} \tag{97}$$

In Eqs. (96) and (97), the matrices are separated into 2×2 and 1×1 minor matrices, which makes the procedures for the operator much easier. Note that the 2 × 2 minor matrix is for the P-SV wave components and that the 1 × 1 minor matrix is for the SH wave component.

4.2 Self-adjointness of the operator $\mathscr{A}_{ij}$

In this section, we discuss the self-adjointness of the operator $\mathscr{A}_{ij}$ and its spectral representation. The domain of the operator $\mathscr{A}_{ij}$ is set by

$$D(\mathscr{A}_{ij}) = \{\psi_i \in L_2(\mathbb{R}_+) \mid \mathscr{A}_{ij}\psi_j \in L_2(\mathbb{R}_+), \mathscr{P}_{ij}\psi_j = 0 \text{ at } x_3 = 0\} \tag{98}$$

with the scalar product

$$\left(u_i, v_i\right)_{L_2(\mathbb{R}_+)} = \int_{\mathbb{R}_+}\left[u_1^*(x_3)v_1(x_3) + u_2^*(x_3)v_2(x_3) + u_3^*(x_3)v_3(x_3)\right]dx_3 \tag{99}$$

for u_i, $v_i \in D(\mathscr{A}_{ij})$. The operation for the differentiation in $\mathscr{A}_{ij}$ is carried out in the sense of the distribution. It is not difficult to show the following:

Lemma 1 *The operator $\mathscr{A}_{ij}$ is symmetric and non-negative.*

[Proof]

Let u_i, $v_i \in D(\mathscr{A}_{ij})$. Then,

$$\begin{aligned}
&\Big(u_i, \mathscr{A}_{ij} v_j\Big)_{L_2(\mathbb{R}_+)} \\
&= -\Big[u_i^* \mathscr{P}_{ij} v_j\Big]_0^\infty + \Big[(\mathscr{P}_{ji} u_i^*) v_j\Big]_0^\infty + \Big(\mathscr{A}_{ji} u_i, v_j\Big)_{L_2(\mathbb{R}_+)} \\
&= \Big(\mathscr{A}_{ji} u_i, v_j\Big)_{L_2(\mathbb{R}_+)}
\end{aligned} \tag{100}$$

$$\begin{aligned}
\Big(u_i, \mathscr{A}_{ij} u_j\Big)_{L_2(\mathbb{R}_+)} &= \int_0^\infty \Big[\partial_3 u_1^* (\lambda_0 + 2\mu_0) \partial u_1 + \partial_3 u_2^* \mu_0 \partial_3 u_2 + \partial_3 u_3^* \mu_0 \partial_3 u_3\Big] dx_3 \\
&+ \int_0^\infty \Big[u_1^* \mu_0 \xi_r^2 u_1 + u_2^* (\lambda_0 + 2\mu_0) \xi_r^2 u_2 + u_3^* \mu_0 \xi_r^2 u_3\Big] dx_3 \\
&+ \int_0^\infty \Big[-\partial_3 u_1^* \lambda_0 \xi_r u_2 - u_2^* \lambda_0 \xi_r \partial_3 u_1 + \partial_3 u_2^* \mu_0 \xi_r u_1 + u_1^* \mu_0 \xi_r \partial_3 u_2\Big] dx_3 \\
&= \int_0^\infty \Big[2\mu_0 |\partial_3 u_1|^2 + \mu_0 |\partial_3 u_3|^2\Big] dx_3 \\
&+ \int_0^\infty \Big[2\mu_0 \xi_r^2 |u_2|^2 + \mu_0 \xi_r^2 |u_3|^2\Big] dx_3 \\
&+ \int_0^\infty \Big[\lambda_0 |\partial_3 u_1 - \xi_r u_2|^2 + \mu_0 |\partial_3 u_2 + \xi_r u_1|^2\Big] dx_3 \geq 0
\end{aligned} \tag{101}$$

□

Next, the following function is defined:

$$\Big(\mathscr{A}_{ij} - \eta^2 \mu_0 \delta_{ij}\Big) g_{jk}(x_3, y_3, \xi_r, \eta) = \delta_{ik} \delta(x_3 - y_3), \quad (\eta \in \mathbb{C}) \tag{102}$$

together with the boundary condition

$$\mathscr{P}_{ij} g_{jk}(x_3, y_3, \xi_r, \eta) = 0, \ \text{ at } x_3 = 0 \tag{103}$$

where $\mathbb{C}$ is a set of complex numbers. The solution of Eq. (102) for for $\eta \in \mathbb{C} \setminus B$ has the following properties:

$$\begin{aligned}
\sup_{x_3 \in \mathbb{R}_+} \int_{\mathbb{R}_+} |g_{ij}(x_3, y_3, \xi_r, \eta)| dy_3 &< \infty \\
\sup_{y_3 \in \mathbb{R}_+} \int_{\mathbb{R}_+} |g_{ij}(x_3, y_3, \xi_r, \eta)| dx_3 &< \infty
\end{aligned} \tag{104}$$

where B is defined by

$$B = B_p \cup B_c \tag{105}$$

in which

$$\begin{aligned}
B_p &= \{\eta \in \mathbb{R} \mid F_R(\xi_r, \eta) = 0\} \\
B_c &= \{\eta \in \mathbb{R} \mid |\eta| \geq \xi_r\}
\end{aligned} \tag{106}$$

Note that F_R in Eq. (106) is the Rayleigh function given by

$$F_R(\xi_r, \eta) = (2\xi_r^2 - \eta^2)^2 - 4\xi_r^2 \gamma\nu \tag{107}$$

where

$$\begin{aligned} \nu &= \sqrt{\xi_r^2 - \eta^2} \\ \gamma &= \sqrt{\xi_r^2 - (c_T/c_L \eta^2} \end{aligned} \tag{108}$$

Lemma 2 *For* $f_i \in L_2(\mathbb{R}_+)$ *and* $\eta \in \mathbb{C} \setminus B$

$$u_i(x_3) = \int_{\mathbb{R}_+} g_{ij}(x_3, y_3, \xi_r, \eta) f_j(y_3) dy_3 \in L_2(\mathbb{R}_+) \tag{109}$$

[Proof]
First, fix i and j and define

$$v_i(x_3) = \int_{\mathbb{R}_+} g_{ij}(x_3, y_3, \xi_r, \eta)\ f_j(y_3) dy_3 \tag{110}$$

Then, the following is obtained by means of the Schwarz inequality:

$$\begin{aligned} |v_i(x_3)| &\leq \left[\int_{\mathbb{R}_+} |g_{ij}(x_3, y_3, \xi_r, \eta)|\ |f_j(y_3)|^2 dy_3\right]^{1/2} \left[\int_{\mathbb{R}_+} |g_{ij}(x_3, y_3, \xi_r, \eta)| dy_3\right]^{1/2} \\ &\leq \left[\int_{\mathbb{R}_+} |g_{ij}(x_3, y_3, \xi_r, \eta)|\ |f_j(y_3)|^2 dy_3\right]^{1/2} M_1 \end{aligned} \tag{111}$$

where

$$M_1 = \sup_{x_3 \in \mathbb{R}_+} \left[\int_{\mathbb{R}_+} |g_{ij}(x_3, y_3, \xi_r, \eta)| dy_3\right]^{1/2} \tag{112}$$

As a result, the following is obtained:

$$\begin{aligned} \int_{\mathbb{R}_+} |v_i(x_3)|^2 dx_3 &\leq M_1^2 \int_{\mathbb{R}_+} \int_{\mathbb{R}_+} |g_{ij}(x_3, y_3, \xi_r, \eta)|\ |f_j(y_3)|^2 dy_3\ dx_3 \\ &\leq M_1^2 M_2 \|f_j\|^2_{L_2(\mathbb{R}_+)} \end{aligned} \tag{113}$$

where

$$M_2 = \sup_{y_3 \in \mathbb{R}_+} \int_{\mathbb{R}_+} |g_{ij}(x_3, y_3, \xi_r, \eta)| dx_3 \tag{114}$$

Equation (113) concludes the proof. □

Theorem 1 *The operator* $\mathscr{A}_{ij}$ *with the domain* $D(\mathscr{A}_{ij})$ *is self-adjoint.*
[Proof]
It is sufficient to prove that $\forall f_i \in L_2(\mathbb{R}_+)$, there exist $u_i^{(+)}, u_i^{(-)} \in D(\mathscr{A}_{ij})$ satisfying

$$\left(\mathscr{A}_{ij} + ip\mu_0\delta_{ij}\right)u_j^{(+)}(x_3) = f_i(x_3) \tag{115}$$

$$\left(\mathscr{A}_{ij} - ip\mu_0\delta_{ij}\right)u_j^{(-)}(x_3) = f_i(x_3) \tag{116}$$

where p is a positive real number. This fact is based on the results of a previous study (Theorem 3.1, Berthier, 1982).
For the construction of $u_i^{(+)}$, define

$$u_i^{(+)}(x_3) = \int_{\mathbb{R}_+} g_{ij}(x_3, y_3, \xi_r, \eta) f_j(y_3) dy_3 \tag{117}$$

where η is chosen such that $\eta^2 = ip$. Note that $\eta \in \mathbb{C} \setminus B$. The following equation:

$$\begin{aligned}
&\int_{\mathbb{R}_+} \varphi_i(x_3)(\mathscr{A}_{ij} + ip\mu_0\delta_{ij})u_j(x_3)dx_3 \\
&= \int_{\mathbb{R}_+} \left[(\mathscr{A}_{ji} + ip\mu_0\delta_{ji})\varphi_i(x_3)\right] u_j(x_3)dx_3 \\
&= \int_{\mathbb{R}_+} (\mathscr{A}_{ji} + ip\mu_0\delta_{ji})\varphi_i(x_3) \int_{\mathbb{R}_+} g_{jk}(x_3, y_3, \xi_r, \eta) f_k(y_3) dy_3 dx_3 \\
&= \int_{\mathbb{R}_+} \left[\int_{\mathbb{R}_+} \left((\mathscr{A}_{ji} + ip\mu_0\delta_{ji})\varphi_i(x_3)\right) g_{jk}(x_3, y_3, \xi_r, \eta) dx_3\right] f_k(y_3) dy_3 \\
&= \left(\varphi_k, f_k\right)_{L_2(\mathbb{R}_+)} , \ (\varphi_i \in \mathscr{D}(\mathbb{R}_+))
\end{aligned} \tag{118}$$

yields Eq. (115), where $\mathscr{D}(\mathbb{R}_+)$ is the Schwartz space. During the derivation of Eq. (118), the following equation:

$$\int_{\mathbb{R}_+} \left[(\mathscr{A}_{ji} + ip\mu_0\delta_{ij})\varphi_i(x_3)\right] g_{jk}(x_3, y_3, \xi_r, \eta) dx_3 = \varphi_k(y_3), \quad (\varphi_i \in \mathscr{D}(\mathbb{R}_+)) \tag{119}$$

is based on the following properties of $g_{ij}(x_3, y_3, \xi_r, \eta)$ at $x_3 = y_3$

$$\begin{aligned}
g_{ik}(y_3 + \epsilon, y_3, \xi_r, \eta) &= g_{ik}(y_3 - \epsilon, y_3, \xi_r, \eta) \\
[\mathscr{P}_{ij} g_{jk}(x_3, y_3, \xi_r, \eta)]_{x_3 = y_3 - \epsilon}^{x_3 = y_3 + \epsilon} &= \delta_{ik}
\end{aligned} \tag{120}$$

In addition, the following is obtained:

$$\begin{aligned}
\mathscr{P}_{ij} u_j(x_3) &= \mathscr{P}_{ij} \int_{\mathbb{R}_+} g_{jk}(x_3, y_3, \xi_r, \eta) f_k(y_3) dy_3 \\
&= \int_{\mathbb{R}_+} \mathscr{P}_{ij} g_{jk}(x_3, y_3, \xi_r, \eta) f_k(y_3) dy_3
\end{aligned} \tag{121}$$

The order of the integral and differential operators of the properties of function g_{ij} are changed such that

$$\lim_{y_3\to\infty} |{y_3}^n \mathscr{P}_{ij} g_{jk}(x_3, y_3, \xi_r, \eta) f_k(y_3)| = 0 \tag{122}$$

for an arbitrary positive integer n. According to Eq. (121), we have

$$\mathscr{P}_{ij} u_j(x_3) = 0, \quad \text{at } x_3 = 0 \tag{123}$$

It has been shown that $u_j \in L_2(\mathbb{R}_+)$ from **Lemma 2**, so that $u_i^{(+)} \in D(\mathscr{A}_{ij})$. The construction of $u_i^{(-)} \in D(\mathscr{A}_{ij})$ is also possible. As a result, the following conclusion is obtained. □

4.3 Generalized Fourier transform for an elastic wave field in a half space

The operator $\mathscr{A}_{ij}$ has been found to be self-adjoint and non-negative, which yields the following spectral representation:

$$\mathscr{A}_{ij} = \int_0^\infty \zeta \, dE_{ij}(\zeta) \tag{124}$$

where E_{ij} is the spectral family. The spectral family is connected with the resolvent by means of the Stone theorem (Wilcox, 1976):

$$\begin{aligned} &\left(\left[(E_{ij}(b) + E_{ij}(b-)) - (E_{ij}(a) + E_{ij}(a-))\right] u_j, v_i\right)_{L_2(\mathbb{R}_+)} \\ &= \lim_{\epsilon \downarrow 0} \frac{1}{\pi i} \int_a^b d\zeta \left(\left[R_{ij}(\zeta + i\epsilon) - R_{ij}(\zeta - i\epsilon)\right] u_j, v_i\right)_{L_2(\mathbb{R}_+)} \end{aligned} \tag{125}$$

for u_i, $v_i \in L_2(\mathbb{R}_+)$. Note that R_{ij} is the resolvent of the operator $\mathscr{A}_{ij}$ and $R_{ij}(\zeta)u_j$ is defined by

$$\left(R_{ij}(\zeta)u_j\right)(x_3) = \int_0^\infty g_{ij}\left(x_3, y_3, \xi_r, \sqrt{\zeta/\mu_0}\right) u_j(y_3) dy_3 \tag{126}$$

Let $0 < a < \mu_0 \eta_R^2$ and $b > \mu \xi_r^2$. Then, the right-hand side of Eq. (125) for the integral becomes

$$\begin{aligned} &\int_a^b d\zeta \left(\left[R_{ij}(\zeta + i\epsilon) - R_{ij}(\zeta - i\epsilon)\right] u_j, v_i\right)_{L_2(\mathbb{R}_+)} \\ &= (-2\pi i) \operatorname*{Res}_{\eta=\eta_R} (2\eta\mu_0) \left(R_{ij}(\zeta) u_j, v_i\right)_{L_2(\mathbb{R}_+)} \\ &\quad + \int_{\xi_r}^{\sqrt{b/\mu}} d\eta \, (2\eta\mu_0) \left(\left[R_{ij}(\zeta + i\epsilon) - R_{ij}(\zeta - i\epsilon)\right] u_j, v_i\right)_{L_2(\mathbb{R}_+)} \end{aligned} \tag{127}$$

where $\zeta = \mu_0\eta^2$ and η_R is defined by $F_R(\xi_r, \eta_R) = 0$. The path of integration in the complex η plane shown in Fig. 5 is used for the evaluation of the integral.

In the following, the relationship between the right-hand side of Eq. (127) and the eigenfunctions is presented. Let $v_i(x_3, \xi_r, \eta) \in D(\mathscr{A}_{ij})$ satisfy

$$(\mathscr{A}_{ij} - \mu_0 \eta^2 \delta_{ij}) v_j(x_3, \xi_r, \eta) = 0 \tag{128}$$

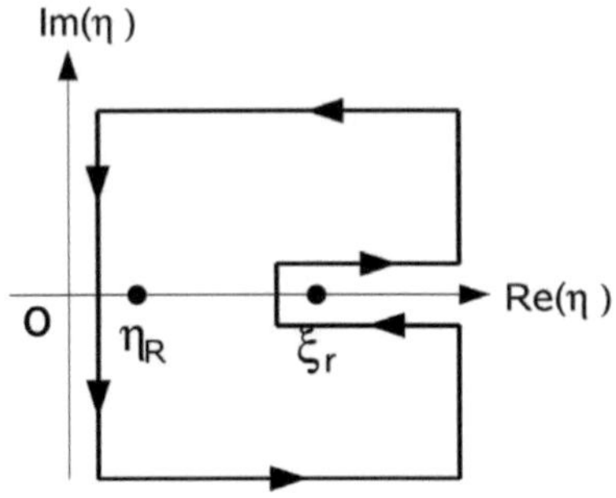

Fig. 5. Path of the integral.

and define the scalar function $W(\eta)$ such that

$$\begin{aligned} W(\eta) &= \left(v_i, (\mathscr{A}_{ij} - \mu_0 \eta^2 \delta_{ij}) v_j\right)_{L_2(\mathbb{R}_+)} \\ &\quad - v_i^*(0, \xi_r, \eta) \mathscr{P}_{ij} v_j(0, \xi_r, \eta) \end{aligned} \tag{129}$$

It is easy to derive the following properties of $W(\eta)$ by means of the boundary conditions for v_i:

$$\begin{aligned} W(\eta) &= 0, \quad (\eta = \eta_R) \\ W(\eta) &\neq 0, \quad (\eta \neq \eta_R) \end{aligned} \tag{130}$$

Note that $v_i(x_3, \xi_r, \eta_R)$ becomes the eigenfunction (Rayleigh wave mode) satisfying the free boundary conditions. Otherwise, $v_i(x, \xi_r, \eta)$, $(\eta \neq \eta_R)$ cannot satisfy the free boundary conditions. As a result, Eq. (130) is established. Integration by parts of Eq. (129) yields

$$W(\eta) = I_1(\eta) - \mu_0 \eta^2 I_2(\eta) \tag{131}$$

where

$$\begin{aligned} I_1(\eta) &= \int_0^\infty \Big[2\mu_0 |\partial_3 v_1|^2 + \mu_0 |\partial_3 v_3|^2 + 2\mu_0 \xi_r^2 |v_2|^2 + \mu_0 \xi_r^2 |v_3|^2 \\ &\quad + \lambda_0 |\partial_3 v_1 - \xi_r v_2|^2 + \mu_0 |\partial_3 v_2 + \xi_r v_1|^2 \Big] dx_3 \\ I_2(\eta) &= \int_0^\infty \Big[|v_1|^2 + |v_2|^2 + |v_3|^2 \Big] dx_3 \end{aligned} \tag{132}$$

The following lemma can then be obtained:

Lemma 3 *The residue of* g_{ij} at $\eta = \eta_R$ *can be expressed in terms of the eigenfunction such that*

$$\underset{\eta=\eta_R}{Res}\; g_{ij}(x_3, y_3, \xi_r, \eta) = -\frac{\psi_{im}(x_3, \xi)\psi_{jm}(y_3, \xi)}{2\mu_0 \eta_R} \tag{133}$$

where $\xi = (\xi_1, \xi_2, \eta_R)$ *and* $\psi_{im}(x_3, \xi)$ *is the eigenfunction defined by*

$$\mathscr{A}_{ij} \psi_{jm}(x_3, \eta_R) = \mu_0 \eta_R^2 \psi_{im}(x_3, \eta_R), \quad \psi_{im}(x_3, \eta_R) \in D(\mathscr{A}_{ij}) \tag{134}$$

[Proof]
Based on a previous study (Aki & Richards, 1980), the function g_{ij} can be constructed by

$$g_{ij}(x_3, y_3, \xi_r, \eta) = w_{ij}(x_3, y_3, \xi_r, \eta) + v_i(x_3, \xi_r, \eta)\Delta_j(\eta) \tag{135}$$

where w_{ij} is defined by

$$\begin{aligned} w_{ij}(x_3, y_3, \xi_r, \eta) &= 0, \quad (x_3 > y_3) \\ \mathscr{P}_{ij} w_{jk}(x_3, y_3, \xi_r, \eta) &= +\delta_{ik}, \quad (x_3 + \epsilon = y_3, \ \epsilon \downarrow 0) \end{aligned} \tag{136}$$

In addition, Δ_j is defined such that g_{ij} satisfies the free boundary condition:

$$\mathscr{P}_{ij} w_{jk}(x_3, y_3, \xi_r, \eta) + \mathscr{P}_{ij} v_j(x_3, \xi_r, \eta)\Delta_j(\eta) = 0, \quad (\text{at } x_3 = 0) \tag{137}$$

The definition of $W(\eta)$ shown in Eq. (129) implies that the expression is valid:

$$W(\eta) = -v_i^*(0, \xi_r, \eta)\mathscr{P}_{ij} v_j(0, \xi_r, \eta) \tag{138}$$

Equations (137) and (138) yield

$$\begin{aligned} & g_{ik}(x_3, y_3, \xi_r, \eta) \\ = \ & w_{ik}(x_3, y_3, \xi_r, \eta) + \frac{v_i(x_3, \xi_r, \eta)}{W(\eta)}[v_l^*(0, \xi_r, \eta)P_{lj} w_{jk}(0, y_3, \xi_r, \eta)] \end{aligned} \tag{139}$$

Now, let η approach η_R. Due to reciprocity, it is found that

$$v_l^*(0, \xi_r, \eta_R)P_{lj} w_j(0, y_3, \xi_r, \eta_R) = v_l^*(y_3, \xi_r, \eta_R)\delta_{lk} \tag{140}$$

Therefore,

$$v_i(x_3, \xi_r, \eta_R)[v_l^*(0, s)P_{lj} w_{jk}(0, y_3, \xi_r, \eta)] = \psi_{im}(x_3, \xi)\psi_{lm}(y_3, \xi)\delta_{ik} \tag{141}$$

The residue of the resolvent kernel is expressed as

$$\operatorname*{Res}_{\eta=\eta_R} g_{ij}(x_3, y_3, \xi_r, \eta) = \frac{\psi_{im}(x_3, \xi)\psi_{jm}(y_3, \xi)}{W'(\eta_R)} \tag{142}$$

For the case in which the eigenfunction is normalized as $I_2(\eta_R) = 1$, we have $W'(\eta_R) = -2\mu\eta_R$, which concludes the proof. □

Next, the function g_{ij}:

$$g_{ij}(x, y, \xi_r, s + i\epsilon) - g_{ij}(x, y, \xi_r, s - i\epsilon) \tag{143}$$

is investigated for the case in which $s = \text{Re}(\eta) > \xi_r$. The function g_{ij} for this case is constructed by

$$g_{ij}(x, y, \xi_r, s \pm i\epsilon) = w_{ij}(x, y, \xi_r, s \pm i\epsilon) + v_{ik}(x, \xi_r, s \pm i\epsilon)\Delta_{kj}(s \pm i\epsilon) \tag{144}$$

where v_{ik} is the definition function of the improper eigenfunction (Touhei, 2002). The definition function v_{ik} satisfies the following:

$$(\mathscr{A}_{ij} - \mu_0 s^2 \delta_{ij}) v_{jk}(x_3, \xi_r, s) = 0 \tag{145}$$

The relationship between the improper eigenfunction and the definition function is given as

$$\psi_{ik}(x, \xi) = v_{ik}(x, y, \xi_r, s + i\epsilon) - v_{ik}(x, y, \xi_r, s - i\epsilon), \quad (\epsilon \to 0) \tag{146}$$

Next, let us define the following function:

$$W_{kl}(\xi_r, s \pm i\epsilon) = -\psi_{ik}(0, \xi) P_{ij} v_{jl}(0, y, \xi_r, s \pm i\epsilon) \tag{147}$$

Substitution of the explicit forms of the eigenfunction and definition function of Eq. (147) yields the following:

$$W_{kl}(\xi_r, s \pm i\epsilon) = -\frac{\mu_0 i}{\pi} s\, \delta_{kl}, \quad (\epsilon \to 0) \tag{148}$$

In addition, note that

$$w_{ik}(x_3, y_3, \xi_r, s + i\epsilon) - w_{ik}(x_3, y_3, \xi_r, s - i\epsilon) \to 0, \quad (\epsilon \to 0) \tag{149}$$

which is obtained from the definition of w_{ik} shown in Eq. (136). Based on Eqs. (148) and (149), the following lemma is obtained.

Lemma 4 *For the region of* $s > \xi_r$*, the function* g_{ij} *satisfies the following equation:*

$$\begin{aligned} & g_{ij}(x_3, y_3, \xi_r, s + i\epsilon) - g_{ij}(x_3, y_3, \xi_r, s - i\epsilon) \\ = \; & \pi i \frac{\psi_{im}(x_3, \xi)\psi_{jm}(y_3, \xi)}{\mu_0 s}, \quad (\epsilon \to 0) \end{aligned} \tag{150}$$

where $\psi_{im}(x_3, \xi)$ *is the improper eigenfunction.*

[Proof]

The requirement of the free boundary condition for g_{ij} yields the following expression of $\Delta_{k\beta}$:

$$\Delta_{kj}(s \pm i\epsilon) \;=\; [W_{lk}(\xi_r, s \pm i\epsilon)]^{-1} \psi_{il}(0, \xi) \mathscr{P}_{im} w_{mj}(0, y_3, \xi_r, s \pm i\epsilon) \tag{151}$$

Incorporating the following reciprocity relation:

$$\psi_{il}(0, \xi) \mathscr{P}_{im} w_{mj}(0, y_3, \xi_r, s \pm i\epsilon) = \psi_{il}(y_3, \xi) \delta_{i\beta} \tag{152}$$

into Eq. (151) yields

$$\Delta_{kj}(s \pm i\epsilon) = [W_{lk}(\xi_r, s \pm i\epsilon)]^{-1} \psi_{jl}(y_3, \xi) \tag{153}$$

Therefore, the following is obtained:

$$\begin{aligned} g_{ij}(x_3, y_3, \xi_r, s \pm i\epsilon) \;=\; & w_{ij}(x_3, y_3, \xi_r, s \pm i\epsilon) \\ & + v_{ik}(x_3, y_3, \xi_r, s \pm i\epsilon)[W_{lk}(\xi_r, s \pm i\epsilon)]^{-1} \psi_{jl}(y_3, \xi) \end{aligned} \tag{154}$$

Thus, Eqs. (146), (148), and (149) conclude the proof. □

Next, let us again consider Eq. (127). Equation (127) holds for an arbitrary $v_i \in L_2(\mathbb{R}_+)$, so that the following equation can be obtained by incorporating the results of Lemmas 3 and 4:

$$\begin{aligned}
&\int_a^b d\zeta \left[R_{ij}(\zeta+i\epsilon) - R_{ij}(\zeta-i\epsilon)\right] u_j \\
&= (2\pi i) \sum_{\xi\in\sigma_p} \psi_{im}(x_3,\xi)\Big(\psi_{jm}(\cdot,\xi), u_j(\xi_r,\cdot\,))\Big)_{L_2(\mathbb{R}_+)} \\
&\quad +(2\pi i) \int_{\xi_r}^{\sqrt{b/\mu}} d\xi_3\, \psi_{im}(x_3,\xi)\Big(\psi_{jm}(\cdot,\xi), u_j(\xi_r,\cdot)\Big)_{L_2(\mathbb{R}_+)}
\end{aligned} \tag{155}$$

where $u_j(\xi_r,\ \cdot) \in L_2(\mathbb{R}_+)$, $\xi = (\xi_1, \xi_2, \xi_3) \in \mathbb{R}^3_+$ and

$$\sigma_p = \{\xi = (\xi_1,\xi_2,\xi_3) \in \mathbb{R}^3_+ \mid F(\xi_r,\xi_3) = 0\} \tag{156}$$

As mentioned earlier, $0 < a < \mu\eta_R^2$ and $b > \mu\xi_r^2$, so that

$$\begin{aligned}
E_{ij}(a) &= E_{ij}(a-) = 0 \\
E_{ij}(b) &= E_{ij}(b-)
\end{aligned} \tag{157}$$

Therefore, Eqs. (125) and (155) yield

$$\begin{aligned}
\Big(E_{ij}(b)u_j\Big)(\xi_r,x_3) &= \sum_{\xi\in\sigma_p} \Big(\psi_{jm}(\,\cdot,\xi), u_j(\xi_r,\cdot\,)\Big)_{L_2(\mathbb{R}_+)} \psi_{im}(x_3,\xi) \\
&\quad + \int_{\xi_r}^{\sqrt{b/\mu}} \Big(\psi_{jm}(\,\cdot,\xi), u_j(\xi_r,\xi_3)\Big)_{L_2(\mathbb{R}_+)} \psi_{im}(x_3,\xi)\, d\xi_3
\end{aligned} \tag{158}$$

Let b in Eq. (158) approach infinity. Then, the following eigenfunction expansion form of u_i is obtained:

$$\begin{aligned}
u_i(\xi_r,x_3) &= \sum_{\xi\in\sigma_p} \Big(\psi_{jm}(\,\cdot,\xi), u_j(\xi_r,\cdot\,)\Big)_{L_2(\mathbb{R}_+)} \psi_{im}(x_3,\xi) \\
&\quad + \int_{\xi_r}^{\infty} \Big(\psi_{jm}(\cdot,\xi), u_j(\xi_r,\cdot\,)\Big)_{L_2(\mathbb{R}_+)} \psi_{im}(x_3,\xi\,)d\xi_3
\end{aligned} \tag{159}$$

Note that the eigenfunction expansion form shown in Eq. (159) is that for $u_i(\xi_r,\ \cdot)$ having the compact support. This result can be extended to all $u_i(\xi_r,\ \cdot\ ) \in L_2(\mathbb{R}_+)$ by a limiting procedure, namely,

$$\Big(\psi_{jm}(\cdot,\xi), u_j(\xi_r,\cdot\,)\Big)_{L_2(\mathbb{R}_+)} = \underset{M\to\infty}{\text{l.i.m.}} \int_0^M \psi_{jm}(x_3,\xi)u_j(\xi_r,x_3)dx_3 \tag{160}$$

where the convergence is in $L_2(\mathbb{R}_+)$. The transform of the function in $L_2(\mathbb{R}_+)$ obtained here can be summarized as follows:

$$\begin{aligned}
\grave{u}_m(\xi) &= \left(\mathscr{F}^{(v)}_{mj} u_j\right)(\xi) = \int_0^\infty \psi_{jm}(x_3,\xi) u_j(\xi_r, x_3) dx_3 \\
u_i(\xi_r, x_3) &= \left(\mathscr{F}^{(v)^{-1}}_{im} \grave{u}_m\right)(\xi_r, x_3) \\
&= \sum_{\xi\in\sigma_p} \psi_{im}(x_3,\xi)\grave{u}_m(\xi) + \int_{\xi_r}^\infty \psi_{im}(x_3,\xi)\grave{u}_m(\xi) d\xi_3
\end{aligned} \tag{161}$$

At this point, the transformation of the elastic wave field in a half space can be presented. Let us define the subset of the wavenumber space as follows:

$$\sigma_c = \{\xi = (\xi_1, \xi_2, \xi_3) \in \mathbb{R}^3_+ \mid \xi_3 > \xi_r\} \tag{162}$$

The following theorem is obtained based on Eqs. (85), (95), and (158):

Theorem 2 *There exists a map satisfying the free boundary condition of the elastic half space of the wave field from* $L_2(\mathbb{R}^3_+)$ to $L_2(\sigma_p) \oplus L_2(\sigma_c)$ *defined by*

$$\begin{aligned}
\hat{u}_j(\xi) &= \left(\mathscr{U}_{ij} u_j\right)(\xi) \\
&= \left(\mathscr{F}^{(v)}_{im}\, T^*_{mj}\, \mathscr{F}^{(h)} u_j\right)(\xi),
\end{aligned} \tag{163}$$

the inverse of which is

$$\begin{aligned}
u_i(x) &= \left(\mathscr{U}^{-1}_{ij} \hat{u}_j\right)(x) \\
&= \left(\mathscr{F}^{(h)^{-1}}\, T_{mi}\, \mathscr{F}^{(v)^{-1}}_{mj}\, \hat{u}_j\right)(x)
\end{aligned} \tag{164}$$

Here, $\mathscr{U}_{ij} u_j$ *and* $\mathscr{U}^{-1}_{ij} u_j$ *are expressed as follows:*

$$\begin{aligned}
\left(\mathscr{U}_{ij} u_j\right)(\xi) &= \int_{\mathbb{R}^3_+} \Lambda^*_{ji}(\xi, x) u_j(x) dx, \quad (\xi \in \sigma_p \cup \sigma_c) \\
\left(\mathscr{U}^{-1}_{ij} \hat{u}_j\right)(x) &= \int_{\mathbb{R}^2} \sum_{\xi\in\sigma_p} \Lambda_{ij}(x,\xi)\hat{u}_j(\xi) d\xi_1 d\xi_2 \\
&\quad + \int_{\mathbb{R}^2} \int_{\xi_r}^\infty \Lambda_{ij}(x,\xi)\hat{u}_j(\xi) d\xi_3 d\xi_1 d\xi_2, \quad (x \in \mathbb{R}^3_+)
\end{aligned} \tag{165}$$

where

$$\Lambda_{ij}(\xi, x) = \frac{1}{2\pi} \exp(i(\xi_1 x_1 + \xi_2 x_2))\psi_{lj}(x_3,\xi) T_{li} \tag{166}$$

Here, $\mathscr{U}_{ij} u_j$ is referred to as the generalized Fourier transform of u_j, and $\mathscr{U}^{-1}_{ij} \hat{u}_j$ is referred to as the generalized inverse Fourier transform of $\hat{u}_j$. Based on the literature (Reed and Simon, 1975), the domain of the operators $\mathscr{U}_{ij}$ and $\mathscr{U}^{-1}_{ij}$ could be extended from L_2 to the space of tempered distributions $\mathscr{S}'$.

4.4 Method for the volume integral equation

We have obtained the transform for elastic waves in a 3-D half space, which is to be applied to the volume integral equation. Preliminary to showing the application of the transform to the volume integral equation, we have to construct the Green's function for the elastic half space based on the proposed transform. The definition of the Green's function for the half space is expressed as

$$\begin{aligned}\left(L_{ij}(\partial_1,\partial_2,\partial_3)+\delta_{ij}\rho\omega^2\right)G_{jk}(x,y) &= -\delta_{ik}\delta(x-y)\\ P_{ij}^{(0)}G_{jk}(x,y) &= 0 \text{ at } x_3=0, \quad (x,y\in\mathbb{R}_+^3)\end{aligned} \tag{167}$$

The application of the generalized Fourier transform to Eq. (167) yields

$$(\mu_0\xi_3^2-\rho\omega^2)\hat{G}_{kj}(\xi,y)=\Lambda_{jk}^*(\xi,y) \tag{168}$$

where $\hat{G}_{ij}$ is the generalized Fourier transform of the Green's function. Therefore, as a result of Eq. (168), the Green's function for a half space can be represented as

$$\begin{aligned}G_{ij}(x,y) &= \int_{\mathbb{R}^2}d\xi_1 d\xi_2\sum_{\xi\in\sigma_p}\frac{\Lambda_{ik}(\xi,x)\Lambda_{jk}^*(\xi,y)}{\mu_0\xi_3^2-\rho\omega^2-i\epsilon}\\ &\quad+\int_{\mathbb{R}^2}d\xi_1 d\xi_2\int_{\xi_r}^{\infty}\frac{\Lambda_{ik}(\xi,x)\Lambda_{jk}^*(\xi,y)}{\mu_0\xi_3^2-\rho\omega^2-i\epsilon}d\xi_3\end{aligned} \tag{169}$$

Next, let the function $w_i(x)$ be given in the following form:

$$w_i(x)=\int_{\mathbb{R}_+^3}G_{ij}(x,y)f_j(y)dy \tag{170}$$

The formal calculation reveals that

$$\begin{aligned}\left(\mathscr{U}_{ki}w_i\right)(\xi) &= \int_{\mathbb{R}_+^3}\Lambda_{ik}^*(\xi,x)\left[\int_{\mathbb{R}_+^3}G_{ij}(x,y)f_j(y)dy\right]dx\\ &= \int_{\mathbb{R}_+^3}\left[\int_{\mathbb{R}_+^3}\Lambda_{ik}^*(\xi,x)G_{ij}(x,y)dx\right]f_j(y)dy\\ &= \int_{\mathbb{R}_+^3}\hat{h}(\xi)\Lambda_{jk}^*(\xi,y)f_j(y)dy\\ &= \hat{h}(\xi)\left(\mathscr{U}_{kj}f_j\right)(\xi)\end{aligned} \tag{171}$$

where $\hat{h}$ denotes

$$\hat{h}(\xi)=\frac{\mathbf{1}(\xi_1)\,\mathbf{1}(\xi_2)}{\mu_0\xi_3^2-\rho\omega^2-i\epsilon} \tag{172}$$

Note that $\mathbf{1}(\cdot)$ in Eq. (172) is defined such that

$$\mathbf{1}(\xi_1)=1, \quad (\forall\xi_1\in\mathbb{R}) \tag{173}$$

At this point, the application of the generalized Fourier transform to the volume integral equation becomes possible and is achieved as follows:

$$\begin{aligned} \hat{v}_i(\xi) &= -\hat{h}(\xi)\mathscr{U}_{ij}N_{jk}\mathscr{U}_{kl}^{-1}\hat{F}_l(\xi) \\ &\quad -\hat{h}(\xi)\mathscr{U}_{ij}N_{jk}\mathscr{U}_{kl}^{-1}\hat{v}_l(\xi) \end{aligned} \tag{174}$$

where $\hat{v}_i$ is the generalized Fourier transform of v_i and $\hat{F}_l$ is the incident wave field due to the point source expressed by

$$\hat{F}_l(\xi) = \hat{h}(\xi)\Lambda^*_{ml}(\xi, x_s)q_m \tag{175}$$

The volume integral equation for the elastic wave equation in the wavenumber domain in a half space has the same structure as that in a full space. Therefore, almost the same numerical scheme based on the Krylov subspace iteration technique is available. Note that the difference in the numerical scheme between that for the elastic full space and that for the half space lies in the discretization of the wavenumber space. The discretization of the wavenumber space for elastic half space is as follows:

$$\begin{aligned} D_\xi &= \left\{(n_1\Delta\xi_1, n_2\Delta\xi_2, \eta_R)\,\middle|\, n_1 \in \mathbb{N}_1, n_2 \in \mathbb{N}_2, F(\xi_r, \eta_R) = 0\right\} \oplus \\ &\quad \left\{(n_1\Delta\xi_1, n_2\Delta\xi_2, \xi_r + n_3\Delta\xi_3)\,\middle|\, n_1 \in \mathbb{N}_1, n_2 \in \mathbb{N}_2, n_3 \in \mathbb{N}_3\right\} \end{aligned} \tag{176}$$

where $\Delta\xi_j$, (j = 1, 2, 3) are the intervals of the grids in the wavenumber space,

$$\xi_r = \sqrt{n_1^2\Delta\xi_1^2 + n_2^2\Delta\xi_2^2} \tag{177}$$

and $\mathbb{N}_1$, $\mathbb{N}_2$, and $\mathbb{N}_3$ compose the set of integers defined by

$$\begin{aligned} \mathbb{N}_1 &= \{n \mid -N_1/2 \le n < N_1/2\} \\ \mathbb{N}_2 &= \{n \mid -N_2/2 \le n < N_2/2\} \\ \mathbb{N}_3 &= \{n \mid 0 \le n < N_3 - 1\} \end{aligned} \tag{178}$$

where (N_1, N_2, N_3) defines the number of grids in the wavenumber space. Note that Eq. (176) corresponds to the decomposition of the Rayleigh and body waves.

4.5 Numerical example

For the numerical analysis of an elastic half space, the Lam´e constants of the background structure is set such that λ_0 = 4 GPa, μ_0 = 2 GPa and the mass density is set at ρ =2 g/cm^3. Therefore, the background velocity of the P and S waves are 2 km/s and 1 km/s, respectively. and that for the Rayleigh wave velocity is 0.93 km/s. In addition, the analyzed frequency is f = 1 Hz.

First, let us investigate the accuracy of the generalized Fourier transform by composing the Green's function. For the calculation of the generalized Fourier transform, $N_1 = N_2 = N_3 = 256$, $\Delta x_1 = \Delta x_2 = 0.25$ km, and Δx_3=0.125 km are chosen to define D_x and D_ξ. The parameter ϵ for the Green's function is set at 0.6.

Figures 6(a) and 6(b) show the Green's function calculated by the generalized Fourier transform and the Hankel transform. The distributions of the absolute displacements are shown is these figures. For the calculation of the Green's function, the point source is set at a depth of 1 km from the free surface. The direction of the excitation force is vertical, and the excitation force has an amplitude of 1.0 × 10^7 kN. Comparison of these figures reveals good agreement between the calculated results, which confirms the accuracy of the generalized Fourier transform.

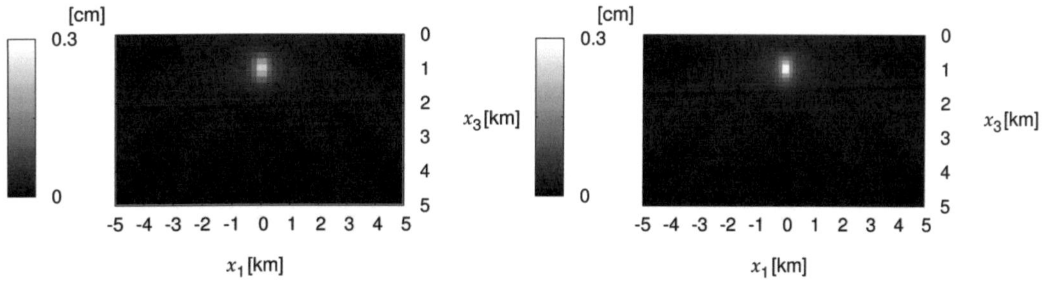

(a) Generalized Fourier transform (b) Hankel transform

Fig. 6. Comparison of the Green's function calculated by the generalized Fourier transform and the Hankel transform.

The following example shows the solution of the volume integral equation. The fluctuation of the elastic wave field is set as follows:

$$\tilde{\lambda}(x) = A_\lambda \exp(-\zeta_\lambda |x - x_c|^2) \tag{179}$$

$$\tilde{\mu}(x) = A_\mu \exp(-\zeta_\mu |x - x_c|^2), \tag{180}$$

where A_λ and A_μ describe the amplitude of the fluctuation, ζ_λ and ζ_μ describe the spread of the fluctuation in the space, and x_c is the center of the fluctuation. These parameters are set at $A_\lambda = A_\mu = 0.6$ GPa, $\zeta_\lambda = \zeta_\mu = 0.3$ km^{-2} and

$$x_c = (0, 0, 1) \text{ km} \tag{181}$$

The fluctuation of the medium in the $x_2 - x_3$ plane at $x_1 = 0$ [km] is shown in Fig. 7.

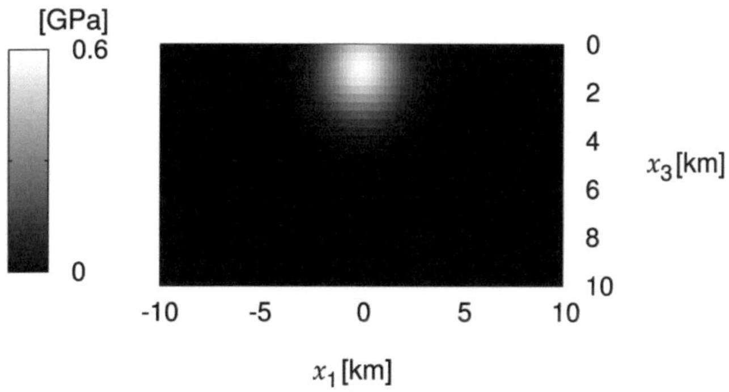

Fig. 7. Fluctuation of the medium

In order to generate the scattered wave field, the location of the point source is set at x_s = (5, 0, 0) km. The direction of the excitation force is vertical, and the excitation force has an amplitude of 1×10^7 kN. Bi-CGSTAB method (Barrett et al., 1994) is used for the Krylov subspace iteration technique. Figures 8 and 9 show the propagation of scattered waves at the free surface and the amplitudes of the scattered waves in the $x_1 - x_3$ plane, respectively. According to Fig. 8, the amplitudes of the scattered waves are larger in the forward region of the fluctuating area, where $x_1 < 0$. Figure 9 shows that the propagation of the Rayleigh waves as the scattered wvaes in the forwrad region. The amplitude of the scattered waves are smaller in the fluctuating area. The scattered waves are found to be reflected at the fluctuating area, thereby generating Rayleigh waves. The above numerical results explain well the propagation of the scattered elastic waves in the half space.

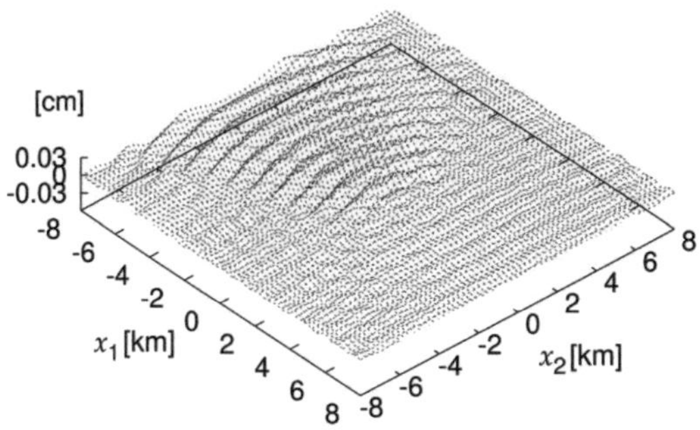

Fig. 8. Scattering of waves at the free surface.

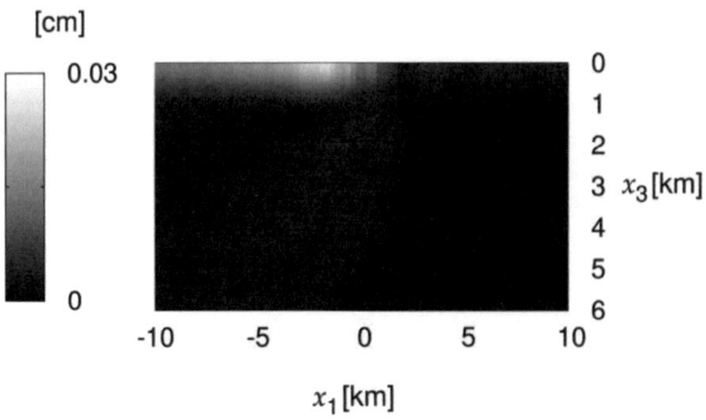

Fig. 9. Distribution of scattered waves in the vertical plane.

The numerical calculations were carried out using a computer with an AMD Opteron 2.4-GHz processor. The CPU time needed for iteration in Bi-CGSTAB was five hours, which is due primarily to the calculation of the generalized Fourier transforms. Note that the 2-D FFT for the horizontal coordinate system was used for the generalized Fourier transform. The transform for the vertical coordinate required a large CPU time. The reduction of this large CPU time requirement should be investigated in the future. The development of a fast algorithm for the generalized Fourier transforms may be required. It is also important to formulate the inverse scattering analysis method and to carry out the analysis.

5. Conclusions

In this chapter, a volume integral equation method was developed for elastic wave propagation for 3-D elastic full and half spaces. The developed method did not require the derivation of a coefficient matrix. Instead, the Fourier transform and the Krylov subspace iterative technique were used for the integral equation. The starting point of the formulation was the volume integral equation in the wavenumber domain. The Fourier transform and the inverse Fourier transform were repeatedly applied during the Krylov iterative process. Based on this procedure, a fast method was realized for both forward and inverse scattering analysis in a 3-D elastic full space via the fast Fourier transform and Bi-CGSTAB method. For example, if the number of iterations was two, the CPU time to obtain accurate solutions was only two minutes. Furthermore, for the inverse scattering problem, the reconstruction of inhomogeneities of the wave field was also successful, even for the multiple scattering problem.

The ordinary Fourier transform is not valid for an elastic half space due to the boundary conditions at the free surface. The generalized Fourier transform and the inverse Fourier transform for elastic wave propagations in a half space were developed for the integral equation based on the spectral theory. The generalized Fourier transform composing the Green's function was also verified numerically. The properties of the scattered wave field in a half space were found to be well explained by the proposed method. At present, the proposed method for an elastic half space requires a large amount of CPU time, which was five hours for the present numerical model. As such, a fast algorithm for the generalized Fourier transforms should be developed in the future.

6. References

Aki, K. and Richards, P.G. (1980). *Quantitative Seismology, Theory and Methods.*, W. H. Freeman and Company.

Barrett, R., Berry, M., Chan, T.F., Demmel, J., Donato, J., Dongarra, J., Eijkhout, V., Pozo, R., Romine, C. and Van der Vorst, H. (1994). *Templates for the solution of Linear Systems: Building Blocks for Iterative Methods*, SIAM.

Bertheir, A.M. (1982). *Spectral theory and wave operators for the Scrödinger equation*, Research Notes in Mathematics, London, Pitman Advanced Publishing Program.

Colton, D. and Kress, R. (1983). *Integral equation methods in scattering theory*, New York, John Wiley and Sons, Inc.

Colton, D. and Kress, R. (1998). *Inverse acoustic and electromagnetic scattering theory*, Berlin, Heidelberg, Springer-Verlag.

De Zaeytijd, J., Bogaert, I. and Franchois, A. (2008). An efficient hybrid MLFMA-FFT solver for the volume integral equation in case of sparse 3D inhomogeneous dielectric scatterers, *Journal of Computational Physics*, 227, 7052-7068.

Fata, S.N. and Guzina, B.B. (2004). A linear sampling method for near-field inverse problems in elastodynamics, *Inverse Problems*, 20, 713-736.

Friedlander, F. G. and Joshi, M. (1998). *Introduction to the theory of distributions*, Cambridge University Press.

Guzina, B.B. and Chikichev, I. (2007). From imaging to material identification: A generalized concept of topological sensitivity, *Journal of Mechanics and Physics of Solids*, 55, 245-279.

Guzina, B. B., Fata, S.N. and Bonnet, M. (2003), On the stress-wave imaging of cavities in a semi-infinite solid, *International Journal of Solids and Structures*, 40, 1505-1523.

Hudson, J. A. and Heritage, J. R. (1981). The use of the Born approximation in seismic scattering problems, *Geophys. J.R. Astr. Soc.*, 66, 221-240.

Hörmander, L. (1983), *The analysis of linear partial differential operators* I, Springer-Verlag, Berlin, Heidelberg, 1983.

Ikebe, T. (1960). Eigenfunction expansion associated with the Schroedinger operators and their applications to scattering theory, *Archive for Rational Mechanics and Analysis*, 5, 1-34.

Kato, T. (1980). *Perturbation Theory for Linear Operators*, Berlin, Heidelberg, Springer-Verlag. Kleinman, R.E. and van den Berg, P.M. (1992). A modified gradient method for two-dimensional problems in tomography, *Journal of Computational and Applied Mathematics*, 42, 17-35.

Reed, M. and Simon, B. (1975). *Method of Modern Mathematical Physics*, Vol. II, Fourier Analysis and Self-adjointness, San Diego, Academic Press.

Touhei, T. (2002). Complete eigenfunction expansion form of the Green's function for elastic layered half space, *Archive of Applied mechanics*, 72, 13-38.

Touhei, T. (2009). Generalized Fourier transform and its application to the volume integral equation for elastic wave propagation in a half space, *International Journal of Solids and Structures*, 46, 52-73.

Touhei,T., Kiuchi, T and Iwasaki (2009). A Fast Volume Integral Equation Method for the Direct/Inverse Problem in Elastic Wave Scattering Phenomena, *International Journal of Solids and Structures*, 46, 3860-3872.

Wilcox, C.H. (1976). Spectral analysis of the Pekeris operator, *Arch. Rat. Mech. Anal.*, 60, pp. 259-300.

Yang, J., Abubaker, A., van den Berg, P.M., Habashy, T.M. and Reitich, F. (2008). A CG-FFT approach to the solution of a stress-velocity formulation of three-dimensional scattering problems, *Journal of Computational physics*, 227, 10018-10039.

2

Uniform Asymptotic Physical Optics Solutions for a Set of Diffraction Problems

Giovanni Riccio
University of Salerno
Italy

1. Introduction

Scattering occurs when an object, inserted in the path of a propagating electromagnetic (EM) wave, modifies the field distribution in the surrounding space. The ability to describe and solve scattering problems is the key to very important applications such as planning radio links, designing antennas and identifying radar targets. In such frameworks one could resort purely numerical methods, f.i., Finite Element Method (FEM) or Method of Moments (MoM), to generate a solution to the considered problem. Unfortunately such techniques have an inherent drawback: they are based on a discrete representation of the system, typically using cell sizes on the order of one-tenth the wavelength of the incident field. Because the number of degrees of freedom increases with frequency, the computation becomes computer-intensive (if not unmanageable) at high frequencies, where asymptotic methods work more efficient. One of these methods has received considerable attention during the last decades because of its peculiarities: the Geometrical Theory of Diffraction (GTD) originated by J.B. Keller (Keller, 1962). What is the essence of this theory? It incorporates the diffraction phenomenon into the ray-oriented framework of the Geometrical Optics (GO), which is based on the assumption that waves can be represented by rays denoting the direction of travel of the EM energy, and that wave fields are characterised mathematically by amplitude, phase propagation factor and polarisation. Diffraction, like reflection and refraction GO mechanisms, is a local phenomenon at high frequencies and is determined by a generalisation of the Fermat principle. It depends on the surface geometry of the obstacle and on the incident field in proximity to the diffraction points creating discontinuities in the GO field at incidence and reflection shadow boundaries.

When using GTD to solve a scattering problem, the first step is to resolve it to smaller and simpler components, each representing a canonical geometry, so that the total solution is a superposition of the contributions from each canonical problem. Accordingly, GTD allows one to solve a large number of real scattering problems by using the solutions of a relatively small number of model problems. In addition, it is simple to apply, it provides physical insight into the radiation and scattering mechanisms from the various parts of the structure, it gives accurate results that compare quite well with experiments and other methods, and it can be combined with numerically rigorous techniques to obtain hybrid methods. The GTD fails, however, in the boundary layers near caustics and GO shadow boundaries. Its uniform version originated by R.C. Kouyoumjian and P.H. Pathak (Kouyoumjian & Pathak, 1974), i.e., the Uniform Theory of Diffraction (UTD), overcomes such limitations and gives useful uniform asymptotic solutions for the calculation of the diffracted field. Because of its characteristics, UTD is usually preferred by researchers and practising engineers for treating real structures with edges, and its formulation for perfectly conducting canonical geometries represents the starting point for heuristic solutions based on (Burnside & Burgener, 1983) and (Luebbers, 1984).

This Chapter is devoted to the Uniform Asymptotic Physical Optics (UAPO) solutions for a set of diffraction problems involving edges in penetrable or opaque planar thin layers (compared to wavelength) illuminated by incident plane waves. Under the assumption that the excited surface waves can be neglected, the analytical difficulties are here attenuated by modelling the truncated layer as a canonical half-plane and by taking into account its geometric, electric and magnetic characteristics into the GO response.

The starting point for obtaining a UAPO solution is that of considering the classical radiation integral with a PO approximation of the electric and magnetic surface currents. By PO current one usually means that representation of the surface current in terms of the incident field, and this dependence is here obtained by assuming the surface currents in the integrand as equivalent sources originated by the discontinuities of the tangential GO field components across the layer. As a matter of fact, by taking into account that the field approaching the layer from the upper side is given by the sum of the incident and reflected fields, whereas the transmitted field furnishes the field approaching the layer from the lower side, the here involved PO surface currents depend on the incident field, the reflection and transmission coefficients. Obviously, this last dependence must be considered only if the transmitted field exists. A useful approximation and a uniform asymptotic evaluation of the resulting radiation integral allow one to obtain the diffracted field, which results to be expressed in terms of the GO response of the structure and the standard transition function of UTD. Accordingly, the UAPO solutions have the same effectiveness and ease of handling of those derived in the UTD framework and, in addition, they have the inherent advantage of providing the diffracted field from the knowledge of the GO field. In other words, it is sufficient to make explicit the reflection and transmission coefficients related to the considered structure for obtaining the UAPO diffraction coefficients. Note that also the heuristic solutions have this advantage, but they do not possess a rigorous analytical justification and therefore should be used with considerable attention.

It must be stressed that the UAPO solutions allow one to compensate the discontinuities in the GO field at the incidence and reflection shadow boundaries, and that their accuracy has been proved by comparisons with purely numerical techniques.

The Chapter is organised as follows. Section 2 is devoted to explain the methodology for obtaining the UAPO expression of the field diffracted by the edge of a penetrable or opaque half-plane when illuminated by an incident plane wave. The solution is given in terms of the UTD transition function and requires the knowledge of the reflection and transmission coefficients, whose expressions are explicitly reported in Section 3 with reference to some test-bed cases. To show the effectiveness of the corresponding UAPO solutions, Section 3 also includes numerical results and comparisons with COMSOL MULTIPHYSICS® simulations. Diffraction by junctions of layers is considered in Section 4. Concluding remarks and future activities are collected in Section 5.

2. UAPO diffracted field by half-planes

The diffraction phenomenon related to a linearly polarised plane wave impinging on a penetrable half-plane surrounded by free-space is considered in the frequency domain. As shown in Fig. 1, the z-axis of a reference coordinate system is directed along the edge and the x-axis is on the illuminated face. The angles (β',ϕ') fix the incidence direction: the first is a measure of the skewness with respect to the edge ($\beta'=90°$ corresponds to the normal incidence) and the latter gives the aperture of the edge-fixed plane of incidence with respect to the illuminated face ($\phi'=0$ corresponds to the grazing incidence). The case $0<\phi'<\pi$ is considered from this point on. The observation direction is specified by (β,ϕ).

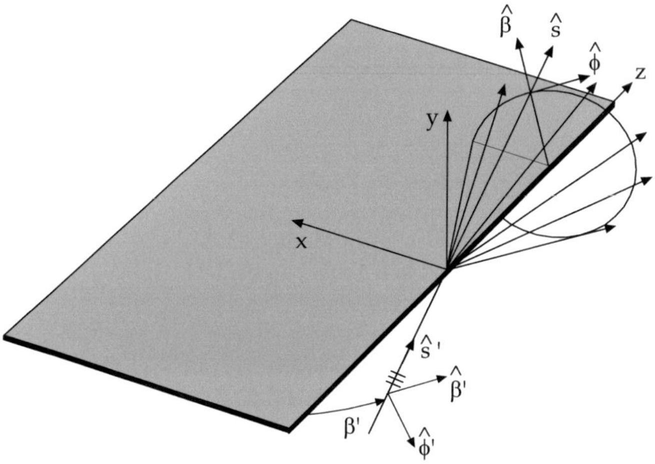

Fig. 1. Diffraction by the half-plane edge.

As well-known, the total electric field at a given observation point P can be determined by the adding the incident field $\underline{E}^i$ and the scattered field $\underline{E}^s$. This last can be represented by the classical radiation integral in the PO approximation:

$$\underline{E}^s = -jk_0 \iint_S \left[(\underline{\underline{I}} - \hat{R}\hat{R})\zeta_0 \underline{J}_s^{PO} + \underline{J}_{ms}^{PO} \times \hat{R} \right] G(\underline{r},\underline{r}')\, dS \tag{1}$$

where ζ_0 and k_0 are the impedance and the propagation constant of the free-space, $\underline{r} = x\hat{x} + y\hat{y} + z\hat{z} = \underline{\rho} + z\hat{z}$ and $\underline{r}' = x'\hat{x} + z'\hat{z} = \underline{\rho}' + z'\hat{z}$ denote the observation and source points, $\hat{R}$ is the unit vector from the radiating element at $\underline{r}'$ to P, and $\underline{\underline{I}}$ is the 3x3 identity matrix. The electric and magnetic PO surface currents $\underline{J}_s^{PO}$ and $\underline{J}_{ms}^{PO}$ induced on S are given in terms of the incident field and can be so expressed:

$$\zeta_0 \underline{J}_s^{PO} = \zeta_0 \underline{J}_s^{*}\, e^{jk_0(x'\sin\beta'\cos\phi' - z'\cos\beta')} \tag{2}$$

$$\underline{J}_{ms}^{PO} = \underline{J}_{ms}^{*}\, e^{jk_0(x'\sin\beta'\cos\phi' - z'\cos\beta')} \tag{3}$$

Moreover,

$$G(\underline{r},\underline{r}')=\frac{e^{-jk_0|\underline{r}-\underline{r}'|}}{4\pi|\underline{r}-\underline{r}'|}=\frac{e^{-jk_0\sqrt{(x-x')^2+y^2+(z-z')^2}}}{4\pi\sqrt{(x-x')^2+y^2+(z-z')^2}}=\frac{e^{-jk_0\sqrt{|\underline{\rho}-\underline{\rho}'|^2+(z-z')^2}}}{4\pi\sqrt{|\underline{\rho}-\underline{\rho}'|^2+(z-z')^2}} \tag{4}$$

is the three-dimensional Green function.
To evaluate the edge diffraction confined to the Keller cone for which $\beta=\beta'$, it is possible to approximate $\hat{R}$ by the unit vector $\hat{s}$ in the diffraction direction, i.e., $\hat{R}=\hat{s}=\sin\beta'\cos\phi\,\hat{x}+\sin\beta'\sin\phi\,\hat{y}+\cos\beta'\,\hat{z}$, Accordingly, it results:

$$\begin{aligned}\underline{E}^s &= -\frac{jk_0}{4\pi}\left[(\underline{\underline{I}}-\hat{s}\hat{s})\zeta_0\underline{J}_s^* + \underline{J}_{ms}^* \times \hat{s}\right]\int_0^\infty\int_{-\infty}^\infty e^{jk_0(x'\sin\beta'\cos\phi'-z'\cos\beta')}\,\frac{e^{-jk_0\sqrt{|\underline{\rho}-\underline{\rho}'|^2+(z-z')^2}}}{\sqrt{|\underline{\rho}-\underline{\rho}'|^2+(z-z')^2}}\,dz'\,dx' \\ &= \left[(\underline{\underline{I}}-\hat{s}\hat{s})\zeta_0\underline{J}_s^* + \underline{J}_{ms}^* \times \hat{s}\right] I_s\end{aligned} \tag{5}$$

2.1 Electric and magnetic PO surface currents

The expressions of the PO surface currents in terms of the incident electric field are here obtained by assuming such currents as equivalent sources originated by the discontinuities of the tangential GO field components across the layer, i.e.,

$$\underline{J}_s^{PO}=\hat{y}\times\left(\underline{H}^+-\underline{H}^-\right)\Big|_S=\hat{y}\times\left(\underline{H}^i+\underline{H}^r-\underline{H}^t\right)\Big|_S=\hat{y}\times\left(\underline{H}_0^i+\underline{H}_0^r-\underline{H}_0^t\right)e^{jk_0(x'\sin\beta'\cos\phi'-z'\cos\beta')} \tag{6}$$

$$\underline{J}_{ms}^{PO}=\left(\underline{E}^+-\underline{E}^-\right)\Big|_S\times\hat{y}=\left(\underline{E}^i+\underline{E}^r-\underline{E}^t\right)\Big|_S\times\hat{y}=\left[\left(\underline{E}_0^i+\underline{E}_0^r-\underline{E}_0^t\right)\times\hat{y}\right]e^{jk_0(x'\sin\beta'\cos\phi'-z'\cos\beta')} \tag{7}$$

where the superscripts i, r and t denote the incident, reflected and transmitted fields, respectively. As well-known, it is convenient to work in the standard plane of incidence and to consider the GO field components parallel ($\|$) and perpendicular ($\perp$) to it. If the ray-fixed coordinate systems sketched in Fig. 2 are used, the EM field can be so expressed:

$$\underline{E}_0^{i,r,t}=E_{\|}^{i,r,t}\,\hat{e}_{\|}^{i,r,t}+E_{\perp}^{i,r,t}\,\hat{e}_{\perp} \tag{8}$$

$$\underline{H}_0^{i,r,t}=\frac{1}{\zeta_0}\left[E_{\|}^{i,r,t}\,\hat{e}_{\perp}-E_{\perp}^{i,r,t}\,\hat{e}_{\|}^{i,r,t}\right] \tag{9}$$

with $\hat{e}_{\|}^t=\hat{e}_{\|}^i$. Accordingly, if $\hat{t}=\hat{n}\times\hat{e}_{\perp}=\hat{y}\times\hat{e}_{\perp}$ and θ^i is the standard incidence angle, it results:

$$\zeta_0\underline{J}_s^*=\left(E_{\perp}^i-E_{\perp}^r-E_{\perp}^t\right)\cos\theta^i\,\hat{e}_{\perp}+\left(E_{\|}^i+E_{\|}^r-E_{\|}^t\right)\hat{t} \tag{10}$$

$$\underline{J}_{ms}^*=\left(E_{\|}^i-E_{\|}^r-E_{\|}^t\right)\cos\theta^i\,\hat{e}_{\perp}-\left(E_{\perp}^i+E_{\perp}^r-E_{\perp}^t\right)\hat{t} \tag{11}$$

where the reflected and transmitted field components can be given in terms of the incident field components by means of the reflection matrix $\underline{\underline{R}}$ and the transmission matrix $\underline{\underline{T}}$:

$$\begin{pmatrix} E_{\|}^{r} \\ E_{\perp}^{r} \end{pmatrix} = \underline{\underline{R}} \begin{pmatrix} E_{\|}^{i} \\ E_{\perp}^{i} \end{pmatrix} = \begin{pmatrix} R_{11} & R_{12} \\ R_{21} & R_{22} \end{pmatrix} \begin{pmatrix} E_{\|}^{i} \\ E_{\perp}^{i} \end{pmatrix} \tag{12}$$

$$\begin{pmatrix} E_{\|}^{t} \\ E_{\perp}^{t} \end{pmatrix} = \underline{\underline{T}} \begin{pmatrix} E_{\|}^{i} \\ E_{\perp}^{i} \end{pmatrix} = \begin{pmatrix} T_{11} & T_{12} \\ T_{21} & T_{22} \end{pmatrix} \begin{pmatrix} E_{\|}^{i} \\ E_{\perp}^{i} \end{pmatrix} \tag{13}$$

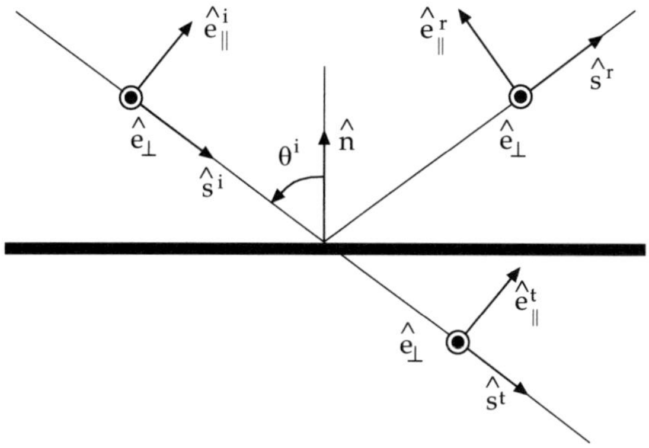

Fig. 2. Ray-fixed coordinate systems for the GO field.

By using the above results, the scattered field can be given in the following matrix formulation:

$$\underline{E}^{s} = \begin{pmatrix} E_{\beta}^{s} \\ E_{\phi}^{s} \end{pmatrix} = \underline{\underline{M}}_0 \begin{pmatrix} E_{\|}^{i} \\ E_{\perp}^{i} \end{pmatrix} I_s = \underline{\underline{M}}_0 \, \underline{\underline{M}}_1 \begin{pmatrix} E_{\beta'}^{i} \\ E_{\phi'}^{i} \end{pmatrix} I_s = \underline{\underline{M}} \begin{pmatrix} E_{\beta'}^{i} \\ E_{\phi'}^{i} \end{pmatrix} I_s \tag{14}$$

with

$$\underline{\underline{M}}_1 = \begin{pmatrix} \hat{e}_{\|}^{i} \cdot \hat{\beta}' & \hat{e}_{\|}^{i} \cdot \hat{\phi}' \\ \hat{e}_{\perp} \cdot \hat{\beta}' & \hat{e}_{\perp} \cdot \hat{\phi}' \end{pmatrix} = \frac{1}{\sqrt{1 - \sin^2\beta' \sin^2\phi'}} \begin{pmatrix} \cos\beta' \sin\phi' & \cos\phi' \\ -\cos\phi' & \cos\beta' \sin\phi' \end{pmatrix} \tag{15}$$

2.2 Matrix elements

The matrix $\underline{\underline{M}}_0$ in (14) allows one to relate the scattered field components along $\hat{\beta}$ and $\hat{\phi}$ to the incident field components along $\hat{e}_{\|}^{i}$ and $\hat{e}_{\perp}$. It accounts for the expressions of the PO surface currents and can be so formulated:

$$\underline{\underline{M}}_0 = \underline{\underline{M}}_7 \left[\underline{\underline{M}}_2 \underline{\underline{M}}_4 \underline{\underline{M}}_5 + \underline{\underline{M}}_3 \underline{\underline{M}}_4 \underline{\underline{M}}_6 \right] \tag{16}$$

where

$$\underline{\underline{M}}_7 = \begin{pmatrix} \hat{\beta} \cdot \hat{x} & \hat{\beta} \cdot \hat{y} & \hat{\beta} \cdot \hat{z} \\ \hat{\phi} \cdot \hat{x} & \hat{\phi} \cdot \hat{y} & \hat{\phi} \cdot \hat{z} \end{pmatrix} = \begin{pmatrix} \cos\beta' \cos\phi & \cos\beta' \sin\phi & -\sin\beta' \\ -\sin\phi & \cos\phi & 0 \end{pmatrix} \tag{17}$$

$$\underline{\underline{M}}_2 = \begin{pmatrix} 1-\sin^2\beta'\cos^2\phi & -\sin\beta'\cos\beta'\cos\phi \\ -\sin^2\beta'\sin\phi\cos\phi & -\sin\beta'\cos\beta'\sin\phi \\ -\sin\beta'\cos\beta'\cos\phi & \sin^2\beta' \end{pmatrix} \tag{18}$$

$$\underline{\underline{M}}_3 = \begin{pmatrix} 0 & -\sin\beta'\sin\phi \\ -\cos\beta' & \sin\beta'\cos\phi \\ \sin\beta'\sin\phi & 0 \end{pmatrix} \tag{19}$$

$$\underline{\underline{M}}_4 = \begin{pmatrix} \hat{x}\cdot\hat{e}_\perp & \hat{x}\cdot\hat{t} \\ \hat{z}\cdot\hat{e}_\perp & \hat{z}\cdot\hat{t} \end{pmatrix} = \frac{1}{\sqrt{1-\sin^2\beta'\sin^2\phi'}} \begin{pmatrix} -\cos\beta' & -\sin\beta'\cos\phi' \\ -\sin\beta'\cos\phi' & \cos\beta' \end{pmatrix} \tag{20}$$

$$\underline{\underline{M}}_5 = \begin{pmatrix} -(R_{21}+T_{21})\cos\theta^i & (1-R_{22}-T_{22})\cos\theta^i \\ 1+R_{11}-T_{11} & R_{12}-T_{12} \end{pmatrix} \tag{21}$$

$$\underline{\underline{M}}_6 = \begin{pmatrix} (1-R_{11}-T_{11})\cos\theta^i & -(R_{12}+T_{12})\cos\theta^i \\ -(R_{21}+T_{21}) & -(1+R_{22}-T_{22}) \end{pmatrix} \tag{22}$$

2.3 Uniform asymptotic evaluation and diffracted field

It is now necessary to perform the evaluation of the following integral:

$$I_s = -\frac{jk_0}{4\pi} \int_0^\infty e^{jk_0 x'\sin\beta'\cos\phi'} \int_{-\infty}^{\infty} e^{-jk_0 z'\cos\beta'} \frac{e^{-jk_0\sqrt{|\underline{\rho}-\underline{\rho}'|^2+(z-z')^2}}}{\sqrt{|\underline{\rho}-\underline{\rho}'|^2+(z-z')^2}}\, dz'\, dx' \tag{23}$$

By making the substitution $z'-z = |\underline{\rho}-\underline{\rho}'|\sinh\zeta$ and using one of the integral representations of the zeroth order Hankel function of the second kind $H_0^{(2)}$, it results (see Appendix D in (Senior & Volakis, 1995) as reference):

$$\int_{-\infty}^{\infty} e^{-jk_0 z'\cos\beta'} \frac{e^{-jk_0\sqrt{|\underline{\rho}-\underline{\rho}'|^2+(z-z')^2}}}{\sqrt{|\underline{\rho}-\underline{\rho}'|^2+(z-z')^2}}\, dz' = -j\pi\, e^{-jk_0 z\cos\beta'} H_0^{(2)}\left(k_0|\underline{\rho}-\underline{\rho}'|\sin\beta'\right) \tag{24}$$

Another useful integral representation of the involved Hankel function can be now considered (Clemmow, 1996):

$$H_0^{(2)}\left(k_0|\underline{\rho}-\underline{\rho}'|\sin\beta'\right) = \frac{1}{\pi}\int_C e^{-jk_0|\underline{\rho}-\underline{\rho}'|\sin\beta'\cos(\theta\mp\alpha)}\, d\alpha \tag{25}$$

where C is the integration path in the complex α-plane as in Fig. 3. The angle θ is between the illuminated face and the vector $\underline{\rho}-\underline{\rho}'$, and the sign $-$ $(+)$ applies if $y>0$ $(y<0)$.

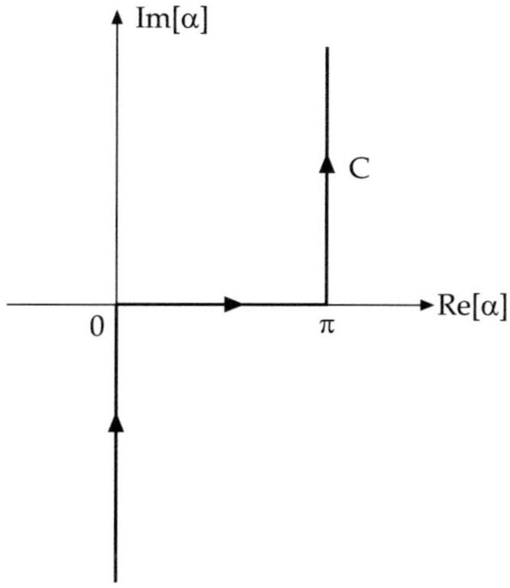

Fig. 3. Integration path C.

If $\rho = |\underline{\rho}|$, according to the geometry shown in Fig. 4, $|\underline{\rho} - \underline{\rho}'|\sin\theta = \rho\sin\phi$ and $|\underline{\rho} - \underline{\rho}'|\cos\theta = \rho\cos\phi - x'$, so obtaining

$$H_0^{(2)}\left(k_0|\underline{\rho} - \underline{\rho}'|\sin\beta'\right) = \frac{1}{\pi}\int_C e^{-jk_0\rho\sin\beta'\cos(\alpha\mp\phi)} e^{jk_0x'\sin\beta'\cos\alpha}\, d\alpha \tag{26}$$

and then

$$\begin{aligned} I_s &= -\frac{k_0}{4\pi} e^{-jk_0z\cos\beta'} \int_0^\infty \int_C e^{-jk_0\rho\sin\beta'\cos(\alpha\mp\phi)} e^{jk_0x'\sin\beta'(\cos\alpha+\cos\phi')}\, d\alpha\, dx' = \\ &= -\frac{k_0}{4\pi} e^{-jk_0z\cos\beta'} \int_C e^{-jk_0\rho\sin\beta'\cos(\alpha\mp\phi)} \int_0^\infty e^{jk_0x'\sin\beta'(\cos\alpha+\cos\phi')}\, dx'\, d\alpha \end{aligned} \tag{27}$$

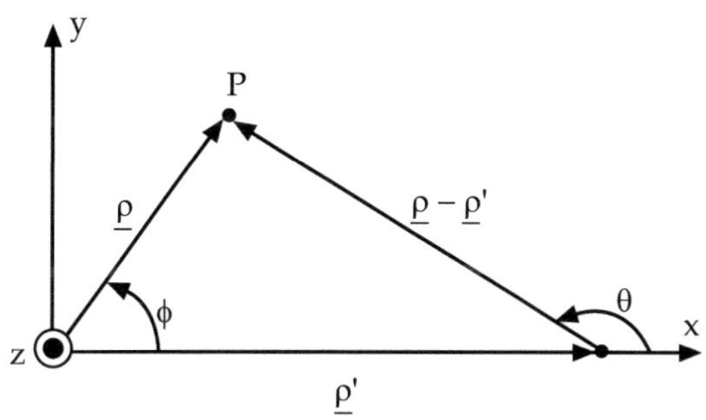

Fig. 4. Geometry in the plane perpendicular to the edge.

By applying the Sommerfeld-Maliuzhinets inversion formula (Maliuzhinets, 1958), it results:

$$\int_0^{\infty} e^{jk_0 x' \sin\beta' (\cos\alpha + \cos\phi')} dx' = \frac{-1}{jk_0 \sin\beta' (\cos\alpha + \cos\phi')} \tag{28}$$

so that

$$I_s = \frac{e^{-jk_0 z \cos\beta'}}{2\sin\beta'} \frac{1}{2\pi j} \int_C \frac{e^{-jk_0 \rho \sin\beta' \cos(\alpha \mp \phi)}}{\cos\alpha + \cos\phi'} d\alpha \tag{29}$$

where the sign $-$ $(+)$ applies in the range $0 < \phi < \pi$ ($\pi < \phi < 2\pi$). Such an integral can be evaluated by using the Steepest Descent Method (see Appendix C in (Senior & Volakis, 1995) as reference). To this end, the integration path C is closed with the Steepest Descent Path (SDP) passing through the pertinent saddle point α_s as shown in Fig. 5. According to the Cauchy residue theorem, the contribution related to the integration along C (distorted for the presence of singularities in the integrand) is equivalent to the sum of the integral along the SDP and the residue contributions $\mathrm{Res}_i(\alpha_p)$ associated with all those poles that are inside the closed path C+SDP, i.e.,

$$I_s(\Omega) = \frac{1}{2\pi j} \int_C g(\alpha) e^{\Omega f(\alpha)} d\alpha = \sum_i \mathrm{Res}_i(\alpha_p) - \frac{1}{2\pi j} \int_{SDP} g(\alpha) e^{\Omega f(\alpha)} d\alpha = \sum_i \mathrm{Res}_i(\alpha_p) + I(\Omega) \tag{30}$$

in which

$$I(\Omega) = -\frac{1}{2\pi j} \int_{SDP} g(\alpha) e^{\Omega f(\alpha)} d\alpha = -\frac{e^{\Omega f(\alpha_s)}}{2\pi j} \int_{SDP} g(\alpha) e^{\Omega [f(\alpha) - f(\alpha_s)]} d\alpha \tag{31}$$

$$\Omega = k_0 \rho \tag{32}$$

$$g(\alpha) = \frac{e^{-jk_0 z \cos\beta'}}{2\sin\beta'} \frac{1}{\cos\alpha + \cos\phi'} = \frac{A}{\cos\alpha + \cos\phi'} \tag{33}$$

$$f(\alpha) = -j \sin\beta' \cos(\alpha \mp \phi) \tag{34}$$

Note that Ω is typically large, $\alpha_p = \pi - \phi'$ and $\alpha_s = \phi$ ($\alpha_s = 2\pi - \phi$) if $0 < \phi < \pi$ ($\pi < \phi < 2\pi$). Moreover, by putting $\alpha = \alpha' + j\alpha''$ and imposing that $\mathrm{Im}[f(\alpha)] = \mathrm{Im}[f(\alpha_s)]$ and $\mathrm{Re}[f(\alpha)] \le \mathrm{Re}[f(\alpha_s)]$, the considered SDP is described by:

$$\alpha' = \alpha_s + \mathrm{sgn}(\alpha'') \cos^{-1}\left(\frac{1}{\cosh\alpha''}\right) = \alpha_s + gd(\alpha'') \tag{35}$$

where $gd(\alpha'')$ is the Gudermann function. By using now the change of variable $f(\alpha) - f(\alpha_s) = -\tau^2 < 0$, (31) can be written as

$$I(\Omega) = \pm \int_{-\infty}^{\infty} G(\tau)\, e^{-\Omega\tau^2} d\tau \tag{36}$$

wherein

$$G(\tau) = \frac{1}{2\pi j} g(\alpha(\tau))\, e^{\Omega f(\alpha_s)} \frac{d\alpha}{d\tau} \tag{37}$$

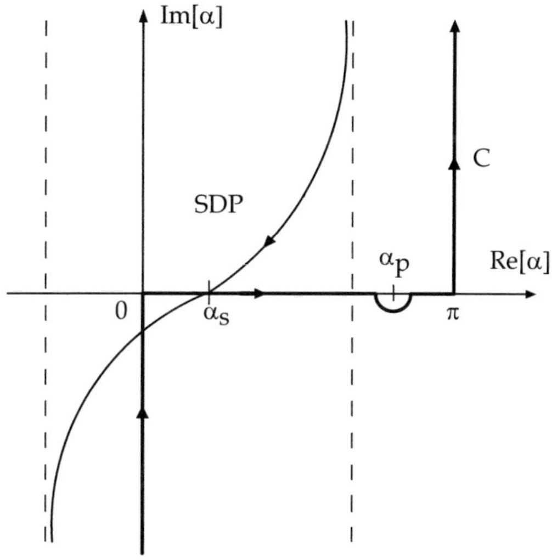

Fig. 5. Integration path.

When α_p is approaching α_s, the function $G(\tau)$ cannot be expanded in a Taylor series. To overcome this drawback it is convenient to regularise the integrand in (36). This procedure (to be referred to as the Multiplicative Method) was described in (Kouyoumjian & Pathak, 1974) as the modified Pauli-Clemmow method (Clemmow, 1950). It requires to introduce the regularised function

$$G_p(\tau) = (\tau - \tau_p) G(\tau) \tag{38}$$

with

$$\tau_p^2 = f(\alpha_s) - f(\alpha_p) = -j \sin\beta'[1 + \cos(\phi \pm \phi')] = -j2 \sin\beta' \cos^2\left(\frac{\phi \pm \phi'}{2}\right) = -j\,\delta \tag{39}$$

and δ is a measure of the distance between α_p and α_s. Accordingly,

$$I(\Omega) = \pm \int_{-\infty}^{\infty} G_p(\tau) \frac{e^{-\Omega\tau^2}}{\tau - \tau_p} d\tau \tag{40}$$

Since $G_p(\tau)$ is analytic near $\tau = 0$, it can be expanded in a Taylor series. By retaining only the first term (i.e., the $\Omega^{-1/2}$- order term) since $\Omega \gg 1$, it results:

$$I(\Omega) \simeq \pm\left[\sqrt{\frac{\pi}{\Omega}}\frac{G_p(0)}{(-\tau_p)}F_t\left(j\Omega\tau_p^2\right)\right] \tag{41}$$

in which

$$\frac{G_p(0)}{(-\tau_p)} = G(0) = \frac{1}{2\pi j}g(\alpha_s)\,e^{\Omega f(\alpha_s)}\left.\frac{d\alpha}{d\tau}\right|_{\tau=0} = \frac{1}{2\pi j}\frac{A\,e^{-j\Omega\sin\beta'}}{\cos\phi+\cos\phi'}\left[\pm\sqrt{\frac{2}{\sin\beta'}}\,e^{j\pi/4}\right] \tag{42}$$

and

$$F_t(\eta) = 2j\sqrt{\eta}\,e^{j\eta}\int_{\sqrt{\eta}}^{\infty} e^{-j\xi^2}d\xi \tag{43}$$

is the UTD transition function (Kouyoumjian & Pathak, 1974). By substituting (32), (39) and (42) in (41), the explicit form of the asymptotic evaluation of $I(\Omega)$ is:

$$\begin{aligned} I(\Omega) &\simeq \frac{e^{-j\pi/4}}{2\sqrt{2\pi k_0\rho\sin\beta'}}\frac{e^{-jk_0(\rho\sin\beta'+z\cos\beta')}}{\sin\beta'(\cos\phi+\cos\phi')}F_t\left(2k_0\rho\sin\beta'\cos^2\left(\frac{\phi\pm\phi'}{2}\right)\right) = \\ &= \frac{e^{-j\pi/4}}{2\sqrt{2\pi k_0}}\frac{e^{-jk_0 s}}{\sqrt{s}}\frac{1}{\sin^2\beta'(\cos\phi+\cos\phi')}F_t\left(2k_0 s\sin^2\beta'\cos^2\left(\frac{\phi\pm\phi'}{2}\right)\right) \end{aligned} \tag{44}$$

where the identities $\rho = s\sin\beta'$ and $z = s\cos\beta'$ are used on the diffraction cone. The above analytic result contributes to the UAPO diffracted field to be added to the GO field and is referred to as a uniform asymptotic solution because $I(\Omega)$ is well-behaved when $\alpha_p \to \alpha_s$. In the GTD framework, the matrix formulation (14) can be rewritten as

$$\underline{\underline{E}}^d = \begin{pmatrix} E_\beta^d \\ E_\phi^d \end{pmatrix} = \underline{\underline{M}}\begin{pmatrix} E_{\beta'}^i \\ E_{\phi'}^i \end{pmatrix} I(\Omega) = \underline{\underline{D}}\begin{pmatrix} E_{\beta'}^i \\ E_{\phi'}^i \end{pmatrix}\frac{e^{-jk_0 s}}{\sqrt{s}} \tag{45}$$

so that the UAPO solution for the 2×2 diffraction matrix $\underline{\underline{D}}$ is given by:

$$\underline{\underline{D}} = \frac{1}{2\sqrt{2\pi k_0}}\frac{e^{-j\pi/4}}{\sin^2\beta'(\cos\phi+\cos\phi')}F_t\left(2k_0 s\sin^2\beta'\cos^2\left(\frac{\phi\pm\phi'}{2}\right)\right)\underline{\underline{M}} \tag{46}$$

Accordingly, the UAPO solutions have the same ease of handling of those derived in the UTD framework and, in addition, they have the inherent advantage of providing the diffracted field from the knowledge of the GO response of the structure. In other words, it is sufficient to make explicit the reflection and transmission coefficients related to the considered structure for obtaining the UAPO diffraction coefficients.

As demonstrated in (Ferrara et al., 2007a), by taking advantage on the fact that the UAPO solutions are UTD-like as regards the frequency if the elements of reflection and transmission matrices are independent on the frequency, it is possible to determine the time domain UAPO diffraction coefficients by applying the approach proposed in (Veruttipong, 1990).

3. Test-bed cases

The effectiveness of the UAPO solution when applied to some test-bed cases is analysed in this Section. As previously stated, the corresponding diffraction coefficients are determined by making explicit the reflection and transmission coefficients to be used in (21) and (22). The numerical results obtained by using the radio frequency module of COMSOL MULTIPHYSICS®, which is a powerful interactive environment for modelling and solving problems based on partial differential equations via FEM, are considered as reference in the case of normal incidence ($\beta' = 90°$).

3.1 Lossy dielectric layer

A lossy dielectric layer having thickness d, relative complex permittivity ε_r and relative permeability $\mu_r = 1$ is now considered. The structure is penetrable and, according to (Ferrara et al., 2007b), the elements of $\underline{\underline{R}}$ and $\underline{\underline{T}}$ can be so expressed:

$$R_{11} = R_{||} = \frac{\overline{R}_{||}\left[1-(P_{da}P_{att})^2 P_a\right]}{1-\overline{R}_{||}^2 (P_{da}P_{att})^2 P_a} \qquad R_{22} = R_{\perp} = \frac{\overline{R}_{\perp}\left[1-(P_{da}P_{att})^2 P_a\right]}{1-\overline{R}_{\perp}^2 (P_{da}P_{att})^2 P_a} \tag{47}$$

$$R_{12} = R_{21} = 0 \tag{48}$$

$$T_{11} = T_{||} = \frac{\left(1-\overline{R}_{||}^2\right)P_{da}P_{att}P_t}{1-\overline{R}_{||}^2 (P_{da}P_{att})^2 P_a} \qquad T_{22} = T_{\perp} = \frac{\left(1-\overline{R}_{\perp}^2\right)P_{da}P_{att}P_t}{1-\overline{R}_{\perp}^2 (P_{da}P_{att})^2 P_a} \tag{49}$$

$$T_{12} = T_{21} = 0 \tag{50}$$

in which

$$\overline{R}_{||} = \frac{\varepsilon_r \cos\theta^i - \sqrt{\varepsilon_r - \sin^2\theta^i}}{\varepsilon_r \cos\theta^i + \sqrt{\varepsilon_r - \sin^2\theta^i}} \tag{51}$$

$$\overline{R}_{\perp} = \frac{\cos\theta^i - \sqrt{\varepsilon_r - \sin^2\theta^i}}{\cos\theta^i + \sqrt{\varepsilon_r - \sin^2\theta^i}} \tag{52}$$

$$P_{da} = e^{-j\beta_{eq} d/\cos\theta^t} \tag{53}$$

$$P_{att} = e^{-\alpha_{eq} d} \tag{54}$$

$$P_a = e^{j2k_0 d \sin\theta^i \tan\theta^t} \tag{55}$$

$$P_t = e^{jk_0 d \cos(\theta^i - \theta^t)/\cos\theta^t} \tag{56}$$

where β_{eq} and α_{eq} are the equivalent phase and attenuation factors relevant to the propagation through the layer (Balanis, 1989), and θ^t is the real angle between the propagation direction and that of attenuation in the layer.
A sample of interesting results is here reported. The first set of figures from Fig. 6 to Fig. 9 refers to a layer characterised by $\varepsilon_r = 4 - j0.23$ and $d = 0.15\,\lambda_0$ (λ_0 is the free-space wavelength) when a plane wave having $E^i_{\beta'} = 1$, $E^i_{\phi'} = 0$ impinges on the structure from $\beta' = 45°$, $\phi' = 60°$. The magnitudes of the electric field β-components of the GO field and the UAPO diffracted field on a circular path with $\rho = 7\,\lambda_0$ are considered in Fig. 6. As expected, the GO pattern presents two discontinuities in correspondence of the incidence and reflection shadow boundaries at $\phi = 240°$ and $\phi = 120°$, respectively. The UAPO field contribution is not negligible in the neighbourhood of such boundaries and guarantees the continuity of the total field across them as shown in Fig. 7. Analogous considerations can be made by considering the electric field ϕ-components as reported in Figs. 8 and 9.
The next four figures relevant to a lossy dielectric layer are useful to assess the accuracy of the UAPO-based approach by means of comparisons with the results obtained by running COMSOL MULTIPHYSICS® in the case of normal incidence. The layer has the same characteristics considered in the first set of figures and the field is observed on the same circular path. An excellent agreement is attained in all the cases, thus confirming the effectiveness of the here described approach.

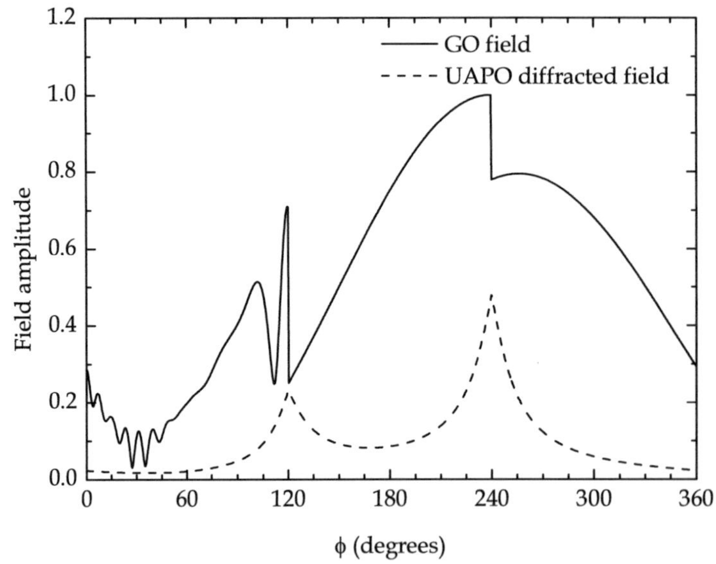

Fig. 6. Amplitude of E_β if $E^i_{\beta'} = 1$, $E^i_{\phi'} = 0$ and $\beta' = 45°$, $\phi' = 60°$. Circular path with $\rho = 7\,\lambda_0$. Layer characterised by $\varepsilon_r = 4 - j0.23$, $\mu_r = 1$ and $d = 0.15\,\lambda_0$.

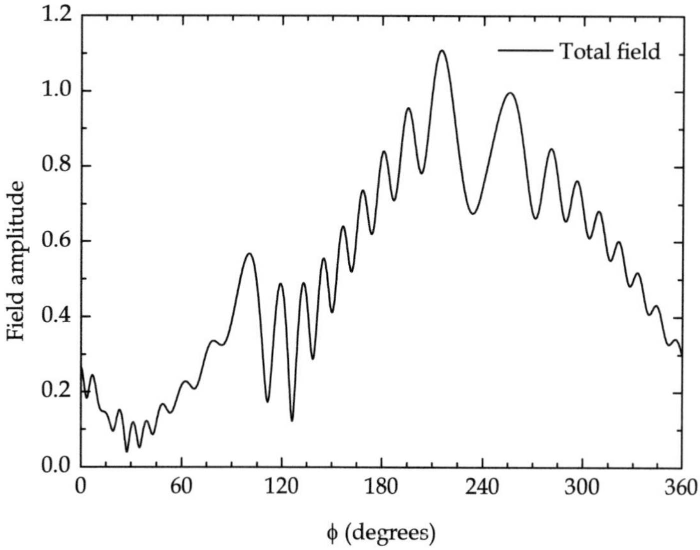

Fig. 7. Amplitude of E_β if $E^i_{\beta'} = 1$, $E^i_{\phi'} = 0$ and $\beta' = 45°$, $\phi' = 60°$. Circular path with $\rho = 7\,\lambda_0$. Layer characterised by $\varepsilon_r = 4 - j0.23$, $\mu_r = 1$ and $d = 0.15\,\lambda_0$.

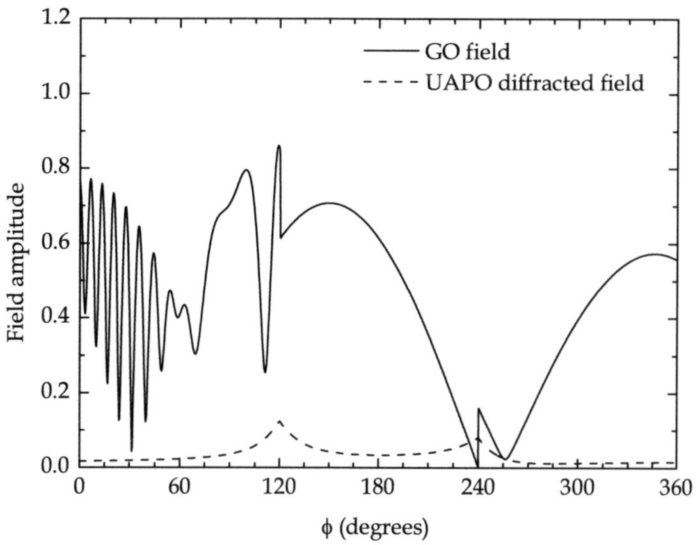

Fig. 8. Amplitude of E_ϕ if $E^i_{\beta'} = 1$, $E^i_{\phi'} = 0$ and $\beta' = 45°$, $\phi' = 60°$. Circular path with $\rho = 7\,\lambda_0$. Layer characterised by $\varepsilon_r = 4 - j0.23$, $\mu_r = 1$ and $d = 0.15\,\lambda_0$.

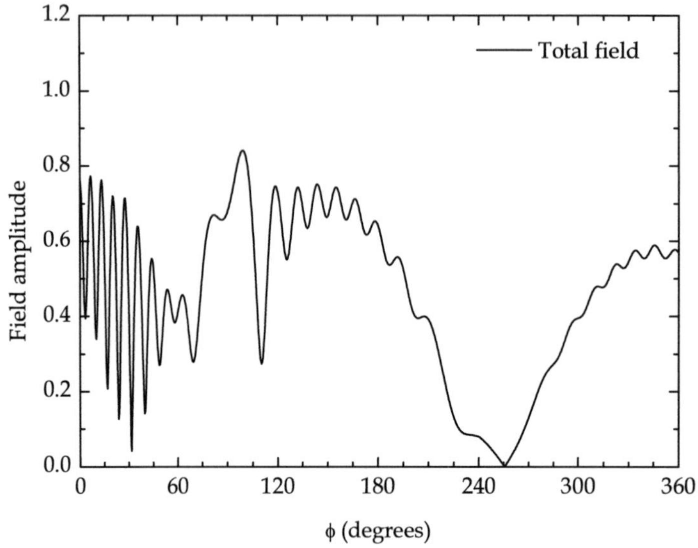

Fig. 9. Amplitude of E_ϕ if $E^i_{\beta'} = 1$, $E^i_{\phi'} = 0$ and $\beta' = 45°$, $\phi' = 60°$. Circular path with $\rho = 7\,\lambda_0$. Layer characterised by $\varepsilon_r = 4 - j0.23$, $\mu_r = 1$ and $d = 0.15\,\lambda_0$.

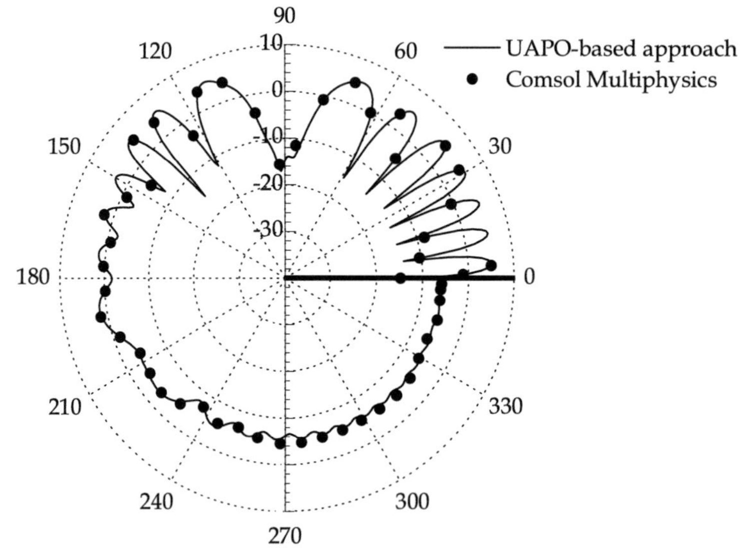

Fig. 10. Amplitude of E_β if $E^i_{\beta'} = 1$, $E^i_{\phi'} = 0$ and $\beta' = 90°$, $\phi' = 30°$. Circular path with $\rho = 7\,\lambda_0$. Layer characterised by $\varepsilon_r = 4 - j0.23$, $\mu_r = 1$ and $d = 0.15\,\lambda_0$.

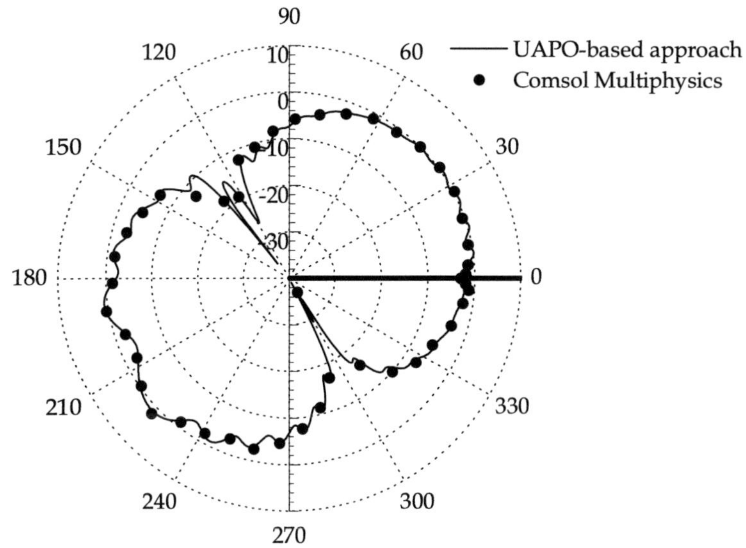

Fig. 11. Amplitude of E_ϕ if $E^i_{\beta'} = 0$, $E^i_{\phi'} = 1$ and $\beta' = 90°$, $\phi' = 30°$. Circular path with $\rho = 7\,\lambda_0$. Layer characterised by $\varepsilon_r = 4 - j0.23$, $\mu_r = 1$ and $d = 0.15\,\lambda_0$.

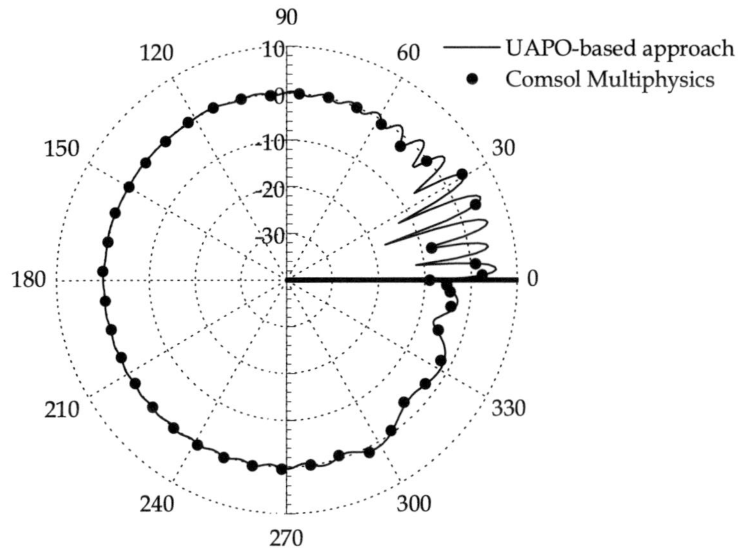

Fig. 12. Amplitude of E_β if $E^i_{\beta'} = 1$, $E^i_{\phi'} = 0$ and $\beta' = 90°$, $\phi' = 140°$. Circular path with $\rho = 7\,\lambda_0$. Layer characterised by $\varepsilon_r = 4 - j0.23$, $\mu_r = 1$ and $d = 0.15\,\lambda_0$.

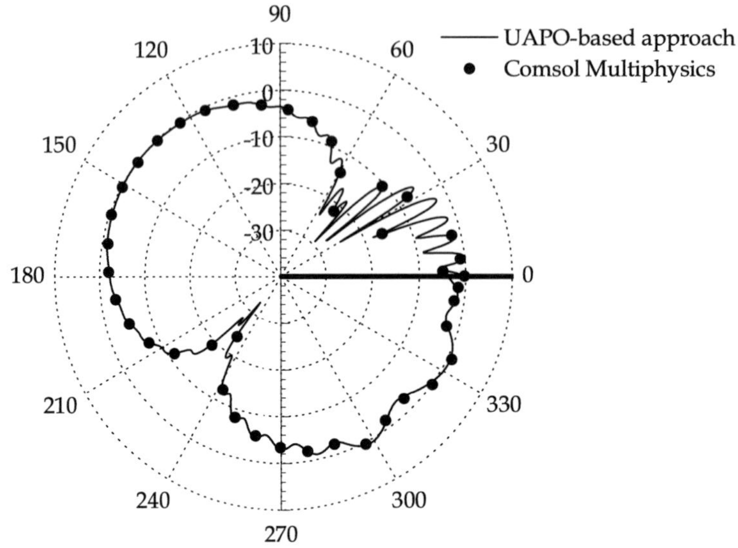

Fig. 13. Amplitude of E_ϕ if $E^i_{\beta'} = 0$, $E^i_{\phi'} = 1$ and $\beta' = 90°$, $\phi' = 140°$. Circular path with $\rho = 7\,\lambda_0$. Layer characterised by $\varepsilon_r = 4 - j0.23$, $\mu_r = 1$ and $d = 0.15\,\lambda_0$.

3.2 Lossless double-negative metamaterial layer

Double-negative metamaterials are unconventional media having negative permittivity and permeability simultaneously, so that they are characterised by antiparallel phase and group velocities, or negative refractive index. Such media can be artificially fabricated by embedding various classes of small inclusions in host media (volumetric structure) or by connecting inhomogeneities to host surfaces (planar structure), and may be engineered to have new physically realizable response functions that do not occur, or may not be readily available, in nature.

A lossless double-negative metamaterial layer having thickness d, relative permittivity $\varepsilon_r = -C$, where C is a positive constant, and relative permeability $\mu_r = -1$ is now considered. The structure is penetrable and, according to (Gennarelli & Riccio, 2009a), the elements of $\underline{\underline{R}}$ and $\underline{\underline{T}}$ are:

$$R_{11} = R_{\parallel} = \frac{\overline{R}_{\parallel}^{12} + \overline{R}_{\parallel}^{23}\, e^{j2k_2 d\cos\theta_2}}{1 + \overline{R}_{\parallel}^{12}\,\overline{R}_{\parallel}^{23}\, e^{j2k_2 d\cos\theta_2}} \qquad R_{22} = R_{\perp} = \frac{\overline{R}_{\perp}^{12} + \overline{R}_{\perp}^{23}\, e^{j2k_2 d\cos\theta_2}}{1 + \overline{R}_{\perp}^{12}\,\overline{R}_{\perp}^{23}\, e^{j2k_2 d\cos\theta_2}} \tag{57}$$

$$R_{12} = R_{21} = 0 \tag{58}$$

$$T_{11} = T_{\parallel} = \frac{\overline{T}_{\parallel}^{12}\,\overline{T}_{\parallel}^{23}\, e^{jk_2 d\cos\theta_2}}{1 + \overline{R}_{\parallel}^{12}\,\overline{R}_{\parallel}^{23}\, e^{j2k_2 d\cos\theta_2}} \qquad T_{22} = T_{\perp} = \frac{\overline{T}_{\perp}^{12}\,\overline{T}_{\perp}^{23}\, e^{jk_2 d\cos\theta_2}}{1 + \overline{R}_{\perp}^{12}\,\overline{R}_{\perp}^{23}\, e^{j2k_2 d\cos\theta_2}} \tag{59}$$

$$T_{12} = T_{21} = 0 \tag{60}$$

in which k_2 is the propagation constant in the double-negative metamaterial, θ_2 is the negative refraction angle, and

$$\overline{R}_{\parallel}^{ij} = \frac{k_j \cos\theta_i - k_i \cos\theta_j}{k_j \cos\theta_i + k_i \cos\theta_j} \quad \overline{R}_{\perp}^{ij} = \frac{k_i \cos\theta_i - k_j \cos\theta_j}{k_i \cos\theta_i + k_j \cos\theta_j} \tag{61}$$

$$\overline{T}_{\parallel}^{ij} = \frac{2k_i \cos\theta_i}{k_j \cos\theta_i + k_i \cos\theta_j} \quad \overline{T}_{\perp}^{ij} = \frac{2k_i \cos\theta_i}{k_i \cos\theta_i + k_j \cos\theta_j} \tag{62}$$

In (61) and (62), $k_1 = k_3 = k_0$, $\theta_1 = \theta_3$ and the superscripts i and j refer to the left and right media involved in the propagation mechanism.
Comparisons with COMSOL MULTIPHYSICS® results are reported in Figs. 14 and 15 with reference to $\phi' = 45°$. As can be seen, the UAPO diffracted field guarantees the continuity of the total field across the two discontinuities of the GO field in correspondence of the incidence and reflection shadow boundaries, and a very good agreement is attained. Accordingly, the accuracy of the UAPO-based approach is well assessed also in the case of a lossless double-negative metamaterial layer.

3.3 Anisotropic impedance layer

A layer characterised by anisotropic impedance boundary conditions on the illuminated face is now considered. Such conditions are represented by an impedance tensor $\underline{\underline{Z}} = Z_{x'} \hat{x}'\hat{x}' + Z_{z'} \hat{z}'\hat{z}'$ having components along the two mutually orthogonal principal axes of anisotropy $\hat{x}'$ and $\hat{z}'$. The structure is opaque so that the transmission matrix does not exist and, according to (Gennarelli et al., 1999), the elements of $\underline{\underline{R}}$ can be so expressed:

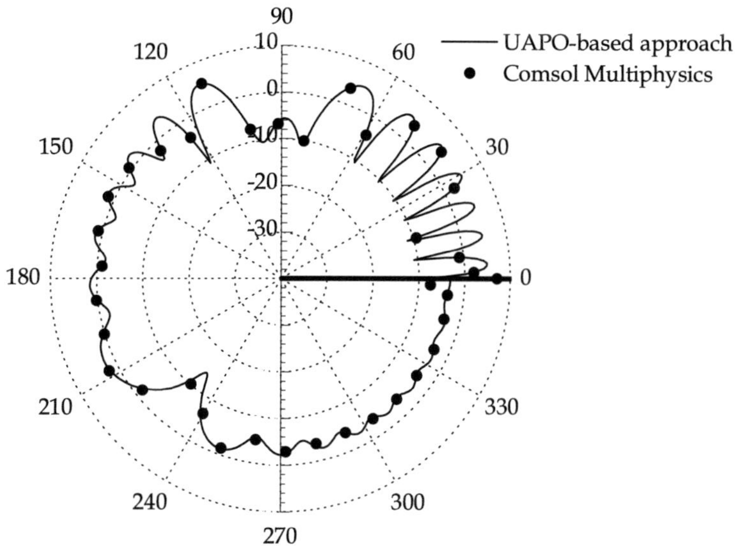

Fig. 14. Amplitude of E_β if $E^i_{\beta'} = 1$, $E^i_{\phi'} = 0$ and $\beta' = 90°$, $\phi' = 45°$. Circular path with $\rho = 5\lambda_0$. Layer characterised by $\varepsilon_r = -4$, $\mu_r = -1$ and $d = 0.125\lambda_0$.

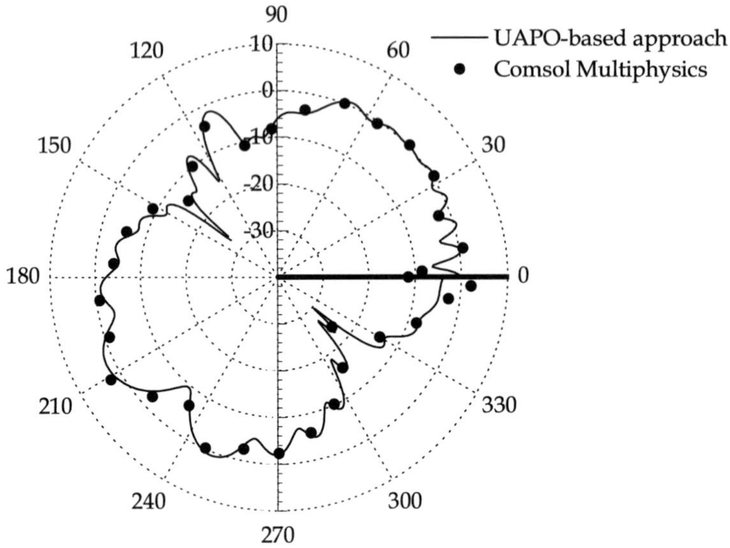

Fig. 15. Amplitude of E_ϕ if $E^i_{\beta'} = 0$, $E^i_{\phi'} = 1$ and $\beta' = 90°$, $\phi' = 45°$. Circular path with $\rho = 5\lambda_0$. Layer characterised by $\varepsilon_r = -4$, $\mu_r = -1$ and $d = 0.125\lambda_0$.

$$R_{11} = \frac{-A + \left(1 + AC + B^2\right)\cos\theta^i - C\cos^2\theta^i}{A + \left(1 - AC - B^2\right)\cos\theta^i - C\cos^2\theta^i} \tag{63}$$

$$R_{12} = -R_{21} = \frac{2B\cos\theta^i}{A + \left(1 - AC - B^2\right)\cos\theta^i - C\cos^2\theta^i} \tag{64}$$

$$R_{22} = -\frac{A + \left(1 + AC + B^2\right)\cos\theta^i + C\cos^2\theta^i}{A + \left(1 - AC - B^2\right)\cos\theta^i - C\cos^2\theta^i} \tag{65}$$

in which, if χ is the angle between $\hat{x}'$ and $\hat{e}_\perp$,

$$A = \frac{Z_{x'}}{\zeta_0}\sin^2\chi + \frac{Z_{z'}}{\zeta_0}\cos^2\chi \tag{66}$$

$$B = \frac{1}{\zeta_0}\left(Z_{x'} - Z_{z'}\right)\sin\chi\cos\chi \tag{67}$$

$$C = -\left[\frac{Z_{x'}}{\zeta_0}\cos^2\chi + \frac{Z_{z'}}{\zeta_0}\sin^2\chi\right] \tag{68}$$

If an isotropic impedance boundary condition is considered (i.,e., the principal axes of anisotropy does not exist and $Z_{x'} = Z_{z'} = Z$), $R_{12} = R_{21} = 0$, whereas R_{11} and R_{22} reduce to the standard reflection coefficients for parallel and perpendicular polarisations.
If the illuminated surface is perfectly electrically conducting, the out diagonal elements are again equal to zero, whereas $R_{11} = 1$ and $R_{22} = -1$ since $Z_{x'} = Z_{z'} = 0$.

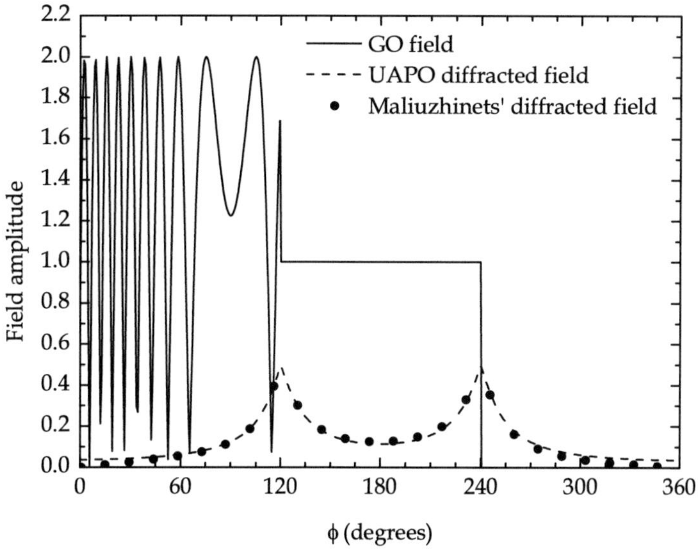

Fig. 16. Amplitude of E_β if $E^i_{\beta'} = 1$, $E^i_{\phi'} = 0$ and $\beta' = 90°$, $\phi' = 60°$. Circular path with $\rho = 5\lambda_0$. Layer characterised by $Z/\zeta_0 = j0.5$.

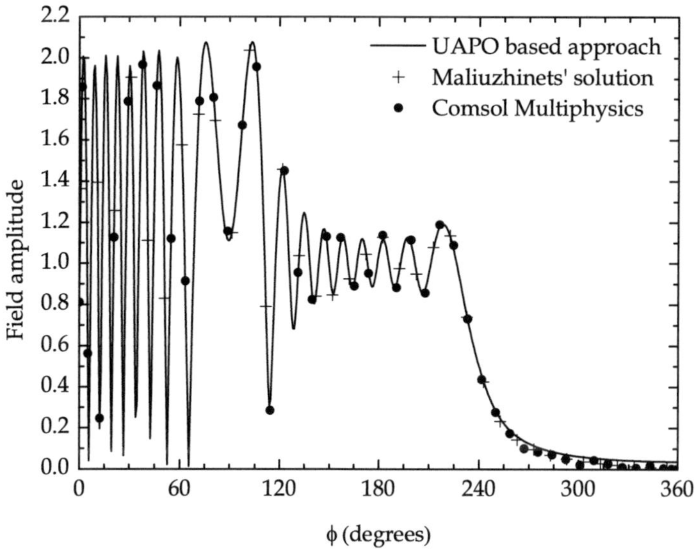

Fig. 17. Amplitude of E_β if $E^i_{\beta'} = 1$, $E^i_{\phi'} = 0$ and $\beta' = 90°$, $\phi' = 60°$. Circular path with $\rho = 5\lambda_0$. Layer characterised by $Z/\zeta_0 = j0.5$.

The magnitudes of the electric field β -components of the GO field and the UAPO diffracted field on a circular path with $\rho = 5\,\lambda_0$ are considered in Fig. 16, where also the diffracted field obtained by using the Maliuzhinets solution (Bucci & Franceschetti, 1976) is reported in the case of an isotropic impedance boundary condition. A very good agreement exists between the two diffracted fields. The accuracy of the UAPO-based approach is further confirmed by comparing the total fields shown in Fig. 17, where also the COMSOL MULTIPHYSICS® results are shown.

4. Junctions of layers

The UAPO solution for the field diffracted by the edge of a truncated planar layer as derived in Section 2 can be extended to junctions by taking into account the diffraction contributions of the layers separately. This very useful characteristic is due to the property of linearity of the PO radiation integral. Accordingly, if the junction of two illuminated semi-infinite layers as depicted in Fig. 18 is considered, the total scattered field in (1) can be so rewritten:

$$\begin{aligned} \underline{E}^s &= -jk_0 \iint_{S_1+S_2} \left[(\underline{\underline{I}} - \hat{R}\hat{R})\zeta_0 \underline{J}_s^{PO} + \underline{J}_{ms}^{PO} \times \hat{R} \right] G(\underline{r},\underline{r}')\, dS = \\ &= -jk_0 \iint_{S_1} \left[(\underline{\underline{I}} - \hat{R}\hat{R})\zeta_0 \underline{J}_{s_1}^{PO} + \underline{J}_{ms_1}^{PO} \times \hat{R} \right] G(\underline{r},\underline{r}')\, dS_1 + \\ &\quad -jk_0 \iint_{S_2} \left[(\underline{\underline{I}} - \hat{R}\hat{R})\zeta_0 \underline{J}_{s_2}^{PO} + \underline{J}_{ms_2}^{PO} \times \hat{R} \right] G(\underline{r},\underline{r}')\, dS_2 = \underline{E}_1^s + \underline{E}_2^s \end{aligned} \qquad (69)$$

and then $\underline{\underline{D}} = \underline{\underline{D}}_1 + \underline{\underline{D}}_2$, with $\underline{\underline{D}}_1$ given by (46). The diffraction matrix $\underline{\underline{D}}_2$ related to the wave phenomenon originated by the edge of the second layer forming the junction can be determined by using again the methodology described in Section 2. If the external angle of the junction is equal to $n\pi$, a $(n-1)\pi$ rotation of the edge-fixed coordinate system must be considered for the second layer. The incidence and observation angles with respect to the illuminated face are now equal to $n\pi - \phi'$ and $n\pi - \phi$, respectively, so that the UAPO solution for $\underline{\underline{D}}_2$ uses $n\pi - \phi'$ instead of ϕ' and $n\pi - \phi$ instead of ϕ. The results reported in (Gennarelli et al., 2000) with reference to an incidence direction normal to the junction of two resistive layers confirm the validity of the approach and, in particular, the accuracy of the solution is well assessed by resorting to a numerical technique based on the Boundary Element Method (BEM).

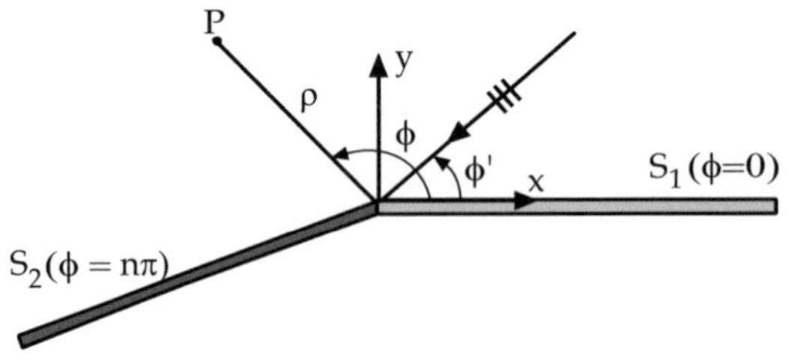

Fig. 18. Junction of two planar truncated layers.

5. Conclusions and future activities

UAPO solutions have been presented for a set of diffraction problems originated by plane waves impinging on edges in penetrable or opaque planar thin layers. The corresponding diffracted field has been obtained by modelling the structure as a canonical half-plane and by performing a uniform asymptotic evaluation of the radiation integral modified by the PO approximation of the involved electric and magnetic surface currents. The resulting expression is given terms of the UTD transition function and the GO response of the structure accounting for its geometric, electric and magnetic characteristics. Accordingly, the UAPO solution possesses the same ease of handling of other solutions derived in the UTD framework and has the inherent advantage of providing the diffraction coefficients from the knowledge of the reflection and transmission coefficients. It allows one to compensate the discontinuities in the GO field at the incidence and reflection shadow boundaries, and its accuracy has been proved by making comparisons with purely numerical techniques. In addition, the time domain counterpart can be determined by applying the approach proposed in (Veruttipong, 1990), and the UAPO solution for the field diffracted by junctions can be easily obtained by considering the diffraction contributions of the layers separately.
To sum up, it is possible to claim that UAPO solutions are very appealing from the engineering standpoint.
Diffraction by opaque wedges has been considered in (Gennarelli et al., 2001; Gennarelli & Riccio, 2009b). By working in this context, the next step in the future research activities may be devoted to find the UAPO solution for the field diffracted by penetrable wedges (f.i., dielectric wedges).

6. Acknowledgment

The author wishes to thank Claudio Gennarelli for his encouragement and helpful advice as well as Gianluca Gennarelli for his assistance.

7. References

Burnside, W.D. & Burgener, K.W. (1983). High Frequency Scattering by a Thin Lossless Dielectric Slab. *IEEE Transactions on Antennas and Propagation*, Vol. AP-31, No. 1, January 1983, 104-110, ISSN: 0018-926X.

Balanis, C.A. (1989). *Advanced Engineering Electromagnetics*, John Wiley & Sons, ISBN: 0-471-62194-3, New York.

Bucci, O.M. & Franceschetti, G. (1976). Electromagnetic Scattering by a Half-Plane with Two Face Impedances, *Radio Science*, Vol. 11, No. 1, January 1976, 49-59, ISSN: 0048-6604.

Clemmow, P.C. (1950). Some Extensions of the Method of Integration by Steepest Descent. *Quarterly Journal of Mechanics and Applied Mathematics*, Vol. 3, No. 2, 1950, 241-256, ISSN: 0033-5614.

Clemmow, P.C. (1996). *The Plane Wave Spectrum Representation of Electromagnetic Fields*, Oxford University Press, ISBN: 0-7803-3411-6, Oxford.

Ferrara, F.; Gennarelli, C.; Pelosi, G. & Riccio, G. (2007a). TD-UAPO Solution for the Field Diffracted by a Junction of Two Highly Conducting Dielectric Slabs. *Electromagnetics*, Vol. 27, No. 1, January 2007, 1-7, ISSN: 0272-6343.

Ferrara, F.; Gennarelli, C.; Gennarelli, G.; Migliozzi, M. & Riccio, G. (2007b). Scattering by Truncated Lossy Layers: a UAPO Based Approach. *Electromagnetics*, Vol. 27, No. 7, September 2007, 443-456, ISSN: 0272-6343.

Gennarelli, C.; Pelosi, G.; Pochini; C. & Riccio, G. (1999). Uniform Asymptotic PO Diffraction Coefficients for an Anisotropic Impedance Half-Plane. *Journal of Electromagnetic Waves and Applications*, Vol. 13, No. 7, July 1999, 963-980, ISSN: 0920-5071.

Gennarelli, C.; Pelosi, G.; Riccio, G. & Toso, G., (2000). Electromagnetic Scattering by Nonplanar Junctions of Resistive Sheets. *IEEE Transactions on Antennas and Propagation*, Vol. 48, No. 4, April 2000, 574-580, ISSN: 0018-926X.

Gennarelli, C.; Pelosi, G. & Riccio, G. (2001). Approximate Diffraction Coefficients of an Anisotropic Impedance Wedge. *Electromagnetics*, Vol. 21, No. 2, February 2001, 165-180, ISSN: 0272-6343.

Gennarelli, G. & Riccio, G. (2009a). A UAPO-Based Solution for the Scattering by a Lossless Double-Negative Metamaterial Slab. *Progress In Electromagnetics Research M*, Vol. 8, 2009, 207-220, ISSN: 1937-8726.

Gennarelli, G. & Riccio, G. (2009b). *Progress In Electromagnetics Research B*, Vol. 17, 2009, 101-116, ISSN: 1937-6472.

Keller, J.B. (1962). Geometrical Theory of Diffraction. *Journal of Optical Society of America*, Vol. 52, No. 2, February 1962, 116-130, ISSN: 0030-3941.

Kouyoumjian, R.G. & Pathak, P.H. (1974). A Uniform Geometrical Theory of Diffraction for an Edge in a Perfectly Conducting Surface. *Proceedings of the IEEE*, Vol. 62, No. 11, November 1974, 1448-1461, ISSN: 0018-9219.

Luebbers, R.J. (1984). Finite Conductivity Uniform UTD versus Knife Diffraction Prediction of Propagation Path Loss. *IEEE Transactions on Antennas and Propagation*, Vol. AP-32, No. 1, January 1984, 70-76, ISSN: 0018-926X.

Maliuzhinets, G.D. (1958). Inversion Formula for the Sommerfeld Integral. *Soviet Physics Doklady*, Vol. 3, 1958, 52-56.

Senior, T.B.A. & Volakis, J.L. (1995). *Approximate Boundary Conditions In Electromagnetics*. The Institution of Electrical Engineers, ISBN: 0-85296-849-3, Stevenage.

Veruttipong, T.W. (1990). Time Domain Version of the Uniform GTD. *IEEE Transactions on Antennas and Propagation*, Vol. 38, No. 11, November 1990, 1757-1764, ISSN: 0018-926X.

3

Differential Electromagnetic Forms in Rotating Frames

Pierre Hillion
Institut Henri Poincaré, 86 Bis Route de Croissy, 78110 Le Vésinet, France

1. Introduction

Differential forms are completely antisymmetric homogeneous r-tensors on a differentiable n-manifold $0 \leq r \leq n$ belonging to the Grasssman algebra [1] and endowed by Cartan [2] with an exterior calculus. These differential forms found an immediate application in geometry and mechanics; introduced by Deschamps [3,4] in electromagnetism, they have known in parallel with the expansion of computers, an increasing interest [5-9] because Maxwell's equations and the constitutive relations are put in a manifestly independent coordinate form.

In the Newton (3+1) space-time, with the euclidean metric $ds^2 = dx^2 + dy^2 + dz^2$ the conventional Maxwell equations in which the **E**, **B**, **D**, **H** fields are 3-vectors have, in absence of charge and current, the Gibbs representation

$$\nabla.\mathbf{B} = 0, \qquad \nabla\wedge\mathbf{E} + 1/c\partial_t\mathbf{B} = 0 \tag{1a}$$

$$\nabla.\mathbf{D} = 0, \qquad \nabla\wedge\mathbf{H} - 1/c\partial_t\mathbf{D} = 0 \tag{1b}$$

and, they also have the differential form representation ($\partial_\tau = 1/c\partial_t$) [5,7]

$$d\wedge \mathtt{E} + \partial_\tau \mathtt{B} = 0, \qquad d\wedge \mathtt{B} = 0 \tag{2a}$$

$$d\wedge \mathtt{H} - \partial_\tau \mathtt{D} = 0, \qquad d\wedge \mathtt{D} = 0 \tag{2b}$$

$d = dx\partial_x + dy\partial_y + dz\partial_z$ is the exterior derivative, $\mathtt{E}$, $\mathtt{H}$ the differential 1-forms

$$\mathtt{E} = E_x\, dx + E_y\, dy + E_z\, dz, \qquad \mathtt{H} = H_x\, dx + H_y\, dy + H_z\, dz \tag{3a}$$

and $\mathtt{B}$, $\mathtt{D}$ the differential 2-forms

$$\mathtt{B} = B_x(dy\wedge dz) + B_y(dz\wedge dx) + B_z(dx\wedge dy)\,,\ \mathtt{D} = -[\,D_x(dy\wedge dz) + D_y(dz\wedge dx) + D_z(dx\wedge dy)] \tag{3b}$$

We are interested here, for reasons to be discussed in Sec.(6) in a Frenet-Serret frame rotating around oz with a constant angular velocity requiring a relativistic processing, We shall prove that this situation leads to an Einstein space-time with a riemannian metric. As an introduction to this problem, we give a succcinct presentation of differential electromagnetic forms in a Minkowski space-time with the metric $ds^2 = dx^2 + dy^2 + dz^2 - c^{-2}\partial_t^2$.

2. Differential forms in Minkowski space-time [7]

In absence of charge and current, the Maxwell equations have the tensor representaation [10, 11]

$$\partial_\sigma F_{\mu\nu} + \partial_\mu F_{\nu\sigma} + \partial_\nu F_{\sigma\mu} = 0 \text{ a)} \qquad \partial_\nu F^{\mu\nu} = 0 \text{ b)} \tag{4}$$

the greek (resp.latin) indices take the values 1,2,3,4 (resp.1,2,3) with $x^1 = x$, $x^2 = y$, $x^3 = z$, $x^4 = ct$, $\partial_j = \partial/\partial x_j$, $\partial_4 = 1/c\partial/\partial_t$ and the summation convention is used. The components of the tensors $F_{\mu\nu}$ and $F^{\mu\nu}$ are with the 3D Levi-Civita tensor ε_{ijk}

$$B_i = \tfrac{1}{2}\, \varepsilon_{ijk} F_{jk}, \quad E_i = -F_{i4}, \qquad H_i = \tfrac{1}{2}\, \varepsilon_{ijk} F^{jk}, \quad D_i = -F^{i4} \tag{5}$$

and in vacuum

$$\mathbf{D} = \varepsilon_0 \mathbf{E}, \qquad \mathbf{H} = \mu_0^{-1}\mathbf{B}, \qquad (\varepsilon_0\mu_0)^{1/2} = 1/c \tag{5a}$$

Let d be the exterior derivative operator

$$d = (\partial_x\, dx + \partial_y\, dy + \partial_z\, dz + \partial_t\, dt)\wedge \tag{6}$$

and $\mathtt{F} = \mathtt{E} + \mathtt{B}$ be the two-form in which :

$$\mathtt{E} = E_x (dx\wedge cdt) + E_y (dy\wedge cdt) + E_z (dz\wedge cdt)$$

$$\mathtt{B} = B_x (dy\wedge dz) + B_y (dz\wedge dx) + B_z (dx\wedge dy) \tag{7}$$

Then the Maxwell equations (4a) have the differential 3-form representation $d\,\mathtt{F} = 0$.
Similarly for $\mathtt{G} = \mathtt{D} + \mathtt{H}$ with :

$$\mathtt{H} = H_x (dx\wedge cdt) + H_y (dy\wedge cdt) + H_z (dz\wedge cdt)$$

$$\mathtt{D} = -[D_x (dy\wedge dz) + D_y (dz\wedge dx) + D_z (dx\wedge dy)] \tag{8}$$

the differential 3-form representation of Maxwell's equations (4b) is $d\,\mathtt{G} = 0$.
To manage the constituive relations (5a) the Hodge star operator [6,9] is introduced

$$*\,(dx\wedge cdt) = c^{-1}\,(dy\wedge dz), \qquad *\,(dy\wedge dz) = c\,(dx\wedge cdt)$$

$$*\,(dy\wedge cdt) = c^{-1}\,(dz\wedge dx), \qquad *\,(dz\wedge dx) = c\,(dy\wedge cdt)$$

$$*\,(dz\wedge cdt) = c^{-1}\,(dx\wedge dy), \qquad *\,(dx\wedge dy) = c\,(dz\wedge cdt) \tag{9}$$

Applying the Hodge star operator to $\mathtt{F}$ gives $*\mathtt{F} = *\mathtt{E} + *\mathtt{B}$ and one checks easily the relation $\mathtt{G} = \lambda_0 *\mathtt{F}$ with $\lambda_0 = (\varepsilon_0/\mu_0)^{1/2}$ so that the Maxwell equations in the Minkowski vacuum, have the diffrerential 3-form representation

$$d\mathtt{F} = 0, \qquad d *\mathtt{F} = 0 \tag{10}$$

3. Electromagnetidsm in a Frenet-Serret rotating frame

We consider a frame rotating with a constant angular velocity Ω around oz. Then, using the Trocheris-Takeno relativistic description of rotation [12, 13], the relations between the cylindrical coordinates R,Φ,Z,T and r,ϕ,z,t in the natural (fixed) and rotating frames are with $ß = \Omega R / c$

$$R = r, \qquad \Phi = \phi \cosh\beta - ct/r \sinh\beta$$

$$Z = z, \qquad cT = ct \cosh\beta - r\phi \sinh\beta \tag{11}$$

and a simple calculation gives the metric ds^2 in the rotating frame

$$ds^2 = c^2dt^2 - dz^2 - r^2d\phi^2 - (1+B^2-A^2)\, dr^2 - 2(A \sinh ß + B \cosh ß)\, cdt\, dr -$$

$$2(A \cosh ß + B \sinh ß)\, r\, dr\, d\phi \tag{12}$$

$$A = ß \sinh ß\, ct/r + ß \cosh ß\, \phi + \sinh ß\, \phi, \quad B = ß \sinh ß\, \phi + ß \cosh ß\, ct/r - \sinh ß\, ct/r \tag{12a}$$

Using the notations $x_4 = ct$, $x_3 = z$, $x_2 = \phi$, $x_1 = r$, we get from (12) $ds^2 = g_{\mu\nu}\, dx^\mu dx^\nu$ with

$$g_{44} = 1, \quad g_{33} = -1, \quad g_{22} = -r^2, \quad g_{11} = -(1+B^2-A^2)$$

$$g_{14} = g_{41} = 2(A \sinh ß + B \cosh ß), \quad g_{12} = g_{21} = 2(A \cosh ß + B \sinh ß) \tag{13}$$

The determinant g of $g_{\mu\nu}$ is

$$g = g_{33}[\, g_{11}g_{22}g_{44} - g_{12}{}^2g_{44} - g_{14}{}^2g_{22}]$$

$$= R\, [g_{11} - g_{12}{}^2r^{-2} - g_{14}{}^2] \; (14)$$

but

$$g_{12}{}^2r^{-2} + g_{14}{}^2 = 4(A^2 - B^2) \tag{14a}$$

and, taking into account the expression (13) of g_{11}, we get finally

$$g = r^2[5(A^2-B^2) - 1], \quad A^2-B^2 = (\phi^2 - c^2t^2/r^2)(ß^2 + \sinh^2 ß + 2ß \sinh ß \cosh ß) \tag{15}$$

So, the rotating Frenet-Serret frame defines an Einstein space-time with the riemannian metric $ds^2 = g_{\mu\nu}\, dx^\mu dx^\nu$, and in this Einstein space-time the Maxwell equations have the tensor representation[14, 15]

$$\partial_\sigma G_{\mu\nu} + \partial_\mu G_{\nu\sigma} + \partial_\nu G_{\sigma\mu} = 0 \quad a) \quad \partial_\nu (|g|^{1/2} G^{\mu\nu}) = 0 \quad b) \tag{16}$$

in which, using the cylindrical coordinates r,ϕ.z ,t with $x_1 = r$, $x_2 = \phi$, $x_3 = z$, $x_4 = ct$; $\partial_1 = \partial_r$, $\partial_2 = \partial_\phi$, $\partial_3 = \partial_z$, $\partial_4 = 1/c\, \partial_t$, the components of the electromagnetic tensors are

$$G_{12} = rB_z, \quad G_{13} = -B_\phi, \quad G_{23} = rB_r; \quad G_{14} = -E_z, \quad G_{24} = -rE_\phi, \quad G_{34} = -E_z$$

$$G^{12} = H_z/r, \quad G^{13} = -H_\phi, \quad G^{23} = H_r/r; \quad G^{14}\; D_z, \quad G_2{}^4 = D_\phi / r, \quad G^{34} = D_z \tag{17}$$

To work with the differential forms, we introduce the exterior derivative

$$\underline{d} = (\partial_r dr + \partial_\phi d\phi + \partial_z dz + \partial_t dt) \wedge \tag{18}$$

(underlined expressions mean that they are defined with the cylindrical coordinates r, ϕ,z,t) and the two-forms $\underline{F} = \underline{E} + \underline{B}$ with

$$\underline{E}= E_r\,(dr\wedge cdt) + E_\phi\,(rd\phi\wedge cdt) + E_z\,(dz\wedge cdt)$$

$$\underline{B} = B_r\,(rd\phi\wedge dz) + B_\phi\,(dz\wedge dr) + B_z\,(dr\wedge rd\phi) \tag{19a}$$

and writing $|g|^{1/2} = rq$, $q = [5(A^2-B^2)-1]^{1/2}$ the two-form $\underline{G} = \underline{D} + \underline{H}$

$$\underline{D} =- q\,[D_r\,(rd\phi\wedge dz) + D_\phi\,(dz\wedge dr) + D_z\,(dr\wedge rd\phi)]$$

$$\underline{H} = q[H_r\,(dr\wedge cdt) + H_\phi\,(rd\phi\wedge cdt) + H_z\,(dz\wedge cdt)] \tag{19b}$$

Then, the Maxwell equatios have the 3-form representation

$$d\underline{F} = 0, \qquad d\,\underline{G} = 0 \tag{20}$$

A simple calculation gives

$$\begin{aligned} d\,\underline{F} = {} & [(\partial_r(rB_r) + \partial_\phi B_\phi + \partial_z(rB_z)]\,(dr\wedge d\phi\wedge dz) + \\ & [\partial_t\,(rB_r) + c\{\partial_\phi E_z - \partial_z(rE_\phi)\}]\,(d\phi\wedge dz\wedge dt) + \\ & [\partial_t B_\phi + c(\partial_z E_r - \partial_r E_z)]\,(dz\wedge dr\wedge dt) + \\ & [\partial_t\,(rB_z) + c\{\partial_r\,(rE_\phi) - \partial_\phi E_r\}]\,(dr\wedge d\phi\wedge dt) \end{aligned} \tag{21a}$$

$$\begin{aligned} d\underline{G} = {} & -\,[\partial_r(qrD_r) + \partial_\phi(qD_\phi) + \partial_z(qD_z)]\,(dr\wedge d\phi\wedge dz) + \\ & [-\partial_t(qrD_r) + c\,\{\partial_\phi(qH_z) - \partial_z(qrH_\phi)\}]\,(d\phi\wedge dz\wedge dt) + \\ & [-\partial_t(qD_\phi) + c\,\{\partial_z(qH_r) - \partial_r(qH_z)\}]\,(dz\wedge dr\wedge dt) + \\ & [-\partial_t(qrD_z) + c\,\{\partial_r(qrH_\phi - \partial_\phi(qH_r)\}]\,(dr\wedge d\phi\wedge dt) \end{aligned} \tag{21b}$$

The Hodge star operator needed to take into account the constitutive relations (5a) in vacuum is defined by the relation

$$\begin{aligned} &*\,(dr\wedge cdt) = -qc^{-1}\,(rd\phi\wedge dz), && *\,(rd\phi\wedge dz) = q^{-1}c\,(dr\wedge cdt) \\ &*\,(rd\phi\wedge cdt) = -qc^{-1}\,(dz\wedge dr), && *\,(dz\wedge dr) = q^{-1}c\,(rd\phi\wedge cdt) \\ &*\,(dz\wedge cdt) = -qc^{-1}\,(dr\wedge rd\phi), && *\,(dr\wedge rd\phi) = q^{-1}c\,(dz\wedge cdt) \end{aligned} \tag{22}$$

Applying (22) to $\underline{F}$ gives $*\underline{\boldsymbol{F}} = *\underline{E} + *\underline{B}$ and it is easily checked that $\underline{G} = \lambda_0\,*\boldsymbol{F}$ with $\lambda_0 = (\varepsilon_0/\mu_0)^{1/2}$ so that in vacuum $d\underline{F} = 0$, $d\,*\underline{\boldsymbol{F}} = 0$.

4. Wave equations in vacuum

4.1 Minkowski space-time

The wave equations satisfied by the electromagnetic field (in absence of charges and currents) are obtained from differential forms with the help of the Laplace-De Rham operator [6,8]

$$L = (d^*\,d^* + *\,d^*\,d)\wedge \tag{23}$$

requiring the Hodge star operators for the n-forms, n = 1,2,3. They are given in Appendix A where, using the exterior derivative (6), and assuming $E_x = E_y = 0$ so that the two-form (7) becomes $\mathrm{E_z} = E_z\,(dz\wedge cdt)$, we get

$$\mathrm{L}\ \mathrm{E_z} = (\Delta - c^{-2}\partial_t^2)\, E_z\,(dz\wedge cdt), \qquad \Delta = \partial_x^2 + \partial_y^2 + \partial_z^2 \tag{24}$$

A similar relation exists for E_x, E_y and for the components of the **B**-field so that, we get finally for the 2-form F:

$$\mathrm{L}\ \mathrm{F} = (\Delta - c^{-2}\partial_t^2)\, F_{\mu\nu}\,(dx^\mu\wedge dx^\nu) \tag{25}$$

so that the wave equation has the 2-form representation $\mathrm{L}\ \mathrm{F} = 0$.

4.2 Einstein space-time

4.2.1 Cartesian frame

In a riemannian cartesian frame, the components of the electromagnetic field tensor $\cdot F_{\mu\nu}$ are solutions of the tensor wave equation [16]

$$g^{\alpha\beta}\nabla_\alpha\nabla_\beta F_{\mu\nu} - 2R_{\mu\nu\alpha\rho}F^{\rho\alpha} + R_\mu{}^\rho F_{\rho\nu} - R_\nu{}^\rho F_{\rho\mu} = 0 \tag{26}$$

$\nabla_\alpha$is the covariant derivative, $R_{\mu\nu\alpha\rho}$ and $R_\mu{}^\rho$ the Riemann curvature and Ricci tensors defined in terms of Christtoffel symbols $\Gamma_{\alpha,\beta\mu}$,$\Gamma^\rho{}_{\alpha\mu}$

$$\Gamma_{\beta,\mu\nu} = \tfrac{1}{2}\,(\partial_\nu g_{\beta\mu} + \partial_\mu g_{\beta\nu} - \partial_\beta g_{\mu\nu}), \qquad \Gamma^\alpha{}_{\mu\nu} = g^{\alpha\beta}\Gamma_{\beta,\mu\nu} \tag{27}$$

by the relations in which $\partial_1 = \partial_x$, $\partial_2 = \partial_y$, $\partial_3 = \partial_y$, $\partial_4 = 1/c\partial_t$,

$$R_{\alpha\beta\mu\nu} = \partial_\nu\Gamma_{\alpha,\beta\mu} - \partial_\mu\Gamma_{\alpha.\beta\nu} + \Gamma^\rho{}_{\alpha\mu}\Gamma_{\rho,\beta\nu} - \Gamma^\rho{}_{\alpha\nu}\Gamma_{\rho,\beta\mu}$$

$$R^{\alpha\beta}{}_{\mu\nu} = g^{\alpha\rho}g^{\beta\sigma}R_{\rho\sigma\mu\nu}, \qquad R_\mu{}^\nu = R^{\nu\alpha}{}_{\alpha\mu}. \tag{28}$$

Now, it is proved [6] that the Laplace-De Rham operator $\mathrm{L} = d{*}d{*} + {*}\mathbf{d}{*}d$ applied to the two form (7) written

$$\mathrm{E} = F_{14}(dx^1\wedge dx^4) + F_{24}(dx^2\wedge dx^4) + F_{34}(dx^3\wedge dx^4) \tag{29}$$

in which $dx^1 = dx$, $dx^2 = dy$, $dx^3 = dz$, $dx^4 = cdt$ gives for the component F_{i4}

$$\mathrm{L}\ \mathrm{E} = \tfrac{1}{2}(g^{\alpha\beta}\nabla_\alpha\nabla_\beta F_{i4} - 2R_{i4\alpha\rho}F^{\rho\alpha} + R_i{}^\rho F_{\rho4} - R_4{}^\rho F_{\rho i})(\,(dx^i\wedge dx^4) \tag{30}$$

$\mathrm{L}\ \mathrm{E} = 0$ gives the 2-form representation of the wave equation in the Einstein space-time with cartesian coordinates

A similar result is obtained for **B** writing $-1/2F_{ij}\,(dx^i\wedge dx^j)$ the B magnetic two-form (7) so that

$$\mathrm{LB} = -1/4(g^{\alpha\beta}\nabla_\alpha\nabla_\beta F_{ij} - 2R_{ij\alpha\rho}F^{\rho\alpha} + R_i{}^\rho F_{\rho j} - R_j{}^\rho F_{\rho i})\,(dx^i\wedge dx^j) \tag{30a}$$

Summing (30) and (30a) gives

$$\mathrm{LF} = \tfrac{1}{2}\,[g^{\alpha\beta}\nabla_\alpha\nabla_\beta F_{\mu\nu} - 2R_{\mu\nu\alpha\rho}F^{\rho\alpha} + R_\mu{}^\rho F_{\rho\nu} - R_\nu{}^\rho F_{\rho\mu}]\,(dx^\mu\wedge dx^\nu) \tag{31}$$

In the Minkowski cartesian frame where $g_{ij} = \delta_{ij}$, $g_{44} = -1$, Eq.(31) reduces to (25).

4.2.2 Frenet-Serret frame

In the Frenet-Serret frame, the Laplace-De Rham operator is defined with the exterior derivative operator (18) and to get a relation such as (30) on the components of the electric field requires some care. First with the greek indices associated to the polar coordinates as previously, one has first to get the Christoffel symbols needed to define the covariant derivative and according to (28), the Riemann curvature and Ricci tensors, a job performed in Appendix B, we are now in position to transpose (30) to a rotating cyindrical frame. To this end, the electric two-form (19a) with

$$\underline{dx}^1\wedge\underline{dx}^4 = dr\wedge cdt, \qquad \underline{dx}^2\wedge\underline{dx}^4 = d\phi\wedge cdt, \quad \underline{dx}^3\wedge\underline{dx}^4 = dz\wedge cdt \tag{32}$$

is written

$$\underline{E} = E_r\,(\underline{dx}^1\wedge\underline{dx}^4) + rE_\phi\,(\underline{dx}^2\wedge\underline{dx}^4) + E_z\,(\underline{dx}^3\wedge\underline{dx}^4) \tag{33}$$

but E_r, rE_ϕ, E_z are the G_{i4} components of the $G_{\mu\nu}$ tensor (17) so that leaving aside a minus sign

$$\underline{E} = G_{14}\,(\underline{dx}^1\wedge\underline{dx}^4) + G_{24}\,(\underline{dx}^2\wedge\underline{dx}^4) + G_{34}\,(\underline{dx}^3\wedge\underline{dx}^4) \tag{34}$$

and we get

$$\underline{L}\ \underline{E} = ½(\,(g^{\alpha\beta}\,\underline{\nabla}_\alpha\underline{\nabla}_\beta G_{i4} - 2\underline{R}_{\,i4\alpha\rho}G^{\rho\alpha} + \underline{R}_{\,i}{}^{\rho}G_{\rho4} - \underline{R}_4{}^{\rho}G_{\rho i})\,)\,(\underline{dx}^i\wedge\underline{dx}^4) \tag{35}$$

so that the components of the electric field are solutions of the two-form equation $\underline{L}\ \underline{E} = 0$ in the Frenet-Serret rotating frame. For the other components of the electromagnetic field, it comes

$$\underline{L}\ \underline{F} = ½(\,(g^{\alpha\beta}\,\underline{\nabla}_\alpha\underline{\nabla}_\beta\,G_{\mu\nu} - 2\underline{R}_{\,\mu\nu\alpha\rho}G^{\rho\alpha} + \underline{R}_{\,\mu}{}^{\rho}G_{\rho\nu} - \underline{R}_\nu{}^{\rho}G_{\rho\mu})\,(\underline{dx}^\mu\wedge\underline{dx}^\nu) \tag{36}$$

We have only considered the two-form $\underline{F}$ because in vacuum $\underline{G} = \lambda_0{}^*\ \underline{F}$.

5. To solve differential form equations

The local 2-form representation (2) of Maxwell's equations follows, as a consequence of the Stokes's theorem, from the Maxwell-Ampère and Maxwell-Faraday integral relations. Then, coming back to these theorems, to solve differential form equations is tantamount to perform the integrals

$$I = \int_M \omega \tag{37}$$

in which ω is a n-form, for instance $\underline{F}$ ou $\underline{G}$, and M an oriented manifold with the same n-dimension as the degree of the ω form [5].

In the 3D-space, the numerical evaluation of (37) is based on the finite element technique, largely used [17, 18] in the simulation of partial differential equations. The manifold M is described by a chain of simplexes made for instance of triangular surfaces, tetrahedral volumes… on which the Whitney forms [5,19,20] gives a manageable description of the n-form ω. A simple example may be found in [19] and a through discussion of the technique in [20]. These solutions may be called weak in opposition to the strong solutions of the Maxwell's equations (1).

The numerical process just described is limited to the 3D-space but in the 4D space-time, in particular for the Frenet-Serret frame, ω depends on dt so that M has to be defined in terms of 2-cells, 3-cells, 4-cells of space-time [21] and the Whitney forms must be generalized accordingly. It does not seem that computational works have been made in this domain.

6. Discussion

Differential electromagnetic forms are usually managed in a Newton space and more rarely in a Minkowski space-time although, in this case, the comparison between tensors and differential forms is very enlightning [7]. This formalism is analyzed here in an Einstein space-time with a Riemann metric, particularly that of a Frenet-Serret frame. From a theoretical point of view, except for some more intricate relations due to Riemann, Ricci tensors and Christoffel symbols there is no difficulty to go from Newton to Einstein differential forms.. The situation is different from a computational point of view, since as mentionned in Sec.5, an important worrk has still to be performed to get the solutions of the differential form equation in an Einstein space-time.

This work may be considered as a first step in a complete analysis of electromagnetic differential forms in an Einstein space-time. The subjects to be discussed go from the presence of charges and currents (left aside here) to boundary conditions with between the introduction of potentials, the energy conveyed by the electromagnetic field and so on. This extension could be performed in the syle used in [7] to analyze the electromagnetic differential forms in a Minkowski space-time. In addition, it would make possible an interesting comparison (al-ready sketched in Sec.4.1) with the electromagnetic tensor formalism of the General Relati-vity [15].

Now, why to take an interest in rotating frames? A first response could have been "Universe" assumed cylindrical. But, although Einstein and Romer (also Levi Civita) have obtained some exact cylindrical wave solutions of the general relativity equations [22], this cosmos has been superseded by a spherical world (nevertheless, because of its particular properties, some works are still devoted to the Levi Civita world [23]. A second response comes from the analysis of the Wilsons' experiments in which was measured the electric potential between the inner and outer surfaces of a cylinder rotating in an external axially directed magnetic field: an analysis with many different approaches [6,23,24,25] (the Trocheris-Takeno des-cription of rotations is used in [25]). Finally, a third response is provided by the increasing at-tention paid to paraxial optical beams with an helicoidal geometrical structure [26], [27] lea-ding to a discussion of light propagation in rotating media: a problem object of some dispu-tes [28-32].The relativistic theory of geometrical optics [15] is still a challenge to which it would be interesting to see what could be the differential form contribution.

Appendix A: Minkowski space-time in vacuum

The four dimensional Hodge operator for Minkowski space-time is defined as follows [8]:

zero-forms and four-forms

$$* (dx\wedge dy\wedge dz \wedge cdt) = -1, \qquad *1 = (dx\wedge dy\wedge dz \wedge cdt) \tag{A.1}$$

one-forms and three-forms

$$* (dx\wedge dy\wedge dz) = -cdt), \qquad * c\,dt = -(dx\wedge dy\wedge dz)$$

$$*(dy\wedge dz\wedge cdt) = -dx, \qquad *dx = -(cdt\wedge dy\wedge dz)$$

$$*(dz\wedge dx\wedge cdt) = -dy, \qquad *dy = -(cdt\wedge dz\wedge dx)$$

$$*(dx\wedge dy\wedge cdt) = -dz, \qquad *dz = -(cdt\wedge dx\wedge dy) \qquad (A.2)$$

two forms

$$*(dy\wedge dz) = -(dx\wedge cdt), \qquad *(dx\wedge cdt) = (dy\wedge dz)$$

$$*(dz\wedge dx) = -(dy\wedge cdt), \qquad *(dy\wedge cdt) = (dz\wedge dx)$$

$$*(dx\wedge dy) = -(dz\wedge cdt), \qquad *(dz\wedge cdt) = (dx\wedge dy) \qquad (A.3)$$

Let us assume $E_x = E_y = 0$, then the electric two-form (7) becomes

$$\mathtt{E} = E_z(dz\wedge cdt) \qquad (A.4)$$

Applying the exterior derivative operator (6) to (B.4) and using (B.2) give

$$*d\,\mathtt{E} = \partial_x E_z\wedge dy - \partial_y E_z\wedge dx \qquad (A.5)$$

and

$$d*d\,\mathtt{E} = \partial_x{}^2E_z\,(dx\wedge dy) + \partial_z\partial_x E_z\,(dz\wedge dy) + \partial_t\partial_x E_z\,(dt\wedge dy) +$$

$$-[\partial_y{}^2E_z\,(dy\wedge dx) + \partial_z\partial_y E_z\,(dz\wedge dx) + \partial_t\partial_y E_z\,(dt\wedge dx)] \qquad (A.6)$$

so that according to (A.3)

$$*d*d\,\mathtt{E} = -\partial_z{}^2E_z\,(dz\wedge cdt) + \partial_z\partial_x E_z\,(dx\wedge cdt) - 1/c\partial_t\partial_x E_z\,(dz\wedge dx) +$$

$$-[\partial_y{}^2E_z\,(dz\wedge cdt) - \partial_z\partial_y E_z\,(dy\wedge cdt) - 1/\,c\partial_t\partial_y E_z\,(dy\wedge dz)] \qquad (A.7)$$

1ow, using (A.3) and (A.2), we also have

$$*d\,*\mathtt{E} = -\partial_z\,E_z\wedge cdt - \partial_t E_z\wedge dz \qquad (A.8)$$

and

$$d*d\,*\mathtt{E} = -\partial_z\partial_x E_z\,(dx\wedge cdt) - \partial_z\partial_y E_z\,(dy\wedge cdt) - \partial_z{}^2E_z\,(dz\wedge cdt) +$$

$$-\,1/c\,[\partial_t{}^2E_z(dt\wedge dz) + \partial_t\partial_x E_z\,(dx\wedge dz) + \partial_t\partial_y E_z\,(dy\wedge dz)] \qquad (A.9)$$

Summing (A.7) and (A.9) gives

$$(*d*d + d*d\,*)\mathtt{E} = -\,(\partial_x{}^2 + \partial_y{}^2 + \partial_z{}^2 - c^{-2}\partial_t{}^2)\,E_z(dz\wedge cdt) \qquad (A.10)$$

Appendix B: Christoffel symbols

The Christoffel symbols are defined in terms of the $g_{\mu\nu}$'s by the well known relations [14-16]

$$\Gamma_{\beta,\mu\nu} = \tfrac{1}{2}\,(\partial_\nu g_{\beta\mu} + \partial_\mu g_{\beta\nu} - \partial_\beta g_{\mu\nu}), \qquad \Gamma^{\alpha}{}_{\mu\nu} = g^{\alpha\beta}\Gamma_{\beta,\mu\nu} \qquad (B.1)$$

In these expressions, the greek indices take the values 1,2,3,4 corresponding in a cylindrical frame to the coordinates $x^1 = r$, $x^2 = \phi$, $x^3 = z$, $x^4 = ct$, while $\partial_1 = \partial_r$, $\partial_2 = 1/r\, \partial_\phi$, $\partial_3 = \partial_z$, $\partial_4 = 1/c\, \partial_t$.

The relations (13) give the components $g_{\mu\nu}$ of the metric tensor for a Frenet-Serret rotating frame and :

$$g_{44} = 1,\ g_{33} = -1,\ g_{22} = -r^2,\ g_{11} = u(r,\phi,t)\ ,\ g_{12} = g_{21} = v(r,\phi,t)\ ,\ g_{14} = g_{41} = w(r,\phi,t) \quad (B.2)$$

the explicit expressions of the functions u,v,w are to be found in (13), no $g_{\mu\nu}$ depends on z. Then, the non-null components of the Christoffel symbols are given for $\mu \leq \nu$ (because of the $\mu\nu$-symmetry)

$$\underline{\Gamma}_{1,11} = \tfrac{1}{2}\partial_r u\ ,\ \underline{\Gamma}_{1,12} = \tfrac{1}{2}\partial_\phi u\ ,\ \underline{\Gamma}_{1,14} = \tfrac{1}{2c}\partial_t u\ ,\ \underline{\Gamma}_{1,22} = 1/r\,\partial_\phi v - r,$$

$$\underline{\Gamma}_{1,24} = \tfrac{1}{2c}\partial_t v + 1/2r\partial_\phi w\ ,\ \underline{\Gamma}_{1,44} = 1/c\partial_t w\ ,\ \underline{\Gamma}_{2,11} = \partial_r v - 1/2r\partial_\phi u$$

$$\underline{\Gamma}_{2,12} = -r\ ,\ \underline{\Gamma}_{2,14'} = 1/2c\,\partial_t v - 1/2\partial_r w\ ,$$

$$\underline{\Gamma}_{4,11} = \partial_r w - 1/2c\,\partial_t u,\quad \underline{\Gamma}_{4,12} = \tfrac{1}{2}\partial_r w - 1/2c\,\partial_t v \quad (B.3)$$

The latin indices taking the values 1,2,3, the covariant derivatives of the components $E_1 = E_r$, $E_2 = E_\phi$, $E_3 = E_z$ of the electric field are

$$\underline{\nabla}_1 E_i = \partial_r E_i - \underline{\Gamma}_{1i}{}^k E_k$$

$$\underline{\nabla}_2 E_i = 1/r\partial_\phi E_i - \underline{\Gamma}_{2i}{}^k E_k$$

$$\underline{\nabla}_3 E_i = \partial_z E_i - \underline{\Gamma}_{3i}{}^k E_k \quad (B.4)$$

Underlined expressions mean they are defined with the cylindrical ccordinates r,ϕ, z,t.
Making in (B.2), u = v = w = 0, gives the metric of the Minkowski frame with polar coordinates and according to (B.3) the only nonnull Christoffeel symbols are

$$\Gamma_{1,22} = -r\ ,\quad \Gamma_{2,12} = \Gamma_{2,21} = -r \quad (B5)$$

7. References

[1] H.Grassmann, *Extension Theory*, (Am. Math.Soc. Providence,2000)

[2] H.Cartan, *Formes différentielles*, (Hermann, Paris, 1967).

[3] G.A.Deschamps, *Exterior Differential Forms* (Springer, Berlin,1970),

[4] G.A.Deschamps, Electromagnetism and differential forms, IEEE Proc. 69 (1981) 676-696.]

[5] A.Bossavit, Differential forms and the computation of fields and forces in electromagnetism . Euro.J.Mech.B,Fluids, 10 (1991) 474-488.

[6] F.W.Hehl and Y.Obhukov, *Foundations of Classical Electrodynamics*, (Birkhauser, Basel, 2003).

[7] I.V.Lindell, *Differential Foms in Electromagnetism*, (Wiley IEEE, Hoboken, 2004).

[8] K.F.Warnick and P.Russer, Two, three and four dimensional electromagnetism using differential forms, Turk.J. Elec.Eng. 14 (2006) 151-172.

[9] F.W.Hehl, Maxwell's equations in Minkowski's world. Ann. der Phys.17 (2008) 691-704.

[10] J.D.Jackson, *Classical Electrodynamics.* (Wiley, New York, 1975).
[11] D.S.Jones, *Acoustic and Electromagnetic waves.*(Clarendon, Oxford, 1956).
[12] M.G.Trocheris, Electrodynamics in a rotating frame of reference, Philo.Mag. 7 (1949) 1143-1155.
[13] H.Takeno, On relativistic theory of rotating disk, Prog. Theor. Phys. 7 (1952) 367-371.
[14] C.Möller, *The Theory of Relativity,* (Clarendon, Oxford, 1952).
[15] J.L.Synge, *Relativity : The General Theory,* (North Holland, Amsterdam, 1960).
[16] A.S.Eddington, *The Mathematical Theory of Relativity.* (University Press, Cambridge, 1951).
[17] R.Dautray and J.L/ Lions (eds), *Analyse mathématique and calcul numérique pour les sciences et les techniques.* (Masson, Paris, 1985).
[18] A.Bossavit , *Computational Electromagnetism.* (Academic Press, San Diego, 1997).
[19] Z.Ren and A.Razeh, Computation of the 3D electromagnetic field using differential based elements and dual formalism. Int.J. Num. Modelling, 9 (1996) 81-96.
[20] A.Stern, Y.Tong, M.Desbrun, J.E.Marsden, Computational electromagnetism with variational integrators and discrete differential forms. ar Xiv :0 707 (2007).
[21] J.L.Synge, *Relativity : The Special Theory,* (North Holland, Amsterdam, 1958).
[22] J.Weber , *Relativity and Gravitation,* (Interscience, New York, 1961).
[23] O.Delice, Kasner generalization of Levi Civita Space-time. Acta Phys. Polo.37 (2006) 2445-2461.
[24] C.T.Ridgely, Applying relativistic electromagnetism to a rotating material medium, Am. J. Phys.66 (1998) 114-121.
[25] P.Hillion, The Wilsons' experiment, Apeiron 6 (1999) 1-8.
[26] G.Rousseaux, On the electrodynamics of Minkowski at low velocities, Eur.Phys.Let. 84 (2008) 2002 p.1-4.
[27] L.Allen, M.V.Beijersbergen, R.J.C.Spreux and J.P.Woerdman, Optical angular momentum and the transformation of Laguerre-Gauss laser modes. Phys.Rev.A 45 (1992) 8185-8189.
[28] I.V.Basistiy, M.S.Soskin and M.V.Vasnetsov, Optical wavefront dislocations and their properties, Opt.Commun. 119 (1995) 604-612.
[29] C.T.Ridgely, Applying covariant versus contravariant electromagnetic tensors in rotating media. Am.J.Phys. 67 (1999) 414-42
[30] G.N.Pellegrini and A.R.Swift, Maxwell's equations in rotating media: is there a problem? Am J.Phys.63 (1995) 694-705.
[31] A.Ya. Bekshaev, M.S.Soskin and M.V.Vasnetsov, Angular momentum of a rotating light beam, Optics Commun. 249 (2005) 347-378.
[32] S.C.Tiwari, Rotating light, OAM paradox and relativistic scalar field, J.Opt .A: Pure Appl.Optics 11 (2009) 065701.

4

Iterative Operator-Splitting with Time Overlapping Algorithms: Theory and Application to Constant and Time-Dependent Wave Equations.

Jürgen Geiser and Asgar Jamneshan
Department of Mathematics, Humboldt Universität zu Berlin,
Unter den Linden 6, D-10099 Berlin,
Germany

1. Introduction

Our study is motivated by wave action models with time dependent diffusion coefficients where the decoupling algorithms are based on iterative splitting methods. The paper is organised as follows. Mathematical models of constant and time dependent diffusion coefficients' wave equations are introduced in Section 2 and we provide analytical solutions as far as possible. In section 3 we give an overview to iterative operator-splitting methods in general, while in section 4 we discuss them with respect to wave equations. For the time dependent case we introduce overlapping schemes. We will do convergence and stability analysis of all methods in use. In section 5 we reformulate the methods for numerical applications, e.g. we give an appropriate discretisation and assembling. We present the numerical results in section 6 and finally, we discuss our future works in the area of splitting and decomposition methods.

2. Mathematical model

Motivated by simulating the propagation of a variety of waves, such as sound waves, light waves and water waves, we discuss a novel numerical scheme to solve the wave equation with time dependent diffusion coefficients, see [21]. We deal with a second-order linearly time dependent partial differential equation. It arises in fields such as acoustics, electromagnetics and fluid dynamics, see [5]. For example, when wave propagation models are physically more complex, due to combined propagations in three dimensions, time dependent equations of such dynamical models become the starting point of the analysis, see [3]. We concentrate on wave propagation models to obtain physically related results for time dependent diffusion parameters, see [5]. For the sake of completion we incorporate the constant case, too.

2.1 Wave equations

In this section we present wave equations with constant and time dependent diffusion coefficients.

Wave equation with constant diffusion coefficients

First we deal with a wave equation that represents a simple model of a Maxwell equation which is needed for the simulation of electro-magnetic fields. We have a linear wave equation with constant coefficients given by:

$$\frac{\partial^2 c}{\partial t^2} = D_1 \frac{\partial^2 c}{\partial x_1^2} + \ldots + D_d \frac{\partial^2 c}{\partial x_d^2} \quad \text{on } \Omega \times [0,T], \tag{1}$$

$$c(x,0) = c_0(x), \text{ and } \frac{\partial c}{\partial t}(x,0) = c_1(x) \quad \text{on } \Omega,$$

$$c(x,t) = c_2(x,t) \text{ on } \partial\Omega_{Dirich} \times [0,T],$$

$$\frac{\partial c}{\partial n} = 0 \text{ on } \partial\Omega_{Neum} \times [0,T],$$

where c_0, c_1 are the initial conditions and c_3 the boundary condition for the Dirichlet boundary. We have $\partial\Omega_{Dirich} \cap \partial\Omega_{Neum} = \partial\Omega$.
For this PDE we can derive an analytical solution:

$$c(x_1,\ldots,x_d,t) = \sin(\frac{1}{\sqrt{D_1}}\pi x_1)\cdot\ldots\cdot\sin(\frac{1}{\sqrt{D_d}}\pi x_d)\cdot\cos(\sqrt{d}\pi t) \tag{2}$$

where d is the spatial dimension.

Wave equation with time dependent diffusion coefficients

Mathematical models often need to have time dependent diffusion coefficients, e.g. hyperbolic differential equations. These are among others the Schrödinger equations or the wave equations with time dependent diffusion coefficients in fluid dynamics. In this paper we shall deal with the uncoupled wave equation with time dependent diffusion coefficients given by:

$$\frac{\partial^2 c}{\partial t^2} = D_1(t) \frac{\partial^2 c}{\partial x_1^2} + \ldots + D_d(t) \frac{\partial^2 c}{\partial x_d^2} \quad \text{on } \Omega \times [0,T], \tag{3}$$

$$c(x,0) = c_0(x), \frac{\partial c}{\partial t}(x,0) = c_1(x) \quad \text{on } \Omega,$$

$$c(x,t) = c_2(x,t) \quad \text{on } \partial\Omega_{Dirich} \times [0,T],$$

$$\frac{\partial c}{\partial n} = 0 \quad \text{on } \partial\Omega_{Neum} \times [0,T],$$

where c_0, c_1 are the initial conditions and c_3 the boundary condition for the Dirichlet boundary. We have $\partial\Omega_{Dirich} \cap \partial\Omega_{Neum} = \partial\Omega$.

In general, we can not derive an analytical solution for arbitrary coefficients' functions. However, given linear diffusion functions, we can deliver an analytical solution with respect to a right hand side (inhomogeneous equation) where we may provide sufficient conditions for the right hand side to vanish in order to obtain an analytical solution for the homogeneous equation. Thus we have

$$\frac{\partial^2 c}{\partial t^2} = D_1(t)\frac{\partial^2 c}{\partial x_1^2} + \ldots + D_d(t)\frac{\partial^2 c}{\partial x_d^2} + f(x_1,\ldots,x_d,t) \quad \text{on } \Omega\times[0,T], \tag{4}$$

$$c(x,0) = c_{anal}(x,t) \;\text{ and }\; \frac{\partial c}{\partial t}(x,0) = c_{anal}(x,t) \quad \text{on } \Omega,$$

$$c(x,t) = c_{anal}(x,t) \quad \text{on } \partial\Omega\times[0,T],$$

where c_{anal} is the assumed analytical solution and $D_j(t) = a_j t + b_j$ with a_j, $b_j \in \mathbb{R}$.

Theorem 1. *We claim to have the following analytical solution for d dimensions:*

$$c(x_1,\ldots,x_d,t) = \sum_{j=1}^{d} \sin(\pi x_j)(\sin(\lambda_j(t)\pi)), \tag{5}$$

while the right hand side $f(x_1, \ldots, x_d, t)$ *is given by*

$$f(x_1,\ldots,x_d,t) = \sum_{j=1}^{d} \pi\frac{a_j}{2}(a_j t + b_j)^{-1/2}\sin(\pi x_j)\cos(\lambda_j(t)\pi), \tag{6}$$

and where

$$\lambda_j(t) = \frac{2}{3a_j}(a_j t + b_j)^{3/2}, \quad j = 1,\ldots,d. \tag{7}$$

Proof. We have the following derivatives

$$\frac{\partial^2 c}{\partial x_j^2} = -\pi^2\sin(\pi x_j)\sin(\lambda_j(t)\pi), \quad j = 1,\ldots,d. \tag{8}$$

$$\frac{\partial^2 c}{\partial t^2} = \sum_{j=1}^{d} sin(\pi x_j)\Big(\pi\lambda_j''(t)\cos(\lambda_j(t)\pi) - \pi^2(\lambda_j'(t))^2\sin(\lambda_j(t)\pi)\Big), \tag{9}$$

where

$$\lambda_j'(t) = (a_j t + b_j)^{1/2},$$

$$\lambda_j''(t) = \frac{a_j}{2}(a_j t + b_j)^{-1/2}, \quad j = 1,\ldots,d.$$

Hence, by employing the derivatives (8)–(9) in (4) we obtian for $f(x_1, \ldots, x_d, t)$

$$f(t) = \sum_{j=1}^{d} \pi \frac{a_j}{2} (a_j t + b_j)^{-1/2} \sin(\pi x_j) \cos(\lambda_j(t)\pi). \tag{10}$$

□

Remark 1. *An analytical solution for the homogeneous equation (3) can be given for $x \in \Omega$ such that $f(x, t) = 0$, i.e.*

$$\Leftrightarrow \sin(\pi x_j) = 0 \quad j = 1, \ldots, d$$

$$\Leftrightarrow x_j \in \mathbb{Z} \quad j = 1, \ldots, d$$

Hence for $x \in \mathbb{Z}^d \cap \Omega$.

2.1.1 Existence of solutions for time dependent wave equations

We assume to have an analytical solution for the following equation.

$$\frac{\partial^2 c}{\partial t^2} = D_1(t) \frac{\partial^2 c}{\partial x_1^2} + \ldots + D_d(t) \frac{\partial^2 c}{\partial x_d^2} + f(x_1, \ldots, x_d, t) \quad \text{on } \Omega \times [0, T], \tag{11}$$

$$c(x, 0) = c_{anal}(x, t) \text{ and } \frac{\partial c}{\partial t}(x, 0) = c_{anal}(x, t) \quad \text{on } \Omega,$$

$$c(x, t) = c_{anal}(x, t) \quad \text{on } \partial\Omega \times [0, T],$$

where the analytical solution is given as $c_{anal}(x_1, \ldots, x_d, t) \in C^2(\Omega) \times C^2([0, T])$ and $f(x_1, \ldots, x_d, t) \in C^2(\Omega) \times C^2([0, T])$.

The equation (11) can be reformulated into a system of first order PDEs. Then we can apply the variation of constants formula which is given by

$$C(x_1, \ldots, x_d, t) = K(x_1, \ldots, x_d, t) + \int_0^t K(x_1, \ldots, x_d, t - s) F(x_1, \ldots, x_d, s) \, ds, \tag{12}$$

where F and C are obtained by the reformulation of a system of first order PDEs. Then we assume that there exists a kernel $K(x_1, \ldots, x_d, t)$ with $C(x_1, \ldots, x_d, 0) = K(x_1, \ldots, x_d, 0)$.

Proof. The variation of constants formula is given by

$$C(x_1, \ldots, x_d, t) = K(x_1, \ldots, x_d, t) + \int_0^t K(x_1, \ldots, x_d, t - s) F(x_1, \ldots, x_d, s) \, ds,$$

Now we assume, given C and F such that we obtain an integral equation, where $K(x_1, \ldots, x_d, t)$ is the unknown.

Based on the rewriting of the Voltera's integral equation there exists a solution when K is bounded, i.e.

$$| K(x_1, \ldots, x_d, t) - K(x_1, \ldots, x_d, t') | \leq L(x_1, \ldots, x_d) \, | t - t' |,$$

for all $(x_1, \ldots, x_d) \in$ and $t, t' \in [0, T]$. This is assumed in solving the solution and that the kernel is bounded, i.e. also for the case $F(x_1, \ldots, x_d, t) \rightarrow 0$. □

Remark 2. *For $F(x_1, \ldots, x_d, t) \equiv 0$ we obtain a solution for the homogeneous equation. Thus, there exists a solution for equation (3).*

3. Splitting methods

Splitting methods have been designed for accelerating solver processes and decomposing them into simpler solvable equation parts, see [24] and [16]. Other ways are to consider the physical behaviour and split it into simpler and solvable equation parts, e.g. symplectic schemes, [22] and [30]. The natural way to decouple a differential equation into simpler parts is done by:

$$\frac{dc(t)}{dt} = A_{\text{full}}\, c(t), \text{ for } t \in (t^n, T), \tag{13}$$

$$\frac{dc(t)}{dt} = (A + B)\, c(t), \quad \text{for } t \in (t^n, T), \tag{14}$$

$$c(t^n) = c^n, \text{ (initial condition)}, \tag{15}$$

where $t, T \in \mathbb{R}^+$ and $t \leq T$. The operator A_{full} can be decoupled into the operators A and B, cf. introduction in [28].

Based on these linear operators the equation (13) can be solved exactly. The solution is given by:

$$c(t^{n+1}) = \exp(\tau A_{\text{full}})\, c(t^n), \tag{16}$$

where the time step is $\tau = t^{n+1} - t^n$ and $t^{n+1} \leq T$.

The simplest operator splitting method is the sequential decoupling into two or more equations. The error for the linear case could be analysed by the Taylor expansion, see [1].

Remark 3. *The introduction for ordinary differential equations presented above can be extended for the abstract Cauchy problem of a parabolic equation by regarding the possibility of defining the operator A_{full} using a Friedrichs' extension. Thus, the mild solutions (or weak solutions) are possible and we can apply the notation of the exp-formulations, see [31].*

3.1 Iterative operator splitting methods for wave equations

In the following we apply the iterative operator-splitting method as an extension to the traditional splitting methods for wave equations. The idea is to repeat the splitting steps with the improved computed solutions. We have to solve a fixed-point iteration and we gain higher-order results.

The iterative splitting method is given in the continous formulation as follows:

$$\frac{\partial^2 c_i(t)}{\partial t^2} = Ac_i(t) + Bc_{i-1}(t) + f(t),$$

$$\text{with } c_i(t^n) = c_{sp}^n, \, c_i'(t^n) = c_{sp}'^n, \tag{17}$$

$$\frac{\partial^2 c_{i+1}(t)}{\partial t^2} = Ac_i(t) + Bc_{i+1}(t) + f(t),$$

$$\text{with } c_{i+1}(t^n) = c_{sp}^n, \, c_{i+1}'(t^n) = c_{sp}'^n, \tag{18}$$

where $c_0(t), c_0'(t)$ are fixed functions for each iteration. Here $c_{\text{sp}}^n, c_{\text{sp}}'^n$ denote the known split approximations at time level $t = t^n$. The split approximation at time level $t = t^{n+1}$ is defined by $c_{sp}^{n+1} = c_{2m+1}(t^{n+1})$.

Remark 4. *The stop criteria is given by:*

$$| c_{k+1} - c_k | \leq \epsilon$$

for $k \in 1, 3, 5, \ldots$ and $\epsilon \in \mathbb{R}^+$. Thus, the solution is given by $c(t^{n+1}) = c_{k+2}$.

For the stability and consistency we can rewrite the equations (17)–(18) in continuous form as follows:

$$\partial_{tt} C_i = \mathcal{A} C_i + \mathcal{F}_i, \tag{19}$$

where $C_i = (c_i, c_{i+1})^t$ and the operators are given by

$$\mathcal{A} = \begin{bmatrix} A & 0 \\ A & B \end{bmatrix}, \mathcal{F}_i = \begin{bmatrix} Bc_{i-1} \\ 0 \end{bmatrix}. \tag{20}$$

We discuss this equation with respect to stability and consistency.

4. Convergence analysis

In the following we present the convergence analysis of the iterative splitting method for wave equations with constant and linear time dependent diffusion coefficients.

4.1 Stability and consistency for the constant case

The stability and consistency results can be done as for the parabolic case. The operator equation with second-order time derivatives can be reformulated into a system of first-order time derivatives.

4.1.1 Consistency

In the following we analyse the consistency and the order of the local splitting error for the linear bounded operators $A, B : \mathbf{X} \rightarrow \mathbf{X}$ where $\mathbf{X}$ is a Banach-space, see [31].

We assume our Cauchy-problem for two linear operators with second-order time derivative.

$$c_{tt} - Ac - Bc = 0 \, , \, t \in (0,T) \, , \tag{21}$$

$$\text{with } \; c(0) = c_0 \, , \, c_t(0) = c_1 \, , \tag{22}$$

where c_0 and c_1 are the initial values, see equation (1).
We rewrite (21)-(22) to a system of first order time derivatives:

$$\partial_t c_1 - c_2 = 0 \text{ in } (0,T), \tag{23}$$

$$\partial_t c_2 - Ac_1 - Bc_1 = 0 \text{ in } (0,T), \tag{24}$$

$$\text{with } \; c_1(0) = c_0, c_2(0) = c_1. \tag{25}$$

where $c_0 = c(0)$ and $c_1 = c_t(0)$ are the initial values. The iterative operator splitting method (17)-(18) is rewritten to a system of splitting methods. The method is given by:

$$\partial_t c_{1,i} = c_{2,i}, \tag{26}$$

$$\partial_t c_{2,i} = Ac_{1,i} + Bc_{1,i-1}, \tag{27}$$

$$\text{with } \; c_{1,i}(t^n) = c_1(t^n), \, c_{2,i}(t^n) = c_2(t^n)$$

$$\partial_t c_{1,i+1} = c_{2,i+1}, \tag{28}$$

$$\partial_t c_{2,i+1} = Ac_{1,i} + Bc_{1,i+1}, \tag{29}$$

$$\text{with } \; c_{1,i+1}(t^n) = c_1(t^n), \, c_{2,i+1}(t^n) = c_2(t^n).$$

We start with $i = 1, 3, 5, \ldots, 2m+1$
We can obtain consistency with the underlying fundamental solution of the equation system.
Theorem 2. *Let $A,B \in \mathcal{L}(\mathcal{X})$ be linear bounded operators. Then the abstract Cauchy problem (21)-(22) has a unique solution and the iterative splitting method (26)-(29) by $i = 1, 3, \ldots, 2m+1$ is consistent with the order of the consistency $\mathcal{O}(\tau_n^{2m})$. The error estimate is given by:*

$$\| e_i \| = K \| B \| \tau_n \| e_{i-1} \| + \mathcal{O}(\tau_n^2)., \tag{30}$$

where $K \in \mathbb{R}^+$, $e_i = \max\{|e_{1,i}|, |e_{i,2}|\}$ and $\|B\|$ is the norm of the bounded operator B. In general, we can do an estimation by recursive arguments:

$$\| e_i \| = \tilde{K} \tau_n^i \| e_0 \| + \mathcal{O}(\tau_n^{i+1}), \tag{31}$$

where $\tilde{K} \in \mathbb{R}^+$ is the growth estimation.

Proof. We derive the underlying consistency of the operator-splitting method. Let us consider the iteration (17)–(18) on the subinterval $[t^n, t^{n+1}]$. For the local error function $e_i(t) = c(t) - c_i(t)$ we have the relations

$$\begin{aligned}
&\partial_t e_{1,i}(t) = e_{2,i}(t), \quad t \in (t^n, t^{n+1}], \\
&\partial_t e_{2,i}(t) = A e_{1,i}(t) + B e_{1,i-1}(t), \quad t \in (t^n, t^{n+1}], \\
&\partial_t e_{1,i+1}(t) = e_{2,i+1}(t), \quad t \in (t^n, t^{n+1}], \\
&\partial_t e_{2,i+1}(t) = A e_{1,i}(t) + B e_{1,i+1}(t), \quad t \in (t^n, t^{n+1}],
\end{aligned} \tag{32}$$

for $m = 0, 2, 4, \ldots$, with $e_0(0) = 0$ and $e_{-1}(t) = c(t)$. We use the notations $\mathbf{X}_4$ for the product space $\mathbf{X}\times\mathbf{X}\times\mathbf{X}\times\mathbf{X}$ endowed with norm $\|(u_1, u_2, u_3, u_4)^t\| = \max\{\|u_1\|, \|u_2\|, \|u_3\|, \|u_4\|\}(u_1, u_2, u_3, u_4 \in \mathbf{X})$.

The elements $\mathcal{E}_i(t), \mathcal{F}_i(t) \in \mathbf{X}^4$ and the linear operator $\mathrm{A} : \mathbf{X}^4 \rightarrow \mathbf{X}^4$ are defined as follows:

$$\mathcal{E}_i(t) = \begin{bmatrix} e_{1,i}(t) \\ e_{2,i}(t) \\ e_{1,i+1}(t) \\ e_{2,i+1}(t) \end{bmatrix}, \quad \mathcal{F}_i(t) = \begin{bmatrix} 0 \\ B e_{1,i-1}(t) \\ 0 \\ 0 \end{bmatrix}, \quad \mathcal{A} = \begin{bmatrix} 0 & I & 0 & 0 \\ A & 0 & 0 & 0 \\ 0 & I & 0 & I \\ A & 0 & B & 0 \end{bmatrix}. \tag{33}$$

Then, using the notations (33), the relations (32) can be written as:

$$\begin{aligned}
&\partial_{tt} \mathcal{E}_i(t) = \mathcal{A}\mathcal{E}_i(t) + \mathcal{F}_i(t), \quad t \in (t^n, t^{n+1}], \\
&\mathcal{E}_i(t^n) = 0.
\end{aligned} \tag{34}$$

Due to our assumptions, $\mathcal{A}$ is a generator of the one-parameter C_0 semi-group $(\exp \mathcal{A}t)_{t\geq 0}$. Hence, using the variations of constants formula, the solution to the abstract Cauchy problem (34) with homogeneous initial conditions can be written as:

$$\mathcal{E}_i(t) = c_0 \int_{t^n}^{t} \exp(\mathcal{A}(t-s))\mathcal{F}_i(s)ds, \tag{35}$$

with $t \in [t^n, t^{n+1}]$ (see, e.g. [6]). Hence, using the denotation

$$\| \mathcal{E}_i \|_\infty = \sup_{t\in[t^n, t^{n+1}]} \| \mathcal{E}_i(t) \|, \tag{36}$$

we have

$$\begin{aligned}
&\| \mathcal{E}_i \| (t) \leq \| \mathcal{F}_i \|_\infty \int_{t^n}^{t} \| \exp(\mathcal{A}(t-s)) \| ds \\
&= \| B \| \| e_{1,i-1} \| \int_{t^n}^{t} \| \exp(\mathcal{A}(t-s)) \| ds, \quad t \in [t^n, t^{n+1}].
\end{aligned} \tag{37}$$

Since $(\mathcal{A}(t))_{t\geq 0}$ is a semi-group, the so-called *growth estimation*

$$\| \exp(\mathcal{A}t) \| \leq K \exp(\omega t), \quad t \geq 0, \tag{38}$$

holds with numbers $K \geq 0$ and $\omega \in \mathbb{R}$, cf. [6].
The estimations (37) and (38) result in

$$\| \mathcal{E}_i \|_\infty = K \| B \| \tau_n \| e_{i-1} \| + \mathcal{O}(\tau_n^2), \tag{39}$$

where $\|e_{i-1}\| = \max\{\|e_{1,i-1}\|, \|e_{2,i-1}\|\}$.

Taking into account the definition of E_i and the norm $\| \cdot \|_\infty$, we obtain

$$\| e_i \| = K \| B \| \tau_n \| e_{i-1} \| + \mathcal{O}(\tau_n^2), \tag{40}$$

and hence

$$\| e_{i+1} \| = K_1 \tau_n^2 \| e_{i-1} \| + \mathcal{O}(\tau_n^3), \tag{41}$$

which proves our statement.
Remark 5. *The proof is aligned to scalar temporal first-order derivatives, see [7]. The generalization for higher-order hyperbolic equations can also be done which are reformulated into first-order systems.*

4.1.2 Stability

The following stability theorem is given for the wave equation done with the iterative splitting method, see (26)–(29).
The convergence is examined in a general Banach space setting and we can prove the following stability theorem.
Theorem 3. *Let us consider the system of linear differential equation used for the spatial discretised wave equation*

$$\partial_t c_1 = c_2, \tag{42}$$

$$\partial_t c_2 = A c_1 + B c_1, \tag{43}$$

$$\text{with } \; c_1(t^n) = c(t^n),\, c_2(t^n) = c_t(t^n),$$

where the operators $A, B : \mathbf{X} \to \mathbf{X}$ are linear and densely defined in the real Banach-space $\mathbf{X}$, see [32]. We can define a norm on the product space $\mathbf{X} \times \mathbf{X}$ with $\|(u, v)^t\| = \max\{\|u\|, \|v\|\}$. We rewrite the equation (42)–(43) and obtain

$$\begin{aligned} &\partial_t \tilde{c}(t) = \tilde{A} c(t) + \tilde{B} c(t), \\ &\tilde{c}(t^n) = \tilde{c}^n , \end{aligned} \tag{44}$$

where $\tilde{c}^n = (c(t^n), c_t(t^n))^T$ *and* $\tilde{A} = \begin{pmatrix} 0 & 1/2I \\ A & 0 \end{pmatrix}$ *and* $\tilde{B} = \begin{pmatrix} 0 & 1/2I \\ B & 0 \end{pmatrix}$. *Let* $\tilde{A}, \tilde{B} : X \to X$ *be linear bounded operators which are generators of the* C_0 *semi-group and* $c_0 \in \mathbf{X}$ *a fixed element. To obtain a scalar estimation for the bounded operators A, B, we assume* $\lambda_{\tilde{A}}$ *as a maximal eigenvalue of*

$\tilde{A}$ and $\lambda_{\tilde{B}}$ as a maximal eigenvalue of $\tilde{B}$. Then the linear iterative operator-splitting method (26) - (29) is stable with the following result:

$$\| \tilde{c}_{i+1}(t^{n+1}) \| \leq \tilde{K} \sum_{j=0}^{i+1} \| c_0 \| \tau^j \lambda_{max}^j , \tag{45}$$

where $\tilde{K} > 0$ is a constant and $\tilde{c}_0 = (c\,(t^n),\ c_t\,(t^n))$ is the initital condition, $\tau = (t^{n+1} - t^n)$ the time step and λ_{max} the maximal eigenvalue of the linear and bounded operators $\tilde{A}$ and $\tilde{B}$.

Proof. Let us consider the iteration (26)–(29) on the subinterval $[t^n, t^{n+1}]$. Then we obtain the eigenvalues of the following linear and bounded operators. Due to the well-posed problem we have: $\lambda_{\tilde{A}}$ eigenvalue of $\tilde{A}$, $\lambda_{\tilde{B}}$ eigenvalue of $\tilde{B}$, see [32] and [13]. Then our iteration methods are given with the eigenvaules as follows:

$$\begin{aligned} &\partial_t \tilde{c}_i(t) = \lambda_{\tilde{A}} \tilde{c}_i(t)) + \lambda_{\tilde{B}} \tilde{c}_{i-1}(t), \quad t \in (t^n, t^{n+1}], \\ &\tilde{c}_i(t^n) = \tilde{c}_0, \end{aligned} \tag{46}$$

and

$$\begin{aligned} &\partial_t \tilde{c}_{i+1}(t) = \lambda_{\tilde{A}} \tilde{c}_i(t) + \lambda_{\tilde{B}} \tilde{c}_{i+1}(t), \quad t \in (t^n, t^{n+1}], \\ &\tilde{c}_{i+1}(t^n) = \tilde{c}_0 , \end{aligned} \tag{47}$$

for $i = 1, 3, 5, \ldots$, with $\tilde{c}_0 = (c(t^n),\ c^t\,(t^n))^t$. The equations can be estimated:

$$\tilde{c}_i(t^{n+1}) = \exp(\lambda_{\tilde{A}} \tau) \tilde{c}_0 + \int_{t^n}^{t^{n+1}} \exp(\lambda_{\tilde{A}}(t-s)) \lambda_{\tilde{B}} \tilde{c}_{i-1}(s) ds,$$

where we can estimate

$$|| \tilde{c}_i(t^{n+1}) || \leq K_1 \, || \tilde{c}_0 || + \tau K_2 \lambda_{\tilde{B}} \, || \tilde{c}_{i-1}(t^{n+1}) || .$$

Further the second equation can be estimated by:

$$\tilde{c}_{i+1}(t^{n+1}) = \exp(\lambda_{\tilde{B}} \tau) \tilde{c}_0 + \int_{t^n}^{t^{n+1}} \exp(\lambda_{\tilde{B}}(t-s)) \lambda_{\tilde{A}} \tilde{c}_i(s) ds,$$

where can be estimated by:

$$|| \tilde{c}_{i+1}(t^{n+1}) || \leq K_3 \, || \tilde{c}_0 || + \tau K_4 \lambda_{\tilde{A}} \, ||| \tilde{c}_i(t^{n+1}) || .$$

With a recursive argument and the maximum of the eigenvalues we can estimate the equations by:

$$|| \tilde{c}_{i+1}(t^{n+1}) || \leq \sum_{j=0}^{i+1} K_j \tau^j \, || \tilde{c}_0 || \, \lambda_{max}^j ,$$

$$|| \tilde{c}_{i+1}(t^{n+1}) || \leq \tilde{K} \sum_{j=0}^{i+1} \tau^j \, || \tilde{c}_0 || \, \lambda_{max}^j ,$$

where $\tilde{K}$ is the maximum of all constants and $\lambda_{max} = \max\{ \lambda_{\tilde{A}}, \lambda_{\tilde{B}} \}$. □

Remark 6. *We have stability for sufficient small time steps τ. Based on the estimation with the eigenvalues we can do the same technique also for unbounded operators that are boundable locally. More accurate estimates can be derived by using the techniques of the mild or weak solutions, see [32].*

4.2 Stability and consistency analysis for time dependent case

We propose the following numerical methods to solve the time dependent wave equation: iterative splitting method with embedded semi-analytical initial conditions and overlapping intervals.

Assumption 1. *For the analysis with the linear time dependent coefficients we assume to have convergence of the semi-analytical solutions to the analytical.*

We give an error estimate for the standard splitting method in Theorem 4. For the iterative operator splitting method with embedded semi-analytical solutions we deal with two time partitions. An inner time partition in which we compute the numerical results and an outer time partition on which we compute the semi-analytical solution.

We have the following definition:

Definition 1. *We part the time interval* $[0, T]$ *as follows*

$$t_{i,j} = i \cdot \tau^{out} + j \cdot \tau^{in}, \quad i = 0,\ldots,M-1 \ and \ j = 0,\ldots,N, \tag{48}$$

$$\tau^{out} = \frac{T}{M}, \quad \tau^{in} = \frac{\tau^{out}}{N} \tag{49}$$

where τ^{out} denotes the outer time step size and τ^{in} the inner.

For all outer time points the semi-analytical solutions are computed to initialize the inner time interval, where we apply the iterative splitting method.

Theorem 4. *The error estimation of the iterative splitting method with embedded semi-analytical solutions is given by*

$$|| u_{num,i}(\tau_n) - u_{semi-anal}(\tau_n) || \leq KN\mathcal{O}(\tau_n^i) \, || \, u(t^m) \, || + \tilde{K} \, || \, e_{out}(\tau) \, ||, \tag{50}$$

$K, N \in \mathbb{R}^+$. $u_{semi-anal}(t)$ is the semi-analytical solution and $u_{num,i}(t)$ is the numerical solution with i-th iterative steps computed and given at the outer time steps. The initial conditions are given at $u(0)$. The overlapping does not cause any significant error in terms of that it would be higher than without overlapping.

We prove theorem 4 for the 2-dimensional equation in the following lemma:

Lemma 1. *The wave equation 3, with parameters $D_1(t) = \tilde{D}_1 \frac{T-t}{T} + \tilde{D}_2 \frac{t}{T}$ and $D_2(t) = \tilde{D}_1 \frac{t}{T} + \tilde{D}_2 \frac{T-t}{T}$, $t \in (0,T)$ and $\tilde{D}_1, \tilde{D}_2 \in \mathbb{R}^+$ is given with the semi-analytical solution by*

$$u_{semi-anal}(x,y,t) = \sin(\frac{1}{\sqrt{\hat{D}_1}}\pi x)\sin(\frac{1}{\sqrt{\hat{D}_2}}\pi x)\cos(\sqrt{2}\pi t),$$

$$t \in (t_m, t_{m+1}), (x,y) \in \Omega,$$

$$\hat{D}_1 = \tilde{D}_1(t_m), \hat{D}_2 = \tilde{D}_2(t_m), \quad m = 0,\ldots,M$$

We apply the iterative splitting scheme (17)–(18) with the η-method in the subtime- intervalls $\tau^{out} \in (t_m, t_{m+1})$, with time step $N\tau_n = \tau$. The local error is given by:

$$|| u_{num,i}(t^{n+1}) - u_{anal}(t^n) || \leq NKO(\tau^i) || u(t^n) || + \tilde{K} || e_{out}(t) ||, \tag{51}$$

Proof. The error analysis is given in two parts:

- Error of the outer intervall $|| u_{anal}(t) - u_{semi-anal}(t) || \leq \tilde{K} || e_{out}(\tau) ||$
- Error of the inner interval $|| u_{num,i}(\tau_n) - u_{semi-anal}(\tau_n) || \leq KNO(\tau_n^i) || u(t^m) || + \tilde{K} || e_{out}(\tau) ||$

Part 1.)
The differential equations are given by:

$$\frac{\partial^2 u_{anal}}{\partial t^2} = D_1(t) \frac{\partial^2 u_{anal}}{\partial x^2} + D_2(t) \frac{\partial^2 u_{anal}}{\partial y^2}, \; t \in (t^m, t^{m+1}) \tag{52}$$

$$\frac{\partial^2 u_{semianal}}{\partial t^2} = \hat{D}_1 \frac{\partial^2 u_{semianal}}{\partial x^2} + \hat{D}_2 \frac{\partial^2 u_{semianal}}{\partial y^2}, \; t \in (t^m, t^{m+1}) \tag{53}$$

We define $e_{out}(t) = u_{anal} - u_{semi-anal}$ and obtain:

$$\frac{\partial^2 e_{out}}{\partial t^2} = (D_1(t^{m+1}) - D_1(t^m)) \frac{\partial^2 e_{out}}{\partial x^2} + (D_2(t^{m+1}) - D_2(t^m)) \frac{\partial^2 e_{out}}{\partial y^2}, \; t \in (t^m, t^{m+1}) \tag{54}$$

We assume $e_{out}(t) \to 0$ for $\tau^{out} \to 0$.

For all other solutions, we assume that the analytical solution $u_{semi-anal}(t^m) \leq u_{anal} \leq u_{semi-anal}(t^{m+1})$ and we estimate

$$|| e_{out}(t) || \leq || \sin\left(\frac{1}{\sqrt{D_1(t^{m+1}) - D_1(t^m)}} \pi x\right) \cdot$$

$$\sin\left(\frac{1}{\sqrt{D_2(t^{m+1}) - D_2(t^m)}} \pi x\right) \cos(\sqrt{2}\pi t) e_{out}(t^m) ||,$$

$$t \in (t_m, t_{m+1}), (x, y) \in \Omega,$$

where $e_{out}(t^m)$ is sufficient small as an initial condition.

Part 2.)
Here we have to use the local error of the iterative splitting method, see previous section 4.1:

$$|| e_i || = K || B || \tau_n || e_{i-1} || + \mathcal{O}(\tau_n^2),$$

where $e_i = \max\{| e_{1,i} |, | e_{i,2} |\}$.

Thus, the combination of both parts reflects the numerical error. □

For the additional overlapping scheme we have only to prove the error of the overlapping regions.

Lemma 2. *In general, we assume that in each time step the error can be bounded by:*

$$\epsilon \geq \epsilon(t) = \sum_{i=1}^{m} | u_{num,i}(t) - u_{ana,i} | \Delta x \Delta y$$

where $m = (x_{steps} + 1)(y_{steps} + 1)$ is the number of the spatial steps in the overlapping area. So that the error for the overlapped time steps t_1 and t_2 can be bounded by:

$$\epsilon_{overlap} \leq \epsilon$$

Proof. The error in t_1 and t_2 is given by

$$\epsilon_{overlap}(t_1) = \sum_{i=1}^{m} | u_{num,i}(t_1) - u_{ana,i}(t_1) | \Delta x \Delta y \leq \epsilon$$

$$\epsilon_{overlap}(t_2) = \sum_{i=1}^{m} | u_{num,i}(t_2) - u_{ana,i}(t_2) | \Delta x \Delta y \leq \epsilon$$

We obtain by overlapping, i.e. averaging the values in t_1 and t_2

$$\frac{\epsilon_{overlap}(t_1) + \epsilon_{overlap}(t_2)}{2} \leq \epsilon$$

Hence, the average error is also in the assumed error bound for the overlapped area. Thus, for a given error bound ϵ, the overlap error is below the error bound for sufficient large m. □

5. Discretization and assembling

We exercise our theory for the two dimensional wave equation. More dimensional equations can be treated analogously, particularly, we did an implementation for the three dimensional case, too. First we present methods (finite difference and iterative methods) to solve the wave equation with constant diffusion coefficients. Besides their benefit for their own, the methods provide basic tools to solve the linearly time dependent equation. By adding overlapping schemes we have all tools at hand to deal with the latter case which we present in a second step.

5.1 Wave equation with constant coefficients

5.1.1 Finite difference discretization

For getting an iterative method for the two dimensional wave equation we first have to apply the η-method together with the mixed forward-backward difference method in time and get:

$$c^{n+1} - 2c^{n} + c^{n-1}$$

$$= \Delta t^2 \left(D_1(t^{n+1}) \eta \frac{\partial^2 c^{n+1}}{\partial x^2} + D_1(t^n)(1 - 2\eta) \frac{\partial^2 c^n}{\partial x^2} + D_1(t^{n-1}) \eta \frac{\partial^2 c^{n-1}}{\partial x^2} \right) \quad (55)$$

$$+\Delta t^2\left(D_2(t^{n+1})\eta\frac{\partial^2 c^{n+1}}{\partial y^2}+D_2(t^{n+1})\left(1-2\eta\right)\frac{\partial^2 c^n}{\partial y^2}+D_2(t^{n-1})\eta\frac{\partial^2 c^{n-1}}{\partial y^2}\right).$$

Now we can apply the mixed forward-backward difference method in space defined as:

$$\frac{\partial^2 c(x_i,y_j)}{\partial x^2}\approx\frac{1}{\Delta x^2}\left(c\left(x_i+\Delta x,y_j\right)-2c\left(x_i,y_j\right)+c\left(x_i-\Delta x,y_j\right)\right),$$

$$\frac{\partial^2 c(x_i,y_j)}{\partial y^2}\approx\frac{1}{\Delta y^2}\left(c\left(x_i,y_j+\Delta y\right)-2c\left(x_i,y_j\right)+c\left(x_i,y_j-\Delta y\right)\right),$$

and get for the uncoupled wave equation (1):

$$\begin{aligned}c^{n+1}&-2c^n+c^{n-1}\\&=\frac{\Delta t^2}{\Delta x^2}D_1(\eta(c^{n+1}_{i+1,j}-2c^{n+1}_{i,j}+c^{n+1}_{i-1,j})+(1-2\eta)(c^n_{i+1,j}-2c^n_{i,j}+c^n_{i-1,j})\\&+\eta(c^{n-1}_{i+1,j}-2c^{n-1}_{i,j}+c^{n-1}_{i-1,j}))\\&+\frac{\Delta t^2}{\Delta y^2}D_2(\eta(c^{n+1}_{i,j+1}-2c^{n+1}_{i,j}+c^{n+1}_{i,j-1})+(1-2\eta)(c^n_{i,j+1}-2c^n_{i,j}+c^n_{i,j-1})\\&+\eta(c^{n-1}_{i,j+1}-2c^{n-1}_{i,j}+c^{n-1}_{i,j-1})).\end{aligned}\tag{56}$$

5.2 Application of the sequential operator-splitting method

In the classical operator-splitting method we part every time step into two smaller substeps to get a better accuracy.

In the first step the partial derivative in x will be discretised implicit. The partial derivative in y will be discretised explicit. In the second step it is the other way around.

We exemplify the idea with the two dimensional wave equation. Thus we get form (55):

$$\begin{aligned}1)\tilde{c}&-2c^n+c^{n-1}\\&=\Delta t^2D_1\left(\eta\frac{\partial^2\tilde{c}}{\partial x^2}+\left(1-2\eta\right)\frac{\partial^2 c^n}{\partial x^2}+\eta\frac{\partial^2 c^{n-1}}{\partial x^2}\right)\\&+\Delta t^2D_2\frac{\partial^2 c^n}{\partial y^2},\end{aligned}\tag{57}$$

$$\begin{aligned}2)c^{n+1}&-2c^n+c^{n-1}\\&=\Delta t^2D_1\left(\eta\frac{\partial^2\tilde{c}}{\partial x^2}+\left(1-2\eta\right)\frac{\partial^2 c^n}{\partial x^2}+\eta\frac{\partial^2 c^{n-1}}{\partial x^2}\right)\\&+\Delta t^2D_2\left(\eta\frac{\partial^2 c^{n+1}}{\partial y^2}+\left(1-2\eta\right)\frac{\partial^2 c^n}{\partial y^2}+\eta\frac{\partial^2 c^{n-1}}{\partial y^2}\right).\end{aligned}\tag{58}$$

5.3 Application of the iterative operator-splitting method

In the two dimensional iterative operator-splitting method we will change the algorithm (57)-(58) in such a way that the result of the second step will be again used in the first step. Thus we get an iterative method with an initial value c^{i-1}:

$$1) c^{i} - 2c^{n} + c^{n-1} = \Delta t^2 D_1 \left(\eta \frac{\partial^2 c^{i}}{\partial x^2} + (1-2\eta) \frac{\partial^2 c^{n}}{\partial x^2} + \eta \frac{\partial^2 c^{n-1}}{\partial x^2} \right) + \Delta t^2 D_2 \left(\eta \frac{\partial^2 c^{i-1}}{\partial y^2} + (1-2\eta) \frac{\partial^2 c^{n}}{\partial y^2} + \eta \frac{\partial^2 c^{n-1}}{\partial y^2} \right), \tag{59}$$

$$2) c^{i+1} - 2c^{n} + c^{n-1} = \Delta t^2 D_1 \left(\eta \frac{\partial^2 c^{i}}{\partial x^2} + (1-2\eta) \frac{\partial^2 c^{n}}{\partial x^2} + \eta \frac{\partial^2 c^{n-1}}{\partial x^2} \right) + \Delta t^2 D_2 \left(\eta \frac{\partial^2 c^{i+1}}{\partial y^2} + (1-2\eta) \frac{\partial^2 c^{n}}{\partial y^2} + \eta \frac{\partial^2 c^{n-1}}{\partial y^2} \right). \tag{60}$$

Now we have an iterative operator-splitting method that stops by achieving a given iteration depth or a given error tolerance

$$\| c^{i} - c^{i-2} \| \leq TOL$$

Hereafter the numerical result for the function c at time point $n+1$ is given by:

$$c^{n+1} := c^{i+1,n+1} = c^{i+1}.$$

For the stability of the function it is important to start the iterative algorithm with a good initial value $c^{i-1,n+1} = c^{i-1}$. Some options for their choice are given in the following subsection.

5.3.1 Initial conditions for the iteration

I.1)

The easiest initial condition for our $c^{i-1,n+1}$ is given by the zero vector, $c^{i-1,n+1} \equiv 0$, but it might be a bad choice, if the stability depends on the initial value.

I.2)

A better variant would be to set the initial value to be the result of the last step, $c^{i-1,n+1} = c^{n}$. Thus the initial value might be next to c^{n+1}, which would be a better start for the iteration.

I.3)

With using the average growth of the function depending on the time, the function at the time point $n+1$ might be even better guessed: $c^{i-1,n+1} = c^{n} + \frac{1}{\Delta t} \cdot (c^{n} - c^{n-1})$

I.4)

A better initial value can be achieved by calculating it with using a method for the first step. The easier one is the explicit method,

$$c^{i-1,n+1} - 2c^{n} + c^{n-1} = \Delta t^2 (D_1 \frac{\partial^2 c^{n}}{\partial x^2} + D_2 \frac{\partial^2 c^{n}}{\partial y^2}).$$

I.5)

The prestepping method might be the best of the ones described in this section because the iteration starts next to the value of c^{n+1}.

5.4 Discretization and assembling

Discretising the algorithm of the iterative operator-splitting method (59)–(60) analogously to (56), we get the following scheme for the two dimensional wave equation:

$$\begin{aligned}
1)\Delta x^2\Delta y^2 c^i_{k,l} &- \Delta t^2\Delta y^2 D_1(t^{n+1})\eta(c^i_{k+1,l} - 2c^i_{k,l} + c^i_{k-1,l}) \\
&= 2\Delta x^2\Delta y^2 c^n_{k,l} + \Delta t^2\Delta y^2 D_1(t^n)(1-2\eta)(c^n_{k+1,l} - 2c^n_{k,l} + c^n_{k-1,l}) \\
&\quad + \Delta t^2\Delta x^2 D_2(t^n)(1-2\eta)(c^n_{k,l+1} - 2c^n_{k,l} + c^n_{k,l+1}) \\
&-\Delta x^2\Delta y^2 c^{n-1}_{k,l} + \Delta t^2\Delta y^2 D_1(t^{n-1})\eta(c^{n-1}_{k+1,l} - 2c^{n-1}_{k,l} + c^{n-1}_{k-1,l}) \\
&\quad + \Delta t^2\Delta x^2 D_2(t^{n-1})\eta(c^{n-1}_{k,l+1} - 2c^{n-1}_{k,l} + c^{n-1}_{k,l-1}) \\
&+\Delta t^2\Delta x^2 D_2(t^{n+1})\eta(c^{i-1}_{k,l+1} - 2c^{i-1}_{k,l} + c^{i-1}_{k,l-1}),
\end{aligned} \tag{61}$$

$$\begin{aligned}
2)\Delta x^2\Delta y^2 c^{i+1}_{k,l} &- \Delta t^2\Delta x^2 D_2(t^{n+1})\eta(c^{i+1}_{k,l+1} - 2c^{i+1}_{k,l} + c^{i+1}_{k,l-1}) \\
&= 2\Delta x^2\Delta y^2 c^n_{k,l} + \Delta t^2\Delta y^2 D_1(1-2\eta)(c^n_{k+1,l} - 2c^n_{k,l} + c^n_{k-1,l}) \\
&\quad + \Delta t^2\Delta x^2 D_2(t^n)(1-2\eta)(c^n_{k,l+1} - 2c^n_{k,l} + c^n_{k,l+1}) \\
&-\Delta x^2\Delta y^2 c^{n-1}_{k,l} + \Delta t^2\Delta y^2 D_1(t^{n-1})\eta(c^{n-1}_{k+1,l} - 2c^{n-1}_{k,l} + c^{n-1}_{k-1,l}) \\
&\quad + \Delta t^2\Delta x^2 D_2(t^{n-1})\eta(c^{n-1}_{k,l+1} - 2c^{n-1}_{k,l} + c^{n-1}_{k,l-1}) \\
&+\Delta t^2\Delta y^2 D_1(t^{n+1})\eta(c^i_{k+1,l} - 2c^i_{k,l} + c^i_{k-1,l}).
\end{aligned} \tag{62}$$

This can be written in a matrix scheme as follows:

$$1)c^i = (Sys^1_i)^{-1}(t^{n+1})\cdot(Sys_{Alt}(t^{n+1})\cdot c^{i-1} + InterB^1(t^n)\cdot c^n + InterC^1(t^{n-1})\cdot c^{n-1}),$$

$$2)c^{i+1} = (Sys_{Neu})^{-1}(t^{n+1})\cdot(Sys^2_i(t^{n+1})\cdot c^i + InterB^2(t^n)\cdot c^n + InterC^2(t^{n-1})\cdot c^{n-1}).$$

With this scheme the sequence c_i can be calculated only with the results of the last steps. It ends when the given error tolerance is achieved. The matrices only have to be calculated once in the program. They do not change during the iteration.

The matrices $Sys^d_i, Sys^d_{Oldj}, Sys^d_{Newj}, InterB^d$ and $InterC^d$ depend on the solutions at different time levels, i.e. $c^{i-1}, c^i, c^{i+1}, c^{n-1}$ and c^n.

5.5 Wave equation with linear time dependent diffusion coefficients

The main idea to solve the time dependent wave equation with linear diffusion functions is to part the time domain $[0,\ T]$ into sub-intervals at which we assume equations with constant diffusion coefficients on each of the sub-intervals. Hence, we reduce the problem of the time depedent wave equation to the one with constant diffusion coefficients.

Mathematically, given:

$$\frac{\partial^2 c}{\partial t^2} = D_1(t)\frac{\partial^2 c}{\partial x^2} + D_2(t)\frac{\partial^2 c}{\partial y^2}, \quad (x,y,t)\in\Omega\times[0,T], \tag{63}$$

$$D_1(t) = \frac{d_2 - d_1}{T} + d_1, \tag{64}$$

$$D_2(t) = \frac{d_1 - d_2}{T} + d_2, \quad d_1, d_2 \in [0,1]. \tag{65}$$

The partition of $[0, T]$ is given by:

$$t_{i,j} = i \cdot \tau^{out} + j \cdot \tau^{in}, \quad i = 0,\ldots,M-1 \ and \ j = 0,\ldots,N, \tag{66}$$

$$\tau^{out} = \frac{T}{M}, \quad \tau^{in} = \frac{\tau^{out}}{N} \tag{67}$$

where τ^{out} denotes the outer time step size and τ^{in} the inner.

We have the following system of wave equations with constant diffusion coefficients on the sub-intervals $[t_{i,0}, t_{i,N}]$ $(i = 0, \ldots, M-1)$:

$$\frac{\partial^2 c^i}{\partial t^2} = D_1(t_{i,0}) \frac{\partial^2 c^i}{\partial x^2} + D_2(t_{i,0}) \frac{\partial^2 c^i}{\partial y^2}, \quad (x,y,t) \in \Omega \times [t_{i,0}, t_{i,N}]. \tag{68}$$

(69)

For each sub-interval $[t_{i,0}, t_{i,N}]$ $(i = 0, \ldots, M-1)$ we can make use of the results in 4.1. In particular, we can give an analytical solution by:

$$c^i_{anal}(x,y,t) = \sin(\frac{1}{\sqrt{D_1(t_{i,0})}} \pi x) \cdot \sin(\frac{1}{\sqrt{D_2(t_{i,0})}} \pi y) \cdot \cos(\sqrt{2}\pi t), \tag{70}$$

$$(x,y,t) \in \Omega \times [t_{i,0}, t_{i,N}], \quad i = 0,\ldots,M-1. \tag{71}$$

Thus we assume for each $i = 0, \ldots, M-1$ following initial and boundary conditions for (68):

$$c^0(x,y,0) = c^0_{anal}(x,y,0), \quad (x,y) \in \Omega, \tag{72}$$

$$c^i(x,y,t) = c^i_{anal}(x,y,t), \quad on \ \partial\Omega \times [0,T]. \tag{73}$$

Furthermore, we can make use of the numeric methods, developed for the wave equation with constant diffusion-coefficients, to give a discretisation and assembling for each sub-interval, see 5.1. We obtain a numerical, resp. semianalytical, solution for the time dependent equation (63) in $\Omega \times [0, T]$ by joining the results c^i of all sub-intervals $[t_{i,0}, t_{i,N}]$ $(i = 0, \ldots, M-1)$. In 4.2 we show that the semi-analytical solution converges to the presumed analytical solution for $\tau^{out} \to 0$. We need the semi-analytical solution as reference solution in order to be able to evaluate the numerical.

In order to reach a more accurate result we propose an interval-overlapping method. Let $p \in \{0,\ldots,[\frac{N}{2}]\}$. We solve the following system:

$$\frac{\partial^2 c^0}{\partial t^2} = D_1(t_{0,0})\frac{\partial^2 c^0}{\partial x^2} + D_2(t_{0,0})\frac{\partial^2 c^0}{\partial y^2}, \quad (74)$$

$$(x,y,t) \in \Omega \times [0, t_{0,N} + p\tau^{in}],$$

$$\frac{\partial^2 c^i}{\partial t^2} = D_1(t_{i,0})\frac{\partial^2 c^i}{\partial x^2} + D_2(t_{i,0})\frac{\partial^2 c^i}{\partial y^2}, \quad (75)$$

$$(x,y,t) \in \Omega \times [t_{i,0} - p\tau^{in}, t_{i,N} + p\tau^{in}], \quad i = 1,\ldots,M-2,$$

$$\frac{\partial^2 c^{M-1}}{\partial t^2} = D_1(t_{M-1,0})\frac{\partial^2 c^{M-1}}{\partial x^2} + D_2(t_{M-1,0})\frac{\partial^2 c^{M-1}}{\partial y^2}, \quad (76)$$

$$(x,y,t) \in \Omega \times [t_{M-1,0} - p\tau^{in}, T],$$

while the initial and boundary conditions are as previously set.

We present the interval-overlapping for the analytical solutions of (74)-(76). Hence, $c_{semi-anal}(x, y, t)$ is

$$\begin{cases} c^0_{anal}(x,y,t), & \text{for } (x,y,t) \in \Omega \times [0, t^0_N - p\tau^{in}]; \\ c^i_{anal}(x,y,t), & \text{for } (x,y,t) \in \Omega \times [t^i_0 + p\tau^{in}, t^i_N - p\tau^{in}], \\ & i = 1,\ldots,M-1; \\ \frac{1}{2}(c^{i-1}_{anal}(x,y,t) + c^i_{anal}(x,y,t)) & \text{for } (x,y,t) \in \Omega \times [t^{i-1}_N - p\tau^{in}, t^i_0 + p\tau^{in}], \\ & i = 1,\ldots,M-1; \\ c^{M-1}_{anal}(x,y,t) & \text{for } (x,y,t) \in \Omega \times [t^{M-1}_0 + p\tau^{in}, T]. \end{cases}$$

The same can be done analogously for the numerical solution.

6. Numerical experiments

We test our methods for the two dimensional wave equation. First we analyse test series for the constant coefficient wave equation. Here, we give some general remarks on how to carry out the experiments, e.g. choise of parameters, and how to interpret the test series correctly, e.g. CFL condition. Moreover, we present a method how to obtain acceptable accuracy with a minimum of cost. In a second step we do an error analysis for the wave equation with linearly time dependent diffusion coefficients. The tables are given at the end of the paper.

6.1 Wave equation with constant diffusion coefficients

The PDE to solve with our numerical methods is given by:

$$\frac{\partial^2 c}{\partial t^2} = D_1\frac{\partial^2 c}{\partial x^2} + D_2\frac{\partial^2 c}{\partial y^2}.$$

We assume Dirichlet boundary conditions:

$$u = u_D \text{ on } \partial\Omega_{Dirich} \quad \text{with}$$

$$u_D(x,y) = sin(\frac{1}{\sqrt{D_1}}\pi x) \cdot sin(\frac{1}{\sqrt{D_2}}\pi y),$$

We can derive an analytical solution which we will use as reference solution for the error estimates:

$$c(x,y,t) = \sin(\frac{1}{\sqrt{D_1}}\pi x_1) \cdot \sin(\frac{1}{\sqrt{D_2}}\pi y) \cdot \cos(\sqrt{2}\pi t),$$

The analytical solution is periodic. Thus it suffices to do the error analysis for the following domain:

$$x \in [0, 2 \cdot D_1]$$

$$y \in [0, 2 \cdot D_2]$$

$$t \in [0, \sqrt{2}]$$

Remark 7. *The analytical solutions for the constant coefficients are given exact solutions for* $t = \frac{n}{\sqrt{2}}$, *for this we obtain the boundary conditons of the solutions. The extrem values are given with respect to* $\cos(\sqrt{2}\pi t) = \pm 0.5$.

We consider stiff and non stiff equations with D_1, $D_2 \in [0, 1]$. In section 5 we gave some options for the initial condition to start the iterative method. In [12] we discussed the optimization with respect to the initialisation process. Here the best initialisation is obtained by a prestep first order method, **I.5**. However, this option needs one more iteration step. Thus we take the explicit method **I.4** for our experiment which delivers almost optimal results.

As already mentioned above we take the analytical solution as reference function and consider an average of L^1-errors over time calculated by:

$$err_{L_1}(t^n) := \sum_{i,j} | u(x^i, y^j, t^n) - u_{anal}(x^i, y^j, t^n) | \cdot \Delta x \cdot \Delta y, \tag{77}$$

$$err_{L_1} := \sum_n err_{L_1}(t^n) \cdot \Delta t, \tag{78}$$

We exercised experiments for non stiff (table (1) and (2)) and stiff (table (3) and (4)) equations while we changed the parameters η and Δt for constant spatial discretisation. Generally, we see that the test series for the stiff equation deliver better results than the one for the non stiff equation. This can be deduced to the smaller spatial grid, see domain restrictions.

In table (1)–(4) we observe that we obtain the best result for $\eta = 0$ and $tsteps = 16$, e.g. for the explicit method. However, for smaller time steps we can always find an η, e.g. implicit method, so that the L^1-error is within an acceptable range. The benefit of the implicit methods is the reduction in computational time, see table (6), with a small loss in accuracy. During our experiments we observed a correlation between η and Δt. It appears that for each given number of time steps there is an η that minimizes the L^1-error indepedently of the equation's stiffness. In tables (1)-(4) we have just listed these numerically computed η's with some additional values to see the movement. We experimented with up to three decimal places for η. We assume, however, that you can minimise the error more if you increase the number of decimal places. This leads us to the idea that for each given time step size there may exist a weight function ω of Δt with which we can obtain a optimal η to reduce the error. We assume that this phenomenon is closely related to the CFL condition and shall give a brief survey on it in the follwing section.

6.2 CFL condition

We look at the CFL condition for the methods in use, see [12], which is given by:

$$\Delta t \leq \frac{1}{\sqrt{D_{max}}} \frac{x_{min}}{\sqrt{2}} \frac{1}{\sqrt{1-2\eta}},$$

where Δt, $D_{max} = \max\{D_1, D_2\}$, $x_{min} = \min\{\Delta x, \Delta y\}$ for $\Delta x = \frac{2 \cdot D_1}{xsteps}$ and $\Delta y = \frac{2 \cdot D_2}{ysteps}$.

Based on the observations in tables (1)-(4) we assume that we need to take an additional value into account to achieve optimal results:

$$\omega(\Delta t) = \frac{x_{min}}{\sqrt{2(\Delta t)^2 D_{max}(1-2\eta)}},$$

where ω may be thought of as a weight function of the CFL condition. In table (5) we calculated ω for the numerically obtained optimal pairs of η and $tsteps$ from the tables (1) and (2). Then, we applied a linear regression to the values in table (5) with respect to Δt and found the linear function

$$\omega(\Delta t) = 9.298\Delta t + 0.2245. \tag{79}$$

With this function at hand, we can determine an ω for every Δt. We can use this ω to calculate an optimal η with respect to Δt in order to minimise the numerical error. Hence, we have a tool to minimise costs without loosing much accuracy. We think that it is even possible to have more accurate ω-functions based on the accuracy of the optimal η with respect to $tsteps$ which we had calculated before to gain ω via linear regression. We will follow this interesting issue in our future work.

Finally, we present test series where we changed the number of iterations in table (7). For different number of time steps we choose the correlated η with the smallest error and exercise on them different types of iteration. We do not observe any significant difference.

Remark 8. *In the numerical experiments we can see the benefit of applying less iterative steps, because of the sufficient accuracy of the method. Thus $i = 2,3$ is sufficient. The optimal iterative steps are realted to the order of the time- and spatial discretisation, see [12]. This means that with time and* *spatial discretisation orders of $O(\Delta t^q)$ and Δx^p the number of iterative steps are i = min p, q, while we assume to have optimal CFL condition. The optimisation in the spatial and time discretisation can be derived from the CFL condition. Here we obtain at least second order methods. The explicit methods are more accurate but need higher computational time, so that we have to balance between sufficient accuracy of the solutions and low computational time achieved by implicit methods, where we can minimise the error using the wight function ω.*

6.3 Wave equation with linearly time dependent diffusion coefficients

We carried out the experiments for the following time dependent PDE:

$$\frac{\partial^2 c}{\partial t^2} = D_1(t)\frac{\partial^2 c}{\partial x^2} + D_2(t)\frac{\partial^2 c}{\partial y^2}, \quad (x,y,t) \in [0,2]\times[0,2]\times[0,\sqrt{2}]$$

$$D_1(t) = \frac{1/1000 - 1}{T} + 1,$$

$$D_2(t) = \frac{1 - 1/1000}{T} + 1/1000.$$

For the experiments we fixe the spatial step sizes Δx and Δy, the iteration depths, η and the inner time step size τ^{in} and change the length of the overlapped region p and the number of outer time steps. We proved that the smaller τ^{out} the closer the numerical (resp. semi-analytical) solution to the assumed analytical. For all subintervals we choose one η and τ^{in} optimally in accordance with our analysis in section 6.2.

We consider L^1-errors over the complete time domain, see (77)–(78), while we take as compare functions the semi-analytical solutions.

In table (8) we compare the L^1-error for different values of p and $tsteps^{out}$. We do not see any significant difference when altering p. This may be a reassurement of what we proved in lemma 2. However, we can observe a considerable decrease of the L^1-error increasing the outer time steps.

Thus, in our next experiment, reflected in table (9), we fixe p = 4, too, and only alter $tsteps^{out}$. We can observe that the error diminshes significantly while raising the number of outer time steps.

Remark 9. *The results show benefits in balancing between time intervals and the optimal CFL number. While implicit methods are less expensive in computations, explicit time discretization schemes are accurate and more expensive. Here we have to taken into account the CFL conditions. Small overlapping and sufficient small iterative steps helps to have an interesting scheme. A balance between time intervalls and iterative steps acchieve the best results in comparison to standard iterative schemes.*

7. Conclusions and discussions

We have presented a new iterative splitting methods to solve time dependent wave equations. Based on a overlapping scheme we could obtain more accurate results of the splitting scheme. Effective balancing of explicit and implicit time-discretization methods, with semi-analytical solutions achieve higher order schemes. Here the delicate problem of time-dependent wave equations are solved with iterative and analytical methods. In future we will continue on nonlinear wave equations and the balancing of time and spatial discretization schemes.

8. References

[1] M. Bjorhus. *Operator splitting for abstract Cauchy problems.* IMA Journal of Numerical Analysis, 18, 419–443, 1998.

[2] W. Cheney, Analysis for Applied Mathematics, Graduate Texts in Mathematics., 208, Springer, New York, Berlin, Heidelberg, 2001.

[3] G. Cohen. *Higher-Order Numerical Methods for Transient Wave Equations.* Series Scientific Computation , Spriner-Verlag, New York, Heidelberg, 2002.

[4] C. N. Dawson, Q. Du, and D. F. Dupont, A finite Difference Domain Decomposition Algorithm for Numerical solution of the Heat Equation, Mathematics of Computation 57 (1991) 63-71.

[5] D.R. Durran. *Numerical methods for wave equations in geophysical fluid dynamics.* Text in applied mathematics, Springer-Verlag, Heidelberg, New York, 1999.

[6] K.-J. Engel and R. Nagel, One-Parameter Semigroups for Linear Evolution Equations, Springer, New York, 2000.

[7] I. Farago and J. Geiser, Iterative Operator-Splitting methods for Linear Problems, Preprint No. 1043 of Weierstrass Institute for Applied Analysis and Stochastics, Berlin, Germany. International Journal of Computational Science and Engineering, accepted September 2007.

[8] M.J. Gander and H. Zhao, Overlapping Schwarz waveform relaxation for parabolic problems in higher dimension, In A. Handlovičová, Magda Komorníkova, and KarolMikula, editors, in: Proc. Algoritmy 14, Slovak Technical University, 1997, pp. 42-51.

[9] E. Giladi and H. Keller, Space time domain decomposition for parabolic problems. Technical Report 97-4, Center for research on parallel computation CRPC, Caltech, 1997.

[10] J. Geiser, Discretisation Methods with embedded analytical solutions for convection dominated transport in porous media, in: Proc. NA&A '04, Lecture Notes in Computer Science, Vol.3401, Springer, Berlin, 2005, pp. 288-295.

[11] J. Geiser, Iterative Operator-Splitting Methods with higher order Time- Integration Methods and Applications for Parabolic Partial Differential Equations, J. Comput. Appl. Math., accepted, June 2007.

[12] J. Geiser and L. Noack, *Iterative Operator-splitting methods for waveequations with stability results and numerical examples,* Preprint 2007-10 of Humboldt University of Berlin, Department of Mathematics, Germany, 2007.

[13] S. Hu, N.S. Papageorgiou. *Handbook of Multivalud Analysis I,II.* Kluwer, Dordrecht, Part I: 1997, Part II: 2000.

[14] W. Hundsdorfer and J.G. Verwer, Numerical Solution of Time-Dependent Advection-Diffusion-Reaction Equations, Springer Series in Computational Mathematics Vol. 33, Springer Verlag, 2003.

[15] J. Kanney, C. Miller, and C.T. Kelley, Convergence of iterative splitoperator approaches for approximating nonlinear reactive transport problems, Advances in Water Resources 26 (2003) 247-261.

[16] K.H. Karlsen and N. Risebro. *An Operator Splitting method for nonlinear convection-diffusion equation.* Numer. Math., 77, 3 , 365–382, 1997.
[17] K.H. Karlsen and N.H. Risebro, Corrected operator splitting for nonlinear parabolic equations, SIAM J. Numer. Anal. 37 (2000) 980-1003.
[18] K.H. Karlsen, K.A. Lie, J.R. Natvig, H.F. Nordhaug and H.K. Dahle, Operator splitting methods for systems of convection-diffusion equations: nonlinear error mechanisms and correction strategies, J. Comput. Phys. 173 (2001) 636-663.
[19] C.T. Kelly. *Iterative Methods for Linear and Nonlinear Equations.* Frontiers in Applied Mathematics, SIAM, Philadelphia, USA, 1995.
[20] P. Knabner and L. Angermann, Numerical Methods for Elliptic and Parabolic Partial Differential Equations, Text in Applied Mathematics, Springer Verlag, Newe York, Berlin, vol. 44, 2003.
[21] J.M. Lees. *Elastic Wave Propagation and Generation in Seismology.* Eos Trans. AGU, 84(50), doi:10.1029/2003EO500012, 2003.
[22] E. Hairer, C. Lubich, and G. Wanner. *Geometric Numerical Integration: Structure-Preserving Algorithms for Ordinary Differential Equations.* SCM, Springer-Verlag Berlin-Heidelberg-New York, No. 31, 2002.
[23] G.I. Marchuk, Some applicatons of splitting-up methods to the solution of problems in mathematical physics, Aplikace Matematiky 1 (1968) 103-132.
[24] R.I. McLachlan, G. Reinoult, and W. Quispel. *Splitting methods.* Acta Numerica, 341–434, 2002.
[25] A.D. Polyanin and V.F. Zaitsev. *Handbook of Nonlinear Partial Differential Equations.* Chapman & Hall/CRC Press, Boca Raton, 2004.
[26] H. Roos, M. Stynes and L. Tobiska. Numerical Methods for Singular Perturbed Differential Equations, Springer-Verlag, Berlin, Heidelberg, New York, 1996.
[27] H.A. Schwarz, Über einige Abbildungsaufgaben, Journal f¨ur Reine und Angewandte Mathematik 70 (1869) 105-120.
[28] B. Sportisse. *An Analysis of Operator Splitting Techniques in the Stiff Case.* Journal of Computational Physics, 161:140–168, 2000.
[29] G. Strang, On the construction and comparision of difference schemes, SIAM J. Numer. Anal. 5 (1968) 506-517.
[30] H. Yoshida, Construction of higher order symplectic integrators, Physics Letters A, Vol. 150, no. 5,6,7, 1990.
[31] E. Zeidler. Nonlinear Functional Analysis and its Applications. II/A Linear montone operators Springer-Verlag, Berlin-Heidelberg-New York, 1990.
[32] E. Zeidler. Nonlinear Functional Analysis and its Applications. II/B Nonlinear montone operators Springer-Verlag, Berlin-Heidelberg-New York, 1990.
[33] Z. Zlatev. Computer Treatment of Large Air Pollution Models. Kluwer Academic Publishers, 1995.

8. Appendix: regression (least square approximation) for extrapolation of functions

Here we have points with values and we assume to have a best approximation with respect to following minimisation:

$$S = \sum_{k=1}^{m} (y_k - L_n(x_k))^2,$$

where $m \geq n$, y_k are the values for the regression and L_n is a function, e.g. polynom, exponential function, etc. that is constructed with the least square algorithm.

9. Tables

η	ErrL1	ErrL1	ErrL1	ErrL1	ErrL1
tsteps	2	3	4	6	8
0.0	3.4461e+01	4.9988e+00	7.1888e-01	1.9701e-01	8.9679e-02
0.02	2.1240e+01	2.9023e+00	4.9343e-01	1.4000e-01	6.1068e-02
0.037	1.4416e+01	1.7632e+00	3.3421e-01	9.4631e-02	3.7458e-02
0.052	1.0303e+01	1.0795e+00	2.1622e-01	5.6866e-02	1.7159e-02
0.065	7.6778e+00	6.5839e-01	1.2939e-01	2.5795e-02	3.5487e-05
0.076	5.9393e+00	3.9440e-01	6.6075e-02	6.7007e-04	1.4458e-02
0.09	4.1978e+00	1.5079e-01	2.3614e-03	3.0836e-02	3.2885e-02
0.102	3.1107e+00	4.7987e-03	5.7883e-02	5.8181e-02	4.8735e-02
0.154	5.0867e-02	5.4006e-01	3.2383e-01	1.7908e-01	1.1781e-01
0.2	1.3615e+00	9.7161e-01	5.6802e-01	2.8732e-01	1.7912e-01
0.3	2.6516e+00	1.6138e+00	1.0523e+00	5.1841e-01	3.1147e-01
0.4	3.2841e+00	1.9514e+00	1.4344e+00	7.3493e-01	4.4047e-01
0.5	3.7528e+00	2.1258e+00	1.7175e+00	9.3132e-01	5.6435e-01

Table 1. $D_1 = 1$, $D_2 = 1$, $\Delta x = \Delta y = \frac{1}{8}$, iter_depth= 2.

η	ErrL1	ErrL1	ErrL1	ErrL1	ErrL1
tsteps	10	12	14	16	32
0.0	4.5568e-02	2.2517e-02	8.8274e-03	1.0083e-15	2.1641e-02
0.02	2.7865e-02	1.0372e-02	5.2088e-05	6.8018e-03	2.3353e-02
0.037	1.3056e-02	1.4938e-04	7.5747e-03	1.2583e-02	2.4807e-02
0.052	1.7048e-04	8.8323e-03	1.4213e-02	1.7683e-02	2.6091e-02
0.065	1.0931e-02	1.6621e-02	1.9967e-02	2.2103e-02	2.7203e-02
0.076	2.0338e-02	2.3214e-02	2.4836e-02	2.5842e-02	2.8144e-02
0.09	3.2324e-02	3.1607e-02	3.1032e-02	3.0600e-02	2.9341e-02
0.102	4.2609e-02	3.8803e-02	3.6342e-02	3.4678e-02	3.0367e-02
0.154	8.7245e-02	6.9988e-02	5.9344e-02	5.2335e-02	3.4811e-02
0.2	1.2674e-01	9.7551e-02	7.9667e-02	6.7935e-02	3.8739e-02
0.3	2.1221e-01	1.5725e-01	1.2371e-01	1.0176e-01	4.7271e-02
0.4	2.9654e-01	2.1645e-01	1.6748e-01	1.3542e-01	5.5789e-02
0.5	3.7915e-01	2.7491e-01	2.1089e-01	1.6887e-01	6.4294e-02

Table 2. $D_1 = 1$, $D_2 = 1$, $\_x = \_y = \frac{1}{8}$, iter_depth= 2.

η	ErrL1	ErrL1	ErrL1	ErrL1	ErrL1
tsteps	2	3	4	6	8
0.0	1.0897e+00	1.5808e-01	2.2733e-02	6.2300e-03	2.8359e-03
0.02	6.7167e-01	9.1778e-02	1.5604e-02	4.4272e-03	1.9311e-03
0.037	4.5587e-01	5.5758e-02	1.0569e-02	2.9925e-03	1.1845e-03
0.052	3.2582e-01	3.4137e-02	6.8376e-03	1.7983e-03	5.4261e-04
0.065	2.4279e-01	2.0820e-02	4.0918e-03	8.1571e-04	1.1222e-06
0.076	1.8782e-01	1.2472e-02	2.0895e-03	2.1189e-05	4.5719e-04
0.09	1.3274e-01	4.7685e-03	7.4675e-05	9.7511e-04	1.0399e-03
0.102	9.5563e-02	1.5175e-04	1.8304e-03	1.8399e-03	1.5412e-03
0.154	1.6086e-03	1.7078e-02	1.0240e-02	5.6631e-03	3.7256e-03
0.2	4.3053e-02	3.0725e-02	1.7962e-02	9.0860e-03	5.6644e-03
0.3	8.3850e-02	5.1033e-02	3.3278e-02	1.6394e-02	9.8494e-03
0.4	1.0385e-01	6.1709e-02	4.5360e-02	2.3241e-02	1.3929e-02
0.5	1.1867e-01	6.7225e-02	5.4311e-02	2.9451e-02	1.7846e-02

Table 3. $D_1 = 1$, $D_2 = 1/1000$, $\Delta x = \frac{1}{8}\Delta y = \frac{1}{8\cdot\sqrt{1000}}$, iter_depth= 2.

η	ErrL1	ErrL1	ErrL1	ErrL1	ErrL1
tsteps	10	12	14	16	32
0.0	1.4410e-03	7.1204e-04	2.7915e-04	1.0391e-16	6.8434e-04
0.02	8.8117e-04	3.2800e-04	1.6472e-06	2.1509e-04	7.3847e-04
0.037	4.1288e-04	4.7238e-06	2.3953e-04	3.9791e-04	7.8448e-04
0.052	5.3910e-06	2.7930e-04	4.4946e-04	5.5919e-04	8.2506e-04
0.065	3.4567e-04	5.2560e-04	6.3142e-04	6.9896e-04	8.6023e-04
0.076	6.4313e-04	7.3408e-04	7.8538e-04	8.1720e-04	8.8998e-04
0.09	1.0222e-03	9.9950e-04	9.8132e-04	9.6767e-04	9.2783e-04
0.102	1.3474e-03	1.2271e-03	1.1492e-03	1.0966e-03	9.6028e-04
0.154	2.7589e-03	2.2132e-03	1.8766e-03	1.6550e-03	1.1008e-03
0.2	4.0079e-03	3.0848e-03	2.5193e-03	2.1483e-03	1.2250e-03
0.3	6.7107e-03	4.9728e-03	3.9121e-03	3.2179e-03	1.4948e-03
0.4	9.3775e-03	6.8447e-03	5.2963e-03	4.2823e-03	1.7642e-03
0.5	1.1990e-02	8.6935e-03	6.6690e-03	5.3403e-03	2.0332e-03

Table 4. $D_1 = 1$, $D_2 = 1/1000$, $\Delta x = \frac{1}{8}\Delta y = \frac{1}{8\cdot\sqrt{1000}}$, iter_depth= 2.

tsteps	η	ω
16	0	1.0000
14	0.02	1.1198
12	0.037	1.2831
10	0.052	1.5145
8	0.065	1.8655
6	0.076	2.4557
4	0.09	3.6222
3	0.102	4.7583
2	0.154	6.6549

Table 5. Calculating ω for different values of dt and η. $D1 = D2 = 1$, $dx = dy = 1/8$, $ttop = sqrt(2)$.

η	tsteps	Comp.Time [msec]
0.154 (implicit)	2	133
0 (explicit)	16	308

Table 6. Computational time of the explicit and implicit schemes.

# iter	ErrL1	ErrL1	ErrL1	ErrL1
tsteps	4	8	12	16
1	1.1990e-03	1.3095e-04	1.3503e-04	1.0083e-15
2	2.3614e-03	3.5487e-05	1.4938e-04	1.0083e-15
4	2.3873e-03	3.5486e-05	1.4938e-04	1.0083e-15
6	2.3873e-03	3.5486e-05	1.4938e-04	1.0083e-15

Table 7. $\Delta x = \Delta y = \frac{1}{8}$. For each tsteps we take the η with the best result from table 1 and 2.

$tsteps^{out}$	ErrL1	ErrL1	ErrL1	ErrL1
p	4	8	16	32
5	0.5551	0.5773	0.5754	0.4877
10	0.3178	0.3453	0.3853	0.4133
20	0.1695	0.1864	0.2169	0.2633
40	0.0865	0.0957	0.1133	0.1446

Table 8. $\Delta x = \Delta y = \frac{1}{64}$, iter_depth= 2, $\eta = 0$ and tsteps= 64.

$tsteps^{out}$	ErrL1
5	0.5551
10	0.3178
20	0.1695
40	0.0865
80	0.0435

Table 9. $\Delta x = \Delta y = \frac{1}{64}$, iter_depth= 2, $\eta = 0$, tsteps= 64 and $p = 4$.

5

Comparison Between Reverberation-Ray Matrix, Reverberation-Transfer Matrix, and Generalized Reverberation Matrix

Jiayong Tian
Institute of Crustal Dynamics, China Earthquake Administration
P.R.China

1. Introduction

Different matrix formulations have been developed to investigate the elastic-wave propagation in a multilayered solid, which have been widely used in the fields of seismology, ocean acoustics(Pao et al., 2000), and non-destructive evaluation (Lowe, 1995), etc. Transfer matrix (TM) method (Haskell, 1953; Thomson, 1950), as one of the most important matrix formulations, yields a simple configuration and efficient computational ability to facilitate its wide application in many research fields. Stiffness matrix (SM) method (Rokhlin & Wang, 2002; Wang & Rokhlin, 2001) has been proposed to resolve the inherent computational instability for the large product of frequency and thickness in TM method. The SM formulation utilizes the stiffness matrix of each sublayer in a recursive algorithm to obtain the stacked stiffness matrix for the multilayered solid. However, the SM formulation is difficult to identify the generalized-ray propagation in the multilayered solid.

In order to evaluate the transient wave propagation in the multilayered solid, Su, Tian, and Pao (Su et al., 2002; Tian & Xie, 2009; Tian et al., 2006) presented reverberation-ray matrix (RRM) formulation. Introducing the local scattering relations at interfaces and the phase relations in sublayers, a system of equations is formulated by a reverberation matrix $\mathbf{R}$, which can be automatically represented as a series of generalized ray group integrals according to the times of reflections and refractions of generalized rays at interfaces. Each generalized ray group integral containing $\mathbf{R}^k$ represents the set of K times reflections and transmissions of source waves arriving at receivers in the multilayered solid, which is very suitable to automatic computer programming for the simple multilayered-solid configuration. However, the dimension of the reverberation matrix will increase as the number of the sublayers increases, which may yield the lower calculation efficiency of the generalized-ray groups in the complex multilayered solid(Tian & Xie, 2009).

In order to increase the calculation efficiency of the generalized-ray groups, Tian presented the reverberation-transfer matrix (RTM) and generalized reverberation matrix (GRM) formulations, respectively. In RTM formulation, RRM formulation is applied to the interested sublayer for the evaluation of the generalized rays and TM formulation to the other sublayers, to construct a RTM of the constant dimension, which is independent of the sublayer number. However, the RTM suffers from the inherent numerical instabilities particularly when the layer thickness becomes large and/ or the frequency is high. GRM formulation is to integrate RRM and SM formulations. The RRM formulation is applied to the interested sublayer for the evaluation of the generalized rays and SM formulation to the other sublayers, to construct a generalized reverberation matrix of the constant dimension, which is independent of the sublayer number. GRM formulation has the higher calculation efficiency and numerical stabilities of the generalized rays in the complex multilayered-solid configuration.

In this chapter, in order to facilitate the wide application of RRM, RTM, and GRM formulations, we compare them clearly to show their difference and applicability.

2. RRM formulation

Here, we only consider in-plane wave propagation in a laminate containing N isotropic sublayers impacted by the vertical force *f(t,x)* on the top surface of the laminate. In RRM formulation, the interfaces between sublayers are expressed by capital letters $I, J, \cdots$. Two local Cartesian coordinate systems $(x,y)^{IJ}$ and $(x,y)^{JI}$ are constructed at two interfaces of sublayer *IJ*, respectively. The thickness of sublayer *IJ* is represented by h^{IJ}, which is shown in Fig.1.

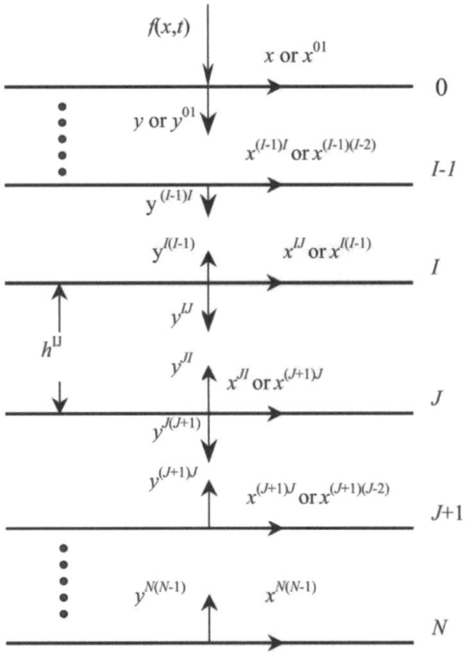

Fig. 1. The schematic diagram of the global and local coordinates for RRM formulation

Introducing the one-side Laplace transform with respect to the time and the double-side Laplace transform with respect to the spatial variable x as(Achenbach, 1973)

$$\hat{f}(\eta,y,p) = \int_{-\infty}^{\infty} e^{-p\eta x} \int_{0}^{\infty} f(x,y,t) e^{-pt} dt dx , \tag{1}$$

the transformed wave equations with related to the displacement potentials φ^{IJ} and ψ^{IJ} in the local Cartesian coordinate system $(x,y)^{IJ}$ are denoted as

$$\left.\begin{aligned}\frac{\partial^2\varphi^{IJ}}{\partial y^2}-\left(\gamma_1^{IJ}\right)^2\varphi^{IJ}=0\\ \frac{\partial^2\psi^{IJ}}{\partial y^2}-\left(\gamma_2^{IJ}\right)^2\psi^{IJ}=0\end{aligned}\right\}, \tag{2}$$

Introduction of the unknown arriving and departing wave-amplitude vectors $\hat{\mathbf{a}}^{IJ}=\left\{a_1^{IJ},a_2^{IJ}\right\}^T$, $\hat{\mathbf{d}}^{IJ}=\left\{d_1^{IJ},d_2^{IJ}\right\}^T$ in the local coordinate system $(x,y)^{IJ}$ yields the displacement and stress vectors $\hat{\mathbf{U}}^{IJ}\left(\eta,y^{IJ},p\right)=\left\{\hat{u}_x^{IJ},\hat{u}_y^{IJ}\right\}^T$, $\hat{\mathbf{F}}^{IJ}\left(\eta,y^{IJ},p\right)=\left\{\hat{\tau}_{yx}^{IJ},\hat{\tau}_{yy}^{IJ}\right\}^T$ from Eq. (2) as

$$\left.\begin{aligned}\hat{\mathbf{U}}^{IJ}\left(\eta,y^{IJ},p\right)&=p\mathbf{A}_u^{IJ}\hat{\mathbf{a}}^{IJ}+p\mathbf{D}_u^{IJ}\hat{\mathbf{d}}^{IJ}\\ \hat{\mathbf{F}}^{IJ}\left(\eta,y^{IJ},p\right)&=\mu^{IJ}p^2\mathbf{A}_f^{IJ}\hat{\mathbf{a}}^{IJ}+\mu^{IJ}p^2\mathbf{D}_f^{IJ}\hat{\mathbf{d}}^{IJ}\end{aligned}\right\}, \tag{3}$$

where $\mathbf{A}_u^{IJ}$ and $\mathbf{D}_u^{IJ}$ are phase-related receiver matrixes for the displacements, $\mathbf{A}_f^{IJ}$ and $\mathbf{D}_f^{IJ}$ for the stresses, respectively. With the definition of the arriving wave amplitude vector $\mathbf{a}^J$ and the departing wave amplitude vector $\mathbf{d}^J$ of interface J as

$$\left.\begin{aligned}\mathbf{a}^J&=\left\{a_1^{J(J-1)},a_2^{J(J-1)},a_1^{J(J+1)},a_2^{J(J+1)}\right\}^T\\ \mathbf{d}^J&=\left\{d_1^{J(J-1)},d_2^{J(J-1)},d_1^{J(J+1)},d_2^{J(J+1)}\right\}^T\end{aligned}\right\}, \tag{4}$$

the application of the boundary conditions yields scattering relation at interface J

$$\mathbf{d}^J=\mathbf{S}^J\mathbf{a}^J+\mathbf{s}^J, \tag{5}$$

where $\mathbf{S}^J$ and $\mathbf{s}^J$ are the scattering matrix and source matrix of interface J, respectively. With the definition of global arriving and departing wave amplitude vectors $\mathbf{a}=\left\{\left\{\mathbf{a}^1\right\}^T,\left\{\mathbf{a}^2\right\}^T,\ldots,\left\{\mathbf{a}^N\right\}^T\right\}^T$ and $\mathbf{d}=\left\{\left\{\mathbf{d}^1\right\}^T,\left\{\mathbf{d}^2\right\}^T,\ldots,\left\{\mathbf{d}^N\right\}^T\right\}^T$, the global scattering matrix can be written in the following form

$$\mathbf{d}=\mathbf{S}\mathbf{a}+\mathbf{s}. \tag{6}$$

Since both vectors **a** and **d** are unknown quantities, an additional equation related to **a** and **d** must be provided. A wave arriving at interface I in the local coordinate $(x,y)^{IJ}$, is also considered as the wave departing from interface J of the same layer in the local coordinate $(x,y)^{JI}$, which yields the other relation between the global arriving and departing wave amplitude vectors

$$\mathbf{a}=\mathbf{P}\mathbf{H}\mathbf{d}, \tag{7}$$

where the phase matrix **P** is a $4N\times4N$ diagonal matrix. **H** is a $4N\times4N$ matrix composed of only one element whose value is one in each line and each row and others are all zero. For example, in vector **d**, if d_i^{JK} and d_i^{KJ} are in the positions p and q respectively, then the elements H_{pq} and H_{qp} in the matrix **H** have the same value one.

3. RTM and GRM formulations

In RTM and GRM formulations, assuming that the receiver is in the sublayer *IJ*, the laminate can be partitioned into three layers, which includes layers *0I*, *IJ*, and *JN* as shown in Fig.2. If the receiver is in the top or bottom sublayer, the laminate will be partitioned into two layers, which includes layers *01*, and *1N*, or 1(*N*-1) and (*N*-1)*N*, respectively.

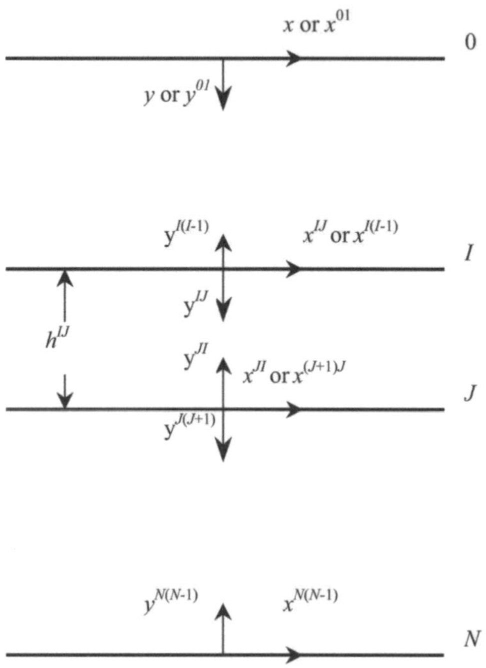

Fig. 2. The schematic diagram of the global and local coordinates for RRM formulation

Considering the continuity conditions at the interfaces *I* and *J* and boundary conditions at the surfaces, the scattering relations at interfaces *0*, *I*, *J*, and *N* yield,

$$\left.\begin{aligned} \mathbf{d}^0 &= \mathbf{S}^0\mathbf{a}^0 + \mathbf{s}^0 \\ \mathbf{d}^I &= \mathbf{S}^I\mathbf{a}^I + \mathbf{s}^I \\ \mathbf{d}^J &= \mathbf{S}^J\mathbf{a}^J + \mathbf{s}^J \\ \mathbf{d}^N &= \mathbf{S}^N\mathbf{a}^N + \mathbf{s}^N \end{aligned}\right\}. \tag{8}$$

If we stack all the local amplitude vectors of the arriving and departing waves as $\mathbf{a} = \left\{\{\mathbf{a}^0\}^T, \{\mathbf{a}^I\}^T, \{\mathbf{a}^J\}^T, \{\mathbf{a}^N\}^T\right\}^T$ and $\mathbf{d} = \left\{\{\mathbf{d}^0\}^T, \{\mathbf{d}^I\}^T, \{\mathbf{d}^J\}^T, \{\mathbf{d}^N\}^T\right\}^T$, the global scattering relation can be denoted as

$$\mathbf{d} = \mathbf{S}\mathbf{a} + \mathbf{s}, \tag{9}$$

where **S** and **s** are the global scattering matrix and source matrix, respectively. Since both **a** and **d** are unknown quantities, an additional equation related to **a** and **d** must be provided.

3.1 Sublayer *IJ*

Similar with the RRM formulation, the local phase matrix for the sublayers *IJ* can be given as

$$\left.\begin{aligned}\mathbf{a}^{IJ} &= \mathbf{P}^{IJ}\mathbf{d}^{JI} \\ \mathbf{a}^{JI} &= \mathbf{P}^{JI}\mathbf{d}^{IJ}\end{aligned}\right\}. \tag{10}$$

3.2 Layers *0I* and *JN*

3.2.1 Reverberation-transfer matrix formulation

In layers *0I* or *JN*, TM formulation is adopted to describe the relations of wave amplitudes between interfaces *0* and *I* or interfaces *J* and *N*. The displacements and stresses at interfaces *0* and *1* in the local coordinate $(x,y)^{01}$ are written as

$$\left.\begin{aligned}\hat{\mathbf{G}}^{01}(0) &= \mathbf{C}^{01}\mathbf{b}^{01} \\ \hat{\mathbf{G}}^{01}\left(h^{01}\right) &= \mathbf{E}^{01}\ \mathbf{b}^{01}\end{aligned}\right\}, \tag{11}$$

where $\mathbf{G}^{01}\left(y^{01}\right)=\left\{\left\{\hat{\mathbf{U}}\left(\eta,y^{01},p\right)\right\}^T,\left\{\hat{\mathbf{F}}\left(\eta,y^{01},p\right)\right\}^T\right\}$, $\mathbf{b}^{01}=\left\{\left\{\mathbf{a}^{01}\right\}^T,\left\{\mathbf{d}^{01}\right\}^T\right\}^T$. Hence, their relations are deduced from the above equations

$$\hat{\mathbf{G}}^{01}\left(h^{01}\right)=\mathbf{E}^{01}\left(\mathbf{C}^{01}\right)^{-1}\hat{\mathbf{G}}^{01}(0). \tag{12}$$

Using the above recursion, we obtain the displacement-stress relation between interfaces *0* and *I*,

$$\hat{\mathbf{G}}^{(I-1)I}\left(h^{(I-1)I}\right)=\left(\prod_{k=1}^{I}\mathbf{E}^{(k-1)k}\left(\mathbf{C}^{(k-1)k}\right)^{-1}\right)\hat{\mathbf{G}}^{01}(0). \tag{13}$$

The displacements and stresses at interface *I* in the local coordinate $(x,y)^{I(I-1)}$ can be denoted as

$$\hat{\mathbf{G}}^{I(I-1)}(0)=\mathbf{C}^{I(I-1)}\mathbf{b}^{I(I-1)}. \tag{14}$$

Here, $\hat{\mathbf{G}}^{I(I-1)}(0)$ and $\hat{\mathbf{G}}^{(I-1)I}\left(h^{(I-1)I}\right)$ represent the displacements and stresses at the same position but in the different local coordinate systems, which have the following relation as,

$$\hat{\mathbf{G}}^{I(I-1)}(0)=\mathbf{T}\hat{\mathbf{G}}^{(I-1)I}\left(h^{(I-1)I}\right), \tag{15}$$

where $\mathbf{T}=diag\{1,-1,-1,1\}$. Substitution of Eqs. (11-14) into Eq.(15) yields

$$\mathbf{b}^{I(I-1)}=\mathbf{L}\mathbf{b}^{01}, \tag{16}$$

where $\mathbf{L}=\left(\mathbf{C}^{I(I-1)}\right)^{-1}\mathbf{T}\left(\prod_{k=1}^{I}\mathbf{E}^{(k-1)k}\left(\mathbf{C}^{(k-1)k}\right)^{-1}\right)\mathbf{C}^{01}$. $\mathbf{b}^{01}$ and $\mathbf{b}^{I(I-1)}$ represent wave amplitudes at interfaces *0* and *I* in the local coordinate $(x,y)^{01}$ and $(x,y)^{I(I-1)}$, respectively. Separation of Eq.(26) into the arriving and departing waves yields

$$\begin{bmatrix} \mathbf{a}^{01} \\ \mathbf{a}^{I(I-1)} \end{bmatrix} = \mathbf{P}^{0I} \begin{bmatrix} \mathbf{d}^{I(I-1)} \\ \mathbf{d}^{01} \end{bmatrix}, \tag{17}$$

where $\mathbf{P}^{0I} = \begin{bmatrix} -\mathbf{L}_{11} & \mathbf{I} \\ -\mathbf{L}_{21} & \mathbf{0} \end{bmatrix}^{-1} \begin{bmatrix} \mathbf{0} & \mathbf{L}_{12} \\ -\mathbf{I} & \mathbf{L}_{22} \end{bmatrix}$ is the phase matrix of layer *0I*, and I is a 2×2 identity matrix. Similarly, the phase relation of the arriving and departing waves in layer *JN* can be denoted as

$$\begin{bmatrix} \mathbf{a}^{J(J+1)} \\ \mathbf{a}^{N(N-1)} \end{bmatrix} = \mathbf{P}^{JN} \begin{bmatrix} \mathbf{d}^{N(N-1)} \\ \mathbf{d}^{J(J+1)} \end{bmatrix}. \tag{18}$$

3.2.2 GRM formulation

In layer *0I* or *JN*, stiffness matrix formulation is adopted to describe the relations of wave amplitudes between interfaces *0* and *I* or interfaces *J* and *N*. The displacements and stresses at interfaces *0* and *1* in the local coordinate system $(x,y)^{01}$ are written as

$$\begin{bmatrix} \hat{\mathbf{U}}(\eta,0^{01},p) \\ \hat{\mathbf{U}}(\eta,h^{01},p) \end{bmatrix} = \mathbf{E}_u^{01}\mathbf{b}^{01}, \tag{19}$$

$$\begin{bmatrix} \hat{\mathbf{F}}(\eta,0^{01},p) \\ \hat{\mathbf{F}}(\eta,h^{01},p) \end{bmatrix} = \mathbf{E}_f^{01}\mathbf{b}^{01}, \tag{20}$$

where

$$\mathbf{E}_f^{01} = \mu^{01}p^2 \begin{bmatrix} \mathbf{A}_f^{01}(0^{01}) & \mathbf{D}_f^{01}(0^{01}) \\ \mathbf{A}_f^{01}(h^{01}) & \mathbf{D}_f^{01}(h^{01}) \end{bmatrix},$$

$$\mathbf{E}_u^{01} = p \begin{bmatrix} \mathbf{A}_u^{01}(0^{01}) & \mathbf{D}_u^{01}(0^{01}) \\ \mathbf{A}_u^{01}(h^{01}) & \mathbf{D}_u^{01}(h^{01}) \end{bmatrix},$$

$$\mathbf{b}^{01} = \left\{ \left[\mathbf{a}^{01}\right]^T, \left[\mathbf{d}^{01}\right]^T \right\}^T.$$

The substitution of $\mathbf{b}^{01}$ from Eq. (19) into Eq. (20) yields the stiffness $\mathbf{K}^{01}$ of sublayer *01*

$$\begin{bmatrix} \hat{\mathbf{F}}(\eta,0^{01},p) \\ \hat{\mathbf{F}}(\eta,h^{01},p) \end{bmatrix} = \begin{bmatrix} \mathbf{K}_{11}^{01} & \mathbf{K}_{12}^{01} \\ \mathbf{K}_{21}^{01} & \mathbf{K}_{22}^{01} \end{bmatrix} \begin{bmatrix} \hat{\mathbf{U}}(\eta,0^{01},p) \\ \hat{\mathbf{U}}(\eta,h^{01},p) \end{bmatrix}. \tag{21}$$

where $\mathbf{K}^{01} = \mathbf{E}_f^{01}\left(\mathbf{E}_u^{01}\right)^{-1}$. Similarly, the stiffness $\mathbf{K}^{12}$ of sublayer *12* can be denoted as

$$\begin{bmatrix} \hat{\mathbf{F}}(\eta,0^{12},p) \\ \hat{\mathbf{F}}(\eta,h^{12},p) \end{bmatrix} = \begin{bmatrix} \mathbf{K}_{11}^{12} & \mathbf{K}_{12}^{12} \\ \mathbf{K}_{21}^{12} & \mathbf{K}_{22}^{12} \end{bmatrix} \begin{bmatrix} \hat{\mathbf{U}}(\eta,0^{12},p) \\ \hat{\mathbf{U}}(\eta,h^{12},p) \end{bmatrix}. \tag{22}$$

The stiffness $\mathbf{K}^{02}$ of layer *02* can be deduced from Eqs. (21) and (22) as

$$\begin{bmatrix} \hat{\mathbf{F}}(\eta,0^{01},p) \\ \hat{\mathbf{F}}(\eta,h^{12},p) \end{bmatrix} = \mathbf{K}^{02} \begin{bmatrix} \hat{\mathbf{U}}(\eta,0^{01},p) \\ \hat{\mathbf{U}}(\eta,h^{12},p) \end{bmatrix}, \tag{23}$$

where

$$\mathbf{K}^{02} = \begin{bmatrix} \mathbf{K}_{11}^{01} + \mathbf{K}_{12}^{01}\left(\mathbf{K}_{11}^{12} - \mathbf{K}_{22}^{01}\right)^{-1}\mathbf{K}_{21}^{01} & -\mathbf{K}_{12}^{01}\left(\mathbf{K}_{11}^{12} - \mathbf{K}_{22}^{01}\right)^{-1}\mathbf{K}_{12}^{12} \\ \mathbf{K}_{21}^{12}\left(\mathbf{K}_{11}^{12} - \mathbf{K}_{22}^{01}\right)^{-1}\mathbf{K}_{21}^{01} & \mathbf{K}_{22}^{12} - \mathbf{K}_{21}^{12}\left(\mathbf{K}_{11}^{12} - \mathbf{K}_{22}^{01}\right)^{-1}\mathbf{K}_{12}^{12} \end{bmatrix}.$$

The stiffness $\mathbf{K}^{0I}$ of layer *0I* can be deduced from the above recursion as

$$\begin{bmatrix} \hat{\mathbf{F}}(\eta,0^{01},p) \\ \hat{\mathbf{F}}(\eta,h^{(I-1)I},p) \end{bmatrix} = \mathbf{K}^{0I} \begin{bmatrix} \hat{\mathbf{U}}(\eta,0^{01},p) \\ \hat{\mathbf{U}}(\eta,h^{(I-1)I},p) \end{bmatrix}, \tag{24}$$

where

$$\mathbf{K}^{0I} = \begin{bmatrix} \mathbf{K}_{11}^{0(I-1)} + \mathbf{K}_{12}^{0(I-1)}\left(\mathbf{K}_{11}^{(I-1)I} - \mathbf{K}_{22}^{0(I-1)}\right)^{-1}\mathbf{K}_{21}^{0(I-1)} & -\mathbf{K}_{12}^{0(I-1)}\left(\mathbf{K}_{11}^{(I-1)I} - \mathbf{K}_{22}^{0(I-1)}\right)^{-1}\mathbf{K}_{12}^{(I-1)I} \\ \mathbf{K}_{21}^{(I-1)I}\left(\mathbf{K}_{11}^{(I-1)I} - \mathbf{K}_{22}^{0(I-1)}\right)^{-1}\mathbf{K}_{21}^{0(I-1)} & \mathbf{K}_{22}^{(I-1)I} - \mathbf{K}_{21}^{(I-1)I}\left(\mathbf{K}_{11}^{(I-1)I} - \mathbf{K}_{22}^{0(I-1)}\right)^{-1}\mathbf{K}_{12}^{(I-1)I} \end{bmatrix}.$$

The displacements and stresses at interface *I* in the local coordinate system $(x,y)^{(I-1)I}$ can be denoted in the local coordinate system $(x,y)^{I(I-1)}$ as

$$\begin{bmatrix} \hat{\mathbf{F}}(\eta,0^{01},p) \\ \hat{\mathbf{F}}(\eta,h^{(I-1)I},p) \end{bmatrix} = \mathbf{T}_f \begin{bmatrix} \hat{\mathbf{F}}(\eta,0^{01},p) \\ \hat{\mathbf{F}}(\eta,0^{I(I-1)},p) \end{bmatrix}, \tag{25}$$

$$\begin{bmatrix} \hat{\mathbf{U}}(\eta,0^{01},p) \\ \hat{\mathbf{U}}(\eta,h^{(I-1)I},p) \end{bmatrix} = \mathbf{T}_u \begin{bmatrix} \hat{\mathbf{U}}(\eta,0^{01},p) \\ \hat{\mathbf{U}}(\eta,0^{I(I-1)},p) \end{bmatrix}. \tag{26}$$

where $\mathbf{T}_f = diag\{1,1,-1,1\}$, $\mathbf{T}_u = diag\{1,1,1,-1\}$. Substitution of Eqs. (25) and (26) into Eq. (24) yields

$$\begin{bmatrix} \hat{\mathbf{F}}(\eta,0^{01},p) \\ \hat{\mathbf{F}}(\eta,0^{I(I-1)},p) \end{bmatrix} = \bar{\mathbf{K}}^{0I} \begin{bmatrix} \hat{\mathbf{U}}(\eta,0^{01},p) \\ \hat{\mathbf{U}}(\eta,0^{I(I-1)},p) \end{bmatrix}, \tag{27}$$

where $\bar{\mathbf{K}}^{01} = \left(\mathbf{T}_f\right)^{-1}\mathbf{K}^{01}\mathbf{T}_u$. Equation (27) can be expressed by the arriving-wave and departing-wave amplitude vectors,

$$\begin{bmatrix} \mathbf{A}_f & \mathbf{D}_f \end{bmatrix} \begin{bmatrix} \mathbf{a}^{01} \\ \mathbf{a}^{I(I-1)} \\ \mathbf{d}^{I(I-1)} \\ \mathbf{d}^{01} \end{bmatrix} = \bar{\mathbf{K}}^{0I} \begin{bmatrix} \mathbf{A}_u & \mathbf{D}_u \end{bmatrix} \begin{bmatrix} \mathbf{a}^{01} \\ \mathbf{a}^{I(I-1)} \\ \mathbf{d}^{I(I-1)} \\ \mathbf{d}^{01} \end{bmatrix}, \tag{28}$$

where

$$\mathbf{A}_f = \begin{bmatrix} \mathbf{A}_f^{01}(0^{01}) & \mathbf{0} \\ \mathbf{0} & \mathbf{A}_f^{I(I-1)}(0^{I(I-1)}) \end{bmatrix},$$

$$\mathbf{D}_f = \begin{bmatrix} \mathbf{0} & \mathbf{D}_f^{01}(0^{01}) \\ \mathbf{D}_f^{I(I-1)}(0^{I(I-1)}) & \mathbf{0} \end{bmatrix},$$

$$\mathbf{A}_u = \begin{bmatrix} \mathbf{A}_u^{01}(0^{01}) & \mathbf{0} \\ \mathbf{0} & \mathbf{A}_u^{I(I-1)}(0^{I(I-1)}) \end{bmatrix},$$

$$\mathbf{D}_u = \begin{bmatrix} \mathbf{0} & \mathbf{D}_u^{01}(0^{01}) \\ \mathbf{D}_u^{I(I-1)}(0^{I(I-1)}) & \mathbf{0} \end{bmatrix}.$$

Separation of Eq. (28) into the wave-amplitude vectors of the arriving and departing waves results in

$$\begin{bmatrix} \mathbf{a}^{01} \\ \mathbf{a}^{I(I-1)} \end{bmatrix} = \mathbf{P}^{0I} \begin{bmatrix} \mathbf{d}^{I(I-1)} \\ \mathbf{d}^{01} \end{bmatrix}, \tag{29}$$

where $\mathbf{P}^{0I} = \left[\mathbf{A}_f - \bar{\mathbf{K}}^{0I}\mathbf{A}_u\right]^{-1}\left[\bar{\mathbf{K}}^{0I}\mathbf{D}_u - \mathbf{D}_f\right]$ is the phase matrix of layer *0I*. Similarly, the phase relation of the arriving and departing waves in layer *JN* can be denoted as

$$\begin{bmatrix} \mathbf{a}^{J(J+1)} \\ \mathbf{a}^{N(N-1)} \end{bmatrix} = \mathbf{P}^{JN} \begin{bmatrix} \mathbf{d}^{N(N-1)} \\ \mathbf{d}^{J(J+1)} \end{bmatrix}. \tag{30}$$

4. Expansion of generalized-ray groups

The unknow amplitude vectors **a** and **d** in RRM, RTM, and GRM formulations yields

$$\left.\begin{aligned} \mathbf{d} &= [\mathbf{I}-\mathbf{R}]^{-1}\mathbf{s} \\ \mathbf{a} &= \mathbf{PH}[\mathbf{I}-\mathbf{R}]^{-1}\mathbf{s} \end{aligned}\right\}, \tag{31}$$

where **R** = **SPH** is reverberation matrix. Once **d** and **a** are known, the transformed displacements can be denoted as

$$\hat{\mathbf{U}}\left(\eta, y^{IJ}, p\right) = p(\mathbf{A}_u\mathbf{PH} + \mathbf{D}_u)(\mathbf{I}-\mathbf{R})^{-1}\mathbf{s}. \tag{32}$$

The transient displacements can be expressed by applying the inverse transforms

$$\mathbf{U}\left(x, y^{IJ}, t\right) = \frac{1}{2\pi i}\int_0^\infty \int_{Br} p^2(\mathbf{A}_u\mathbf{PH} + \mathbf{D}_u)(\mathbf{I}-\mathbf{R})^{-1}\mathbf{s}e^{pt}e^{p\eta x}\, dp d\eta. \tag{33}$$

The replacement of the matrix $\left[\mathbf{I}-\mathbf{R}\right]^{-1}$ by the power series $[\mathbf{I}+\mathbf{R}+\mathbf{R}^2+...+\mathbf{R}^k+...]$ rewrites Eq. (33) as

$$\mathbf{U}\left(x,y^{IJ},t\right)=\sum_{k=0}^{\infty}\frac{1}{2\pi i}\int_0^{\infty}\int_{Br}p^2(\mathbf{A}_u\mathbf{PH}+\mathbf{D}_u)\mathbf{R}^k\mathbf{s}e^{pt}e^{p\eta x}\,dpd\eta\,, \tag{34}$$

where the double integrals can be evaluated by fast Fourier transform (FFT) formulation (Tian & Xie, 2009).

5. Comparisons and discussions

Equation (31) shows that RRM, RTM, and GRM formulations have the same expression of reverberation matrix **R**. In RTM and GRM formulations, the dimension of **R** is in general 12 × 12, which is independent of the sublayer number. If the receiver is in the top or bottom sublayer, reverberation matrix has the order of 8 × 8. However, **R** in RRM formulation is a $4N \times 4N$ matrix, which means that the calculation efficiency of RRM formulation will decrease as the sublayer number increases.

Each term of the double integral in Eq.(34) containing $\mathbf{R}^k$ are defined as a generalized ray group. In RRM formulation, a generalized ray group represents the set of k times reflections and transmissions of the source waves by all interfaces arriving at receivers at (x,y). When k=0, the genralized ray group shows the waves from sources to the receivers directly without any reflection or refraction, which are called as source waves. Here, every generalized ray group contains a series of generalized rays, and the number of generalized rays increases exponentially with the increase of the number of layer and the reflection or refraction times. However, In RTM and GRM formulations, a generalized ray group represents the set of the generalized rays arriving at receiver (x,y) with k times of reflections and transmissions by interfaces 0, I, J, and N.

In RTM formulation, the reverberation matrix **R** has the inherent computational instability for the large product of frequency and thickness, which means that RTM formulation only can be applied to the investigation of low-frequency wave propagation in the multilayered solid. However, RRM and GRM can promise the numerical stability for the large product of frequency and thickness.

In the following, we validate the calculation efficiency of RRM and GRM formulations. Here, we consider the transient vertical displacement at the receiver A ($h^{01}/2$, $h^{01}/2$) in the top subalyer of a laminate containing N sublayers of the same thickness h, density ρ, and Poisson ratio ν impacted by a $F_0\delta(t)\delta(x)$ at the top surface. The Young modulus of the even sublayers is two times of that of the odd sublayers. Table I shows that the calculation time for GRM formulation increases much smaller than that for RRM formulation as the times of reflection or transmission k increases.

The influence of the sublayer number N on the calculation time of the vertical displacement at the receiver A for the eighth generalized ray group is shown in Table II. For the small N, the time consuming for GRM and RRM formulations are almost the same. Compared with GRM formulation, the calculation time for RRM formulation increases remarkably as the sublayer number N increases, which yields the lower calculation efficiency.

	t_{GRM} (s)	t_{RRM} (s)	t_{GRM}/t_{RRM} (%)
$\mathbf{R}^0$	800	1147	69.7
$\mathbf{R}^1$	801	1206	66.4
$\mathbf{R}^2$	808	1271	63.6
$\mathbf{R}^3$	818	1319	62.0
$\mathbf{R}^4$	813	1361	59.7
$\mathbf{R}^5$	811	1414	57.4
$\mathbf{R}^6$	808	1436	56.3
$\mathbf{R}^7$	810	1494	54.2

Table I. Calculation time of the transient vertical displacement at the receiver A (h^{01}, $h^{01}/2$) in the top sublayer of a ten-sublayered laminate

	t_{GRM} (s)	t_{RRM} (s)	t_{GRM}/t_{RRM} (%)
N=4	486	539	90.2
N=6	605	830	72.9
N=8	724	1145	63.2
N=10	810	1494	54.2
N=12	959	2000	48.0
N=14	1077	2565	42.0
N=16	1200	3203	37.5
N=18	1318	4006	32.9
N=20	1439	4886	29.5

Table II. Calculation time of the transient vertical displacement for $\mathbf{R}^7$ at the receiver A (h^{01}, $h^{01}/2$) in the top sublayer of a N-sublayered laminate

6. Conclusions

In conclusion, we present the formulations of the reverberation-ray matrix, reverberation-transfer matrix, and generalized reverberation matrix clearly. Their comparison shows that the application of the RRM formulation to the receiving sublayer and the SM and TM formulations to the other sublayers in the GRM and RTM methods yields a generalized reverberation matrix of the constant dimension, which is independent of the sublayer number. But RTM has the numerical instability for the large product of frequency and thickness, which means that it is only suitable for the low-frequency response in the multilayered solids. The numerical examples show that the calculating time for transient wave propagation in GRM formulation increases much smaller than that for RRM formulation as the times of reflection or transmission k and the sublayer number N increase, which promises the higher calculation efficiency of the generalized rays in the complex multilayered configuration compared with RRM formulation.

7. Acknowledgements

A part of this study was supported by the National Natural Science Foundation of China (No. 10602053 and No. 50808170), research grants from Institute of Crustal Dynamics (No. ZDJ2007-2) and for oversea-returned scholar, Personnel Ministry of China.

8. References

Achenbach, J. D. (1973). "Wave propagation in elastic solids," North-Holland, Amsterdam.

Haskell, N. A. (1953). the dispersion of surface waves on multilayered media. *Bulletin of the Seismological Society of America* 43, 17-34.

Lowe, M. J. S. (1995). Matrix techniques for modeling ultrasonic waves in multilayered media. *IEEE Transactions on Ultrasonics, Ferroelectrics, and Frequency Control* 42, 525-542.

Pao, Y. H., Su, X. Y., and Tian, J. Y. (2000). Reverberation matrix method for propagation of sound in a multilayered liquid. *Journal of Sound and Vibration* 230, 743-760.

Rokhlin, S. I., and Wang, L. (2002). Stable recursive algorithm for elastic wave propagation in layered anisotropic media: Stiffness matrix method. *Journal of the Acoustical Society of America* 112, 822-834.

Su, X. Y., Tian, J. Y., and Pao, Y. H. (2002). Application of the reverberation-ray matrix to the propagation of elastic waves in a layered solid. *International Journal of Solids and Structures* 39, 5447-5463.

Thomson, T. (1950). Transmission of elastic waves through a stratified solid medium. *Journal of Applied Physics* 21, 89-93.

Tian, J., and Xie, Z. (2009). A hybrid method for transient wave propagation in a multilayered solid. *Journal of Sound and Vibration* 325, 161-173.

Tian, J. Y., Yang, W. X., and Su, X. Y. (2006). Transient elastic waves in a transversely isotropic laminate impacted by axisymmetric load. *Journal of Sound and Vibration* 289, 94-108.

Wang, L., and Rokhlin, S. I. (2001). Stable reformulation of transfer matrix method for wave propagation in layered anisotropic media. *Ultrasonics* 39, 413-424.

6

Accelerating Radio Wave Propagation Algorithms by Implementation on Graphics Hardware

Tobias Rick and Torsten Kuhlen
RWTH Aachen University, JARA-SIM
Germany

1. Introduction

Radio wave propagation prediction is a fundamental prerequisite for planning, analysis and optimization of radio networks. For instance coverage analysis, interference estimation or channel and power allocation all rely on propagation predictions. In wireless communication networks optimal antenna sites are determined by either conducting a series of expensive propagation measurements or by estimating field strengths numerically. In order to cope with the vast amount of different configurations to select the best candidate from and to avoid expensive measurement campaigns, numerical predictions have to be both accurate and fast. In this chapter we focus on accelerating techniques for radio wave propagation algorithms in dense urban environments with the target frequency range of common mobile communication systems, i.e., several hundred MHz up to few GHz. One important aspect in radio wave propagation is the prediction of the mean received signal strength which can be simulated by taking complex interactions between radio waves and the propagation environment (see Figure 1) into account. Thus, the simulation of radio waves for propagation predictions becomes a computationally intensive task.
A promising approach is the use of ordinary graphics cards, nowadays available in every personal computer. With over 1000 Gigaflops, modern graphics hardware offers the computational power of a small-sized supercomputer. This is achieved by a strict parallel many-core architecture which can be accessed by a high level of programmability. The main challenge of utilizing graphics hardware for scientific computations is to trick the graphics processors into general purpose computing by casting problems as graphics: Input data is transformed into images and algorithms are turned into image synthesis. However, in the last couple of years a growing support of so-called "General Purpose Computation on Graphics Hardware" has led to recent changes in this architecture, allowing more common ways of parallel programming. Much effort and interest has been put on the acceleration of ray optical approaches, since most ray tracing algorithms tend to be computational intensive and exhibit run times up to hours. Therefore, we focus on the efficient implementation of wave guiding effects on graphics hardware. Among the most time consuming tasks in ray tracing is the problem of visibility between objects, i.e., the identification of all possible interaction sources for diffracted or reflected propagation rays. The algorithms we will present here are specifically designed to reduce the computational cost of the visibility computations by exploiting special features of the graphics card.

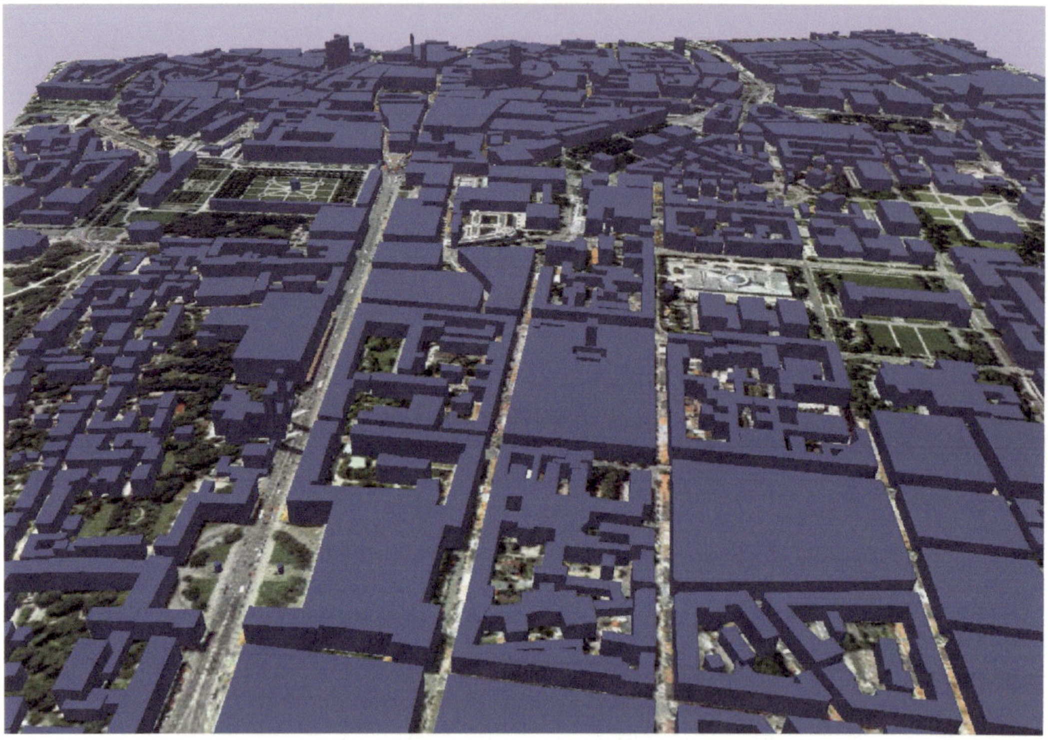

Fig. 1. Typical propagation environment with 2.5 dimensional building structures. The underlying satellite image provides geographical reference.

In principle our method is based on ray tracing. However, interaction sources are found by tracing full line-of-sight beams instead of single rays. Each beam is constructed such that it covers all rays that fall inside its angular opening. Thus, fewer beams than rays have to be processed. Furthermore, we do not actually compute any intersections between propagation environment and radiation source. In fact, we use a discrete sampling approach, i.e., ray tracing is accomplished by traversing discrete points of all rays that lie within a certain beam. Our approach results in an acceleration of wave propagation algorithms by developing a special implementation that exploit the parallel architecture of modern graphics hardware. As prove of concept, we have implemented four basic propagation effects: (1) line-of-sight propagation, (2) wall penetration depth in non-line-of-sight, (3) diffraction into street canyons and (4) diffraction over building rooftops.

The remainder of this chapter is organized as follows. After briefly reviewing previous work in Section 2, we give an overview of the general concept of programmable graphics hardware in Section 3. Then, we cover the basic prerequisites of our algorithm in Section 4 and present our method for the computation of urban propagation phenomena in Section 5. Our path loss model is described in Section 6. We finally give a performance analysis in terms of computation time and accuracy in Section 7 and conclude the chapter in Section 8.

2. Related work

A theoretical foundation of radio wave propagation is given for instance by Rappaport (1995). Since a variety of approaches exists for solving the problem of predicting mean received signal strength we point the reader to Correia (2006) for an overview. In general, we distinguish between empirical (stochastic) channel models and deterministic propagation algorithms. Empirical models, approximate the actual path loss by parametrized functions which are commonly accompanied by extensive measurement campaigns, whereas deterministic approaches are often based on the principle of ray tracing. That is, they identify ray paths through the propagation environment based on wave guiding effects like reflection or diffraction.

Well-known examples of empirical models are the work of Hata (1980) and Ikegami et al. (1984). They propose to model the radio propagation phenomena by approximating the actual propagation loss (path loss) by parametrized functions. Hata determined parameter values by conducting extensive measurement campaigns. Ikegami extended Hata's work by analyzing the dependence of approximate equations of mean field strength in urban propagation environments with respect to height gain, dependence on street width, propagation distance and radio frequency. More recently, Erceg et al. (1999) presented a stochastic channel model which can be applied to frequency ranges above two GHz (WiMAX). Additionally to height gains and propagation distance, Erceg included a parametrization of the environment into his model according to a flat or hilly terrain with either high or low vegetation. Empirical models are typically characterized by short evaluation time but are prone to huge prediction errors and perform especially poor in heterogeneous propagation environments like historically grown cities, cf. Damosso (1999).

Therefore, most deterministic algorithms rely on the computation of actual propagation paths due to wave guiding effects like reflection, diffraction and scattering. Typical approaches are often based on ray tracing which was originally introduced by Whitted (1980) to compute global illumination effects based on geometric optics for image synthesis. Although, global illumination as formulated by Kajiya (1986) and radio wave propagation are similar problem statements, different propagation effects like diffraction or interference become dominant when shifting from visible light to radio waves due to the different size of wavelengths. There are various approaches that focus on acceleration techniques by mapping global illumination algorithms onto the GPU, among them are publications by Horn et al. (2007); Carr et al. (2006) and Dachsbacher et al. (2007). Global illumination techniques have been used for different problems before, for instance for sound rendering. Notable here are the works of Tsingos et al. (2001; 2004) and Funkhouser et al. (1999). Some work on the diffraction effect is described for instance by Stam (1999) for application in computer graphics, whereas Tsingos et al. apply diffraction theory for modeling acoustics in virtual environments.

Ikegami et al. (1991) showed that classical ray tracing can also be applied to the estimation of radio propagation losses. Due to complex interaction of radio waves and geometric structures, this is a time consuming task. However, high prediction accuracy can be achieved. For instance, Schaubach et al. (1992); Erceg et al. (1997); Kim et al. (1999) and Schmitz & Kobbelt (2006) state that their predicted path loss values were generally within 4 to 8 dB of the measured path loss which is considered as a very good result. In Mathar et al. (2007) a ray launching algorithm is used which represents an urban environment as a grid of discrete blocks. Ray-object intersections are found by traversing the blocks by a line sampling method, thereby greatly reducing computation time. In order to further reduce the computational complexity Rick et al. presented an GPU-based approach to radio wave propagation in Catrein et al. (2007) and Rick & Mathar (2007). They trace propagation paths in a discrete fashion by repeated rasterization of line-of-sight regions. Propagation predictions are computed at

interactive rates. However, the accuracy depends on the resolution of the rasterization. Part of their work is presented here. A more general overview of general purpose computations on GPUs and on the evolution of GPU architectures is given for instance by Owens et al. (2005) and Owens et al. (2008), respectively.
The idea of ray tracing can be further extended to the concept of beams, which are a continuum of rays. Beam tracing was first introduced by Heckbert & Hanrahan (1984). The main benefits of beam tracing can be summarized as reduced intersection tests and less sampling problems since after a few iterations ray samples tend to become either too sparse or too dense. Published work in this area includes application of real time rendering by Overback et al. (2007) or audio rendering by Funkhouser et al. (1999). An application of beam tracing to the problem of radio wave propagation can be found in the work by Rajkumar et al. (1996) and more recently by Schmitz et al. (2009). They especially address the issue of delay spread due to multi path propagation.

3. Graphics hardware

Up to 1999 graphics cards had a non-programmable so-called *fixed-function* architecture. Over the last decade they evolved to configurable pipelines and recently into fully programmable floating-point *graphics processing units* (GPUs).
Modern GPUs are extremely powerful computing devices. Figure 2 depicts the evolution of *floating point operations per second* (FLOPS) of the GPU in comparison to the CPU over the last few years, c.f. Owens et al. (2005). The performance gain of graphics cards roughly doubles every half year, clearly outperforming the CPU when competing for Giga FLOPS (GFLOPS). The latest NVIDIA G80 GPU achieves over 300 GFLOPS. The performance of a Quad-Core Intel Xeon processor (3GHz) is roughly 80 GFLOPS Intel Corporation (2008).
The GPU is specialized for computational intensive, highly parallel calculations. Rather than caring for data caching and flow control as the CPU, the GPU is especially designed to support data processing. The GPU architecture follows the *Single Instruction Multiple Data* paradigm. Many processors simultaneously execute the same instructions on different parts of a datastream. The key challenge of programming graphics hardware is to correctly map problems to the graphical rendering context. Input data is transformed into images or geometry and algorithms are turned into image synthesis.

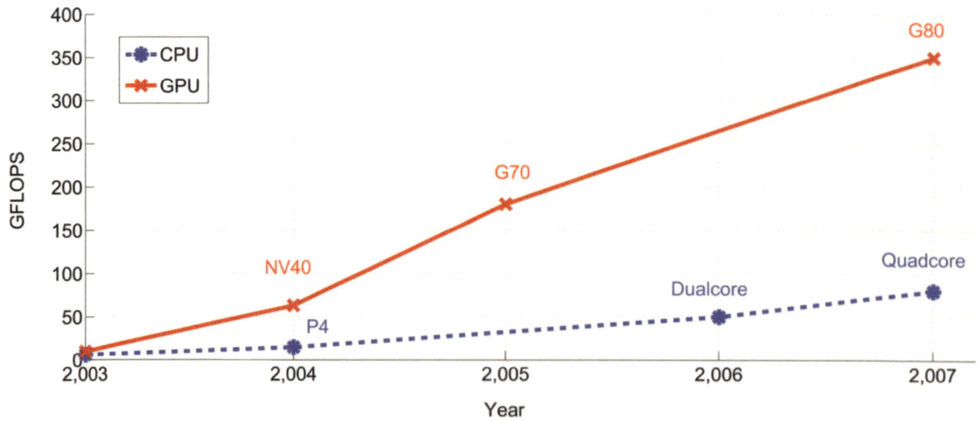

Fig. 2. Floating point operations per second: GPU vs. CPU. The graphics card clearly outperforms the CPU in terms of floating point operations.

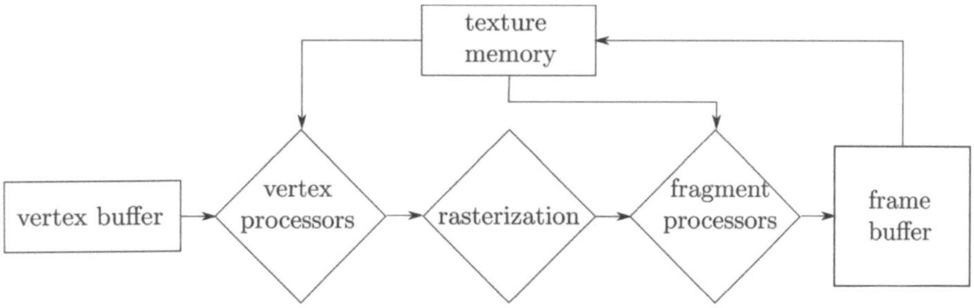

Fig. 3. The graphics rendering pipeline. Geometric primitives are loaded into the vertex buffer and transformed by multiple vertex processors in parallel. The rasterization samples the geometry into pixel positions such that multiple fragment processors can assign colors that are recorded in the frame buffer for the final image.

In general, a computation on the GPU consists of a pass through the various stages of the so-called *graphics rendering pipeline*, see Figure 3. We refer to this pass as a *rendering pass*.
Basic geometrical primitives like points or triangles are loaded into the *vertex buffer*. The primitives are described by their location in space, i.e., coordinates (*vertices*) and associated attributes like material or color. Additionally, data arrays (*textures*) of integer or floating point values can be allocated directly in graphics memory.
First, multiple vertex processors execute in parallel the instructions from a user-written program (*vertex program*). Vertex programs operate on single vertices with access to their attributes and global read-only texture memory. Typically, geometric transformations like translation, rotation and projection are applied.
In the subsequent step, the transformed geometry is sampled (*rasterized*) into discrete points (*fragments*). Each fragment corresponds to a single *pixel* (picture element) position on the screen, and typically has additional information like color and a depth value, i.e., the distance between viewer and the object which originally corresponded to the pixel.
The fragment processors operate analogous to the vertex processors. A user-written *fragment program* is executed on each fragment in a parallel fashion. The fragment program is executed *once per fragment*. Most instructions are floating-point vector operations. Typically, lighting models are evaluated and corresponding fragment colors are assigned (*shaded*).
The *frame buffer* is a two-dimensional array of pixels. The task of the frame buffer is to assemble the final result of the GPU computation by collecting and recording all fragments. *Frame buffer operations* decide how the color from the incoming fragments is combined with the color already stored at the same pixel position. Thus, many fragments can contribute to the final color of a pixel. Commonly, the frame buffer contains color information and hence, it can be displayed on the screen. Information in the frame buffer can also be read back to main memory, which is especially useful for GPU computations. For further readings we point the reader to introductory texts and advanced techniques on graphics and shader rogramming from Akenine-Möller & Haines (2002); Shreiner et al. (2003) or Fernando & Kilgard (2003).

4. Overview

Similar to Wahl et al. (2005), the data requirements for our algorithms are building structures with corresponding building heights. The shape of rooftops is usually omitted

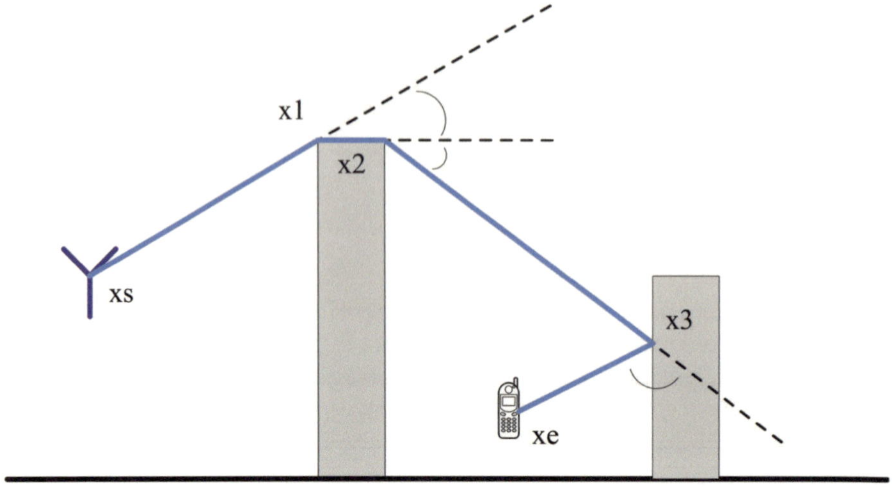

Fig. 4. Example propagation path (side profile). A ray is emanating from the radiation source, diffracted at the rooftop of the first building and reflected the second building into the street: $x_s \rightsquigarrow x_1 \rightsquigarrow x_2 \rightsquigarrow x_3 \rightsquigarrow x_e$.

and hence, a building is described by its polygonal outline and one height value. We refer to this representation as 2.5 dimensional. Building structures ares usually given in vector format with a location accuracy in the order of 1 to 10 meter. In order to produce reliable results, height accuracy should be around 1 to 2 meter. Information about vegetation is not considered and terrain is assumed to be flat. Figure 1 shows a typical propagation environment.

The basic propagation phenomena are reflection, diffraction and scattering. All effects contribute to the radio signal distortions and give rise to signal fluctuations (fading) and additional signal propagation losses. We distinguish propagation effects according to the characteristics of the propagation environment by approximating which propagation paths (see Figure 4) are most likely to occur. In the course of this chapter we will present specialized algorithms for each of the following propagation effects.

If receiver and transmitter antenna have an unobstructed direct path, they are in *line-of-sight* (LOS), otherwise in *non-line-of-sight* (NLOS). An urban *high-rise* scenario consists predominantly of streets lined with tall buildings of several floors each. High building heights make significant contributions from diffraction over multiple rooftops rather unlikely. Therefore, if transmitter antennas are mounted below rooftops, dominant propagation effects are expected from reflection and diffraction into street canyons. See Figure 8(b) for an illustration, more details can be found in Andersen et al. (1995). Furthermore, an urban *low-rise* scenario is characterized by wide streets and buildings with less than three floors. Antennas are usually mounted above average rooftop level and diffraction over rooftops (Figure 9(a)) becomes the dominant propagation effect. Propagation paths due to reflection are not considered in this chapter.

Hence, we formulate the requirements of our propagation algorithm as follows: (1) It is necessary to efficiently distinguish between regions in LOS and in NLOS. (2) In case of NLOS, we require our algorithm to compute different propagations paths, namely wall penetration depth, diffraction into street canyons and diffraction over rooftops.

5. Algorithm

The input for our algorithms is a database of building structures as described in Section 4. In general, the output is a discrete two-dimensional raster image which serves as a role model for the discrete representation of results from GPU computations. An image consists of pixels that are organized in a regular array. Pixels are the data elements of this structure. However, pixels are not restricted to contain only color information. Pixel data is interpreted according to the current context of the GPU computation.

In order to provide a formal description of the presented algorithms we introduce the following notation (see Table 1): Let $s = (x,y,z)$ be a radiation source location, e.g., the transmitter antenna, where the height is referred to as $s.z$. All height values are relative to ground. A wall w of a building structure consists of two points p_0 and p_1 at ground level and two points p_2 and p_3 at rooftop level. The set containing all walls w is denoted by $\mathcal{W}$. Additionally we associate a normal vector $\vec{n}$ with each wall. Normal vectors are perpendicular to their wall and point away from the corresponding building.

Since the terrain is assumed to be rather flat and the receiver points are typically located 1.5 meter above ground we define a *receiver plane* $\mathcal{R}$ to be a discrete set of receiver points at constant height $\mathcal{R}.z$. We restrict ray path calculations to only those paths that intersects the receiver plane.

In the following we will describe algorithms which are explicitly designed to run directly on graphics hardware. First, we present a method for determining line-of-sight regions. Then we will show how this algorithm can be extended to provide additional non-line-of-sight information. Furthermore, we show how the graphics card can be exploited for ray path calculation due to diffraction into street canyons and diffraction over rooftops. GPU implementations are obtained by separating the calculation of these effects into distinct algorithms.

5.1 Line-of-sight

The *line-of-sight* (LOS) algorithm computes a sampling of the receiver plane $\mathcal{R}$ where each point $p \in \mathcal{R}$ is marked whether it is in clear line-of-sight to a source s or not. This algorithm will be one of the main building blocks in the computation of ray paths on graphics hardware. The GPU computation is based on the concept of so-called *shadow volumes*, c.f. Fernando (2004). This technique constructs a polygonal representation of the shadow cone (shadow volume) for 3-dimensional triangular geometry. Since the description of urban propagation environments involves usually just a polygonal outline and one corresponding height value, we propose a specialized algorithm for this particular form of geometry (see Catrein et al. (2007)).

$s = (x,y,z)$	Radiation source location with height $s.z$
$\mathcal{R}$	Discrete set of receiver points with constant ground level $\mathcal{R}.z$
$\mathcal{W}$	Set of all building walls w
$w = (p_0, p_1, p_2, p_3, \vec{n})$	Building wall with p_0, p_1 at ground level and p_2, p_3 at rooftop level. The normal vector $\vec{n}$ is perpendicular to the wall and points away from the associated building.
$\text{NLOS}(s \rightarrow w)$	Shadow polygon cast by wall w when viewed from point s
$\text{LOS}(s) \subseteq \mathcal{R}$	Subset of $\mathcal{R}$ that is in line-of-sight to s

Table 1. Notation

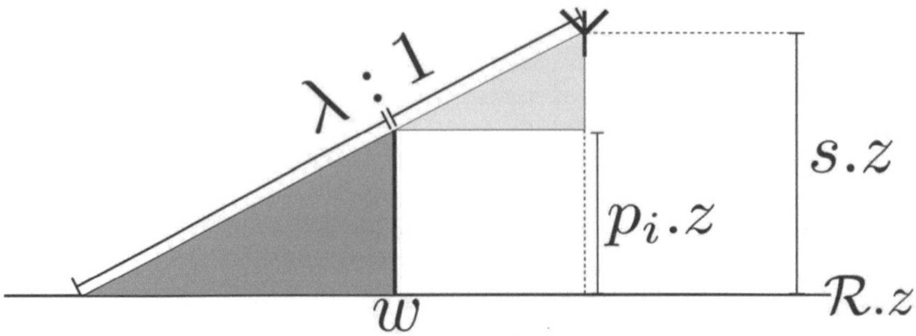

Fig. 5. Size of the shadow polygon (dark gray) according to the intercept theorem for a source height $s.z$ greater than the wall height $p_i.z$ (side profile).

The main idea is to extract the shadow of each building wall directly in the receiver plane. We refer to the intersection of the shadow cone and the receiver plane as *shadow polygon*. Regions are in LOS, if and only if they are in no shadow polygon.
The construction of the shadow polygon of a wall w proceeds as follows: each shadow polygon is a quadrangle with corners (q_0, q_1, q_2, q_3). Points q_0 and q_1 are given by the corners of the wall p_0 and p_1 on the ground. The remaining two corners are determined by the intersection of the receiver plane $\mathcal{R}$ and each of the straight lines through the source point s and the wall point at roof level p_2 and p_3.
According to the intercept theorem (see Figure 5), q_i, $i \in \{2, 3\}$ is then given by

$$q_i = p_i + \lambda \cdot (p_i - s) \tag{1}$$

where

$$\lambda = \begin{cases} \frac{p_i.z - \mathcal{R}.z}{s.z - p_i.z} & \text{,if } s.z - p_i.z > 0 \\ \infty & \text{,otherwise} \end{cases} . \tag{2}$$

The corners of each shadow polygon are computed by a vertex program in parallel for each wall and sampled into discrete points by the subsequent rasterization phase. The result is a two-dimensional pixel array, every (discrete) receiver location lies either in LOS or inside a shadow polygon, hence in NLOS.
With q_2 and q_3 according to equation (1), we refer to the shadow polygon of a single wall w with source point s as

$$\text{NLOS}(s \rightarrow w) = \{p \in \mathcal{R} \mid p \in \text{polygon}(w.p_0, w.p_1, q_2, q_3)\} \tag{3}$$

A *LOS beam* LOS(s) is the set of discrete points in the receiver plane that do not lie in any shadow polygon of the source s

$$\text{LOS}(s) = \bigcap_{w \in \mathcal{W}} \{p \in \mathcal{R} \mid p \notin \text{NLOS}(s \rightarrow w)\} . \tag{4}$$

Figure 6 shows an example of a discretized receiver plane with pixels in LOS (dark grey) and NLOS (light grey) pixels. Note, that although the result is two-dimensional, all shadow computations are performed using all 2.5-dimensional information.

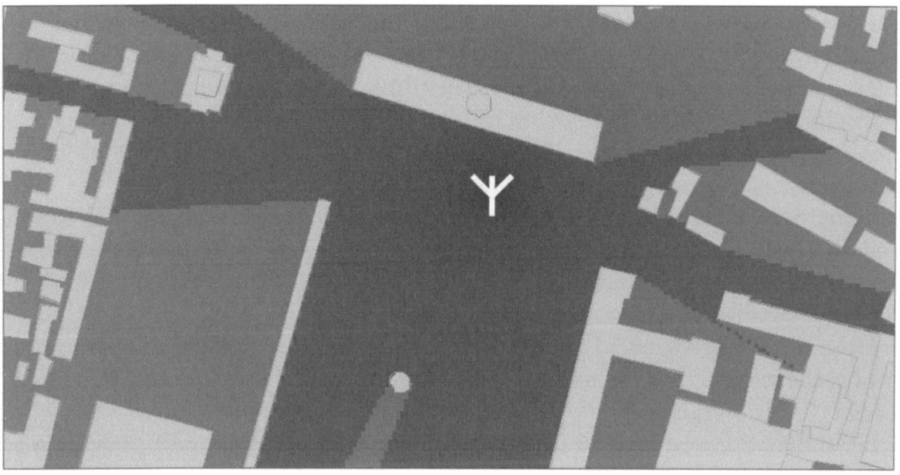

Fig. 6. Top view of the discrete line-of-sight region (dark grey) with the source point in the center (white cross). Building interior is shown in light grey.

5.2 Wall transmission

The algorithm for LOS beams can be extended to provide additional *non*-line-of-sight information. This can include for instance the number of penetrated walls or material which can be achieved by taking all walls into account that intersect the direct ray from the source to a receiver as sketched in Figure 7.

We first provide a more thorough look at GPU frame buffer operations which are an integral part of the algorithm for transmission depth. When fragments are collected and recorded in the frame buffer at the final stage of the rendering pipeline (c.f. Section 3), frame buffer operations decide how fragments that fall on the same pixel position, contribute to the final color of that pixel. Commonly, the fragment with the lowest depth value, i.e., which is nearest to the viewer, determines (replaces) the pixel color. Alternatively, the final color can be a combination (interpolation) of the values of both fragments, the one already in the frame buffer and the new one, which wants to occupy the same pixel position. This technique is called *blending*. In image synthesis, blending is commonly used to draw translucent objects.

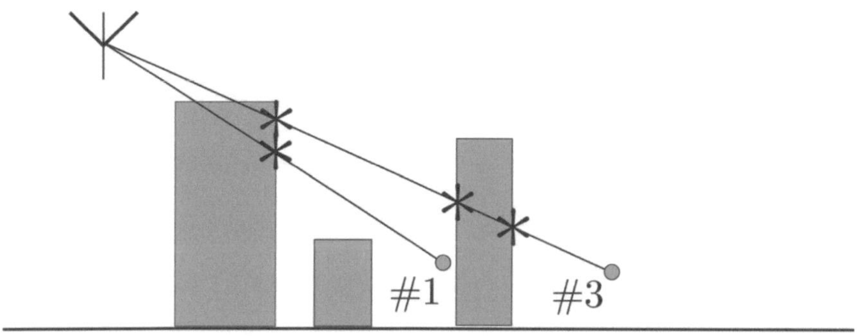

Fig. 7. Wall transmission depth can be taken into account by counting the number of walls in the direct path between the source and the receiver points. On the graphics card, this is achieved by an additive blending of shadow polygons.

Here, we use the blending capability to increase the value in the frame buffer with every rendered shadow polygon. Thus, instead of a solid rendering (replacement of fragments) of the shadow polygons, we apply an additive blending. This effectively counts the number of shadow fragments at each receiver location.

However, frame buffer operations like blending are currently implemented in hardware with a precision of 8 bits. This would means that we could only count up to 255 wall transmissions, which may not be enough for large and complex scenarios. A solution is presented by graphics cards that offer buffers (textures) of higher precision like 32 bit. Blending has to be implemented by a user-written fragment program since high precision blending is not yet directly supported in hardware.

To overcome this drawback and to support propagation environments with an arbitrary number of walls we propose the following hybrid approach between 8 bit hardware and 32 bit fragment shader blending.

Two 32 bit buffers are created and filled with zeros. These buffers will provide a so-called ping-pong scheme because current GPUs do not support a simultaneous read and write access to the same buffer. One buffer is referred to as the *next* buffer, it is the buffer which will be rendered to. The other buffer is referred to as the *current* buffer, it is bound as a texture and read from in the updating step. After each pass, the buffers are swapped, so that the *next* buffer becomes the *current* buffer and vice versa.

Hence, one buffer will keep track of the total transmission depth for every fragment and the other will store intermediate results. The actual rendering of the shadow polygons is done into a third 8 bit frame buffer with hardware supported additive blending. The vertex buffer containing the wall geometry is rendered in separate chunks of 255 walls each. After each chunk rendering, the transmission depth of every fragment is added to *next* buffer. This is done by a fragment shader which reads the *current* buffer and the 8 bit buffer containing the transmission depths of the latest wall chunk. A simple add operation is performed and the result is stored in the *next* buffer. This procedure is repeated until all wall chunks have been processed.

This approach is a multi-pass algorithm, for n walls it requires $\lceil \frac{n}{255} \rceil$ rendering passes. This number can be reduced, if chunks are rendered in parallel whose shadow polygons do not overlap. Roughly the same performance as the original LOS algorithm is achieved since this approach utilizes parts of the rendering pipeline (frame buffer operations) that have been idle before.

5.3 Diffraction into street canyons

Classical ray tracing algorithms commonly model diffraction effects by tracing a multitude of rays into the respective diffraction cone as illustrated in Figure 8(a). This is computationally quite intensive, as one ray (the one hitting the diffraction edge) is split into many secondary rays. Shooting fewer rays, usually results in an under sampling of the propagation environment, leading to regions where no rays arrive. Thus, no path loss calculations can be performed there.

A solution to this problem is presented by the following approach. In order to achieve a high throughput of diffraction computations, we trace full beams instead of single rays. This has the advantage that beams cover a lot more area than single rays and thus, a lot fewer beams than rays have to be processed for a sufficient sampling of the propagation environment.

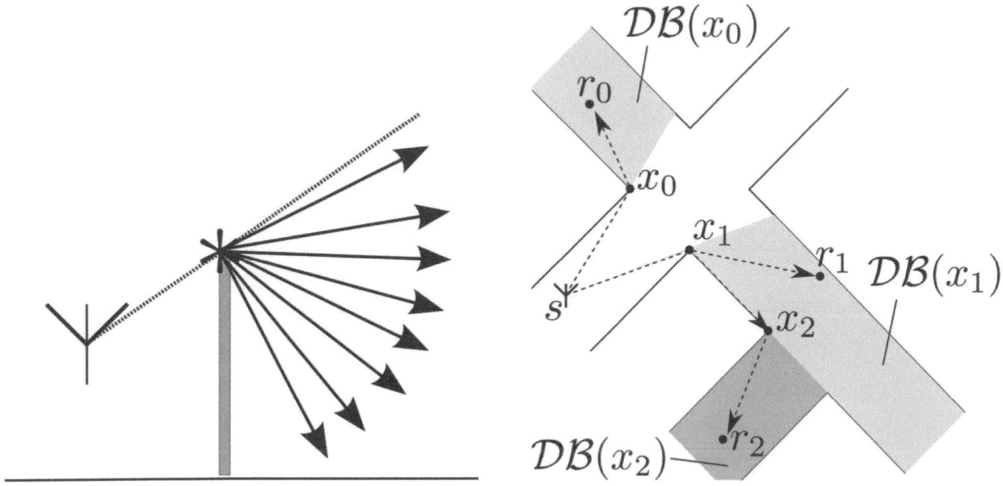

(a) (b)

Fig. 8. (a) Modeling diffraction by shooting a multitude of rays into the diffraction cone. (b) Diffraction beams in nearby street canyons. Diffraction sources are $\mathcal{DS}(s) = \{x_0, x_1\}$ and $\mathcal{DS}(x_1) = \{x_2\}$. Geometric ray paths are for instance $s \rightsquigarrow x_0 \rightsquigarrow r_0$ or $s \rightsquigarrow x_1 \rightsquigarrow x_2 \rightsquigarrow r_2$.

The algorithm consist of three main steps. (1) Identify all potential diffraction sources for a propagation into street canyons. (2) Determine the propagation paths of the secondary wave fronts. (3) Compute geometrical paths in reverse order from the target location towards the diffraction source. This procedure can be applied recursively if propagation paths along multiple street canyons are desired. The idea is sketched in Figure 8(b).

Let s be a radiation source. According to Section 5.1 the set of discrete points that are in line-of-sight to s is defined by LOS(s). We define a *diffraction source* as an end point of a wall w that satisfies the following criteria: (1) The wall is in LOS to the radiation source if one of the end points is in LOS

$$w \in \mathrm{LOS}(s) \Longleftrightarrow w.p_0 \in \mathrm{LOS}(s) \vee w.p_1 \in \mathrm{LOS}(s) \tag{5}$$

and (2) if the normal vector of the wall is facing away from from the radiation source

$$\langle w.\vec{n}, \| w.p_0 - s \|_2 \rangle < 0. \tag{6}$$

For a radiation source s, we collect all possible diffraction sources in the set $\mathcal{DS}(s)$

$$\mathcal{DS}(s) = \bigcup_{w \in \mathrm{LOS}(s)} \begin{cases} \{w.p_j\} & ,\text{if } \langle w.\vec{n}, \| w.p_0 - s \|_2 \rangle < 0 \\ \emptyset & ,\text{otherwise} \end{cases} \tag{7}$$

where each $w.p_j$ is choosen such that

$$j \in \{0,1\} \wedge \| w.p_j - s \|_2 \leq \| w.p_{1-j} - s \|_2 . \tag{8}$$

Hence, a diffraction source is always a point on the wall that is closest to the radiation source.

Now, we construct the diffraction beam $\mathcal{DB}(x)$ as a secondary wave front that originates from the diffraction source $x \in \mathcal{DS}(s)$. With our notation we can write a diffraction beam as

$$\mathcal{DB}(x) = \{r \in \mathcal{R} \mid r \in \text{LOS}(x) \wedge r \notin \text{LOS}(s)\} \tag{9}$$

All points which have been in line-of-sight to the original radiation source s will not be part of the diffraction beam. Thereby, we ignore regions that would result in very large diffraction angles which would in turn not contribute to the overall signal level, significantly.

The final steps consists of the reconstruction of geometric ray paths based on the beam information. For each beam $b \in \mathcal{DB}(x)$, we create a set of ray paths $\mathcal{RP}(b)$

$$\mathcal{RP}(b) = \{s \rightsquigarrow x \rightsquigarrow r \mid r \in \mathcal{DB}(x)\} \tag{10}$$

Thus, every point within the beam travels along its diffraction source towards the original radiation source. Diffraction paths of arbitrary length can be constructed by assigning the start points of the rays as new diffraction sources in a recursive fashion.

The computational bottleneck of this method are the numerous LOS computations. Therefore, we propose to use the LOS implementation on the GPU as described in Section 5.1.

5.4 Propagation over rooftops

The algorithm for diffraction into street canyon (see Section 5.3) always maps one wall edge to one source point for LOS since only ray paths that hit the receiver plane are computed. The calculation of propagation paths over rooftops is different because diffraction source points lie on the whole edge of the roof, see. Figure 9(b).

Therefore, this sections describes how propagation paths due to diffraction over rooftops can be calculated efficiently on the graphics processing unit. The algorithm basically consists of two steps: (1) a discretized version of the diffraction cones is constructed for every rooftop. (2) Ray paths are found by going backwards from each receiver location towards the transmitter. An integral part of our method is the computation of diffraction beams for every rooftops simultaneously. Therefore, no identification of diffraction sources is required, we define the set of roof diffraction sources directly as the set of all building walls $\mathcal{W}$.

Let s be a radiation source. As illustrated in Figure 9(a), a *roof diffraction beam* $\mathcal{RDB}(w)$ of a wall w directly corresponds to the shadow polygon

$$\mathcal{RDB}(w) = \text{NLOS}(s \rightarrow w)\,. \tag{11}$$

Due to the construction of the beam all geometric ray paths from $\mathcal{RDB}(w)$ to the radiation source s have a deflection point somewhere on the wall w at rooftop level, e.g. $s \rightsquigarrow w \rightsquigarrow r$.

For a point $r \in \mathcal{RDB}(w)$ the exact deflection point x_r on the wall can by found be the intersection of the two lines $l_0 = (w.p_2, w.p_3)$ and $l_1 = (r, s)$, hence

$$x_r = \text{intersect}(l_0, l_1)\,. \tag{12}$$

The idea is sketched in Figure 9(b). The geometric ray paths $\mathcal{RP}(b)$ for each beam $b \in \mathcal{RDB}(w)$ are then

$$\mathcal{RP}(b) = \{s \rightsquigarrow x_r \rightsquigarrow r \mid r \in \mathcal{RDB}(w)\}\,. \tag{13}$$

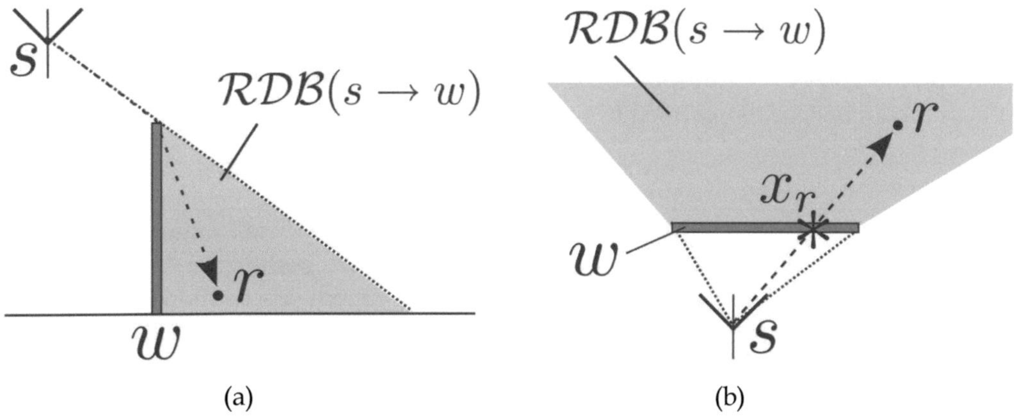

Fig. 9. Illustration of a roof diffraction beam. (a) The diffraction beam of the source s and wall w is depicted in a side profile. (b) The top view reveals the exact deflection point x_r as the intersection of the direct connection between s and r. This yields the ray path $s \rightsquigarrow x_r \rightsquigarrow r$.

The implementation of this method again involves the LOS computation on the GPU. However, some care has to be taken when ray paths are concatenated due to diffraction over multiple rooftops. If two consecutive points of a ray path are in LOS to each other, all points in between must be removed from the final path. This can be achieved by always computing the upper convex hull of all points on a ray path. Further implementation details can be found in Catrein et al. (2007).

6. Path loss calculation

This section depicts how a ray path is mapped to received signal strength. We handle the cases for LOS and NLOS separately. We have an unobstructed clear line-of-sight path if a ray path is of the form

$$\tau^{\rightsquigarrow} = s \rightsquigarrow x \tag{14}$$

for a receiver location x and a radiation source s. We write the corresponding path loss in dB as

$$PL^{\mathrm{dB}}(r) = 10\log_{10}\left(\frac{P_t}{P_r(\| x - s \|_2)}\right) \tag{15}$$

with antenna gains G, wavelength λ, transmitted power P_t and received power P_r

$$P_r(d) = \frac{P_t G \lambda^2}{(4\pi)^2 d^2} \tag{16}$$

according to the free space propagation loss by Rappaport (1995).
The path loss prediction in regions with no direct LOS is known to be more complex. Consider a ray path $\upsilon^{\rightsquigarrow}$ given as

$$\upsilon^{\rightsquigarrow} = x_{i-1} \rightsquigarrow x_i \rightsquigarrow x_{i+1} \tag{17}$$

The change of direction $\alpha_{\upsilon^{\rightsquigarrow},i}$ according to the deflection of the wave front at x_i is given by

$$\alpha_{v^{\sim},i} = \text{acos}\left(\langle \| x_i - x_{i-1} \|_2, \| x_{i+1} - x_i \|_2 \rangle\right). \tag{18}$$

Let α be the change of direction of the corresponding deflection, the angle depended attenuation can then be represented by a polynomial of degree 3

$$PL(\alpha) = a_0 + a_1\alpha + a_2\alpha^2 \tag{19}$$

We retrieve the unknown model coefficients (α_0, α_1, α_2) by a calibration to real-world measurements, thus modeling the stochastic influence of traffic and vegetation. The calibration can be realized by a formulation as a constraint least-square problem. The optimal parameter vector can then be calculated by common solver algorithms like Gauss-Newton or Levenberg-Marquardt (see Levenberg (1944)).

Different types of wave guiding effects are taken into account by introducing distinct attenuation functions, the attenuation functions due to diffraction into street canyons and over rooftops are denoted by $PL^{\text{dB}}_{street}(\alpha)$ and $PL^{\text{dB}}_{roof}(\alpha)$. For the attenuation due to wall transmissions $PL^{\text{dB}}_{wall}(n)$ we use a model similar to the Multi-Wall model by Lott & Forkel (2001) which depends on the number of penetrated walls n.

In logarithmic notation, the overall attenuation of a receiver location r is then the sum of the arriving ray paths and can be written as

$$\begin{aligned} PL^{\text{dB}}_{\text{NLOS}}(r) = & PL^{\text{dB}}(d) + \sum_{i=1}^{N_{street}} PL^{\text{dB}}_{street}\left(\alpha_{street,i}\right) \\ & + \sum_{j=1}^{N_{roof}} PL^{\text{dB}}_{roof}\left(\alpha_{roof,j}\right) + PL^{\text{dB}}_{wall}\left(N_{wall}\right), \end{aligned} \tag{20}$$

where N_{street}, N_{roof} and N_{wall} denote the number of deflection points and $\alpha_{street,i}$, $\alpha_{roof\ ,j}$ the corresponding changes of direction.

7. Results

In this section we present prediction accuracy and calculation time. For the purpose of benchmarking we use building and urban micro cell measurements of downtown Munich, Germany. This data has been created during the COST 231 action as described by Damosso (1999) and is now publicly available at Mannesmann Mobilfunk GmbH, Germany (1999). The scenario comprises 7 km^2 of approximately 18000 building walls. Three different measurement routes are available, referred to as METRO200, METRO201 and METRO202. All predictions were performed on the whole area of 7 km^2 with a resolution of 5 meter. This requires the evaluation of more than $3 \cdot 10^5$ receiver points. The transmitter location is depicted in Figure 10 which shows a typically field strength prediction. Colors are chosen such that diffraction at building edges is clearly visible. In the following, all path loss calculations are based on a calibration on the measurement points along route METRO202.

7.1 Computational performance

The key component of the presented algorithm is the computation of LOS regions and is therefore analyzed in more detail. We have evaluated the performance of our GPU implementation of the LOS computation on different generations of graphics cards: NVIDIA

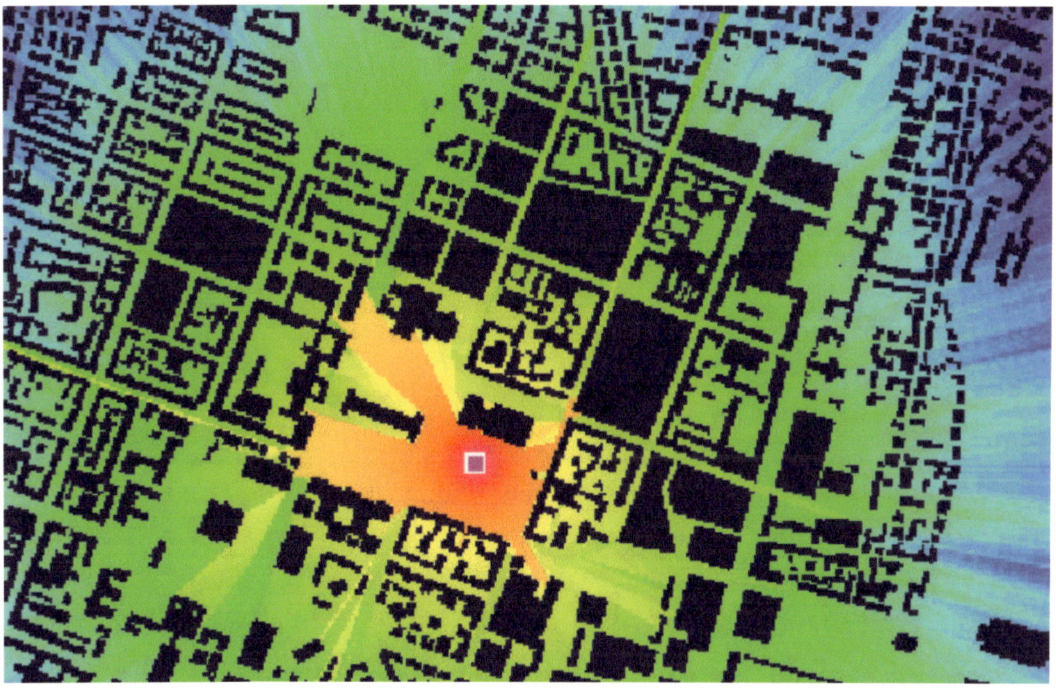

Fig. 10. Field strength prediction in downtown Munich (Germany) with colors according to intensity.

GeForce 6600 GT (May 2004), 7800 GT (August 2005) and a 8800 GTX (May 2007). Figure 11 sketches the computations per second over the size of the underlying discrete grid. At a resolution of 5 meters these cards achieve 100, 200 and over 500 LOS computations per second, respectively. Thus, we observed a speedup by a factor of 2 for every new generation of graphics cards. Hence, we conclude that the implementation of this algorithm in graphics hardware scales well on new generations of graphics hardware. In terms of throughput, the 500 LOS computations per second can also be expressed as roughly $1.6 \cdot 10^8$ receiver points per second. As point of reference, the total number of receiver points is roughly $3 \cdot 10^5$ at a resolution of 5 meter.

The implementation of the transmission depth exhibits similar performance behavior as the LOS computation since there is almost no noticeable increase in rendering load by this approach.

Table 2 gives an overview of different propagation predictions with respect to accuracy and runtime for route METRO201. We use the CPU implementation of the ray launching algorithm CORLA by Mathar et al. (2007) as reference for our GPU implementation. In our test scenario, CORLA exhibits run times of about 8 seconds. When switching from the CPU implementation of CORLA to our GPU method we observed a speedup of roughly 2.5X for the roof diffraction and a speedup of 160X for the diffraction into street canyon with additional transmission depth. The computation time of propagation paths over rooftops is about 3 seconds. The most time (2 seconds) is currently consumed by the computation of upper convex hull, since this has to be executed on the CPU, due to a missing stack implementation on the GPU.

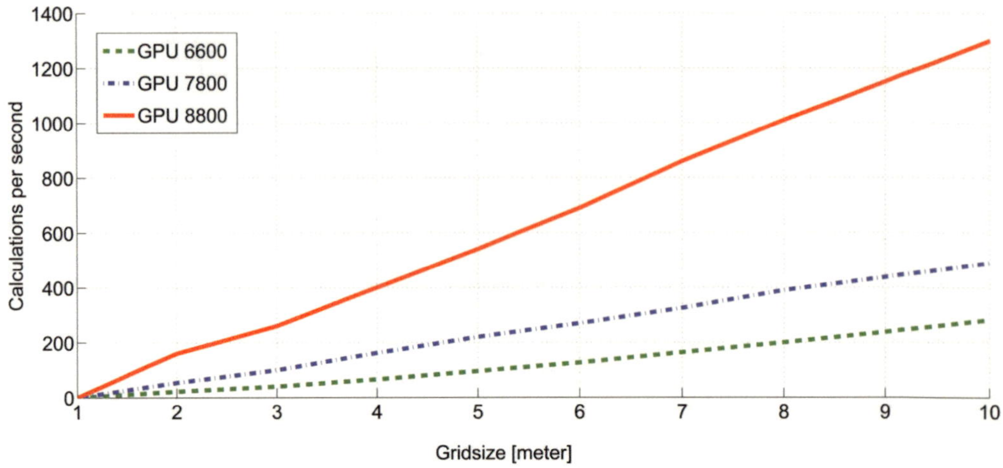

Fig. 11. Line-of-sight calculation time on different generations of graphics cards. Test scenario is COST 231 Munich.

Prediction model	mean error	std. dev.	runtime
CORLA Mathar et al. (2007) (CPU)	0.1 dB	4.2 dB	8 s
Propagation over roofs (GPU)	−0.2 dB	4.7 dB	3.05 s
Propagation into street canyons with transmission depth (GPU)	0.1 dB	4.5 dB	0.05 s

Table 2. Accuracy and runtimes of propagation models in COST 231 Munich along route METRO201.

Furthermore, we neglected the angle deviation in the propagation paths over rooftops and counted only the number of diffractions (which can be performed by the algorithm for transmission depth). In combination with the calculation of diffraction paths into street canyons, the runtime was reduced to approximately 0.05 seconds at roughly the same prediction accuracy.

7.2 Prediction accuracy

We quantify the accuracy of our propagation predictions by the *mean squared error* (mse) and the *standard deviation* (std. dev.) between prediction and measurement data. Let the number of measurements points be N, and r_i the ith measurement point. $M_{dB}(r_i)$ denotes the measured, and $PL_{dB}(r_i)$ the predicted path loss at r_i.
We define the mean squared error as

$$mse = \sqrt{\frac{1}{N}\sum_{i=1}^{N}[M^{dB}(r_i) - PL^{dB}(r_i)]^2} \tag{21}$$

and the standard deviation as

$$std.dev. = \sqrt{\frac{1}{N-1}\sum_{i=1}^{N}[M^{dB}(r_i) - PL^{dB}(r_i) - \overline{x}]^2} \tag{22}$$

with mean error

$$\overline{x} = \frac{1}{N}\sum_{i=1}^{N}\left[M^{\mathrm{dB}}(r_i) - PL^{\mathrm{dB}}(r_i)\right]. \tag{23}$$

A comparison between our prediction and the three available measurement routes according to these performance criteria is given in Table 3. Propagation accuracy according to the mean squared error lies between 4.5 and 6.2 dB. Whether the GPU or CPU was involved in the prediction computation had no influence on the propagation accuray as depicted in Table 2. The exact shape of the predicted and measured path loss is illustrated in Figure 12 for route METRO201. It can be seen that although the path loss is sometimes over- or underestimated, most of the important features in the measurement data are captured quite well by the propagation prediction.

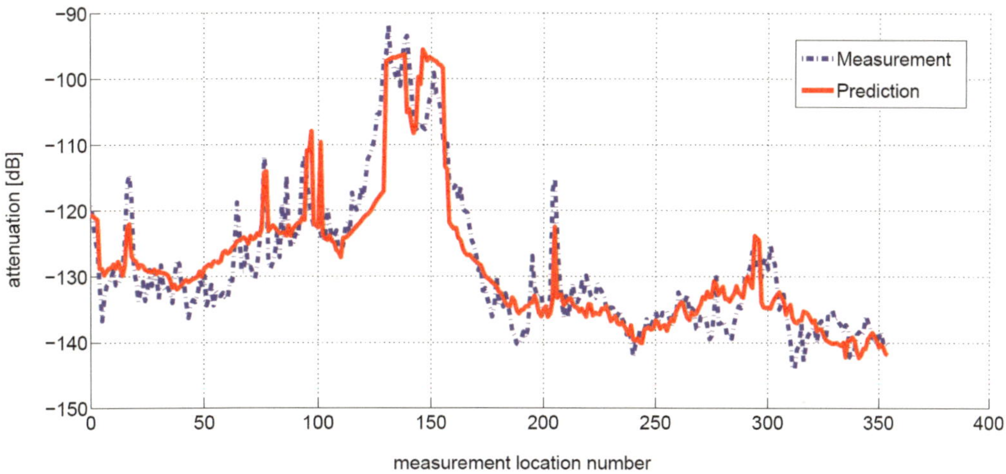

Fig. 12. Comparison between measured and predicted pathloss along route METRO201 in COST 231 Munich.

Measurement Route	mean error	mse	std. dev.
METRO200	2.1 dB	6.2 dB	5.9 dB
METRO201	0.1 dB	4.5 dB	4.5 dB
METRO202	-0.1 dB	5.7 dB	5.7 dB

Table 3. Prediction accuracy in COST 231 Munich.

8. Conclusion

In this chapter, we exploited graphics hardware for accelerating the computation of radio wave propagation predictions. Our method traces discrete line-of-sight beams and reconstructs ray paths based on radiation source, beam origin and all discrete points within a sampled beam.

All algorithms are designed to benefit from the computational power and parallel architecture of modern graphics cards. Core component of the GPU algorithms is a high

throughput LOS computation which leads to computation times of 3 to 0.05 seconds in a scenario of 7 km^2 with roughly 18000 walls. This results in a speedup of 2.5X to 160X times compared to the CPU algorithm CORLA. This is achieved by designing separate algorithms for distinct propagation effects like diffraction over rooftops and diffraction into street canyons. Hence, a proper combination of propagation effects computed on the GPU can deliver propagation predictions at interactive rates. The accuracy of our propagation predictions is quantified by a standard deviation between predictions and measurement data of 4 to 7 dB which is considered as a very good result. Thus, the use of graphics hardware for field strength predictions does not diminish propagation accuracy.
We summarize the main contribution as the development of a run-time efficient algorithm that accurately predicts mean received signal strengths in dense urban propagation environments.

A. Abbreviations and mathematical operators

LOS line-of-sight
NLOS non-line-if-sight
$\| \vec{x} \|_2$ length of vector $\vec{x}$
$\langle \vec{x}, \vec{y} \rangle$ scalar product of two vectors $\vec{x}$ and $\vec{y}$
$\wedge$ logical operator *and*
$\vee$ logical operator *or*
$\cap$ set operator *intersection*
$\cup$ set operator *union*
$\emptyset$ empty set
$x_{i-1} \rightsquigarrow x_i \rightsquigarrow x_{i+1}$ ray path that starts at x_{i-1}, changes direction at x_i and arrives at x_{i+1}

9. References

Akenine-Möller, T. & Haines, E. (2002). *Real-Time Rendering*, Natick, MA: A K Peters, Ltd.
Andersen, J., Rappaport, T. & Yoshida, S. (1995). Propagation measurements and models for wireless communication channels, *IEEE Commun. Mag.* 33: 42–49.
Carr, N. A., Hoberock, J., Crane, K. & Hart, J. C. (2006). Fast gpu ray tracing of dynamic meshes using geometry images, *GI '06: Proceedings of Graphics Interface 2006*, Canadian Information Processing Society, Toronto, Ont., Canada, Canada, pp. 203–209.
Catrein, D., Reyer, M. & Rick, T. (2007). Accelerating radio wave propagation predictions by implementation on graphics hardware, *Proc. IEEE Vehicular Technology Conference*, Dublin, Ireland, pp. 510–514.
Correia, L. M. (ed.) (2006). *COST Action 273: Mobile Broadband Multimedia Networks, Final Report*, Academic Press.
Dachsbacher, C., Stamminger, M., Drettakis, G. & Durand, F. (2007). Implicit visibility and antiradiance for interactive global illumination, *SIGGRAPH '07: ACM SIGGRAPH 2007 papers*, ACM, New York, NY, USA, p. 61.

Damosso, E. (ed.) (1999). *COST Action 231: Digital mobile radio towards future generation systems, Final Report*, Office for Official Publications of the European Communities, Luxembourg.

Erceg, V., Fortune, S. J., Ling, J., Rustako, J. & Valenzuela, R. A. (1997). Comparison of a computer-based propagation prediction tool with experimental data collected in urban microcellular environments, *IEEE J. Sel. Areas Commun.* 15: 677–684.

Erceg, V., Greenstein, L. J., Tjandra, S. Y., Parkoff, S. R., Gupta, A., Kulic, B., Julius, A. A. & Bianchi, R. (1999). An empirically based path loss model for wireless channels in suburban environments, *IEEE J. Sel. Areas Commun.* 17: 1205–1211.

Fernando, R. (2004). *GPU Gems*, Addison-Wesley, chapter Effective Shadow Volume Rendering, pp. 137–166.

Fernando, R. & Kilgard, M. (2003). *The Cg Tutorial: The Definitive Guide to Programmable Real-Time Graphics*, Reading, MA: Addison-Wesley.

Funkhouser, T., Min, P. & Carlbom, I. (1999). Real-time acoustic modeling for distributed virtual environments, *SIGGRAPH '99: Proceedings of the 26th annual conference on Computer graphics and interactive techniques*, ACM Press/Addison-Wesley Publishing Co., New York, NY, USA, pp. 365–374.

Hata, M. (1980). Empirical formula for propagation loss in land mobile radio services, *IEEE Trans. Veh. Technol.* 29: 317–325.

Heckbert, P. S. & Hanrahan, P. (1984). Beam tracing polygonal objects, *SIGGRAPH '84: Proceedings of the 11th annual conference on Computer graphics and interactive techniques*, ACM, New York, NY, USA, pp. 119–127.

Horn, D. R., Sugerman, J., Houston, M. & Hanrahan, P. (2007). Interactive k-d tree gpu raytracing, *I3D '07: Proceedings of the 2007 symposium on Interactive 3D graphics and games*, ACM, New York, NY, USA, pp. 167–174.

Ikegami, F., Takeuchi, T. & Yoshida, S. (1991). Theoretical prediction of mean field strength for urban mobile radio, *IEEE Trans. Antennas Propag.* 39: 299–302.

Ikegami, F., Yoshida, S., Takeuchi, T. & Umehira, M. (1984). Propagation factors controlling mean field strength on urban streets, *IEEE Trans. Antennas Propag.* 32: 822–829. Intel Corporation (2008). URL: *http://www.intel.com*

Kajiya, J. T. (1986). The rendering equation, SIGGRAPH '86: *Proceedings of the 13th annual conference on Computer graphics and interactive techniques*, ACM Press, New York, NY, USA, pp. 143–150.

Kim, S.-C., Guarino, B. J., III, T. M.W., Erceg, V., Fortune, S. J., Valenzuela, R. A., Thomas, L.W., Ling, J. & Moore, J. D. (1999). Radio propagation measurements and prediction using three-dimensional ray tracing in urban environments at 908 mhz and 1.9 ghz, *IEEE Trans. Veh. Technol.* 48: 931–946.

Levenberg, K. (1944). A method for the solution of certain problems in least squares, *Appl. Math.* 2: 164–168.

Lott, M. & Forkel, I. (2001). A multi-wall-and-floor model for indoor radio propagation, *Proc. IEEE Vehicular Technology Conference*, Vol. 1, Rhodes, Greece, pp. 464–468.

Mannesmann Mobilfunk GmbH, Germany (1999). Cost 231 - urban micro cell measurements and building data.
URL: *http://www.ihe.uni-karlsruhe.de/forschung/cost231/cost231.en.html*

Mathar, R., Reyer, M. & Schmeink, M. (2007). A cube oriented ray launching algorithm for 3d urban field strength prediction, *IEEE International Conference on Communications*, pp. 5034 - 5039.

Overback, R., Ramamoorthi, R. & Mark,W. R. (2007). A real-time beam tracer with application to exact soft shadows, *EGSR*.

Owens, J. D., Houston, M., Luebke, D., Green, S., Stone, J. E. & Phillips, J. C. (2008). GPU computing, *Proceedings of the IEEE* 96(5): 879–899.

Owens, J. D., Luebke, D., Govindaraju, N., Harris, M., Krger, J., Lefohn, A. E. & Purcell, T. J. (2005). A survey of general-purpose computation on graphics hardware, *Eurographics 2005, State of the Art Reports*, pp. 21–51.

Rajkumar, A., Naylor, B. F., Feisullin, F. & Rogers, L. (1996). Predicting rf coverage in large environments using ray-beam tracing and partitioning tree represented geometry, *Wirel. Netw.* 2(2): 143–154.

Rappaport, T. S. (1995). *Wireless Communications: Principles and Practice*, Prentice-Hall, Inc.

Rick, T. & Mathar, R. (2007). Fast edge-diffraction-based radio wave propagation model for graphics hardware, *Proc. IEEE 2nd International ITG Conference on Antennas*, Munich, Germany, pp. 15–19.

Schaubach, K. R., IV, N. J. D. & Rappaport, T. S. (1992). A ray tracing method for predicting path loss and delay spread in microcellular environments, *Proc. IEEE Vehicular Technology Conference*, Vol. 2, pp. 932–935.

Schmitz, A. & Kobbelt, L. (2006). Wave propagation using the photon path map, *PE-WASUN '06*, ACM, New York, NY, USA, pp. 158–161.

Schmitz, A., Rick, T., Karolski, T., Kuhlen, T. & Kobbelt, L. (2009). Simulation of Radio Wave Propagation by Beam Tracing, *Eurographics Symposium on Parallel Graphics and Visualization*, pp. 17–24.

Shreiner, D., Neider, J., Woo, M. & Davis, T. (2003). *OpenGL Programming Guide*, fourth edn, Reading MA: Addison-Wesley.

Stam, J. (1999). Diffraction shaders, *SIGGRAPH '99: Proceedings of the 26th annual conference on Computer graphics and interactive techniques*, ACM Press/Addison-Wesley Publishing Co., New York, NY, USA, pp. 101–110.

Tsingos, N., Funkhouser, T., Ngan, A. & Carlbom, I. (2001). Modeling acoustics in virtual environments using the uniform theory of diffraction, *SIGGRAPH '01: Proceedings of the 28th annual conference on Computer graphics and interactive techniques*, ACM, New York, NY, USA, pp. 545–552.

Tsingos, N., Gallo, E. & Drettakis, G. (2004). Perceptual audio rendering of complex virtual environments, *SIGGRAPH '04: ACM SIGGRAPH 2004 Papers*, ACM, New York, NY, USA, pp. 249–258.

Wahl, R., Wlfle, G., Wertz, P., Wildbolz, P. & Landstorfer, F. (2005). Dominant path prediction model for urban scenarios, *14th IST Mobile and Wireless Communications Summit*.

Whitted, T. (1980). An improved illumination model for shaded display, *Commun. ACM* 23(6): 343–349.

II. Wave Propagation in Metamaterials, Micro/nanostructures and Porous Media.

7

Application of Media with Negative Refraction Index to Electromagnetic Imaging. Fundamental Aspects.

Andrey Petrin
Joint Institute for High Temperatures of Russian Academy of Science
Russia

1. Introduction

In the recent decade, there has been growing interest in the electrodynamics of materials characterized by negative refractive index. These materials have been introduced into consideration by V. G. Veselago (1967). The progress in the nanotechnology of composite materials made it possible to design new materials whose properties can be explained by assuming that the refractive index of these materials is negative [Smith (2000), Shelby(2001)]. Some authors [Pendry (2000), Lagarkov (2004)] suggested that these materials with negative refraction allow one to overcome the diffraction limit in optical devices. This thesis has met many objections [Williams (2001)], which are not absolutely indisputable [Pendry (2001).]. Nevertheless, the concept of a superlens (in the form of a layer of medium with negative refraction), which was proposed in [Pendry (2000)] and allows one to completely overcome the diffraction limit in the ideal case, has found many supporters (see, for example, [Cui (2005), Chen (2006), Alitalo (2007), Cheng (2005), Scalora (2007), Podolskiy (2005)] and references therein).

The superlens, as proposed by Pendry (2000), is a plane plate of material with negative refractive index. The geometry of rays originated from a point source *S* and focused in the point *P* for the superlens is shown in Fig. 1. The refraction of the rays takes place according to the Snellius law applied to the material with negative refraction index. Fig.1 shows the case $n = -1$ for which there are no geometrical aberrations. According to Pendry (2000) there is a singularity of the electromagnetic fields at the image point *P* of a point source *S*.

However, the very concept of a superlens with a singularity at the image point *P* seems quite strange. Indeed, it has been known since Fresnel's times that the dimensions of a focal spot are determined only by the angle between the marginal converging rays and by the radiation wavelength in the focusing region (see, for example, the textbook [Sivukhin (1980), Section 55]). The dimensions of the focal spot are a property of a converging wave and do not depend on the optical system that has formed this converging wave. In the case of a superlens, a converging wave propagates in vacuum, outside the layer with negative refractive index, and it seems reasonable that it should focus into a region of standard dimensions.

The goal of the chapter is to examine completely this knotty problem. One approach in strict formulation has been suggested in two papers of A. B. Petrin (2008). It was considered the

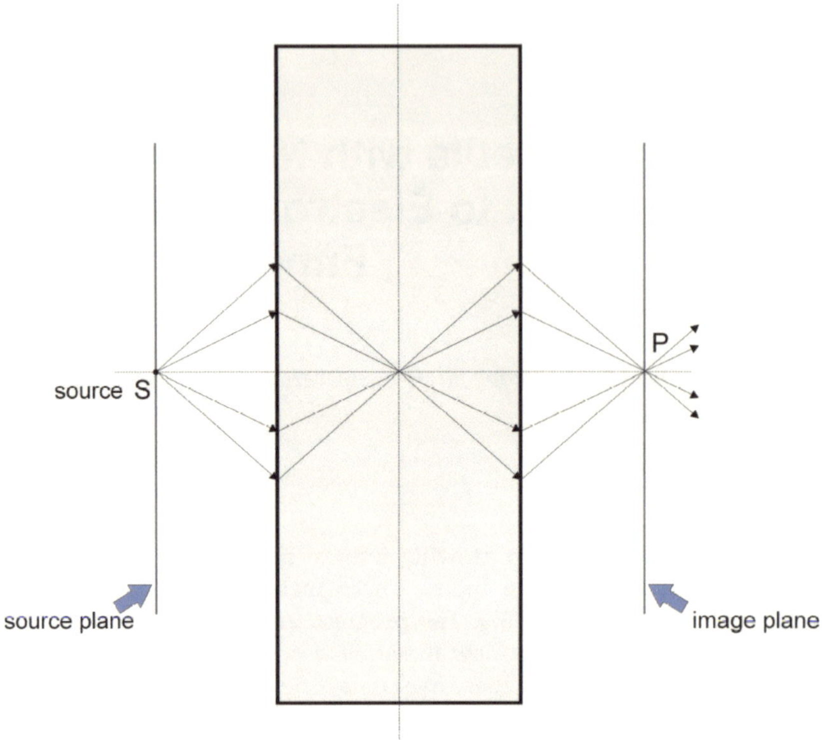

Fig. 1. The geometry of the superlens and the refracted rays from a point source focused into the image point *P*.

propagation of an electromagnetic wave radiated by an elementary electric Hertz dipole located in the air (or vacuum) parallel to the boundary of a layer or, in a particular case, to the boundary of a half-space filled with a material with negative refractive index. It was applied a rigorous approach that dates back to Sommerfeld [Sommerfeld (1926), Wait (1998)]. The scientific discussion around the resolution of superlens has shown that there is some lack of understanding in the theory of diffraction which has a fundamental significance. The main goal of the chapter is to clear the theme of diffraction limit of lens with negative refraction from several points of view.

2. The ultimate resolution of electromagnetic imaging devices and the limits of subwavelength imaging. Fundamental aspects.

Before considering strict solutions of the problem let us investigate the ultimate resolution of electromagnetic devices from common point of view since the propagation of electromagnetic waves is obeyed the Maxwell's equations. Moreover, in the regions outside the superlens the electromagnetic waves propagate in vacuum and it is possible to apply the well-known approaches to the problem of resolution of superlens.

Let us consider the preliminary auxiliary problems.

First, consider a point source of electromagnetic radiation (in particular a point Herzian dipole, see Fig.2). Consider the image plane parallel to the dipole. The distribution of mean electric field on the plane depends on the distance from the dipole to the plane.

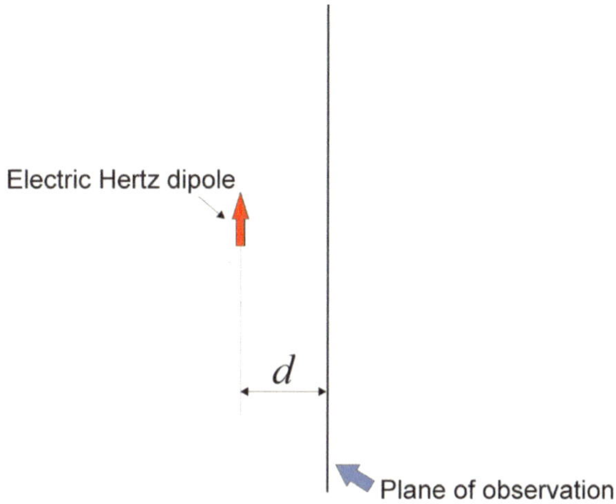

Fig. 2. One elementary Herzian dipole near the image plane (plane of observation)

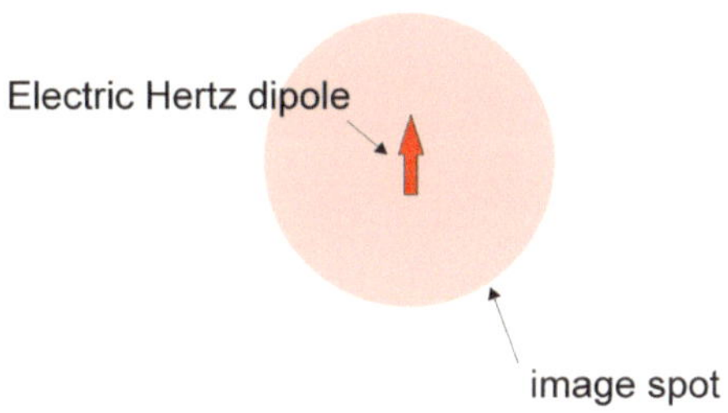

Fig. 3. The geometry of the image spot from a dipole on the image plane.

Fig. 3 shows the image spot (shown approximately by the circle) located on the plane of observation on some level of electric field strength in the near field of the dipole. Fig. 3 shows the case when the distance d from the dipole to the plane of observation is much smaller than the wavelength of the electromagnetic wave λ. In this case of the near field zone ($d << \lambda$) the dimensions of the spot are of the order of d due to the fast decrease of the electric field from the dipole in the near field zone. Thus, formally, *in the near field zone there is subwavelength resolution of the dipole.*

Let's take two dipoles as shown in Fig. 4. In this case it is possible to resolve the mutual position of the two dipoles with subwavelength accuracy in the near field zone. Fig. 5 approximately shows the image spots from these two dipoles on the plane of observation. *So, in principal (when $d/\lambda \to 0$), the subwavelength resolution may be achieved without use of any lenses.* It is necessary to emphasize that such subwavelength resolution in the near field zone is well-known and trivial.

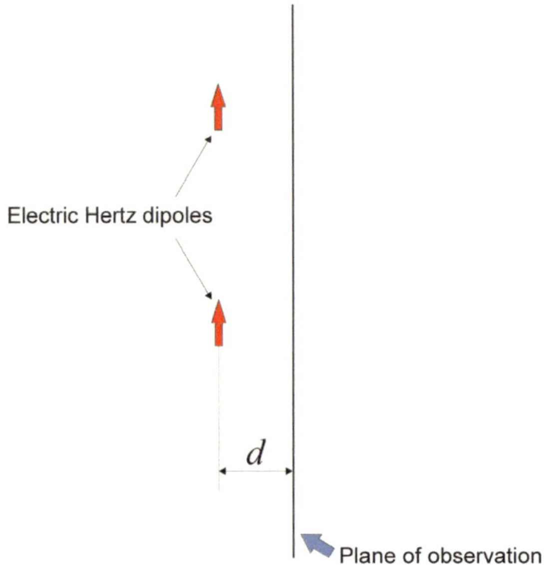

Fig. 4. The geometry of two elementary Herzian dipoles near the image plane (plane of observation).

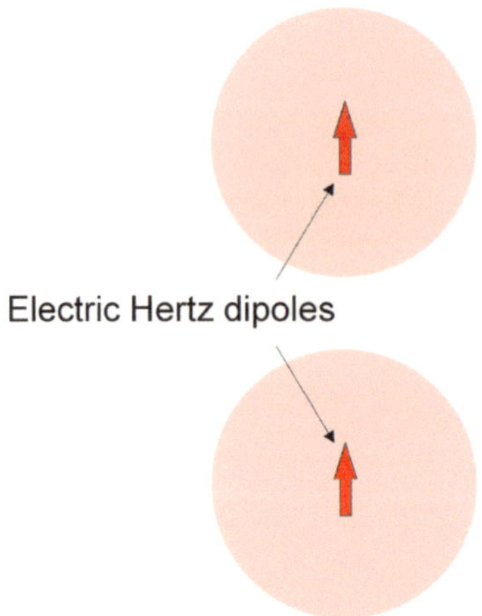

Fig. 5. The geometry of the image spots from two dipoles on the image plane.

Let's again consider the superlens shown in Fig. 1. And let's show that the superresolution (that is the existence of singularity in the image spot from the point source such as dipole as in the work of Pendry (2000)) is impossible. It will be shown below that there is no superresolution neither in near field zone nor in the far field zone ($d >> \lambda$).

The existence of singularity in the image point is contrary to the Huygens principle of wave propagation and to the analogue of the principle in electromagnetism - the formulas of Stratton and Chu (see Stratton (1941)). Indeed, to the right of the plate of negatively refracted material of the superlense in Fig. 1 is the vacuum. So, the image point may be encircled with sphere C as shown in Fig. 6.

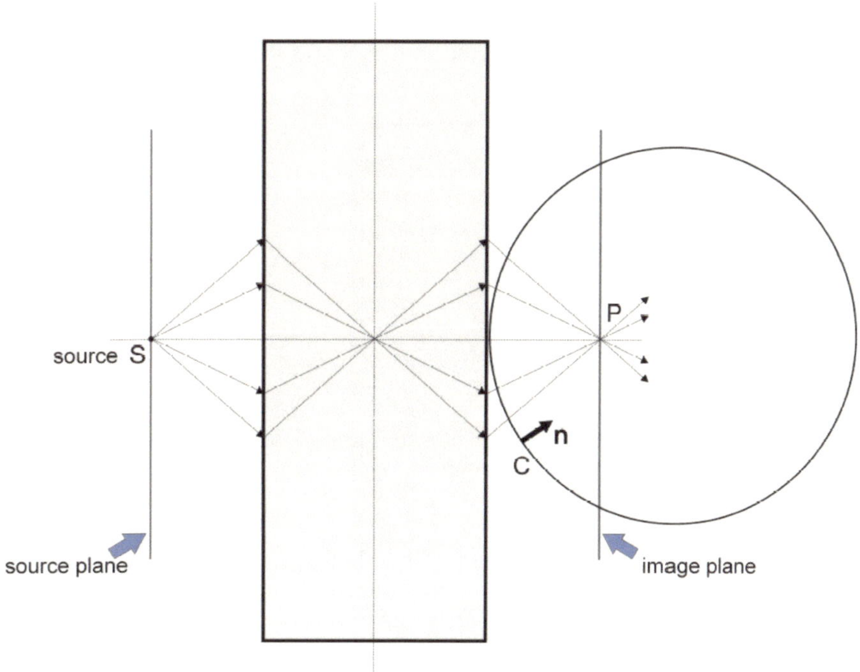

Fig. 6. The sphere surface C around the image point P situated in vacuum outside the refracted material.

Thus, for any point inside the sphere, including the image point P, it is possible to write the formulas of Stratton and Chu (see Stratton (1941)) in the following form:

$$\mathbf{E}_P = i\omega\mu\mathbf{A}_e - \frac{1}{i\omega\varepsilon} graddiv\mathbf{A}_e - rot\mathbf{A}_m, \quad (1)$$

$$\mathbf{H}_P = i\omega\varepsilon\mathbf{A}_m - \frac{1}{i\omega\mu} graddiv\mathbf{A}_m - rot\mathbf{A}_e, \quad (2)$$

where $\mathbf{A}_e = \frac{1}{4\pi}\iint_C [\mathbf{n},\mathbf{H}]\frac{e^{ikr}}{r}dC$ is the electric vector potential; $\mathbf{A}_m = \frac{1}{4\pi}\iint_C [\mathbf{E},\mathbf{n}]\frac{e^{ikr}}{r}dC$ is the magnetic vector potential; $r = |\mathbf{r}_P - \mathbf{r}_C|$ is the distance from the point of integration to the point P.

The formulas of Stratton and Chu show that the electromagnetic fields at the point P are the superposition (the vector sum) of the fields from the sources situated on the surface C. The sources are the surface electric and magnetic dipoles defined by the tangential components of electromagnetic fields. Since the surface C may be of arbitrary form it is possible to deform the surface and transform it into the plane surface as shown in Fig. 7 (the closing semisphere of the surface is depicted in Fig. 7 by doted line). Due to the Zommerfeld's condition of radiation the integral on infinitively distant semisphere is vanished and the integral on C is reduced to the integral over the plane (see Fig. 7).

Thus, the elementary sources of the electromagnetic fields are situated on the plane C and give the fields at the point P by summation. So, it is obvious that the dimensions of the focus region are defined by the two parameters: the wavelength λ and the distance d between the surface of the plate of the negatively refracted material and the image plane.

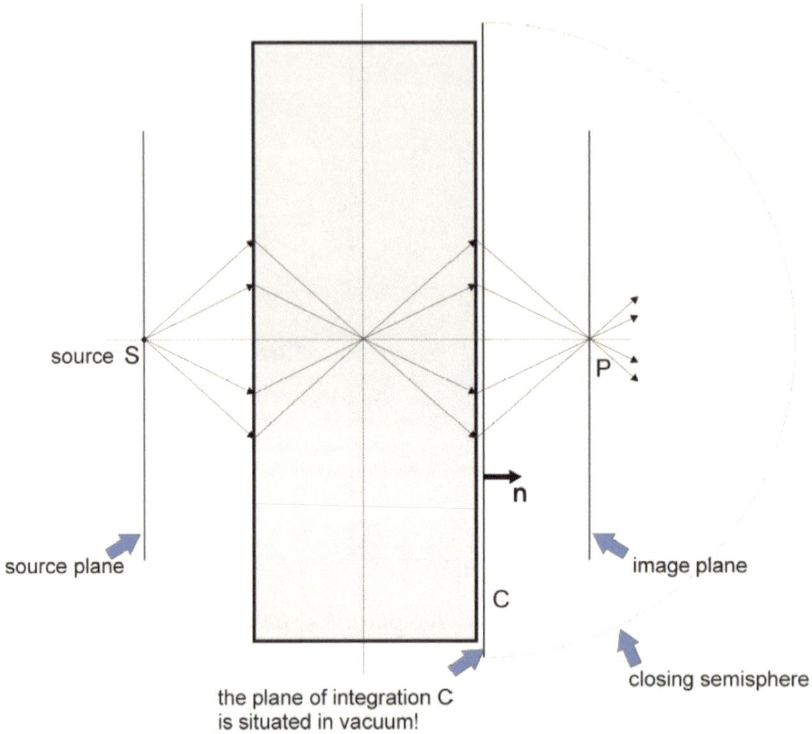

Fig. 7. The transformed surface C (the plane and the closing semisphere).

Let's proove the following: *if the image plane is in the far zone ($d >> \lambda$), the focal region dimensions are of the order of λ and if the image plane is in the near zone ($d << \lambda$) the focal region dimensions are of the order of d*. Consider two arbitrary situated elements on the plane surface C (see Fig. 8). These elements radiate the electromagnetic fields according the formulas (1) and (2) in which the vector potential of the elements are $d\mathbf{A}_e = [\mathbf{n},\mathbf{H}]e^{ikr}dC/4\pi r$ and $d\mathbf{A}_m = [\mathbf{E},\mathbf{n}]e^{ikr}dC/4\pi r$. Consider how the contributions to the fields from the elements change if the point of observation moves from the point P to the point P' (see Fig. 8).

If the image plane is in the far zone ($d >> \lambda$) and if the displacement of the point P is of the order of λ then the distance from any element dC is approximately the same (in relative sence) if the point of observation move from the point P to P' (due to $\lambda/d \rightarrow 0$). So, only phase changing defines the changing in the fields from the element dC. The phase shanges on the distance of the orser of λ. Therefore, the focal region dimensions are of the order of λ. Since the fields from any element dC change noticeably for the displacement of the order of λ the same conclusion may be made for the total or summed fields for the considering case of far zone.

If the image plane is in the near zone ($d << \lambda$) and if the displacement of the point P is of the order of d then the phase difference is neglectively small. So, only the terms without phase exponent give contribution to the fields at the points P and P'. Due to fast decreasing of the near fields of the elementary dipoles the focal region dimensions are of the order of d. Since

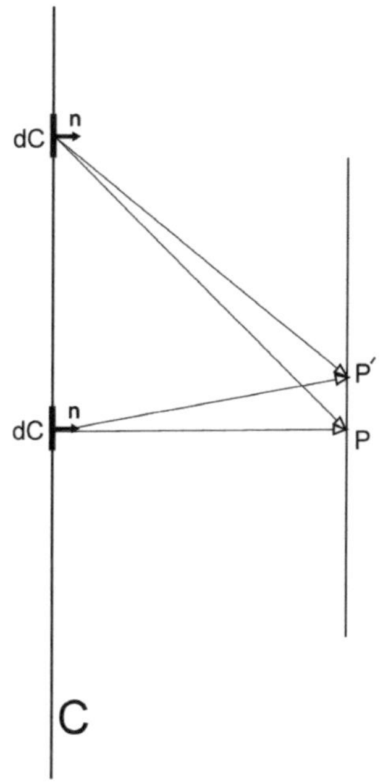

Fig. 8. The geometry of two elementary sources which give contribution to the electromagnetic fields in the points P and P'.

the fields from any element dC change noticeably for the displacement of the order of d the same conclusion may be made for the total or summed fields for the considering case of near field zone. The same dimensions of the spot will be in the case of Fig. 5, so, in this sence the lens with negative material is unnecessary for achieving better resolution.

Thus, the conclusion from the above reasoning is the following: *there is no singularity in principle neither in near field zone nor in the far or intermediate zone.* This conclusion is not depened on the lens disign or the apsorption of the material of the lens. So, the term superresolution has no sence. It is only possible to speak about common subwavelenth resolution in the near field zone.

3. The strict electromagnetic theory of lens with negative refraction (a plane layer filled with a medium with negative refractive index) and the image of a point source of radiation. Correspondence of common media and media with negative refraction.

In this paragraph we consider the strict solution of the problem of focus spot dimension from the point source inside the semispace with negative refraction and in vacuum for the superlens (after refraction in a plane layer filled with a medium with negative refractive index).

3.1 Radiation from a Hertzian dipole parallel to a layer of medium with negative refractive index. Strict theory.

Consider (Fig. 9) an elementary horizontal dipole (a Hertzian dipole) with unit current moment

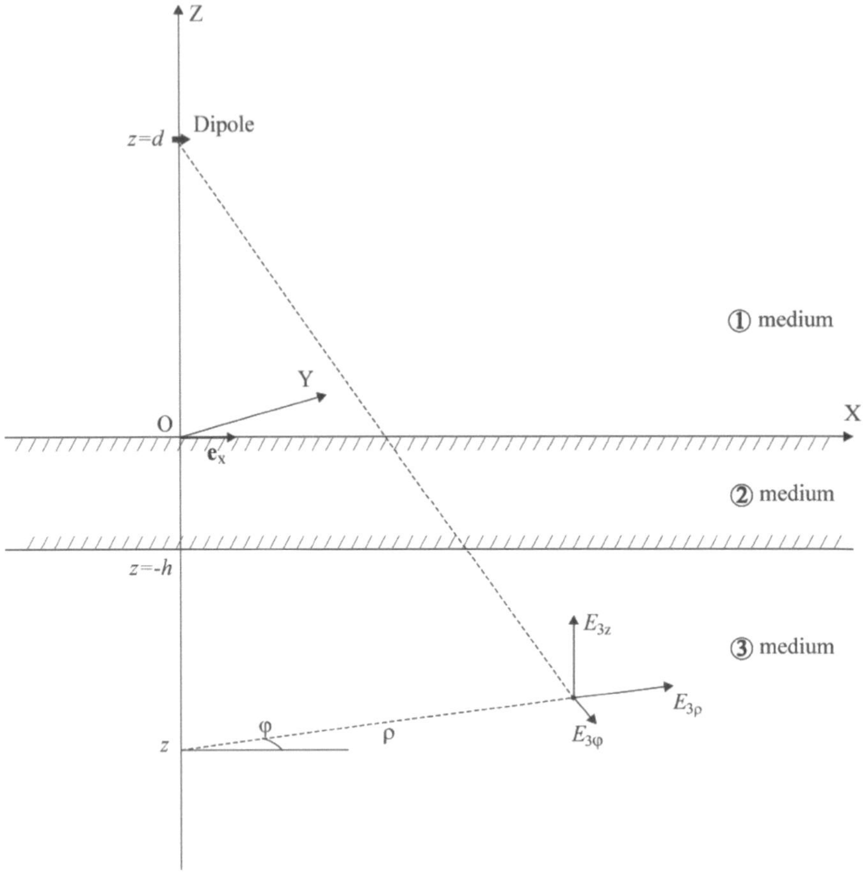

Fig. 9. Geometry of the problem. Media 2 and/or 3 are characterized by negative refractive index. The source of electromagnetic waves, an electric Hertzian dipole, is located in medium 1 (vacuum or air, $\varepsilon_1 = \mu_1 = 1$).

$$\mathbf{J} = \delta(x)\delta(y)\delta(z-d)\mathbf{e}_x , \tag{3}$$

where $\delta(x)$ is the delta function, $\mathbf{e}_x$ is a unit vector directed along the x-axis, and d is the z-coordinate of the dipole. Note that in this paper we use a complex representation of all quantities with the time dependence $e^{-i\omega t}$. A detailed account of the problem of propagation of electromagnetic waves radiated from a Hertzian dipole oriented parallel to the plane boundary of a half-space filled with an absorbing medium with positive real parts of permittivity and permeability was given in King&Smith (1981). Below, we generalize the approach developed in King&Smith (1981) and apply it to solving the problem of radiation from a dipole parallel to a layer with negative refractive index.
We can write Maxwell's equations in the three domains as

$$rot\,\mathbf{E}_j = i\omega\mathbf{B}_j , \tag{4}$$

$$rot\,\mathbf{B}_j = \mu_j\left(-i\omega\varepsilon_j\mathbf{E}_j + \mathbf{J}\right), \tag{5}$$

where j = 1 for domain 1 (z > 0), j = 2 for domain 2 ($-h < z < 0$), and j = 3 for domain 3 ($z < -h$), h is the layer thickness.

The boundary conditions on the surfaces of the layer require that the tangential components of the fields **E** and $\mathbf{H} = \mathbf{B}/\mu$, as well as the normal components of the fields $\mathbf{D} = \varepsilon\mathbf{E}$ and **B**, be continuous (provided that there are no external current sources on the surfaces). However, one can show that, in the case of harmonic fields, it suffices to require that only the tangential components of **E** and $\mathbf{H} = \mathbf{B}/\mu$ should be continuous.

The electric field can be represented as a Fourier expansion:

$$\mathbf{E}(x,y,z) = (2\pi)^{-2} \int_{-\infty}^{+\infty} d\xi \int_{-\infty}^{+\infty} d\eta\, e^{i\xi x + i\eta y} \tilde{\mathbf{E}}(\xi,\eta,z). \tag{6}$$

Similar expressions can be written for **B** and **J**:

$$\mathbf{B}(x,y,z) = (2\pi)^{-2} \int_{-\infty}^{+\infty} d\xi \int_{-\infty}^{+\infty} d\eta\, e^{i\xi x + i\eta y} \tilde{\mathbf{B}}(\xi,\eta,z),$$

$$\mathbf{J}(x,y,z) = (2\pi)^{-2} \int_{-\infty}^{+\infty} d\xi \int_{-\infty}^{+\infty} d\eta\, e^{i\xi x + i\eta y} \tilde{\mathbf{J}}(\xi,\eta,z).$$

Using the Fourier transformation, from (3) we obtain the Fourier transform of the current source:

$$\tilde{\mathbf{J}}(\xi,\eta,z) = \delta(z-d)\mathbf{e}_x. \tag{7}$$

Using the Fourier representations of the fields, we can rewrite Maxwell's equations in the three domains as

$$i\eta\tilde{E}_{jz} - \frac{\partial}{\partial z}\tilde{E}_{jy} = i\omega\tilde{B}_{jx}, \tag{8}$$

$$\frac{\partial}{\partial z}\tilde{E}_{jx} - i\xi\tilde{E}_{jz} = i\omega\tilde{B}_{jy}, \tag{9}$$

$$i\xi\tilde{E}_{jy} - i\eta\tilde{E}_{jx} = i\omega\tilde{B}_{jz}, \tag{10}$$

$$i\eta\tilde{B}_{jz} - \frac{\partial}{\partial z}\tilde{B}_{jy} = -i\omega\varepsilon_j\mu_j\tilde{E}_{jx} + \delta_{1j}\mu_j\delta(z-d), \tag{11}$$

$$\frac{\partial}{\partial z}\tilde{B}_{jx} - i\xi\tilde{B}_{jz} = -i\omega\varepsilon_j\mu_j\tilde{E}_{jy}, \tag{12}$$

$$i\xi\tilde{B}_{jy} - i\eta\tilde{B}_{jx} = -i\omega\varepsilon_j\mu_j\tilde{E}_{jz}, \tag{13}$$

where δ_{sp} is the Kronecker delta.

Using Eqs. (9) and (13), we express $\tilde{E}_{jz}$ and $\tilde{B}_{jy}$ in terms of $\tilde{E}_{jx}$ and $\tilde{B}_{jx}$ and, using (10) and (12), express $\tilde{E}_{jy}$ and $\tilde{B}_{jz}$ in terms of $\tilde{E}_{jx}$ and $\tilde{B}_{jx}$:

$$\tilde{E}_{jy} = \frac{1}{k_j^2 - \xi^2}\left(-\eta\xi\tilde{E}_{jx} + i\omega\frac{\partial}{\partial z}\tilde{B}_{jx}\right), \tag{14}$$

$$\tilde{E}_{jz} = \frac{1}{k_j^2 - \xi^2}\left(i\xi\frac{\partial}{\partial z}\tilde{E}_{jx} + \omega\eta\tilde{B}_{jx}\right), \tag{15}$$

$$\tilde{B}_{jy} = \frac{1}{k_j^2 - \xi^2}\left(-i\omega\varepsilon_j\mu_j\frac{\partial}{\partial z}\tilde{E}_{jx} - \xi\eta\tilde{B}_{jx}\right), \tag{16}$$

$$\tilde{B}_{jz} = \frac{1}{k_j^2 - \xi^2}\left(-\eta\omega\varepsilon_j\mu_j\tilde{E}_{jx} + i\xi\frac{\partial}{\partial z}\tilde{B}_{jx}\right), \tag{17}$$

where we introduced complex wavenumbers for the domains by the formulas $k_j^2 = \omega^2\varepsilon_j\mu_j$, $j = 1, 2, 3$.
Substituting expressions (14)–(17) into (8) and (11), we obtain

$$\left(\frac{d^2}{dz^2} + \gamma_j^2\right)\tilde{E}_{jx} = \frac{k_j^2 - \xi^2}{i\omega\varepsilon_j}\delta(z-d), \tag{18}$$

$$\left(\frac{d^2}{dz^2} + \gamma_j^2\right)\tilde{B}_{jx} = 0. \tag{19}$$

where $\gamma_j^2 = k_j^2 - \xi^2 - \eta^2$.
For fixed values of ξ and η, expressions (18) and (19) are ordinary differential equations in the variable z. The problem consists in solving Eqs. (18) and (19) for $\tilde{E}_{jx}$ and $\tilde{B}_{jx}$ in the three domains. Knowing the components $\tilde{E}_{jx}$ and $\tilde{B}_{jx}$, one can determine the other components of the field from Eqs. (14)–(17).
The solution must satisfy not only the equations but also the boundary conditions on the two interfaces $z = 0$ and $z = -h$ between media.
Formulas (18) and (19) contain the functions $\gamma_j^2 = k_j^2 - \xi^2 - \eta^2$. To find the solutions of these equations, we should single out analytic branches of the functions γ_j versus the complex variable λ, where $\lambda^2 = \xi^2 + \eta^2$. For real wavenumbers, we have the formula (see King & Smith (1981))

$$\gamma_j(\lambda) = \begin{cases} \sqrt{k_j^2 - \lambda^2}, \ \lambda^2 \le k_j^2 \\ i\sqrt{\lambda^2 - k_j^2}, \ \lambda^2 \ge k_j^2. \end{cases} \tag{20}$$

In the general case of an absorbing medium, there exist two branching points of the function $\gamma_j(\lambda)$: the point

$$k_{j,1} = \omega\sqrt{|\varepsilon_j||\mu_j|}\exp\left(i\left(\arg(\varepsilon_j) + \arg(\mu_j)\right)/2\right)$$

and the point $k_{j,2} = e^{i\pi}k_{j,1}$. An analytic branch of the function $\gamma_j(\lambda)$ suitable for describing materials with negative refractive index should be the same as that for ordinary materials, either with or without absorption (with appropriate values of permeability and permittivity). This branch can be determined as

$$\gamma_j(\lambda)=\sqrt{|k_{j,1}-\lambda|}\exp\left(\frac{i\arg(k_{j,1}-\lambda)}{2}\right)\sqrt{|k_{j,2}-\lambda|}\exp\left(\frac{i\arg(\lambda-k_{j,2})}{2}\right), \tag{21}$$

where the functions $|\xi|$ and $\arg(\xi)$ are the modulus and the argument of the complex variable ξ.

Impotante that the considering choise of the analytic branch of the function $\gamma_j(\lambda)$ may be considered as a principle of correspondence of common media and media with negative refraction. The discription of common media and media with negative refraction are based on the same analytic branch. There is no physical reason of jumping from one branch to another.

Figure 10 shows why it is the analytic branch (21) that correctly describes both an ordinary lossy medium (in the limit, with infinitely small losses) and a passive (generally speaking, lossy) medium with negative refractive index. Figure 10a represents the real and imaginary parts of the function $\gamma(\lambda)$ described by formula (21) for an ordinary lossy medium with parameters $\varepsilon=1+i0.01$ and $\mu=1+i0.01$. The imaginary parts of these parameters are chosen solely to make clear the diagrams; the wave frequency is 3 GHz. For such a function $\gamma(\lambda)$, the waves $e^{i\gamma z}$ with positive increasing z will propagate with decay along the z-axis. Figure 10b represents the real and imaginary parts of the function $\gamma(\lambda)$ described by formula (21) for a lossy medium with negative refractive index and with the parameters $\varepsilon=-1+i0.01$ and $\mu=-1+i0.01$. For such a function $\gamma(\lambda)$, the waves $e^{i\gamma z}$ with positive increasing z will propagate opposite to the axis z (Re < 0), but they decay along the z-axis; i.e., they decay while propagating away from the wave source. An incorrect choice of the analytic branch of the function $\gamma(\lambda)$ often leads to unphysical results that violate the cause-and-effect relation.

A solution to Eq. (19) in domain 1 is given by the function

$$\tilde{B}_{1x}=\mu_1\left(C_1'e^{-i\gamma_1 z}+C_1e^{i\gamma_1 z}\right). \tag{22}$$

If k_1 is real, then, for $\lambda^2=\xi^2+\eta^2>k_1^2$, from (20) we obtain

$$\tilde{B}_{1x}=\mu_1\left(C_1'e^{\sqrt{\lambda^2-k_1^2}z}+C_1e^{-\sqrt{\lambda^2-k_1^2}z}\right). \tag{23}$$

When $z>0$, the first term exponentially increases, which contradicts common sense (the field of a source in free space increases at infinity). The integration of the first term by formula (6) yields an infinite integral. Therefore, $C_1'=0$, and for $z>0$ we obtain

$$\tilde{B}_{1x}=\mu_1 C_1 e^{i\gamma_1 z}. \tag{24}$$

Similarly we can show that a solution to Eq. (19) in domain 3 for z < –*h* is given by

$$\tilde{B}_{3x}=\mu_3 C_3 e^{-i\gamma_3 z}. \tag{25}$$

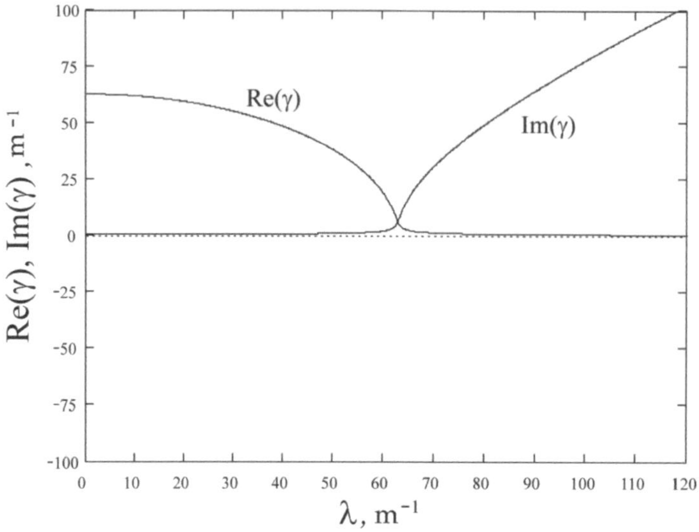

a)

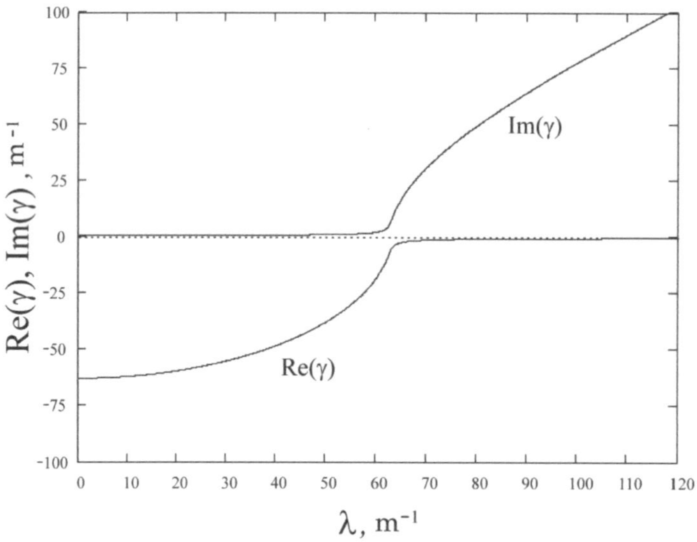

b)

Fig. 10. Real and imaginary parts of the function $\gamma(\lambda)$ described by formula (21) (a) for an ordinary medium with losses and parameters $\varepsilon = 1 + i0.01$ and $\mu = 1 + i0.01$ and (b) for a medium with losses and negative refractive index for $\varepsilon = -1 + i0.01$ and $\mu = -1 + i0.01$.

In domain 3, Eq. (18) for the electric field is also homogeneous; therefore, applying similar arguments, for z < -*h* we obtain

$$\tilde{E}_{3x} = S_3 e^{-i\gamma_3 z} . \tag{26}$$

In domain 2, there exist two linearly independent solutions; therefore,

$$\tilde{B}_{2x} = \mu_2\left(C_2' e^{-i\gamma_2 z} + C_2 e^{i\gamma_2(z+h)}\right), \tag{27}$$

$$\tilde{E}_{2x} = S_2' e^{-i\gamma_2 z} + S_2 e^{i\gamma_2(z+h)}. \tag{28}$$

Equation (18) is inhomogeneous in domain 1. A partial solution to this equation is given by

$$\left(\tilde{E}_{1x}\right)_p = -\frac{k_1^2 - \xi^2}{2\gamma_1 \omega \varepsilon_1} e^{i\gamma_1|z-d|},$$

therefore, the general solution is

$$\tilde{E}_{1x} = S_1' e^{-i\gamma_1 z} + S_1 e^{i\gamma_1 z} - \frac{k_1^2 - \xi^2}{2\gamma_1 \omega \varepsilon_1} e^{i\gamma_1|z-d|}.$$

Again, repeating the arguments that a wave cannot increase as $z \to +\infty$, we find that $S_1' = 0$. As a result, the electric field in domain 1 is represented as

$$\tilde{E}_{1x} = S_1 e^{i\gamma_1 z} - \frac{k_1^2 - \xi^2}{2\gamma_1 \omega \varepsilon_1} e^{i\gamma_1|z-d|}. \tag{29}$$

Let us write the continuity conditions for the tangential components of the electric and magnetic fields, $\tilde{E}_x$ and $\tilde{B}_x/\mu$, on the two boundaries $z = 0$ and $z = -h$:

$$C_1 = C_2' + C_2 e^{i\gamma_2 h} \quad \text{at } z = 0 \tag{30}$$

$$C_2' e^{i\gamma_2 h} + C_2 = C_3 e^{i\gamma_3 h} \quad \text{at } z = -h \tag{31}$$

for the magnetic field and

$$S_1 - \frac{k_1^2 - \xi^2}{2\gamma_1 \omega \varepsilon_1} e^{i\gamma_1 d} = S_2' + S_2 e^{i\gamma_2 h} \quad \text{at } z = 0 \tag{32}$$

$$S_2' e^{i\gamma_2 h} + S_2 = S_3 e^{i\gamma_3 h} \quad \text{at } z = -h. \tag{33}$$

for the electric field. We have four equations and eight unknowns. Therefore, we should add equations for the continuity of $\tilde{E}_y$ and $\tilde{B}_y/\mu$ on the two boundaries. However, first, using relations (30)–(33), we eliminate the coefficients describing the fields in domain 2, i.e., C_2', C_2, S_2', S_2. From (30) and (31) we obtain

$$C_2 = \frac{1}{2i\sin(\gamma_2 h)} C_1 - \frac{e^{i\gamma_3 h - i\gamma_2 h}}{2i\sin(\gamma_2 h)} C_3, \tag{34}$$

$$C_2' = -\frac{e^{-i\gamma_2 h}}{2i\sin(\gamma_2 h)} C_1 + \frac{e^{i\gamma_3 h}}{2i\sin(\gamma_2 h)} C_3, \tag{35}$$

and from (32) and (33) we obtain

$$S_2' = -\frac{e^{-i\gamma_2 h}}{2i\sin(\gamma_2 h)}S_1 + \frac{e^{i\gamma_3 h}}{2i\sin(\gamma_2 h)}S_3 + \frac{k_1^2-\xi^2}{2\gamma_1\omega\varepsilon_1}\frac{e^{i\gamma_1 d - i\gamma_2 h}}{2i\sin(\gamma_2 h)}, \tag{36}$$

$$S_2 = \frac{1}{2i\sin(\gamma_2 h)}S_1 - \frac{e^{i(\gamma_3-\gamma_2)h}}{2i\sin(\gamma_2 h)}S_3 - \frac{k_1^2-\xi^2}{2\gamma_1\omega\varepsilon_1}\frac{e^{i\gamma_1 d}}{2i\sin(\gamma_2 h)}. \tag{37}$$

Substituting expressions (34)-(37) into the boundary conditions for the tangential components of the fields $\tilde{E}_{jy}$ and $\tilde{B}_{jy}/\mu_j$ on the boundary $z=0$, we obtain

$$\tilde{B}_{jy}/\mu_j = \left(-i\omega\varepsilon_j\,\partial\tilde{E}_{jx}/\partial z - \xi\eta\tilde{B}_{jx}/\mu_j\right)/\left(k_j^2-\xi^2\right),$$

$$\frac{\tilde{B}_{1y}}{\mu_1} = \frac{1}{k_1^2-\xi^2}\left(-\xi\eta C_1 + \omega\varepsilon_1\gamma_1 S_1 + \omega\varepsilon_1\gamma_1 Q\right),$$

$$\frac{\tilde{B}_{2y}}{\mu_2} = \frac{1}{k_2^2-\xi^2}\left(-\xi\eta C_1 - i\omega\varepsilon_2\gamma_2\,\mathrm{ctg}(\gamma_2 h)S_1 + i\omega\varepsilon_2\gamma_2\frac{e^{i\gamma_3 h}S_3}{\sin(\gamma_2 h)} + i\omega\varepsilon_2\gamma_2\,\mathrm{ctg}(\gamma_2 h)Q\right),$$

where

$$Q = \frac{k_1^2-\xi^2}{2\gamma_1\omega\varepsilon_1}e^{i\gamma_1 d}.$$

Then, the boundary condition for the magnetic field at z = 0 is expressed as

$$\begin{gathered}\frac{1}{k_1^2-\xi^2}\left(-\xi\eta C_1 + \omega\varepsilon_1\gamma_1 S_1 + \omega\varepsilon_1\gamma_1 Q\right) = \\ \frac{1}{k_2^2-\xi^2}\left(-\xi\eta C_1 - i\omega\varepsilon_2\gamma_2\,\mathrm{ctg}(\gamma_2 h)S_1 + i\omega\varepsilon_2\gamma_2\frac{e^{i\gamma_3 h}}{\sin(\gamma_2 h)}S_3 + i\omega\varepsilon_2\gamma_2\,\mathrm{ctg}(\gamma_2 h)Q\right)\end{gathered}. \tag{38}$$

Proceeding along the same vein, we obtain a boundary condition for the magnetic field on the second boundary z = -*h*:

$$\begin{gathered}\frac{1}{k_2^2-\xi^2}\left(-\xi\eta e^{i\gamma_3 h}C_3 - \frac{i\omega\varepsilon_2\gamma_2 S_1}{\sin(\gamma_2 h)} + i\omega\varepsilon_2\gamma_2\,\mathrm{ctg}(\gamma_2 h)e^{i\gamma_3 h}S_3 + i\omega\varepsilon_2\gamma_2\frac{Q}{\sin(\gamma_2 h)}\right) = \\ \frac{1}{k_3^2-\xi^2}\left(-\xi\eta e^{i\gamma_3 h}C_3 - \omega\varepsilon_3\gamma_3 e^{i\gamma_3 h}S_3\right)\end{gathered}. \tag{39}$$

To obtain the boundary condition for the electric field on the boundary z = 0, we apply the expression

$$\tilde{E}_{jy} = \left(-\eta\xi\tilde{E}_{jx} + i\omega\,\partial\tilde{B}_{jx}/\partial z\right)/\left(k_j^2-\xi^2\right),$$

then the boundary condition is expressed as

$$\begin{aligned}&\frac{1}{k_1^2-\xi^2}\left(-\omega\gamma_1\mu_1 C_1-\eta\xi S_1+\eta\xi Q\right)=\\&=\frac{1}{k_2^2-\xi^2}\left(i\omega\mu_2\gamma_2\operatorname{ctg}(\gamma_2 h)C_1-i\omega\mu_2\gamma_2\frac{e^{i\gamma_3 h}}{\sin(\gamma_2 h)}C_3-\eta\xi S_1+\eta\xi Q\right)\end{aligned}. \tag{40}$$

Finally, the boundary condition for the electric field on the boundary z = -*h* is expressed as

$$\begin{aligned}&\frac{1}{k_2^2-\xi^2}\left(i\omega\mu_2\gamma_2\frac{1}{\sin(\gamma_2 h)}C_1-i\omega\mu_2\gamma_2\operatorname{ctg}(\gamma_2 h)e^{i\gamma_3 h}C_3-\eta\xi e^{i\gamma_3 h}S_3\right)=\\&=\frac{1}{k_3^2-\xi^2}\left(\omega\gamma_3\mu_3 e^{i\gamma_3 h}C_3-\eta\xi e^{i\gamma_3 h}S_3\right)\end{aligned}. \tag{41}$$

The system of equations (38)–(41) can be represented in the matrix form as $\mathbf{A}\cdot\mathbf{T}=\mathbf{B}$, where the matrix of the problem is given by

$$\mathbf{A}=\begin{bmatrix}A_{11}&A_{12}&A_{13}&A_{14}\\A_{21}&A_{22}&A_{23}&A_{24}\\A_{31}&A_{32}&A_{33}&A_{34}\\A_{41}&A_{42}&A_{43}&A_{44}\end{bmatrix},$$

where

$$A_{11}=\xi\eta\left(\frac{1}{k_2^2-\xi^2}-\frac{1}{k_1^2-\xi^2}\right),\ A_{21}=0,\ A_{31}=-\omega\left(\frac{\gamma_1\mu_1}{k_1^2-\xi^2}+\frac{i\mu_2\gamma_2\operatorname{ctg}(\gamma_2 h)}{k_2^2-\xi^2}\right),$$

$$A_{41}=\frac{i\omega\mu_2\gamma_2}{k_2^2-\xi^2}\frac{1}{\sin(\gamma_2 h)},\ A_{12}=0,\ A_{22}=\xi\eta\left(\frac{1}{k_3^2-\xi^2}-\frac{1}{k_2^2-\xi^2}\right)e^{i\gamma_3 h},$$

$$A_{32}=\frac{i\omega\mu_2\gamma_2}{k_2^2-\xi^2}\frac{e^{i\gamma_3 h}}{\sin(\gamma_2 h)},\ A_{42}=-\omega\left(\frac{\gamma_3\mu_3}{k_3^2-\xi^2}+\frac{i\mu_2\gamma_2\operatorname{ctg}(\gamma_2 h)}{k_2^2-\xi^2}\right)e^{i\gamma_3 h},$$

$$A_{13}=\omega\left(\frac{\varepsilon_1\gamma_1}{k_1^2-\xi^2}+\frac{i\varepsilon_2\gamma_2\operatorname{ctg}(\gamma_2 h)}{k_2^2-\xi^2}\right),\ A_{23}=-\frac{i\omega\varepsilon_2\gamma_2}{k_2^2-\xi^2}\frac{1}{\sin(\gamma_2 h)},$$

$$A_{33}=\xi\eta\left(\frac{1}{k_2^2-\xi^2}-\frac{1}{k_1^2-\xi^2}\right),\ A_{43}=0,\ A_{14}=-\frac{i\omega\varepsilon_2\gamma_2}{k_2^2-\xi^2}\frac{e^{i\gamma_3 h}}{\sin(\gamma_2 h)},$$

$$A_{24}=\omega\left(\frac{\varepsilon_3\gamma_3}{k_3^2-\xi^2}+\frac{i\varepsilon_2\gamma_2\operatorname{ctg}(\gamma_2 h)}{k_2^2-\xi^2}\right)e^{i\gamma_3 h},\ A_{34}=0,\ A_{44}=\xi\eta\left(\frac{1}{k_3^2-\xi^2}-\frac{1}{k_2^2-\xi^2}\right)e^{i\gamma_3 h}.$$

The right-hand side of the matrix equation and the vector of unknown coefficients are given by

$$\mathbf{B}^T=\left[\omega\left(\frac{i\varepsilon_2\gamma_2\operatorname{ctg}(\gamma_2 h)}{k_2^2-\xi^2}-\frac{\varepsilon_1\gamma_1}{k_1^2-\xi^2}\right)Q;\ -\frac{i\omega\varepsilon_2\gamma_2}{k_2^2-\xi^2}\frac{Q}{\sin(\gamma_2 h)};\ \xi\eta\left(\frac{1}{k_2^2-\xi^2}-\frac{1}{k_1^2-\xi^2}\right)Q;\ 0\right],$$

$$\mathbf{T}^T=\left[C_1\ \ C_3\ \ S_1\ \ S_3\right].$$

From the equation $\mathbf{A}\cdot\mathbf{T}=\mathbf{B}$ one can determine the column of coefficients $\mathbf{T}=\mathbf{A}^{-1}\cdot\mathbf{B}$ and, as a particular case, $C_3=T_2$ and $S_3=T_4$; then, the components of the electric and magnetic fields in domain 3 are determined by the formulas

$$\tilde{E}_{3x}=S_3 e^{-i\gamma_3 z};$$

$$\tilde{B}_{3x}=\mu_3 C_3 e^{-i\gamma_3 z};$$

$$\tilde{E}_{3y}=\frac{1}{k_3^2-\xi^2}\left(-\eta\xi\tilde{E}_{3x}+i\omega\frac{\partial}{\partial z}\tilde{B}_{3x}\right)=\frac{1}{k_3^2-\xi^2}\left(-\eta\xi S_3+\gamma_3\omega\mu_3 C_3\right)e^{-i\gamma_3 z};$$

$$\tilde{B}_{3y}=\frac{1}{k_3^2-\xi^2}\left(-i\omega\varepsilon_3\mu_3\frac{\partial}{\partial z}\tilde{E}_{3x}-\xi\eta\tilde{B}_{3x}\right)=\frac{\mu_3}{k_3^2-\xi^2}\left(-\gamma_3\omega\varepsilon_3 S_3-\xi\eta C_3\right)e^{-i\gamma_3 z};$$

$$\tilde{E}_{3z}=\frac{1}{k_3^2-\xi^2}\left(i\xi\frac{\partial}{\partial z}\tilde{E}_{3x}+\omega\eta\tilde{B}_{3x}\right)=\frac{1}{k_3^2-\xi^2}\left(\xi\gamma_3 S_3+\omega\eta\mu_3 C_3\right)e^{-i\gamma_3 z};$$

$$\tilde{B}_{3z}=\frac{1}{k_3^2-\xi^2}\left(-\eta\omega\varepsilon_3\mu_3\tilde{E}_{3x}+i\xi\frac{\partial}{\partial z}\tilde{B}_{3x}\right)=\frac{\mu_3}{k_3^2-\xi^2}\left(-\eta\omega\varepsilon_3 S_3+\xi\gamma_3 C_3\right)e^{-i\gamma_3 z}.$$

Knowing the Fourier components of the fields, one can determine the field components themselves by applying the inverse Fourier transformation. As a result, for example, the formula for the electric field component $E_{3x}(x,y,z)$ in domain 3 is given by

$$E_{3x}(x,y,z)=(2\pi)^{-2}\int_{-\infty}^{+\infty}d\xi\int_{-\infty}^{+\infty}d\eta\, e^{i\xi x+i\eta y}\left[\mathbf{A}^{-1}\cdot\mathbf{B}\right]_4 e^{-i\gamma_3 z}. \tag{42}$$

In the special case of a Hertzian dipole parallel to the boundary of a half-space when the media 2 and 3 are filled with the same material ($\varepsilon_2=\varepsilon_3$, $\mu_2=\mu_3$, $k_2=k_3$, and $\gamma_2=\gamma_3$), the system of equations (38)–(41) is simplified and reduces to

$$\begin{cases} C_3=C_1 \\ S_1=S_3+Q \\ -\left\{\frac{\mu_3\gamma_3}{k_3^2-\xi^2}+\frac{\gamma_1\mu_1}{k_1^2-\xi^2}\right\}\omega C_1+\left\{\frac{1}{k_3^2-\xi^2}-\frac{1}{k_1^2-\xi^2}\right\}\eta\xi S_3=0 \\ \left\{\frac{1}{k_3^2-\xi^2}-\frac{1}{k_1^2-\xi^2}\right\}\xi\eta C_1+\left(\frac{\varepsilon_3\gamma_3}{k_3^2-\xi^2}+\frac{\varepsilon_1\gamma_1}{k_1^2-\xi^2}\right)\omega S_3=-2\frac{\omega\varepsilon_1\gamma_1}{k_1^2-\xi^2}Q \end{cases}.$$

The substitution of

$$Q=\frac{k_1^2-\xi^2}{2\gamma_1\omega\varepsilon_1}e^{i\gamma_1 d},\ k_1^2=\omega^2\varepsilon_1\mu_1 \text{ and } k_3^2=\omega^2\varepsilon_3\mu_3$$

into the last two equations, we obtain precisely the system of equations (3.14) from King & Smith (1981):

$$\begin{cases} \left\{\dfrac{\mu_3\gamma_3}{k_3^2-\xi^2}+\dfrac{\gamma_1\mu_1}{k_1^2-\xi^2}\right\}C_1-\left\{\dfrac{1}{k_3^2-\xi^2}-\dfrac{1}{k_1^2-\xi^2}\right\}\eta\xi\left(\dfrac{S_3}{\omega}\right)=0 \\ -\left\{\dfrac{1}{k_3^2-\xi^2}-\dfrac{1}{k_1^2-\xi^2}\right\}\xi\eta C_1-\left(\dfrac{k_3^2\gamma_3}{\mu_3\left(k_3^2-\xi^2\right)}+\dfrac{k_1^2\gamma_1}{\mu_1\left(k_1^2-\xi^2\right)}\right)\left(\dfrac{S_3}{\omega}\right)=e^{i\gamma_1 d} \end{cases}.$$

It can be shown that explicit expressions for the coefficients are given by

$$\begin{cases} C_1=-\dfrac{\xi\eta\left(\mu_1\varepsilon_1-\mu_3\varepsilon_3\right)}{\left(\mu_1\gamma_3+\mu_3\gamma_1\right)\left(\varepsilon_1\gamma_3+\varepsilon_3\gamma_1\right)}e^{i\gamma_1 d} \\ S_3=-\dfrac{\gamma_1\mu_1\left(k_3^2-\xi^2\right)+\mu_3\gamma_3\left(k_1^2-\xi^2\right)}{\omega\left(\mu_1\gamma_3+\mu_3\gamma_1\right)\left(\varepsilon_1\gamma_3+\varepsilon_3\gamma_1\right)}e^{i\gamma_1 d} \end{cases}.$$

Then, in the particular case considered, formula (42) reduces to the following explicit form:

$$E_{3x}(x,y,z)=-(2\pi)^{-2}\int_{-\infty}^{+\infty}d\xi\int_{-\infty}^{+\infty}d\eta\, e^{i\xi x+i\eta y}\frac{\gamma_1\mu_1\left(k_3^2-\xi^2\right)+\mu_3\gamma_3\left(k_1^2-\xi^2\right)}{\omega\left(\mu_1\gamma_3+\mu_3\gamma_1\right)\left(\varepsilon_1\gamma_3+\varepsilon_3\gamma_1\right)}e^{i\gamma_1 d-i\gamma_3 z}. \tag{43}$$

An explicit expression for the y-component of the electric field is as follows:

$$E_{3y}(x,y,z)=-(2\pi)^{-2}\int_{-\infty}^{+\infty}d\xi\int_{-\infty}^{+\infty}d\eta\, e^{i\xi x+i\eta y}\frac{\xi\eta\left(\gamma_1\mu_1+\mu_3\gamma_3\right)}{\omega\left(\mu_1\gamma_3+\mu_3\gamma_1\right)\left(\varepsilon_1\gamma_3+\varepsilon_3\gamma_1\right)}e^{i\gamma_1 d-i\gamma_3 z}. \tag{44}$$

Passing to cylindrical coordinates, from expressions (43) and (44) we can determine the electric field components in cylindrical coordinates King&Smith (1981) (i.e., in the plane of image perpendicular to the z-axis; further analysis will be focused on the field distribution in this plane):

$$E_{3\rho}=-\frac{\cos\varphi}{4\pi}\int_0^{\infty}\left\{\frac{\omega\mu_1\mu_3}{M}\left[J_0(\lambda\rho)+J_2(\lambda\rho)\right]+\frac{\gamma_1\gamma_3}{\omega N}\left[J_0(\lambda\rho)-J_2(\lambda\rho)\right]\right\}e^{i(\gamma_1 d-\gamma_3 z)}\lambda d\lambda \tag{45}$$

$$E_{3\varphi}=\frac{\sin\varphi}{4\pi}\int_0^{\infty}\left\{\frac{\omega\mu_1\mu_3}{M}\left[J_0(\lambda\rho)-J_2(\lambda\rho)\right]+\frac{\gamma_1\gamma_3}{\omega N}\left[J_0(\lambda\rho)+J_2(\lambda\rho)\right]\right\}e^{i(\gamma_1 d-\gamma_3 z)}\lambda d\lambda \tag{46}$$

where $M=\mu_1\gamma_3+\mu_3\gamma_1$, $N=\varepsilon_1\gamma_3+\varepsilon_3\gamma_1$, J_n is a Bessel function of order n, and (ρ,φ,z) are cylindrical coordinates (see Fig. 9).

3.2 Results of calculating the radiation from a dipole parallel to the boundary of a half-space with negative refractive index and discussion of these results.

We calculated radiation from a dipole parallel to the boundary of a half-space filled with a medium with negative refractive index (medium 3; the layer with medium 2 is missing). We investigated radiation transmitted into medium 3 in order to study the focusing properties of the medium, i.e., in order to find the dipole field near the point of the geometric-optics image of the dipole. The electric field components parallel to the plane of image were calculated by formulas (45) and (46). We verified that formulas (43) and (44) give identical results for appropriate components; however, because of the double integration, these calculations take much more time.

The electric field components were calculated for a dipole located at the point with coordinates $x = 0$, $y = 0$, $z = d = 1$ m. The radiation frequency is $f = \omega/2\pi = 3$ GHz. In free space, this frequency corresponds to a wavelength of $\lambda_0 = 0.1$ m. All the calculations were carried out for materials with negative refractive index and small absorption ($\varepsilon_3'' = 10^{-3}$ and $\mu_3'' = 10^{-3}$), because the case of low absorption is of primary importance for possible applications.

First, we calculated the electric field for a material with $\varepsilon_3' = -1$ and $\mu_3' = -1$. According to the laws of geometrical optics [see Veselago (2003)], after refraction, the dipole radiation must be focused at a mirror symmetric point with coordinates $x = 0$, $y = 0$, $z_{im} = -d$. Note that if we substitute the relative permittivity and permeability into the equations for the fields, we should multiply these equations by the permittivity ε_0 and permeability μ_0 of vacuum. The distributions of $|E_{3\rho}|_{\varphi=0}$ in the xz-plane (i.e., in the E plane of the dipole) and of $|E_{3\varphi}|_{\varphi=\pi/2}$ in the yz-plane (i.e., in the H plane of the dipole) are shown in Fig. 11. One can see that the radiation is indeed focused to a spot centered at the point of the geometric-optics image. The dimensions of the spot at the level of 0.707 of amplitude (at the half power level) are $w_E \approx 0.0589$ m in the E plane and $w_H \approx 0.0422$ m in the H plane.

Next, we carried out calculations for a material with $\varepsilon_3' = -2$ and $\mu_3' = -2$. It is reasonable to assume that the focal spot moves to a point with coordinates $x = 0$, $y = 0$, $z_{im} = -2d = -2$m according to the laws of geometrical optics. Since the wavelength in the medium is half that in vacuum, we can assume that the dimensions of the spot are also reduced by a factor of two. Calculations have shown that this is not the case. We have found that, although the position of the image spot on the z-axis, determined by the maximum of the field, approximately corresponds to the geometric-optics position $z_{im} \approx -2.45$ m, the dimensions of the spot in the E and H planes at the level of 0.707 of amplitude are greater than the expected values and are equal to $w_E \approx 0.0675$ m and $w_H \approx 0.0621$ m, respectively.

After that, we carried out calculations for a material with $\varepsilon_3' = -10$ and $\mu_3' = -10$. Again, according to the laws of geometrical optics, it is reasonable to assume that the image point has coordinates $x = 0$, $y = 0$, $z_{im} = -10d = -10$m, and the spot is ten times smaller than that in the first case. Again, calculations have shown that the image plane moves to the point $z_{im} \approx -12.4$ m, and the dimensions of the spot in the E and H planes of the dipole at the level of 0.707 of amplitude are much greater than the expected values, namely, $w_E \approx 0.0701$ m and $w_H \approx 0.0662$ m. In this case ($\varepsilon_3' = -10$ and $\mu_3' = -10$), the distributions of $|E_{3\rho}|_{\varphi=0}$ in the xz-plane (in the E plane of the dipole) and $|E_{3\varphi}|_{\varphi=\pi/2}$ in the yz-plane (in the H plane of the dipole) are shown in Fig. 12. Note that, when the absorption of a medium with negative refractive index is small, as in the case considered, and the modulus $|n|$ of the refractive index is large, the image spot depends not on the relation between $|\varepsilon_3|$ and $|\mu_3|$ but only on the product $|\varepsilon_3\mu_3|$. This fact is illustrated by Eqs. (45) and (46), in which the properties of medium 3 are represented by the product $\varepsilon_3\mu_3$. This property was verified directly by numerical calculation: the dimensions of the image in the case of $\varepsilon_3' = -10$ and $\mu_3' = -10$ are almost the same as in the case of $\varepsilon_3' = -100$ and $\mu_3' = -1$.

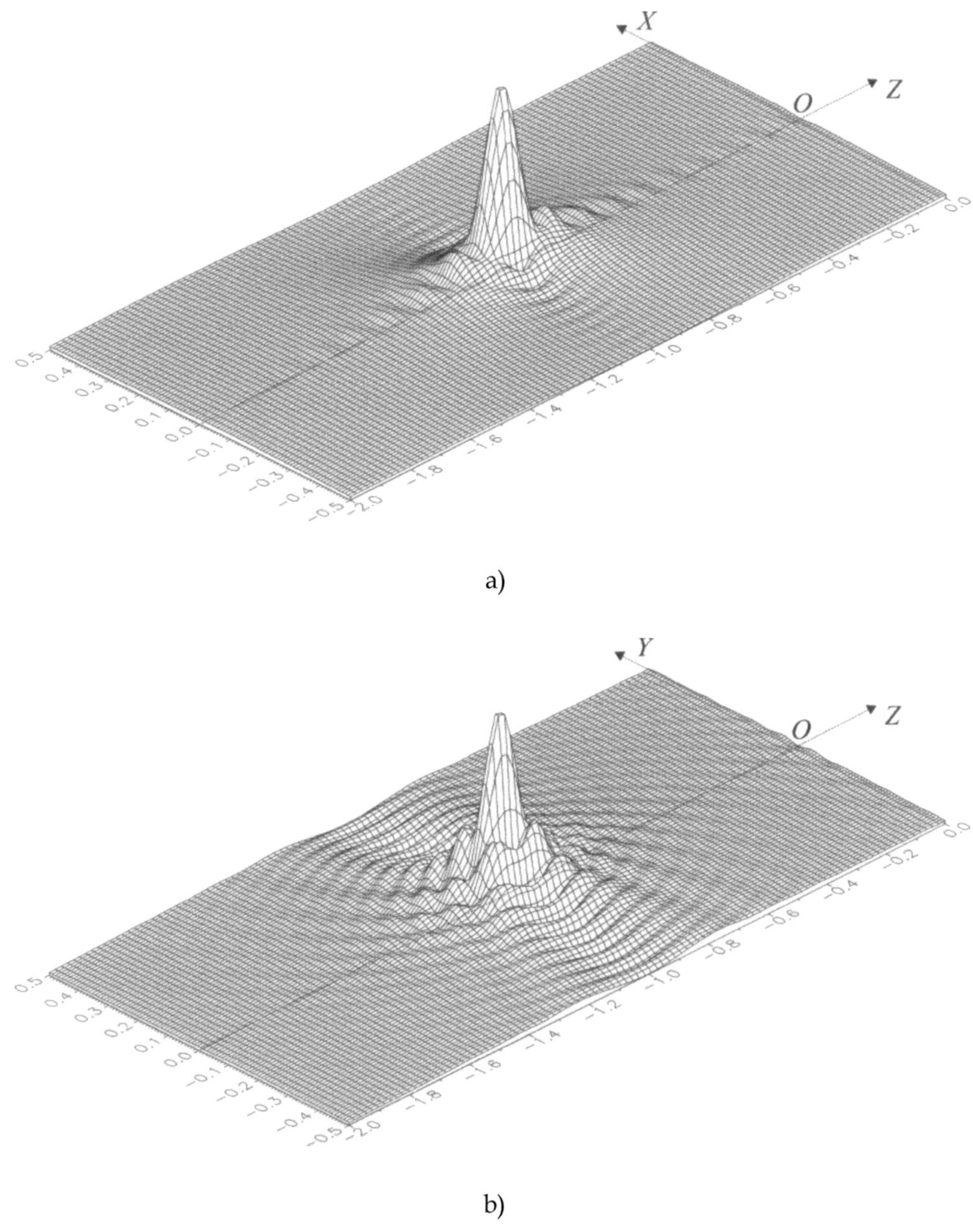

Fig. 11. Distributions of normalized (to maximum) moduli (a) $\left|E_{3\rho}\right|_{\varphi=0}$ in xz-plane and (b) $\left|E_{3\varphi}\right|_{\varphi=\pi/2}$ in the yz-plane in a half-space with negative refraction in the case of $\varepsilon_3'=-1$ and $\mu_3'=-1$.

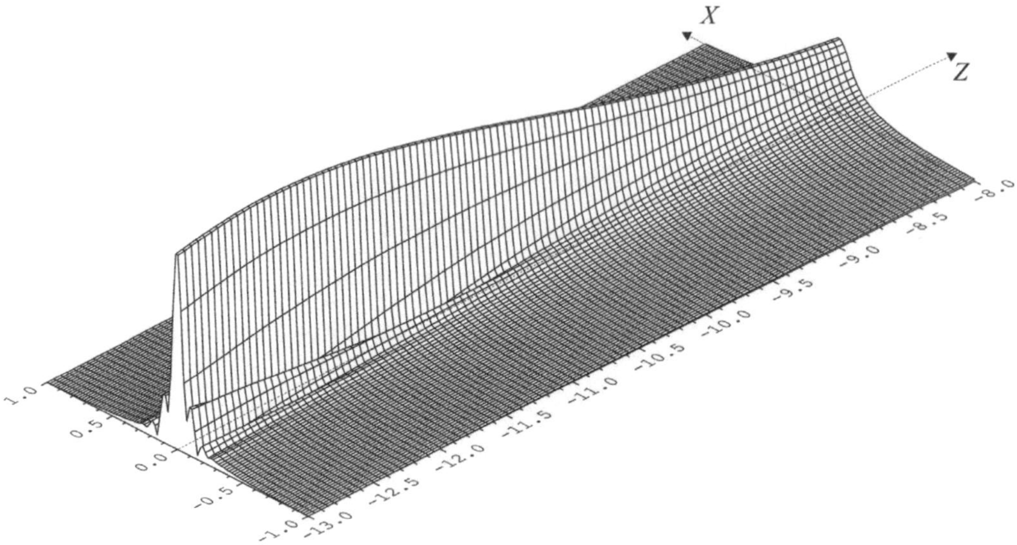

a)

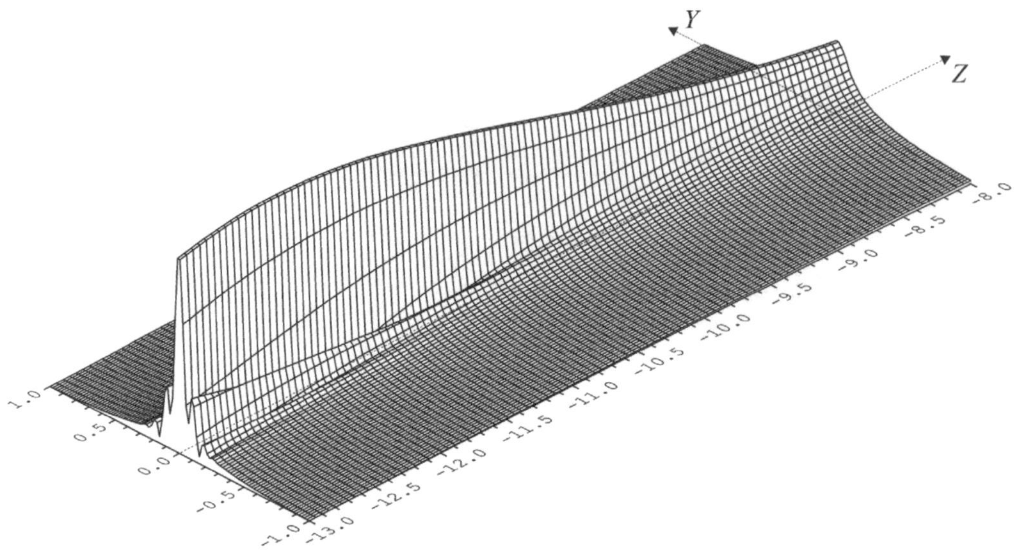

b)

Fig. 12. Same as in Fig. 11 for the case of $\varepsilon_3' = -10$ and $\mu_3' = -10$.

We determined the distance between the boundary and the image plane (where the electric field of the wave attains its maximum) as a function of the modulus of the complex refractive index $|n| \approx \sqrt{|\varepsilon_3||\mu_3|}$ (see Fig. 13). For $|n| \geq 4$, the curve can be approximated by the straight line $z_{im} = -1.243|n|d$.

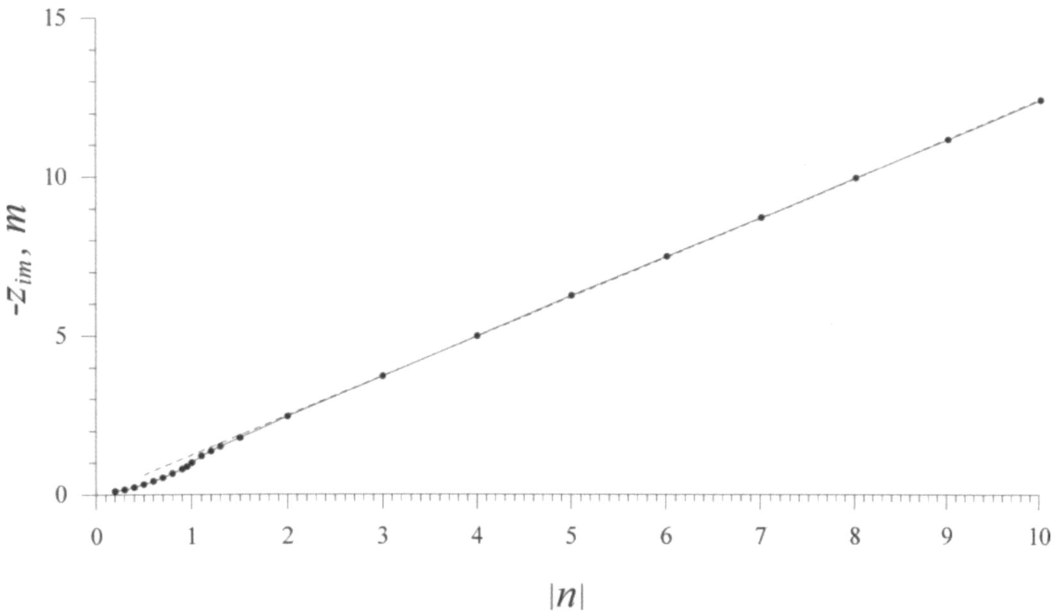

Fig. 13. The distance from the plane boundary to the image (i.e., to the maximum-signal point) as a function of the modulus of the refractive index $|n| \approx \sqrt{|\varepsilon_3||\mu_3|}$. The dashed line represents the function $z_{im} = -1.243|n|d$, which approximates the above function for large values of $|n|$.

Based on the function $z_{im} = z_{im}(|n|)$, we calculated the dimensions w_E and w_H of the spots, normalized by the wavelength λ_0 in free space, as a function of $|n|$. The curves are shown in Fig. 14. The functions $w_E = w_E(|n|)$ and $w_H = w_H(|n|)$ have a minimum at $|n| = 1$ and tend to constant values greater than $\lambda_0/2$ as $|n| \to \infty$. This result is very important. It implies that the use of the plane boundary of a homogeneous material with negative refractive index as an elementary lens does not improve the quality of the image; therefore, such a lens cannot overcome the diffraction limit.

There is a simple explanation for this fact. The dipole radiation excites a certain area of size $D \sim 2d$ on the interface between media. In a medium with negative refraction, this excited surface area works as a lens (antenna) with effective aperture of about D. The wavelength in a material with negative refractive index is $\lambda_{NR} = \lambda_0/|n|$. The distance form this lens to the plane of the image is $|z_{im}| \sim |n|d$ (as found for large values of $|n|$). However, it is well known that the dimensions of the image can be estimated as

$$w \sim |z_{im}|\lambda_{NR}/D \sim (|n|d\lambda_0/|n|)/D \sim d\lambda_0/D \sim \lambda_0/2 = const\ .$$

Thus, for large values of the modulus $|n|$ of the refractive index, the dimensions of the image are nearly constant and do not depend on $|n|$.

The minimum on the curve in the region where $\varepsilon_2' \approx -1$ and $\mu_2' \approx -1$ may be explained by the absence of achromatic aberration of this elementary lens for the particular case.

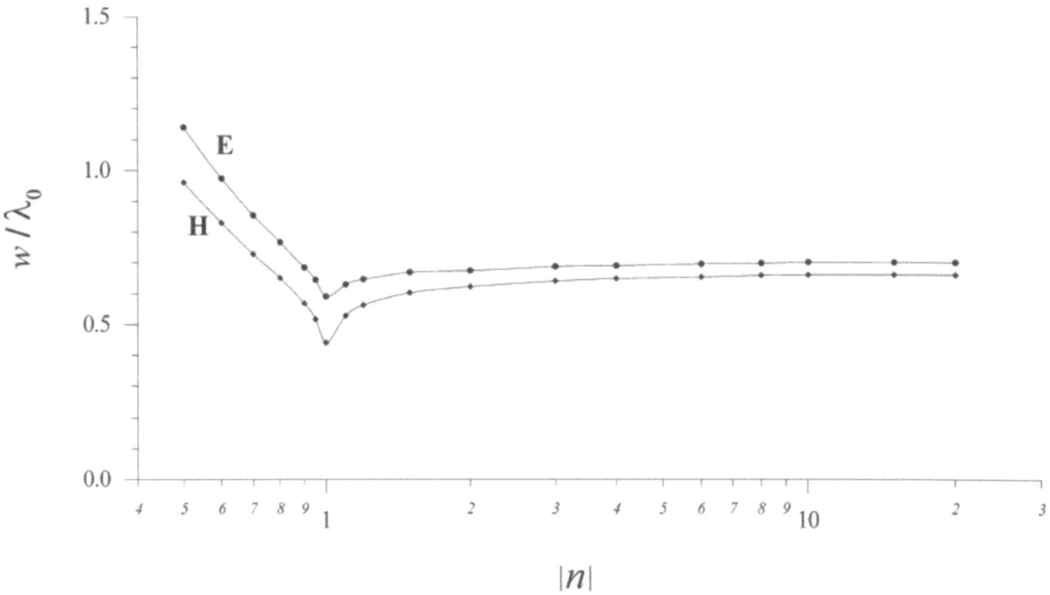

Fig. 14. Normalized half-power widths (or widths at an amplitude level of 0.707) of the image in the E and H planes of the dipole as a function of $|n|$ (λ_0 is the radiation wavelength in free space).

In [Petrin A. B. (2008)], it was shown that, as $\varepsilon_2'' \to 0$ and $\mu_2'' \to 0$, the dimensions of the focusing region vary by at most 30% and this variation occurs only in the domain where $\varepsilon_2' \approx -1$ and $\mu_2' \approx -1$; i.e., our conclusions about the absence of superresolution remain valid even in the limit case of a lossless material with negative refractive index.

3.3 Results of calculating the dipole radiation transmitted through a layer with negative refractive index and discussion of these results.

The results of the previous section show that the focusing of radiation from a point source of electromagnetic radiation due to refraction on the interface between air and a half-space with negative refractive index does not provide a superresolution predicted in [Pendry J. B. (2000)]. To close the question completely, we considered focusing of radiation of a point source transmitted through a layer of medium with negative refractive index. According to [Pendry J. B. (2000)], such a layer represents a lens.

We calculated the electric field for a layer of medium of thickness h = 2m with parameters $\varepsilon_2' = -1$, $\varepsilon_2'' = 10^{-3}$; $\mu_2' = -1$, $\mu_2'' = 10^{-3}$ (see Fig. 9). Media 1 and 3 are air; i.e., $\varepsilon_1' = \varepsilon_3' = 1$, $\varepsilon_1'' = \varepsilon_3'' = 0$; $\mu_1' = \mu_3' = 1$ and $\mu_1'' = \mu_3'' = 0$. The radiation frequency is the same as in the previous section (3 GHz). According to the laws of geometric optics [see Pendry (2000) and Veselago (2003)], after refraction, the dipole radiation should be focused at a point with coordinates x = 0, y = 0, and $z_{im} = -(d+h) = -3\,\text{m}$. The electric field in domain 3 was calculated by formula (42). Figure 15 shows the distribution of the modulus of the electric field $|E_{3x}(0, 0, z)|$ along the z-axis near the point $z_{im} = -3$ m of the geometric-optics image. One can see that the field attains its maximum at the point $z_{im} = -3$ m. However, the size of the distribution is on the order of λ_0. After that, we determined the distributions of the modulus of the electric field in the plane of the image (in the plane passing through the point $z_{im} = -3$ m perpendicular to z-axis) on the intersection lines with the E and H planes of the dipole.

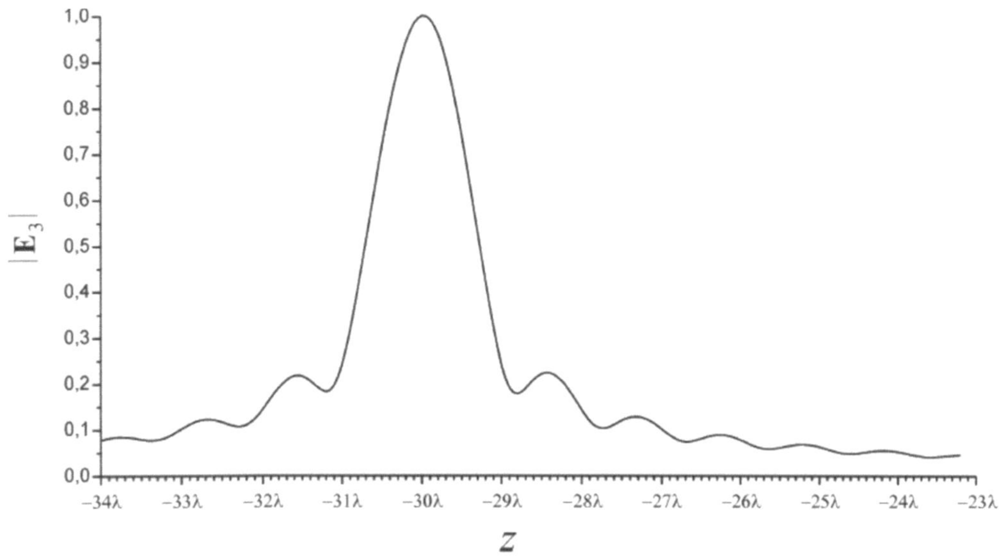

Fig. 15. Normalized (to maximum) distribution of the electric field modulus $|E_{3x}(0, 0, z)|$ along the z-axis near the point of the geometric-optics image $z_{im} = -3\text{m} = -30\lambda$.

The distribution of the modulus of the electric field $|\mathbf{E}_3| = |E_{3x}(x, 0, z_{im})|$ along the intersection line of the xz-plane (i.e., the E plane of the dipole) and the plane of image is shown in Fig. 16a. A similar distribution $|\mathbf{E}_3| = |E_{3x}(0, y, z_{im})|$ along the intersection line of the yz-plane (i.e., the H plane of the dipole) and the plane of image is shown in Fig. 16b. Figures 16 show that radiation is indeed focused to a spot centered at the point of the geometric-optics image. The dimensions of the spot at the amplitude level of 0.707 (at the half-power level) are $w_E \approx 0.0594$ m in the E plane and $w_H \approx 0.0432$ m in the H plane. These results allow us to draw an important conclusion: the distribution of the electric field modulus is slightly expanded after refraction on the second boundary. In other words, superresolution in such a system is not observed.

Thus, the results of the investigation lead to a unanimous conclusion that the use of materials with negative refractive index does not allow one to reach a superresolution of noncoherent objects (to overcome the diffraction limit in far zone).

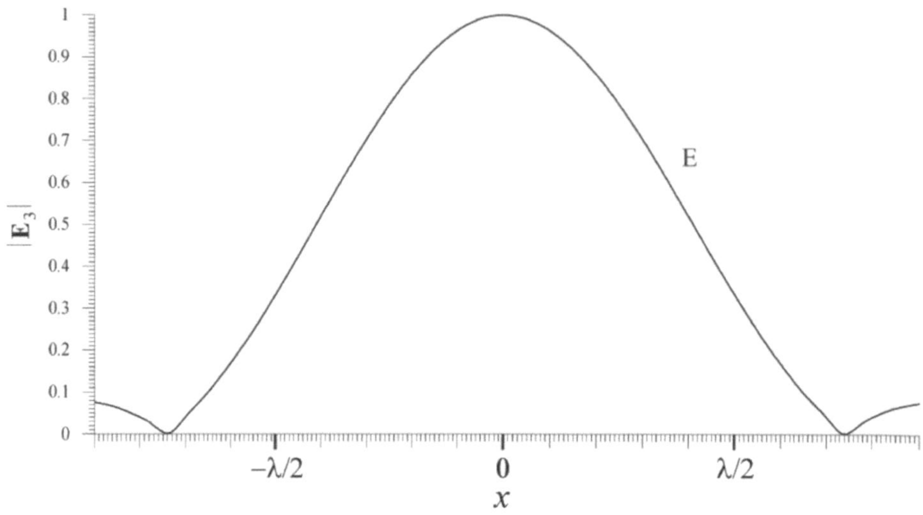

a)

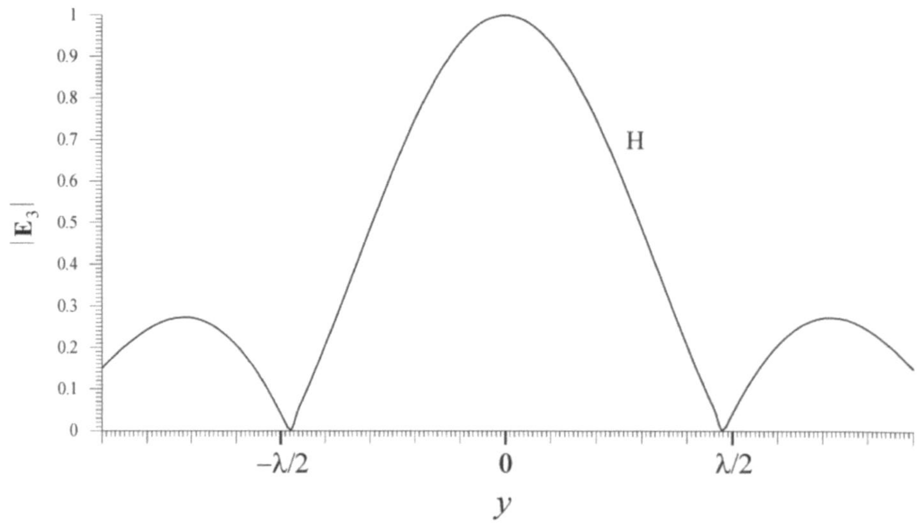

b)

Fig. 16. Normalized (to maximum) distributions of the electric field moduli (a) $|\mathbf{E}_3| = |E_{3x}(x,0,z_{im})|$ and (b) $|\mathbf{E}_3| = |E_{3x}(0,y,z_{im})|$ along the intersection lines of the *xz*-and *yz*-planes, respectively, with a plane passing through the image point $z_{im} = -3\text{m} = -30\lambda$.

4. Conclusion

In this chapter it was proved the fundamental result of diffraction theory: *for any focusing system the existence of singularity at image point of a point source is impossible in principle.* The result was illustrated for the important focusing system consisted of a layer of material with negative refraction (the superlens).

5. References

Chen J. J., Grzegorczyk T. M., Wu B., et al. (2006), *Phys. Rev. E: Stat. Nonlinear, Soft Matter Phys.* Vol. 74, p. 046615.

Cheng Q. and Cui T. J. (2005), *Opt. Lett.* Vol. 30, p. 1216.

Cui T. J., Cheng Q., Lu W. B., et al. (2005), *Phys. Rev. B: Condens. Matter* Vol. 71, p. 045114.

King R. W. P. and Smith G. S. (1981), *Antennas in Matter,* Massachusetts Institute of Technology Press, Cambridge, MA, United States.

Lagarkov A. N. and Kissel V. N. (2004), Near-Perfect Imaging in a Focusing System Based on a Left-Handed-Material Plate, *Phys. Rev. Lett.* Vol. 92, No. 7, p. 077401.

Pendry J. B. (2000), Negative Refraction Makes a Perfect Lens, *Phys. Rev. Lett.* Vol. 85, No. 18, pp. 3966 - 3969.

Pendry J. B. (2001), Reply on the Comments, *Phys. Rev. Lett.* Vol. 87, No. 24, pp. 249704.

Petrin A. B. (2008), Electromagnetic Wave Propagation from a Point Source in Air through a Medium with a Negative Refractive Index, *JETP Letters,* Vol. 87, No. 9, pp. 464–469.

Petrin A. B. (2008), A Point Radiator Parallel to a Plane Layer with Negative Refractive Index, *Journal of Experimental and Theoretical Physics,* Vol. 107, No. 3, pp. 364–373.

Podolskiy V. A., Kuhta N. A., and Milton G. W. (2005), *Appl. Phys. Lett.* Vol. 87, p. 231113.

Scalora M., D'Aguanno G., Mattiucci N., et al. (2007), *Opt. Express* Vol. 15, p. 508.

Shelby R. A., Smith D. R., and Schultz S. (2001), Experimental Verification of a Negative Index of Refraction, *Science,* Vol. 292, p. 77.

Sivukhin D. V. (1980), *General Course of Physics: Optics, Vol. 4,* Nauka, Moscow [in Russian].

Smith D. R., Padilla W. J., Vier D. C., et al. (2000), *Phys. Rev. Lett.* Vol. 84, p. 4184.

Sommerfeld A. (1926), Propagation of Waves in Wireless Telegrapgy, *Ann. Phys.* (Leipzig, Ger.) Vol. 81, pp. 1135-1153.

Stratton J. A. (1941), *Electromagnetic Theory,* McGraw-Hill, New York.

Veselago, V. G. (1967) The electrodynamics of substances with simultaneously negative values of ε and μ, *Sov. Phys. Usp.,* Vol. 10, pp. 509- 518.

Veselago, V. G. (2003) Electrodynamics of materials with negative refractive index, *Usp. Fiz. Nauk,* Vol. 173, No. 7, pp. 790- 794.

Wait J. R. (1998), The Ancient and Modern History of EM Ground-Wave Propagation, *IEEE Antennas Propag. Mag.* Vol. 40, No. 5, pp. 7-24.

Williams J. M. (2001), Some Problems with Negative Refraction, *Phys. Rev. Lett.* Vol. 87, No. 24, p. 249703.

8

Electromagnetic Wave Propagation in Multilayered Structures with Negative Index Material

Mariya Golovkina
Povolzhskiy State University of Telecommunacations and Informatics
Russia

1. Introduction

The emergence of artificially designed subwavelength electromagnetic materials, known as metamaterials, catches an increasing interest of researchers. Metamaterials are artificially structured materials featuring properties that do not or may normally take place and can not be acquired in nature (Engheta & Ziolkowski, 2006). In recent years greater attention has been paid to the metamaterials with negative index of refraction which have quite uncommon electromagnetic properties. The new type of materials with the negative index of refraction were theoretically predicted in 1968 by Veselago (Veselago, 1967). In these materials both the permittivity and the permeability take on simultaneously negative values at certain frequencies. In materials with the negative refractive index the direction of the Pointing vector is antiparallel to the one of the phase velocity, as contrasted to the case of plane wave propagation in conventional media.

The complex refractive index of a medium is defined as the ratio between the speed of an electromagnetic wave in medium and that in vacuum and can thus be expressed as $n^2 = \mu\varepsilon$, where μ is relative magnetic permeability and ε relative dielectric permittivity. If we change simultaneously the signs of ε and μ, the ratio $n^2 = \mu\varepsilon$ will not change. If both ε and μ are positive, this means that $n = \sqrt{\varepsilon\mu}$, if ε and μ are negative in a given wavelength range, this means that $n = -\sqrt{\varepsilon\mu}$. In negative index metamaterials the Poynting vector $\mathbf{S}=[\mathbf{EH}]$ and vectors $\mathbf{E}$ and $\mathbf{H}$ form the left-hand triple, which leads to opposite directions of the group and phase velocity of plane waves propagating in the material. Consequently, metamaterials with simultaneously negative permittivity and permeability are named as "left-handed" metamaterials, "backward-wave media", double negative materials,

Smith, Shelby et al. were the first to demonstrate by means of experiment the existence of metamaterials with simultaneously negative permittivity and permeability at microwave frequencies (Smith et al., 2000), (Shelby et al., 2001). After the experimental demonstration of such materials, the properties and possible applications of various metamaterials with negative index of refraction gained a rapidly increasing interest. Now the negative refractive index metamaterials are demonstrated for near infrared and optical range (Falcone et al. 2004), (Iyer & Eleftheriades, 2002), (Caloz & Itoh, 2002). In metamaterials with negative refractive index many interesting phenomena that do not appear in natural media can be

observed. Among them there are the modification of the Snell's law (Ramakrishna, 2005), reversal of the Doppler effect and Cerenkov radiation (Lu et al., 2003), reformulation of the Fermat principle (Veselago, 2002).
One of the most interesting directions of analysis of the metamaterials unusual properties is the study of periodic structures or photonic crystals composed of metamaterials. Photonic crystals are artificial materials with periodically modulation of the refractive index (Joannopoulos et al., 1997). The simplest one-dimensional photonic crystal consists of alternating layers with different dielectric constant (Yeh, 1988). Such a structure may include partial bandgaps for certain ranges of propagation directions. Photonic crystals containing metamaterials demonstrate a variety of new physical effects, and possess a number of new potentials (Wu et al., 2003 b). In one-dimensional photonic crystals with negative index metamaterial the new zero-index bangaps can be observed (Shadrivov et al., 2003), (Wu et al., 2003 a), (Li et al., 2003). Such novel periodic structures unveil many unusual properties, including substantial suppression of the Anderson localization and long-wavelength resonances (Asatryan et al., 2007). Specially designed one-dimensional structures with negative refraction may include a complete three-dimensional bandgap (Shadrivov et al., 2005). The result is sharply contrasted to the periodic structures with usual dielectrics which lack the complete gap.
In this chapter we study the propagation of electromagnetic waves in one-dimensional periodic structures composed of alternating layers of negative index metamaterial, conventional material (dielectric or semiconductor) and the thin superconducting film in resistive state. If we put the structure with thin superconducting film in external magnetic field, the latter can penetrate into the thickness of superconductor in the form of Abricosov vortex lattice. The presence of the thin superconducting film in the structure leads to attenuation of electromagnetic wave. However, the interaction of electromagnetic waves with moving Abricosov vortex lattice can lead to its amplification (Popkov, 1989), (Glushchenko & Golovkina, 1998). The important feature is that parameters of attenuation and amplification of electromagnetic wave considerably depend on the value of magnetic field. That is why the thin superconducting film can be a control element in considered periodic structures.

2. Periodic structure negative index metamaterial - dielectric

Let us consider a one-dimensional periodic structure shown schematically in Fig. 1, where the layers of usual dielectric material with width d_1 are separated by the layers of the negative index metamaterial with the width d_2.We will describe the variation of the refractive index in the pair of layers in the following way:

$$n(z) = \begin{cases} n_1 = \sqrt{\varepsilon_1 \mu_1}, \; z \in (z_m, z_m + d_1) \\ n_2 = -\sqrt{\varepsilon_2 \mu_2}, \; z \in (z_m + d_1, z_m + d) \end{cases}, \qquad (1)$$

where n_1 and n_2 are the refractive indices of dielectric and metamaterial layers, respectively, $d = d_1 + d_2$ is period of the structure. The structure is uniform in the y direction ($\partial/\partial y = 0$). We consider TE-polarized waves propagating in the (x, z) plane. The electric field can be expressed as $\exp(i\omega t - ik_x x - ik_{z1,2} z)$.The field is assumed to be monochromatic with the frequency ω.

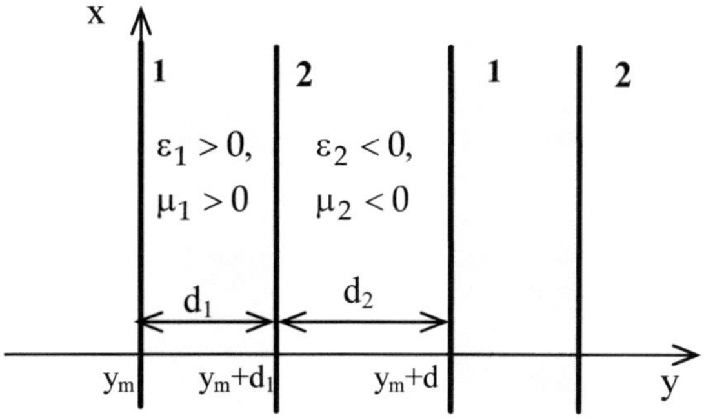

Fig. 1. Geometry of problem: 1 - dielectric, 2 - negative index metamaterial.

The eigenmodes of the problem can be adopted by applying the Bloch theorem for periodic systems and the corresponding Bloch-wave formalism (Yeh, 1988). The Bloch wavenumber K for TE-polarized waves is defined from the dispersion relation:

$$\cos Kd = \frac{1}{2}\mathrm{Tr}(M) = \cos k_{y1}d_1 \cos k_{y2}d_2 - \frac{1}{2}\left(\frac{k_{y1}\mu_2}{k_{y2}\mu_1} + \frac{k_{y2}\mu_1}{k_{y1}\mu_2}\right)\sin k_{y1}d_1 \sin k_{y2}d_2 , \qquad (2)$$

where Tr(M) is trace of the transfer matrix M characterizing the wave scattering in a periodic structure, $k_{y1,2} = \sqrt{\frac{\omega^2}{c^2} n_{1,2}^2 - k_x^2}$ are the longitudinal wave vector components, k_x is the wave vector component along the layers, while the indexes 1, 2 indicate the layer. The value $\cos Kd$ determines the band structure. It should take real values for lossless media and real k_y. Regimes where $|\cos Kd| < 1$ correspond to real K and thus to propagating Bloch waves. In regimes where $|\cos Kd| > 1$, K has an imaginary part; therefore, the Bloch waves are evanescent, and this regime corresponds to forbidden bands of the periodic medium. If the Bloch wavenumber K remains complex for all real k_x the bandgap is complete. The dispersion relation of the TM polarized waves is obtained by replacing ε and μ in Eq. (2):

$$\cos Kd = \cos k_{y1}d_1 \cos k_{y2}d_2 - \frac{1}{2}\left(\frac{k_{y1}\varepsilon_2}{k_{y2}\varepsilon_1} + \frac{k_{y2}\varepsilon_1}{k_{y1}\varepsilon_2}\right)\sin k_{y1}d_1 \sin k_{y2}d_2 . \qquad (3)$$

In periodic structures containing alternating layers of negative refractive metamaterials and usual dielectrics arise the novel partial bandgaps (Nefedov & Tretyakov, 2002), (Li, 2003). These bandgaps appear when the condition $k_{y1}d_1 + k_{y2}d_2 = 0$ is satisfied for certain angles of propagating. This is possible because k_y is positive in usual dielectric and negative in negative refractive metematerial.

Moreover the one-dimensional layered structure with layers of negative index materials can exhibit a complete two-dimensional bandgap (Shadrivov et al., 2005). This bandgap arise only for specified set of material parameters which eliminates the guided modes and transmission resonance. The more complicated one-dimensional periodic structure may possess the absolute bandgap. This structure contain three different kinds of layers in order to suppress the conditions for the existence of the Brewster angle. The dispersion relation for the structure with the parameters $\varepsilon_1 = \mu_1 = 1$, $\varepsilon_2 = \mu_3$, $\mu_2 = \varepsilon_3$ and $d_2 = d_3$ is written in the following way (Shadrivov et al., 2005):

$$\cos Kd = \cos k_{y1}d_1 \cos k_{y2}d_2 - \frac{1}{2}\left(\frac{\varepsilon_2}{\mu_2} + \frac{\mu_2}{\varepsilon_2}\right)\cos k_{y1}d_1 \sin^2 k_{y2}d_2 - \\ -\frac{1}{4}(\varepsilon_2 + \mu_2)\left(\frac{k_{y1}}{k_{y2}} + \frac{1}{\varepsilon_2\mu_2}\frac{k_{y2}}{k_{y1}}\right)\sin(k_{y1}d_1)\sin(2k_{y2}d_2). \tag{4}$$

This dispersion relation remains the same for both TE and TM polarizations. Though, the bandgaps will appear for both polarizations simultaneously. The presence of the bandgaps for both polarizations indicates the existence of an absolute bandgap. The one-dimensional periodic structure, under present consideration, can trap light in three dimensions as opposed to conventional photonic crystals which require the presence of two- and three-dimensional periodicity.

3. Periodic structure negative index metamaterial - semiconductor

The main difference between the periodic structures with semiconductor layers and structures with dielectric layers consists in the frequency dependence of permittivity in semiconductor layers. The dependence of permittivity on frequency leads to emergence of new types of waves and formation of various instabilities. We will now dwell upon the case of absence of losses in the layers. In order to find out the basic features of behaviour of eigenmodes in structures with semiconductors layers we should assume that the permittivity and permeability of negative index material layers are constant.
We will use the hydrodynamic model to describe the movement of charge carriers in the semiconductor (Yu & Cardona, 2005). The application of hydrodynamical model is reasonable for frequencies $\omega >> \nu$, where ν is effective collision frequency, and for frequencies $\omega << \nu$. The typical values of the effective collision frequency for semiconductor make it possible to draw a conclusion that hydrodynamical model is applicable for the description of wave processes with the frequencies $\omega > 10^{10}\,s^{-1}$. In the considered periodic structure the hydrodynamical approach can be used if the period of structure $d = d_1 + d_2$ is much larger than the wavelength of electromagnetic wave. An individual semiconductor layer of thickness d_1 is described in hydrodynamic approach by dielectric permittivity

$$\varepsilon_1 = \varepsilon_{01}\left(1 - \omega_{p1}{}^2 / \omega^2\right), \tag{5}$$

where ε_{01} is the high-frequency dielectric constant, $\omega_{p1} = \left(Ne^2 / m\varepsilon_0\varepsilon_{01}\right)^{1/2}$ is plasma frequency; e, m and N are charge, effective mass and density of the free charge carriers; ε_0 is the permittivity of the free space.
Here, let us turn to the propagation of TM polarized waves. The dispersion relation for TM waves (3) is same as for the structure negative index material - dielectric. We will compare the position of the bandgaps in the structures semiconductor - negative index metamaterial

and semiconductor - dielectric (Golovkina, 2007). Let the left part of the equation (3) be marked as $F(k_x,\omega)$. Fig. 2 shows a graph of function $\cos Kd = F(k_x,\omega)$ for the structure semiconductor - negative index metamaterial for arbitrary fixed value of frequency ω. Comparatively, Fig.3 shows the graph of function F for the periodic structure semiconductor - usual dielectric, which has the same magnitudes of permittivity and permeability as the negative index metamaterial.

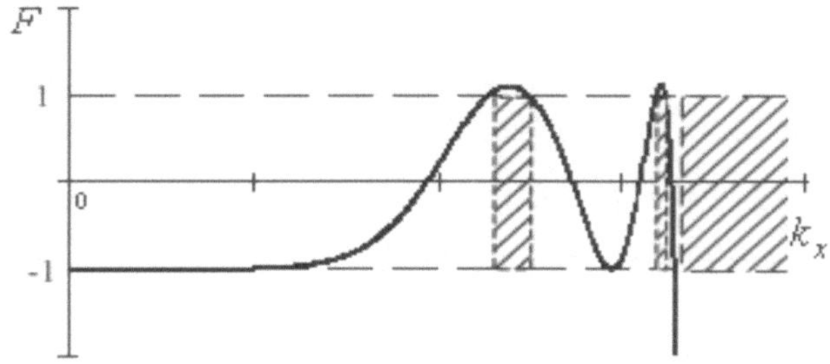

Fig. 2. The graphic detection of bandgaps for structure semiconductor - negative index metamaterial corresponding to the equation (3).

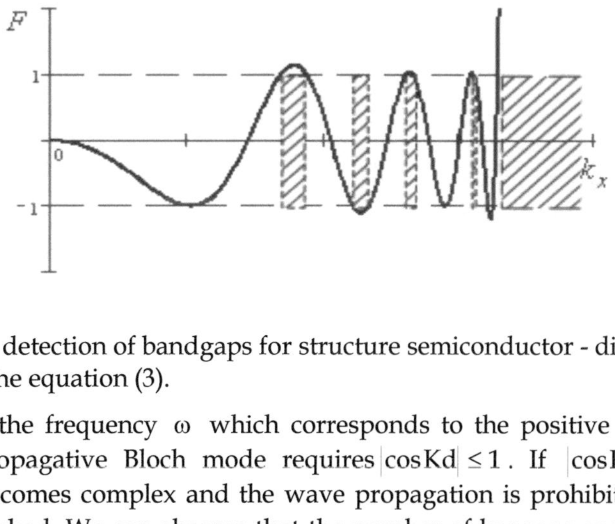

Fig. 3. The graphic detection of bandgaps for structure semiconductor - dielectric corresponding to the equation (3).

We will consider the frequency ω which corresponds to the positive values of ε_1. The existence of a propagative Bloch mode requires $|\cos Kd| \le 1$. If $|\cos Kd| > 1$, the Bloch wavenumber K becomes complex and the wave propagation is prohibited. These areas in the figures are hatched. We can observe that the number of bangaps and their position are different in the structures negative index metamaterial - semiconductor and negative index metamaterial - dielectric.

As is obvious from Fig. 2, 3, with increasing k_x, the bandgaps become narrower. Finally, following further increase of k_x, the permitted bands become so narrow that forbidden bands merge with each other. Though, all this considered, one of the layers is getting opaque. At last, since certain value of k_x both of a layer become opaque and the continuous forbidden band is formed. This value of k_x is the same for the structures semiconductor - negative index metamaterial and semiconductor - dielectric. In the permitted band the Bloch wavenumber varies from $2m\pi/d$ up to $(2m+1)\pi/d$. On each border a direct and reflected waves appear in the equal phase. We can see that the position and quantity of the forbidden bands for structures semiconductor - negative index metamaterial and semiconductor - dielectric is various because the Bloch wavenumbers for these structures have different values.

Now we will try to examine the case when $k_x^2 >> \omega^2 n_{1,2}^2 / c^2$. Then, accordingly,

$$k_{z1,2} = \pm i k_x . \tag{6}$$

The described above inequality imposes constraints on the thickness of layers. The deepest influence of periodicity occurs at $k_x d \approx 1$. Thus, the following restriction should be satisfied $d^2 << c^2 / \omega^2 [\max(\varepsilon_1, \varepsilon_2)]$. The dispersion equation for TM polarized waves can be transformed to the following form (Golovkina, 2009 a):

$$\frac{\varepsilon_1^2(\omega)+\varepsilon_2^2(\omega)}{\varepsilon_1(\omega)\varepsilon_2(\omega)} = -2\frac{\cos Kd - \mathrm{ch}\, k_x d_1 \,\mathrm{ch}\, k_x d_2}{\mathrm{sh}\, k_x d_1 \,\mathrm{sh}\, k_x d_2} . \tag{7}$$

Fig. 4 gives a sketchy idea of the corresponding dispersion curves. As is known, the spectrum of the semiconductor-dielectric structure consists of two different branches. The acoustic branch starts at $k_x = 0$ and $\omega = 0$ and tends to ω_{ps}, where $\omega_{ps} = \left[4\pi e^2 n_{01} / m(\varepsilon_{01} + |\varepsilon_2|)\right]^{1/2}$ is the frequency of surface plasmon, n_{01} is the concentration of carriers. The optic branch starts at $\omega = \omega_{p1}$ and $k_x \to 0$ and tends to ω_{ps} (Fig 1, a).

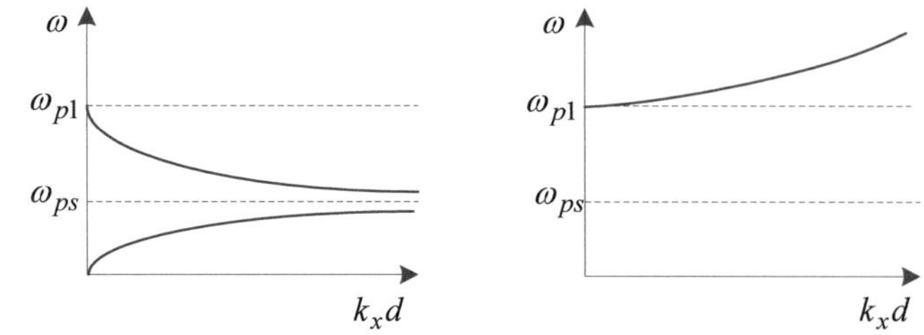

Fig. 4. Dispersion curves of TM-wave for the periodic structure: a) semiconductor-dielectric, b) semiconductor-negative index metamaterial.

When frequencies are highly electromagnetic waves can not propagate. Nevertheless, in structure semiconductor - negative index metamaterial they can propagate on high frequencies $\omega > \omega_{ps}$ (Fig 1, b). This is possible because the energy flux and wave vector have opposite directions in negative index material layer.

4. Periodic structure with thin superconducting film

When in the 1986 high temperature superconductors were discovered the problems of their practical use in microelectronics became actual. New techniques have been developed in recent years to produce superconducting layered systems. Multilayer stackings of different superconducting materials with very small layer thickness intercalated with dielectric layers are obtainable (Silva et al., 1996). Rapid progress in such techniques as molecular beam epytaxy or chemical vapour deposition enables us to grow systems with pre-determined film thickness. Such systems seem to provide a new type of material which does not exist naturally (Wua, 2005). One can create materials with properties distinct from those of any single constituent. The successes of technology have lead to constant increase of use of multilaered structures based on high temperature superconductors in microelectronics and computer hardware (Lancaster, 2006). In most of the experimental and theoretical work on artificially superconducting layered systems, researches have dealt with a study electromagnetic wave propagation along the interfaces (Ghamsari & Majedi, 2007). Nevertheless the problem of electromagnetic wave propagation in such structures in arbitrary directions is practically investigated.

In this section we present the theoretical development for calculating the dispersion relations for electromagnetic wave propagation in an infinite periodic structure. The periodic structure we consider is shown in Fig. 5.

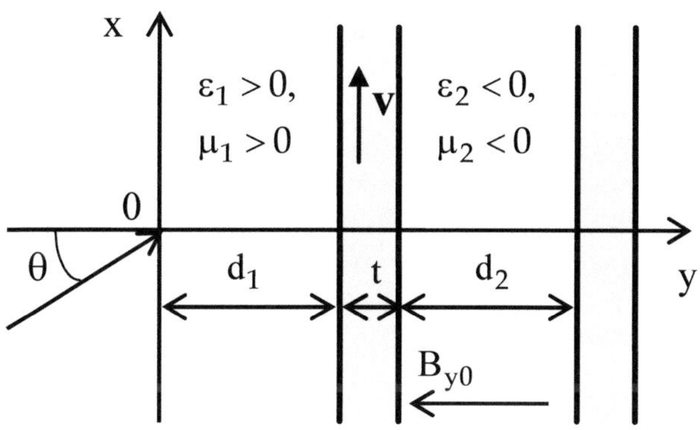

Fig. 5. Geometry of problem. One-dimensional structure dielectric - superconductor - negative index metamaterial.

The structure consists of alternating layers of dielectric with thickness d_1, type-II superconductor with thickness $t \ll \lambda$, where λ is a microwave penetration depth and layers of negative index metamaterial with thickness d_2. We let the interfaces of the superlattice lie parallel to the x-z plane, while the y axis points into the structure. A static magnetic field B_{y0} is applied antiparallel the y axis, perpendicular to the interfaces of the superlattice. The value of magnetic field does not exceed the second critical field for a superconductor. The magnetic field penetrate in the depth of the superconductor in the form of Abrikosov vortex lattice. Under the impact of transport current directed perpendicularly to magnetic field B_{y0} along the 0z axis, the flux-line lattice in the superconductor layers starts to move along the 0x axis. Let's consider the propagation in the given structure p-polarized wave being incident with angle θ in the x0y plane. It can be assumed that $\partial/\partial z = 0$.

The presence of a thin superconductor layer with the thickness of $t \ll l$ is reasonable to be accounted by introduction of a special boundary condition because of a small amount of thickness. Let's consider the superconductor layer at the boundary y=0. At the inertia-free approximation and without account of elasticity of fluxon lattice (the presence of elastic forces in the fluxon lattice at its deformation results in non-linear relation of the wave to the lattice, that is insignificant at the given linear approximation) the boundary condition is written in the following way (Popkov, 1989):

$$\frac{\partial B_y}{\partial t}(y=t)+\frac{j_{z0}\Phi_0}{\eta}\frac{\partial B_y}{\partial t}(y=t)=\frac{B_{y0}\Phi_0}{\eta}\frac{\partial}{t\,\partial x}[H_x(y=t)-H_x(y=0)], \tag{8}$$

where j_{z0} is the current density in the superconducting layer and η is the vortex viscosity. Let's write the boundary condition (8) in the form of matrix M_s, binding fields at the boundaries $y=d_1$ and $y=d_1+t$:

$$\begin{pmatrix} E_z(d_1+t) \\ H_x(d_1+t) \end{pmatrix} = M_s \begin{pmatrix} E_z(d_1) \\ H_x(d_1) \end{pmatrix}, \tag{9}$$

$$M_s = \begin{pmatrix} 1 & 0 \\ \frac{t}{B_{y0}}\left(\frac{\eta}{\Phi_0}-\frac{j_{z0}k_x}{\omega}\right) & 1 \end{pmatrix},$$

where k_x is the projection of the passing wave vector onto the 0x axis and ω is the frequency of the passing wave.

Let us apply the transfer matrix method. This method has been extensively applied to band structure calculations of photonic crystals containing absorptive and frequency dispersive materials. In considered periodic structure the presence of thin superconducting film leads to the dissipation of energy. By usage of matrix method we expressed dispersion relation for TE-wave in the following way:

$$\begin{aligned}\cos Kd = {} & \cos k_{y1}d_1\cos k_{y2}d_2-\frac{1}{2}\left(\frac{k_{y1}\mu_2}{k_{y2}\mu_1}+\frac{k_{y2}\mu_1}{k_{y1}\mu_2}\right)\sin k_{y1}d_1\sin k_{y2}d_2- \\ & -\frac{1}{2}\frac{i\omega\mu_0 t}{B_{y0}}\left(\frac{j_{z0}k_x}{\omega}-\frac{\eta}{\Phi_0}\right)\left(\frac{\mu_1}{k_{y1}}\sin k_{y1}d_1\cos k_{y2}d_2+\frac{\mu_2}{k_{y2}}\cos k_{y1}d_1\sin k_{y2}d_2\right),\end{aligned} \tag{10}$$

where $K=K'+iK''$ is the Bloch wave number and k_y is the projection of passing wave vector onto the 0y axis. The interaction of electromagnetic wave with thin superconducting film leads to emergence of the imaginary unit in the dispersion equation. The presence of imaginary part of the Bloch wave number indicates that electromagnetic wave damps exponentially when passing through the periodic system even if the negative index metamaterial and dielectric layers are lossless. However, when one of the conditions

$$\frac{\mu_1}{k_{y1}}\sin k_{y1}d_1\cos k_{y2}d_2+\frac{\mu_2}{k_{y2}}\cos k_{y1}d_1\sin k_{y2}d_2=0, \tag{11}$$

$$\frac{j_{z0}k_x}{\omega}-\frac{\eta}{\Phi_0}=0 \tag{12}$$

is executed, the Bloch wave vector becomes purely real and electromagnetic wave may penetrate into the periodic structure (Golovkina, 2009 c). The implementation of condition (11) depends on the relation between the parameters of layers and the frequency of electromagnetic wave, while the implementation of condition (12) depends - on parameters of superconducting film, namely on current density j_{z0}. Still, we are able to manage the attenuation and propagation of electromagnetic waves by changing the value of j_{z0}. Moreover, the electromagnetic wave can implement the amplification in such a structure. It is well-known that electromagnetic wave can be amplified in structures with electron flow if the velocity of electromagnetic wave is equal to the velocity of electrons. Such amplification is observed in the traveling-wave tube and backward-wave tube at the interaction between the electromagnetic wave and the electron flow. As is shown in papers (Glushchenko & Golovkina, 1998), (Glushchenko & Golovkina, 2007), the electromagnetic wave can be amplified in layered structures with thin superconducting film in resistive state by means of interaction with moving Abrikosov vortex lattice. Nonetheless, when the layered structure includes the combination of two layers with positive and negative refractive index, the amplification occurs at a lesser velocity of vortex lattice.

The results of numerical calculation concerning the module and the phase of the reflection coefficient R as a function of external magnetic field B_{y0} are given in Fig. 6 and Fig. 7. The structure consists of the layer MgO with thickness $d_1 = 0.1\,\mu m$, the layer of negative index metamaterial with thickness $d_2 = 0.1\,\mu m$ ($\varepsilon_2 = -9 - 0.09i$, $\mu_2 = -1 - 0.01i$), thin superconducting film $YBa_2Cu_3O_7$ on the substrate of $SrTiO_3$. This calculations touched upon the losses in the layers. For the given structure the reflection coefficient is greatly dependent on incident angle θ, thickness of superconducting film t and magnitudes of external magnetic field B_{y0}. We can observe from Fig. 6 that the module of the reflection coefficient reaches the value more than unit at the specified magnitude of the magnetic field. The amplification appears out of energy from moving flux line lattice.

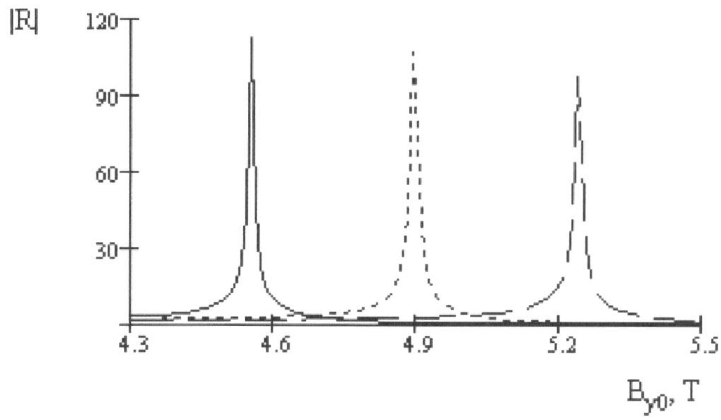

Fig. 6. The module of the reflection coefficient R versus external magnetic field for two layer structure with thin superconducting film $YBa_2Cu_3O_7$ on the substrate of $SrTiO_3$. The parameters are as follows: $\eta = 10^{-8} N/(s \cdot m^2)$, $j_{z0} = 10^8 A/m^2$, $\omega = 10^{13} rad/s$. The solid line: t=40 nm, the short dashed line: t=43 nm, the long dashed line: t=46 nm, where t is thickness of superconducting film.

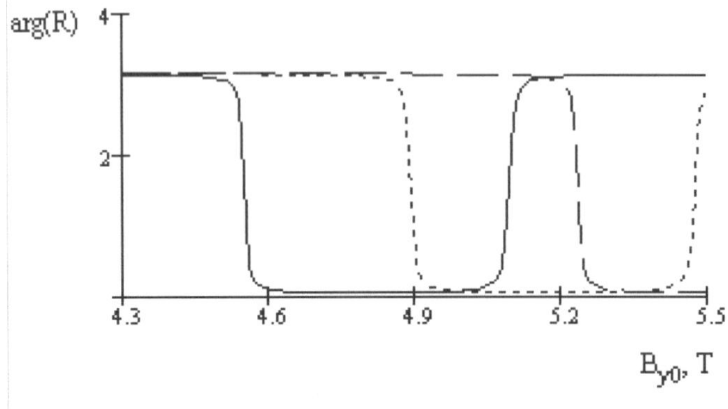

Fig. 7. The phase of the reflection coefficient R versus external magnetic field. The parameters are the same as in Fig. 6.

A high electromagnetic wave amplification coefficient during the reflection from the layered structure with superconducting film by the interaction with a fluxon structure enables to create wide-ranged amplifying devices using the structures analyzed above. Heavy dependence of reflection coefficient on incident wave frequency, incident angle and external magnetic field magnitude makes it possible to create new devices and filters with high parameter selectivity controlled by magnetic field by means using the considered structures.

5. Waveguide structure with thin superconducting film

We have examined the wave propagation in two-layered waveguide divided by a thin superconducting film. On layer of thickness d_1 is a negative index material ($\varepsilon_1 < 0$, $\mu_1 < 0$) and the other one of thickness d_2 is a usual dielectric ($\varepsilon_2 > 0$, $\mu_2 > 0$) (Fig. 8). The thin film of type-II superconductor with thickness t is placed in the plane y0z. The thickness of superconductor $t << \lambda$, where λ is magnetic field penetration depth. A static magnetic field B_0 is applied parallel to the x axis, perpendicular to the plane of the film. The value of magnetic field does not exceed the second critical field for a superconductor. Under the impact of transport current directed along the 0y axis, the flux-line lattice in the superconductor film moves along the 0z axis. We have considered the TE_{0n} modes which effectively interact with flux-line lattice in superconductor.

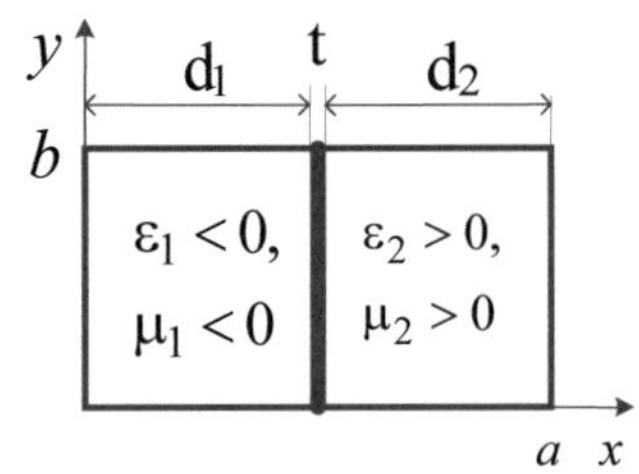

Fig. 8. The two-layered rectangular waveguide with thin superconducting film.

We have received the dispersion relation for the TE_{0n} modes in the following way (Golovkina, 2009 b):

$$\frac{k_{x1}}{\mu_1}\,\mathrm{ctg}\,k_{x1}d_1+\frac{k_{x2}}{\mu_2}\,\mathrm{ctg}\,k_{x2}d_2=-\frac{i\,\mu_0\,\eta t}{B_0\Phi_0}\left(\frac{j_{y0}\Phi_0\beta}{\eta}-\omega\right), \tag{13}$$

where j_{y0} is transport current density, β is propagation constant, ω is angular frequency. The time dependence is $\exp(i\omega t)$. One important difference from the two-layered waveguide without superconducting film is a presence of imaginary part of the propagation constant $\beta = \beta' + i\beta''$. The negative values of imaginary part β correspond to the attenuation of electromagnetic waves, the positive values of β correspond to amplification.

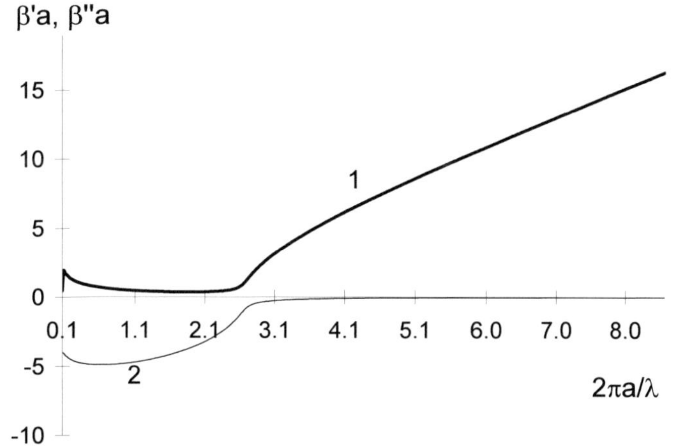

Fig. 9. The real and imaginary part of propagatin constant β versus normalized frequency. Parameters: $B_0 = 0.1\,\mathrm{T}$, $j_{z0} = 10^6\,\mathrm{A/m^2}$, $\varepsilon_1 = -2$, $\mu_1 = -1$, $\varepsilon_2 = 1$, $\mu_2 = 1$, $d_1/d_2 = 3/2$. Curve 1: the real part $\beta' a$, curve 2: the imaginary part $\beta'' a$.

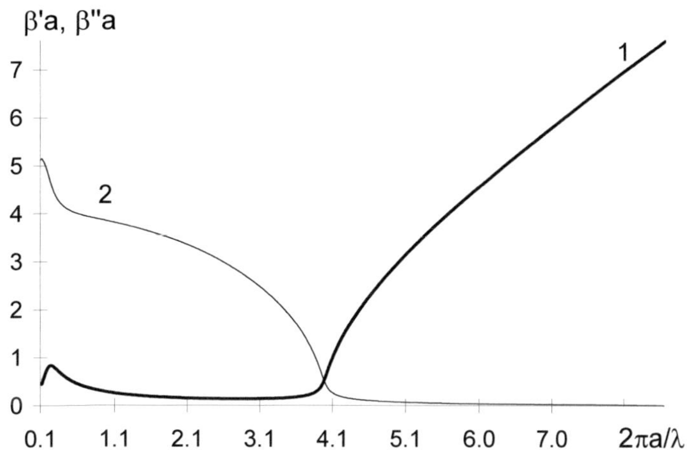

Fig. 10. The real and imaginary part of propagatin constant β versus normalized frequency. Parameters: $B_0 = 0.1\,\mathrm{T}$, $j_{z0} = 10^6\,\mathrm{A/m^2}$, $\varepsilon_1 = -2$, $\mu_1 = -1$, $\varepsilon_2 = 1$, $\mu_2 = 1$, $d_1/d_2 = 1/4$. Curve 1: the real part $\beta' a$, curve 2: the imaginary part $\beta'' a$.

Numerical calculations have resulted in the following. If the velocity of Abricosov vortex lattice is small, the imaginary part of β is negative (Fig. 9). When the frequency is less than cutoff frequency of usual two-layered waveguide, the imaginary part β'' is large, and the electromagnetic wave attenuates. When the frequency is greater than cutoff frequency of usual two-layered waveguide, the imaginary part β'' becomes smaller and the real part β' increases. This means that the electromagnetic wave propagates with smaller attenuation. If the condition of equality of electromagnetic wave velocity and vortex velocity is executed, the imaginary part of propagation constant becomes positive (Fig. 10). The positive values of β'' indicates the presence of amplification.

The amplification depends on frequency, parameters of superconducting film and magnetic field magnitude. It is essential to note that the significant amplification is observed below from cutoff frequency of the two-layered waveguide. In this frequency domain the growing evanescent waves increase as well because of coupling to the moving vortex in superconductor. The dependence of coefficient of amplification (or attenuation) on the external magnetic field allows to create the amplifiers and filters controlled by magnetic field on the basis of the considered above waveguide structure.

6. Conclusion

In this chapter we examined some of the most interesting features of one-dimensional layered structure containing the negative index metamaterials. The greater number of features of these structures has resulted from the point that the waves in the negative index material are backward, since the phase and group velocities are antiparallel. In periodic structures with negative index metamaterials the number and position of bandgaps changes, new type of bandgaps appears. The presence of thin superconducting film in resistive state in one-dimensional periodic structure can cause the attenuation or amplification of electromagnetic waves by means of moving Abricosov vortex lattice in superconducting layer. The heavy dependence of amplification parameters on the value of external magnetic field allows managing the electromagnetic waves propagation in structures with superconducting films. On the basis of periodical structures with negative index material layers and thin superconducting film it is highly possible to create new devices such as filters and amplifiers controlled by magnetic field.

7. References

Asatryan, A.A.; Botten, L.C.; Byrne, M.A.; Freilikher, V.D.; Gredeskul, S.A.; Shadrivov, I.V.; McPhedran, R.C. & Kivshar, Y.S. (2007). Suppression of Anderson localization in disordered metamaterials. *Phys. Rev. Lett.*, Vol. 99, 193902-4.

Caloz, C. & Itoh, T. (2002). Application of the transmission line theory of left-handed (LH) materials to the realization of a microstrip LH transmission line. *Proceedings .of IEEE-AP-S USNC/URSI National Radio Science Meeting*, Vol. 2, 412–415.

Engheta, N. & Ziolkowski, R.W. (2006). Introduction, history, and selected topics in fundamental theories of metamaterials, In: *Metamaterials*, Ed. by Engheta, N. & Ziolkowski, R.W., John Wiley & Sons, ISBN 13978-0-47176102-0, New York.

Falcone, F.; Lopetegi, T.; Laso, M. G.; Baena, J. D.; Bonache, J.; Beruete, M.; Marques, R.; Martin, F. & Sorolla, M. (2004). Babinet principle applied to the design of metasurfaces and metamaterials. *Phys. Rev. Lett.*, Vol. 93, 197401-4.

Ghamsari, B.G. & Majedi, A.H. (2007). Rigorous analysis of superconducting multilayer optical waveguides. *Appl. Superconductivity, IEEE Transactions. on*, Vol. 17, 590-593.

Glushchenko, A.G. & Golovkina, M.V. (1998). Reflection of an electromagnetic wave by a layered superconductor-dielectric structure. *Technical Physics Letters*, Vol. 24, No. 4, 9-12.

Glushchenko, A.G. & Golovkina, M.V. (2007). Propagation of electromagnetic waves in periodic structures with superconducting layers having electrodynamic parameters in the nonlinearity range of the dynamic mixed. *Technical Physics*, Vol. 52, 1366-1368.

Golovkina, M.V. (2007) Electromagnetic wave propagation in a multilayered periodic structure containing negative index material, *Proceedings of IEEE International Siberian Conf. on Control and Communications (SIBCON-2007)*, pp. 174-178, Tomsk, Russia, April 2007.

Golovkina, M.V. (2009 a). Periodic semiconductor structures with metamaterials, *Proceedings of IEEE International Siberian Conference on Control and Communications (SIBCON-2009)*, pp. 133-137, Tomsk, Russia, March 2009.

Golovkina, M.V. (2009 b). Electromagnetic wave propagation in waveguide with thin superconducting film and metamaterial slab, *Progress in Electromagnetics Research Symposium*, Moscow , Russia, August 2009.

Golovkina, M.V. (2009 c). Characteristics of electromagnetic waves propagation in multilayered structure semiconductor - superconductor, *Vestnic Pomorskogo Universiteta*, Vol. 3, 70-75, ISSN 1728-7340.

Iyer, A. K. & Eleftheriades G. V. (2002). Negative refractive index metamaterials supporting 2-D waves. *IEEE MTT-S Int. Microwave Symp. Dig.*, Vol. 2, pp. 1067-1070, Seattle, WA, June 2-7, 2002.

Joannopoulos, J. D.; Villeneuve, P. R. & Fan, S. (1997). Photonic crystals - putting a new twist on light nature. *Nature*, 386, Vol. 386, No. 6621, 143-149.

Lancaster, M.J. (2006). *Passive microwave device applications of high-temperature superconductors.* Cambrige Unuversity Press, ISBN 0521034175, New York.

Li, J. ; Zhou, L.; Chan, C.T. & P. Sheng, (2003). Photonic band gap from a stack of positive and negative index materials. *Phys. Rev. Lett.*, Vol. 90, 083901-4.

Lu, J.; Grzegorczyk, T. M.; Zhang, Y.; Jr., P.J.; Wu, B.-I.; Kong, J. A. & M. Chen (2003). Cerenkov radiation in materials with negative permittivity and permeability. *Opt. Express*, Vol. 11, 723-734.

Nefedov, I. S. & Tretyakov, S. A. (2002). Photonic band gap structure containing metamaterial with negative permittivity and permeability. *Phys. Rev. E*, Vol. 66, 036611-4.

Popkov, A.F. (1989). The magnetostatic wave amplification by magnetic vortex flux in structure ferrite - superconductor. *Technical Physics Letters*, Vol. 15, 9-14.

Ramakrishna, S. A. (2005). Physics of negative refractive index materials. *Rep. Prog. Phys.* , Vol. 68, 449-521.

Shadrivov, I.V.; Sukhorukov, A.A. & Kivshar, Y.S. (2003). Beam shaping by a periodic structure with negative refraction. *Appl. Phys. Lett.*, Vol. 82, 3820-3822.

Shadrivov, I.V.; Sukhorukov, A.A. & Kivshar, Y.S. (2005). Complete bandgaps in one-dimensional left-handed periodic structures, *Phys. Rev. Lett.*, Vol. 95, 193903-4.

Shelby, A.; Smith, D. R. & Schultz, S. (2001). Experimental verification of a negative index of refraction. *Science*, Vol. 292, 77-79.

Silva, E.; Lanucara, M. & Marcon, R. (1996). The effective surface resistance of superconductor/ dielectric/ metal structures. *Supercond. Sci. Technol.*, Vol. 9, 934-941.

Smith, D. R.; Padilla, W. J.; Vier, D. C.; Nemat-Nasser, S.C. & Schultz, S. (2000). Composite medium with simultaneously negative permeability and permittivity. *Phys. Rev. Lett.*, Vol. 84, 4184-4187.

Veselago, V. G. (1968) The electrodynamics of substances with simultaneously negative values of epsilon and mu. *Sov. Phys. Uspekhi*, Vol. 10, 509-514.

Veselago, V.G. (2002). On the formulation of Fermat's principle for light propagation in negative refraction materials. *Uspekhi Phys. Nauk*, Vol. 172, No. 10, 1215-1218.

Wu, L.; He, S. & Shen, L. (2003). Band structure for a one-dimensional photonic crystal containing left-handed materials. *Phys. Rev. B*, Vol. 67, 235103-6.

Wu, L.; He, S. L. & Chen, H. (2003). On unusual narrow transmission bands for a multilayered periodic structure containing left-handed materials. *Optics Express*, Vol. 11, No. 11, 1283–1290.

Wua, C.-J.; Chenb, M.-S. & Yangb, T. (2005). Photonic band structure for a superconductor - dielectric superlattice. *Physica C: Superconductivity*, Vol. 432, 133-139.

Yeh, P. (1988). *Optical Waves in Layered Media*, John Wiley & Sons, ISBN 0471828661, New York.

Yu, P.Y. & Cardona, M. (2005). *Fundamentals of semiconductors: physics and material properties*, Springer, ISBN 3540254706.

9

The Nature of Electromagnetic Waves in Metamaterials and Metamaterial-inspired Configurations

Rui Yang[1,2] and Yongjun Xie[2]
[1] *National Laboratory of Antennas and Microwave Technology, Xidian University, Xi'an, 710071,*
[2] *Department of Electronic Engineering Queen Mary College, University of LondonLondon E1 4NS,*
[1]*China.*
[2]*UK*

1. Introduction

Metamaterials possessing various peculiar features have recently attracted an increasing amount of attention in the electromagnetics community. Their unexpected properties have opened up a number of different research directions that are geared towards the enhancement of the performance of microwave components, and overcoming current limitations. In this Chapter, the fundamental model properties of metamaterials and metamaterial based structures are demonstrated to study the renovated wave propagation.

2. Causality in the resonance behavior of metamaterials

Great of interest has been devoted to split ring resonator (SRR) which composes the essential part of left-handed materials [1-3]. Inherently bianisotropic, SRR metamaterials can be obtained by doping a host isotropic medium with two concentric rings separated by a gap, both having splits at opposite sides. As a result, besides the electric and magnetic coupling, the incident field also induces the magnetoelectric coupling [4,5]. Therefore, this kind of artificial magnetic media needs a careful control of the SRR orientation relative to the incident wave as well as the SRR design. Otherwise, the electromagnetic response is significantly more complicated. Smith et al. explored the electromagnetic characterization of the symmetric and asymmetric SRR plane [6]. Vasundara et al. presented the effects of gap orientation on the properties of SRR metamaterials with measured scattering parameters [7]. Aydin et al. investigated the influence of periodicity, misalignment, and disorder on the magnetic resonance gap of SRRs [8]. Gay-Balmaz et al. studied experimentally and numerically the electromagnetic resonances in individual and coupled SRRs [9]. Katsarakis et al. discovered the electric coupling to the magnetic resonance of SRRs under certain orientation [10]. Correspondingly, several analysis modals are employed to unravel the resonance property in the SRR transmission spectra, such as the physical intuition initiated by Pendry et al. [11], lumped element equivalent circuit model proposed by Martin et al.

[12], and improved by Aznar et al. [13]. Meanwhile, lots of numerical simulations as well as experimental verifications are carried out for the metamaterial design [14-16]. Especially, Simovski et al. clarified the physical meaning of local constitutive parameters of metamaterials [17], and discussed Bloch material parameters of magneto-dielectric metamaterials [18].

A rigorous full wave analysis of bianisotropic SRR metamaterials is presented here for different electromagnetic field polarization and propagation directions. An alternative physical explanation is gained by revealing the fact that imaginary wave number leads to the SRR resonance. The field distribution over SRRs is then expanded into Floquet modes [19,20] to examine the transmission properties through metamaterials under arbitrary incident waves. Evanescent Floquet modes are proved to engender the resonance behavior which accords with the full wave analysis.

2.1 Full wave analysis of the SRR metamterials

To account for the magnetoelectric coupling in Maxwell's equations, SRR metamaterials can be described by the constitutive relations [21]

$$\boldsymbol{D} = \varepsilon_0(\bar{\varepsilon} \cdot \boldsymbol{E} + Z_0 \bar{\kappa} \cdot \boldsymbol{H}) \tag{1a}$$

$$\boldsymbol{B} = \mu_0(-\frac{1}{Z_0}\bar{\kappa}^T \cdot \boldsymbol{E} + \bar{\mu} \cdot \boldsymbol{H}) \tag{1b}$$

where $Z_0 = \sqrt{\mu_0/\varepsilon_0}$, $\bar{\varepsilon}$ and $\bar{\mu}$ are the relative electric permittivity and relative magnetic permeability tensors, $\bar{\kappa}$ is the magnetoelectric coupling dimensionless tensor.

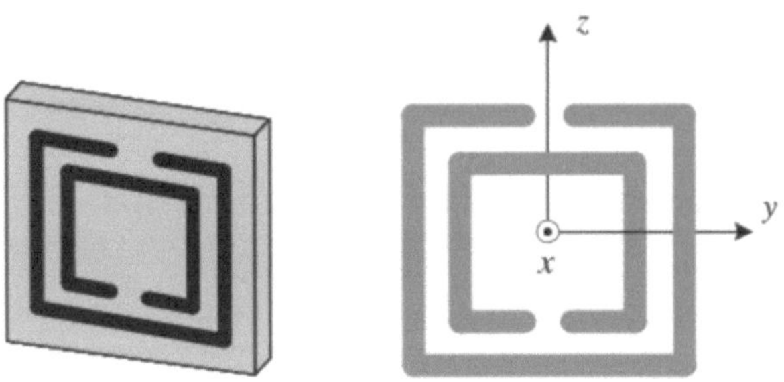

Fig. 1. The SRR unit cell

For axes fixed to the SRR as shown in Fig. 1, only certain components of $\bar{\varepsilon}$, $\bar{\mu}$ and $\bar{\kappa}$ tensors are of significance without losses [4,5]

$$\varepsilon_{xx} = 1\,,\ \varepsilon_{yy} = a + \frac{b\omega^2}{(\omega_0^2 - \omega^2)}\,,\ \varepsilon_{zz} = a \tag{2a}$$

$$\mu_{xx} = 1 + \frac{c\omega^2}{(\omega_0^2 - \omega^2)}\,,\ \mu_{yy} = 1\,,\ \mu_{zz} = 1 \tag{2b}$$

$$\kappa_{yx} = -i\kappa = -\frac{id\omega_0\omega}{(\omega_0^2 - \omega^2)} \tag{2c}$$

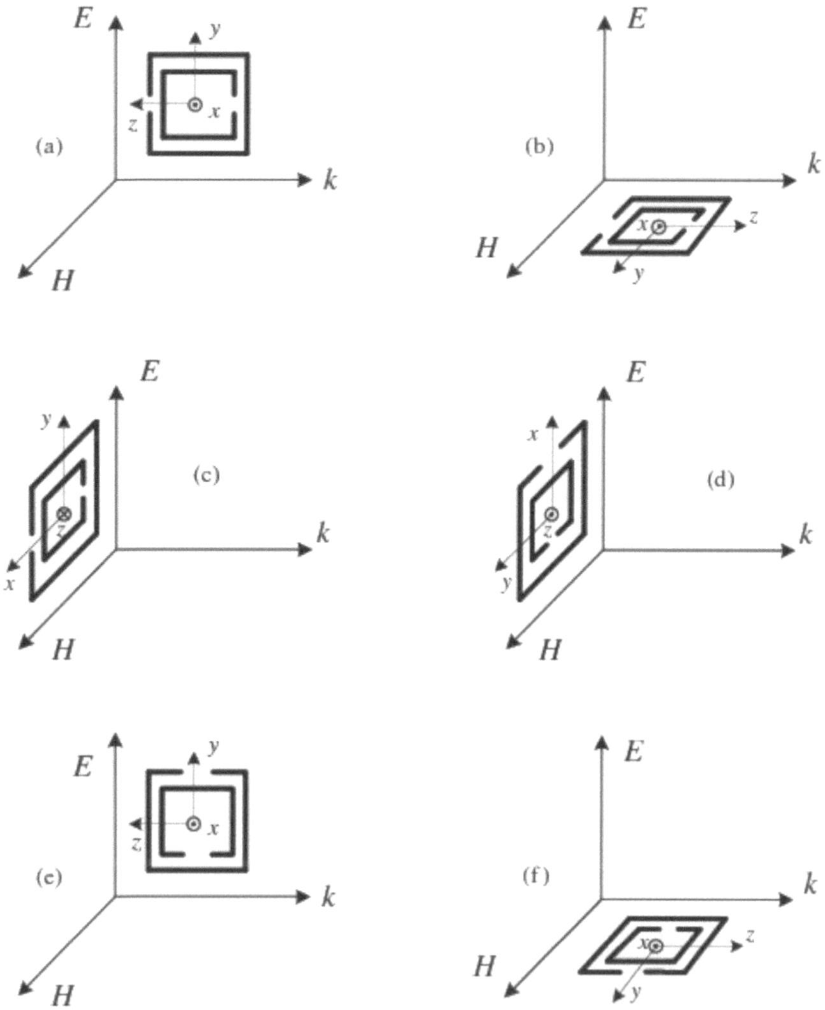

Fig. 2. Six orientations of SRR relative to different electromagnetic field polarization and propagation direction

ω_0 is the resonance frequency, and a, b, c, d in Fig. 1 are related to the geometry of the SRR. For other SRR orientations, the $\bar{\varepsilon}$, $\bar{\mu}$ and $\bar{\kappa}$ tensors just need a coordinates transformation. Introducing a normalized magnetic field $\boldsymbol{h} = Z_0\boldsymbol{H}$, from Maxwell's curl equation for source free regions together with (1), (2), one may write

$$-i\nabla'\times\mathbf{h} = \bar{\varepsilon}\cdot\mathbf{E} + \bar{\kappa}\cdot\mathbf{h} \tag{3a}$$

$$i\nabla'\times\mathbf{E} = -\bar{\kappa}^T\cdot\mathbf{E} + \bar{\mu}\cdot\mathbf{h} \tag{3b}$$

where $\nabla' = \nabla/k_0$.

For the case in Fig. 2a, where magnetic field $\boldsymbol{H}$ is perpendicular to the SRR plane and incident $\boldsymbol{E}$ is parallel to the gapbearing sides of SRR. One obtains

$$(\kappa_{yx} + \beta)h_x = -\varepsilon_{yy}E_y \tag{4a}$$

$$(\kappa_{yx} - \beta)E_y = \mu_{xx}h_x \tag{4b}$$

The normalized wave number of the TEM wave satisfies

$$\beta^2 = \mu_{xx}\varepsilon_{yy} + \kappa_{yx}^2 \tag{5}$$

Ref. [4] concluded the same results by considering the bianisotropy role in SRR metamaterials. At given frequency ω, only those modes having $\mu_{xx}\varepsilon_{yy} + \kappa_{yx}^2 > 0$ will propagate. Those modes with $\mu_{xx}\varepsilon_{yy} + \kappa_{yx}^2 < 0$ will lead to an imaginary β, meaning that all field components will decay exponentially away from the source of excitation. Since $\kappa_{yx}^2 < 0$, this SRR orientation will achieve the resonance stop band when the constitutive parameters are single negative, including $\varepsilon_{yy} > 0$, $\mu_{xx} < 0$ case, as well as $\varepsilon_{yy} < 0$, $\mu_{xx} > 0$ case. When the constitutive parameters are double negative or double positive with the condition $|\mu_{xx}\varepsilon_{yy}| < |\kappa_{yx}^2|$, the resonance stop band will also occur.
For the case in Fig. 2b, where incident ***E*** is perpendicular to the SRR plane, and magnetic field ***H*** is parallel to the gapbearing sides of SRR, one obtains

$$\beta h_y = \varepsilon_{xx}E_x \tag{6a}$$

$$\beta E_x = \mu_{yy}h_y \tag{6b}$$

The normalized wave number satisfies

$$\beta^2 = \varepsilon_{xx}\mu_{yy} = 1 \tag{7}$$

which indicates that metamaterials with this SRR orientation has little influence to do with TEM waves of such electromagnetic field polarization and propagation direction. Meanwhile, there is no resonance stop band.
Through the similar analysis, metamaterials with the six SRR orientations can be re-categorized into three groups according to Maxwell's equations. The ones shown in Fig. 2a, 2b are one group, so do those in Fig. 2c, 2d, as well as those in Fig. 2e, 2f. The wave numbers for the other four cases are listed in Table 1. The case in Fig. 2c has been studied in Ref. [10],

SRR orientation	Fig. 2c	Fig. 2d	Fig. 2e	Fig. 2f
β^2	$\varepsilon_{yy}\mu_{xx}$	$\varepsilon_{xx}\mu_{yy}$	$\varepsilon_{yy}\mu_{xx}$	$\varepsilon_{xx}\mu_{yy}$
ε	$\varepsilon_{yy} = a + \dfrac{b\omega^2}{(\omega_0^2 - \omega^2)}$	$\varepsilon_{xx} = a$	$\varepsilon_{yy} = a$	$\varepsilon_{xx} = 1$
μ	$\mu_{xx} = 1$	$\mu_{yy} = 1$	$\mu_{xx} = 1 + \dfrac{c\omega_0^2}{(\omega_0^2 - \omega^2)}$	$\mu_{yy} = 1$

Table 1. Wave numbers for the SRR metamaterials shown in Fig2. c ~ Fig2. f.

where the authors identified the SRR with its outer ring at low frequencies, and illustrated the simulated currents to explain the resonance phenomenon. Here we can see it more clearly that ε_{yy} becomes less than zero when frequency ω is larger than the resonance frequency ω_0, leading to the imaginary wave number β, thus the resonance stop band is achieved. Also Fig. 2e case has the chance to become resonance when $\mu_{xx} < 0$, and there is no resonance stop band for the Fig. 2d and Fig. 2f case.
From the analysis above we can easily conclude that imaginary wave number actually leads to the SRR resonance. Such result has been noted by Simovski *et al.* in their previous work when studied the metamaterial parameters [17,18]. Here we provide an alternative means of characterizing the resonance of SRR metamaterials.

2.2 Floquet modes analysis of the SRR metamterials

Consider an electromagnetic wave to be incident on the SRR metamaterial plane with each element distributed periodically along $\hat{x}$ and $\hat{y}$ direction as shown in Fig. 3.

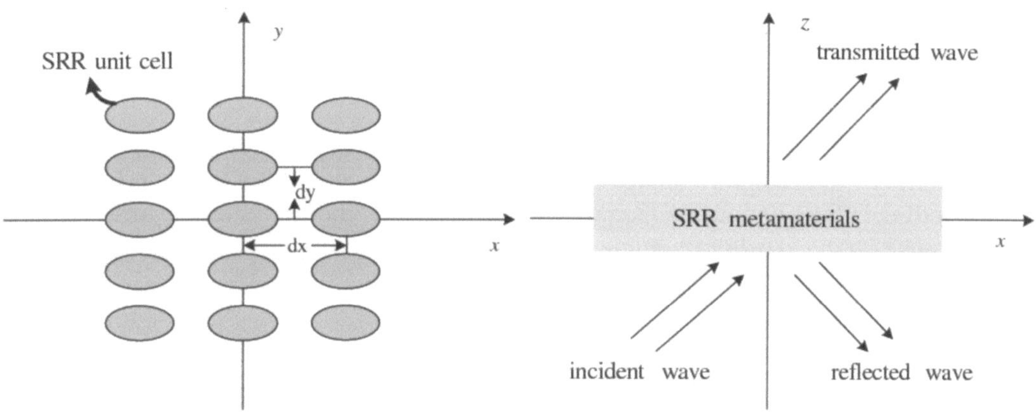

(a) (b)

Fig. 3. Geometry of SRR metamterial plane with incident plane wave (a) Front view (b) Side view

The electromagnetic fields near the SRR elements must satisfy the periodicity requirements imposed by Floquet's theorem. Thus the scattered and the transmitted fields can be expanded as [20]

$$\boldsymbol{E}_{S(T)} = \sum_{m=1}^{2} \sum_{p=-\infty}^{+\infty} \sum_{q=-\infty}^{+\infty} R_{mpq}(B_{mpq}) \boldsymbol{\Psi}_{mpq} \tag{8}$$

where R_{mpq} and B_{mpq} stand for the reflection and the transmission coefficients respectively. The TE and TM vector mode functions ψ_{mpq} can be written as

$$\psi_{1pq} = \frac{1}{(d_x d_y)^{1/2}} \left(\frac{v_{pq}}{t_{pq}} \hat{x} - \frac{u_{pq}}{t_{pq}} \hat{y} \right) \varphi_{pq} \quad \text{TE-Floquet mode} \tag{9a}$$

$$\psi_{2pq} = \frac{1}{(d_x d_y)^{1/2}}\left(\frac{u_{pq}}{t_{pq}}\hat{x} + \frac{v_{pq}}{t_{pq}}\hat{y}\right)\varphi_{pq} \qquad \text{TM-Floquet mode} \tag{9b}$$

The dominant modes have $p = q = 0$ and the higher order modes have $p \neq 0$ or $q \neq 0$. And

$$\varphi_{pq} = \exp(-i(u_{pq}\hat{x} + v_{pq}\hat{y} + \gamma_{pq}\hat{z})) \tag{10}$$

Suppose the incidence wave in the direction of (θ,ϕ) with the wave number k, one has

$$u_{pq} = k\sin\theta\cos\phi + 2\pi p / d_x \tag{11a}$$

$$v_{pq} = k\sin\theta\sin\phi + 2\pi q / d_y \tag{11b}$$

$$\begin{aligned}\gamma_{pq} &= (k^2 - t_{pq}^2)^{1/2}, && \text{for } k^2 > t_{pq}^2 \\ &= -i\left|(k^2 - t_{pq}^2)^{1/2}\right| && \text{for } k^2 < t_{pq}^2\end{aligned} \tag{11c}$$

with

$$t_{pq}^2 = u_{pq}^2 + v_{pq}^2 \tag{12}$$

It is known that a homogenous electromagnetic wave can always be decomposed into a combination of two plane waves with E-field perpendicular or parallel to the incident plane corresponding to the TE- and TM-Floquet modes. Therefore, the effects of any incident wave of either polarization at arbitrary angle (θ,ϕ) will be easily examined for the SRR resonance behavior.

The modal propagation constant γ_{pq} is positive real for the propagating modes and is negative imaginary for the evanescent modes. Since the resonance of SRR metamaterials is often manifested by a dip in the transmitted curves, let's see the *S* parameters for the SRR metamaterial plane.

$$S_{11}^{mpq} = \frac{R_{mpq}(1 - T_{pq}^2)}{1 - R_{mpq}^2 T_{pq}^2}, \quad S_{21}^{mpq} = \frac{T_{pq}(1 - R_{mpq}^2)}{1 - R_{mpq}^2 T_{pq}^2} \tag{13}$$

with the reflection coefficients $|R_{mpq}| \leq 1$ and propagation factor $T_{pq} = \exp(-i\gamma_{pq} z)$. Apparently, S_{21} decreases while T_{pq} gets smaller. When γ_{pq} becomes pure imaginary, all the field components will decay exponentially from the source of excitation, leading to the dip in transmitted (S_{21}) curves. This reveals the fact that evanescent Floquet modes actually engender the resonance behavior which exactly accords with the full wave analysis.

Consider the SRR metamaterials with the orientation in Fig. 2a, the dimensions of the SRR defined in (2) are $a = 0.84$ mm, $b = 1.17$ mm, $c = d = 0.33$ mm, and the dielectric substrate with $\varepsilon_r = 4.8$ is 1.6 mm thick. The SRR metamaterial plane is 3.63 mm in $\hat{z}$ direction with element period $dx = 5.6$ mm, $dy = 5$ mm along $\hat{x}$ and $\hat{y}$ direction. Fig. 4 shows the resonance behavior for a plane wave incident in the XoZ plane ($\phi = 0^\circ$). With E field perpendicular to the incident plane, the resonance only happens in the TE-Floquet modes.

The resonant frequency 8.2 GHz hardly changes, while the resonance intensity is sensitive to the incident angle. One knows that an H component perpendicular to the SRR plane will induce a circular current flow inside the SRRs, which in turn produces just above the resonance frequency a large magnetic dipole moment antiparallel to $H_{\perp} = H\cos\theta$, leading to a negative μ. As the SRR orientation shown in Fig. 2a, the E-field parallel to the gap bearing also introduces the electric resonance, thus the SRR resonance behavior becomes more complicated as the variation of incident angle. As shown in Fig. 4a, when $\theta = 40^{\circ}$, the resonance intensity is of most significance. When $\theta = 0^{\circ}$ and $\theta = 20^{\circ}$, one obtains the almost equal resonance intensity. When $\theta = 60^{\circ}$, the resonance is much weaker. In addition, the bandwidth increases a little as the incidence angle increases. The phases of S_{11} and S_{21} for the SRR metamaterials are given in Fig. 4b. When $\theta = 0^{\circ}$, one finds that the phase of S_{11} goes to zero and S_{21} has an extremely sharp change at the resonance point, which makes the metamaterials can be characterized as a magnetic conductor in this region. However, this does not hold true for the most significant resonance intensity when $\theta = 40^{\circ}$, since it is not merely engendered by H component.

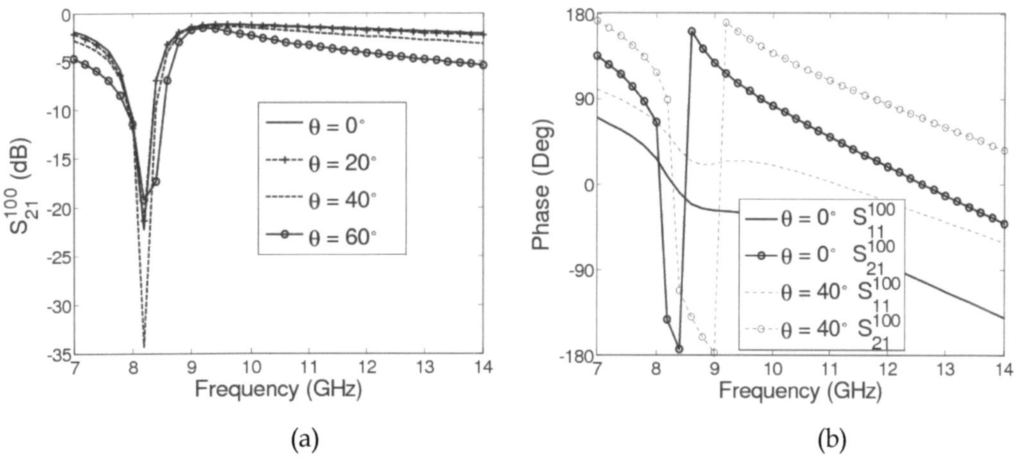

Fig. 4. Transmitted TE-Floquet modes for the SRR metamaterials with the orientation in Fig. 2a (a) magnitudes (b) phases

The graph in Fig. 5 shows the resonance behavior for a plane wave incident in the YoZ plane ($\phi = 90^{\circ}$). With E-field parallel to the incident plane, the resonant behavior this time only happens in the TM-Floquet modes. Similar to the resonance behavior in Fig. 4a, the resonance frequency 8.2 GHz stays the same but the resonance intensity changes with the incidence angle. When $\theta = 60^{\circ}$, the resonance intensity is of most significance. When $\theta = 20^{\circ}$ and $\theta = 40^{\circ}$, one obtains the almost equal and weakest resonance. The intermediate resonance intensity is obtained when $\theta = 0^{\circ}$. The bandwidth decreases a little as the incidence angle increases. In Fig. 5b the phases of S parameters are illustrated. One finds that metamaterials can still be characterized as a magnetic conductor when $\theta = 0^{\circ}$, while the phase of S_{11} and S_{21} are more complex when $\theta = 60^{\circ}$.

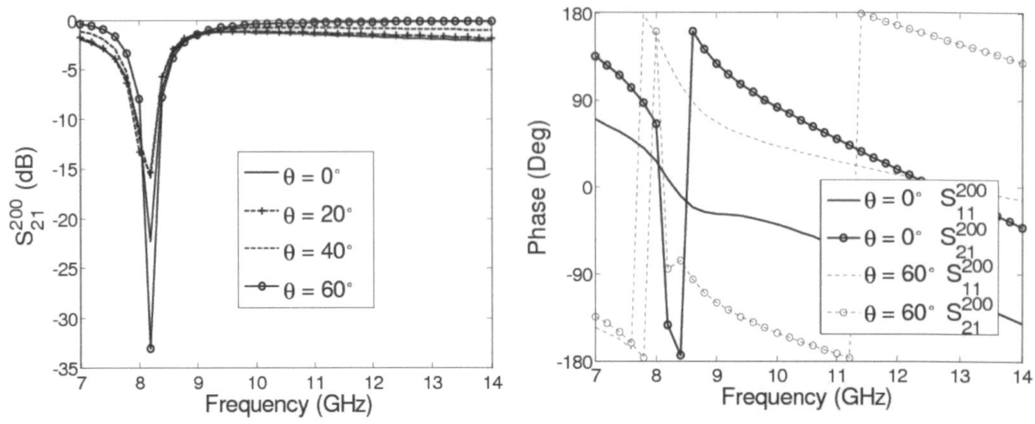

(a) (b)

Fig. 5. Transmitted TM-Floquet modes for the SRR metamaterials with the orientation in Fig. 2a (a) magnitudes (b) phases

Table 2 demonstrates the transmission factor for TE- and TM-Floquet modes at the resonance frequency $8.2\ \text{GHz}$. As is shown, $|T|$ turns smaller when the resonance intensity becomes more significant. And it is the fact that $|T|$ turns smaller because the evanescent modes play more important role, therefore one can conclude that metamaterial resonance is engendered by the evanescent Floquet modes.

TE-Floquet mode $\phi = 0^\circ$	$\theta = 0^\circ$	$S_{11}^{100} = -0.4117 + i(-0.9080)$ $S_{21}^{100} = -0.0614 + i(-0.0466)$	$\lvert T \rvert = 0.0424$
	$\theta = 40^\circ$	$S_{11}^{100} = 0.7535 + i(-0.6571)$ $S_{21}^{100} = 0.0001 + i(0.0193)$	$\lvert T \rvert = 0.0147$
TM-Floquet mode $\phi = 90^\circ$	$\theta = 0^\circ$	$S_{11}^{200} = -0.4117 + i(-0.9080)$ $S_{21}^{200} = -0.0614 + i(-0.0466)$	$\lvert T \rvert = 0.0424$
	$\theta = 60^\circ$	$S_{11}^{200} = 0.6596 + i(-0.8780)$ $S_{21}^{200} = 0.0019 + i(-0.0222)$	$\lvert T \rvert = 0.0126$

Table 2. Transmission factor for TE- and TM-Floquet modes at the resonance frequency 8.2 GHz

For the other two resonance cases in Fig. 2c (1.6 mm thick in $\hat{z}$ direction with $dx = 3.63\text{mm}$, $dy = 5\ \text{mm}$) and Fig. 2e (5 mm thick in $\hat{z}$ direction with $dx = 5.6\ \text{mm}$, $dy = 3.63\ \text{mm}$), Fig. 6 shows the resonance behavior for a plane wave incident in the XoZ plane ($\phi = 0^\circ$) and similar results hold true for the wave incident in the YoZ plane ($\phi = 90^\circ$). The electric resonance for Fig. 2c case shown in Fig. 6a has the most significance resonance intensity when $\theta = 30^\circ$ at $8.6\ \text{GHz}$. On the other hand, the magnetic resonance for Fig. 2e case shown in Fig. 6b demonstrates the general downward trend with the incident angle $8.8\ \text{GHz}$, which has its most significant intensity when $\theta = 0^\circ$, and weakest when $\theta = 60^\circ$.

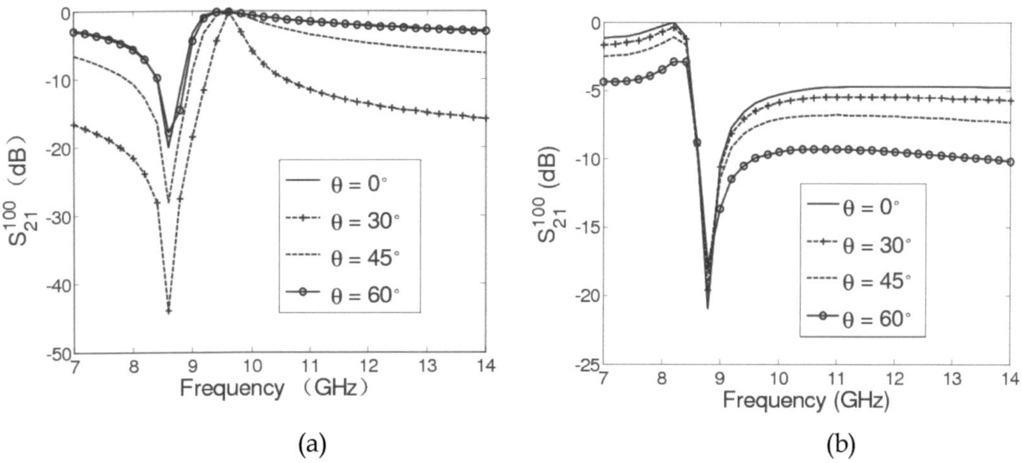

Fig. 6. Transmitted TE-Floquet modes for the SRR metamaterials with (a) the orientation in Fig. 2c (b) the orientation in Fig. 2e

3. Propagation features of waveguides structures with SRR metamaterials

Waveguiding structures based on metamaterial media have recently been considered by several research groups showing how the presence of one or both negative constitutive parameters may give rise to unexpected and interesting propagation properties [22]-[29]. The absence of fundamental mode and sign-varying energy flux in the negative refractive index waveguide are revealed [25]. Rectangular waveguide filled with anisotropic single negative metamaterials are shown to support backward-wave propagation [26]. Moreover, Results for isotropic double negative metamaterial H waveguides are reported, including backward propagation, mode bifurcation and coupling effects [27]. The use of single negative metamaterials as the embedding medium for nonradiative dielectric (NRD) waveguides is examined [28]. Unimodal surface wave propagation in metamaterial NRD waveguides is obtained [29]. However, the presented literatures almost focus on the negative effects of both permittivity and permeability to the metamaterial based waveguides, whereas magnetoelectric coupling of the bianisotropic effects may lead to more dramatically unexpected features in the waveguiding structures.

With the consideration above, metamaterial loaded waveguiding structures are investigated to explore the different dispersion properties of guided waves. It is shown that transverse magnetic and transverse electric waves with non-cutoff frequency and enhanced bandwidth become into existence under certain circumstances in metamaterial parallel plate waveguide and rectangular waveguide. When doping uniaxial bianisotropic SRR metamaterials into NRD waveguides and H waveguides, both longitudinal-section magnetic (LSM) and longitudinal-section electric (LSE) waves are capable of propagating very slowly due to metamaterial bianisotropic effects. Particularly, some abnormal higher-order LSM and LSE modes with negative slope of the phase constant versus frequency may appear when metamaterials are double negative. Such modes will eventually lead to the leakage. Fortunately, for other modes, leakage can be reduced due to the magnetoelectric coupling. Particularly, when the metamaterials are of single negative parameters, leakage elimination can be achieved.

3.1 Parallel plate waveguides and rectangular waveguides

Geometry of parallel plate waveguide and rectangular waveguides filled with SRR metamaterials are shown in Fig. 7. The strip width W in Fig. 7(a) is assumed to be much greater than the separation l between the two plates, so that fringing fields and any x-variation can be ignored. And it is standard convention to have the longest side of the rectangular waveguide along the x -axis, so that $u > v$ in Fig. 7(b).

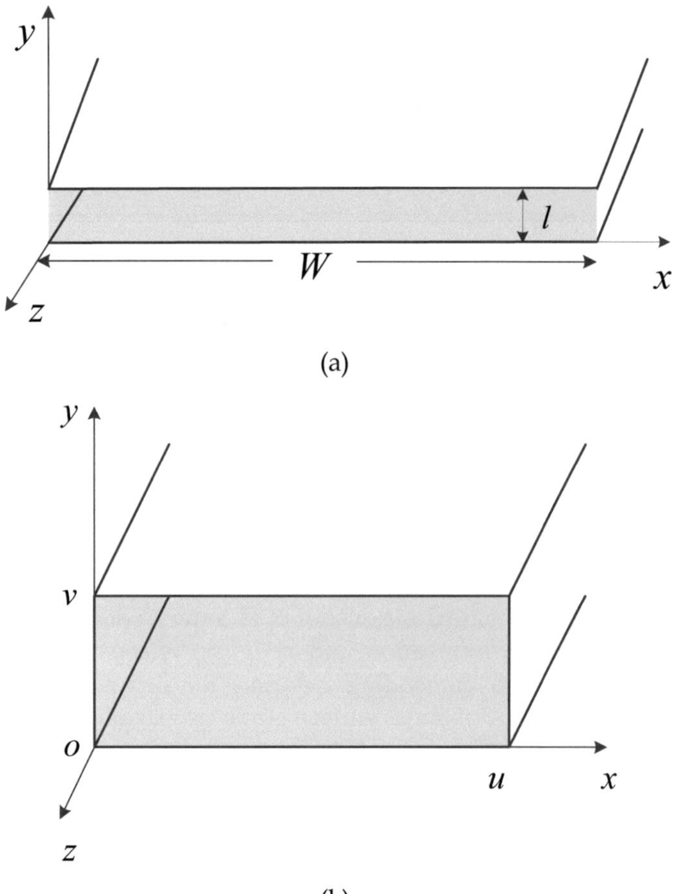

Fig. 7. Geometry of parallel plate waveguide and rectangular waveguide filled with SRR metamaterials

According to Fig. 2, metamaterial waveguide with different SRR orientations are illustrated in Fig. 8. For Fig. 8(a), one can express the following coupled equations for the longitudinal fields:

$$\left(\frac{\varepsilon_{xx}}{\beta^2-\varepsilon_{xx}\mu_{yy}}\partial^2_{x'}+\frac{\varepsilon_{yy}}{\beta^2-\mu_{xx}\varepsilon_{yy}+\kappa^2}\partial^2_{y'}-\varepsilon_{zz}\right)E_z = \left(\frac{-\beta}{\beta^2-\varepsilon_{xx}\mu_{yy}}+\frac{\beta-i\kappa}{\beta^2-\mu_{xx}\varepsilon_{yy}+\kappa^2}\right)\partial^2_{x'y'}h_z \quad (14a)$$

$$(\frac{\mu_{xx}}{\beta^2-\mu_{xx}\varepsilon_{yy}+\kappa^2}\partial_{x'}^2+\frac{\mu_{yy}}{\beta^2-\varepsilon_{xx}\mu_{yy}}\partial_{y'}^2-\mu_{zz})h_z =(\frac{-\beta}{\beta^2-\varepsilon_{xx}\mu_{yy}}+\frac{\beta-i\kappa}{\beta^2-\mu_{xx}\varepsilon_{yy}+\kappa^2})\partial_{x'y'}^2E_z \tag{14b}$$

3.1.1 Non-cutoff frequency modes

For the parallel plate waveguides, TM waves are characterized by $h_z=0$ and a nonzero E_z field which satisfies the reduced wave Eq. (14a), with $\partial x'=0$,

$$[\partial_{y'}^2+\frac{\varepsilon_{zz}}{\varepsilon_{yy}}(\mu_{xx}\varepsilon_{yy}-\kappa^2-\beta^2)]E_z=0 \tag{15}$$

where

$$k_z^2=\beta^2k_0^2=(\mu_{xx}\varepsilon_{yy}-\kappa^2)k_0^2-\frac{\varepsilon_{yy}}{\varepsilon_{zz}}k_c^2 \tag{16}$$

and $k_c=\frac{n\pi}{l}$, $(n=0,1,2\cdots)$ is the cutoff wave number constrained to discrete values.

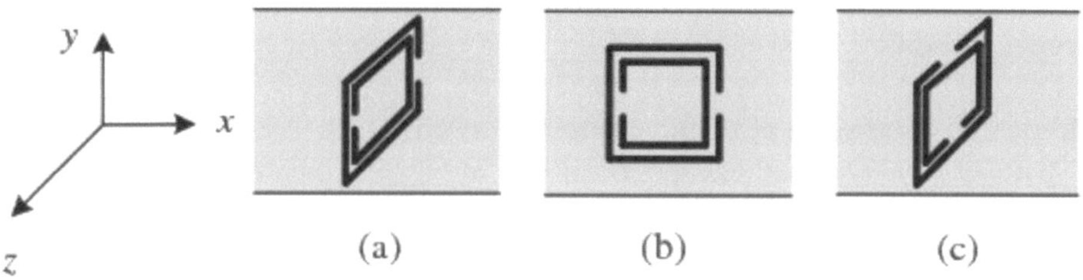

Fig. 8. Metamaterial waveguide with different SRR orientations

Observe that for $n=0$, the TM$_0$ mode is actually identical to the TEM mode shown in Fig. 2(a), therefore, this TM$_0$ mode has a cutoff phenomenon while SRR resonance. However, from (16) one knows that TM mode has the chance to propagate with no cutoff frequency when $\varepsilon_{yy}<0$ and $\mu_{xx}\varepsilon_{yy}-\kappa^2>0$, as shown in Fig. 9. Similar results hold true for the TE modes in Fig. 8(b), and TM modes in Fig. 8(c), which are corresponding to the resonance cases in the Fig. 2.

For the rectangular waveguides, one can see that if $\kappa\neq 0$, decoupling of E_z and h_z occurs only when $\partial_{x'}\equiv 0$ or $\partial_{y'}\equiv 0$, therefore we only consider TE$_{mn}$ modes with $m=0$ or $n=0$, since neither m nor n can be zero for TM modes in a rectangular waveguide. When $\partial_{y'}\equiv 0$, one has the following decoupled equation and boundary condition for TE$_{m0}$ modes from Eq. (14b)

$$(\partial_{x'}^2-\frac{\beta^2-\mu_{xx}\varepsilon_{yy}+\kappa^2}{\mu_{xx}}\mu_{zz})h_z=0 \tag{17}$$

where

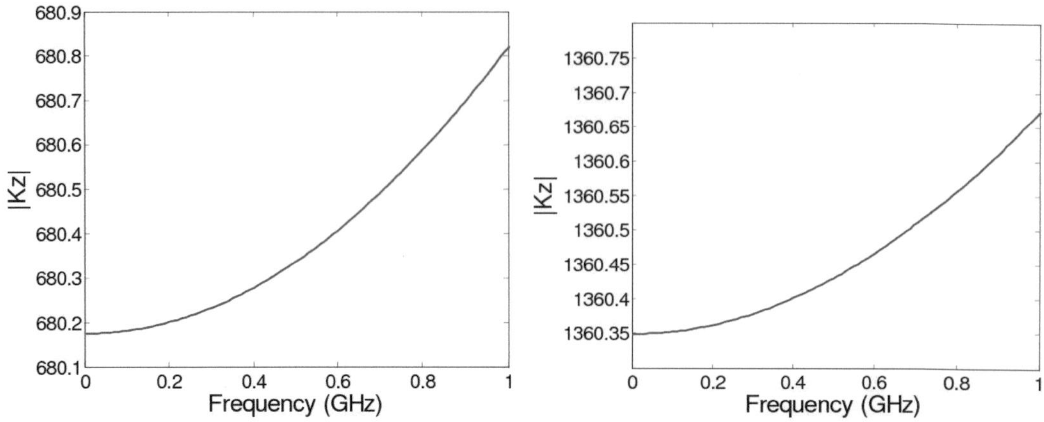

(a) TM_1 (b) TM_2

Fig. 9. The none cutoff frequency TM modes in parallel plate waveguide with SRR metamaterials $\varepsilon_{yy} = -3$, $\varepsilon_{zz} = 1$, $\mu_{xx} = -1$, $\kappa = 1$, $l = 8$ mm

$$k_z^2 = \beta^2 k_0^2 = (\mu_{xx}\varepsilon_{yy} - \kappa^2)k_0^2 - \frac{\mu_{xx}}{\mu_{zz}} k_c^2 \tag{18}$$

Akin to the modes in the parallel plate waveguide, none-cutoff frequency modes also exist under certain condition.

3.1.2 Enhanced bandwidth of single mode operation

For the parallel plate waveguides, the TE modes in Fig. 8(a), characterized by $E_z = 0$ and a nonzero h_z field which satisfies the reduced wave Eq. (14b), with $\partial x' = 0$. Through the similar derivation, one can obtain

$$f_c = \frac{nc}{2l} \qquad (n \geq 1) \tag{19}$$

The TM modes in Fig. 8(b), and TE modes in Fig. 8(c) corresponding to the non-resonance cases in the Fig. 2 achieve the identical cutoff frequencies $f_c = \frac{nc}{2l}$, which is the maximum value for ordinary TM and TE waves, promising a bandwidth enhancement for single-mode operation in material containing waveguide.

For the rectangular waveguides, one can see that TE_{0n} modes in Fig. 8(a) and Fig. 8(c) achieve the cutoff frequency of $f_c = \frac{nc}{2v}$, TE_{m0} modes in Fig. 8(b) obtains the cutoff frequency of $f_c = \frac{nc}{2u}$, which are equal to the ones of air containing rectangular waveguide, promising a bandwidth enhancement for single-mode operation in material containing waveguide.

3.2 Nonradiative dielectric waveguides and H waveguides

Consider the particular case of SRR metamaterials where two sets of SRR microstructures with different orientations are included in NRD waveguides and H waveguides as shown in Fig. 10.

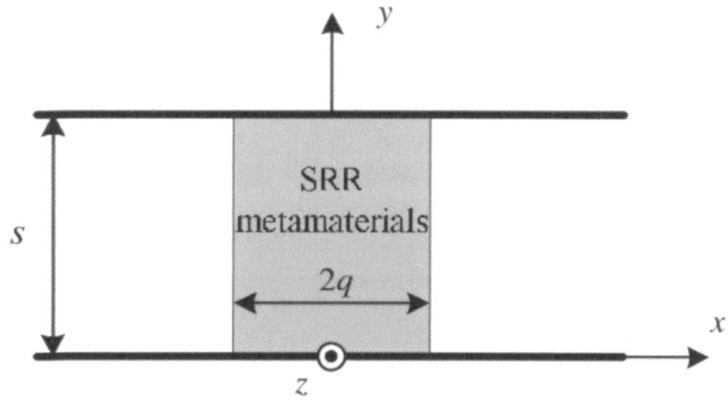

Fig. 10. Configuration of NRD waveguide and H waveguide with SRR metamaterials

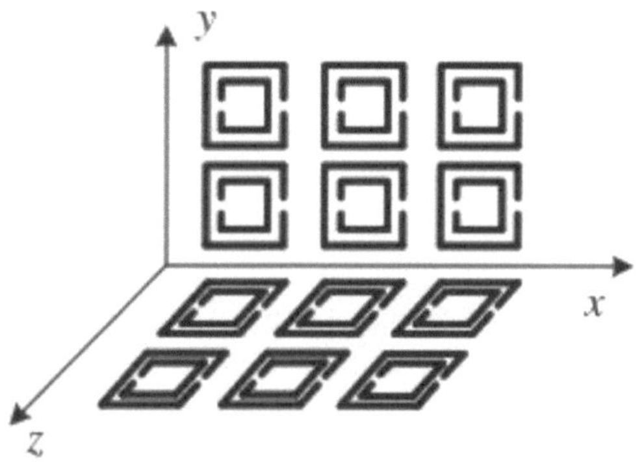

Fig. 11. Spatial orientation of SRRs in the host isotropic medium

The relative orientation of these two ensembles is in Fig. 11 and the $\bar{\varepsilon}$, $\bar{\mu}$ and $\bar{\kappa}$ tensors in Eq. (1) have the following uniaxial form [30]

$$\bar{\varepsilon}=\begin{bmatrix}\varepsilon_1 & 0 & 0\\ 0 & \varepsilon_2 & 0\\ 0 & 0 & \varepsilon_2\end{bmatrix},\ \bar{\mu}=\begin{bmatrix}\mu_1 & 0 & 0\\ 0 & \mu_2 & 0\\ 0 & 0 & \mu_2\end{bmatrix},\ \bar{\kappa}=\mathrm{i}\begin{bmatrix}0 & 0 & 0\\ 0 & 0 & \kappa\\ 0 & -\kappa & 0\end{bmatrix} \tag{20}$$

where

$$\varepsilon_1 \propto a,\ \varepsilon_2 \propto 1+\varsigma\frac{a(\omega_0^2-\omega^2)+b\omega^2}{(\omega_0^2-\omega^2)} \tag{21a}$$

$$\mu_1 \propto 1,\ \mu_2 \propto \xi+\frac{(\omega_0^2-\omega^2)+c\omega^2}{(\omega_0^2-\omega^2)} \tag{21b}$$

$$\kappa \propto \frac{d\omega_0\omega}{(\omega_0^2-\omega^2)} \tag{21c}$$

ς , ξ reflect the different changes of ε and μ components in $\hat{y}\hat{y}$ and $\hat{z}\hat{z}$ direction. Therefore, ε_1 and μ_1 are always positive, whereas ε_2 and μ_2 can be negative in certain frequency band.

For the full wave analysis of the NRD waveguides and H waveguides modal properties, the eigenvalue problem is solved in the complex plane so as to examine the leakage property of H waveguides. The LSM modes are characterized by $h_x = 0$ with nonzero h_y as the supporting field which satisfies

$$\partial_{x'}^2 h_y + \frac{\varepsilon_2}{\varepsilon_1}\partial_{y'}^2 h_y = -(\varepsilon_2\mu_2 - \kappa^2 - \frac{\varepsilon_2}{\varepsilon_1}k_z'^2)h_y \tag{22}$$

where $k_z' = k_z/k_0 = \beta_z - \mathrm{i}\alpha_z$ is the complex normalized longitudinal wave number. All other field components can be expressed as

$$h_z = -\mathrm{i}\frac{1}{k_z'}\partial_{y'}h_y \tag{23a}$$

$$E_x = -\frac{1}{k_z'\varepsilon_1}(\partial_{y'}^2 h_y - k_z'^2 h_y) \tag{23b}$$

$$E_y = \frac{1}{k_z'\varepsilon_2}(\partial_{y'}\partial_{x'}h_y - \kappa\partial_{y'}h_y) \tag{23c}$$

$$E_z = -\mathrm{i}\frac{1}{\varepsilon_2}(\partial_{x'}h_y - \kappa h_y) \tag{23d}$$

Express h_y as a product of two separate variable functions in the form

$$h_y = f(x')g(y')\exp(-\mathrm{i}k_z'z') \tag{24}$$

such that

$$\partial_{x'}^2 f(x') + k_x'^2 f(x') = 0 \tag{25a}$$

$$\partial_{y'}^2 g(y') + k_y'^2 g(y') = 0 \tag{25b}$$

where $k_x' = k_x / k_0$ and $k_y' = k_y / k_0$ are the complex normalized transverse wave numbers. Substituting back into Eq. (22), the normalized wave numbers can be expressed

$$k_{x1}'^2 + \frac{\varepsilon_2}{\varepsilon_1}(k_y'^2 + k_z'^2) = \varepsilon_2\mu_2 - \kappa^2 \tag{26a}$$

$$k_{x0}'^2 + k_y'^2 + k_z'^2 = 1 \tag{26b}$$

One has $k_x' = k_{x1}' = \beta_{x1} - \mathrm{i}\alpha_{x1}$ for $|x'| < q'$, while for $|x'| > q'$, one should take $k_x' = k_{x0}' = \beta_{x0} - \mathrm{i}\alpha_{x0}$. Apply the boundary conditions on the perfectly electric conductor planes to other field components, one can write

$$g(y') = G\sin(k_y'y') \qquad (n = 1, 2, 3, \cdots) \tag{27}$$

where G is the amplitude constant, and $k_y' = n\frac{\pi}{s'}$ with $s' = k_0 s$. The n index gives the number of half waves along y. And

$$f(x') = \begin{cases} F_1 \exp[k_{x0}'(x'+q')] & x' < -q' \\ F_2[\cos(k_{x1}'x') + R\sin(k_{x1}'x')] & -q' < x < q' \\ F_3 \exp[-k_{x0}'(x'-q')] & q' < x' \end{cases} \tag{28}$$

Enforcing the continuity conditions at both $x' = \pm q'$, the modal equation for the LSM modes can be finally derived

$$[k_{x1}' \cot(k_{x1}'q') + k_{x0}'\varepsilon_2][k_{x1}' \tan(k_{x1}'q') - k_{x0}'\varepsilon_2] + \kappa^2 = 0 \tag{29}$$

the order of eigensolution of Eq. (29) gives the m index ($m = 0,1,2,\cdots$) appearing in LSM_{mn}. From the similar derivation, the LSE modes can be defined as

$$[k_{x1}' \cot(k_{x1}'q') + k_{x0}'\mu_2][k_{x1}' \tan(k_{x1}'q') - k_{x0}'\mu_2] + \kappa^2 = 0 \tag{30}$$

and the normalized transverse wavenumber in the slab should be given by

$$k_{x1}'^{\,2} + \frac{\mu_2}{\mu_1}(k_y'^{\,2} + k_z'^{\,2}) = \varepsilon_2\mu_2 - \kappa^2 \tag{31}$$

instead of Eq. (26a). Hereafter, we only consider the LSM modes, and the following results hold true for the LSE modes.

3.2.1 Slow wave propagation

Figure 12(a) presents the operational diagram for LSM modes, the real part of the longitudinal wave number $|\beta_z|$ decreases gradually as the magnetoelectric coupling turns larger in the case that ε_2 and μ_2 are of positive values. Maximum κ is achieved under the cutoff condition $k_z' = 0$.

$$|\kappa_{\max}| = \sqrt{\varepsilon_2\mu_2 - \left|\frac{\varepsilon_2}{\varepsilon_1}\right| k_y'^{\,2} - k_{x1}'^{\,2}} \tag{32}$$

Since the guide wavelength defined as $\lambda_g = \frac{2\pi}{\beta_z}$ becomes smaller when longitudinal wave number increases, the corresponding phase velocity $v = T\lambda$ of the modes will be much slower. From Eq. (21), one can see that in the frequency that ω is far larger than the resonance frequency ω_0, both positive ε_2 and μ_2 as well as smaller absolute value of κ can be obtained, thus slow wave propagation will appear.

In Fig. 12(b), one can see that $|\beta_z|$ shows a general upward trend when the magnetoelectric coupling becomes significant in the case that ε_2 and μ_2 are both negative. Minimum value for κ with the cutoff condition $k_z' = 0$ can be obtained,

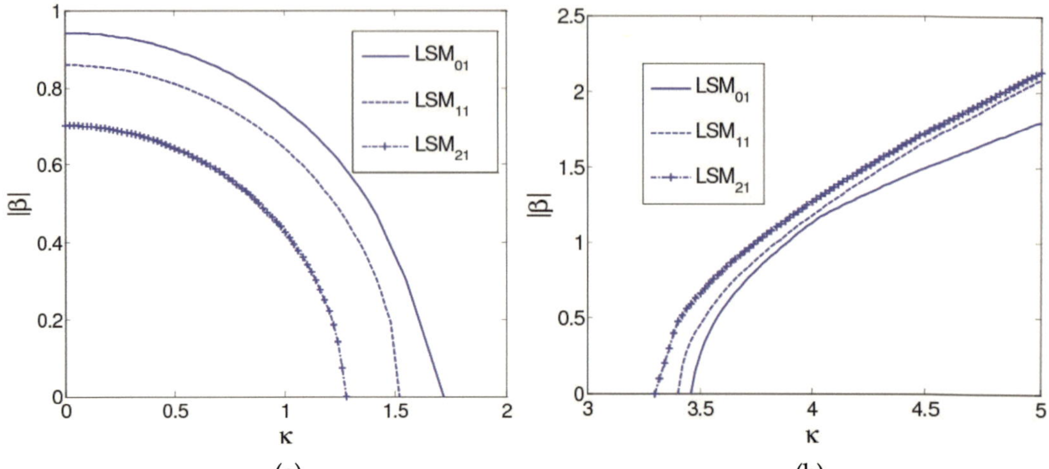

Fig. 12. Relationship of $|\beta_z|$ and magnetoelectric coupling κ in nonradiative dielectric waveguide with SRR metamaterials
(a) $\varepsilon_1 = 1$, $\varepsilon_2 = 3$, $\mu_2 = 2.5$, $f = 35$ GHz, $s = 0.4\lambda_0$, $q = 0.6\lambda_0$
(b) $\varepsilon_1 = 1$, $\varepsilon_2 = -3$, $\mu_2 = -2.5$, $f = 35$ GHz, $s = 0.4\lambda_0$, $q = 0.6\lambda_0$

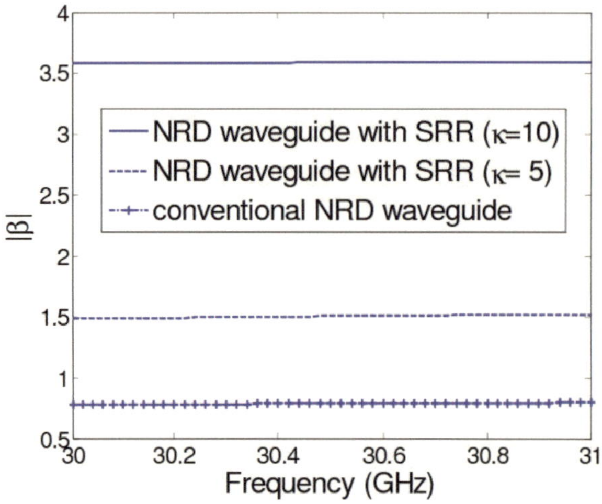

Fig. 13. Variation of the $|\beta_z|$ with frequency f for the dominant LSM01 (conventional NRD waveguide with $\varepsilon_r = 4$, $\mu_r = 1$, $s = 4$ mm, $q = 5$ mm; double negative metamaterial NRD waveguide with $\varepsilon_1 = 1$, $\varepsilon_2 = -4$, $\mu_2 = -1$, $\kappa = 5$ and 10, $s = 4$ mm, $q = 5$ mm)

$$|\kappa_{\min}| = \sqrt{\varepsilon_2\mu_2 + \left|\frac{\varepsilon_2}{\varepsilon_1}\right| k_y'^2 - k_{x1}'^2} \tag{33}$$

From Eq. (21), ε_2 and μ_2 have the chance to become negative when ω is little larger than the resonance frequency ω_0. Meanwhile, the magnetoelectric coupling κ has the absolute value which can be infinitely large within this frequency band. Therefore, the guided waves are able to propagate very slowly, and even approach zero velocity.

Besides, we should stress that when the magnetoelectric coupling vanishes, Eq. (29) and Eq. (30) become a product of two elementary modal equations, which is similar to those of conventional NRD waveguides and H waveguides. Fig. 13 shows that the operational diagram for LSM_{01} modes in a conventional NRD waveguide and a NRD waveguide with SRR metamaterials. Since there exists minimum value for magnetoelectric coupling κ in the double negative case, we choose $\kappa = 5$ and 10 to make sure that LSM_{01} propagates. As can be seen within $[30\text{ GHz},\ 31\text{ GHz}]$, the longitudinal wave number of LSM_{01}mode in NRD waveguide with SRR metamaterials is always larger than that of the conventional one, thus traveling more slowly, which indicates that NRD waveguide with SRR metamaterials allows more number of wavelength to propagate within the same length, providing feasibility of miniaturization for NRD waveguide.

Let's further consider the power flow of LSM_{01} modes in the NRD waveguide with SRR metamaterials. The time-average power passing a transverse cross-section of the NRD waveguide is

$$\begin{aligned} P_{01} &= \frac{1}{2}\mathrm{Re}\int_{x'=-l'}^{l'}\int_{y'=0}^{s'} E\times h^{*}\cdot\hat{z}dy'dx' \\ &= \pm\frac{1}{2}\mathrm{Re}\int_{x'=-l'}^{l'}\int_{y'=0}^{s'} E_x h_y^{*}dy'dx' \\ &= \pm\frac{F_2^2G^2s'l'}{2\varepsilon_1}[\frac{1}{\beta_z}(\frac{\lambda_0}{2s})+\beta_z] \end{aligned} \tag{34}$$

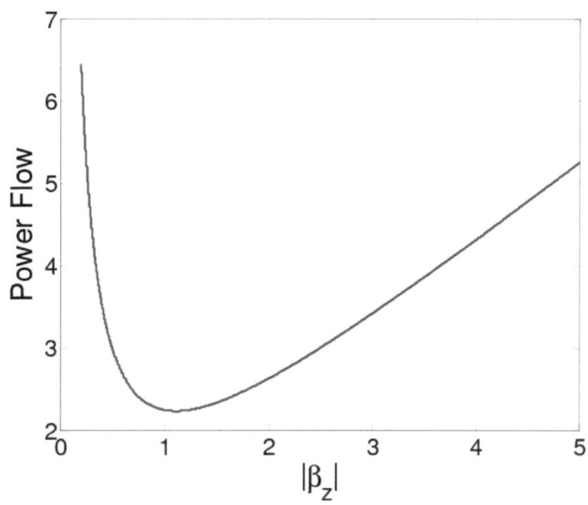

Fig. 14. Energy flow of LSM_{01} mode varied with longitudinal wave number in the NRD waveguide with SRR metamaterials

For ε_2 and μ_2 are both positive, we choose '+', and for the double negative metamaterial case, we choose '-'. As we all know, the double negative metamaterials have the negative wave number, which leads to the positive Poynting vector in Eq. (34). With the choice of

$s = 0.4\lambda_0$, Fig. 14 shows general trend of power flow varied with absolute value of longitudinal wave number by not taking the constant coefficient into account. It can be seen that P_{01} will increase as $|\beta_z|$ becomes larger than $\sqrt{\lambda_0 / 2s} = 1.118$. What is meant by this is that the increasing $|\beta_z|$ will enhance the power flow. From the above analysis, we can discern that metamaterial waveguide with double negative parameters has the chance to achieve infinite large $|\beta_z|$ around the metamaterial resonance frequency. Therefore, inserting metamaterials will strengthen the power flow of the conventional NRD waveguide. It is worth noting that such strengthen trend of power flow is based on the variation of $|\beta_z|$, and the total energy will still be conserved since the electromagnetic wave propagates much more slowly.

In the above results, $s < 0.5\lambda_0$ is assumed throughout which means that the proposed waveguiding structure functions as NRD waveguide. When s increases to $s > 0.5\lambda_0$, the waveguiding structure in Fig. 10 becomes H waveguide, and the results such as slow wave propagation and enhanced energy flow shown in Fig. 12-14 still work.

3.2.2 Abnormal guidance and leakage suppression

Usually, all the NRD waveguide and H waveguide components preserve the vertical symmetry so that a general $n = 1$ dependence may be assumed. Therefore, if the modes can leak power, they must do so in the form of TM_1 or TE_1 mode in the air-filled parallel plate region. Consequently, the condition for leakage can be written as $\beta_z < k_p$, where k_p is the normalized longitudinal wave number of the TM_1 or TE_1 satisfying $k_p^2 = 1 - (\pi/s')^2$. Given $s = 5$ mm, the waveguiding structure in Fig. 10 works as a NRD waveguide when frequency $f < 30$ GHz. No electromagnetic wave can propagate between the parallel plates because of the cutoff property, thus no leaky wave exits. However, when $f > 30$ GHz, it functions as an H waveguide, therefore, leakage may come into being.

Considering the propagation features of the NRD waveguide and H waveguide with bianisotropic SRR metamaterials, the $|\beta_z|$ of most modes become larger when f increases like the modes in conventional waveguides. However, some higher-order LSM modes in the proposed waveguiding structure may operate dramatically differently. From Eq. (26a),

$$k_z'^2 = \frac{\varepsilon_1}{\varepsilon_2}(\varepsilon_2\mu_2 - \kappa^2 - k_{x1}'^2) - k_y'^2 \tag{35}$$

one can conclude that when $\frac{\varepsilon_1}{\varepsilon_2}(\varepsilon_2\mu_2 - \kappa^2) > 0$, $|k_z|$ will demonstrate a general upward trend as f increase, and may have the chance to get a fall while $\frac{\varepsilon_1}{\varepsilon_2}(\varepsilon_2\mu_2 - \kappa^2) < 0$. The similar trend holds true for $|\beta_z|$.

Fig. 15 presents the abnormal falling behavior of LSM higher-order modes in the proposed NRD waveguide and H waveguide with double negative parameters, and TM_1 mode in air filled parallel plate guide. In the region where the dispersion curve of LSM mode is located below the curve of TM_1 mode, one expects the leakage. As can be seen, if the falling behavior continues in the H waveguide, leakage will eventually happens. Fortunately, such higher order modes can propagate only when the parameters are both negative. Moreover, lots of them cannot exit in the H waveguide region since the cutoff of $|\beta_z| = 0$ while f becomes larger.

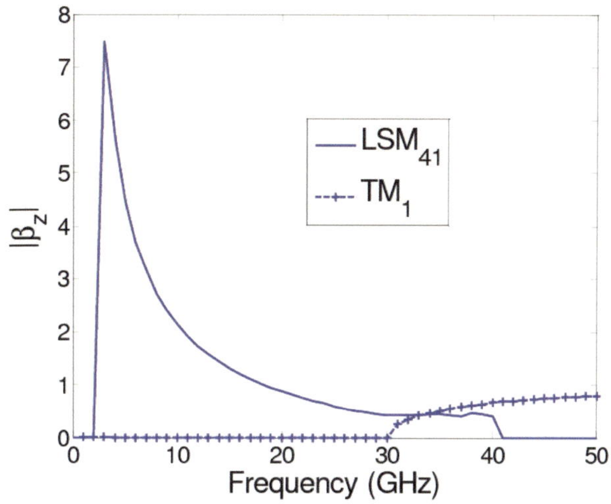

Fig. 15. Abnormal falling behavior of LSM higher modes in the proposed NRD waveguide/H waveguide with double negative parameters ($\varepsilon_1 = 1$, $\varepsilon_2 = -3$, $\mu_2 = -1$, $\kappa = 1$, $s = 5$ mm, $q = 5$ mm) and TM_1 mode in air filled parallel plate guide

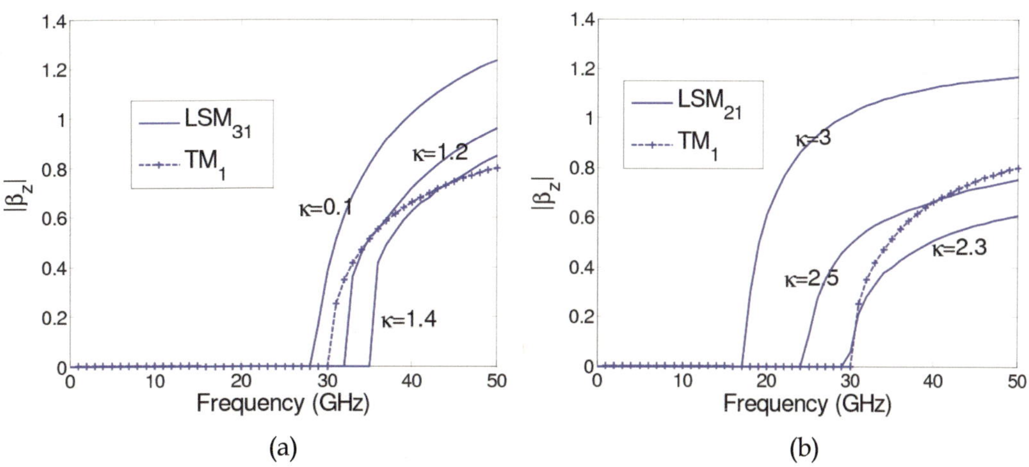

Fig. 16. Variation of the $|\beta_z|$ versus frequency f for LSM modes in the proposed NRD waveguide/H waveguide ((a) for double positive metamaterials with $\varepsilon_1 = 1$, $\varepsilon_2 = 2$, $\mu_2 = 2.5$, $\kappa = 0.1$, 1.2, and 1.4, $s = 5$ mm, $q = 5$ mm (b) for double negative metamaterials with $\varepsilon_1 = 1$, $\varepsilon_2 = -3$, $\mu_2 = -1.5$, $\kappa = 2.3$, 2.5, and 3, $s = 5$ mm, $q = 5$ mm) and TM_1 mode in air filled parallel plate guide

Fig. 16 shows dispersion curve $|\beta_z|$ versus frequency f for LSM modes in proposed H waveguide under the condition $\frac{\varepsilon_1}{\varepsilon_2}(\varepsilon_2\mu_2 - \kappa^2) > 0$. It can be seen in Fig. 16(a), $|\beta_z|$ presents an increase when f becomes lager, but experience a decrease as κ turn larger at the same frequency while ε_2 and μ_2 are positive. Therefore, smaller magnetoelectric coupling κ will reduce the leakage of the propagating modes. To examine further, from Eq. (21) one knows that positive ε_2 and μ_2 as well as minimum κ can exit together in the frequency bands that are far larger than the SRR resonance frequency ω_0, in which the proposed waveguiding structure may exactly works as a H waveguide, thus such reduced leakage scheme can be fulfilled. On the contrary, Fig. 16(b) shows that under the same frequency, $|\beta_z|$ increases when κ becomes more significant in the case that ε_2 and μ_2 are both negative. From Eq. (21), one can see that ε_2 and μ_2 have the chance to become both negative when the frequency is little larger than the SRR resonance frequency ω_0. Meanwhile, the magnetoelectric coupling κ has the absolute value which can be infinitely large within this frequency band. Therefore, the leakage of guided waves is able to be greatly reduced.

For the more special case when ε_2 and μ_2 are of single negative value, Fig. 17 illustrates that when $\mu_2 > 0$ and $\varepsilon_2 < 0$, $|\beta_z|$ of LSM modes become larger as κ increases. Furthermore, the leakage can hardly happen in this case even when $\kappa = 0$, since $\beta_z > k_p$ can be easily satisfied. However, LSE modes will no longer exist under such condition. On the other hand, when $\mu_2 < 0$ and $\varepsilon_2 > 0$, only LSE modes can propagate in the proposed nonradiative dielectric waveguide amd H waveguide, and by choosing proper parameters, the leakage can also be eliminated.

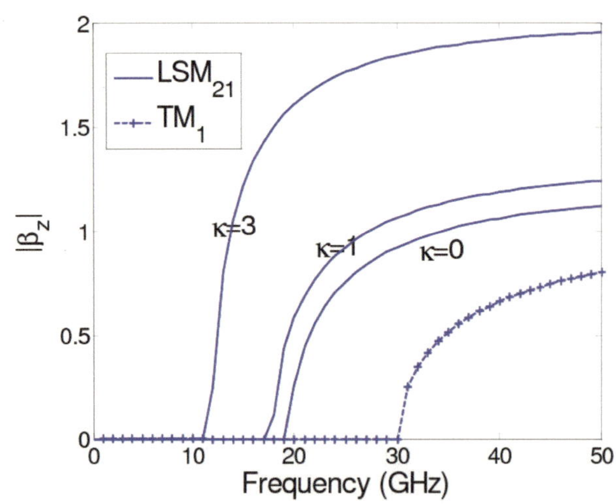

Fig. 17. Variation of the $|\beta_z|$ versus frequency f for LSM modes in proposed NRD waveguide/H waveguide with single negative metamaterials $\varepsilon_1 = 1$, $\varepsilon_2 = -3.5$, $\mu_2 = 1.5$, $\kappa = 0,\ 1,$ and 3, $s = 5$ mm, $q = 5$ mm and TM$_1$ mode in air filled parallel plate guide

4. Metamaterial based multiband frequency selective surfaces

Frequency selective surface (FSS) is a kind of periodically arranged 2D array structure, working like special resonances composed of one or more layers of planar screens and designed to reflect or transmit incident plane waves over specified ranges of frequency. With such unique resonance property, SRR metamaterials are propitious to make frequency selective surfaces (FSSs). Sun et al. [31] investigated a new type of band-pass FSS with metamaterial structures. Baena et al. [32] proposed an isotopic FSS structure based on cubic arrangements of SRR unit, realizing the transmission properties were angle and polarization independent. Marqués et al. [33] illustrated two FSS structures with conventional as well as complementary SRRs and presented ab initio analysis theory.

Typically, the characteristics of an FSS are mainly determined by the shape, size and distribution of the resonant cell. Due to the coupling between the unit cells, the conventional closed element FSSs are often single-band which displays only one dip in the reflection and transmission spectrums. Dual band FSSs, as a breakthrough to the conventional single-band ones, have already realized by fractal elements [34]. And multi-layer structures or complex resonator units are often employed in order to achieve multi-band band-pass or band-stop characteristics of FSS [35,36]. The salient properties of SRR metamaterials have made, and continue to make it possible to fabricate FSSs with enhanced properties that can be rarely obtained by conventional resonance units. Here we demonstrated the multiband FSSs which have several dips in the reflection and transmission spectrums by introducing SRR into the design procedure.

Choosing the FSS structure as shown in Fig. 18, the SRR units are embedded in the dielectric slab with $h = 1$ mm and $\varepsilon_r = 2.2$. The geometric parameters of SRR are $g = 0.4$ mm, $w = 0.5$ mm, $r = 3.9$ mm, $s = 0.6$ mm and repeated in periods of $p = 8$ mm in the x and y direction. It should be noted that such FSS structure will turn into the conventional dual-ring FSS with ring cells when the gaps of the SRR is enclosed.

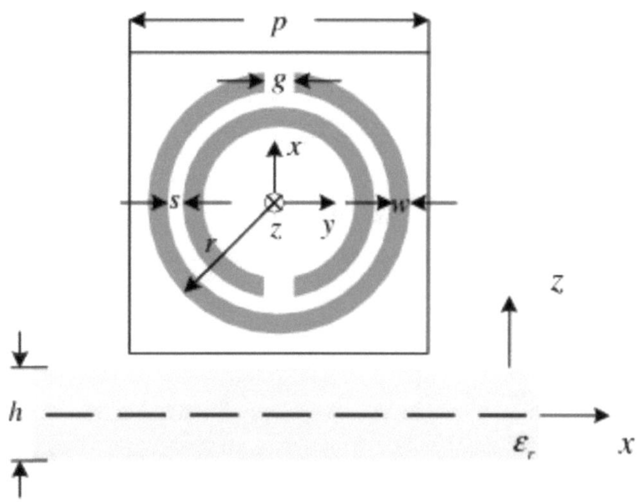

Fig. 18. FSS structure with SRR elements

Fig. 19 shows the resonance response of the FSSs with the SRR and the corresponding dual-ring structure under the normal incident electromagnetic wave. For Fig. 19a ($\theta = 0^\circ$, $\phi = 0^\circ$), the electromagnetic wave propagates along the z-direction with the electric field parallel to the gap bearing sides of the SRR. It can be seen that the transmitted TE-Floquet modes present three resonances when the SRR acts as the resonant cell, while the corresponding FSS with dual-ring structure only demonstrates single resonance. For Fig. 19b ($\theta = 0^\circ$, $\phi = 90^\circ$), the magnetic field is parallel to the gap bearing sides of SRR. The transmitted TE-Floquet modes represent only one resonant point and the frequency is approximately consistent with the FSS with dual-ring unit structure. Meanwhile, Fig. 19 also indicates that the resonant frequency of the FSS with dual-ring structure is insensitive to ϕ at perpendicular incidence. However, the incident angle is crucial for the FSS with the SRR.

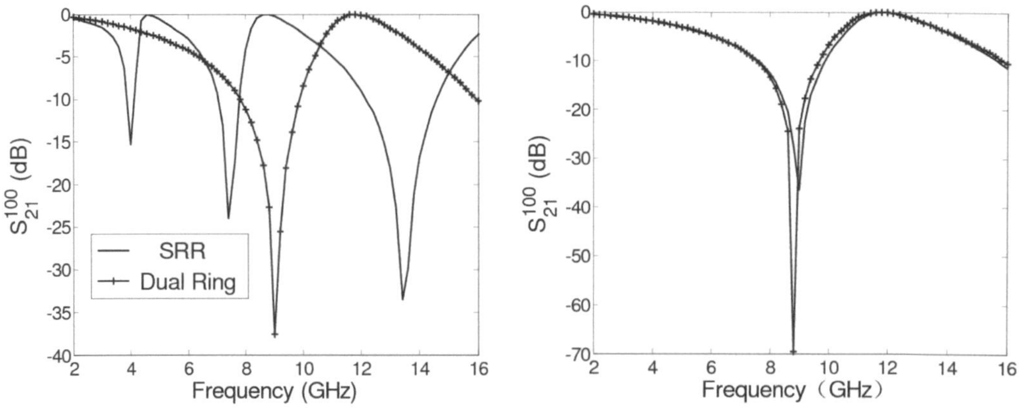

(a) $\theta = 0^\circ$, $\phi = 0^\circ$ (b) $\theta = 0^\circ$, $\phi = 90^\circ$

Fig. 19. Resonance response of the FSS based on SRR and dual-ring structure

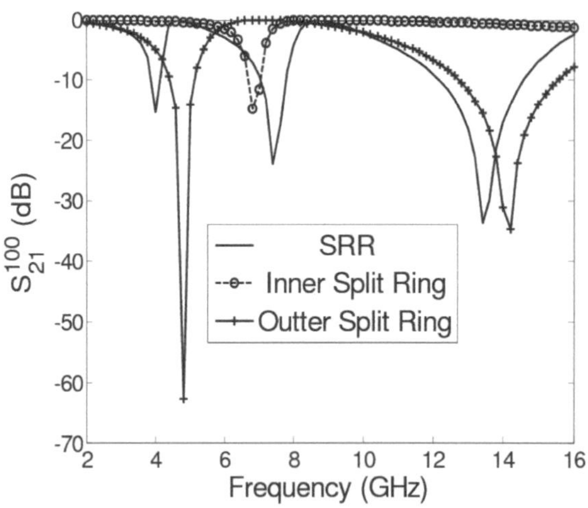

Fig. 20. Resonance response of FSS based on SRR, inner split ring and outer split ring

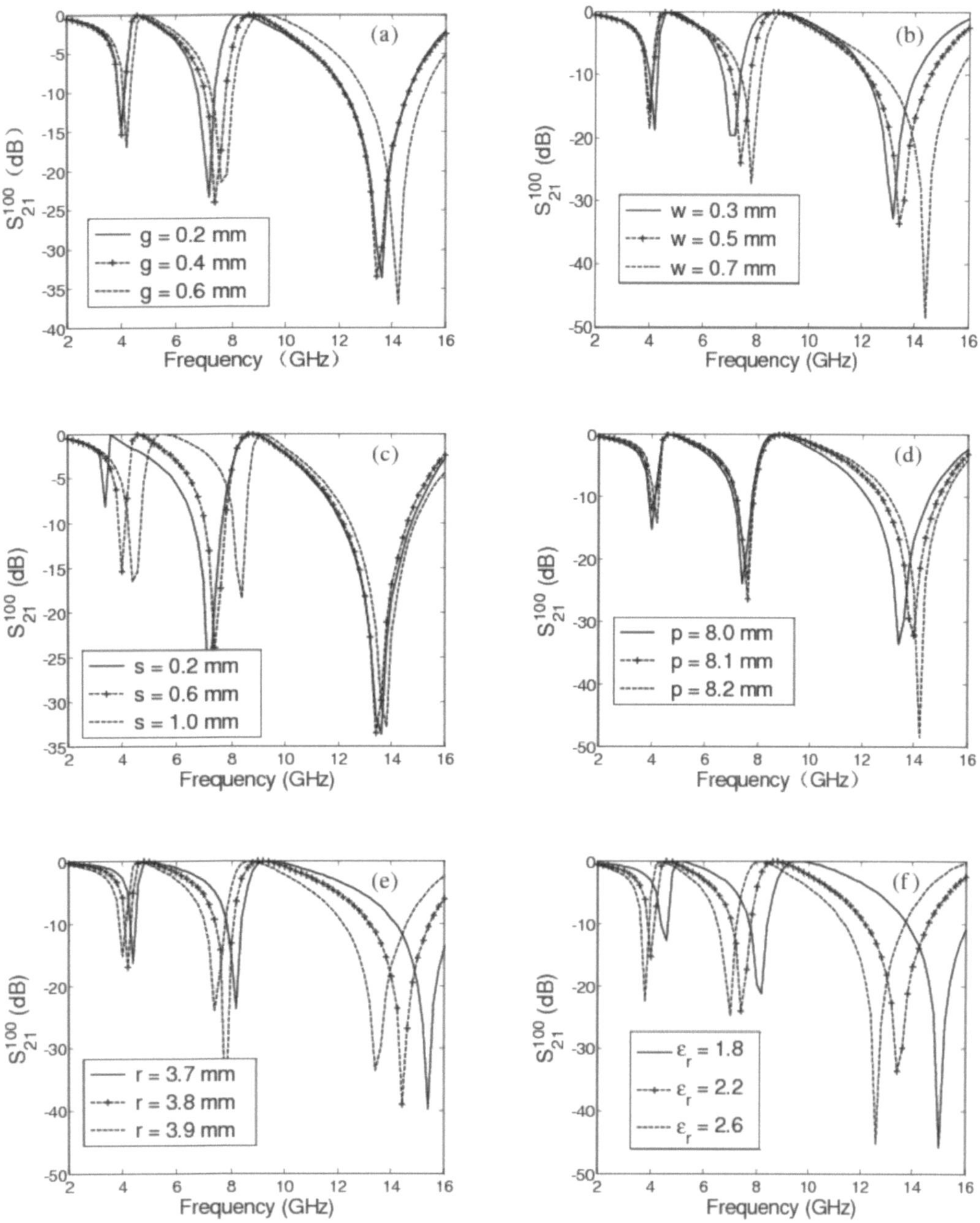

Fig. 21. SRR parameter effects on the resonance properties of the FSSs (a) gap size (b) ring width (c) ring separation (d) unit period (e) outer split ring radius (f) dielectric constant

Fig. 20 displays the resonance response of FSS based on SRR, inner split ring and outer split ring for perpendicular incidence($\theta = 0^\circ$, $\phi = 0^\circ$). It can be concluded that the three resonant points of the FSS with SRR are engendered by the inner split ring and outer split ring respectively. And the inner split ring determines one point, while the outer split ring determines two. The slight shifts between the resonant frequencies are due to the coupling between the two split rings. Similar results have been illustrated by Kafesaki et al. [37] when studied the transmission properties of left-handed materials, here we provide an alternative numerical techniques by introduce Floquet modes into the analysis of metamaterial based FSSs.

Fig. 21 illustrates the influences of SRR parameter on the resonance properties of the FSSs. By making other parameters stay the same, it can be seen that the increase of gap size g, ring width w, ring separation s, and unit period p will lead to the increase of the resonance frequency of the FSS. In contrast, the increase of outer split ring radius r, and dielectric constant ε_r will result in the decrease of the resonance frequency.

Fig. 22 displays the resonance response of the FSS to the electromagnetic wave incident in the z-direction ($\theta = 0^\circ$) from different incidence planes. The orientation of the SRR resonant cell in the FSS to the incident wave changes with the variation of ϕ . Namely, the direction of the E-field turns from parallel to the gap to perpendicular to the gap, which makes the electric resonance of the SRR become weaker. As is shown, when ϕ increases, the resonance response of the TE-Floquet modes weakens and the three resonant points of the FSS finally transit into only one resonant frequency corresponding to $\phi = 0^\circ$ and $\phi = 90^\circ$, respectively. On the other hand, the resonance response of the TM-Floquet modes experience the opposite variation trend, and the single resonance frequency of the FSS becomes into three resonances due to the gradually enhanced electric resonance.

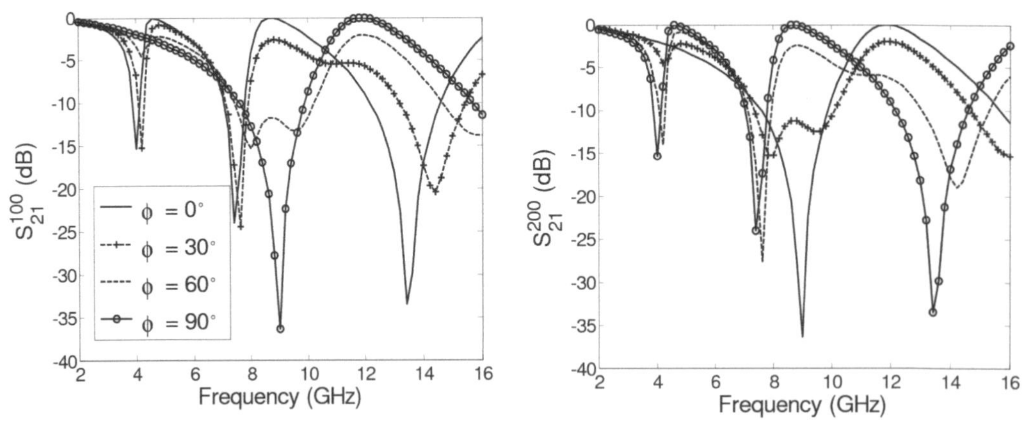

(a) TE-Floquet

(b) TM-Floquet

Fig. 22. The FSS resonance response to the electromagnetic wave in the z-direction from different incident planes

Fig. 23(a) shows the transmission property of FSS with the wave incident in the XoZ plane ($\phi = 0^\circ$). As we have demonstrated, the resonance response only occurs in the transmitted Floquet modes with the same polarization as the incident wave from the dominant plane. When $\theta = 0^\circ$, only electric resonance exits. However, the increase of θ makes H component perpendicularly transmit the SRR, which leads to magnetic resonance. The FSS still keeps three resonances with one point of 7.4 GHz stays the same and the other two points shift slightly. Resonance intensity turns much more complicated because of the combined action of the electric resonance and the magnetic resonance. The resonance point near 4 GHz is of the most significance when $\theta = 30^\circ$, while the resonance points at 7.4 GHz and near 14 GHz are relatively strong when $\theta = 80^\circ$. In addition, the bandwidth of the three resonances turns broader as θ increases. Fig. 23(b) shows the transmission property FSS with the wave incident in YoZ plane ($\phi = 90^\circ$). In this case, electric resonance weakens a little and the bandwidth decreases when θ increases.

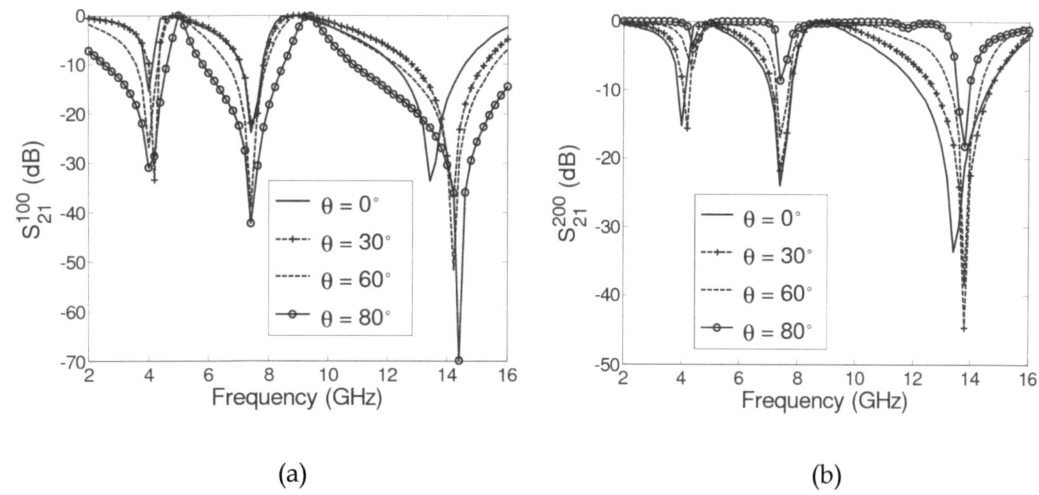

Fig. 23. The FSS resonance response to the wave incident in (a) the XoZ plane (b) YoZ plane

Fig. 24 and Fig. 25 show the TE- and TM-Floquet mode resonance responses of the FSS for the vertical- and parallel-polarized electromagnetic wave incident in the arbitrary planes, respectively. For any incident angle θ, the resonance response of TE-Floquet mode becomes weaker as ϕ increases, while the resonance response for the TM-Floquet mode keeps nearly unchanged as shown in Fig. 20. On the other hand, the TM-Floquet mode resonance response becomes stronger with the increase of ϕ, while the TE-Floquet mode keeps almost unchanged as shown in Fig. 21. We can concluded that both TE- and TM-Floquet modes will share FSS resonance when either polarized wave incident in the none-dominant plane, however, the transmitted modes with the same polarization as the incident wave are much more sensitive to the incident angle.

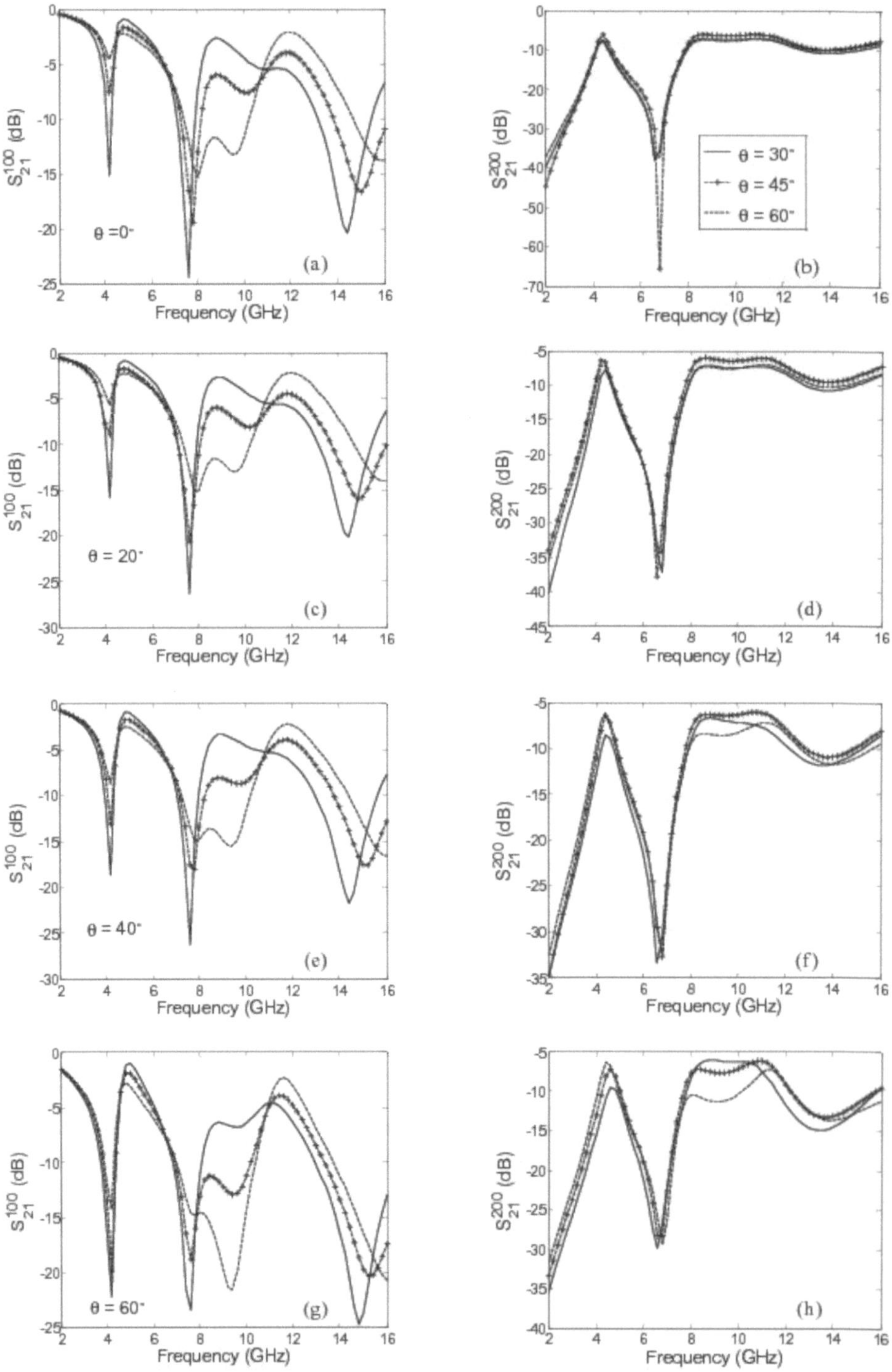

Fig. 24. The resonance responses of the FSS for the vertical-polarized electromagnetic wave in arbitrary plane (a) (c) (e) (g) TE-Floquet modes; (b) (d) (f) (h) TM-Floquet modes

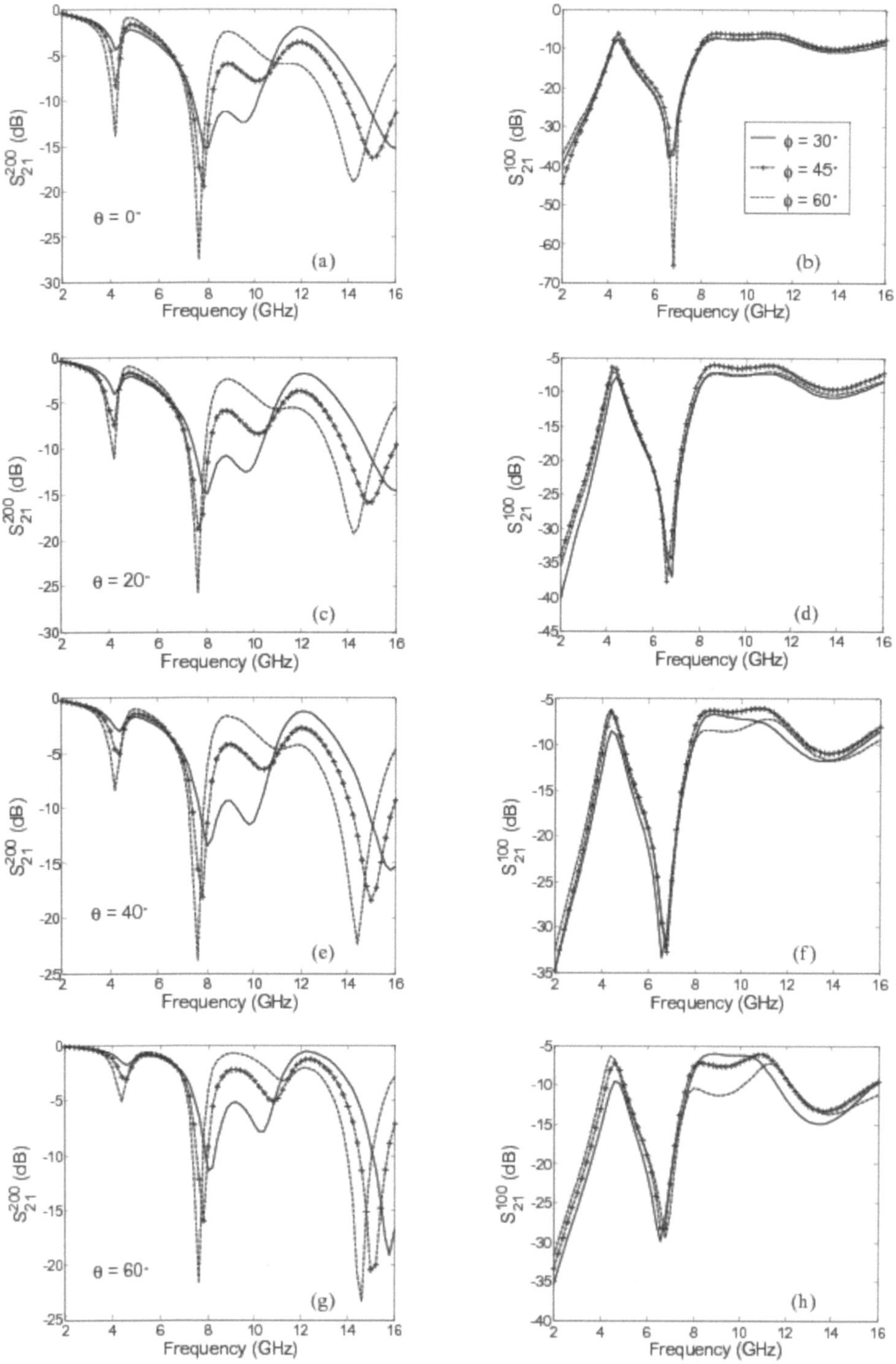

Fig. 25. The resonance responses of the FSS for the parallel-polarized electromagnetic wave in arbitrary plane (a) (c) (e) (g) TM-Floquet modes; (b) (d) (f) (h) TE-Floquet modes

5. Conclusion

The nature of electromagnetic waves in metamaterials and metamaterial-inspired configurations has been studied in this chapter [38-42]. First, causality in the resonance behavior of metamaterials has been developed. Both a rigorous full wave analysis and Floquet mode theory have been employed to examine the various electromagnetic fields effects on metamaterials. Second, propagation features of metamaterial waveguiding structures have been intensively explored. It is shown that different dispersion properties, such as non-cutoff frequency mode propagation and enhanced bandwidth of single mode operation, have been impart to metamaterial based parallel plate waveguide and rectangular waveguide. Meanwhile, slow wave propagation, enhanced energy flow, abnormal higher-order modes, as well as leakage suppression of LSM and LSE modes are illustrated in the NRD waveguide and H waveguide with bianisotropic SRR metamsterials. Finally, multiband transmission property of FSSs with SRR unit cells is illustrated. As is shown, such multi-band spectrum is closely related to the SRR parameters as well as incident wave.

It is worth noting that the electromagnetic responses of SRR metamaterials are in fact much more complicated for practical use. Therefore, restrictive conditions, such as the absence of losses, certain direction of the incident wave, are assumed throughout the full wave analysis. However, we here consider the macroscopical effects of material parameters of the media, providing alternative means of characterizing the SRR metamaterials and its inspired waveguding structures. In addition, the SRR element spacing of the metamaterials are usually small compared to the resonance wavelength, and only the modes of $p=0$ and $q=0$ can propagate in the distant transmitted fields. Therefore, resonance behavior of SRR metamaterials and their transmission properties have been examined only through the dominant TE- and TM-Floquet modes in the Floquet mode analysis.

6. Acknowledgement

This work was supported in part by National Natural Science Foundation of China (Grant No. 60771040) and in part by State Key Laboratory Foundation (Grant No. 9140C0704060804). The work of Dr. Rui Yang was supported by Newton International Fellowship.

7. Reference

[1] V. G. Veselago, *Sov. Phys. Usp.*, 10 (1968) 509.

[2] J. B. Pendry, A. J. Holden, W. J. Stewart, and I. Youngs, *Phys. Rev. Lett.*, 76 (1996) 4773.

[3] R. A. Shelby, Smith D. R., and Schultz S., *Science*, 292 (2001) 77.

[4] R.Marqués, F. Medina, and R.Rafii-El-Idrissi, *Phys. Rev. B*, 65 (2002) 144440.

[5] D. R. Smith, J. Gollub, J. J. Mock, W. J. Padilla, and D. Schurig, *J. Appl. Phys.*, 100 (2006) 024507.

[6] D. R. Smith, D. Schurig, and J. J. Mock, *Phys. Rev. E*, 74 (2006) 036604.

[7] V. V. Varadan, and A. R. Tellakula, *J. Appl. Phys.*, 100 (2006) 034910.

[8] K. Aydin, K. Guven, N. Katsarakis, C. M. Soukoulis, and E. Ozbay, *Opt. Exp.*, 12 (2004) 5896.
[9] P. Gay-Balmaz, and O. J. F. Martin, *J. Appl. Phys.*, 92 (2002) 2929.
[10] N. Katsarakis, T. Koschny, M. Kafesaki, E. N. Economou, and C. M. Soukoulis, *Appl. Phys. Lett.*, 84 (2004) 2943.
[11] J. B. Pendry, *Phys Rev Lett.*, 85 (2000) 3966.
[12] F. Martín, F. Falcone, J. Bonache, R. Marqués, and M. Scrolla, *Appl. Phys. Lett.*, 83 (2003) 4652.
[13] F. Aznar, J. Bonache, and F. Martín, *Appl. Phys. Lett.*, 92 (2008) 043512.
[14] R. W. Ziolkowski, *IEEE Trans. Antennas Propagat.*, 45 (2003) 656.
[15] K. Aydin, I. Bulu, K. Guven, M. Kafesaki, C. M. Soukoulis, and E. Ozbay, *New J. Phys.*, 7 (2005)168.
[16] T. F. Gundogdu, I. Tsiapa, A. Kostopoulos, G. Konstantinidis, N. Katsarakis, R. S. Penciu, M. Kafesaki, E. N. Economou, Th. Koschny, and C. M. Soukoulis, *Appl. Phys. Lett.*, 89 (2006) 084103.
[17] C. R. Simovski, and S. A. Tretyakov, *Phys. Rev. B*, 75 (2007) 195111.
[18] C. R. Simovski, *Metamaterials*, 1 (2007) 62.
[19] N. Amitay, and V. Galindo, *Bell Syst. Tech. J.*, 47 (1968) 1903.
[20] C. C. Chen, *IEEE Trans. Microwave Theory Tech.*, 18 (1970) 627.
[21] C Krowne., *IEEE Trans. Antennas Propag.*, 32 (1984) 1224.
[22] A. Alú and N. Engheta, *IEEE Trans. Microw. Theory Tech.*, 52, (2004) 199.
[23] B.-I. Wu, T. M. Grzegorczyk, Y. Zhang, and J. A. Kong, *J. Appl. Phys.*, 93, (2003) 9386.
[24] Y. S. Xu, *Microwave Opt. Technol. Lett.*, 41, (2004) 426.
[25] I. V. Shadrivov, A. A. Sukhorukov, and Y. S. Kivshar, *Phys. Rev. E*, 67, (2003) 057602.
[26] S. Hrabar, J. Bartolic, Z. Sipus, *IEEE Trans. Antennas Propag.*, 53, (2005) 110.
[27] A. L. Topa, C. R. Paiva, and A. M. Barbosa, *Microwave Opt. Technol. Lett.*, 47, (2005) 185.
[28] P. Yang, D. Lee, and K. Wu, *Microwave Opt. Technol. Lett.*, 45, (2005) 207.
[29] P. Baccarelli, P. Burghignoli, F. Frezza, A. Galli, P. Lampariello, and S. Paulotto, *Microwave Opt. Technol. Lett.*, 48, (2006) 2557.
[30] S. A. Tretyakov, *J. Electromagn. Wave Applic.*, 12, (1998) 821.
[31] Q. H. Sun, Q. Cheng, H. S. Xu, B. Zhou, and T. J. Cui, *META08'Nanjing*, (2008) 267.
[32] J. D. Baena, L. Jelinek, R. Marqués, J. J. Mock, J. Gollub, and D. R. Smith, *Appl. Phys. Lett.*, 91, (2007) 191105.
[33] R. Marqués, J. D. Baena, M. Beruete, F. Falcone, T. Lopetegi, M. Sorolla, F. Martín, and J. Garcia, *J. Opt. A: Pure Appl. Opt.*, 7, (2005) 38.
[34] J. Romen, and Y. Rahmat-Samii, Electron. Lett., 35, (1999) 702.
[35] R. Mittra, C. H. Chan, and T. Cwik, Proc. IEEE, 76, (1998) 1593.
[36] B. A. Munk, New York: Wiley, 2000.
[37] M. Kafesaki, Th. Kosohny, R. S. Penciu, T. F. Gundogdu, E. N. Economou, and C. M. Soukoulis, *J. Opt. A: Pure Appl. Opt.*, 7, (2005) 12.
[38] R. Yang, Y. J. Xie, X. F. Li, Y. Y. Wang, J. Jiang, Europhys. Lett., 84, (2008), 34001.

[39] R. Yang, Y. J. Xie, X. D. Yang, R. Wang, B. T. Chen, Opt. Express, 17, (2009), 6101.
[40] R. Yang, Y. J. Xie, J. Jiang, Y. F. Gong, IEEE Antennas Wireless Propag. Lett., 7, (2008), 481.
[41] R. Yang, Y. J. Xie, X. F. Li, J. Jiang, Infrared Phys. Techn., 51, (2008), 555.
[42] R. Yang, Y. J. Xie, META08'Nanjing, Invited Paper.

10

Electromagnetic Propagation Characteristics in One Dimensional Photonic Crystal

Arafa H Aly
Physics department, Faculty of Sciences, Beni-suef University, Egypt

1. Introduction

Photonic crystals (PCs) are structures with periodically modulated dielectric constants whose distribution follows a periodicity of the order of a fraction of the optical wavelength. Since the first pioneering work in this field, many new interesting ideas have been developed dealing with one-dimensional (1D), two-dimensional (2D), and three-dimensional (3D) PCs. Researchers have proposed many new and unique applications of photonic devices which may revolutionize the field of photonics in much the same way as semiconductors revolutionized electronics. They can generate spectral regions named photonic band gaps (PBGs) where light cannot propagate in a manner analogous to the formation of electronic band gaps in semiconductors [1,2]. There are several studies of metallic [3-7] and superconducting photonic crystals [7,8] which are mostly concentrated at microwave, millimeterwave, and far-infrared frequencies. In those frequencies, metals act like nearly perfect reflectors with no significant absorption problems.

Yablonovitch [1] main motivation was to engineer the photonic density of states in order to control the spontaneous emission of materials embedded with photonic crystal while John's idea was to use photonic crystals to affect the localization and control of light. However due to the difficulty of actually fabricating the structures at optical scales early studies were either theoretical or in the microwave regime where photonic crystals can be built on the far more reading accessible centimeter scale. This fact is due to the property of the electromagnetic fields known as scale invariance in essence, the electromagnetic fields as the solutions to Maxwell's equations has no natural length scale and so solutions for centimeter scale structure at microwave frequencies as the same for nanometer scale structures at optical frequencies.

The optical analogue of light is the photonic crystals in which atoms or molecules are replaced by macroscopic media with different dielectric constants and the periodic potential is replaced by a periodic dielectric function. if the dielectric constants of the materials is sufficiently different and also if the absorption of light by the material is minimal then the refractions and reflections of light from all various interfaces can produce many of the same phenomena for photons like that the atomic potential produced for electrons[9].

The previous details can guide us to the meaning of photonic crystals that can control the propagation of light since it can simply defined as a dielectric media with a periodic modulation of refractive index in which the dielectric constant varies periodically in a

specific directions. Also it can be constructed at least from two component materials with different refractive index due to the dielectric contrast between the component materials of the crystal .it's characterized by the existence of photonic band gap (PBG) in which the electromagnetic radiation is forbidden from the propagation through it.
One of the most important properties of photonic crystal is controlling propagation of light in small submicron scale volumes. We will try to mention another example that may develop our sense about photonic crystals. Metallic wave guides and cavities are related to photonic crystals. They are widely used to control microwave propagation. The metallic cavity prohibits the propagation of electromagnetic waves with frequency below the threshold frequency while the metallic waveguides allow propagation only along its axis [9]. Today a huge number of researchers go ahead to enhance and widen the circle of applications through photonic crystals by using a composite structure which is composed of (metal / dielectric) structure instead of (dielectric / dielectric) structure. Since the metallic - dielectric photonic crystals (MDPCs) can be useful for a large number of applications and also will open the door to new discoveries that depend mainly on the optical properties of the metal. MDPCs have many advantages over dielectric based photonic crystals since:

1. Metal components can be used for wiring of future optoelectronic merger circuits this is due to that, nanowire materials are characterized by the existence of surface plasmon polaritons (spp) and also possess negative refractive index then it can act as left-handed materials[6,7].
2. Metals have stronger non linearity than other optical materials due to that most of metals possess negative refractive indices.
3. The photonic band gap (PBG) of the metallic dielectric photonic crystals (MDPCs) can be more robust against structural disorders.
4. High temperature operating may be possible.
5. Excitation of surface plasmon polaritons (spp) is also possible.

There are certain advantages to introducing 1D MDPCs. These include reduced size and weight, easier fabrication methods, and lower costs. Here, we use the transfer matrix method to study the effect of absorption of metal at the near-infrared and visible frequencies. In particular, we study a simple-one dimensional metallic-dielectric structure. Silver, Gold, Copper, Lithium and Aluminum have been used in order to study the effect of different metals on optical properties.

2. Numerical methods

In order to study the electromagnetic waves propagating through photonic crystal structures and determine the optical properties of these periodic structures. There are six main methods are used to study the propagation of electromagnetic waves through photonic crystals numerically and they called them[10]; the plane wave method(PW), finite time domain method (FDTD), finite element method.(FEM), transfer matrix method (TMM), scattering matrix method (SMM), and study of diffraction gratings (DG). The advantages and defects of these methods were obtained in ref.10. We will now explain in brief a mathematical treatment with a simple one dimensional photonic crystal structure (1DPC) (see fig.1) which is composed of two materials with thicknesses (d_2 and d_3) and refractive indices (n_2 and n_3) respectively. The analysis of the incident electromagnetic radiation on this structure will be performed using the transfer matrix method (TMM).

Fig. 1. Stack of 1 D MDPC ,where n_2 is the refractive index of the metal, d_2 is the thickness of the metal , n_3 is the refractive index of the dielectric material and d_3 is the thickness of the dielectric material.

We also use n_1 to be the refractive index of vacuum while n_4 is that of the substrate material as in fig.1. The electric field that satisfies Maxwell's equations is [11].

$$E(x)=R\exp(-ik_x.x)+L\exp(ik_x.x) = A(x)+\beta(x) \quad (1)$$

Where β the Z-component of the wave is vector and ω is the angular frequency. Also the electric field consists of right and left traveling waves since:-
With $\pm k_x$ are the components of the wave vector and (R, L) being constants in each homogeneous layer.
We will go to mention the mathematical form of the dynamical matrices and for the propagation matrix to obtain an expressions for the reflection and transmission, the dynamical matrices take the form [11]:

$$D_\alpha = \begin{pmatrix} 1 & 1 \\ n_\alpha Cos\theta_\alpha & -n_\alpha Cos\theta_\alpha \end{pmatrix} \text{ for S - wave} \quad (2)$$

$$D_\alpha = \begin{pmatrix} Cos\theta_\alpha & Cos\theta_\alpha \\ n_\alpha & -n_\alpha \end{pmatrix} \text{ for P - wave} \quad (3)$$

with

$$\beta = n_\alpha \frac{\omega}{c} Sin\theta_\alpha \text{, and } k_{\alpha x} = n_\alpha \frac{\omega}{c} Cos\theta_\alpha$$

while the propagation matrix take the form:

$$P_\eta = \begin{pmatrix} \exp(i\phi_\alpha) & 0 \\ 0 & \exp(-i\phi_\alpha) \end{pmatrix} \quad (4)$$

Since the number of the propagation matrix depend on the number of materials which build our structure [11]. Finally the transfer matrix method can take the form:

$$M = \begin{pmatrix} M_{11} & M_{12} \\ M_{21} & M_{22} \end{pmatrix} \quad (5)$$

$$P_2 = \begin{pmatrix} Cos\phi_1 + i\,Sin\phi_1 & 0 \\ 0 & Cos\phi_1 - i\,Sin\phi_1 \end{pmatrix}$$

$$P_3 = \begin{pmatrix} Cos\phi_2 + i\,Sin\phi_2 & 0 \\ 0 & Cos\phi_2 - i\,Sin\phi_2 \end{pmatrix}$$

Since;

$$\phi_1 = \frac{2\pi d_2}{\lambda} n_2\,Cos\theta_2, \text{ and } \quad \phi_2 = \frac{2\pi d_3}{\lambda} n_3\,Cos\theta_3 \;.$$

The components of the transfer matrix method can be written in a detailed form for an S - wave as:

$$M_{11} = \frac{1}{2}\left[\left(1 + \frac{n_4\,Cos\,\theta_4}{n_1\,Cos\,\theta_1}\right)Cos\,\varphi_2 + \left(\frac{n_3\,Cos\,\theta_3}{n_1\,Cos\,\theta_1} + \frac{n_4\,Cos\,\theta_4}{n_3\,Cos\,\theta_3}\right)(i\,Sin\,\phi_2)\right]Cos\,\phi_1 + \frac{1}{2}\left[\left(\frac{n_2\,Cos\,\theta_2}{n_1\,Cos\,\theta_1} + \frac{n_4\,Cos\,\theta_4}{n_2\,Cos\,\theta_2}\right)Cos\,\phi_2 + \left(\frac{n_3\,Cos\,\theta_3}{n_2\,Cos\,\theta_2} + \frac{n_2 n_4 Cos\,\theta_2\,Cos\,\theta_4}{n_1 n_3\,Cos\,\theta_1\,Cos\,\theta_3}\right)(i\,Sin\,\phi_2)\right](i\,Sin\,\phi_1) \tag{6}$$

$$M_{12} = \frac{1}{2}\left[\left(1 - \frac{n_4\,Cos\,\theta_4}{n_1\,Cos\,\theta_1}\right)Cos\,\phi_2 + \left(\frac{n_3\,Cos\,\theta_3}{n_1\,Cos\,\theta_1} - \frac{n_4\,Cos\,\theta_4}{n_3\,Cos\,\theta_3}\right)(i\,Sin\,\phi_2)\right]Cos\,\phi_1 + \frac{1}{2}\left[\left(\frac{n_2\,Cos\,\theta_2}{n_1\,Cos\,\theta_1} - \frac{n_4\,Cos\,\theta_4}{n_2\,Cos\,\theta_2}\right)Cos\,\phi_2 + \left(\frac{n_3\,Cos\,\theta_3}{n_2\,Cos\,\theta_2} - \frac{n_2 n_4 Cos\,\theta_2\,Cos\,\theta_4}{n_1 n_3\,Cos\,\theta_1\,Cos\,\theta_3}\right)(i\,Sin\,\phi_2)\right](i\,Sin\,\phi_1) \tag{7}$$

$$M_{21} = \frac{1}{2}\left[\left(1 - \frac{n_4\,Cos\,\theta_4}{n_1\,Cos\,\theta_1}\right)Cos\,\phi_2 - \left(\frac{n_3\,Cos\,\theta_3}{n_1\,Cos\,\theta_1} - \frac{n_4\,Cos\,\theta_4}{n_3\,Cos\,\theta_3}\right)(i\,Sin\,\phi_2)\right]Cos\,\phi_1 - \frac{1}{2}\left[\left(\frac{n_2\,Cos\,\theta_2}{n_1\,Cos\,\theta_1} - \frac{n_4\,Cos\,\theta_4}{n_2\,Cos\,\theta_2}\right)Cos\,\phi_2 - \left(\frac{n_3\,Cos\,\theta_3}{n_2\,Cos\,\theta_2} - \frac{n_2 n_4 Cos\,\theta_2\,Cos\,\theta_4}{n_1 n_3\,Cos\,\theta_1\,Cos\,\theta_3}\right)(i\,Sin\,\phi_2)\right](i\,Sin\,\phi_1) \tag{8}$$

$$M_{22} = \frac{1}{2}\left[\left(1 + \frac{n_4\,Cos\,\theta_4}{n_1\,Cos\,\theta_1}\right)Cos\,\phi_2 - \left(\frac{n_3\,Cos\,\theta_3}{n_1\,Cos\,\theta_1} + \frac{n_4\,Cos\,\theta_4}{n_3\,Cos\,\theta_3}\right)(i\,Sin\,\phi_2)\right]Cos\,\phi_1 - \frac{1}{2}\left[\left(\frac{n_2\,Cos\,\theta_2}{n_1\,Cos\,\theta_1} + \frac{n_4\,Cos\,\theta_4}{n_2\,Cos\,\theta_2}\right)Cos\,\phi_2 - \left(\frac{n_3\,Cos\,\theta_3}{n_2\,Cos\,\theta_2} + \frac{n_2 n_4 Cos\,\theta_2\,Cos\,\theta_4}{n_1 n_3\,Cos\,\theta_1\,Cos\,\theta_3}\right)(i\,Sin\,\phi_2)\right](i\,Sin\,\phi_1) \tag{9}$$

Where the reflectance and transmittance can be written as:

$$R = \left|r^2\right| = \left|\frac{M_{21}}{M_{11}}\right|^2 \tag{10}$$

$$T = \frac{n_4 \, Cos \, \theta_4}{n_1 \, Cos \, \theta_1}\left|t\right|^2 = \frac{n_4 \, Cos \, \theta_4}{n_1 \, Cos \, \theta_1}\left|\frac{1}{M_{11}}\right|^2 \tag{11}$$

Where r and t is the reflection and transmission and we can also obtain by the same method the components of the transfer matrix method (TMM) for P - wave. Finally the dispersion relation for the metallic -dielectric photonic band gap (MDPBG) which describes the relation between the forbidden band (K) and the frequency (ω) can be written as [11]:

$$Cos \, K\Lambda = Cos\left(k_2 d_2\right) Cos\left(k_3 d_3\right) - \frac{1}{2}\left(\frac{n_3}{n_2} + \frac{n_2}{n_3}\right) Sin\left(k_2 d_2\right) Sin\left(k_3 d_3\right) \tag{12}$$

with $k_1 = \frac{\omega}{c} n_2 \; , \quad k_2 = \frac{\omega}{c} n_3 \; , \quad \Lambda = d_2 + d_3$

Where is Λ the lattice constant and K is the forbidden band.

3. Results and discussions

The periodicity of the permittivity plays the same role for the photons that propagate inside the structure than the atomic potential for the electrons. Leading further this analogy, the thicknesses and the index contrast of the photonic crystal determinate many of its optical properties as it does for conduction properties of semiconductors. Playing on these two parameters, we can obtain frequency ranges for which light propagation is forbidden in the material and others ranges for which light can propagate. These frequency ranges are also scale dependent. Reducing the size of the elementary cell of the periodic lattice shifts the whole frequency range to higher values. The consequence of this property is the possibility

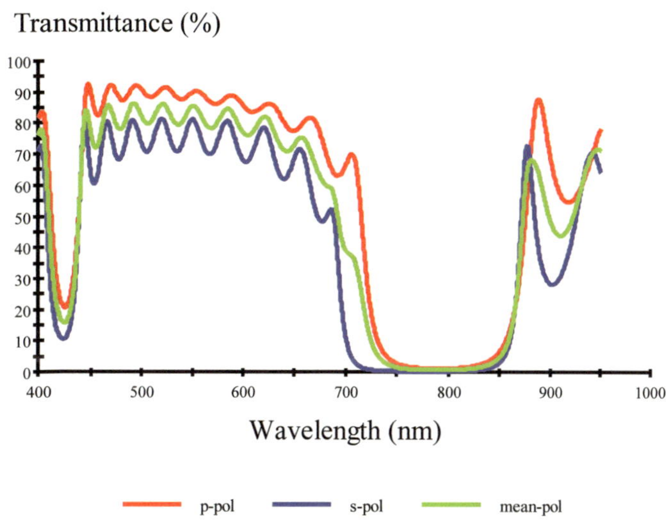

Fig. 2. The transmittance of p-pol, s-pol and the mean pol to Ag-SiO2 structure.

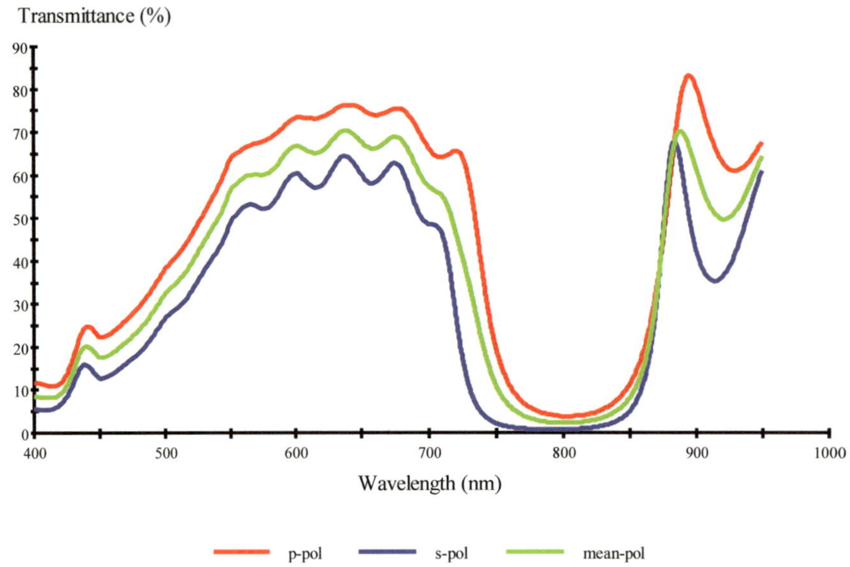

Fig. 3. The transmittance of p-pol, s-pol and the mean pol to Au-SiO2 structure.

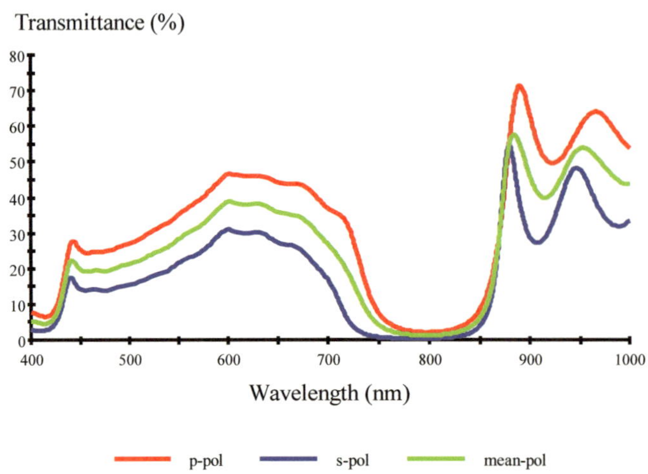

Fig. 4. The transmittance of p-pol, s-pol and the mean pol to Cu-SiO2 structure.

to transpose a photonic crystal design from the microwave domain to infrared or visible range. In our results we have studied one dimensional metallic (different metals) dielectric (SiO2) photonic crystals (MDPC's). In all our figures we have used the thickness of SiO2 layer is 340nm and the thickness layer of metal is 4nm for all results. Also we have used 10 periods and the incidence of angle is 50° for the all our results (Fig's 2-6). In the case of Ag-SiO2 (fig.2) we can see the magnitude of transmittance near the unity with visible light and there is a small photonic band gap (PBG) appeared within the short IR (750-850 nm). The magnitude of transmittance in the case of Au-SiO2 (fig.3) is 10% at 400 nm and increased to ~80 % within 550-700nm with the same PBG range 750-850 nm with s-polarized.

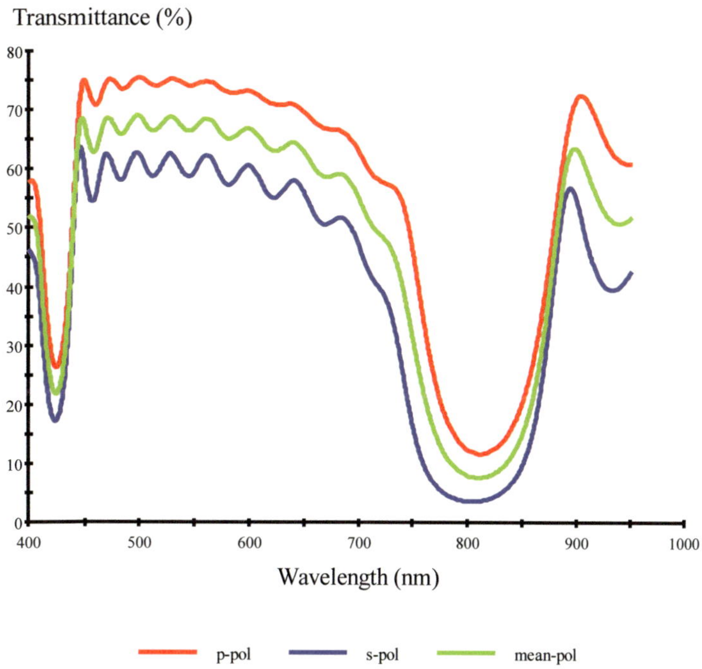

Fig. 5. The transmittance of p-pol, s-pol and the mean pol to Li-SiO2 structure.

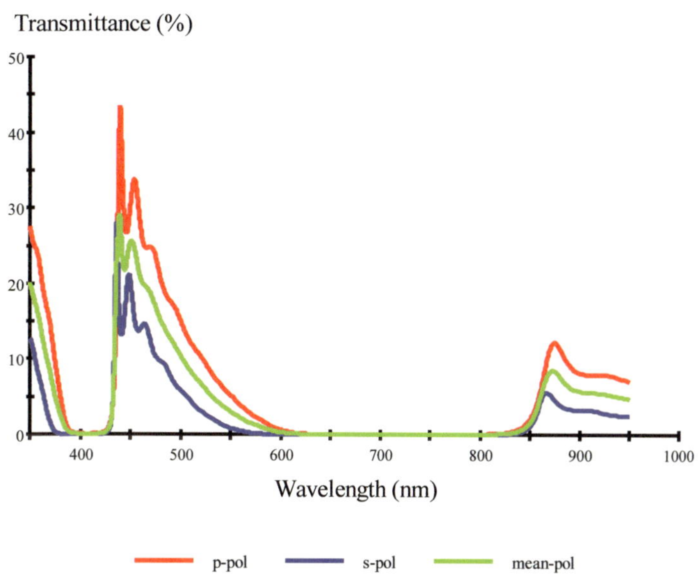

Fig. 6. The transmittance of p-pol, s-pol and the mean pol to Al-SiO2 structure.

In figure (4) the magnitude of transmittance in the case of Cu-SiO2 is go downing under 40% within the range of visible light (400-700) and we got the same PBG between 750-850 nm. But in fig.(5) with Li-SiO2, the magnitude of transmittance going up to around 70% within the visible light and the PBG disappeared.
In the last figure (6) we have examined Al-SiO2 with the same conditions and we found aluminum that gives the lowest possible transmittance within the visible and near IR frequencies.

4. Conclusion

We studied metallic photonic crystals at visible and near-infrared wavelengths with the transfer matrix method. We focused on the transmittance of these structures, thus we studied the simple 1D structure. We expect that our conclusions regarding the transmittance will hold to any other metallic structures. By comparing the results for different metals, we found that copper gives the lowest possible transmittance within the visible frequencies. Silver gives slightly higher transmittance within the visible with small PBG near IR. Gold gives low possible transmittance within short visible and going to near 70% between 550-700nm with small PBG within 750-850nm. Lithium gives higher transmittance. Aluminum is very lossy and is not recommended for optical photonic crystals. Silver, gold and Lithium are acceptable although slightly lower in performance. We also found our results are very sensitive to the thickness of layers as well as PBG is sensitive to the kind of polarizations-pol.

5. Acknowledgements

The author thankful Prof Macleod for his offers me the Essential Macleod program to achieve this work and my thanks extended to Mr H.Abdel Rahman for his efforts in this work.

6. References

[1] E. Yablonovitch, Phys. Rev. Lett. 58, 2059, (1987).
[2] S. John, Phys. Rev. Lett. 58,2487, (1987).
[3] A. A. Maradudin and A. R. McGurn, Phys. Rev. B 48, 17 576, (1993).
[4] M. M. Sigalas, C. T. Chan, K. M. Ho, and C. M. Soukoulis, Phys. Rev. B 52, 11 744, (1995).
[5] J. S. McCalmont, M. M. Sigalas, G. Tuttle, K. M. Ho, and C. M. Soukoulis, Appl. Phys. Lett. 68, 2759, (1995).
[6] Arafa H Aly and Sang-Wan Ryu, J. of Computational and Theoretical Nannoscience, Vol 5, 1-15, (2008).
[7] Arafa H Aly, Materials Chemistry and Physics, 115, 391, (2009).
[8] Arafa H. Aly, Heng-Tung Hsu, Tzong-Jer Yang, Chien-Jang Wu, and C-K, J. of Applied Physics, 105, 083917, (2009).
[9] John D. Joannopoulos, Steven G. Johnson, Joshua N. Winn, and, Robert D. Meade, "Photonic crystals moleding the flow of light. 2nd Edition. Princeton University Press, (2008).
[10] G. Guida, A. de Lustrac, and A. Priou, PIER 41, 1–20, (2003).
[11] P. Yeh., "Optical Waves in Layered Media", J. Wiley a& Sons, Inc., Hoboken, New Jersey, (2005).

11

Wave Propagation in Elastic Media with Micro/Nano-Structures

G. L. Huang[1,3], C.T. Sun[2] and F. Song[3]
[1]*Department of Systems Engineering, University of Arkansas at Little Rock*
[2]*School of Aeronautics and Astronautics, Purdue University*
[3]*Department of Applied Science, University of Arkansas at Little Rock*
USA

1. Introduction

Micro- and nano-scale materials and structures such as plate- or beam-like structures with submicron or nano thicknesses have attracted considerable interest from the scientific community due to the increasingly strong demands of miniaturization in the fields of microelectronics and nanotechnology. More and more nano-structures, e.g. ultra-thin films, nanowires and nanotubes, have been fabricated and served as the basic building blocks for microelectromechanical systems (MEMS) and nanoelectromechanical systems (NEMS) (Jin et al., 1998; Craighead, 2000; Husain et al., 2003; Feng et al., 2007). For long-term stability and reliability of various devices at nanoscale, researchers should possess a deep understanding and knowledge of mechanical properties of nano materials and structures, especially for dynamic properties.

Among many techniques, high-frequency acoustic wave technique has been regarded as one of very efficient nondestructive methods to characterize elastic media with micro- or nano-structures. (Hernaandez et al., 2002) used high-frequency laser-excited guided acoustic waves to estimate the in-plane mechanical properties of silicon nitride membranes. Mechanical properties and residual stresses in the membranes were evaluated from measured acoustic dispersion curves. The mean values of the Young's modulus and density of three nanocrystalline diamond films and a free standing diamond plate were determined by analyzing the dispersion of laser-generated surface waves (Philip et al., 2003). Recently, growing interest of using terahertz (THz) waves in nanoscale materials and nano-photonic or nano-phononic devices has opened a new topic on the wave characteristics of nanomaterials (Schneider et al., 2000, Vollmannn et al., 2004; Ramprasad & Shi, 2005; Sampathkumar et al., 2006). As dimensions of the material become smaller, however, their resistance to deformation is increasingly determined by internal or external discontinuities (such as surfaces, grain boundary, strain gradient, and dislocation). Although many sophisticated approaches for predicting the mechanical properties of nanomaterials have been reported, few addressed the challenges posed by interior nanostructures such as the surfaces, interfaces, structural discontinuities and deformation gradient of the nanomaterials under extreme loading conditions. The use of atomistic simulation may be a potential solution in the long run. However, it is well known that the capability of this approach is much limited by its need of prohibitive computing time and an astronomical amount of data generated in the calculations.

It has been well recognized that classical continuum models become inadequate to describe the response of solids when the characteristic length (or wave length) of deformation becomes comparable to or smaller than the characteristic length of microstructures in the solid. The main reason for this deficiency of the classical continuum model can be attributed to its inability to account for the local motion of the microstructure. One common way to solve the above mentioned problem is to employ additional kinematic variables to describe the nonhomogeneous local deformation in the micro- or nano- structure of the solid. This approach leads to Cosserat continuum models (Cosserat & Cosserat, 1909) or micropolar models (Mindlin, 1964) or the like (Achenbach et al., 1968). However, the large number of material constants were left undetermined, which would require rather prohibitive experiments. Recently, (Chen & Lee, 2003a,b) have attempted to determine the material constants in the micromorphic theory (Eringen, 1999) by relating the micromorphic theory to atomistic models. To overcome those challenges, another approach toward developing a high order continuum theory was taken by considering the exact local configurations of the media. By employing several kinematic variables to describe the local motion in addition to the macro kinematic variables, the microstructure continuum theory was derived (Sun et al., 1968; Huang & Sun, 2006a,b, 2007, 2008; Huang & Song, 2008) for periodically layered and ealstic media with nanostructures. The main advantage of the approach is that the microstctural material constants can be obtained directly from the original material system without ambiguity.

The presence of surface stresses is another challenge of understanding the mechanical and physical properties of nanostructures espectially for such nanostructures that have small size and thus large surface area to volume (SAV) ratio. Much effort has been made to understand surface stress effects on the dynamic behavior of nanowires. Using the newly-developed surface Cauchy-Born model, (Park & Klein 2008) analyzed the surface stress effects on the nanowire resonant frequencies for varying boundary conditions. It was observed that if finite deformation kinematics are considered, the strain independent surface stresses will substantially alter the resonant frequencies of the nanowires. Wu and Dzenis (2006) investigated the longitudinal and flexural wave propagation in nanowires/fibers within the framework of conventional continuum mechanics. In the study, the surface effects were considered by introducing the strain-independent surface tension/stress in the conventional love's rod theory and Timoshenko beam theory. However, in those approaches, heterogeneous nanostructure effects upon the wave propagation in the nanowire were not considered. It should be mentioned that the heterogeneous nanostructure effects should be addressed when the characteristic length of deformation (or wave-length) becomes comparable to or smaller than the characteristic length of the heterogeneous nanostructures. Therefore, to capture heterogeneity of the nanowires for high frequency nanowire-based devices, (Song et al., 2009; Song & Huang, 2009) developed a high-order continuum model to study the surface stress influences on high-frequency longitudinal and flexural wave propagation in nanowires, from which the size dependent wave information were also observed. In the proposed model, additional kinematic variables were introduced to account for the nanostructure heterogeneity and the local motion in the nanowires. Surface effects were first incorporated into the current model by using the incremental deformation approach (Biot, 1965; Sun, 1972). The equations of motion including both strain-independent and –dependant surface stress effects were derived for

longitudinal and flexural wave motion. Some new physical wave phenomena related to the surface stress effects were analyzed and discussed.
The objective of this chapter is to introduce the microstructure continuum theory to study the wave propagation in elastic media with different micro- or nano- structures. First, wave propagation in the one-dimensional nanophononic crystal heterostructure will be investigated when the frequency is in range of THz. It is found that the current theory can give very good prediction of wave dispersion in nonphononic crystal structures. Then, the microstructure continuum theory will be used to study the high-frequency wave propagation in the two-dimensional ultra-thin films. The nanostructural effects upon the wave propagation in ultra-thin films are investigated. Both the in-plane and out-of-plane waves are considered. Finally, surface stress effects on dynamic begavior of nanowires are studied by using the new-developed high-order continuum theory.

2. Wave propagation in nanophononic crystal heterostrucutre

In this section, we introduce the microstrucuture continuum model to study wave propagation in a one-dimensional nanophononic crystal heterostructure. For simplicity, the heterostrucutre is represented by a one-dimensional lattice model system. We employ the microstructure continuum model to describe the mechanical behavior of the discrete lattice system. The accuracy of the microstructure continuum model is examined. Rather than the transverse wave as considered by (Ramprasad & Shi, 2005), the longitudinal wave propagation is investigated.

2.1 Microstructure continuum model for one-dimensional nanostructures

Consider a discrete heterostructure system consisting of two alternating layers of two different materials as shown in Fig. 1 (a). It is assumed that the material (material 1) in the first layer is comprised of a number of unit cells of atoms each consisting of two masses m_1 and m_2 (Fig. 1(a) only shows one unit cell in the each layer). Similarly, the unit cell in the other layer consists of masses m_3 and m_4. The pairwise interactions within each cell are represented by elastic springs with spring constants k_1, k_2 , k_3 and k_4, respectively, as shown in Fig. 1(a). The lattice constant between adjacent masses is denoted by a. It should be noted that the pairwise interaction parameters k_2 and k_4 are different from those in the corresponding bulk materials, since the interaction effects between different layers are assumed to be included in these parameters. However, it should be noted that the thickness of the interfacial layer between the two materials is neglected in this study.

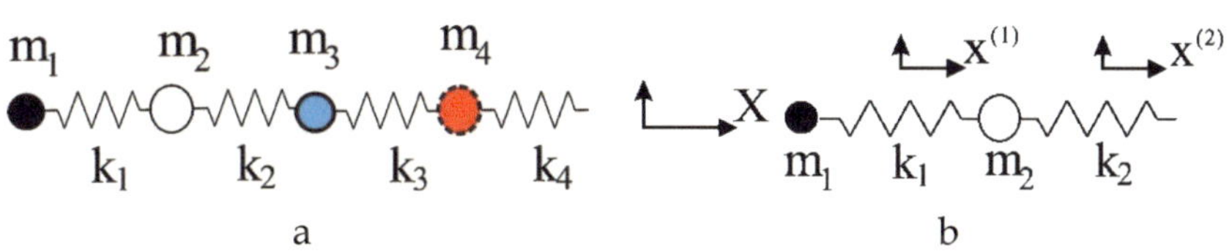

Fig. 1. (a) The lattice system for a nanophononic crystal with two-layered strucuture, and each layer has one unit cell consisting of two atoms. (b) The unit cell of material in layer 1

A microstructure continuum model is now developed to study longitudinal harmonic wave propagation in the nano heterostructure. A unit cell in layer 1 is shown in Fig. 1(b), where $x^{(1)}$ and $x^{(2)}$ indicate the local coordinate axes and X the macro coordinate axis. It is noted that the origin of the local coordinate system is located at the mid-point between the two masses. The local displacement fields in the two sub-regions in the unit cell are approximated by the global kinematic variables as

$$u^{(1)}(x^{(1)},X)=U\,(X)+\phi^{(1)}(X)x^{(1)},\quad u^{(2)}(x^{(2)},X)=U\,(X+a)+\phi^{(2)}(X)x^{(2)} \tag{1}$$

where $u^{(1)}(x^{(1)},X)$ and $u^{(2)}(x^{(2)},X)$ are the local displacements in the two sub-regions, and the global displacement U and the kinematic variables $\phi^{(i)}(i=1,2)$ are functions of the global coordinate variable X. By using the continuity condition, $u^{(1)}(a/2,X)=u^{(2)}(-a/2,X)$, $\phi^{(2)}$ can be eliminated. From these approximate displacements given by Eq. (1), we can obtain the strain energy and kinetic energy densities, respectively, for the equivalent continuum:

$$W=\frac{1}{2a}\left[\frac{k_1}{2}(a\Phi)^2+\frac{k_2}{2}(2a\frac{\partial U}{\partial X}-a\Phi)^2\right] \tag{2}$$

and

$$T=\frac{1}{2a}\left[\frac{m_1}{2}(\dot{U}-\frac{a}{2}\dot{\Phi})^2+\frac{m_2}{2}(\dot{U}+\frac{a}{2}\dot{\Phi})^2\right] \tag{3}$$

where $\Phi\equiv\phi^{(1)}$. Subsequently, by using the Hamilton's principle (Achenbach, 1973), the equations of motion for the microstructure continuum medium are obtained as

$$(m_1+m_2)\ddot{U}-\frac{a}{2}(m_1-m_2)\ddot{\Phi}=k_2\left(4a^2\frac{\partial^2 U}{\partial X^2}-2a^2\frac{\partial\Phi}{\partial X}\right) \tag{4}$$

$$\frac{a^2}{4}(m_1+m_2)\ddot{\Phi}-\frac{a}{2}(m_1-m_2)\ddot{U}=-(k_1+k_2)a^2\Phi+2k_2a^2\frac{\partial U}{\partial X} \tag{5}$$

with the stress boundary condition along X direction,

$$t=2k_2a\frac{\partial U}{\partial X}-k_2a\Phi \tag{6}$$

where t is traction at the boundary. Eqs. (2)-(6) give a nonconventional continuum model with microstructure which is used to provide a better representation of the discrete lattice system.

The microstructure continuum model for layer 2 with one unit cell can be obtained by replacing m_1 and m_2 with m_3 and m_4, and k_1 and k_2 with k_3 and k_4, respectively. The equations of motion for the classical continuum model that represents this discrete system can be derived by ignoring microstructure effects. We have

$$k\frac{\partial^2 U}{\partial X^2}=\rho_0\ddot{U} \tag{7}$$

where $k\equiv 2ak_1k_2/(k_1+k_2)$ and $\rho_0\equiv(m_1+m_2)/(2a)$, from which the phase velocity of the representative classical continuum medium can be obtained as $c_0=2a\sqrt{(k_1k_2)/(k_1+k_2)(m_1+m_2)}$.

2.2 Numerical simulation

The accuracy of the microstructure continuum model for material 1 as described by Eqs. (2) and (3) or equivalently by Eqs. (4) and (5) can be evaluated by studying harmonic wave propagation in material 1 of infinite extent. Basically, this material is composed of identical unit cells shown in Fig. 1(b). Fig. 2(a) shows the comparison of the lowest dispersion curves (the acoustic mode) plotted with nondimensional phase velocity $c^*=c_p/c_0$ vs nondimensional wave number KL for m_2/m_1=0.4, k_2/k_1=0.5. The solution (labeled as "micro structure model") for the microstructure continuum model is obtained by assuming the wave form $U = A_1 e^{iK(X-c_p t)}$, $\Phi = A_2 e^{iK(X-c_p t)}$, where K and c_p denote wave number and phase velocity, respectively. The curve labeled as "exact solution" is calculated based on the discrete lattice system (Huang & Sun, 2006b). It is found that the micro structure continuum model is reasonably accurate for predicting wave dispersion at least up to KL=1.5. In contrast, the classical continuum model predicts a nondispersive behavior.

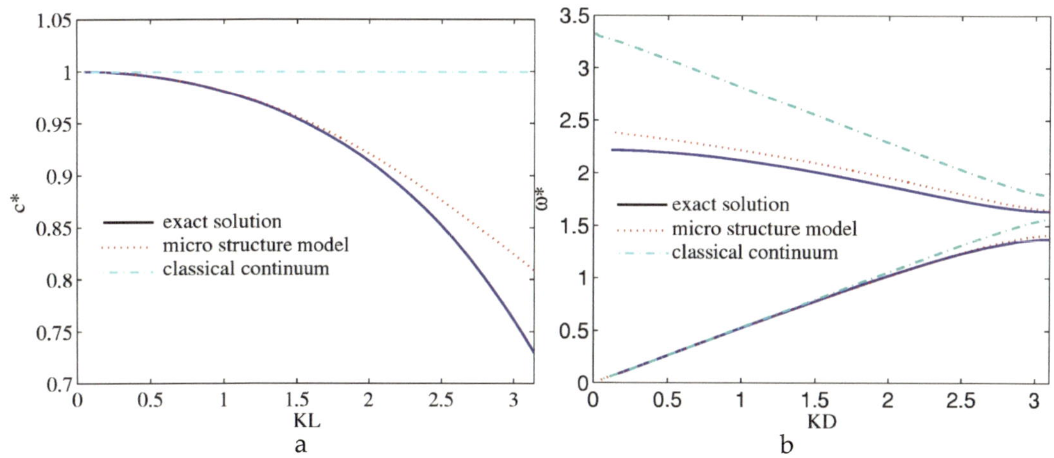

Fig. 2. (a) Comparison of the dispersive curves for bulk material one obtained with the lattice model and the microstructure continuum model (L =2a). (b) Comparison of the dispersive curves for nanophononic crystal obtained with the lattice model, the microstructure continuum model, and classical continuum model (D =4a)

Consider now longitudinal wave propagation in the nano-phononic structure consisting of alternating layers of materials 1 and 2 as shown in Fig. 1(a). Each layer has only one unit cell. Based on the microstructure continuum model, the two discrete layers of materials 1 and 2 can be represented by the respective continua with microstructures. The resulting system is a layered medium of two alternating layers of materials 1 and 2, respectively. Dispersion curves for longitudinal wave propagating though this layered medium can be solved using the Floquet theory (Sun et al., 1968). The numerical result for nondimensional angular frequency $\omega^* = \omega / \omega_0$, with $\omega = Kc_p$, $\omega_0 = c_0 / D$ and $D = 4a$, vs nondimensional wave number KD as shown in Fig. 2(b) is obtained using the following material properties for the discrete nano-phononic system: m_2/m_1=0.4, m_3/m_1=0.6, m_4/m_1=0.3, k_2/k_1=0.5, k_3/k_1=0.5 and k_4/k_2=2.2. The ratio of the densities for the two materials is $\rho_1/\rho_2 = (m_1 + m_2)/(m_3 + m_4) = 0.64$. These material properties are selected to match those considered by (Ramprasad & Shi, 2005). For the purpose of comparison, the dispersion curves for the nano-phononic crystal obtained from the full discrete lattice system (labeled as "exact solution") and from the classical continuum model are also shown in Fig. 2(b). It is evident that the present microstructure continuum model yields an almost perfect prediction for the acoustic mode and a very reasonable prediction for the optic mode. On the other hand, the classical continuum model is much less accurate especially for shorter wavelengths and for the optic mode.

In summary, this investigation has demonstrated that the microstructure continuum model can be used to describe with excellent accuracy wave propagation in nano-phononic crystals. It is also concluded that the conventional continuum model is adequate for certain heterostructures in which the dimensions of the constituent materials are large relative to those of the unit cells.

3. Wave propagation in ultra-thin films

In this section, we employ a procedure to develop a microstructure continuum model with microstructures based on the atomic structure of the ultra-thin film. The atomistic crystal structure of the thin film, for the sake of simplicity, is represented by a lattice model. The dimensions of the crystal structure naturally appear in the constitutive equations and the equations of motion of the representative continuum. The accuracy of the present model is evaluated by comparing dispersions of free harmonic waves predicted by the continuum model and exact analysis based on the lattice model.

3.1 Microstructure continuum model for ultra-thin films

A thin film of cubic structure as shown in Fig. 3 is considered. The thin film has a uniform thickness h and is assumed to be in a state of plane strain parallel to the X_1-X_3 plane. The discrete solid dots denote atoms m_1 and m_2. The spacing between two adjacent atoms is a. Only the interactions between the nearest and next-nearest neighbors are considered. By assuming infinitesimal deformations, the interactions are represented by linearly elastic springs with spring constants α_1 and α_2, respectively (Ghatak et al., 1972). Although the cubic structure is chosen for the sake of mathematical simplicity, the proposed approach can be readily applied for thin films with other crystal structures.

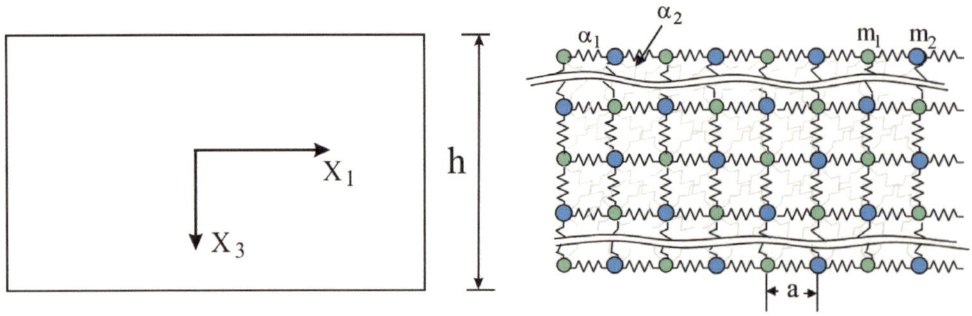

Fig. 3. Thin film structure and its lattice system

A representative unit cell (see Fig. 4) of the cubic lattice model is considered in this study. The representative cell is composed of four sub-cells. In each subcell, a local coordinate system is set up as shown in Fig. 4. Note that in this representative element, atoms 3, 6, 7, 8, and 9 are not included in the representative cell because they are included in the adjacent cells. The four local coordinate systems $(x_1^{(k)}, x_3^{(k)})$ with origins located at the geometrical centers of the four sub-cells (k = 1-4), respectively, are set up so that x_1 and x_3 are parallel to the global (macro) coordinates X_1 and X_3, respectively.

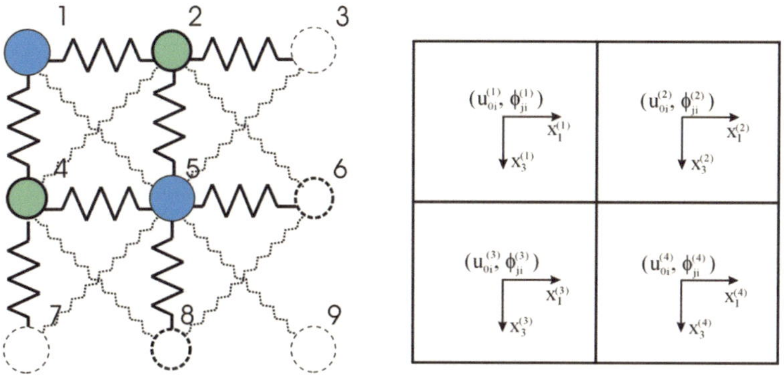

Fig. 4. Representative unit cell and the equivalent continuum

The local displacements $u_i^{(k)}$ in each sub-element are expanded in power series with respect to the respective local coordinates as

$$u_i^{(k)} = u_{0i}^{(k)} + \phi_{1i}^{(k)} x_1^{(k)} + \phi_{3i}^{(k)} x_3^{(k)}, \; k = 1, 2, 3, 4 \tag{8}$$

where the expansion is truncated at the linear terms. Greater accuracy may be achieved by adding higher order terms in Eq. (8) (Huang & Sun, 2008). We assume that local translational displacements $u_{0i}^{(k)}$ of the four sub-cells are values of macro-displacement U_i at the four sub-cells, respectively, i.e.,

$$\begin{aligned}
u_{0i}^{(1)}(X_1, X_3) &= U_i(X_1, X_3), \\
u_{0i}^{(2)} &= U_i(X_1 + a, X_3) \approx U_i(X_1, X_3) + a\frac{\partial U_i}{\partial X_1}, \\
u_{0i}^{(3)} &= U_i(X_1, X_3 + a) \approx U_i(X_1, X_3) + a\frac{\partial U_i}{\partial X_3}, \\
u_{0i}^{(4)} &= U_i(X_1 + a, X_3 + a) \approx U_i(X_1, X_3) + a\frac{\partial U_i}{\partial X_1} + a\frac{\partial U_i}{\partial X_3}
\end{aligned} \tag{9}$$

The kinematic variables $u_{0i}^{(k)}, \phi_{1i}^{(k)}$ and $\phi_{1i}^{(k)}$ are required to satisfy the displacement continuity conditions along the boundaries shared by the adjacent pairs of the sub-cells. These boundary conditions lead to the following relations:

$$[\phi_{ji}^{(2)}] = \begin{bmatrix} 2\frac{\partial U_1}{\partial X_1} - \Phi_{11} & 2\frac{\partial U_3}{\partial X_1} - \Phi_{13} \\ \Phi_{31} & \Phi_{33} \end{bmatrix} \tag{10}$$

$$[\phi_{ji}^{(3)}] = \begin{bmatrix} \Phi_{11} & \Phi_{13} \\ 2\frac{\partial U_1}{\partial X_3} - \Phi_{31} & 2\frac{\partial U_3}{\partial X_3} - \Phi_{33} \end{bmatrix} \tag{11}$$

$$[\phi_{ji}^{(4)}]=\begin{bmatrix} 2\dfrac{\partial U_1}{\partial X_1}-\Phi_{11} & 2\dfrac{\partial U_3}{\partial X_1}-\Phi_{13} \\ 2\dfrac{\partial U_1}{\partial X_3}-\Phi_{31} & 2\dfrac{\partial U_3}{\partial X_3}-\Phi_{33} \end{bmatrix} \tag{12}$$

in which

$$\Phi_{ji}(X_1,X_3)\equiv\phi_{ji}^{(1)}(X_1,X_3) \tag{13}$$

The deformation energy in each sub-cell can be written as

$$W^{(1)}=W_{1-2}+W_{1-4}+W_{1-5}+W_{2-4} \tag{14}$$

$$W^{(2)}=W_{2-3}+W_{2-5}+W_{2-6}+W_{3-5} \tag{15}$$

$$W^{(3)}=W_{4-5}+W_{4-7}+W_{4-8}+W_{5-7} \tag{16}$$

where superscripts 1, 2, 3, and 4 represent the sub-cells, and W_{1-2}, $W_{1-5}\cdots$ denote the total deformation energies due to stretching of the springs between atoms 1 and 2, and 1 and5..., respectively. The total deformation energy in sub-cell 1 is

$$W^{(1)}=\frac{1}{2}\alpha_1 a^2\left[\left(\phi_{11}^{(1)}\right)^2+\left(\phi_{33}^{(1)}\right)^2\right]+\frac{1}{4}\alpha_2\left(\phi_{11}^{(1)}+\phi_{31}^{(1)}+\phi_{13}^{(1)}+\phi_{33}^{(1)}\right)^2+\frac{1}{4}\alpha_2\left(\phi_{11}^{(1)}-\phi_{31}^{(1)}-\phi_{13}^{(1)}+\phi_{33}^{(1)}\right)^2 \tag{18}$$

The deformation energies for other sub-cells have a similar expression as Eq. (18).
Based on the continuity conditions of displacements, Eqs. (10)-(12), the kinematic variables $\phi_{ji}^{(k)}(j,i=2,3,4)$ in sub-cells 2, 3, and 4 can be eliminated and expressed in terms of the kinematic variables of sub-cell 1. Thus, the total deformation energy stored in the representative unit cell (that consists of 4 sub-cells) can be expressed in terms of U_i and Φ_{ji} and their derivatives. By dividing this total energy with the planar area $4a^2$, we obtain the strain energy density W of the representative cell after some manipulations. We have

$$\begin{aligned} W&=\frac{1}{4a^2}\left(W^{(1)}+W^{(2)}+W^{(3)}+W^{(4)}\right)\\ &=\frac{1}{2}\alpha_1(E_{11}^2+E_{33}^2+\gamma_{11}^2+\gamma_{33}^2)+\frac{1}{2}\alpha_2\left[\left(E_{11}+E_{33}\right)^2+\gamma_{11}^2+\gamma_{33}^2+\gamma_{13}^2+\gamma_{31}^2+4E_{13}^2\right] \end{aligned} \tag{19}$$

where

$$E_{ij}=\frac{1}{2}\left(U_{i,j}+U_{j,i}\right),\quad i,j=1,3 \tag{20}$$

is the macro strain and

$$\gamma_{ji}=U_{j,i}-\Phi_{ji},\; i,j=1,3 \tag{21}$$

is the relative strain. These two deformation variables resemble those in Mindlin's microstructure theory (Mindlin, 1964). This strain energy density function forms the base of the continuum model that represents the discrete lattice system.

To establish the constitutive relations we define Cauchy stress and relative stress, respectively

$$\Sigma_{ij} = \frac{\partial W}{\partial E_{ij}} \tag{22}$$

$$\sigma_{ij}^{R} = \frac{\partial W}{\partial \gamma_{ij}} \tag{23}$$

It is noted that if γ_{ij} is absent, then the relative stress σ_{ij}^{R} vanishes, and the strain energy density function reduces to that of the classical continuum. This reduced model will be referred to as the "effective modulus" theory.

The kinetic energy density function for the representative cell of the thin film can be derived from the discrete system based on the local displacements given by Eq. (8). We obtain

$$\begin{aligned} T = {} & \frac{m_1}{4a^2}\left[\dot{U}_1^2 + \dot{U}_3^2 + \frac{a^2}{4}\left(\dot{\Phi}_{11} + \dot{\Phi}_{31}\right)^2 + \frac{a^2}{4}\left(\dot{\Phi}_{13} + \dot{\Phi}_{33}\right)^2\right] + \\ & \frac{m_2}{4a^2}\left[\dot{U}_1^2 + \dot{U}_3^2 + \frac{a^2}{4}\left(\dot{\Phi}_{11} - \dot{\Phi}_{31}\right)^2 + \frac{a^2}{4}\left(\dot{\Phi}_{13} - \dot{\Phi}_{33}\right)^2\right] \end{aligned} \tag{24}$$

where a dot represents the derivative with respect to time.

To complete the continuum model, the Hamilton's principle is conducted to obtain the equations of motion:

$$\begin{aligned} & -\frac{1}{2a^2}(m_1 + m_2)\ddot{U}_1 + \alpha_1\left(2\frac{\partial^2 U_1}{\partial X_1{}^2} - \frac{\partial \Phi_{11}}{\partial X_1}\right) \\ & +\alpha_2\left(2\frac{\partial^2 U_1}{\partial X_1{}^2} + 2\frac{\partial^2 U_1}{\partial X_3{}^2} + 2\frac{\partial^2 U_3}{\partial X_1 \partial X_3} - \frac{\partial \Phi_{11}}{\partial X_1} - \frac{\partial \Phi_{31}}{\partial X_3}\right) = 0 \end{aligned} \tag{25}$$

$$\begin{aligned} & -\frac{1}{2a^2}(m_1 + m_2)\ddot{U}_3 + \alpha_1\left(2\frac{\partial^2 U_3}{\partial X_1{}^2} - \frac{\partial \Phi_{33}}{\partial X_3}\right) \\ & +\alpha_2\left(2\frac{\partial^2 U_3}{\partial X_1{}^2} + 2\frac{\partial^2 U_3}{\partial X_3{}^2} + 2\frac{\partial^2 U_1}{\partial X_1 \partial X_3} - \frac{\partial \Phi_{13}}{\partial X_1} - \frac{\partial \Phi_{33}}{\partial X_3}\right) = 0 \end{aligned} \tag{26}$$

$$\frac{1}{8}(m_1 + m_2)\ddot{\Phi}_{11} + \frac{1}{8}(m_1 - m_2)\ddot{\Phi}_{31} + (\alpha_1 + \alpha_2)\left(\Phi_{11} - \frac{\partial U_1}{\partial X_1}\right) = 0 \tag{27}$$

$$\frac{1}{8}(m_1 + m_2)\ddot{\Phi}_{33} + \frac{1}{8}(m_1 - m_2)\ddot{\Phi}_{13} + (\alpha_1 + \alpha_2)\left(\Phi_{33} - \frac{\partial U_3}{\partial X_3}\right) = 0 \tag{28}$$

$$\frac{1}{8}(m_1+m_2)\ddot{\Phi}_{31}+\frac{1}{8}(m_1-m_2)\ddot{\Phi}_{11}+\alpha_2\left(\Phi_{31}-\frac{\partial U_1}{\partial X_3}\right)=0 \tag{29}$$

$$\frac{1}{8}(m_1+m_2)\ddot{\Phi}_{13}+\frac{1}{8}(m_1-m_2)\ddot{\Phi}_{33}+\alpha_2\left(\Phi_{13}-\frac{\partial U_3}{\partial X_1}\right)=0 \tag{30}$$

The boundary conditions are

$$T_j=\left(\Sigma_{ji}+\sigma_{ji}^R\right)n_i\,,\; P_{ji}=0 \tag{31}$$

where n_i the unit vector normal to the boundary surface, T_i is external traction, and P_{ji} the external couple. It is noted that, for present continuum model developed based on the linear displacement expansion of Eq. (8), no couple stress is present.

The equations of motion for the reduced effective modulus theory are readily obtained by requiring $\gamma_{ij}=0$ in the energy density function. We have

$$\alpha_1\frac{\partial^2 U_1}{\partial X_1{}^2}+\alpha_2\left(\frac{\partial^2 U_1}{\partial X_1{}^2}+\frac{\partial^2 U_1}{\partial X_3{}^2}+2\frac{\partial^2 U_3}{\partial X_1\partial X_3}\right)=\frac{1}{2a^2}(m_1+m_2)\ddot{U}_1 \tag{32}$$

$$\alpha_1\frac{\partial^2 U_3}{\partial X_1{}^2}+\alpha_2\left(\frac{\partial^2 U_3}{\partial X_1{}^2}+\frac{\partial^2 U_3}{\partial X_3{}^2}+2\frac{\partial^2 U_1}{\partial X_1\partial X_3}\right)=\frac{1}{2a^2}(m_1+m_2)\ddot{U}_3 \tag{33}$$

3.2 Wave propagation

The continuum theory with microstructures previously presented is now employed to study propagation of harmonic waves in a thin film of thickness h and of infinite in-plane dimensions. The thin film is assumed to consist of a number of atom layers, and the surfaces $X_3=\pm h/2$ are free of tractions. Harmonic waves propagating in the X_1-direction can be expressed as

$$U_1=f_1(X_3)e^{ik(X_1-ct)}\,,\; U_3=f_2(X_3)e^{ik(X_1-ct)} \tag{34}$$

$$\Phi_{11}=f_3(X_3)e^{ik(X_1-ct)}\,,\; \Phi_{33}=f_4(X_3)e^{ik(X_1-ct)} \tag{35}$$

$$\Phi_{31}=f_5(X_3)e^{ik(X_1-ct)}\,,\; \Phi_{13}=f_6(X_3)e^{ik(X_1-ct)} \tag{36}$$

where k denotes wave number, c is phase velocity, and $f_i(X_3)$ are unknown functions. Substituting Eqs. (34)-(36) in the equations of motion (25)-(30), and eliminating f_2, f_3, f_4, f_5, and f_6, we obtain

$$\begin{aligned}&C_1C_5f_1^{''''}+(C_5C_2+C_6C_1)f_1^{'''}-(C_4C_8-C_5C_3-C_6C_2-C_1C_7)f_1^{''}\\&+(C_3C_6+C_2C_7)f_1^{'}+C_7C_3f_1=0\end{aligned} \tag{37}$$

where

$$C_1 = (2 - D_8)\alpha_2 \,,\; C_2 = -ik(\alpha_1 + \alpha_2)D_2 - \alpha_2 D_7$$

$$C_3 = \frac{\omega^2}{2a^2}(m_1 + m_2) - 2k^2(\alpha_1 + \alpha_2) - ik(\alpha_1 + \alpha_2)D_1$$

$$C_4 = C_8 = 2ik\alpha_2 \,,\; C_5 = (\alpha_1 + \alpha_2)(2 - D_4)$$

$$C_6 = -(\alpha_1 + \alpha_2)D_3 - ik\alpha_2 D_6 \,,\; C_7 = \frac{\omega^2}{2a^2}(m_1 + m_2) - 2k^2\alpha_2 - ik\alpha_2 D_5$$

$$D_1 = -ik(\alpha_1 + \alpha_2)\left[\frac{\omega^2}{2a^2}(m_1 + m_2) - \alpha_2\right],\; D_2 = \frac{\omega^2}{8a^2}(m_1 - m_2)\alpha_2$$

$$D_3 = ik\frac{\omega^2}{8a^2}(m_1 - m_2)\alpha_2 \,,\; D_4 = -(\alpha_1 + \alpha_2)\left[\frac{\omega^2}{2a^2}(m_1 + m_2) - \alpha_2\right]$$

$$D_5 = -ik\alpha_2\left[\frac{\omega^2}{2a^2}(m_1 + m_2) - \alpha_1 - \alpha_2\right],\; D_6 = \frac{\omega^2}{8a^2}(m_1 - m_2)(\alpha_1 + \alpha_2)$$

$$D_7 = ik\frac{\omega^2}{8a^2}(m_1 - m_2)(\alpha_1 + \alpha_2) \,,\; D_8 = -\alpha_2\left[\frac{\omega^2}{2a^2}(m_1 + m_2) - \alpha_1 - \alpha_2\right]$$

The general solutions of Eq. (37) for the function f_1 depend on the type of the roots of the characteristic equation

$$\begin{aligned} &C_1C_5\beta^4 + (C_5C_2 + C_6C_1)\beta^3 - (C_4C_8 - C_5C_3 - C_6C_2 - C_1C_7)\beta^2 \\ &+(C_3C_6 + C_2C_7)\beta + C_7C_3 = 0 \end{aligned} \tag{38}$$

By solving the above biquadratic equation, the general solution for f_1 can be expressed as

$$f_1 = E_1 e^{\beta_1 X_3} + E_2 e^{\beta_2 X_3} + E_3 e^{\beta_3 X_3} + E_4 e^{\beta_4 X_3} \tag{39}$$

where β_1, β_2, β_3 and β_4 are the four roots of Eq. (38). The solutions for other functions f_i, i=2-6 can be obtained in a similar manner.

The traction-free boundary conditions at $X_3 = \pm h/2$ lead to

$$S_1E_1e^{(\beta_1 h/2)} + S_2E_2e^{(\beta_2 h/2)} + S_3E_3e^{(\beta_3 h/2)} + S_4E_4e^{(\beta_4 h/2)} = 0 \tag{40}$$

$$S_5E_1e^{(\beta_1 h/2)} + S_6E_2e^{(\beta_2 h/2)} + S_7E_3e^{(\beta_3 h/2)} + S_8E_4e^{(\beta_4 h/2)} = 0 \tag{41}$$

$$S_1E_1e^{(-\beta_1 h/2)} + S_2E_2e^{(-\beta_2 h/2)} + S_3E_3e^{(-\beta_3 h/2)} + S_4E_4e^{(-\beta_4 h/2)} = 0 \tag{42}$$

$$S_5E_1e^{(-\beta_1 h/2)} + S_6E_2e^{(-\beta_2 h/2)} + S_7E_3e^{(-\beta_3 h/2)} + S_8E_4e^{(-\beta_4 h/2)} = 0 \tag{43}$$

where

$$S_m = (2-D_8)F_m + ikH_m - D_7\,,\; S_{m+4} = (\alpha_1+\alpha_2)\left[(2-D_4)H_m - D_3F_m\right] + ik\alpha_2$$

$$F_m = \beta_m^3\frac{C_5C_1}{C_4C_7} - \beta_m\frac{C_8C_4 - C_5C_3}{C_4C_7}\,,\; H_m = -\frac{1}{C_4}\left(C_3 + \beta_m^2 C_1\right),\; m=1,2,3,4$$

Eqs. (40)-(43) have nontrivial solutions for E_1, E_2, E_3, and E_4 only if the determinant of the coefficients vanishes. This leads to the dispersion equation:

$$\begin{vmatrix} S_1e^{\beta_1h/2} & S_2e^{\beta_2h/2} & S_3e^{\beta_3h/2} & S_4e^{\beta_4h/2} \\ S_5e^{\beta_1h/2} & S_6e^{\beta_2h/2} & S_7e^{\beta_3h/2} & S_8e^{\beta_4h/2} \\ S_1e^{-\beta_1h/2} & S_2e^{-\beta_2h/2} & S_3e^{-\beta_3h/2} & S_4e^{-\beta_4h/2} \\ S_5e^{-\beta_1h/2} & S_6e^{-\beta_2h/2} & S_7e^{-\beta_3h/2} & S_8e^{-\beta_4h/2} \end{vmatrix} = 0 \qquad (44)$$

Surface wave propagation and anti-plane wave propagation in ultra-thin films were investigated following a similar manner, and more details can be found in (Huang & Sun 2006a; Huang & Song, 2008), respectively.

3.3 Numerical simulation

Fig. 5 shows the comparison of dispersion curves for the lowest symmetric mode and anti-symmetric mode obtained according to Eq. (44) of the continuum model with microstructures, the effective modulus theory (Eqs. (32) and (33)), and the exact solution for the lattice system (Huang & Sun, 2006a), respectively. The parameters used in the calculation for the cubic crystal structure are $a = 1.74\times10^{-10}m$, $\alpha_1 = 2.02N/m$, $\alpha_2 = 1.10N/m$, $m_2/m_1 = 10$ and the film thickness is $h = 8a$ (i.e., 7 atom layers). The normalized wave velocity $c^* = c/\sqrt{\alpha_2/\rho_0}$, with $\rho_0 = (m_1+m_2)/(2a^2)$, is used in the figure. The results of Fig. 5(a) clearly shows that both the effective modulus theory and the continuum model with microstructures yield very good predictions for the anti-symmetric wave for long waves with *ka*<0.5. This is expected because at lower frequencies wave dispersion is mainly caused by the finite thickness of the film and not by the effect of microstructures. However, as wave length decreases, the dispersion curve computed from the effective modulus theory deviates substantially from the exact curve, while the curve predicted by the continuum theory with microstructures is in good agreement with the exact curve. On the other hand if one employs the effective modulus theory and waves with *ka*>0.5 to determine elastic constants of a thin film with the aid of the dispersion relations, the values of these constants may be significantly overestimated.

Fig. 5 (b) shows the dispersion curves for the lowest symmetric mode. The dispersion curve predicted by the present continuum model with microstructures shows a better agreement with the exact solution than that by the effective modulus theory. It is of interest to note that the results by both continuum theories show some difference relative to the exact phase velocity at long wave lengths. This discrepancy is attributed to the fact both the present continuum theory and the effective modulus theory are not capable of accounting for the nano scale surface effect on the thin film. However, this surface effect can be neglected in thin films consisting of 20 or more atom layers (Sun & Zhang, 2003). Consequently, if one wants to use the lowest symmetric wave mode to determined thin film material properties, some data corrections should be exercised when the thickness of thin film is less than 6 - 7 nano-meters.

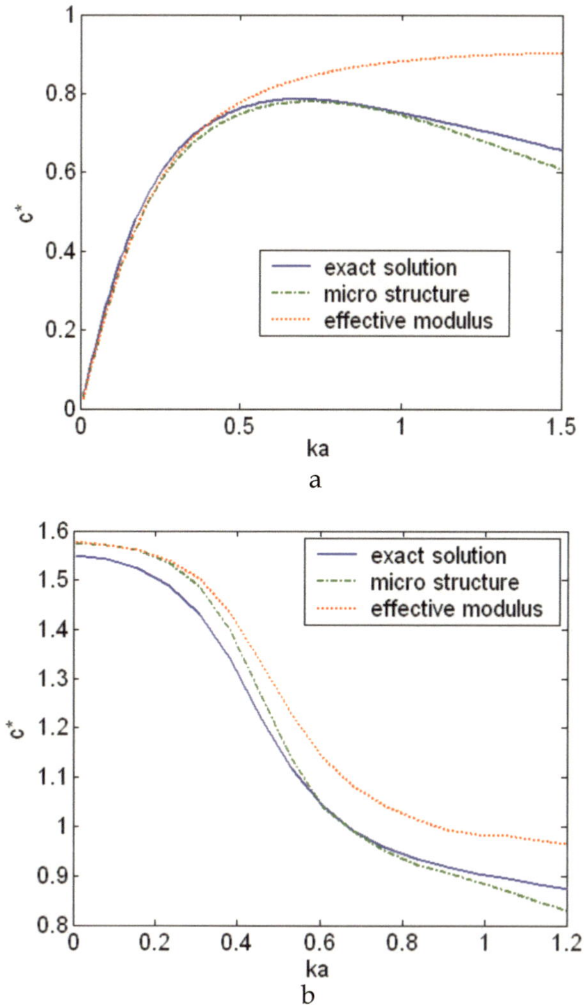

Fig. 5. Dispersive curves obtained with the lattice model and continuum models for a thin film with 7 atom layers. (a) Antisymmetric mode. (b) Symmetric mode

It is well known that in a classical elastic solid, the Rayleigh surface wave is nondispersive. However, surface waves with wavelengths that are comparable to the atomic spacing must be carefully examined if the surface wave technique is to be used in measuring material constants. It is noteworthy that, especially in electronic device applications, surface wave frequencies on the order of GHz -THz are now possible for this type of measurement. We now consider the two-dimensional problem of waves propagating along the surface of a semi-infinite continuum with microstructure as schematically shown in Fig. 6(a). This continuum is assumed to represent a material of the cubic structure and X_1-axis is parallel to the [100] crystal direction. The exact solution for surface waves in a half-space lattice can be refered to (Huang & Sun, 2006a) for the lattice system. To simulate the semi-infinite medium, lattice systems of 90 and 110 layers of atoms are both used with the bottom layer fixed. The results indicate that there is little difference between the two solutions.

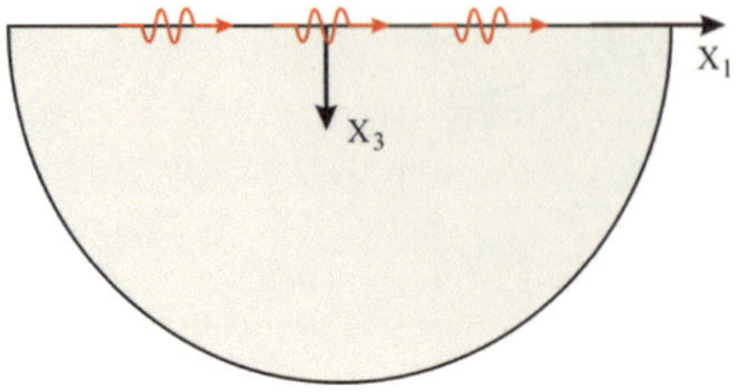

a

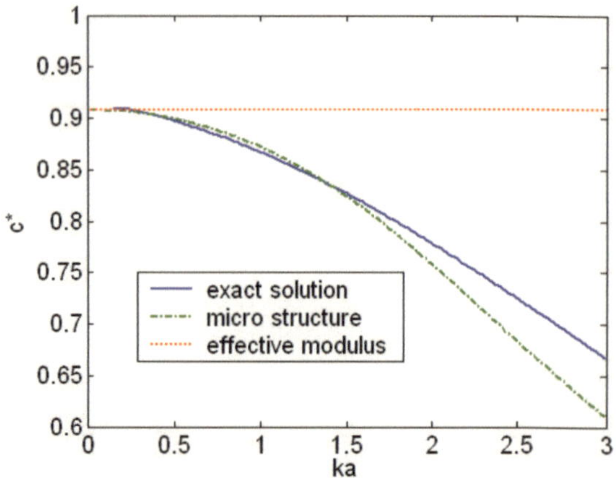

b

Fig. 6. (a) Surface wave propagation in the generalized elastic medium. (b) Dispersive curves for surface waves obtained with the lattice model, the continuum theory with micro structures, and the effective modulus theory

Fig. 6(b) shows the dispersion curves for surface waves, effective modulus theory, and the corresponding lattice system, respectively. The parameters of the cubic crystal structure used in the calculations are $a = 1.74 \times 10^{-10} m$, $\alpha_1 = 2.02N/m$, $\alpha_2 = 1.10N/m$, $m_2/m_1 = 10$. For this set of material parameters, the roots are $\beta_1 = -4.2 \times 10^7 + 1.01 \times 10^7 i$ and $\beta_2 = -4.2 \times 10^7 - 1.01 \times 10^7 i$ for *ka*=0.015. The resulting displacement components in the X_3 direction are found to decay rapidly along the X_3 axis. The rate of decay depends on wave length. In general, the displacement components diminish within about two times the wave length. From Fig. 6(b), it is evident that, unlike the classical continuum, surface waves in the lattice system are dispersive and the present continuum model with microstructures describes this dispersive behavior pretty well. In contrast, the effective modulus theory fails to capture it. We have examined the validity in using the classical continuum (effective modulus) theory to analyze high frequency/short wavelength harmonic waves in thin films. It was observed that the effective modulus theory is inadequate for describing waves of short wavelengths propagating in thin films. It was also found that the continuum theory with microstructures provides much more accurate predictions of dispersive wave velocities of short wavelengths.

4. Wave propagation in nanowires with surface effects

To study high-frequency wave propagation in the nanowire with surface effects, a high-order continuum model is necessary and needed to capture heterogeneous nanostructure effects. In the model, additional kinematic variables are introduced to describe the local motion of nanostructures. Moreover, effects of the surface stress upon the wave propagation will be first considered by using the incremental deformation approach.

4.1 High-order continuum model for nanwires with surface effects

4.1.1 Longitudinal wave motion

Consider first a longitudinal wave propagating in an elastic nanowire with constant circular cross section as illustrated in Fig. 7. A cylindrical coordinate system (r, θ, x) is adopted with the origin of the coordinate at the center of the cross section, where r, θ, x are the radial coordinate, azimuthal coordinate and longitudinal coordinate, respectively. In the figure, the symbol τ represents the surface stress acting along the axis direction of the nanowire, and R denotes the radius of the nanowire.

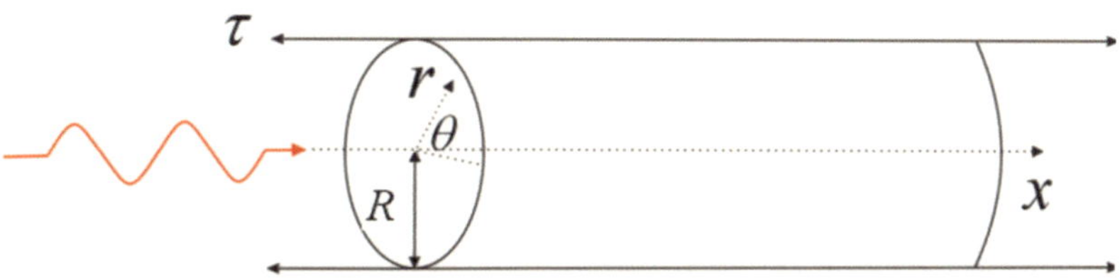

Fig. 7. Longitudinal wave propagation in the nanowire with uniform surface stresses

Based on the surface elasticity theory, the general linear constitutive relation of the surface can be written as

$$\tau_{ij} = \tau_0 \delta_{ij} + \lambda_s \varepsilon_{vv}^s \delta_{ij} + 2G_s \varepsilon_{ij}^s \tag{45}$$

where the subscripts (i and j) denote the stress and strain components along r, θ and x directions, δ_{ij} denotes the Kronecker delta, repeated subscripts follow index summation, τ_{ij} represents the total surface stress, τ_0 represents the strain-independent surface stress, ε_{ij}^s is the surface strain, and λ_s and G_s denote surface moduli, respectively. For the longitudinal wave propagation problem, the constitutive relation of the surface stress can be simplified as

$$\tau = \tau_0 + \tau_e \tag{46}$$

where $\tau_e = (\lambda_s + 2G_s)\varepsilon_{xx}^s = E_s \varepsilon_{xx}^s$ is the strain -dependant stress, and $\varepsilon_{xx}^s = \varepsilon_{xx}\big|_{r=R}$.
To capture heterogeneous nanostructure effects, a generalized high-order continuum model is used to describe local displacements in the nanowire as

$$u_x = U_x(x,t), u_r = \xi_1(x,t)r + \xi_2(x,t)r^2 \tag{47}$$

where u_x and u_r are the local displacements in the longitudinal and radial directions respectively, and $\Phi_1(x,t)$ and $\Phi_2(x,t)$ are the kinematic variables to capture local motion due to the heterogeneity along radial direction. Based on Eq. (47), linearized local strain components can be obtained as

$$\begin{aligned} \varepsilon_{xx} &= \partial U_x(x,t)/\partial x, \ \varepsilon_{\theta\theta} = r\xi_1(x,t) + r\xi_2(x,t) \\ \varepsilon_{rr} &= r\xi_1(x,t) + 2r\xi_2(x,t) \end{aligned} \tag{48}$$

and the isotropic constitutive equations can be expressed as

$$\sigma_{xx} = \lambda(\varepsilon_{xx} + \varepsilon_{rr} + \varepsilon_{\theta\theta}) + 2G\varepsilon_{xx} \tag{49}$$

where λ and G are the Lame constants of the nanowire.
To consider surface stress τ, the formulation according to Trefftz's theory will be taken (Biot, 1965; Sun, 1972). Using Trefftz's incremental stress components which are assumed to

be related linearly to the accompanying deformation, we can obtain an expression for the incremental strain energy density for the surface stressed nanowire as

$$\Delta U = \left\langle \int_0^{\varepsilon_{xx}} \sigma_{xx} d\varepsilon_{xx} + \int_0^{\varepsilon_{xx}} \tau_e d\varepsilon_{xx} + \int_0^{\hat{\varepsilon}_{xx}} \tau_0 d\hat{\varepsilon}_{xx} - \int_0^{\varepsilon_{xx}} \tau_0 d\varepsilon_{xx} \right\rangle \tag{50}$$

in which $\langle \cdot \rangle = \frac{1}{V} \int_V (\bullet) dV$, V is volume of the nanowire, and $\hat{\varepsilon}_{xx} = \frac{\partial u_x}{\partial x} + \frac{1}{2}\left(\frac{\partial u_x}{\partial x}\right)^2 + \frac{1}{2}\left(\frac{\partial u_r}{\partial x}\right)^2$ is the Lagrangian stain component.

The kinetic energy density for the longitudinal motion can be expressed by

$$T = \left\langle \frac{1}{2} \rho \left(\dot{u}_x^2 + \dot{u}_r^2 \right) \right\rangle \tag{51}$$

where ρ is the mass density of the nanowire.

By applying the Hamilton's principle, the equations of motion in the nanowire with surface stresses can be obtained as

$$\rho \ddot{U}_x - 2\lambda \left(\frac{\partial \xi_1}{\partial x} + R \frac{\partial \xi_2}{\partial x} \right) - \left[\lambda + 2G + (1 - \upsilon) \frac{2\tau_0}{R} + \frac{2E_s}{R} \right] \frac{\partial^2 U_x}{\partial x^2} = 0 \tag{52}$$

$$\rho R^2 \ddot{\xi}_1 + \frac{4}{5} \rho R^3 \ddot{\xi}_2 + (8\lambda + 2G)(\xi_1 + R\xi_2) + 4\lambda \frac{\partial U_x}{\partial x} - 4R\tau_0 \left(\frac{\partial^2 \xi_1}{\partial x^2} + R \frac{\partial^2 \xi_2}{\partial x^2} \right) = 0 \tag{53}$$

$$\begin{aligned} &\frac{4}{5} \rho R^2 \ddot{\xi}_1 + \frac{2}{3} \rho R^3 \ddot{\xi}_2 + (8\lambda + 2G)\xi_1 + R(9\lambda + 10G)\xi_2 \\ &+ 4\lambda \frac{\partial U_x}{\partial x} - 4R\tau_0 \left(\frac{\partial^2 \xi_1}{\partial x^2} + R \frac{\partial^2 \xi_2}{\partial x^2} \right) = 0 \end{aligned} \tag{54}$$

It should be mentioned that Poisson's ratio effects are also included in derivation of the above governing equations. Moreover, different from the conventional high-order continuum theory, the current high-order model contains the parameters related to the strain-independent, the strain-dependant surface stress and intrinsic length scales, and thus can reflect size and surface dependent wave responses in the nanowire. As shown in Eq. (52), the effects of the strain-dependent surface stress upon the wave propagation can be predicted by considering different surface moduli E_s. For simplicity and clarity, we will focus on effects of the strain-independent surface stress upon the wave propagation by setting $E_s = 0$ in the study.

4.1.2 Flexural wave motion

Following a similar manner, a high-order continuum model will be formulated to analyze flexural wave propagation in nanowires to capture heterogeneous nanostructure effects. A coordinate system (x, y, z) will be adopted with the origin of the coordinate at the center of the cross section, where x is the coordinate along the axis direction, and y and z are the coordinates along the cross-section directions. Based on the current high-order model for the flexural wave motion, local displacements can be approximated as

$$u_x = -z\Psi_1(x,t) - z^3\Psi_2(x,t),\ u_z = W_z(x,t) \tag{55}$$

where $\Psi_1(x,t)$ and $\Psi_2(x,t)$ are kinematic variables to capture local motion due to heterogeneity. The linearized strain-displacement relations are given by

$$\varepsilon_{xx} = -z\partial\Psi_1(x,t)/\partial x - z^3\partial\Psi_2(x,t)\partial x \tag{56}$$

$$\gamma_{xz} = -\Psi_1(x,t) - 3z^2\Psi_2(x,t) + \partial W_z(x,t)/\partial x \tag{57}$$

in which ε_{xx} and γ_{xz} are the normal strain and the transverse shear strain respectively. The isotropic constitutive equations of the nanowire are

$$\sigma_{xx} = E\varepsilon_{xx},\ \sigma_{xz} = \kappa G\gamma_{xz} \tag{58}$$

where E and G are the Young's modulus and shear modulus, respectively, and κ is the shear correction coefficient with circular cross section $\kappa = (6+12\nu+6\nu^2)/(7+12\nu+4\nu^2)$ (Kaneko, 1975), in which ν is the Poisson's ratio.

To consider the surface stress effects and follow the Trefftz's theory, the incremental deformation energy density ΔU can be obtained as

$$\Delta U = \left\langle \int_0^{\varepsilon_{xx}} \sigma_{xx} d\varepsilon_{xx} + \int_0^{\gamma_{xz}} \sigma_{xz} d\gamma_{xz} + \int_0^{\varepsilon_{xx}} \tau_e d\varepsilon_{xx} + \int_0^{\hat{\varepsilon}_{xx}} \tau_0 d\hat{\varepsilon}_{xx} - \int_0^{\varepsilon_{xx}} \tau_0 d\varepsilon_{xx} \right\rangle \tag{59}$$

where $\hat{\varepsilon}_{xx} = -z\frac{\partial\Psi_1}{\partial x} - z^3\frac{\partial\Psi_2}{\partial x} + \frac{1}{2}\left(z\frac{\partial\Psi_1}{\partial x}\right)^2 + z^4\frac{\partial\Psi_1}{\partial x}\frac{\partial\Psi_2}{\partial x} + \frac{1}{2}z^6\left(\frac{\partial\Psi_2}{\partial x}\right)^2 + \frac{1}{2}\left(\frac{\partial W_z}{\partial x}\right)^2$ is the Lagrangian stain component for the flexural wave motion. The kinetic energy density for flexural motion can be obtained as

$$T = \left\langle \frac{1}{2}\rho\left(\dot{u}_x^2 + \dot{u}_z^2\right) \right\rangle \tag{60}$$

Similarly, by applying the Hamilton's principle, equations of the flexural motion can be readily obtained as

$$\rho\ddot{W}_z + \kappa G\left(\frac{\partial\Psi_1}{\partial x} - \frac{\partial^2 W_z}{\partial x^2} + \frac{3}{4}R^2\frac{\partial\Psi_2}{\partial x}\right) - \frac{2\tau_0}{R}\frac{\partial^2 W_z}{\partial x^2} = 0 \tag{61}$$

$$\begin{aligned} &\rho R^2\ddot{\Psi}_1 + 4\kappa G\left(\Psi_1 - \frac{\partial W_z}{\partial x} + \frac{3}{4}R^2\Psi_2\right) - ER^2\left(\frac{\partial^2\Psi_1}{\partial x^2} + \frac{1}{2}R^2\frac{\partial^2\Psi_2}{\partial x^2}\right) \\ &+ \frac{1}{2}\rho R^2\ddot{\Psi}_2 - 4R(\tau_0 + E_s)\frac{\partial^2\Psi_1}{\partial x^2} - 3R^3(\tau_0 + E_s)\frac{\partial^2\Psi_2}{\partial x^2} = 0 \end{aligned} \tag{62}$$

$$\begin{aligned} &\rho R^4\ddot{\Psi}_2 + \frac{48}{5}\kappa G\left(\Psi_1 - \frac{\partial W_z}{\partial x} + \frac{3}{2}R^2\Psi_2\right) - \frac{8}{5}ER^2\left(\frac{\partial^2\Psi_1}{\partial x^2} + \frac{5}{8}R^2\frac{\partial^2\Psi_2}{\partial x^2}\right) \\ &+ \frac{8}{5}\rho R^2\ddot{\Psi}_1 - \frac{48}{5}R(\tau_0 + E_s)\frac{\partial^2\Psi_1}{\partial x^2} - 8R^3(\tau_0 + E_s)\frac{\partial^2\Psi_2}{\partial x^2} = 0 \end{aligned} \tag{63}$$

Similarly, the impacts of the strain-independent surface stress on the wave responses will be the focus for simplicity and clarity in the study by setting $E_s = 0$.

4.2 Wave propagation and numerical simulation

By assuming the waveform $U = \bar{A}e^{ik(x-ct)}$, $\xi_1 = \bar{B}e^{ik(x-ct)}$, $\xi_2 = \bar{C}e^{ik(x-ct)}$, and $W_z = \tilde{A}e^{ik(x-ct)}$, $\Psi_1 = \tilde{B}e^{ik(x-ct)}$, $\Psi_2 = \tilde{C}e^{ik(x-ct)}$ for corresponding longitudinal and flexural wave motion, where k and c denote wave number and phase velocity, respectively, the harmonic dispersion realtions of corresponding wave motion can be obtained (Song et al., 2009).

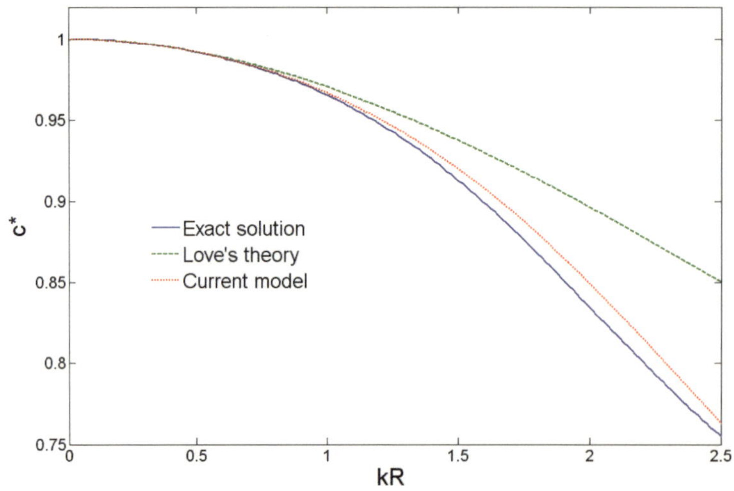

Fig. 8. The comparison of the nondimensional wave dispersion curves

The accuracy of the proposed model is discussed in Fig. 8 by the comparison of the longitudinal wave dispersion relations predicted from the current model and the love's theory (Achenbach, 1973) with that obtained from the exact solution, respectively. The exact solution is based on wave solution from the elasticity theory (Achenbach, 1973). In the example, surface effects are ignored by setting $\tau_0 = 0$. The material parameters for the calculation are $E = 68.5GPa$, $\nu = 0.35$, $\rho = 2700Kg/m^3$. The nondimensional wave velocity is defined as $c^* = c/c_0$, with $c_0 = \sqrt{E/\rho}$, and non-dimensional wave number kR are used in this figure. It can be found that all models can give very good predictions of the longitudinal wave dispersion for the frequency range within kR<1.0. However, the dispersion curve from the love's theory deviates substantially from the exact curve for kR>1.0. The current model can still give a reasonable prediction even when the frequency becomes extremely high for example kR=2.5, which shows the accuracy of the proposed model to capture the heterogeneous nanostructure effects.

Surface stress effects upon longitudinal wave propagation in nanowire are depicted in Fig. 9. In the following simulation, the nondimensional strain-independent surface stress parameter is introduced by $\tau_0^* = \beta\tau_0/E$, where β is a constant coefficient for normalization taken as $1\times10^{9}m^{-1}$ for the current case, and positive and negative values of surface stress parameter τ_0^*, which physically denote surface tension and compression respectively, are selected. The material parameters used for the calculation are assumed to be $E=68.5GPa$, $\nu=0.35$, $\rho=2700Kg/m^3$ and R=3nm. It can be found that the longitudinal wave phase velocity increases with the increase of surface tension, however, wave phase velocity decreases with the increase of surface compression. As a result, if one employs the wave dispersion relation to determine nanowire's material properties ignoring surface effects ($\tau_0^*=0$), their values may be significantly underestimated or overestimated. It should be mentioned that the phase velocity size dependence can be different if different material and surface properties are under investigation.

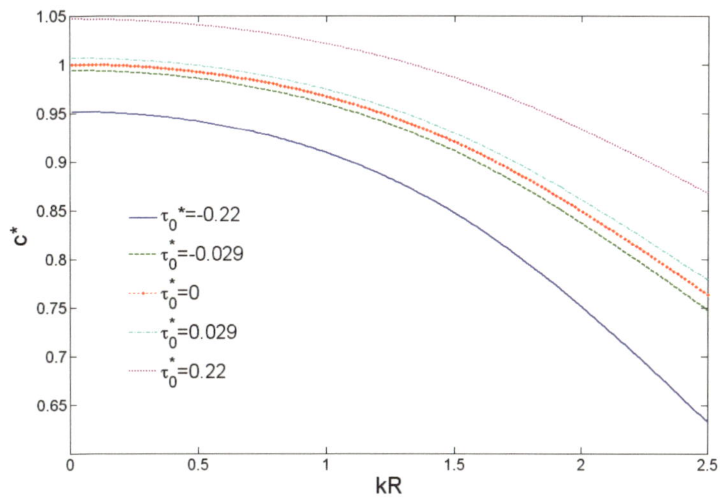

Fig. 9. Nondimensional longitudinal wave dispersion curves of the nanowire with different surface stresses

Size dependent effects of the nanowire upon longitudinal wave propagation are shown in Fig. 10 with the radius of nanowires being of 3, 10, 50 and 100nm respectively. The material parameters for the calculation are assumed to be $E=68.5GPa$, $\nu=0.35$, $\rho=2700Kg/m^3$ and $\tau_0^*=0.22$. For the comparison, the wave dispersive result for $\tau_0^*=0$ is also depicted. It is well known that the dispersion relation does not alter with respect to the radius of the nanowire R if surface effect is neglected ($\tau_0^*=0$), since the continuum mechanics cannot capture size effect of nanomaterials (Lim & He, 2004). However, by considering the surface stress effects in the current model, we can find that the phase wave velocity increases with decrease of the size of the nanowire under surface tension. It should be mentioned that the increase tendency of quasi-static phase velocity (kR approaches zero) predicted by the proposed model has a very good agreement with the theoretical and experimental predictions about the Young's modulus of nanowires under surface tension (Wang & Li, 2008; Cuenot, et al. 2004). It is also noticed that the prediction about the phase velocity, in which the Poisson's ratio effect related to the surface stress is only considered(Wu & Dzenis, 2006), is not in agreement with the results by the current model. It is also worth to point out that the size dependence of the longitudinal motion becomes even more evident with the increase of the wave frequency.

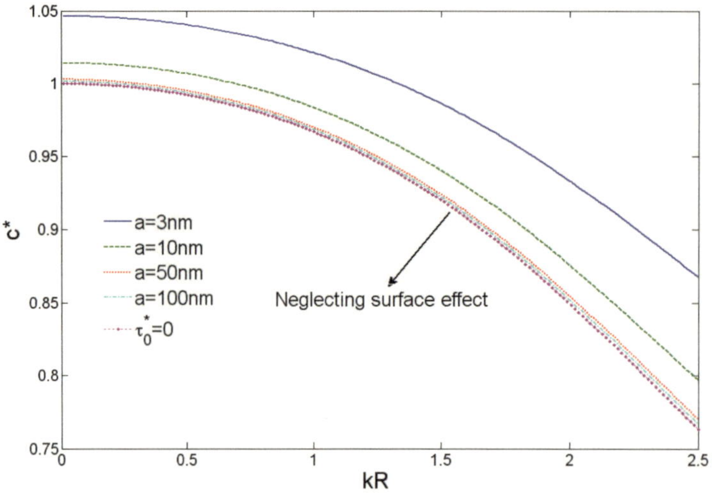

Fig. 10. Nondimensional longitudinal wave dispersion curves of the nanowire with different sizes

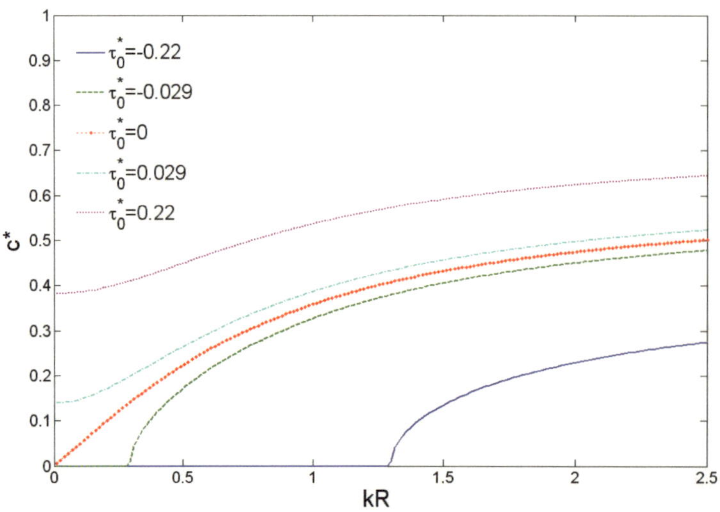

Fig. 11. Nondimensional flexural wave dispersion curves of the nanowire for different surface stresses

Fig. 11 shows surface stress effects upon flexural wave dispersion relations of nanowires. The material parameters are $E = 68.5GPa$, $\nu = 0.35$, $\rho = 2700Kg/m^3$ and R=3*nm*. Similarly, it can be found that the flexural wave phase velocity of the nanowire increases with the increase of the surface tension, while the phase velocity of the nanowire decreases with the increase of the surface compression. It is also of interest to notice that flexural stopping wave band can be observed in the nanowire subject to surface compression ($\tau_0^* < 0$), for example, the stopping wave band is 0<*kR*<0.28 for the nanowire under surface compression $\tau_0^* = -0.029$ and the stopping wave band is 0<*kR*<1.3 for $\tau_0^* = -0.22$.

Fig. 12 shows size dependent effects upon flexural wave propagation in the nanowire with radius of nanowires being of 3, 10, 50 and 100*nm*, respectively. The material parameters used in the calculation are $E = 68.5GPa$, $\nu = 0.35$, $\rho = 2700Kg/m^3$ and $\tau_0^* = 0.22$. For the comparison, wave dispersive curve for the nanowire without surface stress $\tau_0^* = 0$ is also included in the figure. Comparing the results with those in Fig. 10, similar size effects upon the flexural wave propagation can be observed with those in the longitudinal wave propagation. For example, the phase velocity for the nanowire with the radius being 3*nm* is four times bigger than that for the nanowire with the radius being 50*nm* when *kR*=0.

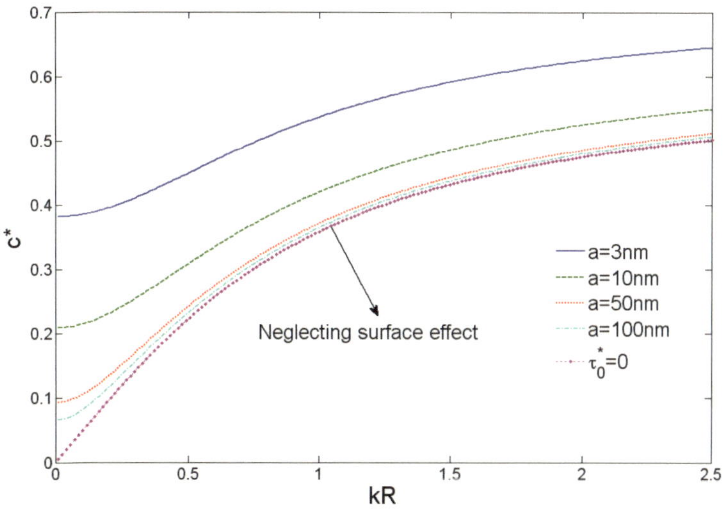

Fig. 12. Nondimensional flexural wave dispersion curves of the nanowire with different sizes

5. Conclusions

In this chapter, we develop a microstructure continuum theory to study wave propagation in the elastic media with different micro- or nano- structures. One of the main advantages of this approach is that the constitutive relations are derived explicitly from the properties and configuration of the microstructure and, theoretically, no additional experiments are necessary. For different nanomaterials, we have demonstrated that microstructure continuum theories can be a good tool for modeling high-frequency wave propagation in nano-phononic crystals and ultra-thin films. A high-order continuum model including surface stress effects is also developed to study wave propagation in nanowires. In this model, surface stress effects are considered by using the incremental deformation theory. The accuracy of the proposed models is validated by numerical simulation. Some new physical wave phenomena related to the heterogeneous effects and surface stress effects on high-frequency wave propagation is discussed.

6. References

Achenbach, J. D. (1973). *Wave Propagation in Elastic Solids,* Elsevier, New York.

Achenbach, J. D.; Sun, C.T. & Herrmann, G. (1968). On the vibrations of a laminated body. *J. Appl. Mech. (Trans. ASME),* Vol. 35, pp. 689-696.

Biot, M. A. (1965). *Mechanics of Incremental Deformations,* Wiley, New York.

Chen, Y. & Lee, J. D. (2003a). Connecting molecular dynamics to micromorphic theory. (I). instantaneous and averaged mechanical variables. *Phys. A,* Vol. 322, pp. 359-376.

Chen, Y. & Lee, J. D. (2003b). Determining material constants in micromorphic theory through phonon dispersion relations. *Int. J. Eng. Sci.,* Vol. 41, pp. 871-886.

Cosserat, E. & Cosserat, F. (1909). *Theorie des Corps Deformables,* A. Hermann & Fils, Paris.

Craighead, H. G. (2000). Nanoelectromechanical system. *Science,* Vol. 290, No. 5496, pp. 1532-1535.

Cuenot, S.; Fre´tigny, C.; Demoustier-Champagne, S. & Nysten, B. (2004). Surface tension effect on the mechanical properties of nanomaterials measured by atomic force microscopy. *Phys. Rev. B,* Vol. 69, pp. 165410.

Eringen, A. C. (1999). *Microcontinuum field theories I: foundations and solids,* Springer Verlag, New York.

Feng, X. L.; He, R. R.; Yang, P. D. & Roukes, M. L. (2007). Very high frequency silicon nanowire electromechanical resonators. Nano Lett., Vol. 7, No. 7, pp. 1953-1959.

Ghatak, A. & Kothari, L. (1972). *An introduction to lattice dynamics,* Addison-Wesley.

Hernaandez, C. M.; Murray, T. W. & Krishnaswarmy S. (2002). Photoacoustic characterization of the mechanical properties of thin films. *Appl. Phys. Lett.,* Vol. 80, No. 4, pp. 691-693.

Huang, G. L. & Sun C. T. (2006a). A continuum model with microstructure for wave propagation in ultra-thin films. *Int. J. Solids Struct.,* Vol. 43, pp. 7104-7127.

Huang, G. L. & Sun C. T. (2006b). Modeling heterostructures of nano-phononic crystals by continuum model with microstructures. *Appl. Phys. Lett.,* Vol. 88, pp. 261908.

Huang, G. L. & Sun C. T. (2007). Continuum modeling of heterogeneous media with microstructures or nanostructures. *Phil. Mag. A,* Vol. 87, pp. 3689-3707.

Huang, G. L. & Sun C. T. (2008). A Higher-order continuum model for elastic media with multiphased microstructures. *Mech. Adv. Mater. Struct.,* Vol. 15, pp. 1-8.

Huang, G. L. & Song, F. (2008). High-frequency antiplane wave propagation in ultra-thin films with nanostructures. *Int. J. Solids Struct.*, Vol. 45, pp. 5368-5380.

Husain, A.; Hone, J.; Henk, W.; Postma, Ch.; Huang, X. M. H.; Drake, T.; Barbic, M.; Scherer, A. & Roukes, M. L. (2003). Nanowire-based very-high-frequency electromechanical resonator. *Appl. Phys. Lett.*, Vol. 83, No. 6, pp. 1240-1242.

Jin, Z. H.; Zhou, H. J.; Jin, Z. L.; Savinell, R. F. & Liu, C. C. (1998). Application of nano-crystalline porous tin oxide thin film for CO sensing. *Sensors and Actuators B: Chemical*, Vol. 52, pp. 188-194.

Kaneko, T. (1975). On Timoshenko's correction for shear in vibrating beams. *J. Phys. D Appl. Phys.*,Vol. 8, pp. 1927-1936.

Lim, C. W. & He, L. H. (2004). Size-dependent nonlinear response of thin elastic films with nano-scale thickness. *Int. J. Mech. Sci.*, Vol. 46, pp. 1715-1726.

Mindlin, R. D. (1964). Micro-Structure in linear elasticity. *Arch. Rat. Mech. Anal.*, Vol. 16, pp. 51-78.

Park, H. S. & Klein, P. A. (2008). Surface stress effects on the resonant properties of metal nanowires: The importance of finite deformation kinematics and the impact of the residual surface stress. *J. Mech. Phys. Solids*, Vol. 56, pp. 3144-3166.

Philip, J.; Hess, P.; Feygelson, T.; Butler, J. E.; Chattopadhyay, S.; Chen, K. H. & Chen, L. C. (2003). Elastic mechanical and thermal properties of nanocrystalline diamond films. *J. Appl. Phys.*, Vol. 93, No. 4, pp. 2164-2171.

Ramprasad, R. & Shi, N. (2005). Scalability of phononic crystal heterostructures. *Appl. Phys. Lett.*, Vol. 87, pp. 111101.

Sampathkumar, A.; Murray, T. W. & Ekinci, K. L. (2006). Photothermal operation of high frequency nanoelectormechanical systems. *Appl. Phys. Lett.*, Vol. 88, pp. 223104.

Schneider, D.; Witke, T.; Schwarz, T.; Schoneich, B. & Schultrich, B. (2000). Testing ultra-thin films by laser-acoustics. *Surf. Coat. Technol.*, Vol. 126, pp. 136-141.

Song, F.; Huang, G. L. & Varadan, V. K. (2010). Study of wave propagation in nanowires with surface effects by using a high-order continuum theory. *Acta Mech.*, Vol. 209, pp.129-139.

Song, F. & Huang, G. L. (2009). Modeling surface stress effects on bending behavior of nanowires: incremental deformation theory, *Phys. Lett. A*, Vol. 373, pp. 3969-3973.

Sun, C. T.; Achenbach, J. D. & Herrmann, G. (1968). Continuum theory for a laminated medium. *J. Appl. Mech. (Trans. ASME)*, Vol. 35, pp. 467-475.

Sun, C. T. (1972). On the equations for a Timoshenko beam under initial stress. *J. Appl. Mech. (Trans. ASME)*, Vol. 39, pp. 282-285.

Sun, C. T. & Zhang, H. T. (2003). Size-dependent elastic moduli of platelike nanomaterials. *J. Appl. Phys.*, Vol. 93, pp. 1212-1218.

Vollmann, J.; Profunser, D. M.; Meier, A. H.; Dobeli, M. & Dual, J. (2004). Pulse laser acoustics for the characterization of inhomogeneities at interfaces of microstructures. *Ultrasonics*, Vol. 42, pp. 657-663.

Wang, G. F. & Li, X. D. (2008). Predicting the Young's modulus of nanowires from first-principles calculations on their surface and bulk materials. *J. Appl. Phys.*, Vol. 104, pp. 113517.

Wu, X. F. & Dzenis, Y. A. (2006). Wave propagation in nanofibers. *J. Appl. Phys.*, Vol. 100, pp. 124318.

12

Wave Propagation in Carbon Nanotubes

Lifeng Wang, Haiyan Hu and Wanlin Guo
Nanjing University of Aeronautics and Astronautics
China

1. Introduction

Interest in carbon nanotubes has grown rapidly since their discovery (Iijima, 1991). A single-walled carbon nanotube can be considered as a single layer graphite sheet rolled into cylinder. Three examples of single-walled carbon nanotubes, armchair, zigzag and chiral, are shown in Figure 1 (Qian *et al.*, 2002). A multi-walled carbon nanotube is similar to a single-walled carbon nanotube, but with many layers of graphite sheets in the cylinder structure.

Recent studies have indicated that carbon nanotubes exhibit superior mechanical and electronic properties over any known materials, and hold substantial promise for new super-strong composite materials, among others. For instance, carbon nanotubes have an exceptionally high elastic modulus (Treacy *et al.*, 1996), and sustain large elastic strain and failure strain (Yakobson *et al.*, 1996; Wong *et al.*, 1997). Apart from an extensive experimental study to characterize the mechanical behavior of carbon nanotubes, theoretical or computational modeling of carbon nanotubes has received considerable attention. The current computational modeling approaches include both atomistic modeling and continuum modeling.

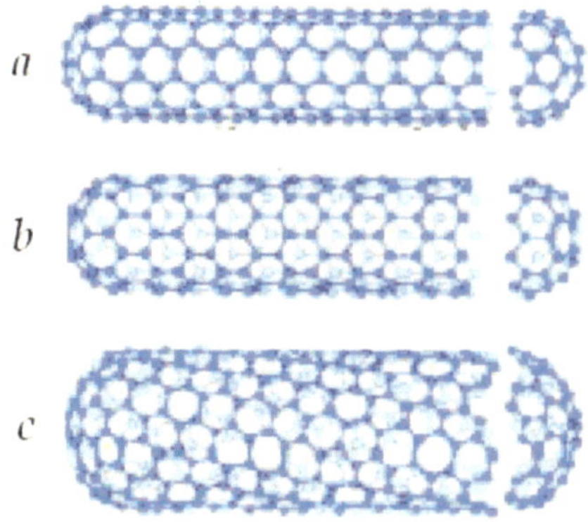

Fig. 1. Examples of carbon nanotubes, (a) armchair, (b) zigzag and (c) chiral ((Qian *et al.*, 2002))

Among the researches of continuum modeling of carbon nanotubes, numerous studies concentrated on the static mechanical behavior, such as buckling, of carbon nanotubes by using the elastic models of beam and cylindrical shell (Qian *et al.*, 2002). Several research teams implemented the elastic models of beam to study the dynamic problems, such as vibration and wave propagation (Poncharal *et al.*, 1999; Popov *et al.*, 2000; Yoon *et al.*, 2002, 2003a,b), of carbon nanotubes. Some other teams used the elastic models of cylindrical shell to study the vibration and wave dispersion relations of carbon nanotubes (Wang *et al.*, 2005a; Dong & Wang, 2006; Natsuki *et al.* 2005, 2006). Furthermore, Chakraborty (2007) modeled a multi-walled carbon nanotube as an assemblage of cylindrical shell elements connected throughout their lengths by distributed springs to investigate the elastic waves of very high frequency in carbon nanotubes, and to give the dispersion relation between the group velocity and the wave number. Their studies showed that both elastic models of beam and cylindrical shell are valid to describe the vibration or wave propagation of carbon nanotubes in a relatively low frequency range.

In recent years, the need to describe the microstructure-dominated behavior of a mechanical component in an averaged sense, without modeling detailed microstructures and deformation processes at the micro-scale, has stimulated numerous studies on enriched continuum models. In these models, the nonstandard deformation and force quantities account for the influences of the microstructures on mechanical behaviors. The description of the deformation behavior of solids is not limited to relations between elastic stresses and strains. The consideration of microstructure, which is usually neglected in the classical theory of elasticity, results in the constitutive law related to the strain gradient (Mindlin, 1964). This constitutive law, different from the generalized Hook's law, enables one to predict many new phenomena, such as the dispersive waves in non-local elastic continuum, owing to the microstructures in a solid. As a consequence, the wave propagation and dispersion in granular media have also drawn considerable attention (Manolis, 2000; Chang & Gao, 1995). For instance, the wave propagation in a rod exhibiting non-local elasticity has become an interesting topic (Nowinski, 1984; Mühlhaus & Oka, 1996; Askes et al, 2002). Several researchers (Sudak, 2003; Zhang et al., 2004, 2005; Wang, 2005b; Wang *et al.*, 2006a; Xie *et al.* 2006) established the non-local elastic models with the second order gradient of stress taken into account so as to describe the vibration and wave propagation of both single- and multi-walled carbon nanotubes in a higher frequency range. The molecular dynamics simulations showed those models worked much better than the elastic models in a relatively high frequency range. However, the molecular dynamics simulations also showed that the micro-structures of a carbon nanotube might have a so significant influence on the waves of very high frequency such that non-elastic models could not predict the wave dispersion well (Wang & Hu, 2005; Wang et al. 2006b).

The primary objective of this work is to study the phase velocity and group velocity of the flexural and longitudinal wave propagations in carbon nanotubes so as to examine the effect of micro-structures of a carbon nanotube on the wave dispersion. This work deals with the dispersion relations of both longitudinal and flexural waves in single-walled and multi-walled carbon nanotubes via the non-local elastic models of both Timoshenko beams and cylindrical shells. The study focuses on the comparison between the non-local elastic models and the elastic models in predicting the dispersion relation. For this purpose, Section 2 presents the molecular dynamics modeling of carbon nanotubes. Section 3 presents the dispersion relation of longitudinal waves in a single-walled carbon nanotube from a non-local elastic model of cylindrical shell, which includes the second order gradient of strain in order to characterize the micro-structures of the carbon nanotube. Section 4 gives the dispersion relation of flexural waves in a single-walled carbon nanotube by using a non-local elastic model of Timoshenko beam. Section 5 turns to the dispersion relation of flexural waves in a multi-walled carbon nanotube from a non-local elastic model of multi-Timoshenko beams, which also takes the second order gradient of strain into account. Similarly, Section 6 gives the dispersion relation of longitudinal waves in a multi-walled carbon nanotube on the basis of a non-local elastic model of multi-cylindrical shells. Finally, the chapter ends with some concluding remarks made in Section 7.

2. Molecular dynamics model for carbon nanotubes

This section presents the molecular dynamics models for wave propagation in a carbon nanotube, respectively, for a wide range of wave numbers. Molecular dynamics simulation consists of the numerical solution of the classical equations of motion, which for a simple atomic system may be written

$$m_i \ddot{\mathbf{r}}_i = \mathbf{F}_i = -\frac{\partial V}{\partial \mathbf{r}_i}. \tag{1}$$

For this purpose the force $\mathbf{F}_i$ acting on the atoms are derived from a potential energy, $V(r_{ij})$ where r_{ij} is the distance from atom *i* to atom *j*.

In the molecular dynamics models of this chapter, the interatomic interactions are described by the Tersoff-Brenner potential (Brenner, 1990), which has been proved applicable to the description of mechanical properties of carbon nanotubes. The structure of the Tersoff-Brenner potential is as follows

$$V(r_{ij}) \equiv \sum_i \sum_{j(>i)} [V_R(r_{ij}) - \bar{B}_{ij} V_A(r_{ij})], \tag{2}$$

$V_R(r_{ij})$ and $V_A(r_{ij})$ are the repulsive and attractive terms given by

$$V_R(r_{ij}) \equiv f_{ij}(r_{ij}) \frac{D_{ij}}{S_{ij}-1} \exp[-\sqrt{2S_{ij}}\beta_{ij}(r - r_0)], \tag{3a}$$

$$V_A(r_{ij}) \equiv f_{ij}(r_{ij}) \frac{S_{ij} D_{ij}}{S_{ij}-1} \exp[-\sqrt{2/S_{ij}}\beta_{ij}(r - r_0)]. \tag{3b}$$

Here $S_{ij} = 1.29$, $D_{ij} = 6.325\text{eV}$, $\beta_{ij} = 15\text{nm}^{-1}$, $r_0 = 0.1315\text{nm}$, f_{ij}, D_{ij}, S_{ij}, β_{ij} are scalars, $f_{ij}(r_{ij})$ is a switch function used to confine the pair potential in a neighborhood with radius of r_2 as following

$$f_{ij}(r_{ij}) \equiv \begin{cases} 1, & r_{ij} < r_1, \\ \dfrac{1}{2}\left[1 + \cos\left(\dfrac{\pi(r_{ij} - r_1)}{r_2 - r_1}\right)\right], & r_1 \le r_{ij} \le r_2, \\ 0, & r_{ij} > r_2, \end{cases} \tag{3c}$$

where $r_1 = 1.7\mathring{A}$, $r_2 = 2.0\mathring{A}$. In Equation (2), $\bar{B}_{ij}$ reads

$$\bar{B}_{ij} \equiv \frac{1}{2}(b_{ij} + b_{ji}), \tag{3d}$$

$$b_{ij} \equiv [1 + \sum_{k \neq i,j} G(\theta_{jik}) f_{ik}(r_{ik})]^{-\delta}, \qquad b_{ji} \equiv [1 + \sum_{k \neq i,j} G(\theta_{ijk}) f_{jk}(r_{jk})]^{-\delta}, \tag{3e}$$

$$G(\theta_{ijk}) \equiv a_0[1 + \frac{c_0^2}{d_0^2} - \frac{c_0^2}{d_0^2 + (1 + \cos\theta_{ijk})^2}], \tag{3f}$$

where θ_{ijk} is the angle between bonds *i-j* and *i-k*, δ = 0.80469, a_0 = 0.011304, $c_0 = 19$ and $d_0 = 2.5$. In addition, the C-C bond length in the model is 0.142nm.

The Verlet algorithm in the velocity form (Leach, 1996) with time step 1 fs is used to simulate the atoms of carbon nanotubes.

$$\begin{aligned} R(t + \delta t) &= R(t) + \delta t V(t) + \frac{1}{2}\delta t^2 a(t) \\ V(t + \delta t) &= V(t) + \frac{1}{2}\delta t[a(t) + a(t + \delta t)] \end{aligned} \tag{4}$$

where, *R* represents the position, *V* is the velocity, *a* denotes the acceleration of atoms, δt is the time step.

3. Flexural wave in a single-walled carbon nanotube

3.1 Non-local elastic Timoshenko beam model

This section starts with the dynamic equation of a non-local elastic Timoshenko beam of infinite length and uniform cross section placed along direction x in the frame of coordinates (x, y, z), with $w(x,t)$ being the displacement of section x of the beam in direction y at the moment t.

In order to describe the effect of microstructure of carbon nanotubes on their mechanical properties, it is assumed that the beam of concern is made of the non-local elastic material, where the stress state at a given reference point depends not only on the strain of this point, but also on the higher-order gradient of strain so as to take the influence of microstructure into account. The simplest constitutive law to characterize the non-local elastic material in the one-dimensional case reads

$$\sigma_x = E(\varepsilon_x + r^2 \frac{\partial^2 \varepsilon_x}{\partial x^2}), \tag{5}$$

where E represents Young's modulus, and ε_x the axial strain. As studied in Askes *et al.* (2002), r is a material parameter to reflect the influence of the microstructure on the stress in the non-local elastic material and yields

$$r = \frac{d}{\sqrt{12}}, \tag{6}$$

where d, referred to as the inter-particle distance, is the axial distance between two rings of particles in the material. For the armchair single-walled carbon nanotube, d is just the axial distance between two rings of carbon atoms.

To establish the dynamic equation of the beam, it is necessary to determine the bending moment M, which reads

$$M = \int_A y\sigma_x \mathrm{d}A \,, \tag{7}$$

where A represents the cross section area of the beam, σ_x the axial stress, y the distance from the centerline of the cross section. It is well known from the theory of beams that the axial strain yields

$$\varepsilon_x = \frac{y}{\rho'} \,, \tag{8}$$

where ρ' is the radius of curvature of beam. Let φ denote the slope of the deflection curve when the shearing force is neglected and s denote the coordinate along the deflection curve of the beam, then the assumption upon the small deflection of beam gives

$$\frac{1}{\rho'} = \frac{\partial\varphi}{\partial x}\frac{\partial x}{\partial s} \approx \frac{\partial\varphi}{\partial x} \,. \tag{9}$$

Substituting Equations (8) and (9) into Equations (5) and (7) gives the following relation between bending moment M and the curvature and its second derivative when the shearing force is neglected

$$M = EI\left(\frac{\partial\varphi}{\partial x} + r^2\frac{\partial^3\varphi}{\partial x^3}\right), \tag{10}$$

where $I = \int y^2 \mathrm{d}A$ represents the moment of inertia for the cross section.

To determine the shear force on the beam, let γ be the angle of shear at the neutral axial in the same cross section. Then, it is easy to see the total slope

$$\frac{\partial w}{\partial x} = \varphi - \gamma \,. \tag{11}$$

For the torsional problem of one dimension, the constitutive law of the non-local elastic material reads

$$\tau = G\left(\gamma + r^2\frac{\partial^2\gamma}{\partial x^2}\right), \tag{12}$$

where τ is the shear stress and G is the shear modulus. Then, the shear force Q on the cross section becomes

$$Q = \beta AG\left[\left(\varphi - \frac{\partial w}{\partial x}\right) + r^2\left(\frac{\partial^2\varphi}{\partial x^2} - \frac{\partial^3 w}{\partial x^3}\right)\right], \tag{13}$$

where β is the form factor of shear depending on the shape of the cross section, and β =0.5 holds for the circular tube of the thin wall (Timoshenko & Gere, 1972).
Now, it is straightforward to write out the dynamic equation for the beam element of length dx subject to bending M and shear force Q as following

$$\begin{cases} \rho A \dfrac{\partial^2 w}{\partial t^2}\mathrm{d}x + \dfrac{\partial Q}{\partial x}\mathrm{d}x = 0, \\ \rho I \dfrac{\partial^2 \varphi}{\partial t^2}\mathrm{d}x + Q\mathrm{d}x - \dfrac{\partial M}{\partial x}\mathrm{d}x = 0. \end{cases} \tag{14}$$

Substituting Equations (10) and (13) into Equation (14) yields the following coupled dynamic equation for the deflection and the slop of non-local elastic Timoshenko beam

$$\begin{cases} \rho \dfrac{\partial^2 w}{\partial t^2} + \beta G[(\dfrac{\partial \varphi}{\partial x} - \dfrac{\partial^2 w}{\partial x^2}) + r^2(\dfrac{\partial^3 \varphi}{\partial x^3} - \dfrac{\partial^4 w}{\partial x^4})] = 0, \\ \rho I \dfrac{\partial^2 \varphi}{\partial t^2} + \beta AG[(\varphi - \dfrac{\partial w}{\partial x}) + r^2(\dfrac{\partial^2 \varphi}{\partial x^2} - \dfrac{\partial^3 w}{\partial x^3})] - EI(\dfrac{\partial^2 \varphi}{\partial x^2} + r^2\dfrac{\partial^4 \varphi}{\partial x^4}) = 0. \end{cases} \tag{15}$$

3.2 Flexural wave dispersion in different beam models

To study the flexural wave propagation in an infinitely long beam, let the dynamic deflection and slope be given by

$$w(x,t) = \hat{w}e^{i\tilde{k}(x-ct)}, \quad \varphi(x,t) = \hat{\varphi}e^{i\tilde{k}(x-ct)}, \tag{16}$$

where $i \equiv \sqrt{-1}$, $\hat{w}$ represents the amplitude of deflection of the beam, and $\hat{\varphi}$ the amplitude of the slope of the beam due to bending deformation alone. In addition, c is the phase velocity of wave, and $\tilde{k}$ is the wave number related to the wave length λ via $\lambda\tilde{k} = 2\pi$. Substituting Equation (16) into Equation (15) yields

$$\begin{cases} (-i\beta AG\tilde{k} + i\beta AGr^2\tilde{k}^3)\hat{w} + (-\rho I\tilde{k}^2c^2 + \beta AG - \beta AGr^2\tilde{k}^2 + EI\tilde{k}^2 - EIr^2\tilde{k}^4)\hat{\varphi} = 0, \\ (-\rho\tilde{k}^2c^2 + \beta G\tilde{k}^2 - \beta Gr^2\tilde{k}^4)\hat{w} + (i\beta G\tilde{k} - i\beta Gr^2\tilde{k}^3)\hat{\varphi} = 0. \end{cases} \tag{17}$$

From the fact that there exists at least one non-zero solution $(\hat{w},\hat{\varphi})$ of Equation (17), one arrives at

$$\frac{\rho^2 I}{\beta G}\tilde{k}^2c^4 - [\rho A + \rho I(1 + \frac{E}{\beta G})\tilde{k}^2](1 - r^2\tilde{k}^2)c^2 + EI\tilde{k}^2(1 - r^2\tilde{k}^2)^2 = 0. \tag{18}$$

Solving Equation (18) for the phase velocity c gives two branches of wave dispersion relation

$$c = \sqrt{\frac{-b_1 \pm \sqrt{b_1^2 - 4a_1c_1}}{2a_1}}, \tag{19}$$

where $a_1 = \rho^2 I\tilde{k}^2 / \beta G$, $b_1 = [\rho A + \rho I(1 + E/\beta G)\tilde{k}^2](r^2\tilde{k}^2 - 1)$ and $c_1 = EI\tilde{k}^2(1 - r^2\tilde{k}^2)^2$. Here, the lower branch represents the dispersion relation of the flexural wave, and the upper branch determines the dispersion relation of the transverse shearing out of interest.

If $r=0$, Equation (15) leads to

$$\rho A\frac{\partial^2 w(x,t)}{\partial t^2}+EI\frac{\partial^4 w(x,t)}{\partial x^4}-\rho I(1+\frac{E}{\beta G})\frac{\partial^4 w(x,t)}{\partial x^2\partial t^2}+\frac{\rho^2 I}{\beta G}\frac{\partial^4 w(x,t)}{\partial t^4}=0\,. \tag{20}$$

This is the dynamic equation of a traditional Timoshenko beam (Timoshenko & Gere, 1972). In this case, the relation of wave dispersion takes the form of Equation (19), but with $a_1=\rho^2 I\tilde{k}^2/\beta G$, $b_1=-[\rho A+\rho I(1+E/\beta G)\tilde{k}^2]$ and $c_1=EI\tilde{k}^2$.

If neither the rotary inertial nor the shear deformation is taken into account, Equation (15) leads to the dynamic equation of a non-local elastic Euler beam as following

$$\rho A\frac{\partial^2 w(x,t)}{\partial t^2}+EI[\frac{\partial^4 w(x,t)}{\partial x^4}+r^2\frac{\partial^6 w(x,t)}{\partial x^6}]=0\,. \tag{21}$$

The condition of non-zero solution $\hat{w}$ of Equation (21) gives the dispersion relation

$$c=k\sqrt{\frac{EI}{\rho A}(1-r^2\tilde{k}^2)}\,. \tag{22}$$

In this case, $r=0$ results in the dispersion relation in the traditional Euler beam

$$c=k\sqrt{\frac{EI}{\rho A}}\,. \tag{23}$$

When $\tilde{k}^2<1/r^2$, there implies a cut off frequency in Equations (19) and (22).

3.3 Flexural wave propagation in a single-walled carbon nanotube

To predict the flexural wave dispersion from the theoretical results in Section 3.2, it is necessary to know Young's modulus E and the shear modulus G, or Poisson's ratio υ. The previous studies based on the Tersoff-Brenner potential gave a great variety of Young's moduli of single-walled carbon nanotubes from the simulated tests of axial tension and compression. When the thickness of wall was chosen as 0.34nm, for example, 1.07TPa was reported by Yakobson *et al* (1996), 0.8TPa by Cornwell and Wille (1997), and 0.44-0.50TPa by Halicioglu (1998). Meanwhile, the Young's modulus determined by Zhang *et al.* (2002) on the basis of the nano-scale continuum mechanics was only 0.475 TPa when the first set of parameters in the Tersoff-Brenner potential (Brenner, 1990) was used. Hence, it becomes necessary to compute Young's modules and Poisson's ratio again from the above molecular dynamics model for the single-walled carbon nanotubes under the static loading.

For the same thickness of wall, the Young's modulus that we computed by using the first set of parameters in the Tersoff-Brenner potential (Brenner, 1990) was 0.46TPa for the armchair (5,5) carbon nanotube and 0.47TPa for the armchair (10,10) carbon nanotube from the molecular dynamics simulation for the text of axial tension. Furthermore, the simulated test of pure bending that we did gave the product of effective Young's modulus $E=0.39$TPa and Poisson's ratio $\upsilon=0.22$ for the armchair (5,5) carbon nanotube, $E=0.45$TPa and $\upsilon=0.20$ for the armchair (10,10) carbon nanotube. Young's moduli and Poisson's ratios obtained from the simulated test of pure bending for those two carbon nanotubes were

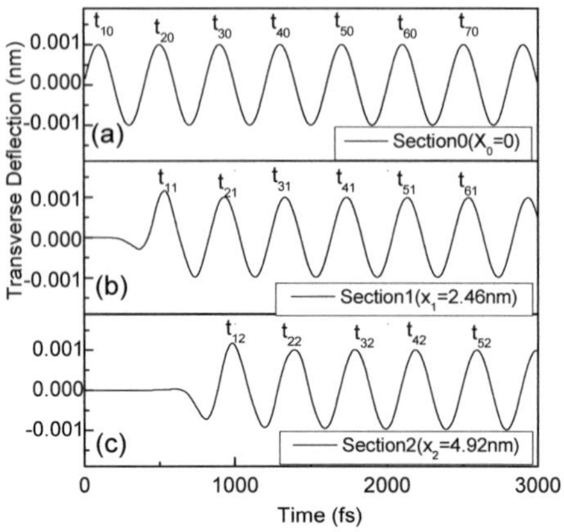

Fig. 2. Time histories of the deflection of different sections of the armchair (5,5) carbon nanotube, where subscripts i and j in t_{ij} represent the number of wave peak and the number of section, respectively. (a) The sinusoidal wave of period $T = 400\text{fs}$ input at Section 0. (b) The deflection of Section 1, 2.46nm ahead of Section 0. (c) The deflection of Section 2, 4.92nm ahead of Section 0. (Wang & Hu, 2005)

used. In addition, Equation (6) gives $r = 0.0355\text{nm}$ when the axial distance between two rings of atoms reads $d = 0.123\text{nm}$. For the single-walled carbon nanotubes, the wall thickness is $h = 0.34\text{nm}$ and the mass density of the carbon nanotubes is $\rho = 2237\text{kg/m}^3$.
It is quite straightforward to determine the phase velocity and the wave number from the flexural vibration, simulated by using molecular dynamics, of two arbitrary sections of a carbon nanotube. As an example, the end atoms denoted by Section 0 at $x_0 = 0$ of the armchair (5,5) carbon nanotube was assumed to be subject to the harmonic deflection of period $T = 400\text{fs}$ as shown in Figure 2(a). The corresponding angular frequency is $\omega = 2\pi/T \approx 1.57 \times 10^{13}\text{rad/s}$. The harmonic deflection was achieved by shifting the edge atoms of one end of the nanotube while the other end was kept free. Figures 2(b) and 2(c) show the flexural vibrations of Section 1 at $x_1 = 2.46\text{nm}$ and Section 2 at $x_2 = 4.92\text{nm}$, respectively, of the carbon nanotube simulated by using molecular dynamics. If the transient deflection of the first two periods is neglected, the propagation duration Δt of the wave from Section 1 to Section 2 can be estimated as below

$$\Delta t \approx \frac{(t_{32} - t_{31}) + (t_{42} - t_{41}) + \cdots + (t_{n2} - t_{n1})}{n-2}. \tag{24}$$

There follow the phase velocity and wave number

$$c = \frac{x_2 - x_1}{\Delta t}, \quad \tilde{k} = \frac{2\pi}{\lambda} = \frac{\omega T}{\lambda} = \frac{\omega}{c}. \tag{25}$$

Figure 3 illustrate the dispersion relations between the phase velocity c and the wave number of flexural wave in the armchair (5,5) and (10,10) carbon nanotubes, respectively. Here, the symbol E represents the traditional Euler beam, T the traditional Timoshenko beam, NE the non-local elastic Euler beam, NT the non-local elastic Timoshenko beam, and MD the molecular dynamics simulation, respectively. In Figures 3, when the wave number

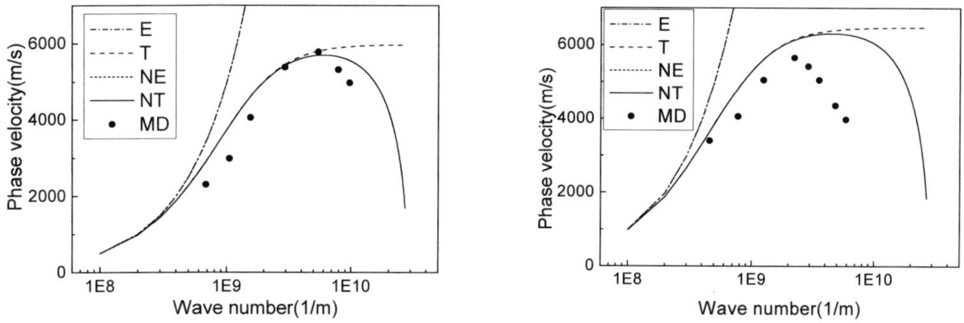

(a) an armchair (5,5) carbon nanotube (b) an armchair (10,10) carbon nanotube

Fig. 3. Dispersion relation of longitudinal wave in single-walled carbon nanotubes. (Wang & Hu, 2005)

$\tilde{k}$ is smaller than $1\times10^{9}m^{-1}$, or the wave length is $\lambda > 6.28\times10^{-9}\text{m}$, the phase velocities given by the four beam models are close to each other, and they all could predict the result of the molecular dynamics well. The phase velocity given by the traditional Euler beam, however, is proportional to the wave number, and greatly deviated from the result of molecular dynamics when the wave number became larger than $1\times10^{9}m^{-1}$. Almost not better than the traditional Euler beam, the result of the non-local elastic Euler beam greatly deviate from the result of molecular dynamics too when the wave number became large. Nevertheless, the results of both traditional Timoshenko beam and non-local elastic Timoshenko beam remain in a reasonable coincidence with the results of molecular dynamics in the middle range of wave number or wave length. When the wave number $\tilde{k}$ is larger than $6\times10^{9}\text{m}^{-1}$ (or the wave length is $\lambda < 1.047\times10^{-9}\text{m}$) for the armchair (5,5) carbon nanotube and $3\times10^{9}\text{m}^{-1}$ (or the wave length is $\lambda < 2.094\times10^{-9}\text{m}$) for the armchair (10,10) carbon nanotube, the phase velocity given by the molecular dynamics begin to decrease, which the traditional Timoshenko beam failed to predict. However, the non-local elastic Timoshenko beam is able to predict the decrease of phase velocity when the wave number is so large (or the wave length was so short) that the microstructure of carbon nanotube significantly block the propagation of flexural waves.

3.4 Group velocity of flexural wave in a single-walled carbon nanotube

The concept of group velocity may be useful in understanding the dynamics of carbon nanotubes since it is related to the energy transportation.

From Equation (19), with $\omega = c\tilde{k}$ considered, the angular frequency ω gives two branches of the wave dispersion relation (Wang *et al.* 2008)

$$\omega = \sqrt{\frac{-b_1 \pm \sqrt{b_1^2 - 4a_1c_1}}{2a_1}}, \tag{26}$$

where $a_1 = \rho^2 I / \beta G$, $b_1 = [-\rho A - \rho I(1+E/\beta G)\tilde{k}^2](1-r^2\tilde{k}^2)$ and $c_1 = EI\tilde{k}^4(1-r^2\tilde{k}^2)^2$. The group velocity reads

$$c_g = \frac{d\omega}{d\tilde{k}} = \frac{1}{2\omega}\left(\frac{-\frac{db_1}{d\tilde{k}} \pm (b_1\frac{db_1}{d\tilde{k}} - 2a_1\frac{dc_1}{d\tilde{k}})\left(b_1^2 - 4a_1c_1\right)^{-1/2}}{2a_1}\right), \tag{27}$$

where

$$\frac{db_1}{dk} = -2\tilde{k}\rho I(1+E/\beta G)(1-r^2\tilde{k}^2) - 2r^2\tilde{k}(-\rho A - \rho I(1+E/\beta G)\tilde{k}^2), \tag{28}$$

$$\frac{dc_1}{dk} = 4EI\tilde{k}^3(1-r^2\tilde{k}^2)^2 - 4EIr^2\tilde{k}^5(1-r^2\tilde{k}^2). \tag{29}$$

Figure 4 shows the dispersion relations between the group velocity and the wave number of flexural waves in an armchair (5,5) single-walled carbon nanotube and in an armchair (10,10) single-walled carbon nanotube. Here the results were not compared with molecular dynamics results, the Young's modulus used the common value. The product of Young's modulus and the wall thickness is $Eh = 346.8\text{Pa}\cdot\text{m}$ and Poisson's ratio is $\upsilon = 0.20$. There follows $G = E/(2(1+\upsilon))$. In addition, the material parameter $r = 0.0355\text{nm}$. The product of the mass density and the wall thickness yields $\rho h \approx 760.5\text{kg/m}^3\cdot\text{nm}$. For the (5,5) single-walled carbon nanotube, the product of the mass density and the section area yields $\rho A = 1.625\times10^{-15}\text{kg/m}$, the product of the mass density and the moment of inertia for the cross section yields $\rho I = 3.736\times10^{-35}\text{kg}\cdot\text{m}$, and there follows $EI = 1.704\times10^{-26}\text{Pa}\cdot\text{m}^4$. For the (10,10) single-walled carbon nanotube, the product of the mass density and the section area yields $\rho A = 3.25\times10^{-15}\text{kg/m}$, the product of the mass density and the moment of inertia for the cross section yields $\rho I = 2.541\times10^{-34}\text{kg}\cdot\text{m}$, and there follows $EI = 1.159\times10^{-25}\text{Pa}\cdot\text{m}^4$. For both lower and upper branches of the dispersion relation, the results of the elastic Timoshenko beam remarkably deviate from those of the non-local elastic Timoshenko beam with an increase in the wave number. Figure 4(a) and (b) show again the intrinsic limit of the wave number $\tilde{k}<2\times10^{10}\text{ m}^{-1}$, instead of $\tilde{k}<\sqrt{12}/d\approx2.82\times10^{10}\text{m}^{-1}$. This fact explains the difficulty that the cut-off flexural wave predicted by the non-local elastic cylindrical shell is $\tilde{k}<\sqrt{12}/d\approx2.82\times10^{10}\text{m}^{-1}$, but the direct molecular dynamics simulation only gives the dispersion relation up to the wave number $\tilde{k}\approx2\times10^{10}\text{ m}^{-1}$ (Wang & Hu, 2005).

4. Longitudinal wave in a single-walled carbon nanotube

4.1 Wave dispersion predicted by a non-local elastic shell model

This section studies the dispersion of longitudinal waves from a thoughtful model, namely, the model of a cylindrical shell made of non-local elastic material. For such a thin cylindrical shell, the bending moments can be naturally neglected for simplicity in theory. Figure 5(a) shows a shell strip cut from the cylindrical shell, where a set of coordinates $(x,\theta,\tilde{r})$ is defined, and Figure 5(b) gives the forces on the shell strip of unit length when the bending moments are negligible (Graff 1975). The dynamic equations of the cylindrical shell in the longitudinal, tangential, and radial directions $(x,\theta,\tilde{r})$ read

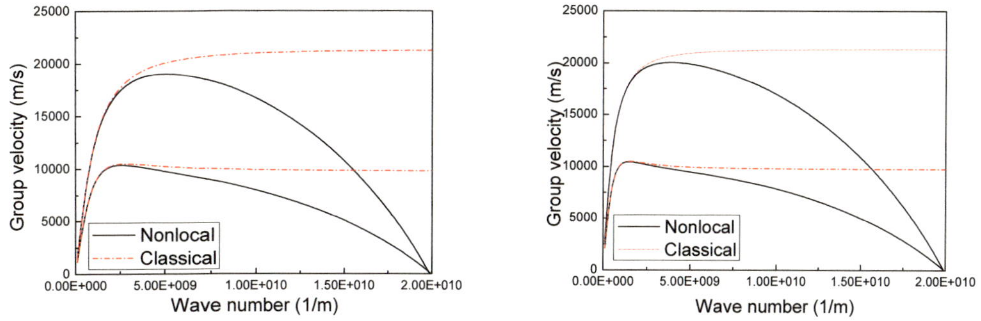

(a) an armchair (5,5) carbon nanotube (b) an armchair (10,10) carbon nanotube

Fig. 4. Dispersion relations between the group velocity and the wave number of flexural waves in single-walled carbon nanotubes (Wang *et al.* 2008a)

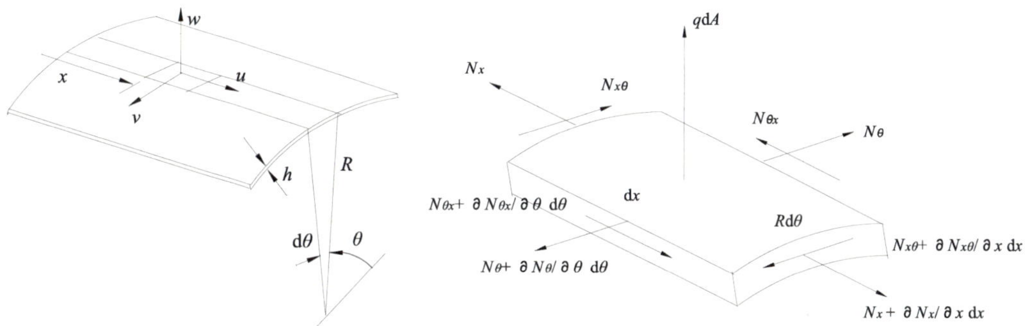

(a) (b)

Fig. 5. The model of a cylindrical shell made of non-local elastic material
(a) A strip from the cylindrical shell,
(b) A small shell element and the internal forces (Wang *et al.* 2006b)

$$\rho h\frac{\partial^2 u}{\partial t^2}-\frac{\partial N_x}{\partial x}-\frac{1}{R}\frac{\partial N_{\theta x}}{\partial \theta}=0\,, \tag{30a}$$

$$\rho h\frac{\partial^2 v}{\partial t^2}-\frac{1}{R}\frac{\partial N_\theta}{\partial \theta}-\frac{\partial N_{x\theta}}{\partial x}=0\,, \tag{30b}$$

$$\rho h\frac{\partial^2 w}{\partial t^2}+\frac{N_\theta}{R}=0\,, \tag{30c}$$

where h presents the thickness of the shell, R the radius of the shell, ρ the mass density, (u,v,w) the displacement components $(x,\theta,\tilde{r})$. $N_x, N_\theta, N_{x\theta}, N_{\theta x}$, the components of the internal force in the shell, can be determined by integrating the corresponding stress components $\sigma_x, \sigma_\theta, \tau_{x\theta}, \tau_{\theta x}$ across the shell thickness as following

$$N_x, N_\theta, N_{x\theta}, N_{\theta x} = \int_{-h/2}^{h/2} (\sigma_x, \sigma_\theta, \tau_{x\theta}, \tau_{\theta x}) dz \, , \tag{31}$$

where z is measured outward from the mid surface of the shell.
The constitutive law of the two-dimensional non-local elastic continuum for the cylindrical shell under the load of axial symmetry as follows,

$$\sigma_x = \lambda(\varepsilon_x + \varepsilon_\theta) + 2\mu\varepsilon_x + \lambda(r_x^2 \frac{\partial^2 \varepsilon_x}{\partial x^2} + r_x^2 \frac{\partial^2 \varepsilon_\theta}{\partial x^2}) + 2\mu r_x^2 \frac{\partial^2 \varepsilon_x}{\partial x^2} \, , \tag{32a}$$

$$\sigma_\theta = \lambda(\varepsilon_x + \varepsilon_\theta) + 2\mu\varepsilon_\theta + \lambda(r_x^2 \frac{\partial^2 \varepsilon_x}{\partial x^2} + r_x^2 \frac{\partial^2 \varepsilon_\theta}{\partial x^2}) + 2\mu r_x^2 \frac{\partial^2 \varepsilon_\theta}{\partial x^2} \, , \tag{32b}$$

$$\tau_{x\theta} = 2\mu\varepsilon_{x\theta} + 2\mu r_x^2 \frac{\partial^2 \varepsilon_{x\theta}}{\partial x^2} \, . \tag{32c}$$

Let $\gamma = 2\varepsilon_{x\theta}$ and γ be the shear strain of the element with $\gamma = \gamma_{x\theta} = \gamma_{\theta x}$. Substituting $\mu = E/(2+2\upsilon) = G$ and $\lambda = E\upsilon/(1-\upsilon^2)$ into Equation (32) yields

$$\sigma_x = \frac{E}{1-\upsilon^2}[\varepsilon_x + r_x^2 \frac{\partial^2 \varepsilon_x}{\partial x^2} + \upsilon(\varepsilon_\theta + r_x^2 \frac{\partial^2 \varepsilon_\theta}{\partial x^2})] \, , \tag{33a}$$

$$\sigma_\theta = \frac{E}{1-\upsilon^2}[\varepsilon_\theta + r_x^2 \frac{\partial^2 \varepsilon_\theta}{\partial x^2} + \upsilon(\varepsilon_x + r_x^2 \frac{\partial^2 \varepsilon_x}{\partial x^2})] \, , \tag{33b}$$

$$\tau_{x\theta} = \tau_{\theta x} = G(\gamma + r_x^2 \frac{\partial \gamma^2}{\partial x^2}) \, , \tag{33c}$$

where $r_x = r = d/\sqrt{12}$ characterizes the influence of microstructures on the constitutive law of the non-local elastic materials, and d, referred to as the inter-particle distance (Askes et al, 2002), is the axial distance between rings of carbon atoms when a single walled carbon nanotube is modeled as a non-local elastic cylindrical shell.
Under the assumption that only the membrane stresses play a role in the thin cylindrical shells, the stress components $\sigma_x, \sigma_\theta, \tau_{x\theta}, \tau_{\theta x}$ are constants throughout the shell thickness such that Equation (31) yields

$$N_x = \frac{Eh}{1-\upsilon^2}[\varepsilon_x + r^2 \frac{\partial^2 \varepsilon_x}{\partial x^2} + \upsilon(\varepsilon_\theta + r^2 \frac{\partial^2 \varepsilon_\theta}{\partial x^2})] \, , \tag{34a}$$

$$N_\theta = \frac{Eh}{1-\upsilon^2}[\varepsilon_\theta + r^2 \frac{\partial^2 \varepsilon_\theta}{\partial x^2} + \upsilon(\varepsilon_x + r^2 \frac{\partial^2 \varepsilon_x}{\partial x^2})] \, , \tag{34b}$$

$$N_{x\theta} = N_{\theta x} = Gh(\gamma + r\frac{\partial \gamma^2}{\partial r^2}) = \frac{Eh}{2(1+\upsilon)}(\gamma + r^2 \frac{\partial \gamma^2}{\partial x^2}) \, . \tag{34c}$$

The geometric relation under the condition $\partial/\partial\theta = 0$ leads to $\varepsilon_x = \partial u/\partial x$, $\varepsilon_\theta = w/R$, $\gamma = \partial v/\partial x$. Substituting Equation (34) into Equation (30) gives a set of dynamic equations of the non-local elastic cylindrical shell

$$\frac{\rho(1-\upsilon^2)}{E}\frac{\partial^2 u}{\partial t^2}-\frac{\partial^2 u}{\partial x^2}-r^2\frac{\partial^4 u}{\partial x^4}-\frac{\upsilon}{R}(\frac{\partial w}{\partial x}+r^2\frac{\partial^3 w}{\partial x^3})=0, \tag{35a}$$

$$\frac{2\rho(1+\upsilon)}{E}\frac{\partial v^2}{\partial t^2}-\frac{\partial^2 v}{\partial x^2}-r^2\frac{\partial^4 v}{\partial x^4}=0, \tag{35b}$$

$$\frac{\rho(1-\upsilon^2)}{E}\frac{\partial^2 w}{\partial t^2}+\frac{1}{R^2}(w+r^2\frac{\partial^2 w}{\partial x^2})+\frac{\upsilon}{R}(\frac{\partial u}{\partial x}+r^2\frac{\partial^3 u}{\partial x^3})=0. \tag{35c}$$

Obviously, Equation (35b) is not coupled with Equations (35a) and (35c) such that the torsional wave in the cylindrical shell is independent of the longitudinal and radial waves. Now consider the motions governed by the coupled dynamic equations in u and w. Let

$$u=\hat{u}e^{i\tilde{k}(x-ct)},\quad w=\hat{w}e^{i\tilde{k}(x-ct)} \tag{36}$$

where $i\equiv\sqrt{-1}$, $\hat{u}$ is the amplitude of longitudinal vibration, $\hat{w}$ the amplitude of radial vibration. c and $\tilde{k}$ are the same as previous definition. Substituting Equation (36) into Equations (35a) and (35c) yields

$$\begin{bmatrix}\frac{\tilde{k}^2c^2}{c_p^2}-\tilde{k}^2+r^2\tilde{k}^4 & i(\tilde{k}-r^2\tilde{k}^3)\frac{\upsilon}{R}\\ -i(\tilde{k}-r^2\tilde{k}^3)\frac{\upsilon}{R} & \frac{\tilde{k}^2c^2}{c_p^2}-\frac{1}{R^2}(1-r^2\tilde{k}^2)\end{bmatrix}\begin{bmatrix}\hat{u}\\ \hat{w}\end{bmatrix}=0. \tag{37}$$

where $c_p\equiv\sqrt{E/((1-\upsilon^2)\rho)}$, which is usually referred to as 'thin-plate' velocity. The existence of non-zero solution $[\hat{u}\ \hat{w}]^T$ of Equation (37) requires

$$c^4-c_p^2(1+\frac{1}{R^2\tilde{k}^2})(1-r^2\tilde{k}^2)c^2+\frac{c_p^4(1-\upsilon^2)}{R^2\tilde{k}^2}(1-r^2\tilde{k}^2)^2=0. \tag{38}$$

Solving Equation (38) for the dimensionless phase velocity c/c_p gives the two branches of the wave dispersion relation as following

$$\frac{c}{c_p}=\sqrt{\frac{(1+\frac{1}{R^2\tilde{k}^2})\mp\sqrt{(1+\frac{1}{R^2\tilde{k}^2})^2-4\frac{1-\upsilon^2}{R^2\tilde{k}^2}}}{2}(1-r^2\tilde{k}^2)}. \tag{39}$$

Equation (39) shows again the intrinsic limit $1-r^2\tilde{k}^2>0$ or $\tilde{k}<\sqrt{12}/d$ for the maximal wave number owing to the microstructure. That is, the longitudinal wave is not able to propagate in the non-local elastic cylindrical shell if the wave length is so short that $\lambda<\pi d/\sqrt{3}\approx1.814d$ holds. If $r=0$ in Equation (39), one arrives at

$$\frac{c}{c_p}=\sqrt{\frac{(1+\frac{1}{R^2\tilde{k}^2})\mp\sqrt{(1+\frac{1}{R^2\tilde{k}^2})^2-4\frac{1-\upsilon^2}{R^2\tilde{k}^2}}}{2}}. \tag{40}$$

This is just the wave dispersion relation of the traditional elastic cylindrical shell.

4.2 Wave propagation simulated by molecular dynamics

This section presents the longitudinal wave dispersion from the theoretical results in Section 4.1 compared with the molecular dynamics results for the longitudinal wave propagation in an armchair (5,5) carbon nanotube and an armchair (10,10) carbon nanotube, respectively, for a wide range of wave numbers.

In the molecular dynamics models, the interatomic interactions are also described by the Tersoff-Brenner potential (Brenner, 1990). It is quite straightforward to determine the phase velocity and the wave number from the longitudinal vibration, simulated by using the molecular dynamics model, of two arbitrary sections of the carbon nanotube. The Young's moduls was 0.46TPa for the armchair (5,5) carbon nanotube and 0.47TPa for the armchair (10,10) carbon nanotube from the molecular dynamics simulation for the text of axial tension. Furthermore, Poisson's ratio $\upsilon = 0.22$ for the armchair (5,5) and $\upsilon = 0.20$ for the armchair (10,10) carbon nanotube.

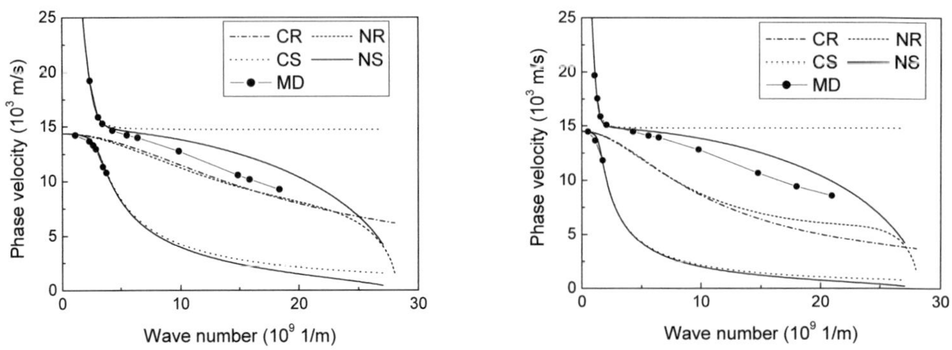

(a) an armchair (5,5) carbon nanotube (b) an armchair (10,10) carbon nanotube.

Fig. 6. The wave dispersion relations between phase velocity versus wave number given by the models of elastic cylindrical shell and non-local elastic cylindrical shell in comparison with molecular dynamics simulations. (Wang *et al.* 2006b)

Figure 6 shows the dispersion relation between the phase velocity c and the wave number $\tilde{k}$, and the dispersion relation between the angular frequency ω and the wave number $\tilde{k}$ given by the models of both elastic cylindrical shell and non-local elastic cylindrical shell in comparison with the numerical simulations of molecular dynamics for the two carbon nanotubes. In Figure 6, the symbol NS represents the model of non-local elastic cylindrical shell as in Equation (40), the symbol CS the model of elastic cylindrical shell as in Equation (41), the symbol CR the rod model of Love theory, the symbol NR the non-local elastic rod model with lateral inertial taken into consideration (Wang, 2005), and MD the molecular dynamics simulation. Obviously, only the results from the model of non-local elastic cylindrical shell coincide well with the two branches of dispersion relations given by molecular dynamics simulations.

4.3 Group velocity of longitudinal wave in a single-walled carbon nanotube

From Equation (39), with $\omega = c\tilde{k}$ considered, the two branches of the wave dispersion relation as following

$$\omega = c_p \sqrt{\frac{(\tilde{k}^2 + \frac{1}{R^2}) \mp \sqrt{(\tilde{k}^2 + \frac{1}{R^2})^2 - 4\frac{(1-\upsilon^2)\tilde{k}^2}{R^2}}}{2}(1 - r^2\tilde{k}^2)} \,. \tag{41}$$

Then, the group velocity reads

$$\begin{aligned} c_g &= \frac{\mathrm{d}\omega}{\mathrm{d}\tilde{k}} \\ &= \frac{c_p}{2\sqrt{2}} \Bigg\{ \left(2\tilde{k} \mp \{[2\tilde{k}(\tilde{k}^2 + \frac{1}{R^2})] - 4\tilde{k}\frac{1-\upsilon^2}{R^2}\}[(\tilde{k}^2 + \frac{1}{R^2})^2 - 4\frac{(1-\upsilon^2)\tilde{k}^2}{R^2}]^{-\frac{1}{2}} \right)(1 - r^2\tilde{k}^2) \\ &+ \left[(\tilde{k}^2 + \frac{1}{R^2}) \mp \sqrt{(\tilde{k}^2 + \frac{1}{R^2})^2 - 4\frac{(1-\upsilon^2)\tilde{k}^2}{R^2}} \right] \left(-2r^2\tilde{k} \right) \Bigg\} \\ & \left\{ \left[(\tilde{k}^2 + \frac{1}{R^2}) \mp \sqrt{(\tilde{k}^2 + \frac{1}{R^2})^2 - 4\frac{(1-\upsilon^2)\tilde{k}^2}{R^2}} \right](1 - r^2\tilde{k}^2) \right\}^{-\frac{1}{2}} . \end{aligned} \tag{42}$$

Figure 7 shows the dispersion relation between the group velocity and the wave number of longitudinal waves in an armchair (5,5) single-walled carbon nanotube and in an armchair (10,10) single-walled carbon nanotube. Now, the product of Young's modulus E and the wall thickness h is $Eh = 346.8\text{Pa}\cdot\text{m}$, and Poisson's ratio is $\upsilon = 0.20$ for the (5,5) single-walled carbon nanotube and the (10,10) single-walled carbon nanotube. In addition, one has $r = 0.0355\text{nm}$ when the axial distance between two neighboring rings of atoms is $d = 0.123\text{nm}$. The product of mass density ρ and the wall thickness h yields $\rho h \approx 760.5$ $\text{kg/m}^3\cdot\text{nm}$. The radii of the (5, 5) and (10, 10) single-walled carbon nanotubes are 0.34nm and 0.68nm, respectively. There is a slight difference between the theory of non-local elasticity and the classical theory of elasticity for the lower branches. The group velocity decreases rapidly with an increase in the wave number. When the wave number approaches to $\tilde{k}$=5×10^9 m^{-1} or so, the group velocity goes to zero for the (5,5) single-walled carbon nanotube. Similarly, the group velocity approaches to zero for the (10, 10) single-walled carbon nanotube when the wave number goes to $\tilde{k}$=3×10^9 m^{-1} or so. This may explain the difficulty that the direct molecular dynamics simulation can only offer the dispersion relation between the phase velocity and the wave number up to $\tilde{k}\approx 5\times10^9$ m^{-1} for the (5,5) single-walled carbon nanotube and up to $\tilde{k}\approx 3\times10^9$ m^{-1} for the (10,10) single-walled carbon nanotube. For the upper branches of the dispersion relation, the difference can hardly be identified when the wave number is lower. However, the results of the elastic cylindrical shell remarkably deviate from those of the non-local elastic cylindrical shells with an increase in the wave number. Figure 7 (a) and (b) show the intrinsic limit $\tilde{k}<2\times10^{10}$ m^{-1}, instead of $\tilde{k}<\sqrt{12}/d\approx 2.82\times10^{10}\text{m}^{-1}$ for the maximal wave number owing to the micro-structures. This can explain the contradiction that the cut-off longitudinal wave predicted by the non-local elastic cylindrical shell is $\tilde{k}<\sqrt{12}/d\approx 2.82\times 10^{10}\mathrm{m}^{-1}$, but the molecular dynamics simulation offers the dispersion relation up to the wave number $\tilde{k}\approx 2\times 10^{10}\ \mathrm{m}^{-1}$ only (Wang et al. 2006b).

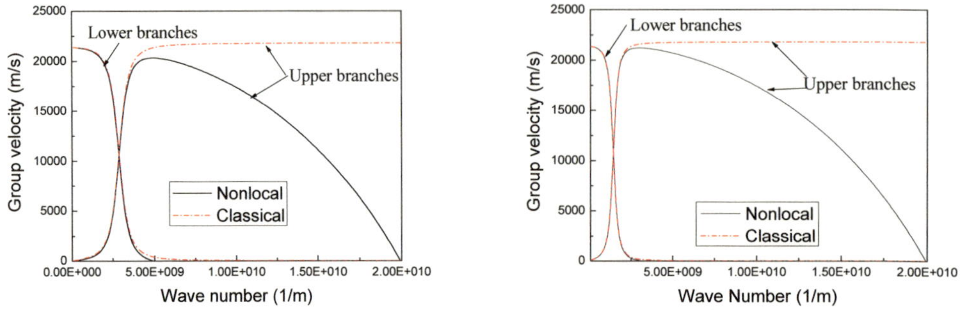

(a) an armchair (5,5) carbon nanotube (b) an armchair (10,10) carbon nanotube

Fig. 7. Dispersion relations between the group velocity and the wave number of longitudinal waves in armchair single-walled carbon nanotubes (Wang *et al.* 2008a)

5. Flexural waves in a multi-walled carbon nanotube

5.1 Multi Timoshenko beam model

Here, a nonlocal multiple-elastic Timoshenko beam model with second order strain gradient taken into consideration is developed, in which each of the nested, originally concentric nanotubes of a multi-walled carbon nanotube is described as an individual elastic beam, and the deflections of all nested tubes are coupled through the van der Waals interaction between any two tubes. Since all nested tubes of a multi-walled carbon nanotube are originally concentric and the van der Waals interaction is determined by the interlayer spacing, the net van der Waals interaction pressure remains zero for each of the tubes provided they deform coaxially. Thus, for small-deflection linear vibration the interaction pressure at any point between any two adjacent tubes linearly depends on the difference of their deflections at that point. Here, we assume that all nested individual tubes of the multi-walled carbon nanotube vibrate in the same plane. Thus, coplanar transverse vibration of N nested tubes of an N wall is described by N coupled equations. The dynamics equations of the j th layer for N-walled carbon nanotube,

$$\rho A_k \frac{\partial^2 w_k}{\partial t^2}+\beta A_k G\left[\left(\frac{\partial \varphi_k}{\partial x}+r^2\frac{\partial^3 \varphi_k}{\partial x^3}\right)-\left(\frac{\partial^2 w_k}{\partial x^2}+r^2\frac{\partial^4 w_k}{\partial x^4}\right)\right]=w_k\sum_{j=1}^{n}C_{kj}-\sum_{j=1}^{n}C_{kj}w_j \tag{43a}$$

$$\rho I_k \frac{\partial^2 \varphi_k}{\partial t^2}+\beta A_k G\left[(\varphi_k-\frac{\partial w_k}{\partial x})+r^2\left(\frac{\partial^2 \varphi_k}{\partial x^2}-\frac{\partial^3 w_k}{\partial x^3}\right)\right]-EI_k(\frac{\partial^2 \varphi_k}{\partial x^2}+r^2\frac{\partial^4 \varphi_k}{\partial x^4})=0 \tag{43b}$$

where E, G ,υ , r, ρ , h represents the same meaning as previous sections. A_k represents the cross section area of the beam, $I_k=\int y^2 dA_k$ represents the moment of inertia for the cross section , w_k being the displacement of section, φ_k denote the slope of the deflection curve when the shearing force is neglected.

For small-deflection linear vibration, the van der Waals pressure at any point between two tubes should be a linear function of the jump in deflection at that point.

$$p_k = \sum_{k=1}^{n} C_{kj}(w_k - w_j) = w_k \sum_{j=1}^{n} C_{kj} - \sum_{j=1}^{n} C_{kj} w_j \tag{44}$$

where N is the total number of layers of the multi-walled carbon nanotube. C_{kj} is the van der Waals interaction coefficients for interaction pressure per unit axial length can be estimated based on an effective interaction width (Ru, 2000)

$$C_{kj} = 2R_k c_{kj} \tag{45}$$

The van der Waals interaction coefficients can be obtained through the Lennard-Jones pair potential (Jones, 1924; Girifalco & Lad, 1956)

$$V(r) = 4\varepsilon[(\sigma / \tilde{r})^{12} - (\sigma / \tilde{r})^{6}]\,. \tag{46}$$

where $\varepsilon = 2.968 \times 10^{-3}\,\text{eV}$, $\sigma = 3.407\,\text{Å}$, where $\tilde{r}$ is the distance between two interacting atoms.

Note that the attractive van der Waals force that is obtained from the Lennard-Jones pair potential is negative, the repulsive van der Waals force is positive, and the downward pressure is assumed to be positive.

Here, van der Waals interaction coefficients c_{jk} obtained through the Lennard-Jones pair potential by He et al. (2005) are used.

$$c_{kj} = -\left[\frac{1001\pi\varepsilon\sigma^{12}}{3a^4}E_{kj}^{13} - \frac{1120\pi\varepsilon\sigma^{6}}{9a^4}E_{kj}^{7}\right]R_j\,, \tag{47}$$

where

$$E_{kj}^{m} = (R_k + R_j)^{-m} \int_0^{\frac{\pi}{2}} \frac{d\theta}{\left[1 - K_{kj}\cos^2\theta\right]^{\frac{m}{2}}}\,, \tag{48}$$

and

$$K_{kj} = \frac{4R_k R_j}{(R_k + R_j)^2}\,. \tag{49}$$

5.2 Dispersion of flexural wave in multi-walled carbon nanotubes

To study the flexural wave propagation, let us consider the deflection and the slope given by

$$w_k = \hat{w}_k e^{i\tilde{k}(x-ct)}, \quad \varphi_k = \hat{\varphi}_k e^{i\tilde{k}(x-ct)}, \quad (k = 1 \cdots n) \tag{50}$$

where $i \equiv \sqrt{-1}$, $\hat{w}_k$ represent the amplitudes of deflections of the k th tube, and $\hat{\varphi}_k$ the amplitudes of the slopes of the j th tube due to bending deformation alone. In addition, c and $\tilde{k}$ are the same as previous definition. Substituting Equation (50) into Equation (43) yields

$$\begin{cases}(\rho\tilde{k}^2c^2-\beta G\tilde{k}^2+\beta Gr^2\tilde{k}^4)\hat{w}_k-(\mathrm{i}\beta G\tilde{k}-\mathrm{i}\beta Gr^2\tilde{k}^3)\hat{\varphi}_k+\hat{w}_k\sum_{j=1}^{n}C_{kj}\Big/A_k-\sum_{j=1}^{n}\frac{C_{kj}}{A_k}\hat{w}_j=0\\(\mathrm{i}\beta A_kG\tilde{k}-\mathrm{i}\beta A_kGr^2\tilde{k}^3)\hat{w}_k+(\rho I_k\tilde{k}^2c^2-\beta A_kG+\beta A_kGr^2\tilde{k}^2-EI_k\tilde{k}^2+EI_kr^2\tilde{k}^4)\hat{\varphi}_k=0\end{cases} \tag{51}$$

From the fact that there exists at least one non-zero solution $(\hat{w},\hat{\varphi})$ of Equation (51), one arrives at

$$\left[\rho\tilde{k}^2c^2\mathbf{I}_{2n\times 2n}+\mathbf{H}_{2n\times 2n}\right]\begin{bmatrix}\hat{w}_1\\\hat{\varphi}_1\\\hat{w}_2\\\hat{\varphi}_2\\\vdots\\\hat{w}_n\\\hat{\varphi}_n\end{bmatrix}=0, \tag{52}$$

where $\mathbf{I}_{2n\times 2n}$ is an identity matrix and the elements in the matrix $\mathbf{H}_{2n\times 2n}$ are

$$\mathbf{H}(2k-1,2j-1)=-\frac{C_{kj}}{A_k}\quad (j\neq k) \tag{53a}$$

$$\mathbf{H}(2k-1,2k-1)=-\beta G\tilde{k}^2+\beta Gr^2\tilde{k}^4+\sum_{j=1}^{n}C_{kj}\Big/A_k\ ; \tag{53b}$$

$$\mathbf{H}(2k-1,2k)=-(\mathrm{i}\beta G\tilde{k}-\mathrm{i}\beta Gr^2\tilde{k}^3)\ ; \tag{53c}$$

$$\mathbf{H}(2k,2k-1)=(\mathrm{i}\beta A_kG\tilde{k}-\mathrm{i}\beta A_kGr^2\tilde{k}^3)/I_k\ ; \tag{53d}$$

$$\mathbf{H}(2k,2k)=(-\beta A_kG+\beta A_kGr^2\tilde{k}^2-EI_k\tilde{k}^2+EI_kr^2\tilde{k}^4)/I_k\ ; \tag{53e}$$

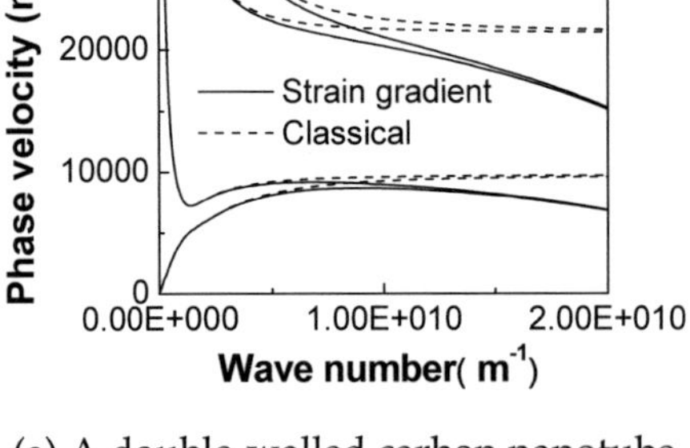

(a) A double-walled carbon nanotube ((5,5),(10,10))

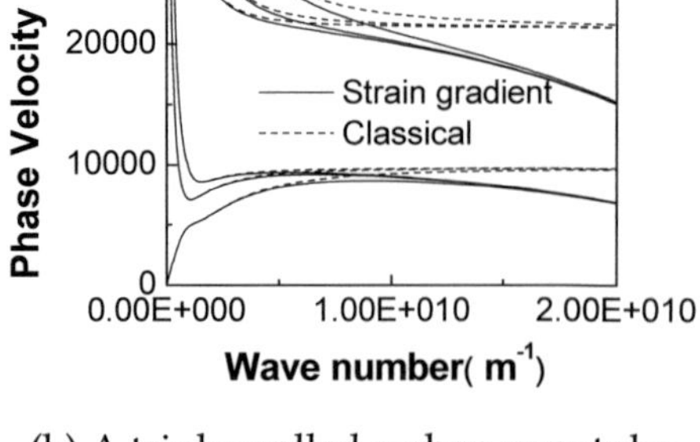

(b) A triple-walled carbon nanotube ((5,5),(10,10),(15,15))

Fig. 8. Flexural wave dispersion of multi-walled carbon nanotubes given by strain gradient Timoshenko model compared with classical Timoshenko model

Figure 8 show the comparison of strain gradient Timoshenko beam model and classical Timoshenko model in deal with the flexural wave dispersion of a double-walled carbon nanotubes ((5,5),(10,10)) and a triple-walled carbon nanotubes ((5,5),(10,10),(15,15)). Where in Equations (52) and (53), Young's modulus $E = 1.02\text{TPa}$, shear modulus $G = E/2(1+\upsilon)$ and poisson ratio $\upsilon = 0.2$, the shear form factor $\beta = 0.5$, the material parameter $r = 0.0355\text{nm}$, the mass density $\rho = 2237\text{kg/m}^3$, thickness $h = 0.34\text{nm}$. In all these figures, it can be seen, the difference of the dispersion relation between phase velocity and wave number given by strain gradient shell model and classical shell model is more and more obviously with the increase of wave number. This is similar as single walled carbon nanotubes' results (Wang & Hu, 2005).

5.3 Group velocity of flexural waves in a multi-walled carbon nanotube

Under the assumption that there exists at least one non-zero solution $(\hat{w}_k, \hat{\varphi}_k)$ of Equation (51), with $\omega = c\tilde{k}$ considered, one arrives at

$$\left[\rho\omega^2 \mathbf{I}_{2n\times 2n} + \mathbf{H}_{2n\times 2n}\right]\begin{bmatrix}\hat{w}_1\\ \hat{\varphi}_1\\ \hat{w}_2\\ \hat{\varphi}_2\\ \vdots\\ \hat{w}_n\\ \hat{\varphi}_n\end{bmatrix} = 0, \tag{54}$$

where $\mathbf{I}_{2n\times 2n}$ is an identity matrix and the entries in matrix $\mathbf{H}_{2n\times 2n}$ are

$$\mathbf{H}(2k-1, 2j-1) = -\frac{C_{kj}}{A_k} \quad (j \neq k), \tag{55a}$$

$$\mathbf{H}(2k-1, 2k-1) = -\beta G\tilde{k}^2 + \beta G r^2 \tilde{k}^4 + \sum_{j=1}^{n} C_{kj} \Big/ A_k, \tag{55b}$$

$$\mathbf{H}(2k-1, 2k) = -(\mathrm{i}\beta G\tilde{k} - \mathrm{i}\beta G r^2 \tilde{k}^3), \tag{55c}$$

$$\mathbf{H}(2k, 2k-1) = (\mathrm{i}\beta A_k G\tilde{k} - \mathrm{i}\beta A_k G r^2 \tilde{k}^3)/I_k, \tag{55d}$$

$$\mathbf{H}(2k, 2k) = (-\beta A_k G + \beta A_k G r^2 \tilde{k}^2 - E I_k \tilde{k}^2 + E I_k r^2 \tilde{k}^4)/I_k. \tag{55e}$$

From Equation (54), the dispersion relation between the group velocity and the wave number of flexural waves in a multi-walled carbon nanotube can be numerically obtained by using $c_g = \mathrm{d}\omega/\mathrm{d}\tilde{k}$.

Number of tube k (radius)	ρA_k (kg/m)	ρI_k(kg·m)	EI_k(Pa·m^4)
k=1 (0.34nm)	1.625×10^{-15}	3.737×10^{-35}	1.704×10^{-26}
k=2 (0.68nm)	3.250×10^{-15}	2.541×10^{-34}	1.158×10^{-25}

Table 1. Parameters of a double-wall carbon nanotube

Number of tube k (radius)	ρA_k (kg/m)	ρI_k(kg·m)	EI_k(Pa·m^4)
k=1 (0.34nm)	1.625×10^{-15}	3.737×10^{-35}	1.704×10^{-26}
k=2 (0.68nm)	3.250×10^{-15}	2.541×10^{-34}	1.158×10^{-25}
k=3 (1.02nm)	4.874×10^{-15}	8.296×10^{-34}	3.782×10^{-25}
k=4 (1.36nm)	6.466×10^{-15}	1.943×10^{-33}	8.860×10^{-25}

Table 2. Parameters of a quadruple-wall carbon nanotube

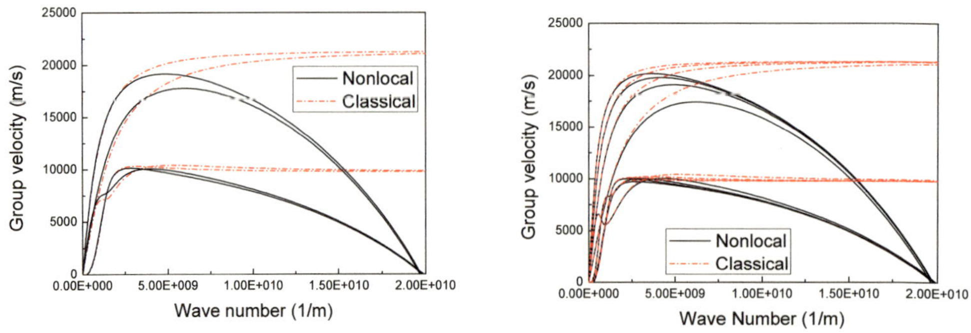

(a) A double-walled carbon nanotube with the radii 0.34nm and 0.68nm.

(b) A quadruple -walled carbon nanotube with the radii 0.34nm, 0.68nm, 1.02nm and 1.36nm

Fig. 9. Dispersion relations between the group velocity and the wave number of flexural wave in multi-walled carbon nanotubes. (Wang *et al.* 2008a)

Figure 9 presents the dispersion relation between the group velocity and the wave number of flexural waves in a double-walled carbon nanotube with the radii $R_1 = 0.34\text{nm}$ and $R_2 = 0.68\text{nm}$ respectively, and in a quadruple-walled carbon nanotube with the radii $R_1 = 0.34\text{nm}$, $R_2 = 0.68\text{nm}$, $R_3 = 1.02\text{nm}$ and $R_4 = 1.36\text{nm}$ respectively. The product of Young's modulus and the wall thickness is $Eh = 346.8\text{Pa}\cdot\text{m}$, and Poisson's ratio is $\upsilon = 0.20$. There follows $G = E/(2(1+\upsilon))$. The product of mass density and the wall thickness yields $\rho h \approx 760.5\text{kg/m}^3\cdot\text{nm}$, and the form factor of shear $\beta = 0.5$. Table 1 shows the parameters $\rho A_k, \rho I_k, EI_k$ of the double-walled carbon nanotube. Table 2 gives the parameters $\rho A_k, \rho I_k, EI_k$ of the quadruple-walled carbon nanotube. For all these lower branches and the upper branches of the dispersion relation, the difference is almost invisible when the wave number is lower. However, the results of the elastic model of Timoshenko beams remarkably deviate from those of non-local elastic model of Timoshenko beams with an increase in the wave number. Figures 9 (a) and (b) show again the intrinsic limit

$\tilde{k}<2\times10^{10}$ m^{-1} , rather than $\tilde{k}<\sqrt{12}/d\approx2.82\times10^{10}$m^{-1} for the maximal wave number owing to the micro-structures.

6. Longitudinal wave in a multi-walled carbon nanotube

6.1 Strain gradient multi shell model

This section studies dispersion of longitudinal wave in a multi-walled carbon nanotube with a multi-cylindrical shell model that takes into consideration the second order strain gradient. For such a thin cylindrical shell, the bending moments can be naturally neglected for simplicity. The dynamics equations of the k th layer for an N-walled carbon nanotube are

$$\frac{\rho(1-\upsilon^2)}{E}\frac{\partial^2 u_k}{\partial t^2}-\frac{\partial^2 u_k}{\partial x^2}-r^2\frac{\partial^4 u_k}{\partial x^4}-\frac{\upsilon}{R_k}(\frac{\partial w_k}{\partial x}+r^2\frac{\partial^3 w_k}{\partial x^3})=0, \tag{56a}$$

$$\frac{2\rho(1+\upsilon)}{E}\frac{\partial^2 v_k}{\partial t^2}-\frac{\partial^2 v_k}{\partial x^2}-r^2\frac{\partial^4 v_k}{\partial x^4}=0, \tag{56b}$$

$$\frac{\rho(1-\upsilon^2)}{E}\frac{\partial^2 w_k}{\partial t^2}+\frac{1}{R_k{}^2}(w_k+r^2\frac{\partial^2 w_k}{\partial x^2})+\frac{\upsilon}{R_k}(\frac{\partial u_k}{\partial x}+\frac{\partial^3 u_k}{\partial x^3})=\frac{q_k(1-\upsilon^2)}{Eh}. \tag{56c}$$

As only infinitesimal vibration is considered, the net pressure due to the van der Waals interaction is assumed to be linearly proportional to the deflection between two layers, i.e.,

$$q_k=\sum_{j=1}^{N}c_{kj}(w_k-w_j)=w_k\sum_{j=1}^{N}c_{kj}-\sum_{j=1}^{N}c_{kj}w_j, \tag{57}$$

where N is the total number of layers of the multi-walled carbon nanotube and c_{kj} is the van der Waals interaction coefficient.

The van der Waals interaction coefficients can be obtained through the Lennard-Jones pair potential (Jones, 1924). Here, van der Waals interaction coefficients obtained through the Lennard-Jones pair potential by He et al. (2005) are used.

Obviously, Equation (56b) is uncoupled from the other two equations in Equation (56) such that the torsional waves in the shell are not coupled with the longitudinal and radial waves.

6.2 Longitudinal wave dispersion in multi-walled carbon nanotubes

In this section, the wave propagations of multi-walled carbon nanotubes are investigated via the second-order strain gradient shell theory.

Consider the motions governed by the coupled dynamic equations in u and w , let

$$u_k=\hat{u}_k e^{i\tilde{k}(x-ct)},\ w_k=\hat{w}_k e^{i\tilde{k}(x-ct)},\ (k=1,2\cdots N) \tag{58}$$

where $i\equiv\sqrt{-1}$, $\hat{u}_k$ is the amplitude of longitudinal vibration, and $\hat{w}_k$ is the amplitude of radial vibration. Substituting Equation (58) into Equations (56a) and (56c) yields

$$\left[\frac{\tilde{k}^2c^2}{c_p^2}-\tilde{k}^2+r^2\tilde{k}^4\right]\hat{u}_k+\left[i(\tilde{k}-r^2\tilde{k}^3)\frac{\upsilon}{R}\right]\hat{w}_k=0, \tag{59a}$$

$$\left[-\mathrm{i}(\tilde{k}-r^2\tilde{k}^3)\frac{\upsilon}{R}\right]\hat{u}_k+\left[\frac{\tilde{k}^2c^2}{c_p^2}-\frac{1}{R^2}(1-r^2\tilde{k}^2)\right]\hat{w}_k+\frac{1-\upsilon^2}{Eh}\left(\hat{w}_k\sum_{j=1}^{n}c_{kj}-\sum_{j=1}^{n}c_{kj}\hat{w}_j\right)=0\,. \qquad (59b)$$

The dispersion relation can then be obtained by solving the eigenvalue equation,

$$\left[\frac{\tilde{k}^2}{c_p^2}c^2\mathbf{I}_{2N\times 2N}+\mathbf{H}_{2N\times 2N}\right]\begin{bmatrix}\hat{u}_1\\ \hat{w}_1\\ \hat{u}_2\\ \hat{w}_2\\ \vdots\\ \hat{u}_N\\ \hat{w}_N\end{bmatrix}=0\,, \qquad (60)$$

where $c_p\equiv\sqrt{E/((1-\upsilon^2)\rho)}$, and $\mathbf{I}_{2N\times 2N}$ is an identity matrix and the elements in the matrix $\mathbf{H}_{2N\times 2N}$ are

$$\mathbf{H}(2k,2j)=-\frac{(1-\upsilon^2)c_{kj}}{Eh}\quad (k\neq j)\,, \qquad (61a)$$

$$\mathbf{H}(2k-1,2k-1)=-\tilde{k}^2+r^2\tilde{k}^4\,, \qquad (61b)$$

$$\mathbf{H}(2k-1,2k)=\mathrm{i}(\tilde{k}-r^2\tilde{k}^3)\frac{\upsilon}{R_k}\,, \qquad (61c)$$

$$\mathbf{H}(2k,2k-1)=-\mathrm{i}(\tilde{k}-r^2\tilde{k}^3)\frac{\upsilon}{R_k}\,, \qquad (61d)$$

$$\mathbf{H}(2k,2k)=-\frac{1}{R_k^2}(1-r^2\tilde{k}^2)+\frac{1-\upsilon^2}{Eh}\sum_{j=1}^{n}c_{kj}\,. \qquad (61e)$$

The parameters of Equation (60) are as follows: $Eh=347\mathrm{Pa{\bullet}m}$, $\rho h=761\,\mathrm{kg{\bullet}nm/m^3}$, $r=0.0355\mathrm{nm}$, $\upsilon=0.2$.

To see the effect of the strain gradient in multi-walled carbon nanotubes clearly; two double-walled carbon nanotubes are using as examples. Equation (60) becomes

$$\begin{bmatrix}\frac{\tilde{k}^2c^2}{c_p^2}-\tilde{k}^2+r^2\tilde{k}^4 & \mathrm{i}(\tilde{k}-r^2\tilde{k}^3)\frac{\upsilon}{R_1} & 0 & 0\\ -\mathrm{i}(\tilde{k}-r^2\tilde{k}^3)\frac{\upsilon}{R_1} & \frac{\tilde{k}^2c^2}{c_p^2}-\frac{1}{R_1^2}(1-r^2\tilde{k}^2)+\frac{1-\upsilon^2}{Eh}c_{12} & 0 & -\frac{(1-\upsilon^2)c_{12}}{Eh}\\ 0 & 0 & \frac{\tilde{k}^2c^2}{c_p^2}-\tilde{k}^2+r^2\tilde{k}^4 & \mathrm{i}(\tilde{k}-r^2\tilde{k}^3)\frac{\upsilon}{R_1}\\ 0 & \frac{(1-\upsilon^2)c_{12}}{Eh} & -\mathrm{i}(\tilde{k}-r^2\tilde{k}^3)\frac{\upsilon}{R_1} & \frac{\tilde{k}^2c^2}{c_p^2}-\frac{1}{R_1^2}(1-r^2\tilde{k}^2)-\frac{1-\upsilon^2}{Eh}c_{12}\end{bmatrix}\begin{bmatrix}\hat{u}_1\\ \hat{w}_1\\ \hat{u}_2\\ \hat{w}_2\end{bmatrix}=0 \qquad (62)$$

Figures 10 compare the strain gradient shell model and the classical shell model in terms of the longitudinal wave dispersion of two double-walled carbon nanotubes (with diameters of 0.68nm and 1.36nm, 6.8nm and 7.48nm, respectively). The difference between the dispersion relation given by the strain gradient shell model and the classical shell model becomes more and more obvious with increasing wave number. This is similar to the results obtained from single-walled carbon nanotubes (Wang & Hu, 2005).

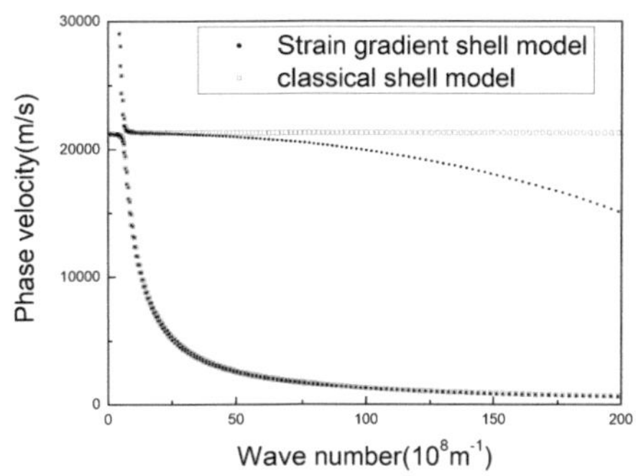

(a) A double-walled carbon nanotube with diameter inner 6.8nm, outer 7.48nm

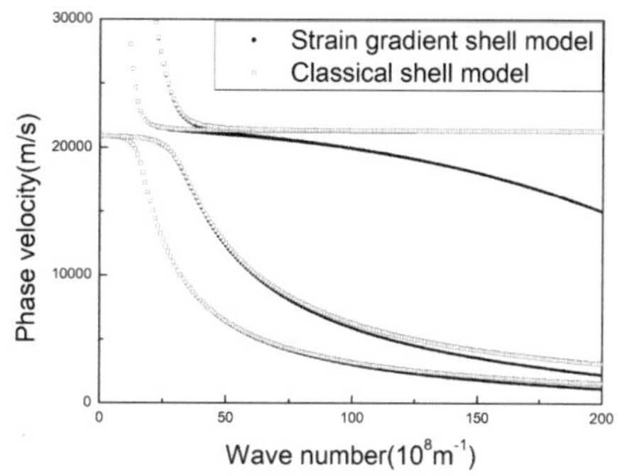

(b) A double-walled carbon nanotube with diameter inner 0.68nm, outer 1.36nm

Fig. 10. Comparison of the strain gradient shell model and the classical shell model on the dispersion relation of double-walled carbon nanotubes(Wang *et al.* 2008b)

6.3 Group velocity of longitudinal wave in a multi-walled carbon nanotube

From Equation (59), with $\omega = c\tilde{k}$ considered, one arrives at

$$\left[\frac{\omega^2}{c_p^2}\mathbf{I}_{2N\times 2N} + \mathbf{H}_{2N\times 2N}\right]\begin{bmatrix}\hat{u}_1\\ \hat{w}_1\\ \hat{u}_2\\ \hat{w}_2\\ \vdots\\ \hat{u}_N\\ \hat{w}_N\end{bmatrix} = 0, \qquad (63)$$

where $c_p \equiv \sqrt{E/((1-\upsilon^2)\rho)}$, $\mathbf{I}_{2N\times 2N}$ is an identity matrix and the entries in matrix $\mathbf{H}_{2N\times 2N}$ are

$$\mathbf{H}(2k,2j) = -(1-\upsilon^2)c_{kj}/Eh, \qquad k \neq j, \qquad (64a)$$

$$\mathbf{H}(2k-1,2k-1) = -\tilde{k}^2 + r^2\tilde{k}^4, \qquad (64b)$$

$$\mathbf{H}(2k-1,2k) = \mathrm{i}(\tilde{k} - r^2\tilde{k}^3)\upsilon/R_k, \qquad (64c)$$

$$\mathbf{H}(2k,2k-1) = -\mathrm{i}(\tilde{k} - r^2\tilde{k}^3)\upsilon/R_k\,, \tag{64d}$$

$$\mathbf{H}(2k,2k) = -\frac{1}{R_k^2}(1 - r^2\tilde{k}^2) + \frac{1-\upsilon^2}{Eh}\sum_{j=1}^{n} c_{kj}\,. \tag{64e}$$

From Equation (63), the dispersion relation between the group velocity and the wave number of longitudinal waves in a multi-walled carbon nanotube can be numerically determined through $c_g = \mathrm{d}\omega/\mathrm{d}\tilde{k}$.

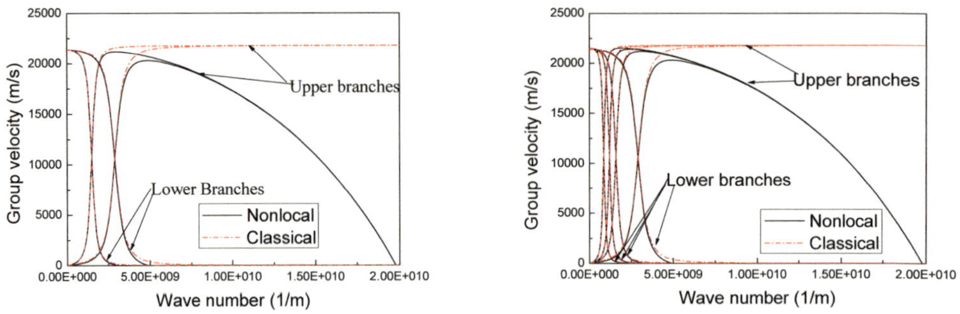

(a) A double-walled carbon nanotube with radii 0.34nm and 0.68nm

(b) A quadruple-walled carbon nanotube with radii 0.34nm, 0.68nm, 1.02nm and 1.36nm

Fig. 11. Dispersion relation between the group velocity and the wave number of longitudinal wave in multi-walled carbon nanotubes(Wang *et al.* 2008a)

Figure 11 presents the dispersion relation between the group velocity and the wave number of longitudinal waves in a double-walled carbon nanotube with the radii $R_1 = 0.34\text{nm}$ and $R_2 = 0.68\text{nm}$ respectively, and in a quadruple-walled carbon nanotube with the radii $R_1 = 0.34\text{nm}$, $R_2 = 0.68\text{nm}$, $R_3 = 1.02\text{nm}$ and $R_4 = 1.36\text{nm}$, respectively. The product of Young's modulus and the wall thickness is $Eh = 346.8\text{Pa}\cdot\text{m}$, and Poisson's ratio is $\upsilon = 0.20$ for both the double-walled carbon nanotube and the quadruple-walled carbon nanotube. In addition, one has $r = 0.0355\text{nm}$ when the axial distance between two neighboring rings of atoms reads $d = 0.123\text{nm}$. The product of mass density and the wall thickness yields $\rho h \approx 760.5\text{kg/m}^3 \cdot \text{nm}$. Now, there is a slight difference between the theory of non-local elasticity and the classical theory of elasticity for the lower branches. The group velocity decreases rapidly with an increase in the wave number. When the wave number approaches to $\tilde{k}=5\times10^9\ \text{m}^{-1}$ or so, the group velocity of all these lower branches goes to zero. For the upper branches of the dispersion relation, the difference is almost invisible when the wave number is lower. However, the results of the elastic cylindrical shells remarkably deviate from those of non-local elastic cylindrical shells with an increase in the wave number. Figures 11 (a) and (b) show again the intrinsic limit of wave number yielding $\tilde{k}<2\times10^{10}\ \text{m}^{-1}$, instead of $\tilde{k}<\sqrt{12}/d\approx2.82\times10^{10}\text{m}^{-1}$.

7. Concluding remarks

This chapter presents a detailed study on the dispersion relation between the phase velocity, group velocity and the wave number for the propagation of flexural and longitudinal waves in single- and multi-walled carbon nanotubes on the basis of the elastic and the non-local elastic Timoshenko beams and cylindrical shells. The study indicates that both elastic and non-local elastic models can offer the right prediction when the wave number is lower. However, the results of the elastic model remarkably deviate from those given by the non-local elastic model with an increase of the wave number. As a result, the micro-structures play an important role in the dispersion of both longitudinal and flexural waves in single- and multi-walled carbon nanotubes.

The cut-off wave number of the dispersion relation between the group velocity and the wave number is about $2\times10^{10}\ 1/\mathrm{m}$ for both flexural and longitudinal waves in both single- and multi-walled carbon nanotubes. This fact interprets the contradiction that the direct molecular dynamics simulation can not give the dispersion relation between the phase velocity and the wave number when the wave number approaches to $2\times10^{10}\ 1/\mathrm{m}$, which is much lower than the cut-off wave number of the dispersion relation between the phase velocity and the wave number predicted by the continuum mechanics.

Propagation of waves in infinite long carbon nanotubes is discussed in this chapter. The effects of boundary on the wave propagation are not considered. It is of great interest to carry out the research on the finite length carbon nanotube, with the effects of boundary conditions taken into consideration. There is little experiment work on the wave propagation in carbon nanotube. Future experiment studies should be conducted to explore this properties and potential applications.

8. Acknowledgements

This work was supported by the National Natural Science Foundation of China under Grants 10702026.

9. References

Askes, H.; Suiker, A. S. J. & Sluys, L. J. (2002). A classification of higher-order strain-gradient models - linear analysis. *Archive of Applied Mechics,* 72, 171-188.

Brenner D. W. (1990) Empirical potential for hydrocarbons for use in simulating the chemical vapor deposition of diamond films. *Physical Review B,* 42, 9458.

Chakraborty, A. (2007). Shell element based model for wave propagation analysis in multi-wall carbon nanotubes. *International Journal of Solids and Structures,* 44, 1628-1642

Chang C. S.; Gao J. (1995). Second-gradient constitutive theory for granular material with random packing structure. *International Journal Solids and Structures,* 32(16), 2279-2293.

Cornwell C. F.; Wille L. T. (1997). Elastic properties of single-walled carbon nanotubes in compression. *Solid State Communications,* 101, 555.

Craig R. R., (1981). *Structural Dynamics* (New York: John Wiley & Sons,).

Dong, K. & Wang, X. (2006). Wave propagation in carbon nanotubes under shear deformation. *Nanotechnology,* 17, 2773-2782.

Graff, K. F. (1975). *Wave Motion in Elastic Solids*,Ohio: Ohio State University.

Girifalco L.A. & Lad, R. A. (1956). Energy of Cohesion, Compressibility, and the Potential Energy Functions of the Graphite System. *The Journal of Chemical Physics,* 25, 693

Halicioglu T. (1998). Stress Calculations for Carbon Nanotubes, *Thin solid films,* 312, 11

He X. Q.; Kitipornchai S; & Liew, K. M. (2005). Buckling analysis of multi-walled carbon nanotubes: a continuum model accounting for van der Waals interaction. *Journal of the Mechanics and Physics of Solids*, 53, 303-326.

Iijima, S. (1991). Helical microtubules of graphitic carbon. *Nature*, 354, 56-58.

Jones, J.E. (1924). The determination of molecular fields: from the variation of the viscosity of a gas with temperature. *Proceedings of the Royal Society of London A* , 106, 441.

Leach, A. R. (1996). *Molecular Modeling*, London: Addison Wesley Longman Limited

Manolis, G. D. (2000). Some basic solutions for wave propagation in a rod exhibiting non-local elasticity. *Engineering Analysis with Boundary Elements,* 24(6), 503-508.

Mindlin, R. D. (1964). Micro-structure in linear elasticity. *Archive of Rational Mechanics and Analysis,* 16(1), 51-78.

Mühlhaus, H. B. & Oka F. (1996). Dispersion and wave propagation in discrete and continuous models for granular materials. *International Journal of Solids and Structures,* 33(19), 2841-2858.

Natsuki, T.; Hayashi, T. & Endo, M. (2005). Wave propagation of carbon nanotubes embedded in an elastic medium. *Journal of Applied Physics,* 97, 044307.

Natsuki, T.; Endo, M. & Tsuda, H. (2006). Vibration analysis of embedded carbon nanotubes using wave propagation approach. *Journal of Applied Physics,* 99, 034311.

Nowinski, J. L. (1984). On a nonlocal theory of longitudinal waves in an elastic circular bar. *Acta Mechanica,* 52(3-4), 189-200.

Poncharal, P.; Wang, Z. L.; Ugarte, D. & de Heer W. A. (1999). Electrostatic Defections and Electromechanical Resonances of Carbon Nanotubes. *Science,* 283, 1513-1516.

Popov, V. N.; Van Doren, V. E. & Balkanski, M. (2000). Elastic properties of single-walled carbon nanotubes. *Physical Review B,* 61, 3078-3084.

Qian, D.; Wagner, G. J.; Liu, W. K.; Yu, M. F. & Ruoff, R. S. (2002). Mechanics of carbon nanotubes. *Applied Mechanics Review,* 55, 495-533.

Ru, C. Q. (2000). Column buckling of multiwalled carbon nanotubes with interlayer radial displacements. *Physical Review B,* 62, 16962.

Sudak, L. J. (2003). Column buckling of multiwalled carbon nanotubes using nonlocal continuum mechanics. *Journal of Applied Physics*, 94, 7281-7287.

Timoshenko, S. & Gere, J. (1972). *Mechanics of Material,* New York: Van Nostrand Reinhold Company.

Treacy, M. M. J.; Ebbesen, T. W. & Gibson J. M. (1996). Exceptionally high Young's modulus observed for individual carbon nanotubes. *Nature*, 381, 678-680.

Wang, C. Y.; Ru, C. Q. & Mioduchowski, A. (2005a). Free vibration of multiwall carbon nanotubes. *Journal of Applied Physics*, 97, 114323.

Wang, C. M.; Zhang, Y. Y.; Ramesh, S. S. & Kitipornchai, S. (2006a). Buckling analysis of micro- and nano-rods/tubes based on nonlocal Timoshenko beam theory. *Journal of Physics D: Applied Physics*, 39, 3904-3909.

Wang, L. F. (2005). *On some mechanics problems in one dimensional nanostructures.* Doctoral dissertation of Nanjing University of Aeronautics and Astronautics (in Chinese).

Wang, L. F. & Hu, H. Y. (2005). Flexural wave propagation in single-walled carbon nanotubes. *Physical Review B*, 71, 195412.

Wang, L. F.; Hu, H. Y., & Guo, W. L. (2006b). Validation of the non-local elastic shell model for studying longitudinal waves in single-walled carbon nanotubes. *Nanotechnology*, 17, 1408-1415.

Wang L. F.; Guo W. L.; Hu H. Y. (2008a). Group velocity of wave propagation in carbon nanotubes, *Proceedings of the Royal Society A*, 464: 1423-1438

Wang, L. F.; Liew, K. M.; He, X. Q.; Hu, Y. G.; Wang, Q.; Guo, W. L. & Hu, H. Y., (2008b), Using Model of Strain Gradient Membrane Shell to Characterize Longitudinal Wave Dispersion in Multi-Walled Carbon Nanotubes. *Journal of Computational and Theoretical Nanoscience*, 5: 1980–1988

Wang, Q. (2005b). Wave propagation in carbon nanotubes via nonlocal continuum mechanics. *Journal of Applied Physics*, 98, 124301.

Wong, E. W.; Sheehan, P.E. & Lieber, C.M. (1997). Nanobeam Mechanics: Elasticity, Strength, and Toughness of Nanorods and Nanotubes. *Science*, 277, 1971-1975.

Xie, G. Q.; Han, X. & Long, S. Y. (2006). Effect of small size on dispersion characteristics of wave in carbon nanotubes. *International Journal of Solids and Structures*, 44, 1242-1255.

Yakobson, B. I.; Brabec C. J. & Bernholc J. (1996). Nanomechanics of Carbon Tubes: Instabilities beyond Linear Response. *Physical Review Letters*, 76, 2511-2514.

Yoon, J.; Ru, C. Q. & Mioduchowski, A. (2002). Non-coaxial resonance of an isolated multiwall carbon nanotube. *Physical Review B*. 66, 233402.

Yoon, J.; Ru, C. Q. & Mioduchowski, A. (2003a). Vibration of embedded multi-wall carbon nanotubes. *Composites Science and Technology*, 63, 1533-1542.

Yoon, J.; Ru, C. Q. & Mioduchowski, A. (2003b). Sound wave propagation in multi-wall carbon nanotubes. *Journal of Applied Physics*, 93, 4801-4806.

Zhang P.; Huang Y.; Geubelle P. H.; Klein P. A. &Hwang K. C., (2002). The elastic modulus of single-walled carbon nanotubes: a continuum analysis incorporating interatomic potentials. *International Journal of Solids and Structures*, 39, 3893.

Zhang, Y. Q.; Liu, G. R. & Wang, J. S. (2004). Small-scale effects on buckling of multiwalled carbon nanotubes under axial compression. *Physical Review B*, 70, 205430.

Zhang, Y. Q.; Liu, G. R. & Xie X. Y. (2005). Free transverse vibrations of double-walled carbon nanotubes using a theory of nonlocal elasticity. *Physical Review B*, 71, 195404.

13

Differential Quadrature Method for Linear Long Wave Propagation in Open Channels

Birol Kaya and Yalcin Arisoy
Dokuz Eylul University, Engineering Faculty, Civil Engineering Department, Izmir, Turkey

1. Introduction

The St. Venant equations which are written for the calculation of free surface, gradually varied unsteady flows can only be solved analytically in very special limited problems. Apart from this they can be solved by numerical solution methods. Depending on the problem's characteristics, St. Venant equations can be arranged in different forms such as kinematic wave, dynamic wave, gravity wave, quasi-steady dynamic wave, and non-inertia wave may emerge. In the solution of non-linear St. Venant equations, different numerical solution techniques may be used by linearization of the equations by mathematical approaches. In this study the St. Venant equations made linear have been solved by using the Differential Quadrature Method (DQM), and the results have been compared by the Finite Difference Method (FDM) and Finite Volume Method's (FVM) results.

2. Background

The St Venant equations are used in the calculation of one-dimensional free surfaced flow. This equations, which is gained by writing the equations of continuity and motion, is semi-linear and hyperbolic. Apart from some very special problems, it cannot be directly integrated. For this reason, it can be solved by numerical solution techniques. The numerical solution techniques, which are generally used, are Finite Difference Method, Finite Volume Method and Finite Element Method.
In Finite Differences Method, such as Preismann scheme (Cunge,1975), Holly-Preismann scheme (Yang et al.,1992), McCormack scheme (Aguirre-Pe et al.,1995; Garcia-Navarro & Saviron, 1992; Garcia & Kahawita, 1986; Fennema & Chaudhry, 1986), Lambda scheme (Fennema & Chaudhry, 1986), Gabutti scheme (Fennema & Chaudhry, 1986), Beam-Warming scheme (Jha et al., 1996;1994; Fennema & Chaudhry, 1987; Mingham & Causon, 1998) and more different schemes are used (Glaister, 1988; 1993; Abbot, 1979; Cunge et al., 1980; Fread, 1983; Wang et al., 2000). Hsu and Yeh (2002) are investigated that is iterative explicit simulation of 1D surges and dam-break flows, and, has been summarized studies use different schemes. In addition to these studies, that have different boundary and initial conditions, Kiladze (2009) has investigated the study of the Stability of Finite Difference Schemes to Solve Saint-Venant Equations, Crossley and Wright (2005) has studied the time accurate local time stepping for the unsteady shallow water equations. Chambers (2000) and Hsu and Yeh (2002) are given studies that Finite Element Method has been used.

Additionally, Zhou et al. (2007) are proposed a split-characteristic Finite Element Model for 1-D unsteady flows. In the solution of this method, such as RDKG2 scheme (Kesserwani et al.; 2009), the modified Godunov method (Savic & Holly, 1993) and the Petrov-Galerking finite element scheme (Yang et al. 1993; Khan 2000) has been used. In recently, use of Finite Volume Method is increasing. Very studies are existing that are use of Finite Volume Method with different approximations (Bradford & Sanders, 2002; Mingham & Causon, 1998; Ying, et al., 2004; Erduran et al., 2002; Goutal & Maurel, 2002; Ghidaoui et al., 2001; Lee & Wright, 2009). Besides that conventional methods, in recently, such as Transfer Matrix Method (Daneshfaraz & Kaya, 2008), Incremental Differential Quadrature Method (Hashemi et al., 2006; 2007) and Differential Quadrature Method (Kaya et al., 2010) has been used. In parallel development of computer technology, the numerical methods can be used more effectively.

St. Venant equations can be written in different forms by omitting some terms and approaches such as kinematic wave, dynamic wave, gravity wave, quasi-steady dynamic wave and inertial wave may emerge. A considerable extent of publications regarding the solution of equations using different wave approaches are summarized in Yen and Tsai (2001) and Fan and Li (2006).

DQM has been developed by Richard Bellman (Bellman et al, 1971). This method offers the solution of gained equations in differential form of any given system, including existing boundary/initial conditions in the equations. Shu ve Richards (1992) have studies using the DQM in the area of some applications of fluid mechanics and the bending and twisting of beams. A recent study (Fung, 2001) has emphasized the application principles of DQM in problems we may face in the area of fluid mechanics and heat transfer. Kaya (2010) has investigated the use of DQM on the solution of Advection Diffusion Equation.

Shu et al. (2003) are presented local radial basis function-based differential quadrature method. Shu et al. (2004) examined the DQM was used to simulate the eccentric Couette–Taylor vortex flow in an annulus between two eccentric cylinders with rotating inner cylinder and stationary outer cylinder. Lo et al. (2005) are presented natural convection in a differentially heated cubic enclosure is studied by solving the velocity–vorticity form of the Navier–Stokes equations by a generalized differential quadrature method. Ding et al. (2006) examined, the local multi-quadric differential quadrature method is applied on three-dimensional incompressible flow problems.

3. Differential quadrature method

In the DQM, the partial derivative of a function with respect to a variable at a discrete point is approximated as a weighted linear sum of the function values at all discrete points in the region of that variable. The approximation of the partial derivative can be written as (Civalek, 2004):

$$u_x^{(r)}(x_i) = \left.\frac{\partial^r u}{\partial x^r}\right|_{x=xi} = \sum_{j=1}^{N} A_{ij}^{(r)} u(x_j) \qquad i = 1,2,...,N \tag{1}$$

where $u_x^{(r)}$ is the rth order derivative of the function, x_j are the discrete points of the variable x, $u(x_j)$ are the function values at points x_j and $A^{(r)}_{ij}$ are the weight coefficients for the rth order derivative of the function. The DQM is an alternative approach to the standard numerical solution methods such as finite differences and finite elements, for the initial and

boundary value problems encountered in physics and mathematics (Shu, 1992; Shu et al., 2003; Shu & Chew, 1997).
The overall sensitivity of the model especially depends on the location and number of sampling grid points. However, Civalek (2003) points out that the determination of the effective choice of sampling grid points for any problem reduces the analysis time (Civalek, 2003). For instance, previous studies show that for the solution of linear equations with homogeneous boundary conditions, selecting equal intervals between the adjacent grid points are adequate. On the other hand, for the vibration problems, the choice of grid points through the Chebyshev-Gauss-Lobatto method is more reasonable. In time-bound equations and initial value problems, selection of unequal intervals for sampling grid points produces the appropriate solutions (Civalek, 2003). For boundary value problems, DQM performance is highly dependent on the boundary conditions and sampling grid points. The boundary conditions can be easily implemented to DQ system and the common type of boundary conditions, which are Dirichlet, Neumann and/or mixed type function, do not create any difficulty in this implementation process (Bert, 1996; Civalek, 2003).
Determining the weight coefficients is the most crucial step in the use of DQM. Shu and Xue (1997) worked on the selection of the weight coefficients and proposed several solutions in their studies. The weight coefficients change upon approximation function and according to the chosen approximation function, the method takes different names such as Polynomial Differential Quadrature, Fourier Expansion Base Differential Quadrature and Harmonic Differential Quadrature (Civalek, 2004; Shu et al., 2002).

4. St.Venant equations and linearization

For unsteady one-dimensional free surface flows, the continuity (Eq.2) and momentum (Eq.3) equations are to be called the Saint-Venant equations. In use of the Saint-Venant equations, some basic assumptions are made that are (1) the streamline curvature is very small and the pressure distributions are hydrostatic; (2) the flow resistance are the same as for a steady uniform flow for the same depth and velocity; (3) the bed slope is small enough to satisfy : $\cos\theta \approx 1$ and $\sin\theta \approx \tan\theta \approx \theta$; (4) the water density is a constant; and (5) the channel has fixed boundaries, and air entrainment and sediment motion are neglected (Chanson, 2004).

$$\frac{\partial A}{\partial t}+\frac{\partial Q}{\partial x}=q_l \tag{2}$$

$$\frac{\partial Q}{\partial t}+\left(\frac{\partial (QV)}{\partial x}-V_x q\right)+gA\frac{\partial y}{\partial x}\text{-}gA(S_0-S_f)=0 \tag{3}$$

In this equations, the velocities and water depths change with time and longitudinal position, and, Q is flow discharge [L3/T]; y, flow depth [L]; A, flow cross sectional area [L2]; q_l, lateral outflow per unit length along the channel [L2] (infiltration, seepage, precipitation, etc); V_x, x component of velocity of lateral flow [L/T]; S_0, channel bed slope [L/L]; S_f friction slope [L/L]; x, distance in longitudinal direction [L/L]; and t is time [T]. In the solution of St. Venant equations different approximations are used namely dynamic wave, quasi-steady dynamic wave, non-inertia wave, kinematic wave and gravity wave (Yen & Tsai, 2001).
In practical terms, lateral outflows (seepage, infiltration etc) are comparatively low and can be neglected ($q=0$). The linearizing of the basis equation of unsteady and gradually varied flows

in prismatic channels can be done by the transformation $Q=Q_0+Q'$ and $A=A_0+A'$. Here Q_0 and A_0 show the steady initial conditions values. When the linearized St. Venant equations have been arranged and the 3rd degree terms have been neglected from the equation

$$\frac{\partial Q'}{\partial t}+C\frac{\partial Q'}{\partial x}=\left((1-F_0^2)+2\frac{C}{u_0^2}F_0^2-\left(\frac{C}{u_0^2}\right)^2F_0^2\right)m\frac{\partial^2 Q'}{\partial x^2} \tag{4}$$

is obtained (Yen & Tsai, 2001). In this equation,

$$m=\frac{u_0 y_0}{2S_0} \tag{5}$$

$$F_0=\frac{u_0}{\sqrt{g y_0}} \tag{6}$$

$$y_0=\frac{A_0}{B_0} \tag{7}$$

$$C=-\frac{\partial Sf/\partial A}{\partial Sf/\partial Q}\bigg|_0 u_0 \tag{8}$$

happens.

$$D_h=\left((1-F_0^2)+2\frac{C}{u_0}F_0^2-\left(\frac{C}{u_0}\right)^2F_0^2\right)m \tag{9}$$

impending, as seen this equation can be written as

$$\frac{\partial Q'}{\partial t}+C\frac{\partial Q'}{\partial x}=D_h\frac{\partial^2 Q'}{\partial x^2} \tag{10}$$

This equation is used in the case of dynamic wave approach. In the other wave approaches because some waves are neglected; in the quasi-steady dynamic wave approach

$$D_h=\left[1-(1-\frac{C}{u_0^2})F_0^2\right]m \tag{11}$$

in the non-inertia wave approach

$$D_h=m \tag{12}$$

and, in the kinematic wave approach $D_h=0$. Instead of the Q yield St. Venant equations can be arranged by the y value by changing $y=y_0+y'$. In this case

$$\frac{\partial y'}{\partial t}+C\frac{\partial y'}{\partial x}=D_h\frac{\partial^2 y'}{\partial x^2} \tag{13}$$

equation is gained (Yen & Tsai, 2001).

The celerity used in Eq. (10) depends on the channel cross-section geometry, the flow area and the resistance formula (S_f). For a trapezoidal channel, C is defined as follows:

$$C=\left[1+\frac{\alpha}{2}\left(1-\frac{2\sqrt{1+z^2}y_0}{b+2\sqrt{1+z^2}y_0}\frac{b+zy_0}{b+2zy_0}\right)\right]u_0 \tag{14}$$

Here, $\alpha=1$ for Chezy formula, $\alpha=4/3$ for the Manning's formula, b is bottom width [L], and z is the side slope [L/L].

5. Application of differential quadrature method

The equation (10) obtained as a result of an arrangement of of the linearized St. Venant equations, can be rewritten for the solution of DQM as seen in Eq. 15

$$\sum_{r=1}^{R}A_{r,s}Q_{i,r}+\left(C\sum_{j=1}^{N}B_{j,i}\text{-}D_h\sum_{j=1}^{N}B^{(2)}_{j,i}\right)Q_{j,s}=0 \quad i=1,2,...,N;\ s=1,2,...,R \tag{15}$$

where N, is the number of sampling grid points in the x direction; R is the number of sampling grid points in time direction; and A, B, $B^{(2)}$ are weight matrix coefficients.
Determining the boundary conditions, Eq. 15 is solved for the $Q(i,s)$ values. For example, when $Q(x,0)=f_1(t)$ and $Q(0,t)=f_2(x)$, Eq. 15 yields

$$\sum_{r=2}^{R}A_{r,s}Q_{i,r}+\left(C\sum_{j=2}^{N}B_{j,i}\text{-}D_h\sum_{j=2}^{N}B^{(2)}_{j,i}\right)Q_{j,s}=-A_{1,s}Q_{i,1}\text{-}\left(cB_{1,i}\text{-}D_hB^{2}_{1,i}\right)Q_{1,s} \tag{16}$$

where, $i=2,3,\ldots N$ and $s=2,3,\ldots R$.
Using the properties of Legendre polynomials, the weight coefficient matrix can be written as (Shu et al. 2004):

$$L_i=\prod_{j=1}^{N}(x_j-x_i) \tag{17}$$

$$B_{i,k}=\frac{L_i}{L_k(x_i-x_k)} \qquad i\neq k \tag{18a}$$

$$B_{i,i}=-\sum_{k=1}^{N}B_{i,k} \qquad i\neq k \tag{18b}$$

$$B^{(2)}_{i,k}=2\left[B_{i,k}B^{)}_{i,k}-\frac{B_{i,k}}{x_i-x_k}\right] \qquad i\neq k \tag{19a}$$

$$B^{(2)}_{i,i}=-\sum_{k=1}^{N}B^{(2)}_{i,k} \qquad i\neq k \tag{19b}$$

For the numerical discretization, several different methods are available. It can be selected with equal intervals, or non-equal intervals like Chebyshev-Gauss-Lobatto grid points, or with the normalization of the routes of Legendre polynomials (Shu, 2000). In the present study, several different approximations have been tested, and, Chebyshev-Gaus-Lobatto grid points have been selected. In the Chebyshev-Gaus-Lobatto approximation, by writing

$$r_i = \frac{1}{2}\left(1 - \cos\frac{i-1}{N-1}\pi\right) \tag{20}$$

the locations as the calculation points can be calculated as follows:

$$x_i = \frac{r_i - r_1}{r_N - r_1} \tag{21}$$

In time domain, this yields

$$t_i = \frac{r_i - r_1}{r_R - r_1} \tag{22}$$

by considering R, in place of N, in Eq.(21) and Eq.(22). A matrix can be written like B matrix as follows.

$$L_i = \prod_{j=1}^{R}(t_j - t_i) \tag{23}$$

$$A_{i,k} = \frac{L_i}{L_k(t_i - t_k)} \qquad i \neq k \tag{24a}$$

$$A_{i,i} = -\sum_{k=1}^{R} A_{i,k} \qquad i \neq k \tag{24b}$$

6. Numerical examples

In the Murumbidgee River in South Wales, Austria, in 1978 and 1983, two serious floods happen (Sivapalan et al. 1997). In the present study, the numerical calculations were performed by use of the flood propagation between Gundagai and Wagga Wagga stations over the river. Gundagai is located upstream of the river with 120 km length and 0.032% longitudinal slope. The Gundagai measurements were undertaken as initial data, and then the flood hydrograph at Wagga Wagga was determined. A comparison was conducted taking into account the DQM, FVM, FDM, actual measurements, and nonlinear FDM solution performed by Sivapalan et al. (1997). Here, the first three analyses (DQM FVM and FDM) were performed for constant C and D_h values.

In the first example, Gundagai station's measurements in 1978 were accepted as boundary data, and the flow rate values at Wagga Wagga station were calculated by use of Implicit Finite Difference Method (IFDM), Explicit Finite Difference Method (EFDM), Implicit Finite Volume Method (IFVM), Explicit Finite Volume Method (EFVM) and DQM. Gundagai and Wagga Wagga stations measurements and IFDM, IFVM and DQM results are given in Fig. 1. An additionally, nonlinear solution at given Sivapalan et al (1997) was examined. The solution of Sivapalan et al. (1997), EFDM, EFVM and DQM results are given in Fig. 2.

The accuracy of these different numerical methods was obtained by using the Root Mean Square (RMS) equation as follows:

$$\|e\|_2 = \sqrt{\frac{1}{R}\sum_{i=1}^{R}|e_i|^2} \qquad e_i = Q_{observed} - Q_{num.method} \tag{25}$$

where $Q_{observed}$ is the observed value and $Q_{num.method}$ is the numerical solution i.e., DQM, EFDM, IFDM, EFVM, IFVM and Sivapalan et al.(1997) values and R is the number of points in time direction.

RMS error values of FDM, FVM and DQM were given in Fig. 3 and in Fig. 4. RMS errors in Sivapalan et al. (1997) for a nonlinear solution was calculated as approximately 170 m^3/sec. RMS error values in FVMs, FDMs and DQM are converged to approximately 165, 137 and 122, respectively.

In the watershed in 1983, a flood had occurred again. Gundagai and Wagga Wagga stations measurements and solution of Sivapalan et al. (1997) at Wagga Wagga were given in Fig. 5. In addition, the results of IFDM, IFVM and DQM are seen in this Figure. EFDM, EFVM results are given in Fig. 6. In Fig. 7 and 8, FDMs, FVMs and DQM RMS error values are given as if in 1978 flood.

RMS error values are converged to 73, 76 and 71 in FDM, FVM and DQM, respectively. This number is 41 in Sivapalan's solution.

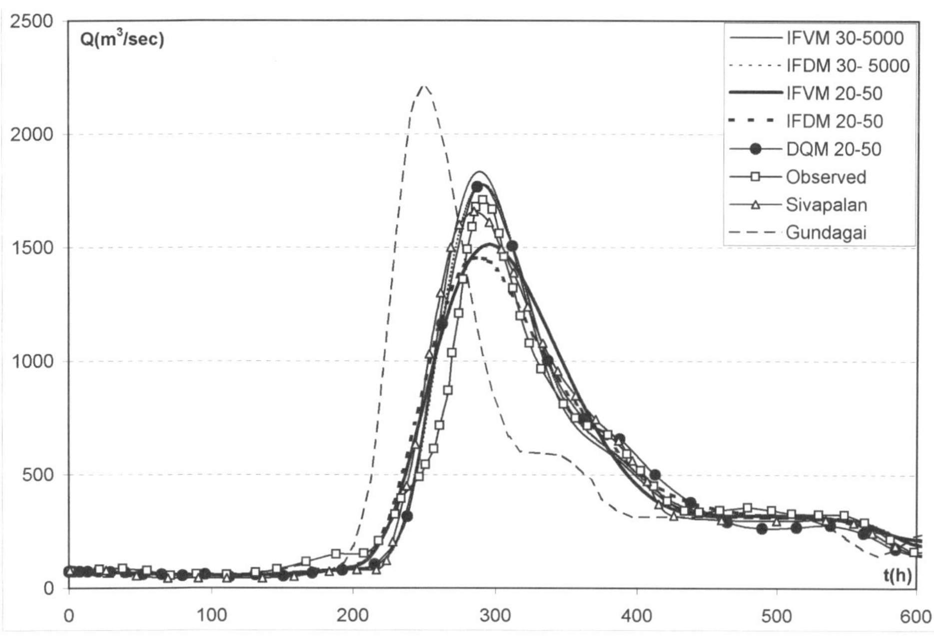

Fig. 1. IFVM, IFDM and DQM results and Sivapalan et al.(1997) solution at Wagga Wagga and actual measurements at Gundagai (upstream boundary) and Wagga Wagga (1978 flood)

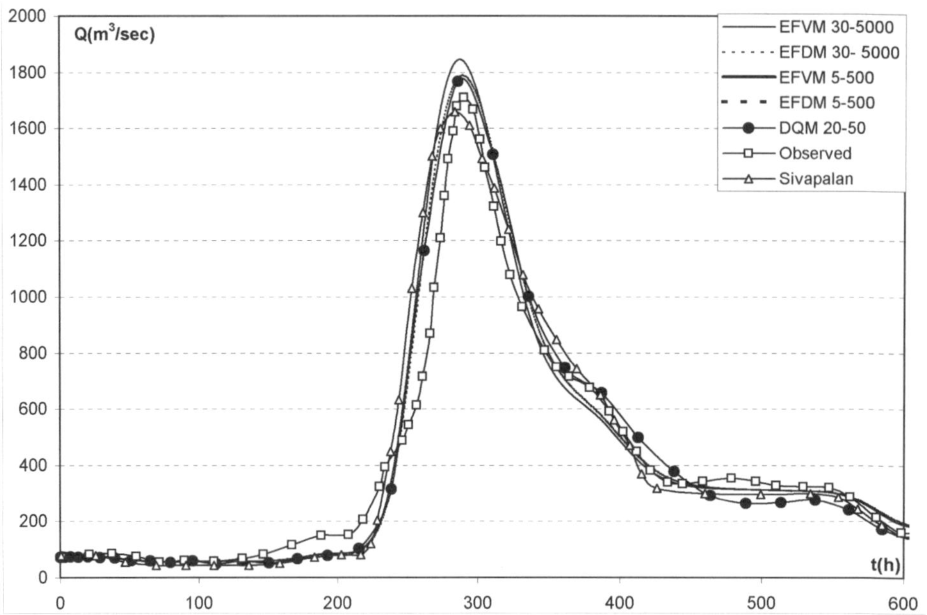

Fig. 2. EFDM, EFVM and DQM results, observed values and Sivapalan et al. (1997) solution at Wagga Wagga (1978 flood)

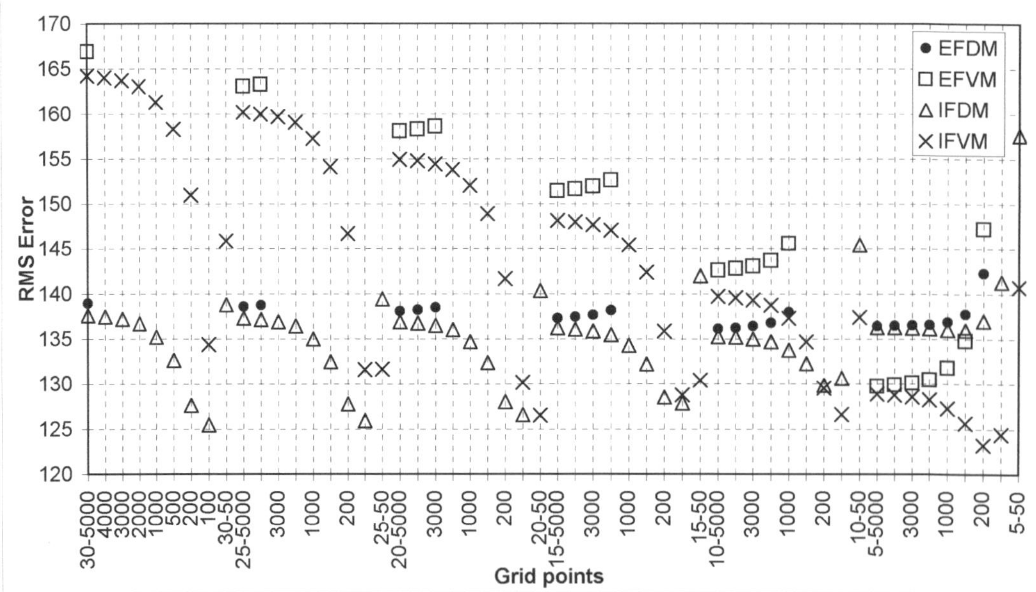

Fig. 3. FDMs and FVMs RMS error values for different sampling grid points (1978 flood)

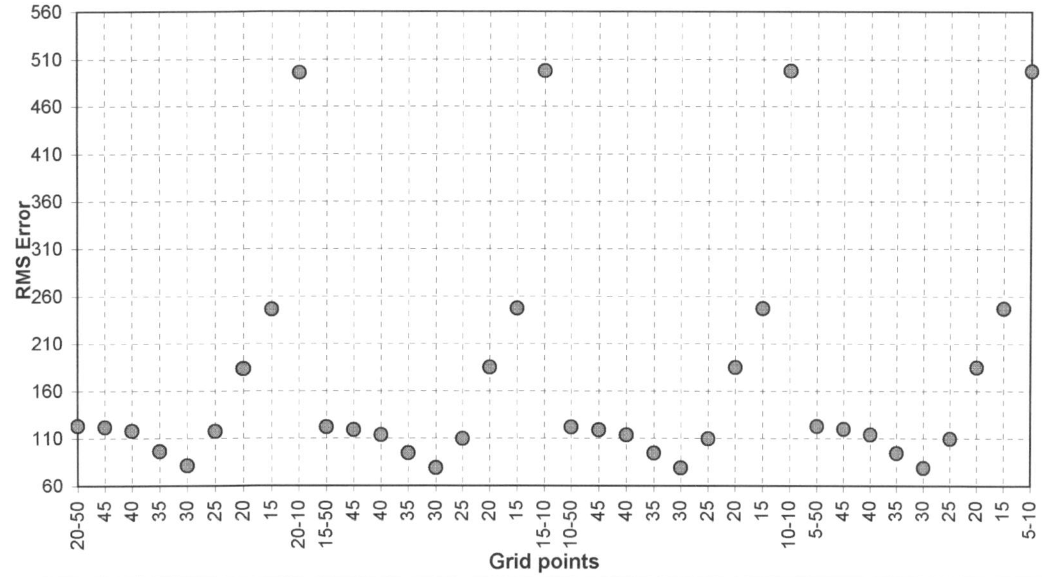

Fig. 4. DQM RMS error values for different sampling grid points (1978 flood)

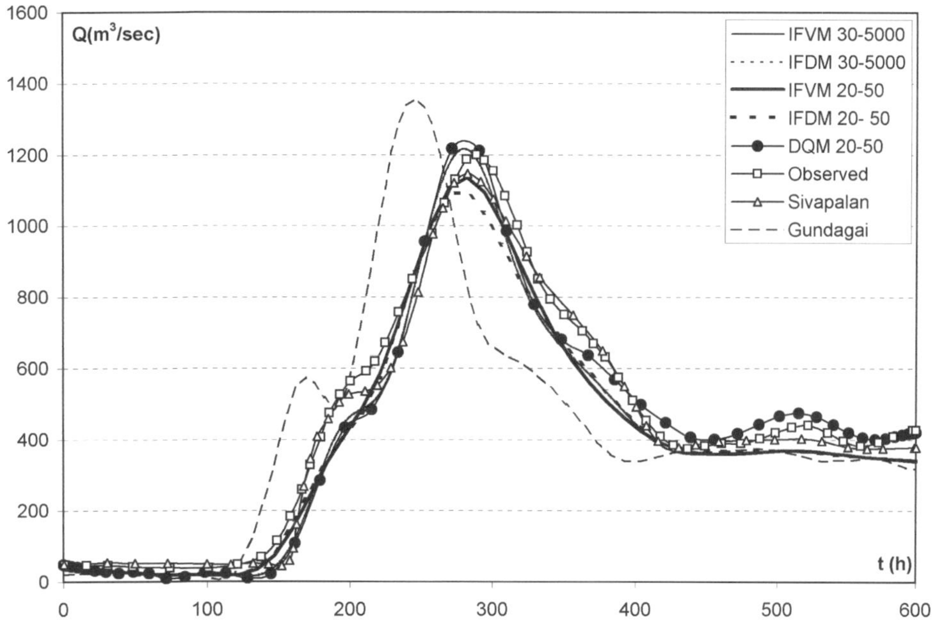

Fig. 5. IFVM, IFDM and DQM results and Sivapalan et al.(1997) solution at Wagga Wagga and actual measurements at Gundagai (upstream boundary) and Wagga Wagga (1983 flood)

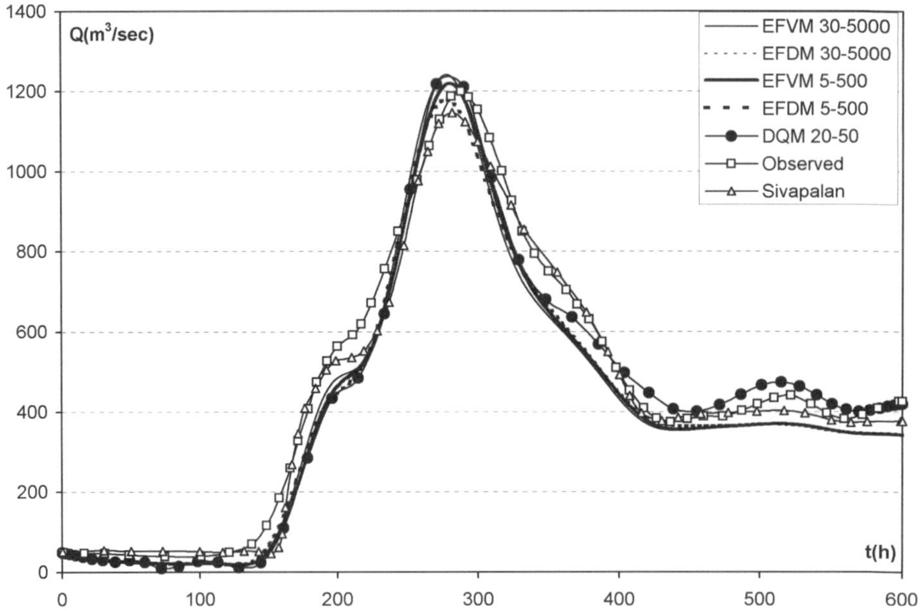

Fig. 6. EFDM, EFVM and DQM results, observed values and Sivapalan et al. (1997) solution at Wagga Wagga (1983 flood)

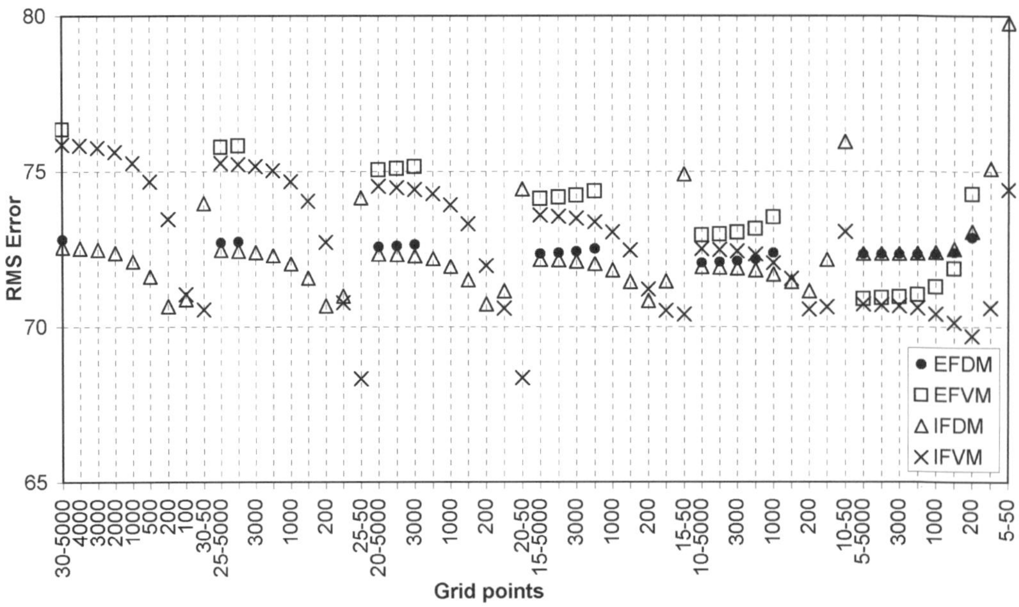

Fig. 7. FDMs and FVMs RMS error values for different sampling grid points (1983 flood)

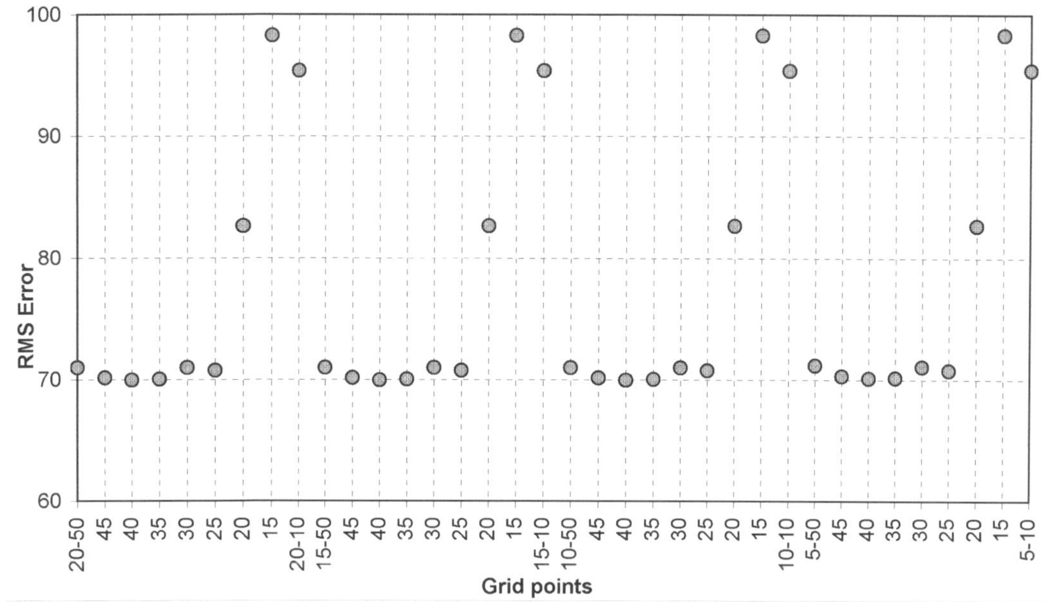

Fig. 8. DQM RMS error values for different sampling grid points (1983 flood)

7. Results and conclusion

DQM has found increasing use in recent years in Hydraulic Engineering, because it is an alternative approach to the standard methods such as Finite Difference Method, Finite Element Method and Finite Volume Method. From the previous applications of DQM in different engineering fields, it is seen that the results of DQM are converged rapidly and closer to analytical solutions than other numerical solutions. A similar consequence is observed in an application of Hydraulic Engineering (Kaya et al., 2010).

The St. Venant equations that are written for an unsteady open channel flow under some acceptance can be translated to linear form, and then can be solved by numerical solution methods. In this study, DQM were employed for one-dimensional, gradually varied, unsteady open-channel flows. The DQM results were compared with the results using the FDMs and FVMs. It is obvious from the examples given above that the DQM results are closer to the measurements than the other numerical methods.

In calculations making use of 1978 Murumbidgee River flood data, the DQM's RMS error values are less than nonlinear solution results in Sivapalan et al. (1997); On the contrary, in calculations of 1983 flood, that is bigger. Yet, for a linear solution performed in this study, the DQM's RMS error values are less than other numerical methods.

In contrast to the FDM and FVM, in the DQM, a small number of grid points are sufficient for a stable solution. The results of FDMs and FVMs are converged from 2000 at t direction, but this value is 25 in DQM.

Determining the weight coefficients is the most crucial step in the use of DQM. In wave propagation problems in open channels, by making use of Legendre polynomials for weight coefficients and Chebyshev-Gauss-Lobatto points for the numerical discretization, the results can be obtained closer to an analytical solution (Kaya, 2010).

8. References

Abbott, M.B. (1979). *Computational Hydraulics: Elements of the Theory of Free Flow*, Pitman, London, 1979.

Aguirre-Pe, J.; Quisca, S. & Plachco, F.P. (1995). Tests and numerical one-dimensional modelling of a high-viscosity fluid dam-break wave, *Journal of Hydraulics Research*, Vol.33, No.1, 17–26.

Bellman, R. & Casti, J. (1971). Differential quadrature and long-term integration, *Journal of Mathematical Analysis And Applications*, Vol.34, 235-238.

Bert, C.W. & Malik, M. (1996). Differential quadrature method in computational mechanics: A review, *Applied Mechanics Review*, Vol.49, No.1, 1-28.

Chambers, Z. (2000). *A Characteristics Finite Element Algorithm for Computational Open Channel Flow analysis*, PhD.Thesis, The University of Tennessee, Knoxville.

Chanson, H. (2004). *The Hydraulics of Open Channel Flow: An Introduction*, Elsevier Butterworth-Heinemann, Oxford, 585 p.,ISBN 0 7506 5978 5.

Civalek, Ö. (2003). *Linear and Non-Linear Dynamic Analysis of Multi-Degree-of-Freedom Systems by Harmonic Differential Quadrature Method*, PhD Thesis (in Turkish), Dokuz Eylul University, The Graduate School of Natural and Applied Science, Izmir, Turkey.

Civalek, Ö. (2004). Application of differential quadrature (DQ) and harmonic differential quadrature (HDQ) for buckling analysis of thin isotropic plates and elastic columns, *Engineering Structures*, Vol.26, No.2, 171-186.

Crossley, A.J. & Wright, N.G. (2005). Time accurate local time stepping for the unsteady shallow water equations, *Int. J. Numer. Meth. Fluids*, Vol.48, No.7, 775–799

Cunge, J.; Holly, F.M. & Verwey, A. (1980). *Practical aspects of computational river hydraulics*, Pitman Advanced Publishing Program, London, 420 pages, ISBN 0-273-08442-9.

Cunge, J.A. (1975). *Rapidly Varied Flow in Power and Pumping Canals in Unsteady Flow in Open Channel*, Water Resources Publications, Fort Collins, CO, Ch. 14.

Daneshfaraz, R. & Kaya, B. (2008) Solution of the propagation of the waves in open channels by the transfer matrix method, *Ocean Engineering*, Vol.35, No.11-12, 1075-1079.

Ghidaoui, M.S., Deng, J.Q., Gray, W.G. & Xu, K. (2001) A Boltzmann based model for open channel flows, *Int. J. Numer. Meth. Fluids*, Vol.35, 449–494.

Ding, H.; Shu, C.; Yeo, K.S. & Xu, D. (2006). Numerical computation of three-dimensional incompressible viscous flows in the primitive variable form by local multiquadric differential quadrature method, *Comput. Methods Appl. Mech. Engrg.*, No.195, 516–533.

Erduran, K.S.; Kutija, V. & Hewett, C.J.M. (2002). Performance of finite volume solutions to the shallow water equations with shock-capturing schemes, *Int. J. Numer. Meth. Fluids*, Vol.40, No.10, 1237–1273

Fan, P. & Li, J.C. (2006). Diffusive wave solutions for open channel flows with uniform and concentrated lateral inflow, *Advances in Water Resources*, No.29, 1000-1019.

Fennema, R.J. & Chaudhry, M.H. (1986). Explicit numerical schemes for unsteady free-surface flows with shocks, *Water Resources Research*, Vol.22, No.13, 1923–1930.

Fennema, R.J. & Chaudhry, M.H. (1987). Simulation of one-dimensional dam-break flows, *Journal of Hydraulics Research*, Vol.25, No.1, 41–51

Fread D. (1985). *Flood routing*. In: Anderson MG, Burt TP, editors. *Hydrological forecasting*, Chichester: Wiley.

Fung, T.C. (2001). Solving initial value problems by differential quadrature method - part I: first-order equations, *Int. Journal For Numerical Methods in Eng.*, No.50, 1411-1427.

Garcia, R. & Kahawita, R.A. (1986). Numerical solution of the St. Venant equations with the MacCormack finite difference scheme, *Int. J. Numer. Meth. Fluids*, Vol.6, 59–274.

Garcia-Navarro, P. & Saviron, J.M. (1992). McCormack's method for the numerical simulation of one-dimensional discontinuous unsteady open channel flow, *Journal of Hydraulics Research*, Vol.30, No.1, 95–105.

Glaister, P. (1988). Approximate Riemann solutions of the shallow water equations, *Journal of Hydraulics Research*, Vol.26, No.3, 293–306.

Glaister, P. (1993). Flux difference splitting for open-channel flows, *Int. J. Numer. Meth. Fluids*, Vol.16, 629–654.

Goutal, N. & Maurel, F. (2002). A finite volume solver for 1D shallow-water equations applied to an actual river, *Int. J. Numer. Meth. Fluids*, Vol.38, 1–19

Hashemi, M.R. ; Abedini, M.J. & Malekzadeh, P. (2006). Numerical modelling long waves in shallow water using incremental differential quadrature method, *Ocean Engineering*, Vol.33, 1749-1764.

Hashemi, M.R. ; Abedini, M.J. & Malekzadeh, P. (2007). A differential quadrature analysis of unsteady open channel flow, *Applied Mathematical Modelling*, No.31, 1594-1608.

Hsu, C.T. & Yeh, K.C. (2002). Iterative explicit simulation of 1D surges and dam-break flows, *Int. J. Numer. Meth. Fluids*, Vol.38, 647–75

Jha, A.K.; Akiyama, J. & Ura, M. (1994). Modeling unsteady open-channel flows—modification to Beam and Warming scheme, *Journal of Hydraulics Engineering, ASCE*, Vol.120, No.4, 461–476.

Jha, A.K.; Akiyama, J. & Ura, M. (1996). A fully conservative Beam and Warming scheme for transient open channel flows, *Journal of Hydraulics Research*, Vol.34, No.5, 605–621.

Kaya, B. (2010) Solution of the advection-diffusion equation using the differential quadrature method, *KSCE Journal of Civil Engineering*, Vol.14, No.1., 69-75.

Kaya, B. ; Arisoy, Y. & Ulke, A. (2010). Differential Quadrature Method (DQM) for Numerical Solution of the Diffusion Wave Model, *Journal of Flood Engineering* (accepted)

Kesserwani, G., Liang, Q. Vazquez, J., & Mose, R. (2009). Well-balancing issues related to the RKDG2 scheme for the shallow water equations, *Int. J. Numer. Meth. Fluids*, Published online in Wiley InterScience (www.interscience.wiley.com). DOI: 10.1002/fld.2027

Khan, A.A. (2000). Modeling flow over an initially dry bed, *J. Hydraul. Res.*, Vol.38, No.5, 383–388.

Kiladze, R. (2009). Study of the Stability of Finite Difference Schemes to Solve Saint-Venant Equations, *Bulletin of the Georgian National Academy of Sciences*, Vol.3, No.1, 96-99

Lo, D.C. ; Young, D.L. & Murugesan, K. (2005). GDQ method for natural convection in a cubic cavity using velocity-vorticity formulation, *Numerical Heat Transfer, Part B*, Vol.48, 363–386.

Lee, S.H. & Wright, N.G. (2009). Simple and efficient solution of the shallow water equations with source terms, *Int. J. Numer. Meth. Fluids*, Published online in Wiley InterScience (www.interscience.wiley.com). DOI: 10.1002/fld.2071

Mingham, C.G. & Causon, D.M. (1998). A fully conservative Beam and Warming scheme for transient open channel flows, *Journal of Hydraulics Engineering, ASCE,* Vol.124, No.6, 605–614.

Shu, C. (2000). *Differential Quadrature and Its Application in Engineering,* Springer-Verlag London Limited.

Shu, C. & Chew, Y.T. (1997). Fourier expansion-based differential quadrature and its application to Helmholtz eigenvalue problems, *Communications in Numerical Methods in Engineering,* Vol.13, 643–653.

Shu, C.; Ding, H. & Yeo, K.S. (2003). Local radial basis function-based differential quadrature method and its application to solve two-dimensional incompressible Navier–Stokes equations, *Comput. Methods Appl. Mech. Engineering.,* No.192, 941–954.

Shu, C. & Richards, B.E. (1992). Application of generalized differential quadrature to solve two-dimensional incompressible Navier-Stokes equations, *International Journal For Numerical Methods In Fluids,* Vol.15, 791-798.

Shu, C. ; Wang, L. ; Chew, Y.T. & Zhao, N. (2004). Numerical study of eccentric Couette–Taylor flows and effect of eccentricity on flow patterns, *Theoret. Comput. Fluid Dynamics,* No.18, 43–59.

Shu, C. & Xue, H. (1997). Explicit computations of weighting coefficients in the harmonic differential quadrature, *Journal of Sound and Vibration,* Vol.204, No.3, 549-555.

Shu, C.; Yao, Q.; Yeo, K.S. & Zhu, Y.D. (2002). Numerical analysis of flow and thermal fields in arbitrary eccentric annulus by differential quadrature method, *Heat and Mass Transfer,* Vol.38, 597-608.

Wang, J.S. ; Ni, H.G. & He, Y.S. (2000). Finite Difference TVD Scheme for Computatin of Dam-Break Problems, *Journal of Hydraulic Engineering,* Vol.126, 153-262

Yang, J.C.; Chen, K.N. & Lee, H.Y. (1992). An accurate computation for rapidly varied flow in an open channel, *International Journal for Numerical Methods in Fluids,* Vol.14, 361–374.

Yen, B.C. & Tsai, C.W. (2001). On noninertia wave versus diffusion wave in flood routing, *Journal of Hydrology,* Vol.244, 97-104.

Ying, X., Khan, A.A. & Wang, S.S.Y. (2004). Upwind conservative scheme for Saint Venant equations, *Journal of Hydraulic Engineering ASCE,* Vol.130, No.10, 977-987.

Zhou, Y.L., Tang, H.W. & Liu X.H. (2007). A split-characteristic finite element model dor 1-D unsteady flows, *Journal of Hydrodynamics, Ser.B,* Vol.19, No.1, 54-61.

14

A Parabolic Equation for Wave Propagation over Porous Structures

Tai-Wen Hsu and Jen-Yi Chang
Department of Hydraulic and Ocean Engineering, National Cheng Kung University, Tainan, Taiwan

1. Introduction

Porous structures such as seawalls, detached breakwaters, or submerged breakwaters are frequently used to protect shorelines from the effects of waves. The effectiveness of these structures is due to the fact that they are able to reflect, absorb, and dissipate wave energy. Wave energy reduction on the leeside of porous structures increases so that only a small part of the wave energy is transmitted to the nearshore. Consequently, the wave field on the leeside of such structures becomes quiet and the intensity of the wave action on the shoreline decreases; as a result, coastal erosion and the corresponding coastal disasters are mitigated. Wave transformation over porous structures must be understood in order to determine the stability of these porous structures and to evaluate their usefulness with regard to wave energy reduction.

Several numerical models have been developed to study the wave transformation over porous structures. Based on the mild-slope assumption, Rojanakamthorn et al. (1989) adopted Sollitt and Cross's (1972) theory to derive a modified mild-slope equation (MSE) for describing non-breaking waves traveling over a general finite porous bed. Later Rojanakamthorn et al. (1990) extended their model to the case of wave breaking on a submerged permeable breakwater. The validity of this model has been verified by comparing its numerical results with those of experiments and an analytical solution of a rectangular submerged breakwater.

The MSE is suitable for explaining the wave deformation of an unsteady flow in a porous structure that is linearized using an approximation of the nonlinear friction forces. The wave energy dissipation due to resistance in the porous structure is considered in the model and the depth-averaged equation yields an elliptic-type MSE on a permeable bed. Losada et al. (1996) presented the results of studies similar to that of Rojanakamthorn et al. (1990) for the kinematics and dynamics of wave interaction with permeable breakwaters that encounter non-breaking obliquely incident regular waves and directional random waves. The influence of structural geometry, properties of porous materials, and wave characteristics over and inside the breakwaters was also investigated. Following Rojanakamthorn et al. (1990) and Losada et al. (1996), Mèndez et al. (2001) derived a theory to analyze the influence of wave reflection and energy dissipation in wave breaking and porous flow, as induced by a porous submerged structure. The analytical expressions for the second-order

mean quantities mass flux, energy flux, radiation stress, and mean water level are given in terms of the shape functions of the structure.
With increases in the number of available design possibilities, engineers want to work with increasingly larger coastal regions; however, it is still necessary for a simpler model to be applied in these situations—one that can provide reliable results to the practical questions posed. An evolution equation of mild slope equation (EEMSE) usually selected to simulate the combined wave refraction-diffraction in the vicinity of porous structures. The main reason for developing such a model is that it can predict coastal processes on a regional scale so that coastal changes resulting from driving forces and man-made structures can be determined. This wave information can be used as forcing functions to drive models that calculate alongshore and on-offshore sediment transports. Numerical wave models such as RCPWAVE (Ebersole et al., 1986), REF/DIF-1 (Kirby and Dalrymple, 1991), and the Parabolic Wave Model (Mordane et al., 2004), all of which neglect the wave reflection, are frequently used as tool for predicting coastal erosion and deposition.
The main purpose of this paper is to extend the EEMSE addressed by Hsu and Wen (2001) and Hsu et al. (2008b) for waves propagating over submerged permeable structures in the surf zone. The governing equation is a parabolic formulation of the EEMSE including the breaking and energy dissipation effect of porous structures. The approximate MSE enables a more accurate description of combined wave refraction, diffraction and reflection in the computational domain. Following Rojanakamthorn et al. (1990), wave breaking and energy dissipation of porous structures are accounted for in a relatively straightforward manner by adding an energy dissipation coefficient to the EEMSE. An implicit finite-difference approximation scheme is implemented in the numerical solution. Several numerical cases were computed and compared through experiments to examine the validity of the present model. This model was also applied to practical cases of wave-height evolution in the vicinity of the submerged permeable structures over a complicated bathymetry.

2. Theoretical formulation

2.1 Governing equation

The definition sketch for linear periodic waves propagating over a submerged permeable structure in the surf zone is shown in Fig. 1, where h is the local water depth of the pure water region and h_p, the thickness of the porous layer. Following Rojanakamthorn et al. (1989), the MSE for describing the wave deformation over porous structures is derived by multiplying the Laplace equation using corresponding vertical eigenfunctions solved by Sollitt and Cross (1972). In this paper, wave breaking, energy dissipation, and large-angle incidence are included in the MSE. Based on Rojanakamthorn et al. (1989), the MSE is formulated to consider the wave transformation over porous structures, as given by

$$\nabla_h \cdot \left(\alpha_p \nabla_h \phi\right) + k_0^2 \alpha_p \left(1 + i f_d\right) \phi = 0 \tag{1}$$

where $\phi(x,y)$ is the complex velocity potential; $\nabla_h = \left(\partial / \partial x, \partial / \partial y\right)$, is the horizontal gradient operator; k_0, is the complex wavenumber; f_d, is the energy dissipation factor in the surf zone after wave- breaking; and α_p, is a parameter expressed by

$$\alpha_p = \alpha_1 + n_0\left(S - i f_p\right)\alpha_2 \tag{2}$$

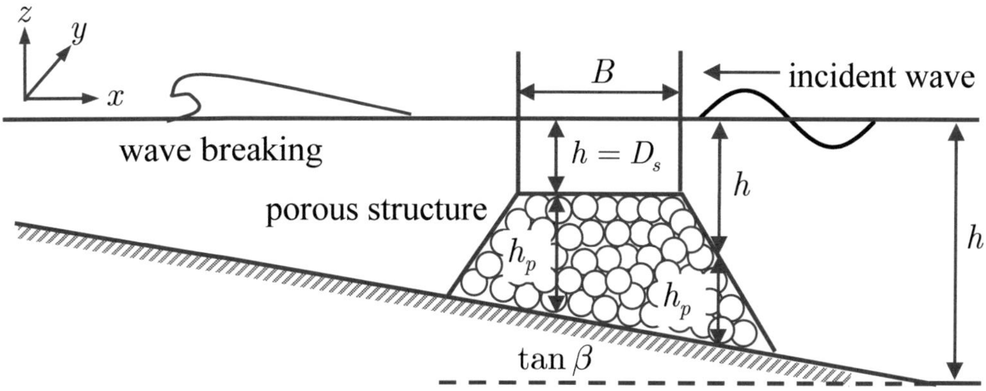

Fig. 1. Waves propagating over a submerged permeable structure.

where α_1 and α_2 are respectively given by

$$\alpha_1 = h\beta_1^2\left[\frac{\beta_2^2\left(1-e^{-2k_0h}\right)}{2k_0h} - \frac{\beta_3^2\left(1-e^{2k_0h}\right)}{2k_0h} - 2\beta_2\beta_3\right] \tag{3}$$

$$\alpha_2 = \frac{1}{2}h\beta_1^2\left[1+\frac{\sinh\left(2k_0h_p\right)}{2k_0h_p}\right] \tag{4}$$

$$\beta_1 = \frac{1}{n_0e^{k_0h}\sinh\left(k_0h_p\right)-\delta_p\cosh\left(k_0h\right)} \tag{5}$$

$$\beta_2 = n_0e^{k_0h}\sinh\left(k_0h_p\right)-\frac{1}{2}\delta_pe^{k_0h} \tag{6}$$

$$\beta_3 = \frac{1}{2}\delta_pe^{-k_0h} \tag{7}$$

$$\delta_p = n_0\sinh\left(k_0h_p\right)-\left(S-if_p\right)\cosh\left(k_0h_p\right) \tag{8}$$

in which n_0 is the porosity of the permeable material; $S = n_0 + (1-n_0)C_M$, the inertial coefficient; C_M, the virtual mass coefficient; f_p, a linear friction coefficient; and $i=\sqrt{-1}$, a unit complex number. Following Radder (1979), a scale factor

$$\varphi = \sqrt{\alpha_p}\phi \tag{9}$$

is introduced into Eq. (1) and the mathematical manipulation yields the Helmholtz equation

$$\nabla_h^2\varphi + k_c^2\varphi = 0 \tag{10}$$

where k_c is a pseudo-wavenumber written in the form

$$k_c^2 = k_0^2\left(1 + if_d\right) - \frac{\nabla_h^2\sqrt{\alpha_p}}{\sqrt{\alpha_p}} \tag{11}$$

The complex wavenumber can be determined through the following dispersion relation:

$$\omega^2 = gk_0\frac{n_0 e^{k_0 h}\sinh\left(k_0 h_p\right) - \delta_p \sinh\left(k_0 h\right)}{n_0 e^{k_0 h}\sinh\left(k_0 h_p\right) - \delta_p \cosh\left(k_0 h\right)} \tag{12}$$

where ω is the angular frequency and g, the gravitational acceleration. The wave motion is assumed to be purely harmonic in time. Equation (10) is easily split into two equations by using the commuting operators $\partial / \partial x$ and $\left(1 / k_c^2\right)\partial^2/\partial^2 y$:

$$\left(\frac{\partial}{\partial x} - ik_c\sqrt{1+\mathrm{X}}\right)\left(\frac{\partial}{\partial x} + ik_c\sqrt{1+\mathrm{X}}\right)\phi = 0 \tag{13}$$

where X is the orthogonal operator given by

$$\mathrm{X} = \frac{1}{k_c^2}\frac{\partial^2}{\partial y^2} \tag{14}$$

In Eq. (13), it is noted that the wave field is split into forward- and backward-scattering fields. As mentioned, solving the MSE incurs a significant computational cost whenever large coastal areas are considered. In these circumstances, a simpler method can provide an efficient and reliable tool to solve practical problems. The combined effect of wave refraction and diffraction in the vicinity of shoals, islands, and coastal structures can be simulated by a parabolic approximation in which the wave reflection is neglected. Although the wave reflection might be important for submerged structures, it is usually ignored in the model for solving practical coastal engineering problems such as nearshore current, sediment transport, and ocean environment pollution. The inference of reflection will be considered in section 6.

Eq. (13) is thus simplified as

$$\left(\frac{\partial}{\partial x} - ik_c\sqrt{1+\mathrm{X}}\right)\phi = 0 \tag{15}$$

According to Sollitt and Cross (1972), the linear friction factor f_p is evaluated from Lorentz's condition of equivalent work, and this relation can be expressed as

$$f_p = \frac{1}{\omega}\frac{\int_\forall \int_0^T \left(\frac{n_0^2 \upsilon}{k_p}\left|\vec{u}_s\right|^2 + \frac{n_0^3 C_p}{\sqrt{k_p}}\left|\vec{u}_s\right|^3\right) dt d\forall}{\int_\forall \int_0^T n_0 \left|\vec{u}_s\right|^2 dt d\forall} \tag{16}$$

where $\forall$ is the volume of the flow field; T the wave period; υ the kinematic viscosity; k_p the intrinsic permeability; C_p the turbulent friction coefficient; and $\vec{u}_s = -\nabla_h \phi$ the seepage velocity vector. The value of f_p is obtained through an iterative technique suggested by Sollitt and Cross (1972). The detailed procedure will be briefly presented in the following section. The values of k_p and C_p are usually determined through laboratory tests; however, it is difficult to estimate their proper values in practical applications without experiments. Some empirical formulae may be applicable in the computation. Following Furukawa and McDougal (1991), the value of the intrinsic permeability k_p is determined by the empirical formula:

$$k_p = 1.643 \times 10^{-7} \left(\frac{d_{50}}{10} \right)^{1.57} \frac{n_0^3}{\left(1-n_0\right)^2} \tag{17}$$

where d_{50} is the grain or gravel diameter in millimeters. If the porous materials are stones or armor units, it is plausible that one could calculate the diameter using the equivalent volume of the sphere. The coefficient C_p that describes the resistance properties of porous media is generally estimated from experiments under steady flow conditions. Arbhabhiramar and Dinoy (1973) conducted laboratory tests and proposed a formula through experimental data:

$$C_p = 3.162 \times 10^6 \left[d_{50} \sqrt{\frac{n_0}{k_p}} \right]^{-1.50} \tag{18}$$

It is noted that Eqs. (11) and (12) are reduced to the expressions of the pseudo-wavenumber $k_c^2 = k_0^2 - \nabla_h^2 \sqrt{CC_g} / \sqrt{CC_g}$ and the dispersion relation $\omega^2 = g k_0 \tanh k_0 h$ for the case of impermeable structures without wave breaking and energy dissipation, respectively. These results are identical to those of Mordane et al.'s (2004) theoretical formulations.

2.2 Wave energy dissipation

In order to describe the wave breaking and energy dissipation of waves propagating over a porous structure, many empirical formulae can be used in the MSE. For energy dissipation after wave breaking over an impermeable bed, Dally et al. (1985) used the hydraulic jump model to account for the energy dissipation. Battjes and Janssen (1978) estimated the energy dissipation by means of periodic bores. Isobe (1987) adopted a turbulence model to represent the energy losses of breaking waves; this model was later modified by Watanabe and Dibajnia (1988) from comparisons of model calculations and experimental data. In order to define the region of the surf zone, many criteria have been utilized to determine the breaking point; these include criteria based on theoretical formulations and empirical relationships such as those of Miche (1951), Goda (1970), and Isobe (1987) and others.

It is desirable to find a more suitably coupled formula for describing the wave breaking and energy dissipation for porous structures in MSE. When studying the possible combinations of these empirical formulae, experimental data from permeable beds are rare. Rojanakamthorn et al. (1990) adopted Watanabe and Dibajnia's (1988) model to derive a new equation for describing the processes of wave decay and recovery due to porosity and wave breaking; therein, a similar mechanism of energy loss between impermeable and permeable beds is assumed. Based on their equation, the energy dissipation formulae mentioned above are selected to account for wave deformation on the porous structures across the surf zone. The energy dissipation coefficients related to wave breaking and porosity are finally expressed as

$$f_d = \frac{k_r}{C_g}\frac{K_2}{k_r h'}\left(1-\frac{K_1^2}{\gamma^2}\right) \tag{19a}$$

$$f_d = \frac{k_r}{C_g}\frac{\alpha' Q_b}{n\pi}\left(\frac{H_{\max}}{H_{rms}}\right)^2 \tag{19b}$$

$$f_d = \frac{k_r}{C_g}\alpha_D \tan\beta\sqrt{\frac{g}{h'}}\sqrt{\frac{\gamma-\gamma_r}{\gamma_s-\gamma_r}} \tag{19c}$$

where K_1, K_2, α', and α_D are calibrated parameters; k_r the real part of the progressive complex wavenumber k_0; $\gamma = H/h'$ the ratio of wave height to the effective water depth h'; $\gamma_r = 0.4(H/h')_b$ the critical value of a recovery wave on h' ; $\gamma_s = 0.8(0.57+5.3\tan\beta)$ a ratio of wave height to h'; $H_{\max} = (0.88/k_r)\tanh(\gamma_0 kh/0.88)$ the maximum wave height; H_{rms} = the root mean square value of wave height; $Q_b = \exp\left[(1-Q_b)/(H_{rms}/H_{\max})^2\right]$; $\gamma_0 = 0.5+0.4\tanh(33H_0/L_0)$; and $\tan\beta$ the average bottom slope calculated from the breaking point to an offshore distance $5h_b'$. The subscript "b" denotes the value at the breaking point. For regular waves, the value of Q_b equals 1 and H_{rms} is replaced by H_0. Based on Losada et al. (1997), the effective water depth h' is used to be instead of h in the energy dissipation model in order to consider the porosity of the structure which is given by

$$h' = \frac{1}{k_r}\tanh^{-1}\left(\frac{\omega^2}{gk_r}\right) \tag{20}$$

A breaking index is used as a criterion to define the incipient breaking point of breaking waves traveling over a permeable structure. The following breaking criteria of an impermeable bed as presented by Miche (1951), Goda (1970), and Isobe (1987), respectively are modified in the wave calculations for porous structures:

$$\gamma_b = 0.142\frac{L_b}{h_b'}\tanh\frac{2\pi h_b'}{L_b} \tag{21a}$$

$$\gamma_b = 0.17\left\{1-\exp\left[-1.5\frac{\pi h_b'}{L_0}\left(1+15(\tan\beta)^{4/3}\right)\right]\right\} \tag{21b}$$

$$\gamma_b = 1.06-0.6\exp\left(-3\sqrt{h_b'/L_0}\right)+10(\tan\beta)^{3/2}\cdot\exp\left[-45\left(\sqrt{h_b'/L_0}-0.1\right)^2\right] \tag{21c}$$

where γ_b is the wave breaking index defined by $\gamma_b = H_b/L_b$ and H_b and L_b, are the breaking wave height and wavelength, respectively. Eqs. (19) and (21) are respectively selected to form a coupled equation to solve the energy dissipation due to porosity and wave breaking.

In practical applications, the parameters α_D, K_1, K_2, and γ' are determined experimentally. According to the analysis of Mèndez et al. (2001), $\alpha_D = 1.0$, $K_1 = 0.4$, $K_2 = 0.35$, and $\alpha' = 1.0$ were suggested in the numerical simulation. Nine sets of coupled equations were examined in the computation. In order to find a more suitable coupled equation for solving the wave breaking and energy dissipation on a porous structure, two sets of experimental data (Cruz et al. 1997; Lara et al. 2006) were used by Hsu et al. (2008a) to examine the applicability of the nine coupled equations. A comparison of the model results and experimental data is presented in Table 1. It is observed that the combination of Eqs. (19c) and (21b) achieves the best agreement with observations in which a higher correlation coefficient C_R is obtained, where C_R is defined by

$$C_R = 1 - \sqrt{\frac{1}{N}\sum_{i=1}^{N}\frac{\left[(H_m)_i - (H_c)_i\right]^2}{\left[(H_m)_i\right]^2}} \tag{22}$$

where N is the total number of observed data points; H_m the value measured from laboratory experiments; and H_c the value calculated from the model. As a result, this coupled equation is used in the study to calculate wave breaking and energy dissipation on permeable structures.

Eq.(21) / Eq.(19)	Miche(1951) (Eq.(21a))	Goda(1970) (Eq.(21b))	Isobe(1987) (Eq.(21c))
Dally et al.(1985) (Eq.(19a))	0.845	0.831	0.853
Battjes and Janssen (1978) (Eq.(19b))	0.799	0.818	0.788
Watanabe and Dibajnia (1988) (Eq.(19c))	0.859	0.871	0.846

Table 1. Comparisons of the correlation coefficients C_R using different coupled equations of breaking criteria and energy dissipation formulae.

2.3 Parabolic approximations

In Eq. (15), it should be noted that different approximations of the function $\sqrt{1+X}$ will lead to different orders of the parabolic equations. The most classical approximation is an expansion of the Taylor series, wherein a series with a greater number of terms could lead to a better approximation of the incident wave angle. However, this is a time-consuming process due to the large matrix dimension inherent in a large coastal environment; this would results in a disadvantage of the numerical calculation. An alternative to the rational function approximation permits us to improve the angular capacity of the incident angle as well as the computational effort by summing a relatively fewer number of terms related to the Taylor series. In this investigation, following Mordane et al. (2004), we selected a quadratic rational function of the Padé [2,2] approximation, given by

$$\sqrt{1+X} = \frac{P_0 + P_1X + P_2X^2}{Q_0 + Q_1X + Q_2X^2} + O\left(X^5\right) \tag{23}$$

where P_i and Q_i $(i=0,1,2)$ are Padé coefficients. Several methods can be used to obtain the Padé coefficients. According to St. Mary (1985), these coefficients are determined using an ad-hoc technique of Chebychev referred to as Pkni = { $P_0 = 1.628909$, $P_1 = 2.428289$, $P_2 = 0.8308198$, $Q_0 = 1$, $Q_1 = 1.615038$, $Q_2 = 0.235499$ }. In the present model, Pkni is used to model a wave traveling in very large coastal areas of several wavelengths. Mordane et al. (2004) showed that the maximal propagation wave angle of Pkni coefficients is around 77°. Substituting Eq. (23) into Eq. (15) yields the following expression:

$$\left(Q_0 + Q_1 X + Q_2 X^2\right)\frac{\partial \varphi}{\partial x} = ik_c\left(P_0 + P_1 X + P_2 X^2\right)\varphi \tag{24}$$

Eq. (24) can be further derived using the expressions $X = \left(1/k_c^2\right)\partial^2/\partial^2 y$ and $X^2 = \left[\left(1/k_c^2\right)\partial^2/\partial^2 y\right]^2$; the resulting equation is

$$\left(\lambda_1 + \lambda_2\frac{\partial^2}{\partial y^2} + \lambda_3\frac{\partial^3}{\partial y^3} + \lambda_4\frac{\partial^4}{\partial y^4}\right)\frac{\partial \varphi}{\partial x} = ik_c\left(\vartheta_1 + \vartheta_2\frac{\partial^2}{\partial y^2} + \vartheta_3\frac{\partial^3}{\partial y^3} + \vartheta_4\frac{\partial^4}{\partial y^4}\right)\varphi \tag{25}$$

where the corresponding coefficients are given by $\lambda_1 = Q_0$, $\lambda_2 = \left(Q_1/k_c^2\right) + \left(2Q_2/k_c^5\right)\left[\left(3/k_c\right)\left(\partial k_c/\partial y\right)^2 - \left(\partial^2 k_c/\partial y^2\right)\right]$, $\lambda_3 = -\left(4Q_2/k_c^5\right)\left(\partial k_c/\partial y\right)$, $\lambda_4 = Q_2/k_c^4$, $\vartheta_1 = P_0$, $\vartheta_2 = \left(P_1/k_c^2\right) + \left(2P_2/k_c^5\right)\left[\left(3/k_c\right)\left(\partial k_c/\partial y\right)^2 - \left(\partial^2 k_c/\partial y^2\right)\right]$, $\vartheta_3 = -\left(4P_2/k_c^5\right)\left(\partial k_c/\partial y\right)$ and $\vartheta_4 = P_2/k_c^4$. Equation (25) is a fifth-order partial differential equation, and it can be applied to simulate wave transformation over porous structures, including wave breaking and energy dissipation. We notice that the finite difference equations can be obtained by making the governing equation discrete; they are solved by a standard Crank-Nicholson scheme for traditional partial differential equations. The details of the numerical scheme are provided in the paper of Mordane et al. (2004).

It follows that by differentiating the Helmholtz equation with respect to y and using the rigid boundary condition, the boundary conditions of Eq. (25) are thus given by

$$\frac{\partial \phi}{\partial y}(x, y_0) = 0,\ \frac{\partial \phi}{\partial y}(x, y_m) = 0 \tag{26}$$

$$\frac{\partial^3 \phi}{\partial y^3}(x, y_0) = 0,\ \frac{\partial^3 \phi}{\partial y^3}(x, y_m) = 0 \tag{27}$$

for all x, where y_0 and y_m are the initial and terminal points of the lateral boundary, respectively. By using a central finite difference to discretize the boundary conditions of Eqs. (26) and (27) and by introducing the fictitious points $\varphi_{i,-1}$, $\varphi_{i,-2}$, $\varphi_{i,m+1}$, and $\varphi_{i,m+2}$, the boundary conditions in this model can be reduced to $\phi_{i,-1} = \phi_{i,1}$, $\phi_{i,m+1} = \phi_{i,m-1}$, $\phi_{i,-2} = \phi_{i,2}$, and $\phi_{i,m+2} = \phi_{i,m-2}$.

For a large wave angle incidence, the radiation boundary condition is also specified by Eq. (15) and approximated by Eq. (25). However, it is too complicated to achieve the matrix formulation by using Eq. (25) as the boundary conditions in the numerical solution. An alternative method is to rotate the computational domain, in order to ensure that the given boundary is always normal to the incident wave direction. This method enables us to change the radiation boundary conditions as the rigid boundary conditions at the lateral boundary.

3. Model verifications

The applicability and validity of the present model were verified through experiments in which waves propagated over impermeable composite uniform slope, submerged permeable breakwaters and sloping beaches. In the first case, Nagayama (1983) measured wave deformation on a bar-type beach profile. It consists of three sections of slope 1/20 and the wave deformation includes shoaling, first breaking, wave decay, wave recovery and the of second breaking. The experimental conditions are $H_0 = 0.07$ m and $T = 0.94$ s, respectively. Fig. 2. shows the comparison of wave height between the numerical simulation of the present model and experimental data. It can be seen from the figure that the numerical results follow the experimental data fairly well.

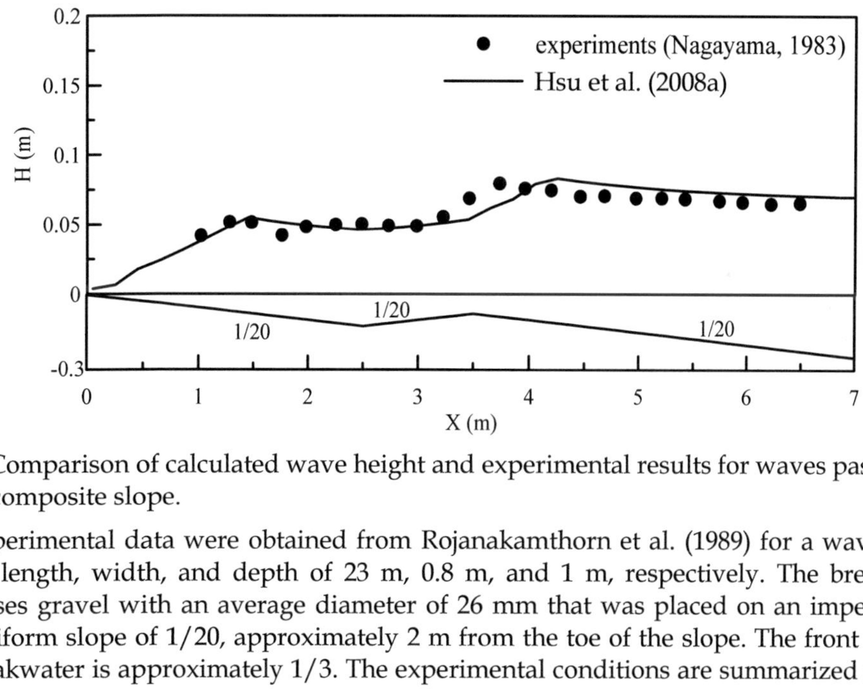

Fig. 2. Comparison of calculated wave height and experimental results for waves passing over a composite slope.

The experimental data were obtained from Rojanakamthorn et al. (1989) for a wave flume with a length, width, and depth of 23 m, 0.8 m, and 1 m, respectively. The breakwater comprises gravel with an average diameter of 26 mm that was placed on an impermeable and uniform slope of 1/20, approximately 2 m from the toe of the slope. The front slope of the breakwater is approximately 1/3. The experimental conditions are summarized in Table 2, where H_0 is the incident wave height; D_s, the water depth above the structure; and B, the width of the structure (Fig. 1).

According to Hsu et al. (2008a) the key parameters $n_0 = 0.39$, $k_p = 3.77 \times 10^{-7}$ m^2, and $C_p = 0.332$ were used in the computation; these are identical to those used by Rojanakamthorn et al. (1989). The added mass coefficient arose due to the unsteady convergence and divergence of streamlines around a solid body. Since the unsteady flow around a body with a complicated shape has not been fully clarified, the coefficient C_M is not readily determined for randomly and densely packed solid materials. Therefore, based on to Sollitt and Cross (1972) and Madsen (1974), the inertial coefficient S is approximately regarded as unity for this computation.

The numerical calculation in the Hsu et al.'s (2008a) model includes an iteration procedure for f_p. Following Rojanakamthorn et al. (1989) and for convenience, the initial value of f_p is first estimated as unity. For a given incident wave condition, the local complex wavenumber k_0 is estimated from Eq. (12) through the iteration technique. The coefficient α_p is next computed from Eqs. (2) to (8); the scale factor φ and the seepage velocity vector $\vec{u}_s$ are then determined at every grid point. The value of f_p is again estimated from Eq. (16). The iteration procedure is repeated until the change in f_p is less than a tolerance of the relative error $\varepsilon = 10^{-2}$. Details of the numerical scheme are referred to Hsu et al. (2008a).

A comparison of the model results with those of the laboratory experiments is shown in Fig. 3; these results are found to be in good agreement. It is observed from Fig. 3 that the wave amplitude rapidly decreases while waves propagate over the submerged permeable breakwater due to energy decay on the porous structures. We note that in the free surface of the experiments, the model results were unable to catch the modulations. On the other hand, the values of f_p are different from those given in Rojanakamthorn et al. (1989) (see Table 2). The difference in values may be caused by neglecting certain physical processes such as wave reflection or different numerical schemes (parabolic or elliptic type of MSE) in the model.

Items / Authors	slopes	H_0 (m)	d_{50} (mm)	T (sec)	h (m)	B (cm)	D_s (cm)	n_0	k_p (m^2)	C_p	f_p
Rojanakamthorn et al. (1989)	1/20	0.0471	20-35	0.93	0.375	10	6.5	0.39	3.77E-7	0.332	(1.155)0.412
	1/20	0.0321	20-35	0.93	0.375	30	6.5	0.39	3.77E-7	0.332	(1.039)0.384
	1/20	0.0472	20-35	1.08	0.375	235	8.0	0.39	3.77E-7	0.332	(1.507)0.871
Cruz et al. (1997)	1/20*; 1/6.67+	0.0220	6.7	1.02	0.176	—	—	0.44	2.5E-8	0.40	(2.576)
	1/20*; 1/6.67+	0.0430	6.7	1.00	0.1748	—	—	0.44	2.5E-8	0.40	(2.557)
Lara et al. (2006)	1/20	0.15	—	2.0	0.4	—	—	—	—	—	—
	1/20	0.15	19	2.0	0.4	—	—	0.49	(2.036E-7)	(0.624)	(5.726)
	1/20	0.15	39	2.0	0.4	—	—	0.49	(6.296E-7)	(0.495)	(2.495)
	1/20	0.15	—	3.0	0.4	—	—	—	—	—	—
	1/20	0.15	19	3.0	0.4	—	—	0.49	(2.036E-7)	(0.624)	(6.520)
	1/20	0.15	39	3.0	0.4	—	—	0.49	(6.296E-7)	(0.495)	(1.829)

Table 2. Experimental conditions for wave propagating over submerged permeable breakwaters (* : the front slope; + : the rear slope. Values in brackets were calculated by the present model.)

The model is also applied to simulate wave transformation on a porous sloping bottom. Experiments were performed by Cruz et al. (1997) in a 0.30 × 0.20 × 11.0 m wave flume at the University of Tokyo, Japan. The bathymetry of the sloping bottom is shown in Fig 3. The triangular porous bar side slopes of are 1:20 and 1:6.67, respectively. The key parameters $n_0 = 0.44$, $k_p = 2.5 \times 10^{-8}\,\mathrm{m}^2$, and $C_p = 0.4$ were used in the computation, as suggested by Cruz et al. (1997). The experimental conditions are listed in Table 2. A comparison of the numerical results and experiments for the two cases (Table 2) is shown in Fig. 4. Fig 4(a) shows the measured and computed wave heights for an incident wave with small damping without breaking. It is evident that the general trend of the damped wave height is approximated quite well by the model. In Fig 4(b), the propagating wave breaks at around $x = 5.8$ m. The computed and observed wave heights are in good agreement.

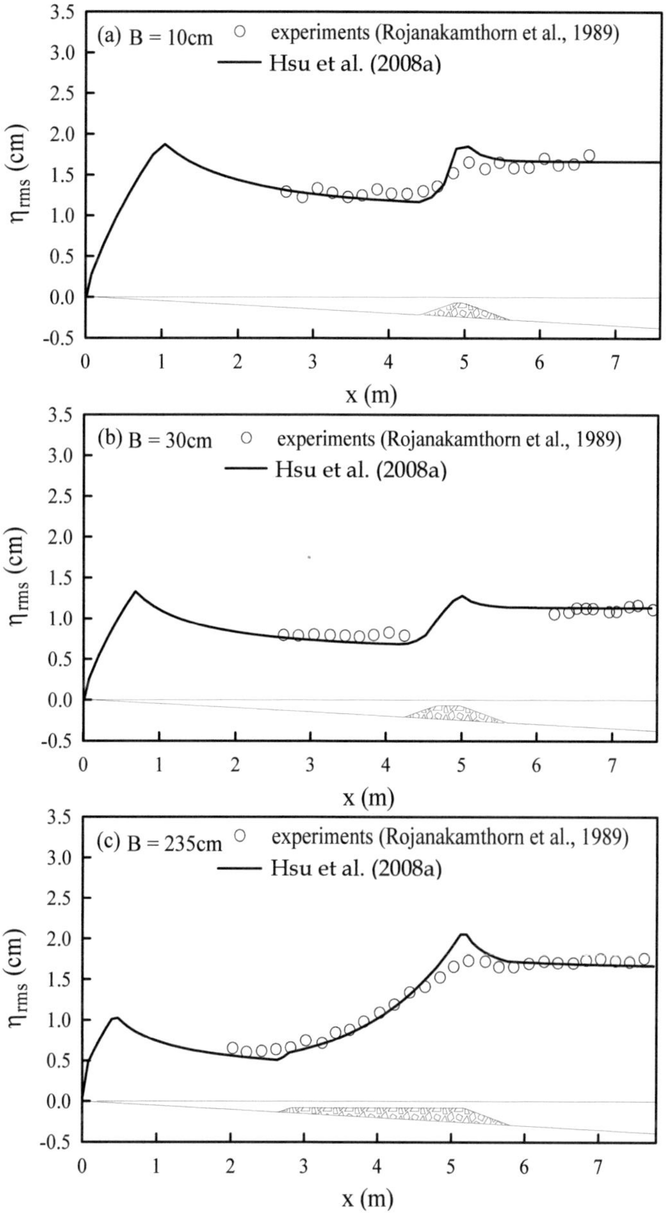

Fig. 3. Distribution of root mean square value of water surface elevations. (a) $B = 10$ cm; (b) $B = 30$ cm; and (c) $B = 235$ cm.

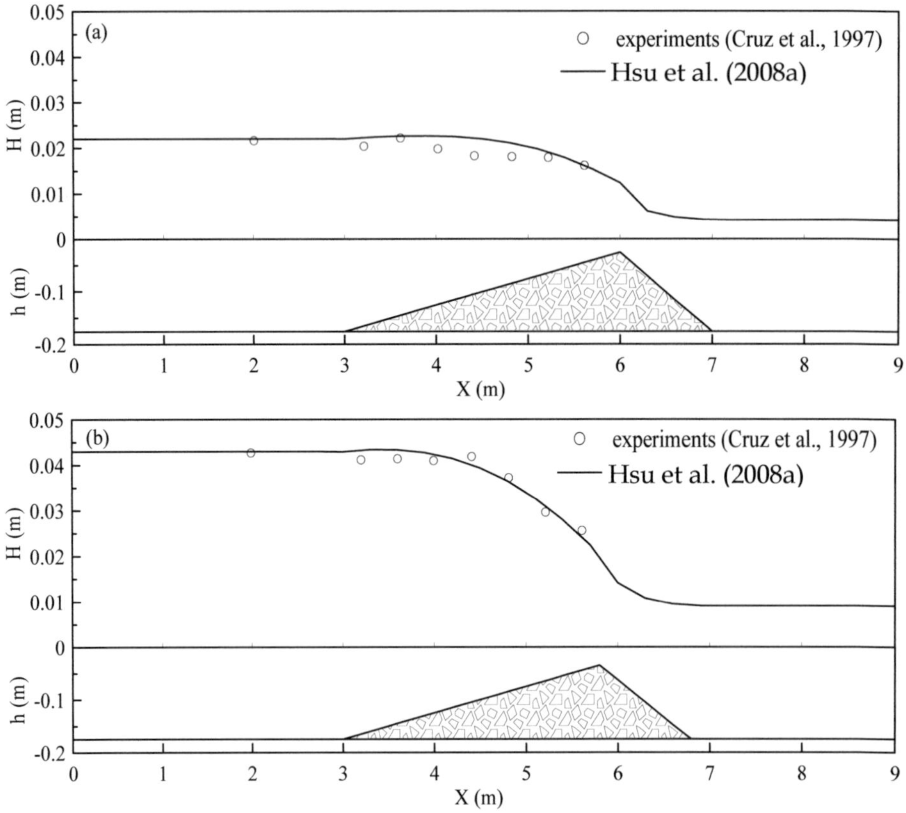

Fig. 4. Wave transformation on plane porous slope. (a) H_0 = 0.022 m; T = 1.02 s and (b) H_0 = 0.043 m, T = 1.00 s.

Numerical simulations of the wave breaking processes over a gravel sloping bottom were also performed by Hsu et al. (2008a) and compared with the experimental data provided by Lara et al. (2006). Experiments were conducted in a wave flume with a length, width, and depth of 24 m, 0.6 m, and 0.8 m, respectively. Two sloping layers with a porosity of $n_0 = 0.49$ were constructed with different mean gravel diameters— $d_{50} = 19$ mm and 30 mm, respectively. These key parameters were used to estimate the values of k_p and C_p by Eqs. (17) and (18); f_p is obtained through an iteration procedure. Two different incident wave conditions $H_0 = 0.15$ m, $T = 2.0$ s and $H_0 = 0.15$ m, $T = 3.0$ s (Table 2) were selected to verify the capability of the present model. The types of wave breaking in these two cases are spilling breaker ($T = 2.0$ s) and plunging breaker ($T = 3.0$ s), respectively. The experimental conditions are summarized in Table 2. Fig. 5 shows the spatial variations of the wave heights from a shoaling zone to the still water level. The numerical results of Hsu et al. (2008a) model can reproduce the wave transformation over a permeable sloping bottom. Notably, under incident wave conditions, the breaking wave height, location of the breaking point, and wave decay caused by energy dissipation against breaking and porous gravel slopes were accurately predicted.

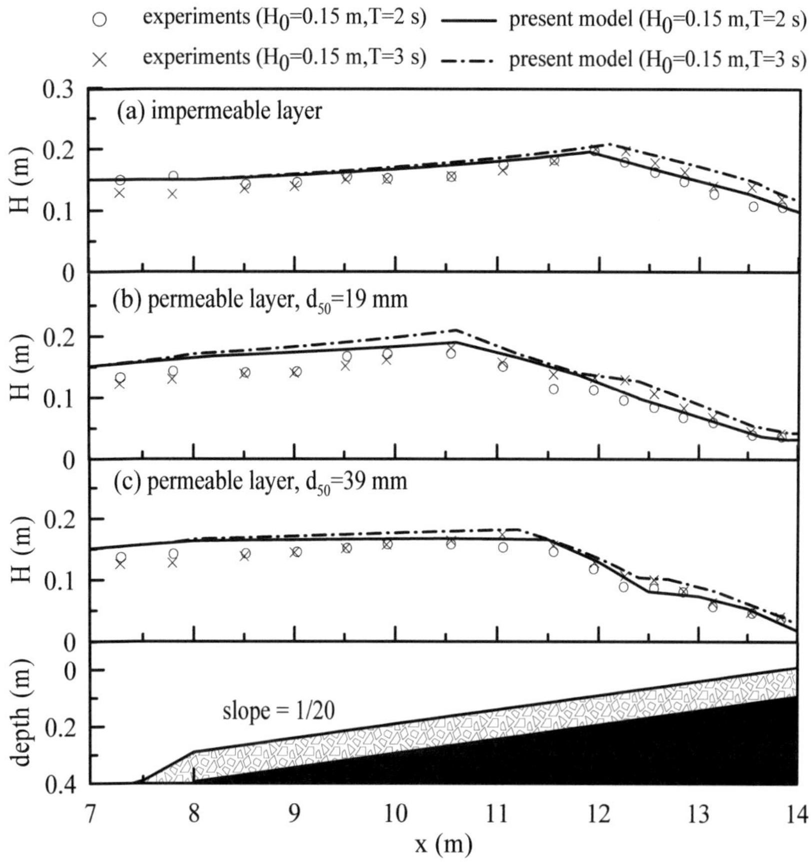

Fig. 5. Wave height evolution on permeable gravel slopes predicted by Hsu et al.'s (2008a) model. (a) impermeable layer; (b) permeable layer (d_{50} =19 mm); and (c) permeable layer (d_{50} =39 mm). The thickness of the porous layer is 11 cm. The experiments were carried out by Lara et al. (2006).

4. Experimental verification

In order to examine the validity of the present model, a large-scale experiment was conducted by Hsu et al. (2008a) in a three-dimensional wave basin to create waves propagating over a permeable submerged circular pile that is rested on a gravel sloping bottom with normal incidence. This model test was carried out in a laboratory of the National Cheng Kung University, Taiwan. As show in Fig. 6, the wave basin has a length, width, and depth of 40 m, 40 m, and 1.2 m. Waves were generated by two piston-type wavemakers, and a wave absorption system was installed on all the boundaries. The sloping bottom was carefully constructed in concrete with straight and parallel contours that made an angle of 15° with the wave generator; the bottom slope was $1/29$. The porous circular pile was constructed with gravel with $d_{50} = 35.5$ mm, a diameter of 3 m, $D_s = 0.1$ m, and $n_0 = 0.419$; it is located at the center of the test area, which has a water depth varying from 0.175 m to 0.275 m.

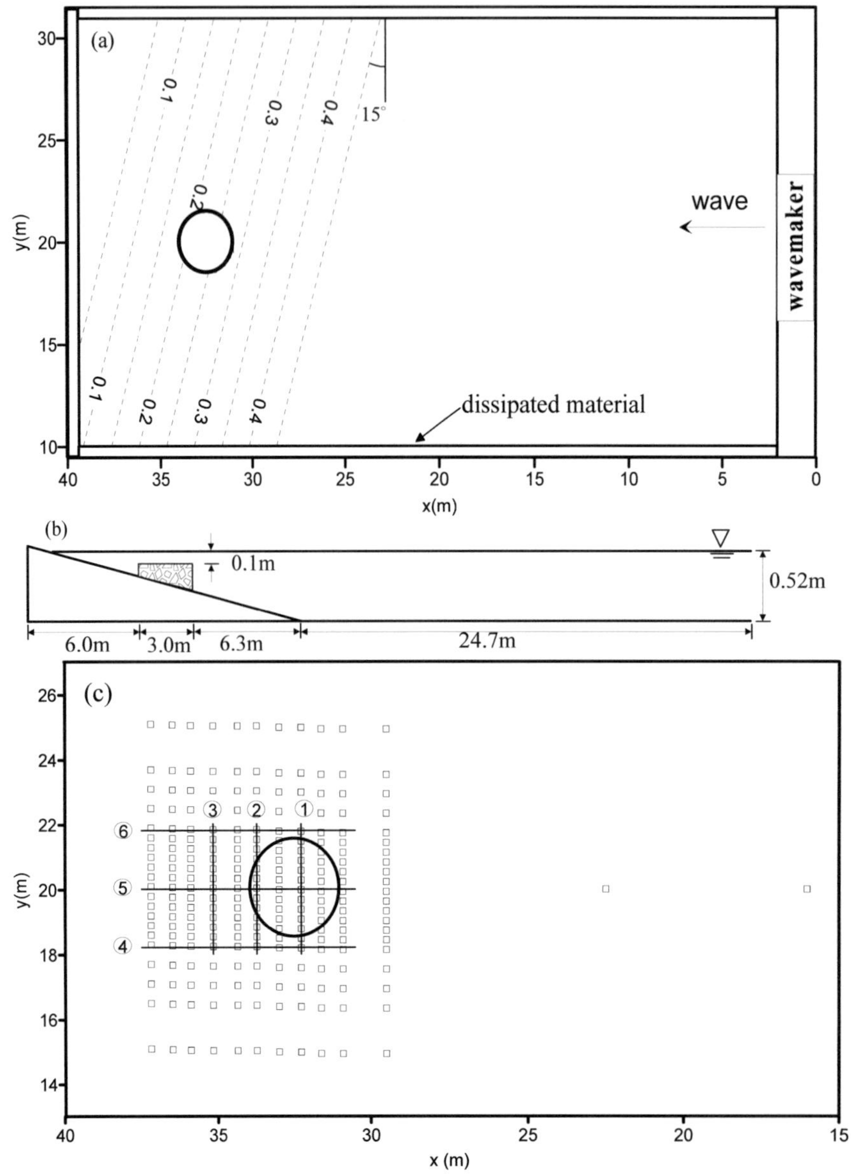

Fig. 6. Laboratory experiment in a wave basin with a submerged permeable circular pile placed on a sloping bottom. (a) plane view; (b) section view; and (c) measuring points of wave gauges.

The still water level in the constant-depth region was maintained at 0.52 m for all experiments. The free surface elevation was measured by 233 capacity-type wave gauges installed at different locations (Fig. 6). The values of intrinsic permeability and turbulent drag coefficient for the circular pile were obtained from water-head experiments, and the results were $k_p = 1.9 \times 10^{-7}$ m^2 and $C_p = 0.1048$, respectively. The incident wave conditions at deep water depth were $H_0 = 0.04$ m, $T = 0.9$ s with normal incident waves.

The porous submerged circular pile acts as a lens and focuses the incoming wave energy into a strong convergence zone. This experiment provides a comprehensive data set that can be used to verify the capability of Hsu et al.'s (2008a) model to predict accurate wave transformation including refraction, diffraction, and energy dissipation due to porous structures on a plane beach. Wave height data were analyzed and collected at many locations within the test zone, along 21 transections (see Fig. 6). The simulated wave heights were compared with the observed data at six selected sections.

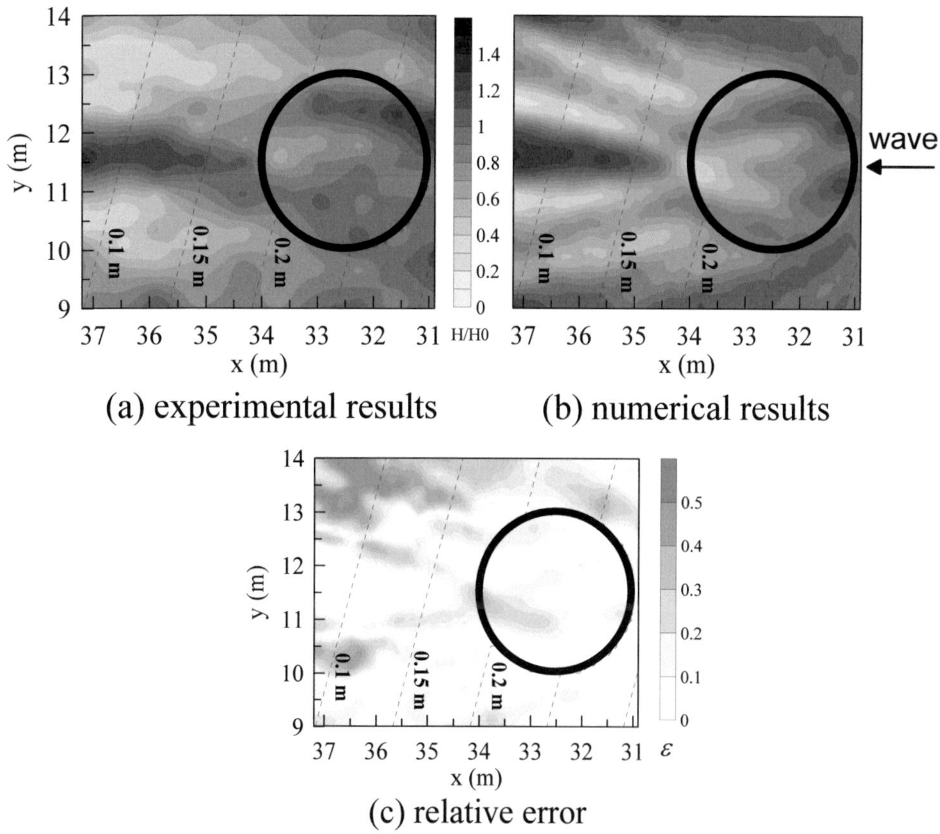

Fig. 7. A comparison of wave patterns around a submerged permeable cylinder. $H_0 = 0.04$ m, $T = 0.9$ s. (a) experimental results; (b) numerical results; and (c) relative error $\varepsilon = |(H_m - H_c)/H_m|$ between the experimental and numerical results.

Fig. 7(a) and Fig. 7(b) shows the results of relative wave heights of experimental data and numerical data in a planar wave pattern. The comparison between experimental data and numerical data demonstrates that the model is capable of describing the combined effect of wave refraction, diffraction, and porosity. Note that the model predicts a favorable focusing wave pattern. The difference between the numerical calculations and experiments as shown in Fig. 7(c) may be caused by wave reflection from the side walls in the wave basin. Fig. 8 presents comparisons of the wave height variations for all six profiles; it is interesting to note that the numerical results are in good agreement with the experimental observations.

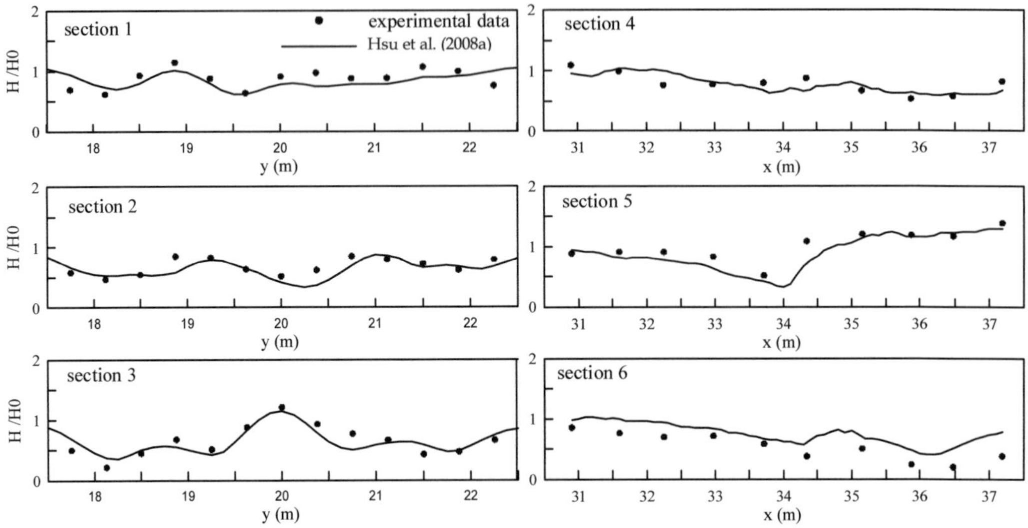

Fig. 8. Comparisons of wave heights at different transactions for waves passing over a submerged circular pile.

5. Model applications

In addition to the laboratory verification, the proposed by Hsu et al.'s (2008a) model was also verified using two typical real cases. The first case simulates wave transformation over a porous elliptical shoal superimposed on a plane beach. The second case tested waves propagating over a permeable submerged breakwater using field data obtained during a field experiment at the CERC Field Research Facility (FRF) in Duck, North Carolina, U.S.A.

In coastal engineering, a porous shoal is generally constructed with gravel or stones for ecological recovery and coastal protection at offshore regions. In the numerical run, the shoal was placed on a sloping bottom with a slope of 1:20, which was rotated clockwise at an angle of 20° from a straight wave paddle and had major and minor radii of 4.0 m and 3.0 m, respectively. The experiment for an impermeable elliptical shoal was conducted by Berkhoff et al. (1982) in a wave basin where the deep mean water depth was h=0.45 m and the incident wave height and period were H_0 = 0.0464 m and T = 1s, respectively. Since the water depth near the wave board was constant, the wave crests generated by the wave paddle traveled in uniform wave trains. According to Eqs. (17) and (18), the typical key parameters of k_p and C_p used in the numerical computation for the porous shoal constructed with gravel of d_{50} = 50 mm and n_0 = 0.20, n_0= 0.39 are summarized in Table 3. For the parabolic MSE model, wave rays and wave height contours for both impermeable (n_0= 0) and permeable shoals (n_0 = 0.20, n_0= 0.39) are shown in Fig. 8. Based on Eqs. (17) and (18), it is noted that different values of the porosity n_0 and different gravel sizes (d_{50} = 50 mm) could yield different values for k_p, C_p, and f_p in the calculation, as listed in Table 3. The numerical result indicates that the wave height continues to increase behind the shoal and decay as the waves reach to the shoreline. The focus of wave trajectories (caustics) is clearer for the impermeable shoal due to the focal phenomenon of wave refraction-diffraction. The comparison in Fig. 9 implies that the combined wave refraction-diffraction effects are rapidly reduced to some extent by the porous structures. A higher porosity of a shoal with a constant grain, gravel or stone diameter would produce a higher attenuation of transmitted waves.

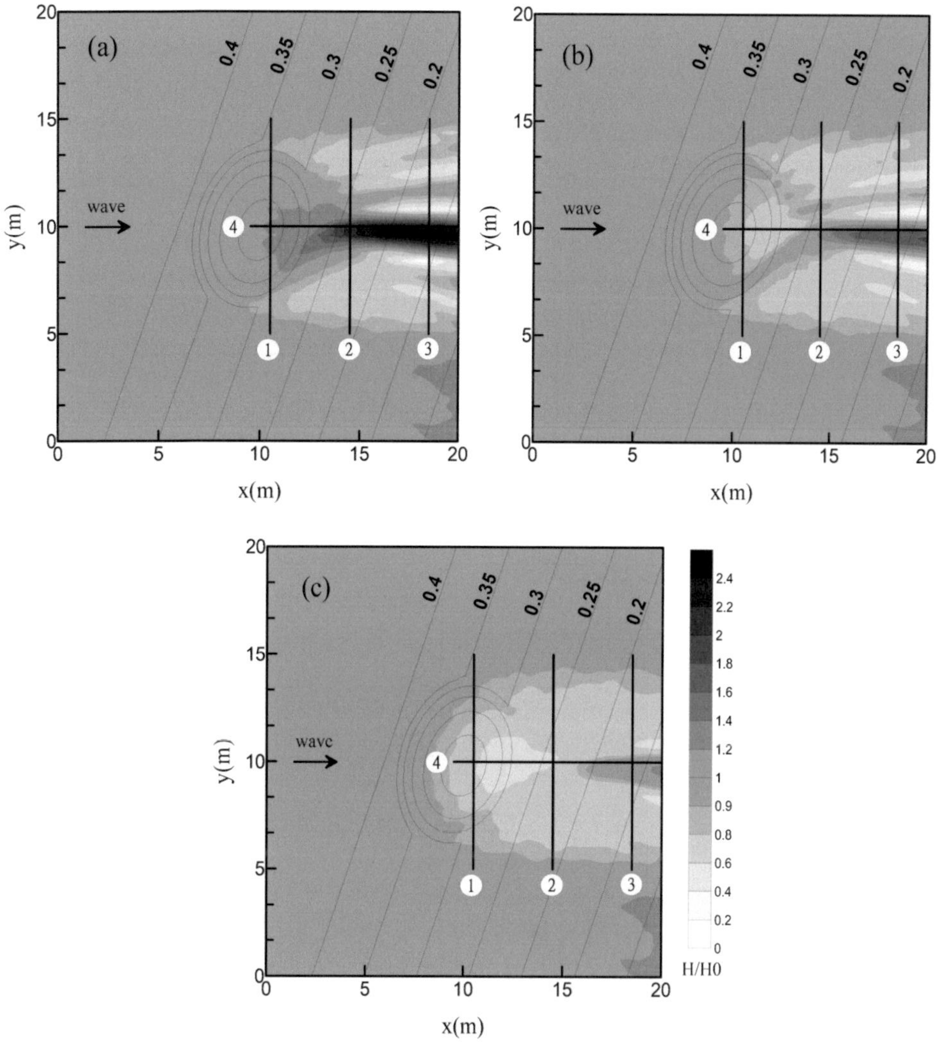

Fig. 9. Calculated wave fields for (a) impermeable ($n_0 = 0.0$), (b) permeable ($n_0 = 0.20$, $d_{50} = 50$ mm), and (c) permeable ($n_0 = 0.39$, $d_{50} = 50$ mm) elliptical shoal on a sloping bottom. The thin solid lines indicate bathymetric contours. The bold solid lines labeled with numbers are transections 1~4 of the wave height distribution, provided for the purpose of comparison. (Hsu et al., 2008a)

When verifying the proposed by Hsu et al.'s (2008a) model, the wave heights are compared at four selected sections, labeled as transections 1~4. A comparison of the computational results with the observations is shown in Fig. 10. It is observed from all the figures that the present model results are in much closer agreement with the experimental data than those of the linear model (Dalrymple et al., 1989) for an impermeable shoal. For any case that involves wave refraction, diffraction, and energy dissipation, it is evident that the present model is capable of accomplishing this job. Notably, Fig. 10 shows that the focus is clearly reduced by the porosity of the shoal.

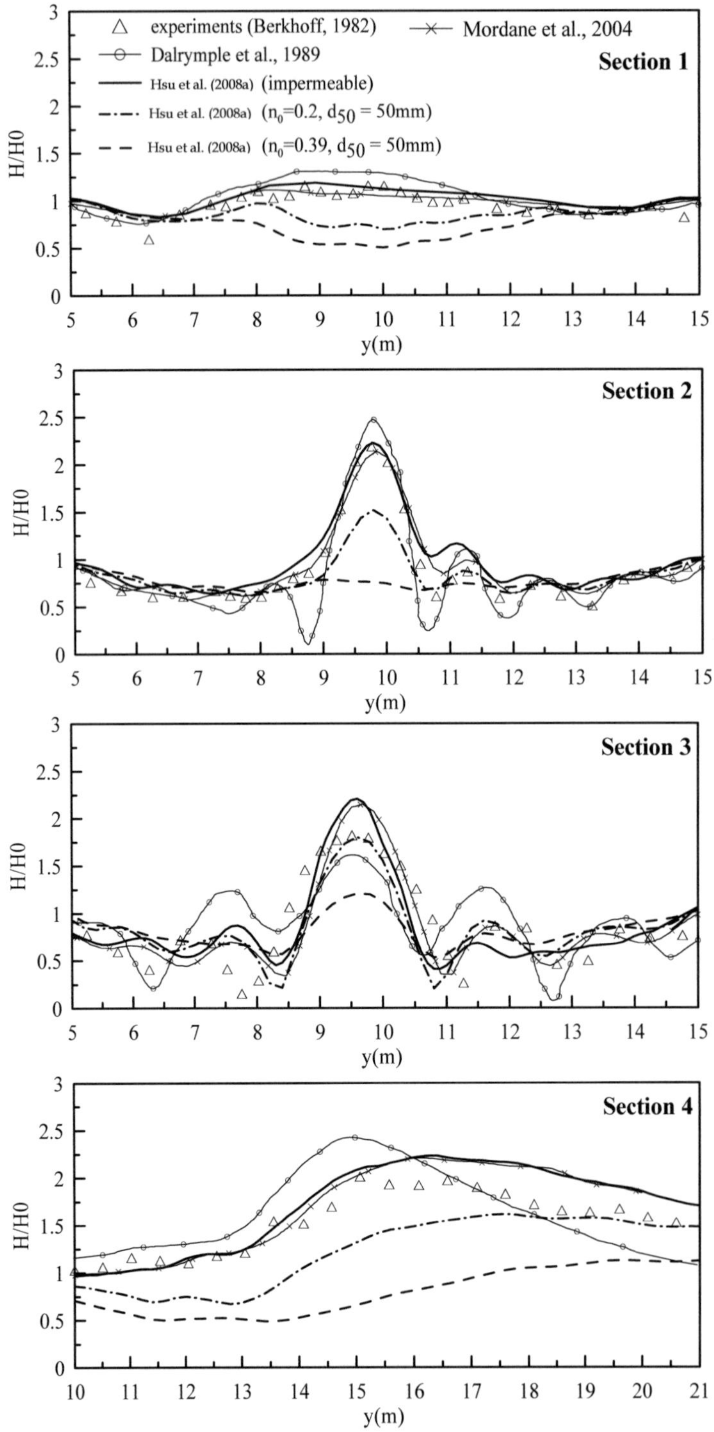

Fig. 10. Comparison of the model results ($n_0 = 0$, 0.2, 0.39) against experimental data by Berkhoff et al. (1982) along sections 1~4, respectively (see Fig. 9).

n_0	d_{50} (mm)	k_p (m^2)	C_p	f_p
0.20	50	2.6E-8	0.06	0.372
0.39	50	6.0E-7	0.15	0.565

Table 3. Key parameters used in the computation of elliptical shoal

The bathymetry in the FRF coastal area is generally straight and parallel to the coastline, except in the vicinity of the research pier, as shown in Fig. 11. The pier was built in order to collect field data. Two cases with and without a submerged breakwater were designed for simulating a wave field using the available wave data in the numerical calculation. The submerged breakwater hass a length, width, and length of 200 m, 30 m, and 2.5 m, respectively, and it comprised large stones; it was placed at a water depth of $h = 4$ m. For a practical engineering design, the porosity and stone size of the submerged breakwater were taken as $n_0 = 0.49$ and $d_{50} =$ 1,000 mm. The values of $k_p = 1.025 \times 10^{-4}\, m^2$ and $C_p = 0.173$ were then obtained from Eqs. (17) and (18). The input monochromatic wave conditions are $H_0 = 1.56$ m, $T = 6.87$ s, and $\theta_0 = 43^\circ$, based on field observations.

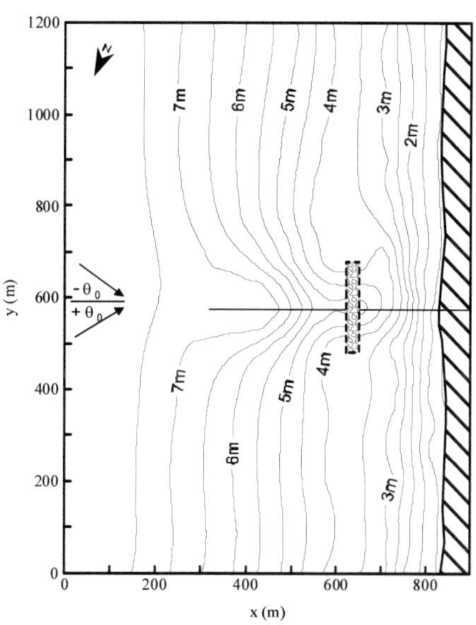

Fig. 11. Application of model for a complicated bathymetry constructed using a permeable submerged breakwater in the FRF pier, Duck, North Carolina, U.S.A.

From Fig. 12(a), we notice that the scouring hole causes the wave energy to diverge, inducing a reduction in the wave height along the pier without a structure. For the construction of the submerged breakwater, Fig. 12(b) demonstrates that the model reproduces the general wave height variation due to wave refraction, diffraction, wave breaking, and energy dissipation. Comparisons between the model results and observed data for field verification without a structure are presented in Fig. 13. The results show that the proposed model accurately predicts wave propagation for these types of wave conditions over a complicated bathymetry. It is also noted that the model does decay the wave height on account of refraction, diffraction, wave breaking, and porosity behind the submerged detached breakwater.

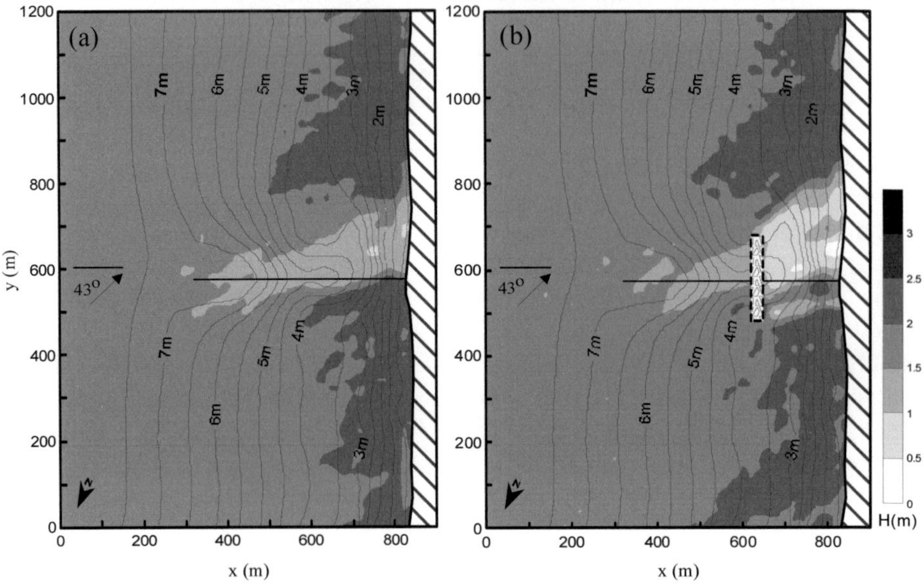

Fig. 12. Planar wave patterns (a) with and (b) without a permeable submerged breakwater on a complicated bathymetry at FRF coastal area, Duck, North Carolina, U.S.A. The line at the center represents the FRF pier.

6. Evolution Equation of Mild-Slope Equation (EEMSE)

To further consider the effect of wave reflection, the EEMSE was developed by Hsu et al. (2008b) to simulate waves propagation over a porous structure. Following Hsu et al. (2008b) a numerical model which includes wave-porous media interaction, wave reflection, wave breaking and dissipation was presented. Eq.(10) can be expressed as

$$-\left[(2i\omega)/\alpha_p\right]\varphi_t = \nabla_h^2\varphi + k_c^2\varphi \tag{28}$$

The radiation boundary condition (RBC) in the model is specified so that waves are forced to go out radically along the bounder of a computational domain. There are two types of RBC: a full or partial reflection radiation boundary condition and a given boundary condition which are specified along the boundary of x and y directions as follows, respectively.

$$\phi_x = \pm(-1)^m i\alpha\phi k\cos\theta + 2i\phi_i k\cos\theta \tag{29}$$

$$\phi_y = \pm(-1)^m i\alpha\phi k\sin\theta \tag{30}$$

where $\alpha = (1-R)/(1+R)$ is an absorption coefficient; R is a reflection coefficient, θ is the approaching wave angle to the boundary; and ϕ_i denotes a given velocity potential at the boundary of incident waves. For full reflection, $\phi_i = 0$, $m = 0$, and $\alpha = 0$; for partial reflection, $\phi_i = 0$, $m = 0$, and $0 < \alpha \le 1$; and for a given boundary condition, $m = 1$ and $\alpha = 1$. For this case, the velocity potential is assumed in the form

$$\phi = A(x,y)e^{i(kx\cos\theta + ky\sin\theta - \omega t)} \tag{31}$$

where $A(x,y)$ is the amplitude of the velocity potential. From Eq. (31), we have

$$\frac{1}{k^2}\frac{\partial^2 \phi}{\partial^2 y} = \phi \sin^2\theta \tag{32}$$

For full transmitted waves along the boundary, Eq. (30) is reduced to Eq.(15). In this model wave diffraction, refraction, reflection and breaking from porous structures were considered. The model is validated for wave propagating over a single cylindrical impermeable pile rest on a protecting cylindrical impermeable or permeable step. A typical case was performed by Hsu et al., (2009). Fig. 13 presents the bathymetry of a single cylindrical pile rest on a protecting cylindrical step. The impermeable cylinder has a radius of $r_a = L/4$ and the radius of the cylindrical protection is $r_b = L/2$, where L is the wavelength. The wave incidence is equal to 45°. The incident wave height is $H_0 = 0.03$ m. The results of analytical solutions presented by Silva et al. (2003) and predicted dimensionless wave amplitude fields are shown in Fig. 14. Fig. 15 shows the result of wave propagation over a single cylindrical impermeable pile on a protecting cylindrical permeable step ($n_0 = 0.4$, $f_p = 1.0$).

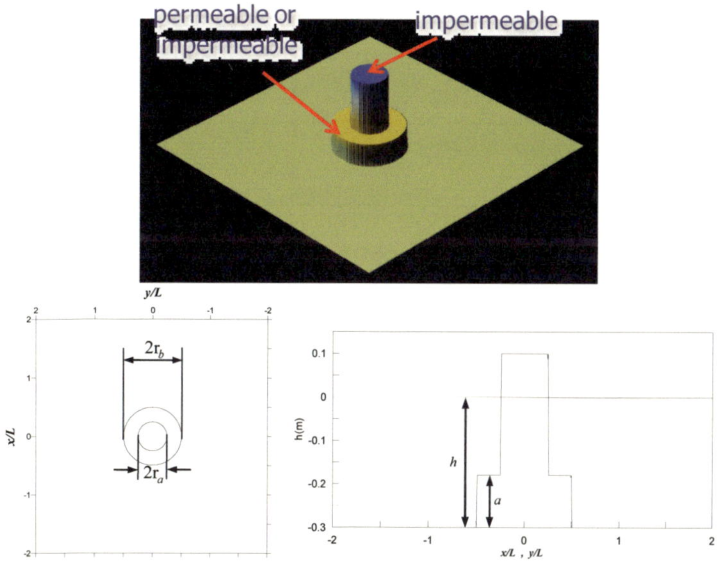

Fig. 13. Bathymetry of a single cylindrical pile on a protecting cylindrical step

In this two figures, the compared profile is along the x/L= 0 and y/L= 0 axes. Comparing with theoretical results and present model, it is clear that the numerical prediction is in good agreement with the analytical solution.

The model is also applied to simulate wave transformation on a group of five vertical porous cylinders. Each cylinder has a radius = L/4. In Fig. 16, the four cylinders are located at the vertices of a square of side dimension which is equal to 2L and the fifth one is located at the centre. The incident wave condition is the same with Fig.15. Fig. 17 shows the wave pattern for waves propagating over a group of five vertical permeable cylinders. The results of analytical and numerical dimensionless wave amplitude fields are compared with in Fig. 17 and good agreement is found.

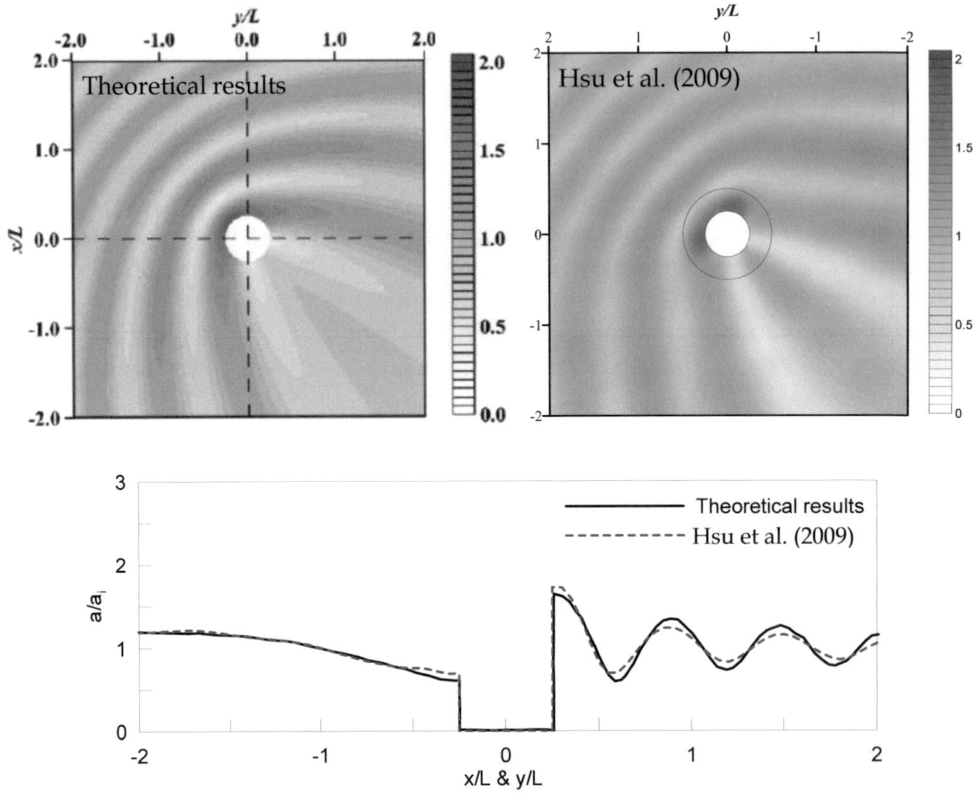

Fig. 14. Wave propagation over a single cylindrical impermeable pile on a protecting cylindrical impermeable step ($n_0 = 0,\ f_p = 0$)

7. Conclusions

A numerical wave model has been developed by Hsu et al. (2008a; 2008b; 2009) for wave propagation over porous structures on the basis of an evolution equation of mild-slope equation (EEMSE), including wave breaking and energy dissipation. The model is simple and it incurs a lower computational cost even when modeling large coastal areas. For progressive waves, the validity of the present model is verified through comparisons with the experimental data and analytical solutions for waves propagating over porous structures. The model is further applied to the practical cases of a permeable elliptical shoal and a submerged permeable breakwater on a complicated bathymetry. Since a higher-order quadratic rational function of the Padé [2,2] approximation is employed in the finite difference discretization, the accuracy of model prediction can be improved.

The wave reflection is also considered in the present parabolic approximation. The MSE developed for describing wave transformation over porous structures is initially well suited

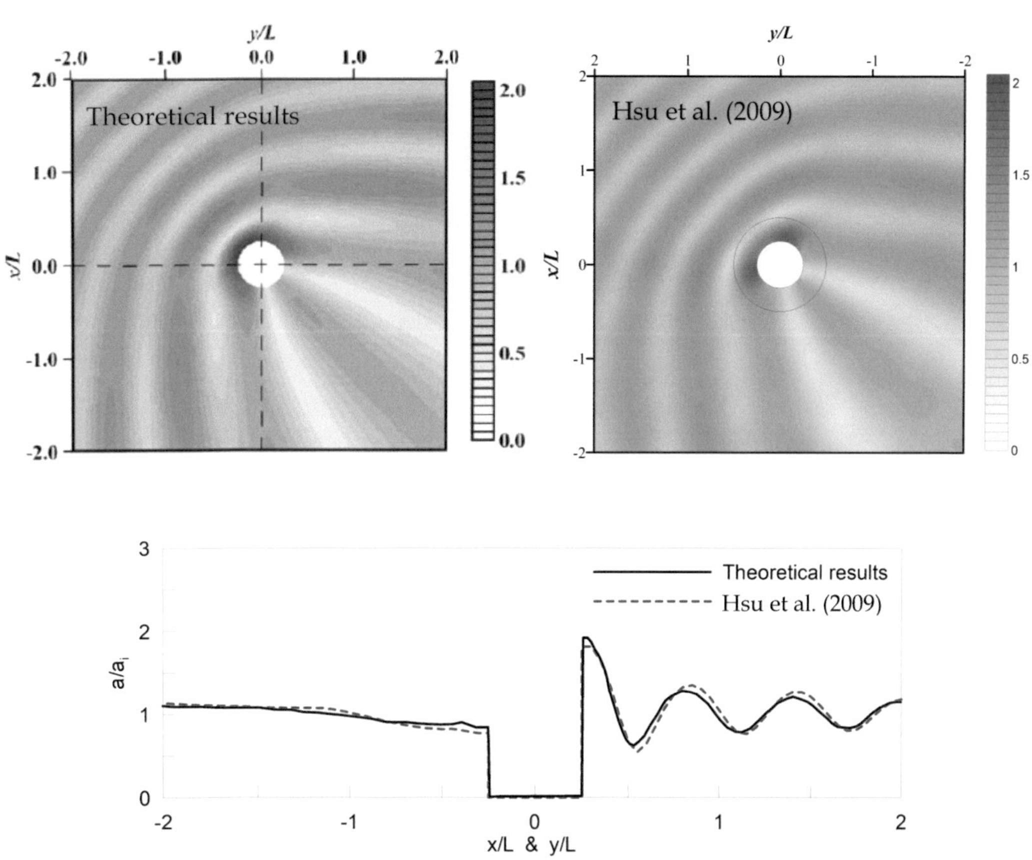

Fig. 15. Wave propagation over a single cylindrical impermeable pile on a protecting cylindrical permeable step ($n_0 = 0.4$, $f_p = 1.0$)

for examining wave interaction with submerged structures. In some cases, reflection might be important for submerged porous structures or waves traveling on a steep sloping beach. However, the combined effect of wave refraction and diffraction for simulating coastal processes is the focus of the parabolic model. For this reason, the parabolic MSE of Mordane et al. (2004) is extended to account for wave propagation over submerged permeable structures. A further development of the parabolic MSE for porous structures including reflection is also carried out by Hsu et al. (2008b).

The other restriction of the parabolic MSE is that it is applicable only to regular waves. In real situations, the sea state is random; therefore, it would be more realistic if the present model could be used to describe wave transformation of porous structures for irregular waves. Following Panchang et al. (1990) and Suh et al. (1997), the present MSE can be modified to derive a wave spectral calculation.

Furthermore, the model can be applied to submerged structures with multiple permeable layers. The concept of equivalent work may be utilized to set a monoporous layer to represent a single layer. The validity of this approach should be examined through experimental data.

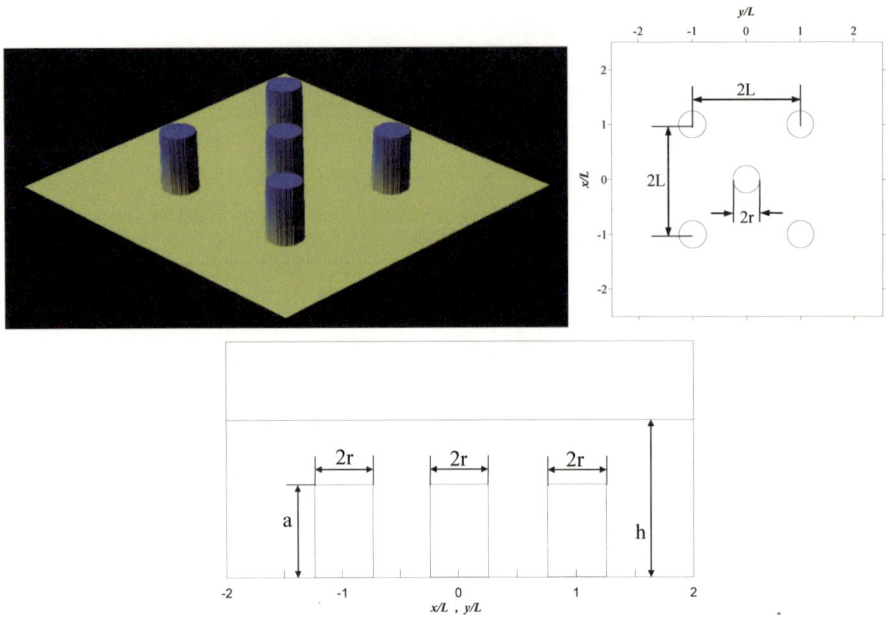

Fig. 16. Bathymetry of a group of five vertical cylinders

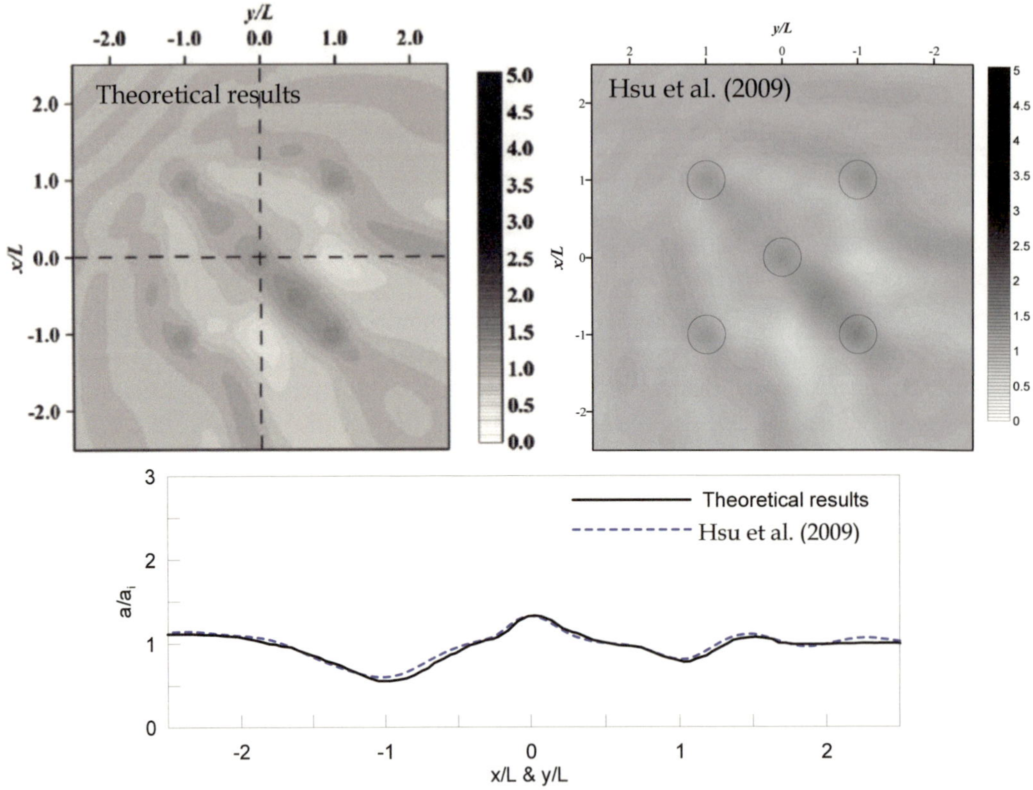

Fig. 17. Wave propagation over a group of five vertical permeable cylinders ($n_0 = 0.4$, $f_p = 1.0$)

8. References

Arbhabhiramar, A.M., Dinoy, A.A. (1973). Friction factor and Reynolds number in porous media flow. *Journal of Hydraulic Division 99 (HY6)*, pp. 901-911

Battjes, J.A., Janssen, J. P. F. M. (1978). Energy loss and set-up due to breaking in random waves. *Proceedings of 16th International Conference on Coastal Engineering*, Hamburg, pp. 569-578.

Berkhoff, J.C.W., Booij, N., Radder, A.C. (1982). Verification of numerical wave propagation models for simple harmonic linear water waves. *Coastal Engineering 6*, pp.255- 279.

Cruz, E.C., Isobe, M., Watanabe, A. (1997). Boussinesq equation for wave transformation on porous beds. *Coastal Engineering 30*, pp.125- 156.

Dally, W.R., Dean, R.G., Dalrymple, R.A. (1985). Wave height variation across beaches of arbitrary profile. *Journal of Geophysical Research 90 (C6)*, pp.11917-11927.

Dalrymple, R.A., Suh, K.D., Kirby, J.T., Chae, J.W., 1989. Models for very wide-angle water waves and wave diffraction. Part 2 : Irregular bathymetry. *Journal of Fluid Mechanics 201*,pp. 299-322.

Ebersole, B.A., Cialone, M.A., Prater, M.D. (1986). Regional coastal processes numerical modeling system. Report 1: RCPWAVE, Department of the Army, Waterway Experiment Station, Corps of Engineers.

Furukawa, K., McDougal, W.G. (1991). Wave diffraction due to trenches and rubble. Report No. 166, Department of Civil Engineering Report, Oregon State University, Corvallis, OR.

Goda, Y. (1970). A synthesis of breaker indices. *Proceedings of 17th International Conference on Coastal Engineering, Vancouver*, Sydney , pp. 471- 490.

Hsu, T.W. and Wen, C.C., (2001). A parabolic equation extended to account for rapidly verying topography. *Ocean Engineering* 28, pp.1479-1498.

Hsu, T.W., Chang, J.Y., Lan, Y.J. and Ou S.H. (2008a). A parabolic equation for wave propagation over porous structures. *Coastal Engineering* 55, pp. 1148-1158.

Hsu T.W., Chang, J.Y., Lin, T.Y. (2008b). An Evolution Equation of Mild-Slope Equation for Water Waves Propagation over Porous Media with Large Angle Incidence. *Proceedings of 31th International Conference on Coastal Engineering, Hamburg, Germany.*

Hsu, T.W., Chang, J.Y., Lan,Y.J. (2009). Modeling of water waves over submerged permeable structures with larger angle. A paper submitted to Coastal Engineering.

Isobe, M. (1987). A parabolic equation model for transformation of irregular waves due to refraction, diffraction and breaking. *Coastal Engineering in Japan* 30, pp.33-47.

Kirby, J.T., Dalrymple, R.A. (1991). User's manual, combined refraction/diffraction model. Center for Applied Coastal Research, Department of Civil Engineering, University of Delaware, Newark, De19716, REF/DIF 1, 2.3.

Lara, J.L., Losada, I.J., Liu, P.L.F. (2006). Breaking waves over a mild gravel slope: Experimental and numerical analysis. *Journal of Geophysical Research* 111, C11019.

Losada, I.J., Silva, R., Losada, M.A. (1996). Interaction of non–breaking directional random waves with submerged breakwaters. *Coastal Engineering* 28, 249-266.

Losada, I.J., Patterson, M.D., Losada, M.A. (1997). Harmonic generation past a submerged porous step. *Coastal Engineering* 31, 281-304.

Madsen, O.S. (1974). Wave transmission through porous structure. *Journal of Waterway, Port, Coastal, and Ocean Engineering* 100, 169-188.

Mèndez, F.J., Losada, I.J, Losada, M.A. (2001). Mean magnitudes induced by regular waves in permeable submerged breakwaters. *Journal of Waterway, Port, Coastal, and Ocean Engineering* 127(1), 7-15.

Miche, R. (1951). Le povoir réfléchissant des ouvrages maritime exposés à l'action de la houle, *Annales Ponts et Chausseess*, 121 Annee, 285-319.

Mordane, S., Mangoub, G., Maroihi, K.L., Chagdali, M. (2004). A parabolic equation based on a rational quadratic approximation for surface gravity wave propagation. *Coastal Engineering* 50, 85-95.

Nagayama, S. (1983) Study on the change of wave height and energy in the surf zone. Bachelor Thesis, Yokohama National University, 80p, Japan.

Panchang, V.G., Ge, W., Pearce, B.G., Briggs, M.J. (1990). Numerical simulation of irregular wave propagation over shoal. *Journal of Waterway, Port, Coastal, and Ocean Engineering* 116(3), 324-340.

Radder, A.C. (1979). On the parabolic equation method for water wave propagation. Journal of Fluid Mechanics 95, 159-176.

Rojanakamthorn, S., Isobe M., Watanabe A. (1989). A mathmatical model of wave transformation over a submerged breakwater. *Coastal Engineering in Japan* 32 (2), 209-234.

Rojanakamthorn, S., Isobe, M., Watanabe, A. (1990). Modeling of wave transformation on submerged breakwater. *Proceedings of 22th International Conference on Coastal Engineering*, Delft, The Netherlands, pp. 1060-1073.

Silva, R., Losada, M.A., Salles, P. (2006) Modelling linear wave transformation induced by dissipative structures—Regular waves. Ocean Engineering 33, pp.2174–2194

Sollitt, C.K., Cross, R.H. (1972). Wave transmission through permeable breakwaters. *Proceedings of 13th International Conference on Coastal Engineering*, Vancouver, pp. 1827-1846.

St. Mary, D.F. (1985). Analysis of an implicit finite difference scheme for very wide angle underwater acoustic propagation. *Proceedings of 11th IMACS World Congress on System Simulation and Scientific Computation*, Oslo, pp. 153-156.

Suh, K.D., Lee, C., Park, W.S. (1997). Generation of waves in Boussinesq models using a source function method, *Coastal Engineering* 32, 91-117.

Watanabe, A., Dibajnia M. (1988). A numerical model of wave deformation in surf zone. *Proceedings of 21th International Conference on Coastal Engineering*, Malaga, Spain, pp. 578-587.

III. Wave Propagation in Elastic and Piezoelectric Media.

15

Propagation of Ultrasonic Waves in Viscous Fluids

Oudina Assia and Djelouah Hakim
Facutly of Physics, University of Sciences and Technology Houari Boumedienne, P.O. 32, El Alia, Bab Ezzouar, Algiers, Algeria

1. Introduction

Viscosity is a property of fluids which can have a great importance in many fields of science and industry. It can be measured by conventional viscometers or by ultrasonic methods (Hertz & Al, 1991). Ultrasonic methods are based either on the determination of the characteristics of propagation (velocity or attenuation) or by the measurement of reflection coefficient (Hertz & Al, 1991) (Malcom J.W Povey, 1997). In the ultrasound based experimental devices, the measurements of the propagation velocity and the attenuation during the propagation allow to determine the physical properties of the medium (He & Zheng, 2001).

When an ultrasonic pulse is propagating into a viscous fluid, the pulse waveforms change because of the attenuation and the dispersion of the propagation medium. After its travelling through a medium, the transmitted is not a simply delayed and attenuated waveform copy of the pulse injected at the input. A modelling of the phenomenon of propagation is necessary for an adequate interpretation of the pulse waveforms.

During the 20 to 30 last years, a special attention has been devoted to the modelling of the ultrasonic wave propagation in viscous media. In almost these models, the phenomenon of propagation is represented by a slightly dispersive linear system for which the phase is a linear function of frequency (Hertz & Al, 1991). Unfortunately the response of such systems is not causal (D.T, 1995). Hilbert transform or Fractional Calculus are some of the main mathematical tools which have been used to develop theoretical models respecting causality (Hertz & Al, 1991) (D.T, 1969) (He, 1999). In the case of viscous media for which the attenuation of the waves is proportional to the frequency squared, the methods based on Laplace transform allow the derivation of impulse responses respecting the causality (Thomas.L, 1995) (D.T, 1969) (Norton & Purrington, 2009).

2. Theory

When an ultrasonic wave travels through a fluid, the phenomenon of attenuation results from several mechanisms of absorption. At the beginning of the propagation, the absorption is mainly due to the viscosity since the thermal agitation did not have enough time to exert its influence. Each layer of the fluid tends to slow down the displacement of the adjacent layers causing thus the damping of the wave as it penetrates into the fluid. Besides viscosity,

there is appearance of the conduction thermal phenomenon which results from the heat transfer between the regions of dilatation and compression. The wave propagation causes a thermal agitation within the fluid characterized by the collisions of the atoms: it is the Brownian motion which causes the coupling or energy exchange between the wave motion and the internal motion (translation, vibration, rotation).
The Brownian motion is a random motion of microscopic particles with different velocities resulting from the increase in temperature and caused by the molecular shocks. The local periodic variation of the pressure due to the wave propagation involves a molecular displacement (rotation + translation). This displacement is caused by the increase in the internal energy of the fluid. The change of the potential energy causes a change of the structure of the fluid because of the modifications of the distances between the different atoms. To pass from one energy level to another, each atom acquires a certain energy, leaving a disorder explained by the presence of holes in the energy levels. These displacements are the same as those which leads to the relaxation mechanism, i.e. the return to equilibrium after the modification of the positions and the structure of the fluid molecules. The return to the equilibrium is achieved after a certain time named relaxation time. The entropy of the fluid increased and any increases in entropy means the establishment of an irreversible equilibrium state accompanied by energy dissipation and dispersion in the frequency domain.

3. Problem formulation

3.1 Theoretical model

The propagation of a plane wave in a viscous fluid can be interpreted by considering that the fluid can be regarded as a set of particles which undergo during the wave propagation a succession of dilatations and compressions. Each particle transmits the vibration which it receives from its neighbors and behaves as a secondary source. The model is represented by a shematic system made up of masses m, separated by a distance a at the equilibrium, and assembled in series with a spring and a piston coupled in parallel model of Voigt-Kelvin. Under the action of the wave, the system oscillates from its equilibrium position (Thomas.L, 2004).

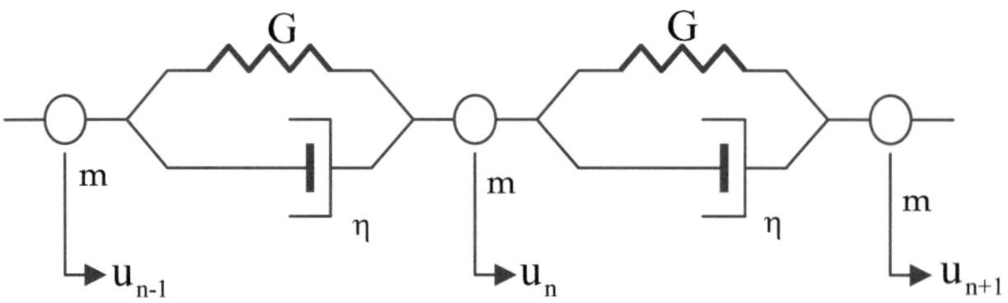

Fig. 1. Representation of the linear model of the discrete chain

3.2 Equation of propagation

The wave propagation equation of the acoustic pressure in a viscous medium is given by (Ludwig & Levin, 1995):

$$\frac{\partial^2 p}{\partial x^2} - \frac{1}{v_0^2}\frac{\partial^2 p}{\partial t^2} + \tau \frac{\partial^3 p}{\partial t \partial^2 x} = 0 \tag{1}$$

where $p(x,t)$, and $\tau = [\eta_V + 4\eta_S / 3] / \rho v_0^{\,2}$ are respectively the acoustic pressure and the relaxation time. η_S and η_V are the shear and bulk viscosity and v_0 is the propagation velocity.

Equation (1) can be solved by using Laplace transform. By taking account of the homogeneous initial conditions, the application of the transform of Laplace to the equation of propagation (1) gives:

$$\frac{\partial^2 P\,(x,s)}{\partial x^2} - \frac{s^2}{v_0^2(1+\tau s)} P(x,s) = 0 \tag{2}$$

$P(x,s)$ represents Laplace transform of $p(x,t)$. The solution that satisfies the boundary condition $\lim_{x\to\infty} \; P(x,s) = 0$ is:

$$P(x,s) = A \quad \exp^{\frac{-sx}{v_0\sqrt{1+s\tau}}} \tag{3}$$

A is a constant determined from the boundary conditions at $x = 0$. If the pressure at $x = 0$ is given by $p(0,t)$, equation (3) can be written as:

$$P(x,s) = P(0,s)\, H(x,s) \tag{4}$$

In this latter expression, $P(0,s)$ is the Laplace transform of $p(0,t)$ and $H(s)$ is a complex valued function defined by:

$$H(x,s) = e^{\frac{-xs}{v_0\sqrt{1+s\tau}}} \tag{5}$$

The inverse transform allows writing the pressure $p(x,t)$ as a convolution product:

$$p\,(x,t) = p\,(0,t) \otimes h\,(x,t) \tag{6}$$

where $h\,(x,t)$ is the inverse Laplace transform of $H(x,s)$, symbol $\otimes$ denotes the convolution in time. The impulse response $h\,(x,t)$ is defined by:

$$h\,(x,t) = \frac{1}{j\,2\pi} \int_{c-j\infty}^{c+j\infty} H(x,s)\; e^{st}\, ds \tag{7}$$

Integral (7) is evaluated in the complex plane along the Bromwich contour. This integral has to be carried out in the complex plan, along the straight line $\mathrm{Re}\,(s) = c$, with $s = c + jy$ where y varies from $-\infty$ à $+\infty$. The real number c is selected in such a way that c is on the right of all the singularities (poles, points of connections, or essential singularities).

The research of singularities shows that equation 7 presents a pole at: $s = -1/\tau$. To carry out the calculation of this integral, the following change for variable is made: $p = \frac{1}{\tau} + s$, $\varsigma = \frac{x}{\tau^{\frac{1}{2}} v_0}$ and $m = \frac{k}{\tau^{\frac{3}{2}} v_0}$.

Thus equation 7 becomes:

$$h(x,t) = e^{\frac{-t}{\tau}} \; TL^{-1}\{H1(p)H_2(p)\} \tag{8}$$

Where $H1(p)$ and $H2(p)$ are defined by:

$$H_1(p) = e^{-k\sqrt{\varsigma}} \text{ et } H_2(p) = e^{\frac{m}{\sqrt{\varsigma}}} \tag{9}$$

Accounting of the convolution theorem, one obtains:

$$h(x,t) = e^{\frac{-t}{\tau}} \; h_1(x,t) \otimes h_2(x,t) \tag{10}$$

$h_1(x,t)$ and $h_2(x,t)$ being respectively the Laplace inverse transforms of $H_1(p)$ and $H_2(p)$. The inverse transform $h_1(x,t)$ of $H_1(p)$ is:

$$h_1(x,t) = \frac{\varsigma}{2\sqrt{\pi t^3}} e^{\frac{-\varsigma^2}{4t}} \tag{11}$$

The inverse Laplace transform of the second term is defined by:

$$h_2(x,t) = \frac{1}{j2\pi} \int_{\gamma-i\infty}^{\gamma+i\infty} e^{\frac{m}{\sqrt{\varsigma}}} e^{\varsigma t} d\varsigma \tag{12}$$

This integral has a pole located at $\varsigma = 0$; γ must thus be a positive real number. The evaluation of this integral is made by considering the Bromwich contour.

The figure (2-a), represents this closed contour C of Bromwitch made up of the segments at the right-hand side Γ and of the half rings C_1 of center $(\gamma, 0)$ and of radius R .This contour does not contain any pole, the integral on C is null according to Cauchy theorem:

$$\int_{\Gamma} = -\int_{C_1} \tag{13}$$

According to the Jordan Lemma, the integral on the half-circle C_1 cancels when the radius R tends towards the infinite. Then:

$$h_2(x,t) = \lim_{R\to\infty} \int_{\Gamma} = 0 \text{ pour } t \le 0 \tag{14}$$

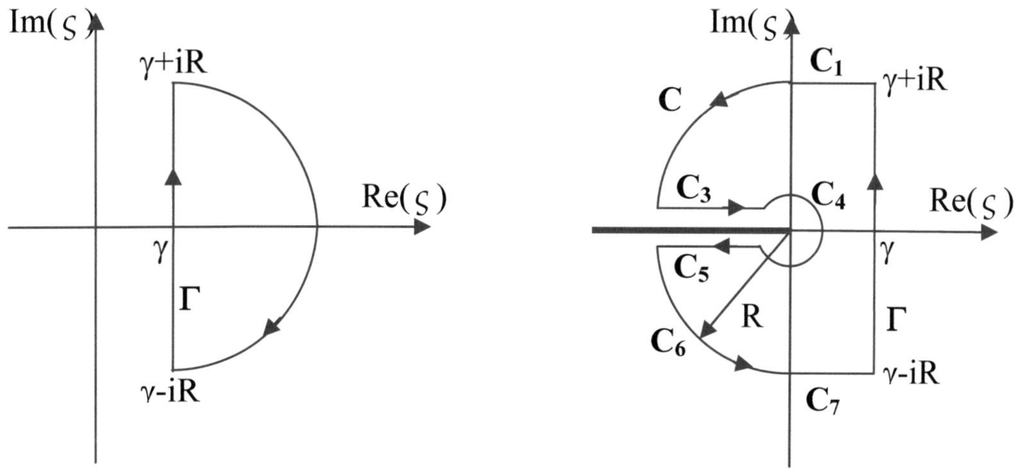

(a) (b)

Fig. 2. (a) Contour of Bromwich for $t \leq 0$ (b) Contour of Bromwich for $t \geq 0$

Figure (2-b), represents a closed contour C made up of the segments on the right-hand side of Γ, C_1 and C_7 , arcs of a circle C_2 and C_6 of radius R centered at the origin, of an arc of a circle C_4 of radius unit and segments on the right-hand side C_3 and C_5. The half-line defined by Im(p)=0 and Re(p)≤0 constitutes a cut of this contour and does not contain any pole, so the integral on C is null according to Cauchy theorem , so :

$$\int_{\Gamma} = -\int_{C_1} - \int_{C_2} - \int_{C_3} - \int_{C_4} - \int_{C_5} - \int_{C_6} - \int_{C_7} \tag{15}$$

Integrals on the segments C_1 and C_7 cancel when R tends towards the infinite. And according to Jordan Lemma, integrals along circle arcs C_2 and C_6 cancel when the radius R tends towards the infinite, so ones obtain:

$$h_2(x,t) = \lim_{R\to\infty} \int_{\Gamma}$$

$$h_2(x,t) = -\lim_{R\to\infty} \left[\int_{C_3} + \int_{C_4} + \int_{C_5} \right] \tag{16}$$

On the segment and contour C_3, the integral I_3 is written under the form :

$$I_3(x,t) = \int_{-\infty}^{-1} e^{\frac{m}{\sqrt{\varsigma}}} e^{\varsigma t} d\varsigma \tag{17}$$

$$\varsigma = re^{j\pi} = r(\cos\pi + j\sin\pi) = -r$$

r being a positive real number. So I_3 is now given by:

$$I_3(x,t) = \int_{+1}^{+\infty} e^{-\frac{i\,m}{\sqrt{r}}}\, e^{-rt} dr \tag{18}$$

On the segment and C_5 contour, the integral I5 and ς can be written as:

$$I_5(x,t) = \int_{-1}^{-\infty} e^{\frac{m}{\sqrt{\varsigma}}}\, e^{\varsigma t} d\varsigma \qquad \varsigma = re^{-j\pi} = r\,(\cos\pi + j\sin\pi) = -r \tag{19}$$

where r is a positive real number. Then, one obtains the expression of I_5 :

$$I_5(x,t) = -\int_{+1}^{+\infty} e^{\frac{i\,m}{\sqrt{r}}}\, e^{-rt} dr \tag{20}$$

The sum of the integrals on the segments C_3 and C_5 gives:

$$I_3 + I_5 = -2i \int_{+1}^{+\infty} \sin\left(\frac{m}{\sqrt{r}}\right) e^{-rt} dr \tag{21}$$

On C_4 contour $\varsigma = e^{-i\theta}$, so the expression of I_4 is written under the following form :

$$I_4(x,t) = -i \int_{+\pi}^{-\pi} e^{[m\ \cos(\frac{\theta}{2}) + t\cos(\theta)]} \times e^{i\,[t\sin(\theta) - m\ \sin(\frac{\theta}{2}) + \theta]} d\theta \tag{22}$$

The integrand is an odd function of θ

$$I_4(x,t) = -2i \int_0^{\pi} e^{[m\ \cos(\frac{\theta}{2}) + t\cos(\theta)]} \times \cos(t\sin(\theta) - m\ \sin(\frac{\theta}{2}) + \theta) d\theta \tag{23}$$

$h_2(x,t)$ can be written under the form:

$$h_2(x,t) = \frac{1}{\pi}\int_{+1}^{+\infty} \sin(\frac{m}{\sqrt{r}}) e^{-rt}\ dr + \frac{1}{\pi}\int_0^{\pi} e^{m\cos(\frac{\theta}{2}) + t\cos\theta} \times \cos(t\sin\theta - m\sin\frac{\theta}{2} + \theta) d\theta \tag{24}$$

To carry out the numerical calculation of the impulse response, one makes the following variable change for the integral going from 1 to $+\infty$:

$$r = \frac{1}{\sin\varphi},\quad dr = \frac{-\cos\varphi}{\sin^2\varphi}\ d\varphi \tag{25}$$

$$\int_{+1}^{+\infty} \sin(\frac{m}{\sqrt{r}}) e^{-rt} dr = \int_0^{\pi/2} \sin(m\sqrt{\sin\varphi}) \times e^{-\frac{t}{\sin\varphi}} \frac{\cos^2\varphi}{\sin\varphi} d\varphi \tag{26}$$

$h_2(x, t)$ becomes then:

$$h_2(x,t) = \int_0^{\pi/2} \sin(m\sqrt{\sin\varphi}) \times e^{-\frac{t}{\sin\varphi}} \frac{\cos^2\varphi}{\sin\varphi} d\varphi + \frac{1}{\pi}\int_0^{\pi} e^{[m\cos(\frac{\theta}{2}) + t\cos(\theta)]} \times \cos(t\sin(\theta) - m\sin(\frac{\theta}{2}) + \theta) d\theta \tag{27}$$

The integrals of the equation (27) are evaluated numerically by a Gauss quadrature method. The impulse response h(x,t) can then be calculated using the equation (12) and, knowing the pressure at the output of the source p(0,t), one can deduce the pressure at any point in the propagation medium by using equation (6).

4. Numerical simulation of the propagation

To represent the pressure injected at the input of the system, we use a damped sinusoid of the form:

$$p(0.t) = f(t) = e^{-\delta t}\sin(2\pi f t) \tag{28}$$

This function represents with a quite good precision the waveform of the pulses delivered by the ultrasonic generators. The damping ratio δ controls the frequency bandwidth of the input pulse around the nominal frequency f which is about several megahertz for current applications.

The output pressure at a given position is obtained by calculating the convolution of p(0,t) with the impulse response of filter propagation defined by equation 7.

The various functions used for the modeling of the propagation process (input pulse, impulse response and output pulse) are given on figure 3. By analyzing figure 3, we quote that the amplitude of the output signal is attenuated but one also notices a spreading in the time of the output pressure which arrives with a delay corresponding to the travel time. The comparison of the spectrum of the output pressure with that of the pressure at the input, shows a shift to lower frequencies. This results from the fact that the attenuation increases with frequency. One also notes the presence of a precursor such as it has been mentioned by other authors (Thomas.L, 1995) (He, 1999) (Ludwig & Levin, 1995). The presence of this precursor appears in the form of the spreading at the beginning of the waveform pulses which, in the absence of viscosity, would arrive at the point of x coordinate x at the moment T = x/cL.

5. Variation of the output pressure due to the propagation phenomena

5.1 Influence of the penetration distance

One notes the progressive reduction in the amplitude and a time-scale of the output signal as the distance from the beginning of the propagation increases figure 4. The pressure decrease of the output is due to the absorption caused by the viscous effects which attenuate the amplitude and cause a progressive filtering depending on the distance travelled by the wave. The frequency components are the beginning attenuated because of the dispersive nature of the medium.

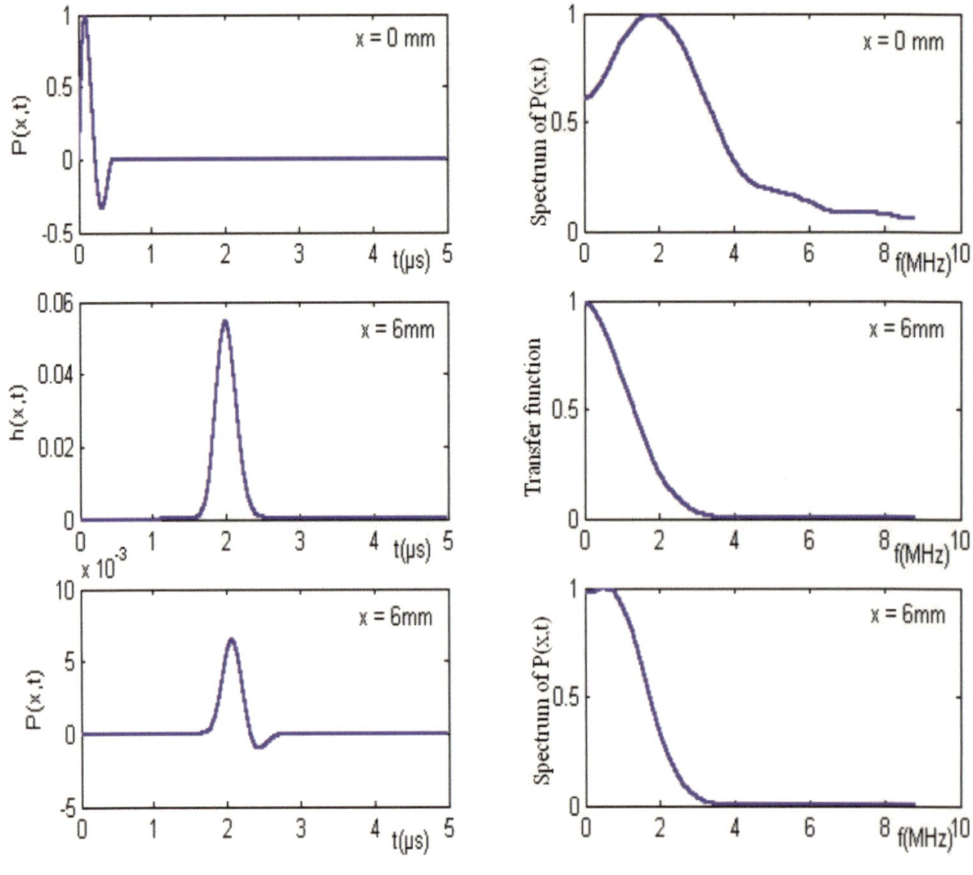

Fig. 3. Graphical representation of the input pulse, the impulse response of the output pulse at x = 6mm and the respective spectra for $\tau = 0.005$s. The nominal frequency is f = 2.25 MHz

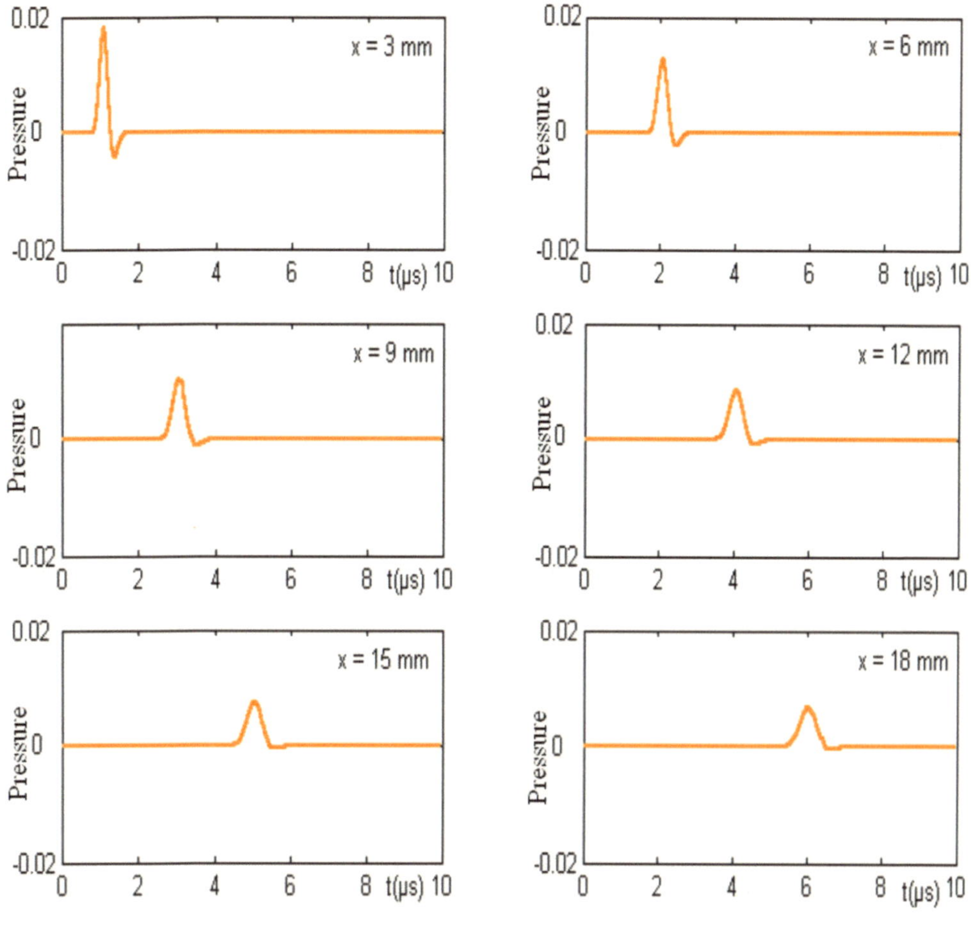

Fig. 4. Graphical representation of the output pulse for different distances for a wave having a center frequency f = 2.25MHz and for τ = 0.005s.

5.2 Influence of frequency variation

Figure 5 represents the variation of the pressure output versus frequency at a given position and for a fixed relaxation time. The increase in the frequency causes a more important reduction in the amplitude at the output. Besides this considerable reduction, we notice a contraction in time and a change of the waveform of the pulses.

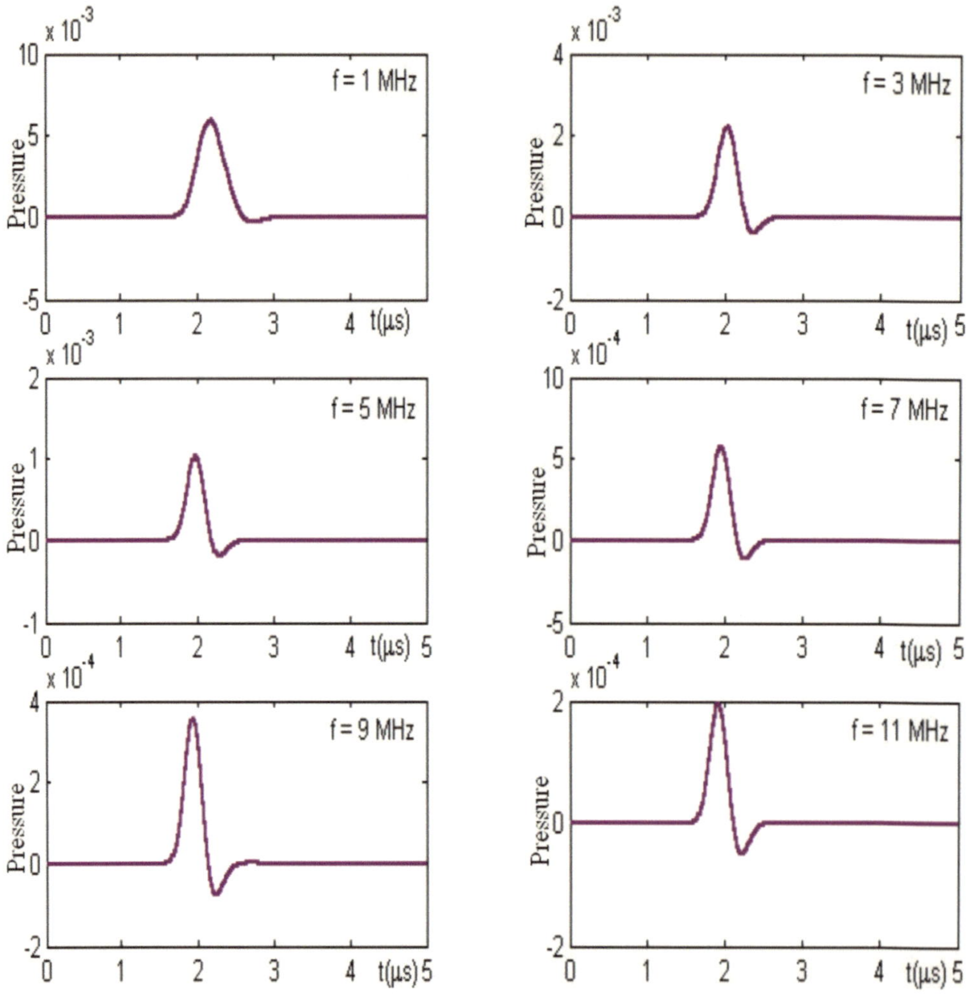

Fig. 5. Graphical representation of the output pulse at x = 6mm for different frequencies of the input pulse with $\tau = 0.005$s

5.3 Influence of relaxation time

The effect of the relaxation time is represented on figure 6, we notice a rapid decrease of amplitude and distortion of the output pulses when the relaxation time increases and a spread in time of the output pressure.

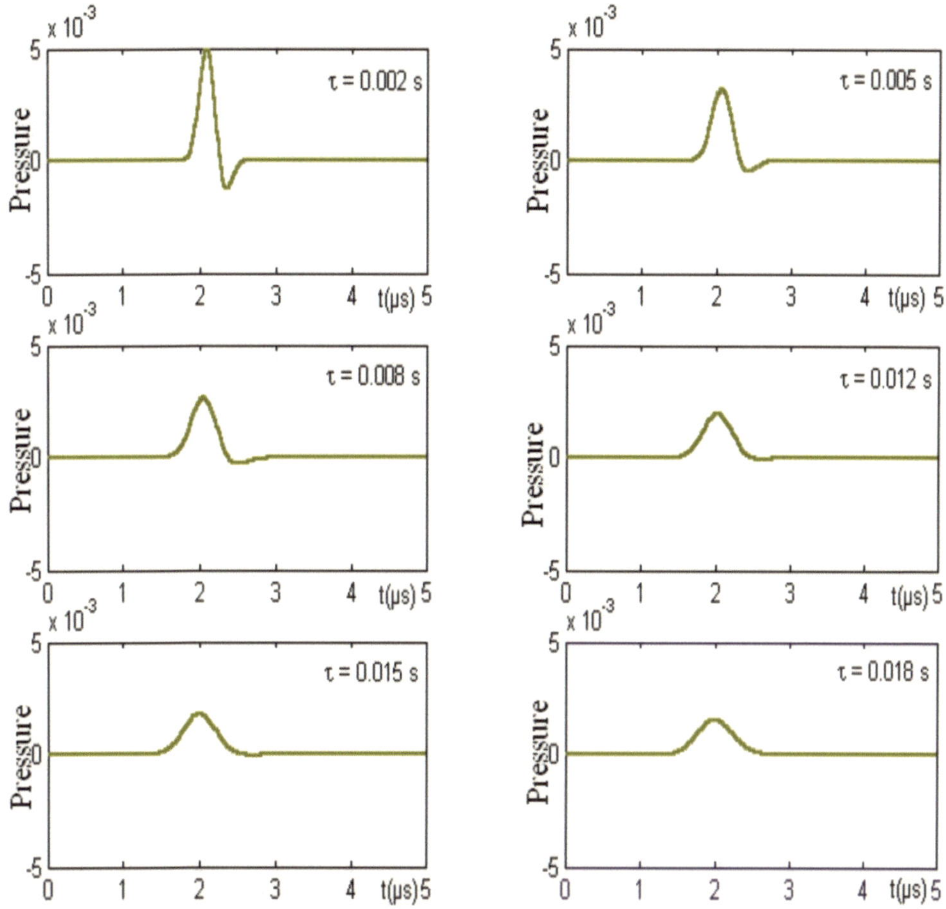

Fig. 6. Graphical representation of the output pulse at x = 6mm for different frequencies of the input pulse with $\tau = 0.005$s

6. Validation of the theoretical model

The influence of the various mechanisms of absorption which contribute to the reduction in the amplitude during the wave propagation, have been checked in experiments by studying the variation of the propagation velocity ultrasonic and the parameter of attenuation versus the temperature in glycerine (Oudina & Djelouah, 2008). The variation of the propagation velocity versus the temperature is represented on figure 8 which shows that the propagation velocity decreases when the temperature increases; this decreasing result from the thermal agitation within the fluid because of the wave motion, and characterized by the collisions of the atoms. In the case of glycerol, the predominant molecular bindings are the OH bindings which are particularly sensitive to the temperature. In fact, a sudden change of temperature induced by the successive collision of atoms causes a break of the chemical bonds of OH type whose energy cohesion is the lowest, leading thus to a decreasing velocity when temperature increases. So it can be said that an increase in the entropy means the establishment of an irreversible steady state.

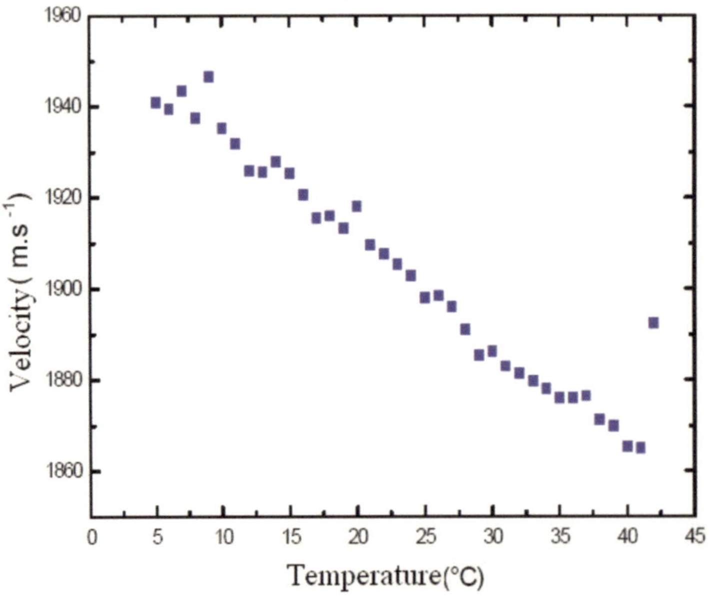

Fig. 8. Variation of the propagation velocity versus the temperature.

The study of the attenuation versus the temperature is represented on figure 9. It is noted that the attenuation decreases with increasing temperature, knowing that the attenuation parameter is closely related to the viscosity and the viscosity of a fluid strongly depends on the temperature; in particular it decreases when the temperature increases.

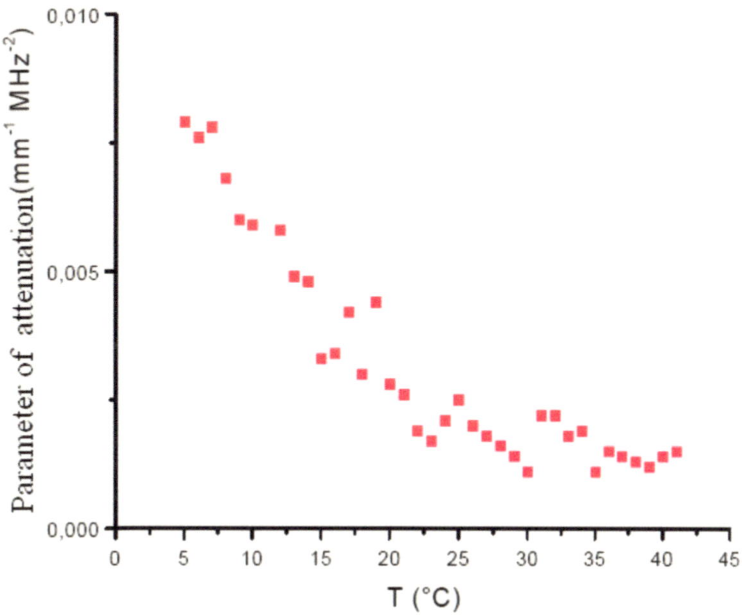

Fig. 9. Variation of the parameter of attenuation versus temperature.

7. Conclusion

The modeling of the equation of propagation in a viscous fluid by the transform of Laplace made it possible to highlight the phenomena of absorption by developing a theoretical model which allows to study the influence of these various parameters on the propagation in the viscous fluids.

8. References

Blackstock, D.T. (1969).Transient solution for sound radiated into a viscous fluid , *J. Acoust. Soc. Am*, Vol.41, No.5, (124-130).

He, Ping.; Zheng. J; (2001). Acoustic dispersion and attenuation measurement using both transmetted and reflected pulses, *Ultrasonics*, Vol.39, (27-32).

He, Ping. (1999). Experimental verification of models for determining dispersion from attenuation, *IEEE Trans, Ultrason, Ferroelect, Freq Contr*, Vol.46, No.3, (706-714).

Hertz, T. G.; Dymling, S. O.; Lindstrom, K. L.; Persson .H.(1991). Viscosity measurement of an enclosed liquid using ultrasound, *Rev. Sci. instrum.* Vol 62., No 2., (Feb 1991), (457-462).

Ludwig, R.; Levin ,P.L. (1995). Analytical and numerical treatement of pulsed wave propagation into a viscous fluid. *IEEE Trans. Ultrason., Ferroelect., Freq. Contr.*, Vol.42, No.4, (July 1995), (789-792).

Malcolm J.W. Povey (1997). *Ultrasonic techniques for fluids characterization,* Academic press, ISBN 978-0-12-563730-5, United States of America.

Norton, G. V.; Purrington . R. D. (2009). The westervelt equation with viscous attenuation versus a causal propagation operator: A numerical comparison, *Journal. of. sound. and. vibration.* Vol 62., No 2., (May 2009), (163-172).

Oudina, A.; Djelouah, H. (2008). Temperature effects on the propagation of transient ultrasonic waves in viscous fluids. *Acta. Acustica., United., With. Acustica.,* Vol.94, (43-47), ISSN 1610-1928.

Szabo, T.L. (1995). Causal theories and data for acoustic attenuation obeying a frequency power law, *J. Acoust. Soc. Am,* Vol.97, No.1, (January 1995), (14-24).

Szabo, T.L. (2004). *Diagnostic ultrasound imaging,* Academic press series in biomedical engineering, ISBN 0-12-680145-2, United States of America.

16

A Multiplicative Method and A Correlation Method for Acoustic Testing of Large-Size Compact Concrete Building Constructions

Vladimir K. Kachanov, Igor V. Sokolov, and Sergei L. Avramenko
Moscow Power Engineering Institute (Technical University)
Russia

1. Introduction

New methods for acoustic testing of large-size (>1.5 m) concrete building structures (foundations, walls of buildings, airfield pavements, bridge pillars, slabs, blocks, beams, etc.) based on the use of methods of free and forced vibrations are reviewed.
Problems of inspection of large-size concrete building structures by the resonance and the impact-echo method are considered. These methods are the only possible acoustic methods for testing large concrete building structures, that cannot be inspected by other (conventional) ultrasonic testing techniques. However, the resonance and impact echo methods for testing large-size concrete building constructions can be used only for testing extended building structures (in extended building structures the inspected thickness h is much smaller than the other dimensions). The resonance and impact echo methods cannot be used in compact building structures (in which the tested dimension, e.g., the thickness h, is comparable with at least one of the other dimensions), because of the influence of geometrical effects (the noise of the article shape does not allow unambiguous determination of the desired maximum in the article's spectral characteristic).
A new multichannel acoustic method for testing large-size compact concrete building constructions is considered. The method is based on the use of the resonance and impact-echo methods with the subsequent multiplication of partial spectral characteristics. The multichannel multiplicative method allows performance of acoustic testing of large-size compact concrete building constructions (blocks, beams, columns, supports, and other standard articles).
The second problem of inspection of large-size compact concrete building structures determined the necessity of calculating the acoustic velocity in compact articles. It is impossible to determine the acoustic-vibration velocity C_l in a compact article because of the effect of geometrical dispersion of the sound velocity. The resonance and impact echo methods can be used only at a known value of the correlation coefficient of the velocity of longitudinal vibrations in a particular compact article. This can be done using the technique of numerical simulation of acoustic fields. A new correlation method for determining the velocity in a compact article with known dimensions is described. It allows monitoring of the strength of arbitrarily shaped large-size compact concrete building constructions.

2. Problems of acoustic testing of large-size concrete building structures

Unfortunately, failures in residential and industrial buildings due to aging of concrete, which destroys the strength of building constructions (BCs), have occurred more frequently in recent years. To prevent such failures, special attention should be given to the problems of inspecting large-size concrete objects of BCs.

Unfortunately, conventional ultrasonic nondestructive testing (NDT) methods allow inspection of concrete articles of only limited thicknesses. The most informative ultrasonic echo method allows one to test BCs only to a maximum depth of 1–1.5 m even if low-frequency signals are used ($f_0 \leq 100$ kHz).

To solve the problem of testing large-size concrete BCs by acoustic methods, inspection methods based on the analysis of eigenfrequencies of an article (the impact method and, less frequently, the resonance method) were developed in the United States approximately 20 years ago (Carino, N.J., 2001; Sansalone, M. & Carino, N.J., 1986 ; Carino, N.J.; Sansalone, M. and Hsu, N.N., 1986; Sansalone, M. and Streett, W.B., 1997). The essence of the impact-echo (IE) method is illustrated in Fig. 1, which shows a diagram of testing of an extended concrete article (hereinafter, the term "extended article" is defined by the condition that the inspected thickness h is much smaller than the other dimensions). Using a small steel ball or a special impactor device, a short but strong mechanical impact is delivered to the BC surface (Fig. 1). This impact initiates free decaying acoustic oscillations in the tested extended article. These oscillations are detected by a broadband receiving PET and then by a spectrum analyzer (SA). The free-oscillation spectrum is the informative parameter for analyzing BCs. The form of the spectral characteristic allows determination of the eigenfrequency f at which a BC thickness resonance is observed. The frequency of the resonance peak allows calculation of the thickness h at a known propagation velocity C_l of a longitudinal acoustic wave: $h \approx C_l / 2f$ (Figs. 1, 2).

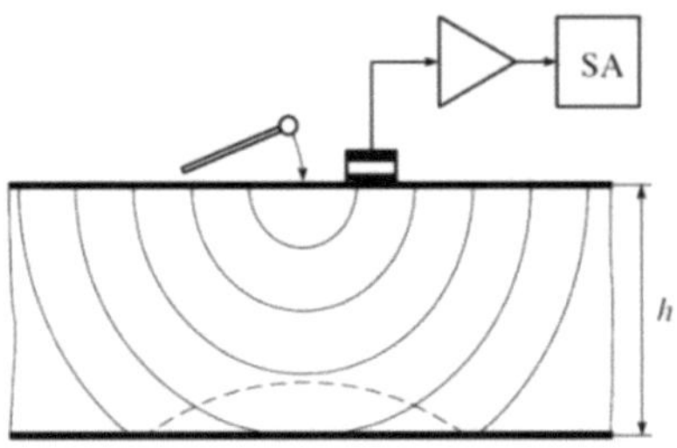

Fig. 1. Schematic diagram of testing of a flaw-free concrete article (h = 0.5 m) by the acoustic IE method.

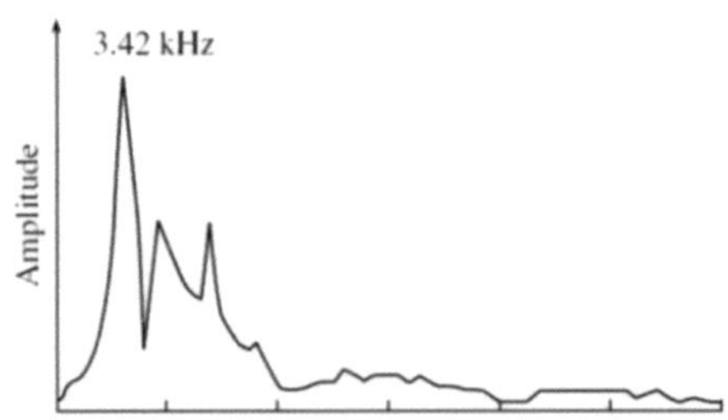

Fig. 2. AFC of an extended flaw-free concrete article: f = 3.42 kHz and h = $C/2f = C\lambda/2$.

The IE method, based on the excitation and measurement of natural oscillations in an article, qualitatively differs from conventional ultrasonic testing methods, based on location principles.

First, when the IE method is applied, a resonance of the article itself arises; the resonance frequency f is determined by the article size ($h \approx \lambda/2$), and, for large-size articles, this frequency may be within the range of tens of hertz to several kilohertz. At such frequencies, the attenuation of acoustic signals in concrete is negligibly low; therefore, it is possible to test concrete BCs with a thickness of up to 10–20 m by the IE method.

Second, the analysis of the resonance characteristics yields only indirect information on the presence of flaws. The presence of a flaw in a BC (Fig. 3) can be determined in comparing the spectral characteristics of a flaw-free (Fig. 2) and a defective article (Fig. 4).

Third, during testing of large-size concrete BCs, the most important task is to determine neither the BC dimensions nor flaws of the article's internal structure, but the structure strength, which is determined mainly by the concrete strength (grade). For this purpose, in some cases, the problem of determining the propagation velocity of acoustic vibrations inside a concrete article becomes predominant because the velocity of longitudinal acoustic vibrations C_l in concrete is directly related to the concrete strength characteristics (Ermolov, I.N. & Lange, Yu.V., 2004).

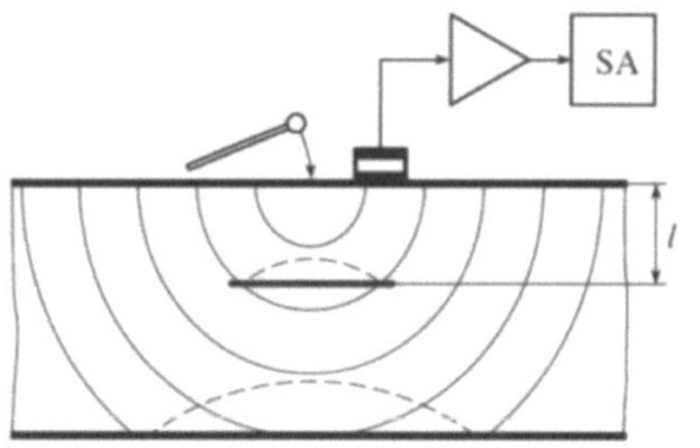

Fig. 3. Schematic diagram of testing of an extended concrete article (h = 0.5 m) with a flaw (l = 0.25 m) by the acoustic IE method.

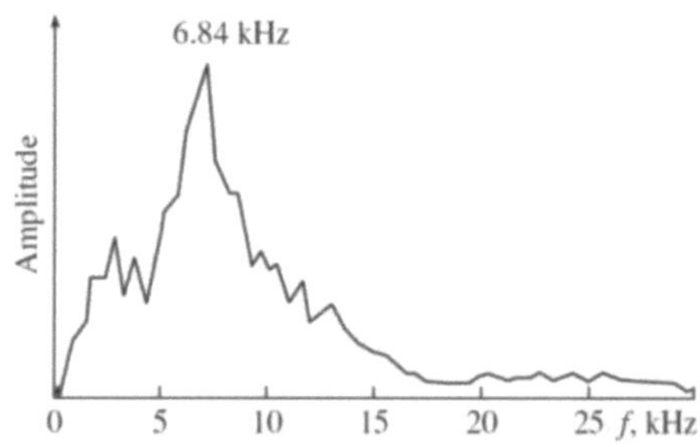

Fig. 4. AFC of an extended concrete article with a flaw: f = 6.84 kHz and l = 0.25 m.

Note that, at present, the problem of measuring the velocity of acoustic vibrations in large-size concrete BCs is far from being solved. The known methods for determining the velocity C_l in concrete from the surface velocity C_{sur} of acoustic vibrations do not always yield precise information on the actual velocity C_l inside a large-size BC because the strength of the concrete surface layer often does not correspond to the strength of its deeper layers. The velocity C_l of an acoustic wave calculated through the surface velocity C_{sur} is usually 10–20% lowers than the actual velocity in the volume of a concrete structure (Ferraro, C.C., 2003).

This phenomenon, which is explained by a change in the concrete properties near the surface due to constant contact with the atmosphere, is additionally aggravated by the presence of steel reinforcement bars inside reinforced-concrete constructions. The velocity of acoustic waves in steel is higher than that in concrete; therefore, the wave propagation velocity in reinforced concrete is higher than that in plain concrete.
The velocity of ultrasonic vibrations in large BCs cannot always be measured by the echo-pulse (shadow) method. Moreover, the velocity of longitudinal ultrasonic vibrations measured by the echo-pulse (shadow) testing method is not equal to the velocity of longitudinal acoustic vibrations deter-mined using the IE method because of the effect of geometrical dispersion in articles at $\lambda \sim h$ (Ermolov, I.N. & Lange, Yu.V., 2004; Bolotin, V.V. 1999). In fact, the velocity C_l is measured using the ultrasonic echo-pulse (shadow) testing method provided that $\lambda << h$; as a result, $C_l = 2h/T$. The velocity of longitudinal vibrations measured by the IE method is measured under the condition of $\lambda \sim h$; as a result, $C_l = (2hf/k)$, where k is the velocity correction factor ($k \neq 1$). For slabs and piles, k = 0.96 and 0.95, respectively (Carino, N.J., 2001). Hence, the BC strength can be measured by the IE method only in articles where the velocity correction factor k is strictly defined.
Proceeding from the above, we can conclude that eigenfrequency methods are the only possible acoustic methods for testing large BCs that cannot be inspected by other (conventional) acoustic testing techniques. However, the IE method allows testing of only extended BCs (for which the factor k is known) and excludes testing of compact concrete BCs.
Hereinafter, we define a compact article as an object for which the ratio of the measured thickness to the two other dimensions (width and length) is <1:5 or >5:1. Note that, according to this definition, both a large concrete article (2x3x4 m) and a small object (2x3x4 cm) are compact.
Figure 5 shows examples of (a) extended and (b) compact BCs; arrows indicate surfaces accessible to testing and determining surfaces of the impactor location. Figure 6 illustrates the problem of testing compact articles. The article has a limited width (a) and a thickness-to-length ratio $h/l \approx 1:1.5$.

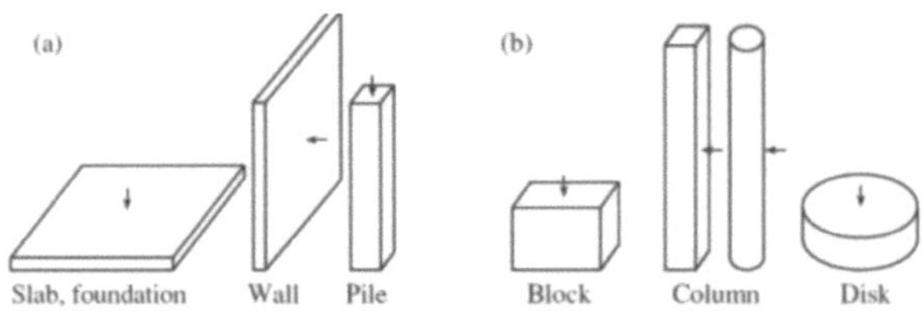

Fig. 5. Examples of (a) extended and (b) compact BCs.

As a result, in addition to a thickness resonance, numerous supplementary resonance peaks caused by geometrical effects appear in the spectral characteristic. Against the background of numerous resonances, it is impossible to unambiguously determine the main thickness-resonance peak at the desired frequency $f \sim 1/h$ (Fig. 6b). In addition, the velocity $C_l = (2hf/k)$ measured by the IE method in a compact article differs from the velocity C_l in an extended article: in a compact article, the velocity correction factor k requires special determination for each article. It is precisely for these reasons (ambiguity of the spectral characteristic and uncertainty of C_l) that, to date, compact concrete BCs are not tested by acoustic methods. Hence, the IE method, which is widespread abroad, is used to test only extended concrete BCs.

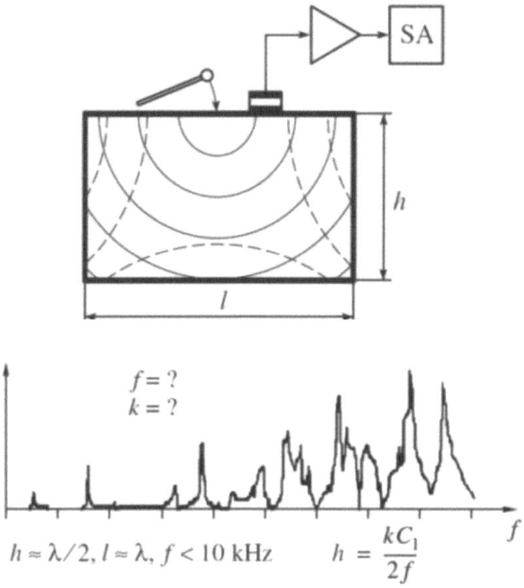

Fig. 6. (a) Schematic diagram of the acoustic IE method for a compact concrete article; (b) AFC of a compact concrete arti-cle. Parameters of the IE method in testing of compact articles.

The IE method is one of the oldest NDT methods. The first paper devoted to the IE method (Mary Sansalone, United States) allowing determination of the thickness and the presence of flaws in extended BCs was published in 1986 (Sansalone, M. & Carino, N.J. 1986). In 1998, the American Society adopted this method for Testing and Materials (ASTM) as a standard (ASTM C1383. Standard Test Method for Measuring the P-Wave Speed and the Thickness of Concrete Plates Using the Impact-Echo Method). The simplicity of the impact method, the high capacity, and the relative cheapness of testing devices has made it quite widespread in industrially advanced countries.

A standard IE complex consists of a computer, an amplifier, a PET, and a set of steel balls of different diameters used to excite free oscillations in BCs.

In some cases, steel balls are replaced by a special device – an impactor, a mechanism for delivering a force- and duration-normalized impact on a surface and situated in one unit with the receiving PET. The latter is built as a transducer with a point contact ensuring a good "dry" acoustic contact with a rough concrete surface. The impactor ensures better stability of the exciting-signal spectrum, thereby contributing to the improvement of the reproducibility of the spectral-characteristic measurement results. In addition, the impactor solves the problem of synchronizing the onset of measurement with the moment of striking.

The resonance method, which involves excitation of forced vibrations in an article using an external generator of a linearly changing voltage, is similar to the IE method based on the analysis of free vibrations in a tested article. In this case, a ball or an impactor is replaced by a broadband emitting PET and signals are received with a broadband receiving probe. The amplitude-frequency characteristic (AFC) of forced vibrations in an article is similar to the spectral characteristic obtained in the IE method. However, the resonance method has a number of advantages over the IE method.

First, a Fourier transform of the received signal is unnecessary. To obtain the resonance characteristic of a test object, it is sufficient to record the received-signal amplitude at each frequency. This allows a detailed study of the AFC in the frequency ranges of interest at as small a step as possible, thus increasing the measurement accuracy.
Second, the resonance method is more sensitive because, in the IE method, excitation is performed by a short impact whose energy is distributed over the entire spectrum (in the resonance method, the signal energy can be concentrated at each individual frequency).
Third, if there is a good acoustic contact between the PET and the article, the reproducibility of measurements is ensured.
Fourth, the AFC of the electroacoustic channel can be corrected by setting the required amplitude of the emitted signal or selecting the gain of the input signal for each frequency.
The resonance method has also not found practical application in the quality control of actual concrete BCs, but the resonance method is indispensable for laboratory studies (when it is necessary to carefully study the AFC of an article in some frequency range, to find the optimal arrangement of probes, to perform identical measurements many times, or to conduct other investigations). For this reason, we actively used the resonance method for detailed studies of the characteristics of concrete articles and the development of new testing techniques.
The IE method is used mainly to determine the thickness of large-size (>1.5 m) concrete BCs. In this case, reliable measurement results are obtained in testing of extended slabs (foundations, walls, floors, etc.) whose lengths and widths exceed the tested thickness by a factor of 5 or more. As a rule, testing of such extended articles allows one to unambiguously determine the BC thickness according to the frequency of the maximum resonance peak in the article's spectral characteristic.
The value of the velocity of acoustic vibrations necessary for the subsequent determination of the BC thickness is determined according to the American standard ASTM C1383 through the surface-wave velocity despite all the aforementioned drawbacks of this method. Note that it is the method of eigenfrequencies that allows determination of the longitudinal-wave velocity in the bulk of an article. However, the field of application of these methods is limited by the shape of tested articles: only extended slabs (with lengths and widths far exceeding the thickness) and long piles (with lengths far exceeding other dimensions) may undergo testing. For compact articles, the value of the coefficient k is unknown; therefore, it is impossible to use the methods of eigenfrequencies for monitoring the wave velocity in such articles.
The analysis of publications on the application of the IE and resonance methods allows us to draw the following conclusions.
(1) To date, the IE method (in some cases, the resonance method) is a quite widespread technique for acoustic testing of concrete BCs (abroad) and is virtually the only acoustic method enabling testing of extended articles thicker than 1.5 m in the case of one-way access.
(2) The IE method has certain limitations: the testing techniques and devices allow inspection by the IE method of virtually any thickness but only for extended articles (foundations, walls, floors, and piles) whose lengths and widths exceed their thicknesses by a factor of >5. The existing testing techniques are unsuitable for testing compact articles (for which at least one dimension differs from the thickness by a factor of <5, a feature typical of supports, columns, blocks, etc.).

2. A correlation method for determining the propagation velocity of an acoustic wave in large-size compact concrete articles

A review of the methods for testing large-size concrete BCs shows that the impact echo method helps to inspect extended concrete BCs (in extended BCs, the tested dimension differs from the other dimensions by a factor of at least 5) (Carino, N.J., 2001). Thus, it becomes possible to measure the thickness h of (at a known velocity C_l) or the velocity C_l in (at a known thickness h) foundations, walls, building floors, bridge supports, piles, etc. However, it is shown in (Carino, N.J., 2001). that one cannot achieve an unambiguous result in compact articles. In fact, in compact articles (in which the tested dimension, e.g., the thickness h, is comparable with at least one of the other dimensions), numerous amplitude peaks (resonances) appear in the spectral characteristics and a shift of the resonance-peak frequency is observed due to the effect of geometrical dispersion of the wave velocity, which leads to ambiguous testing results.

These problems determined the necessity of calculating the acoustic fields in compact articles appearing under the action of a driving force. As is known, either free vibrations (when the action of a driving force has a short-term pulsed character) or forced vibrations (initiated by continuous action of a driving force) arise in an elastic body (Skuchik, E., 1971). Analytical calculation of the vibration spectra for elastic solids of different shapes involves intricate mathematical calculations. It is relatively simple to obtain analytical expressions for the spectral characteristics of natural vibrations only for the simplest geometrical forms—rods and extended slabs (plates) (Ermolov, I.N. & Lange, Yu.V., 2004). For more complex shapes, a solution leads to differential equations of the fifth and higher orders that cannot be solved analytically. Meanwhile, an analytical solution allows one to more deeply understand the essence of the processes proceeding in a tested object affected by an external driving force.

The analytical solution for the simplest shape — a thin rod of finite length l — is known (Fig. 7). In our context, "thin" means that the thickness of the rod is many times smaller than the wavelength and its length l is comparable with the wavelength ($l \approx \lambda$). In the simplest variant, we assume that only longitudinal waves propagate in the rod; flexural and torsional waves are not considered here. In the case of forced vibrations, the behavior of the rod is analyzed under the assumption that a constant harmonic force $F = F_0 e^{i\omega t}$ is applied to one end of the rod ($x = 0$) and its other end ($x = l$) is free. As a result of this analysis, the values of the rod resonance frequencies are determined: $f = nC_l/2l$, where n = 1,2,3,.. .(Korobov, A.I. 2003).

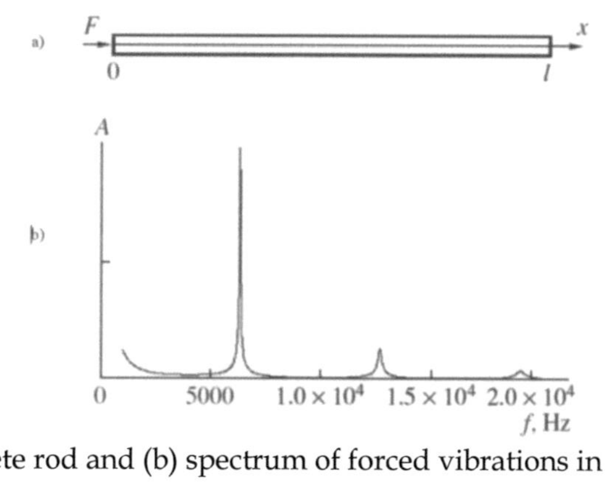

Fig. 7. (a) Thin concrete rod and (b) spectrum of forced vibrations in the rod.

Hence, the rod is a distributed vibratory system with many degrees of freedom (modes), each of which has an eigenfrequency ω. Fig. 7b shows the calculated spectral characteristic of a thin rod of length l = 0.3 m. The following parameters typical of concrete were used in the calculation: the Young modulus E = 3.456x1010 N/m^2, the viscosity E = 5000 $(N_s)/m^2$, and the density ρ = 2400 kg/m^3. The propagation velocity C_r of an acoustic wave in such a rod calculated taking into account the above values of the Young modulus and density is C_r = 3795 m/s. The resonance frequency f_{1a} of the first mode of a longitudinal wave obtained analytically is 6325 Hz.

In practice, we most frequently deal with concrete articles of a more complex shape than rods. Because it is difficult to analytically calculate the vibration spectra of such articles, it becomes necessary to simulate the physical processes occurring in compact BCs by numerical methods. Simulating physical processes is necessary for determining the character of the acoustic-field distribution in BCs, the optimal testing algorithm, and the optimal arrangement of probes on the article surface. In other words, the possibility of constructing a numerical model of a BC with specified dimensions, boundary conditions, and material properties is a necessary condition for studies.

The finite-element method (FEM) is most frequently used for this purpose (Bolotin, V.V. 1999). This method is based on the approximation of a continuous function by a discrete model, which is constructed on a set of piecewise-continuous functions defined on a finite number of subdomains called finite elements. The geometrical domain under study is partitioned into elements so that, on each of them, the unknown function is approximated by a trial function (as a rule, a polynomial). These trial functions must satisfy the continuity boundary conditions coinciding with the boundary conditions imposed by the problem itself. The choice of the approximating function determines the corresponding type of element.

Partition of the geometrical domain into a large number of finite elements and solution of the main equation of motion for each element allow calculation of the spectra of free (transitional analysis) and forced (modal analysis) vibrations. The main equation of motion has the form

$$[M]\{\ddot{u}\}+[C]\{\dot{u}\}+[K]\{U\}=\{F^a\}, \tag{1}$$

where $[M]$ - is the mass matrix, $[C]$ - is the damping matrix, $[K]$ - is the elasticity matrix, $\{\ddot{u}\}$- is the nodal acceleration vector, $\{\dot{u}\}$- is the nodal velocity vector, $\{u\}$- is the nodal displacement vector, and $\{F^a\}$ is the applied-force vector.

In the modal analysis, it is considered that the displacements of all elements obey the harmonic law

$$\{u\}=\{u_{max}e^{i\Phi}\}e^{i\Omega t} \tag{2}$$

where u_{max} is the displacement amplitude, Ω is the circular frequency of forced vibrations, and Φ is the displacement phase shift.

The use of complex form of representation allows reduction of expression (2) to a more compact form:

$$\{u\}=\{u_{max}(\cos\Phi+i\sin\Phi)\}e^{i\Omega t}$$

or

$$\{u\} = (\{u_1\} + i\{u_2\})e^{i\Omega t}, \tag{3}$$

where $\{u_1\} = \{u_{\max} \cos\Phi\}$ and $\{u_2\} = \{u_{\max} \sin\Phi\}$.
Similarly, the force vector can be represented as

$$\{u\} = (\{F_1\} + i\{F_2\})e^{i\Omega t} \tag{4}$$

Substituting (3) and (4) into (1), we obtain

$$([K] - \Omega^2[M] + i\Omega[C])(\{u_1\} + i\{u_2\}) = \{F_1\} + i\{F_2\} \tag{5}$$

Note that, for forced vibrations, this equation does not include the time *t* (only stationary vibrations are considered in the modal analysis).
The solution of Eq. (5) allows calculation of the displacement amplitude of each element of the model at the current frequency of the driving force. Multiple solution of this equation for each frequency in the range of interest makes it possible to construct the spectral characteristic of forced vibrations of the model in this frequency range for the element of interest.
To date, several software packages allowing FEM-based computer calculations have become widespread. In this study, the FEM calculations were performed according to the ANSYS program with the Multiphysics package (Chigarev, A.V.; Kravchuk, A.S., & Smalyuk, A.F., 2004; Kaplun, A.B.; Morozov, E.M., & Olfer'eva, M.A., 2003; Basov, K.A., 2002). The ANSYS graphical environment facilitates constructing a 3D model, specifying the material properties and the boundary conditions, and visualizing the calculation results. In this study, the ANSYS package was used to calculate the spectra of free and forced vibrations in concrete articles with different shapes. The calculation of the spectra of free and forced vibrations for a thin rod similar to that considered above is presented below and is intended for primary verification of the simulation results.
The simulation procedure consists of several stages. A geometrical model of the rod is created — a cylinder with a diameter $d = 0.01$ m and a length $l = 0.3$ m (Fig. 8a). The elastic properties of the material are specified — the density $\rho = 2400$ kg/m^3, the Young modulus $E = 3.456 \times 10^{10}$ N/m^2, and the Poisson ratio $\sigma = 0.2$. A grid is imposed on the geometrical model. In this case, the cylinder is partitioned into 100 elements along the length.

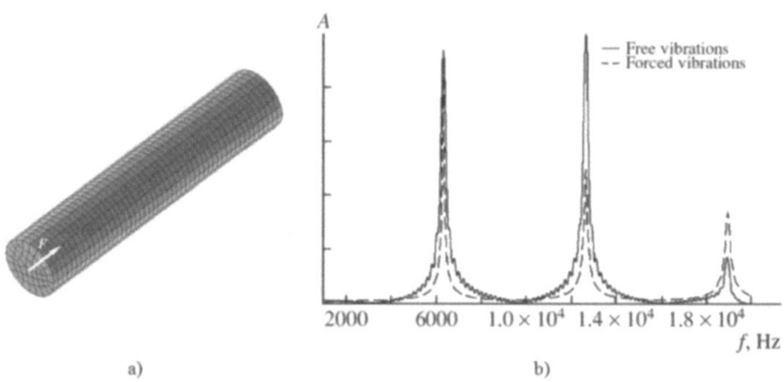

Fig. 8. (a) Model of a thin rod and (b) calculated spectra of free and forced vibrations in the rod.

The boundary conditions are set. An external force is applied to all nodes of the grid positioned at one end of the cylinder. In simulating free vibrations, a short impulse of force in the form of a sine-vibration half-period with a duration of 50 μs is used. A harmonic oscillator is used as an external force for modeling forced vibrations. Subsequently, the type of analysis is chosen — modal and transitional for forced and free vibrations, respectively. In addition, at this stage, the acoustic-wave damping in the rod material is specified.
(In this case, the coefficient of modal damping is 0.005, which corresponds to a damping α = 0.04 dB/m at a frequency of 5 kHz.) The modal analysis also requires that the frequency range and the frequency measurement step of the oscillator used as the source of the driving force be specified. Finally, the analysis is performed and the results are deduced. Because the result of the transitional analysis is the signal shape in the time domain, it is additionally necessary to calculate the spectrum.
The modal analysis results in a spectral characteristic; therefore, no additional calculations are required. Figure 8b shows the spectra of free and forced vibrations obtained as a result of the ANSYS simulation. Comparison of the spectra in Figs. 7b and 8b shows that the results of the numerical simulation for both types of vibration coincide with the analytical solution for the rod. The resonance frequency f_{1m} of the longitudinal wave's first mode resulting from the simulation is f_{1m} = 6250 Hz, $f_{1m} \approx f_{1a}$.
The application of the FEM also allows calculation of the spectra of articles with more complex shapes (Avramenko, S.L. & Kachanov, V.K., 2007). In such articles, first, one has to deal with a large number of degrees of freedom and, second, it is impermissible to disregard the existence of other wave modes, as was done in the analytical calculation for the rod. Below, we present the results of calculating the spectral characteristics of a homogeneous concrete slab with dimensions of 300×300×30 cm (Fig. 9). This slab can be considered an extended article because its length and width far exceed its thickness.

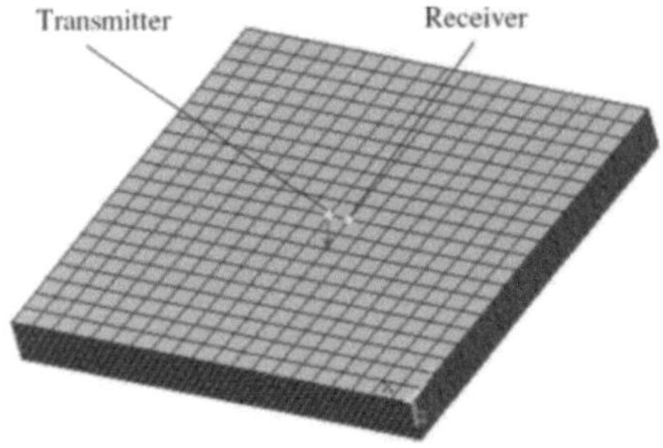

Fig. 9. Model of a slab with dimensions of 300×300×30 cm.

Let us divide the slab's faces into 40×40×10 elements. The elastic properties of concrete in this and all subsequent calculations are the same as those used in the calculation of the rod spectrum. Let us place the source of the external force at the center of the slab and the receiver (the element in which the spectral characteristic is calculated) at a distance of 7.5 cm from it. Figure 10 shows the spectra of free and forced vibrations in the slab obtained with the FEM. A resonance peak at a frequency f = 6400 Hz is clearly pronounced in both spectral characteristics. This resonance peak corresponds to the first mode of a longitudinal wave. The impact echo method for measuring the thickness of concrete slabs implies calculation of the slab thickness h from the formula $h = C_{sl}/2f$, where f is the frequency of the longitudinal wave's first mode and C_{sl} is the velocity of longitudinal waves in the slab. In our case, the calculation of the slab thickness from this formula yields h = 0.3 m, which exactly corresponds to the model thickness.

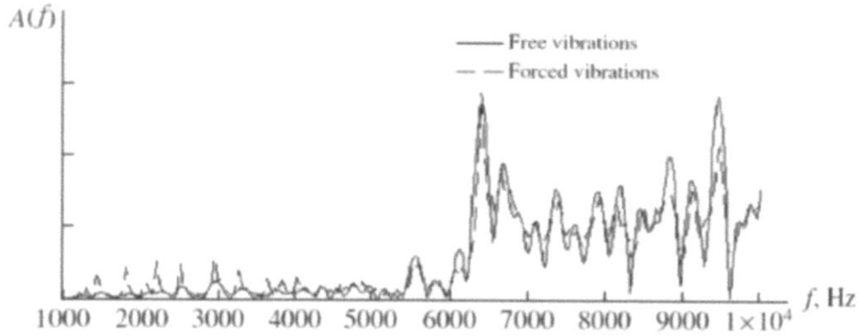

Fig. 10. Spectra of free and forced vibrations for a slab with dimensions of 300×300×30 cm.

Note that the resonance frequency of the longitudinal wave's first mode in the 0.3-m-thick concrete slab (f_{1sl} = 6400 Hz) differs from the resonance frequency of the 0.3-m-long rod of the same material (f_{1r} = 6250 Hz). This is because the velocity of a longitudinal wave C_{sl} in an extended slab differs from the velocity of a longitudinal wave C_r of acoustic vibrations in a rod. The formula $C_{sl} = \sqrt{\frac{E}{\rho}\frac{1}{1-\sigma^2}}$ is valid for a slab. In contrast to the velocity in a rod, the velocity in an extended slab depends on (in addition to E and ρ) the Poisson ratio σ. Calculating the velocity C_{sl} at σ = 0.2 typical of concrete yields C_{sl} = 3872 m/s (compare to C_r = 3795 m/s).

Comparing the two characteristics in Fig. 10 shows that the spectra of free and forced vibrations are identical. Slight differences in the quality factors and the peak amplitudes in the characteristics are related to the error in calculating the spectrum of free vibrations. This indicates that the spectral characteristic of a BC does not depend on the technique of vibration excitation (both a short impulse of force and a harmonic signal with a smoothly increasing frequency can be used as sources of external actions).

The fact that the simulation results obtained with the ANSYS package do not contradict the main formula of the impact echo method confirms the consistency of the chosen simulation technique. This circumstance makes it possible to apply simulation as an efficient tool for searching for optimal techniques for testing various concrete articles (including compact objects). The practical meaning of this conclusion is that spectral characteristics obtained using the impact echo and resonance methods will be identical. At the same time, in contrast to the impact echo method, the resonance method does not require calculation of the signal spectrum but takes a longer time. For this reason, all further calculated spectral characteristics were obtained as a result of modal analysis (forced vibrations).

Two main factors can be distinguished among the reasons for which the thickness of compact BCs cannot be successfully tested by the eigenfrequency methods: the effects of "noise of form" on the amplitude–frequency characteristics (AFCs) of compact articles and the geometrical dispersion of the longitudinal - wave velocity.

Let us consider the influence of the shape of a compact article. Here, it should be noted that a model based on the principles of geometrical acoustics is unacceptable for calculating the resonance characteristic of a compact article. In fact, determination of the signal profile on the surface of a BC from the interference of several echo signals reflected from the BC boundaries can be used only when the wavelength of a probing signal is many times smaller than the BC size. If the wavelength is of the same order of magnitude as the BC size, such an interference model cannot be applied for precise determination of the article's spectral characteristic. Instead of it, we should speak about the resonance properties of the article itself.

The above can be explained by the following example. It is known that, upon an impact on the surface of an infinitely long slab of a certain thickness h, a receiver positioned at a small distance from the point of action registers damped harmonic vibrations at a frequency $f = C_{sl}/2h$. It should be noted that vibrations obey exactly a harmonic law. It follows from this fact that the considered slab is a resonator and the frequency f its eigenfrequency. At the same time, being an elementary signal, a sinusoid (including a damped sinusoid) cannot be divided into simpler components and, consequently, cannot be obtained via any summation (interference) of any other simpler components. Hence, in considering free vibrations of a compact elastic article, the resonance nature of its spectral characteristic should not be explained as resulting from the reflection of wave fronts from boundaries. It is more correct to treat the test object as a body possessing a certain set of eigenfrequencies (Glikman A.G., 2009).

The infinite slab considered above is certainly an idealized example. A real extended slab always has borders and, as a result, can be represented in the form of a thin plate (whose dimensions far exceed its thickness). As shown in (Ermolov, I.N. & Lange, Yu.V., 2004; Bolotin, V.V., 1999), apart from a longitudinal wave, flexural and planar waves propagate in the plate and initiate resonances at frequencies $f_a(m, n)$ and $f_s(m, n)$, respectively. In these expressions, m and n indicate the numbers of wavelengths along the plate length and width, respectively.

Figure 11 shows several modes of flexural and planar vibrations of the plate. Note that the frequencies of the lowest modes of flexural and planar vibrations in extended concrete plates are much lower than the frequency of the first mode of a longitudinal wave because the BC length and width, determining the vibration frequency, are many times larger than the thickness. For the same reason, owing to the high acoustic damping in concrete, the amplitudes of both low and high modes of these vibrations are insignificant. However, in compact articles, whose dimensions are comparable with their thickness, the frequencies of flexural and planar vibrations lie within the same range as the frequency of the longitudinal wave's first mode. In addition, their amplitudes are substantially higher. Thus, a large number of resonance peaks with comparable amplitudes are present in the resonance characteristic.

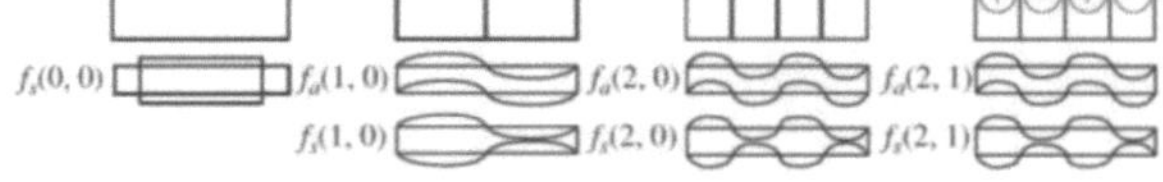

Fig. 11. Some forms of flexural and planar vibrations in a plate.

Let us demonstrate this by an example. To adequately characterize the compactness of a tested article, let us introduce the compactness factor m, equal to the ratio of one of its overall dimensions to its thickness. Figure 12 shows the spectral characteristics of concrete parallelepipeds with dimensions of 150×150×30, 120×120×30, 90×90×30, and 60×60×30 cm resulting from the simulation at an arbitrarily specified velocity of acoustic vibrations in the slab. The compactness factors m of these blocks are equal to 5, 4, 3, and 2, respectively.

Figure 12 shows that, as m decreases, the spectral characteristic becomes more complex. At $m = 4$, it is rather difficult to unambiguously determine the frequency of the first mode of the longitudinal wave, from which the thickness of the compact article should be determined.

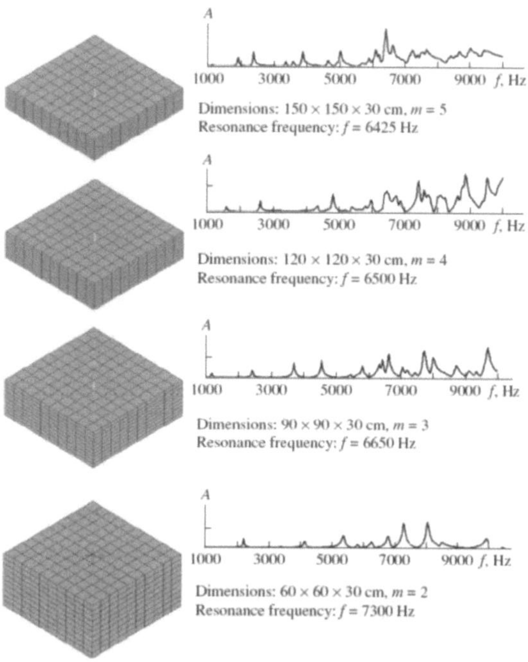

Fig. 12. Spectral characteristics of a compact slab with a thickness of 30 cm at different values of the compactness m.

As a rule, an unambiguous interpretation of the spectrum becomes absolutely impossible at $m < 3$. Thus, as m decreases, interpreting the spectrum on the basis of the feature that the "amplitude of the longitudinal wave's first mode is maximal", becomes problematic. The error in finding the frequency of the longitudinal wave's first mode increases, thereby reducing the reliability of the measurement results because the desired peak cannot be distinguished against a background of numerous other resonances forming a sort of noise disguising the useful signal. This noise can be called "noise of form" (in terms of noiseimmune ultrasonic testing of articles) (Kachanov, V.K. and Sokolov, I.V., 2007) because, on the one hand, it is the shape of a compact article that determines its AFC and, on the other hand, such an AFC with many peaks hinders determination of the sought resonance, i.e., is a sort of interference (noise) disguising the required article thickness. As follows from (Carino, N.J., 2001), the geometrical effects leading to the appearance of noise of form of a compact article do not allow reliable thickness measurements of compact articles with $m < 5$, as was confirmed by the simulation results

Let us now consider the effect of geometrical dispersion of the longitudinal-wave velocity during testing of compact BCs. As was mentioned above, the propagation velocity C_l of a longitudinal wave in an elastic body depends on the geometrical shape and dimensions of this body with respect to the wavelength. The table 1 presents the known analytical formulas for calculating the longitudinal-wave velocity for some very simple geometrical shapes.

The velocity C_l in compact articles can also be determined by the shadow or echo-pulse method, but this is possible only under the obligatory condition that the wavelength λ of the probing signal is much smaller than the size of the article. However, in controlling large-size concrete articles, these methods cannot always be used because of relatively high damping of ultrasonic signals at the testing frequency.

Geometric form	Condition	Formula for velocity calculation
Infinite space	$\lambda <<$ medium dimensions	$C_l = \sqrt{\frac{E(1-\sigma)}{\rho(1+\sigma)(1-2\sigma)}}$
Thin rod of length *l* and diameter *d*	$\lambda \approx l, \lambda >> d$	$C_r = \sqrt{\frac{E}{\rho}}$
Extended slab of length *a*, width *b* and thickness *h*	$\lambda \approx h, \lambda << a, \lambda << b$	$C_r = \sqrt{\frac{E}{\rho}\frac{1}{1-\sigma^2}}$

Table 1. Propagation velocity of acoustic waves in bodies of different geometric forms

The testing methods based on the use of the eigenfrequencies of inspected articles do not allow direct determination of C_l. The value of C_l is calculated using the velocity correction coefficient $k = C/C_l$, where C is the velocity of a longitudinal wave in an article of a certain shape determined by one of the eigenfrequency methods. For example, if the tested article is an extended slab, $C = C_{sl}$. In this case,

$$k = C_{sl}/C_l = \sqrt{\frac{1-2\sigma}{(1-\sigma)^2}}\ .$$

Hence, the coefficient k for an extended slab depends only on the Poisson ratio; i.e., $k = f(\sigma)$. At $\sigma = 0.2$, we obtain the known velocity correction coefficient $k = 0.96$. This value of the velocity correction coefficient is used in the calculation of C_l in an extended slab at a known thickness h: $C_l = 2fh/k$, or in the calculation of h at a known velocity C_l: $h = kC_l/2f$.
If the tested article has the shape of a thin rod, $C = C_r$. In this case, the velocity correction coefficient is determined by the formula:

$$k = C_r/C_l = \sqrt{\frac{(1+\sigma)(1-2\sigma)}{1-\sigma}}\ .$$

At $\sigma = 0.2$ this coefficient in the rod is $k = 0.95$.
A general approach to the problem of determining the velocity of a longitudinal wave in compact articles can be found in studies by I.N.Ermolov (Ermolov, I.N. & Lange, Yu.V., 2004). In addition, according to (Bolotin, V.V., 1999), the propagation velocity of longitudinal waves in a rod generally (when the condition $\lambda >> d$ is not satisfied) depends on the ratio d/λ; consequently, $k = C/C_l = f(\sigma, d/\lambda)$. This phenomenon is called the geometrical dispersion of the velocity. For thin rods with $d/\lambda << 1$ the dispersion is insignificant and $k = C_r/C_l = f(\sigma)$. Similarly, the velocity of a longitudinal wave in a compact slab for which the conditions $\lambda << a$ and $\lambda << b$ are not satisfied, depends on the ratios a/λ and

b/λ. In this case $k = C/C_l = f(\sigma, a/\lambda)$. Attempts to obtain an analytical expression for the function $f(\sigma, a/\lambda)$ at least for some ranges of the ratios d/λ and a/λ yield a result with an error that may exceed 15% relative to experimental results.

Hence, in a general case, the form of the function $f(\sigma, a/\lambda)$ for a compact article of an arbitrary shape is completely unknown. This impedes determination of the velocity C_l by the eigenfrequency methods. The above is confirmed by the simulation results (Fig. 12) showing that, in some cases, it is impossible to unambiguously determine the thickness of a compact article by the impact echo method from the spectral characteristic because of its ambiguous character even if the velocity of acoustic vibrations in the compact article is known. It should be noted that, despite a constant article thickness, the frequency of the resonance peak corresponding to the longitudinal wave's first mode (diagrams in Fig. 12) increases with a decrease in m. Note that, for $m < 4$, the frequency increases so significantly that the geometrical-dispersion effect cannot be disregarded.

So, the problem of determining the wave velocity in compact concrete BCs is very important. It is obvious that the velocity of acoustic vibrations in a compact article determined by the shadow and echo-pulse ultrasonic methods or by measurement of the surface-wave velocity (in view of all limitations of such velocity measurements) is not equal to the true value of the velocity C_l observed in an actual compact concrete article.

Hence, one of the tasks of testing large-size compact concrete BCs by the impact echo method (thickness-measurement problem) remains unrealized because of the unknown velocity C_l in a particular compact article.

In addition, there exists another problem in testing large-size compact concrete BCs—determining the velocity C_l in a large-size compact article all of whose dimensions are known. This problem is aimed at the subsequent determination of the concrete strength and the strength of a concrete BC and the prediction of the failure-free service life of BCs. A similar problem of measuring the propagation velocity of acoustic vibrations in BCs with known dimensions is also necessary for determining the quality of concrete during construction of calibration characteristics for concrete specimens, in which the time-dependent velocity in a particular solidifying concrete block is measured. Thus, such a problem of determining the wave velocity in compact concrete BCs with known dimensions is independent and very important.

To test the propagation velocity of an acoustic wave in arbitrarily shaped compact BCs whose dimensions are known, a new correlation method based on the use of the spectral characteristics of BCs is proposed.

This method consists of the following stages.

I. The experimental spectral characteristic of an arbitrarily shaped compact article is measured, but it does not allow unambiguous determination of the desired resonance frequency.
II. The spectral characteristic of an article similar to a real tested object is calculated by simulation; the velocity C_l is selected arbitrarily.
III. Then, the value of the velocity C_l at which the calculated and experimental characteristics are maximally alike is selected.
IV. The cross-correlation function (CCF) of both characteristics is calculated; the degree of similarity of these characteristics is determined from the CCF maximum.
V. The desired value of C_l is determined from the characteristic at which the CCF maximum is observed.

Note that, when the velocity is selected, it is unnecessary to perform simulation for each new velocity value – as C_l increases, the characteristic proportionally and linearly stretches along the frequency axis, i.e., the frequencies of all resonance peaks proportionally increase. It follows from the above that it is sufficient to perform only one simulation at the minimum selected velocity of a longitudinal wave.

An example of determining C_l according to the proposed method is considered below for a compact concrete block with dimensions of 80×50×30 cm.

First, we preliminarily calculate (simulate) the spectral characteristic of the model of this compact block in the frequency range 1–10 kHz. The initial "base" value of the longitudinal-wave velocity was selected equal to the minimum possible value of C_l for concrete articles: $C_l = 3000$ m/s.

Figure 13 shows the spectral characteristic of the block resulting from simulation and experiments on an actual block. In calculating the spectral characteristic, we used a velocity different from the velocity in the actual block (which is unknown); therefore, the spectral characteristics do not coincide. The problem is how to select a velocity at which the spectral characteristics coincide, thereby maximizing the value of their CCF. For this purpose, it is necessary to calculate the CCF of the experimental spectral characteristic with a set of calculated characteristics, each of which must correspond to a certain value of C_l.

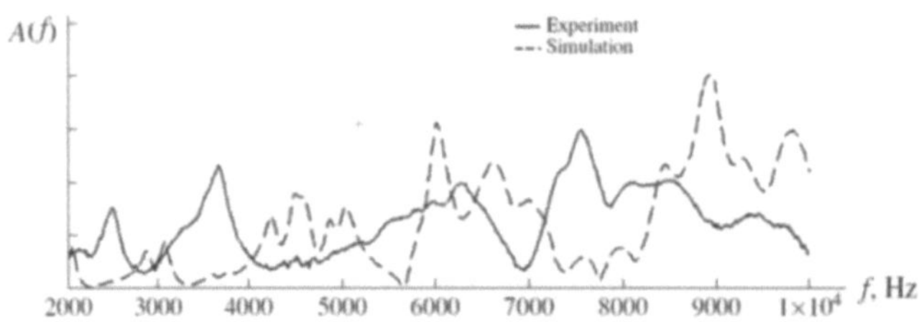

Fig. 13. Experimental and calculated spectral characteristics for a block with dimensions of 80 × 50 × 30 cm (simulation at $C_l = 3000$ m/s).

The calculated spectral characteristics are obtained by stretching the base characteristic along the frequency axis, i.e., via multiplication of the frequency axis by a coefficient equal to the ratio of the desired velocity to the base velocity (3000 m/s). For example, to obtain the characteristics corresponding to velocities of 3000, 3010, 3020, 3030, m/s, etc., the frequency axis of the initial characteristic should be multiplied by factors of 1, 1.0033, 1.0066, 1.0100, etc. As a result of calculating the CCF of the experimental characteristic with 150 calculated characteristics corresponding to the velocity range 3000–4500 m/s with a step of 10 m/s, we obtain the CCF depicted in Fig. 14a. Figure 14b shows the experimental characteristic and the calculated characteristic corresponding to a velocity of 3765 m/s, which coincide quite well. In this case, the obvious CCF maximum determines the velocity of acoustic vibrations in this compact object: $C_l = 3765$ m/s.

Hence, the proposed method allows measurements of the longitudinal-wave velocity in compact large-size arbitrarily shaped concrete articles all dimensions of which are known. The velocity is measured in the entire BC volume and not in some region or, especially, on the surface. In this case, the velocity is measured only using a preliminarily measured spectral characteristic of the compact article, which, owing to its compactness, has no clearly pronounced resonance peak.

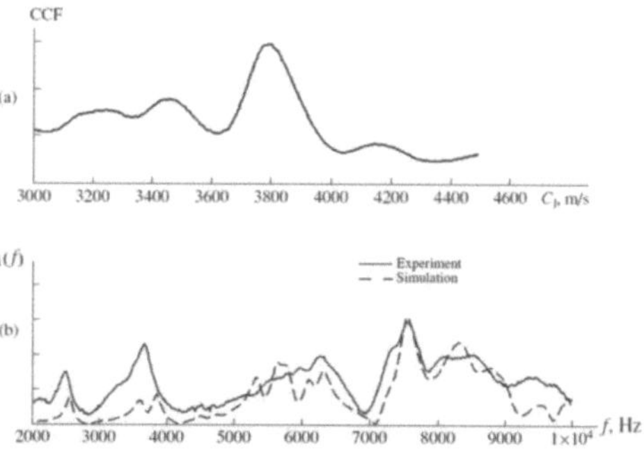

Fig. 14. (a) Dependence of the CCF on the velocity and (b) coincidence of the experimental and calculated spectral characteristics.

Another advantage of the proposed method is that it is less sensitive to the reinforcing structure and large-grained filler than ultrasonic methods because the wavelength of elastic vibrations is of the same order of magnitude as the BC dimensions. In addition, this method has no fundamental limitations on the maximum testing depth – as the BC dimensions increase, the frequency band of the spectral characteristic proportionally shifts toward lower frequencies and the vibration amplitude decreases, but the shape of the characteristic remains constant. Thus, this method is much more sensitive than the methods based on the use of the shadow or echo-pulse methods, in which the dimensions of an article limit the sensitivity. For this reason, the proposed correlation method allows measurements of the velocity of acoustic vibrations in large-size concrete BCs paneled with boards, tiles, etc., that cannot be accessed for measurements and selection of an optimal contact point. An analogous situation arises when the velocity is measured in such concrete BCs with known dimensions as supporting elements of bridges, foundations, piles driven into the ground, etc. Regular measurements of the velocity of acoustic vibrations aimed at the monitoring of strength characteristics of concrete constructions are of fundamental importance for such BCs, parts of which may be submerged, buried in the ground, etc.

One of the limitations of the proposed method for determining the velocity of acoustic vibrations is that the velocity measurement requires knowledge of all dimensions of a tested article and preliminary simulation of the spectral characteristic for each particular article.

3. A multichannel multiplicative method for acoustic testing of large-size compact concrete building constructions

In previous parts it was shown that the IE method for acoustic testing of large-size concrete BCs allows thickness measurements of only extended articles. It is impossible to measure the thicknesses of large-size compact concrete BCs by the IE or resonance methods because of the influence of geometrical effects (the noise of the article shape does not allow unambiguous determination of the desired maximum in the article's spectral characteristic) and the effect of geometrical dispersion of the sound velocity in a compact article (it is impossible to determine the acoustic-vibration velocity C_l).

The correlation method for determining the longitudinal-wave velocity C_l in compact articles with known dimensions requires preliminary simulation of the spectral characteristic of a BC, thereby limiting to a certain degree the acoustic velocity measurements in compact concrete BCs by this method. These problems can be solved by the multichannel multiplicative method for testing compact articles (Kachanov, V.K., Sokolov, I.V., & Avramenko, S.L., 2008; Avramenko, S.L., 2008), which made it possible for the first time to use both the resonance and IE methods for testing the thicknesses of compact BCs.

3.1 Taking into Account the Influence of the Article Shape on the Longitudinal-Wave Velocity

In testing of compact structures, the resonance frequency of the first mode of a longitudinal wave depends not only on the article thickness but also on its compactness *m*. To study this dependence, we calculated the spectral characteristics of blocks with the same thickness h = 30 cm and different values of *m* between 1 and 5. Then, the resonance frequencies *f* of the first mode of a longitudinal wave were determined for each of these characteristics and the correction coefficient of the longitudinal-wave velocity was calculated from the formula $k=2f\,h/C_l$. The dispersion characteristic for parallelepipeds corresponding to the dependence of *k* on *m* was obtained via calculation (Fig. 15). The result of a similar calculation performed for a set of articles in the form of 30-cm-thick disks with diameters in the range 30–150 cm is also shown in Fig. 15.

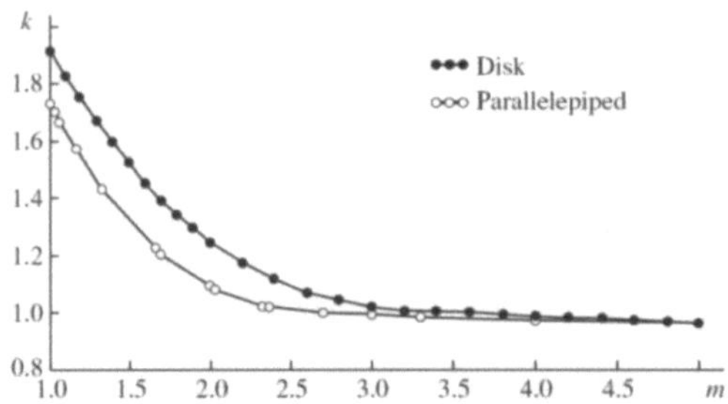

Fig. 15. Dispersion characteristics for a disk and a parallelepiped.

Note that both characteristics for parallelepipeds and disks coincide if the *m* axis of the characteristic is multiplied by a factor of 1.22. The analysis of the obtained dispersion characteristics allows us to draw a number of important conclusions that must be used in calculating compact concrete BCs.

1. The dispersion characteristics for articles of the most widespread standard shapes (parallelepipeds and disks) show that, regardless of the shape of an article at $m > 5$, the velocity-correction coefficient *k* tends to 0.96. This is a quite expected result in view of the fact that, as *m* increases, the shape of the tested article becomes more similar to an extended slab, for which $k = 0.96$.
2. The obtained dispersion characteristics show that, in articles in the form of a disk and a parallelepiped at $m < 3$, the velocity-correction coefficient *k* changes (increases to an appreciable degree).
3. It is of importance that the obtained dispersion characteristics are independent of the scale of BCs; i.e., a proportional change in all dimensions of a BC does not affect the dependence of *k* on *m*. In fact, when the scale increases by a factor *a*, the compactness *m* remains constant, the resonance frequency *f* of a longitudinal wave decreases by a factor of *a*, and the thickness *h* increases by the same factor. Substituting these results into formula $k=2f\,h/C_l$, we obtain that the value of the velocity-correction coefficient in the article remains:

$$k' = \frac{2(f/a)(ha)}{C_1} = \frac{2fh}{C_1}k$$

(k' is the velocity-correction coefficient for an article that is larger than the article with the coefficient k by a factor of a). Owing to this property, the obtained dispersion characteristics can be used for articles of a given shape and an arbitrary thickness.

4. Similarly, the dispersion characteristic is also independent of the longitudinal-wave velocity in a BC. For example, when the velocity C_l increases by a times, the resonance frequency f also increases by the the same factor, while the velocity-correction coefficient remains unchanged:

$$k' = \frac{2(af)h}{(aC_1)} = \frac{2fh}{C_1}k.$$

5. In addition, k is independent of the attenuation of an acoustic wave in concrete. Figure 16 shows the calculated spectral characteristics of a concrete block with dimensions of 80×50×30 cm at values of attenuation α_1 = 0.05 dB/m, α_2 = 0.1 dB/m, and α_3 = 0.2 dB/m. As is seen, a change in α results in a change in the Q factors of resonance peaks, but not in the resonance frequencies.

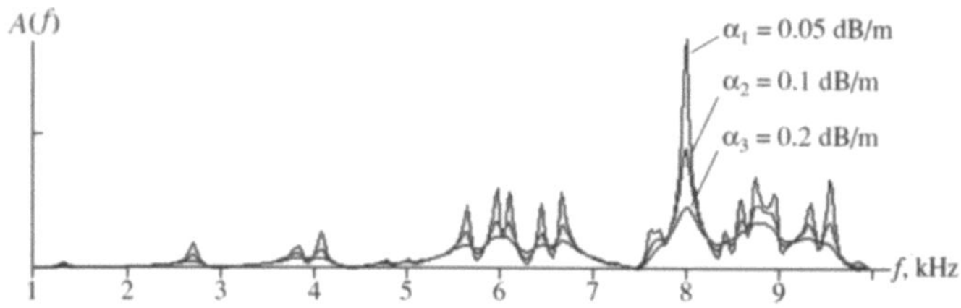

Fig. 16. Dependence of the spectral characteristics for a block with dimensions of 80×50×30 cm on acoustic attenuation.

The practical value of these characteristics is that they make it possible to recalculate the longitudinal-wave velocity C_l in articles of these shapes (in compact articles) into a longitudinal-wave velocity in an infinite space and vice versa. The use of these characteristics makes it possible to calculate the thickness of a compact article if C_l is known or, vice versa, to calculate C_l if the thickness of a compact article is known. In other words, the dispersion characteristics obtained in this study allow testing of large-size compact concrete BCs by eigenfrequency methods.

3.2 Selecting the optimal positions of transducers

It should be noted that, in existing foreign techniques of applying the IE method, the question of selecting the positions of the transmitter and receiver on the BC surface is not considered. This is probably associated with the fact that, for extended articles, for whose testing the IE method is intended, the positions of the transmitter and receiver are not so important as for compact articles. As was shown above, the spectral characteristic of a compact article contains many resonance peaks, which impede the determination of the resonance frequency of the first mode of a longitudinal wave. However, the results of the performed calculations and experiments on actual concrete structures have shown that, when a source–receiver pair moves over the surface of a BC in the form of a parallelepiped, all resonance peaks remain at constant frequencies, but their amplitudes undergo substantial changes. It has been established that, at the surfaces of compact articles, there are areas subjected to ultrasound where the first-mode peak amplitude of a longitudinal wave reaches a maximum and, in most cases, predominates over the amplitudes of other peaks. Thus, such areas should be preferred for placing transducers on the surface of a compact article.

On the basis of the analysis of a large number of calculated and experimental characteristics, it was concluded that the resonance amplitude of a longitudinal wave in compact articles is maximized when the transmitter and receiver lie on one of the axes of symmetry parallel to the BC sides. To confirm this inference, let us consider the calculated spectral characteristics of a symmetric compact block (60×60×30 cm) obtained for different positions of the transmitter and receiver. Special attention will be given to the amplitude of a peak at a frequency of 730 Hz (this is the resonance frequency of the first mode of a longitudinal wave).

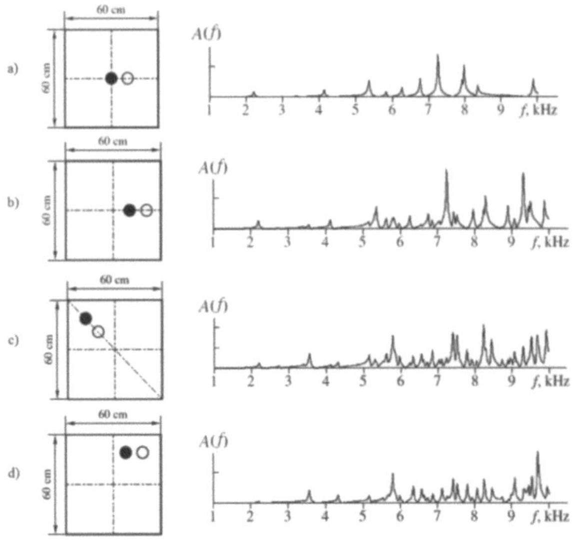

Fig. 17. Dependence of the spectral characteristics for a block with dimensions of 60×60×30 cm on the position of the transmitter–receiver pair.

Figures 17a and 17b show the spectral characteristics corresponding to the positions of the transmitter and receiver on an axis of symmetry parallel to two sides of the block. The characteristic in Fig. 17a was obtained for the transmitter positioned at the point of intersection of both axes of symmetry. As is seen, in this case, the amplitude of the resonance peak for the first mode of a longitudinal wave has a maximum. In the characteristic in Fig. 17b, this peak is still distinguished unambiguously, but the amplitudes of other peaks are higher than those in Fig. 17a. The characteristics in Figs. 17c and 17d were obtained at arbitrary positions of the transmitter and receiver that were not related to the axes of symmetry parallel to the block faces. When the characteristics in Figs. 17c and 17d were calculated, the transducers were positioned along the diagonal axis of symmetry and beyond any axes of symmetry, respectively. In both cases, a useful resonance peak is present in the characteristic but its amplitude is hardly discernable against the background of other

peaks. Hence, as a result of the performed studies, it has been established that a necessary condition for obtaining a good spectral characteristic of a compact article is a correct position of the transmitter-receiver pair on the surface of the tested object: the transmitter-receiver pair must be placed on one of the axes of symmetry.
The dispersion characteristic for parallelepipeds shown in Fig. 15 was obtained as a result of calculating the spectral characteristics of symmetric blocks with identical lengths and widths. Let us determine the possibility of using this characteristic for testing asymmetric parallelepipeds with a length unequal to their width. To do this, let us consider the spectral characteristics of an asymmetric compact block with dimensions of 80×50×30 cm (Fig. 18).

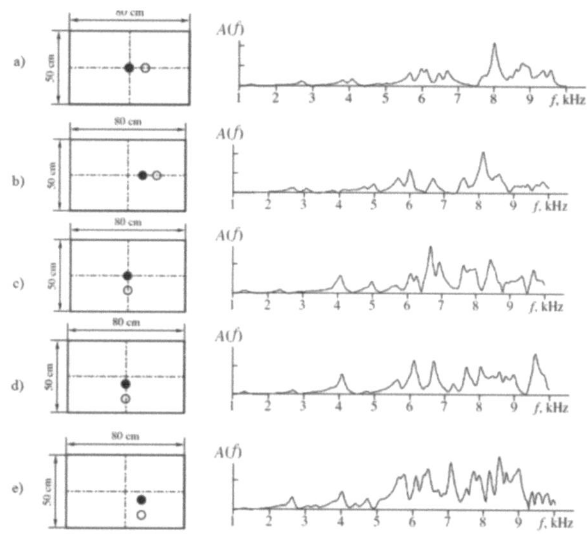

Fig. 18. Spectral characteristics for a block with dimensions of 80×50×30 cm as functions of the position of the transmitter- receiver pair.

In this case, let us place the transmitter and receiver on both axes of symmetry parallel to the sides of the block. In positions (a) and (b), the transmitter and receiver are on the longitudinal axis of symmetry of the compact article perpendicular to the block width. The peak at a frequency f_w = 8025 Hz predominates in the spectral characteristics corresponding to these positions. In positions (c) and (d), the transmitter and receiver are on the transverse axis of symmetry of the compact asymmetric block perpendicular to the block length and the peak at a frequency f_l= 6730 Hz predominates in the spectral characteristics.
The peak at a frequency of 8025 Hz is also observed, but its amplitude is lower. In position (e), the transmitter-receiver pair lies not on the axes of symmetry. The complex form of the spectral characteristic, containing a large number of resonance peaks, once more suggests the problem of correct arrangement of transducers on the surface of a compact article. As a result, it can be concluded that the resonance peak at a frequency f_w = 8025 Hz (the subscript indicates the side that is perpendicular to the axis of symmetry) predominates on the longitudinal axis, perpendicular to the block width, whereas the resonance peak at a frequency f_l = 6730 Hz predominates on the transverse axis, perpendicular to the block length.
Note that two compactness factors can be calculated at once for an asymmetric parallelepiped: the width m_w = 50/30 = 1.67 and length m_l = 80/30 = 2.67 factors. Two

velocity-correction coefficients, k_w = 1.22 and k_l = 1.01, can correspondingly be determined from the dispersion characteristic. In order to find out whether it is possible to use the dispersion characteristic of a symmetric parallelepiped, obtained earlier, for an asymmetric block, let us calculate twice the thickness *h* according to the sets of values *f*, *m*, and *k* for each side (as in all previous cases, the elastic properties of the material during simulation are set such that C_l = 4000 m/s).
As is seen, the calculated thickness exactly coincided with the model thickness:

$$h_w = \frac{k_w C_1}{2f_w} = 0{,}30m\,,\ h_l = \frac{k_l C_l}{2f_l} = 0{,}30m.$$

First, this means that the dispersion characteristic calculated for a symmetric object is also suitable for asymmetric compact parallelepipeds. Second, this means that the frequency of the resonance peak that predominates in the article's spectral characteristics depends on the axis on which the transmitter and receiver are positioned. In our case, when the transducers are positioned along the longitudinal axis, we determined a resonance corresponding to the width of the studied block. For the transducers lying on the transverse axis of symmetry, we determined a resonance corresponding to the length of the studied block. In other words, the frequency of each of these peaks allows us to determine either the thickness (if the velocity is known) or the velocity (if the thickness is known). The possibility of calculating these values by two methods in asymmetric compact blocks allows one to additionally verify the results of thickness measurements at a known velocity C_1.
Let us consider the following example of calculating the thickness of a compact article under the assumption that the block thickness is unknown but the velocity is known, C_l = 4000 m/s. For definitiveness, the spectral characteristics are taken from the previous example. The frequency of the first mode of a longitudinal wave f_w = 8025 Hz is determined from the spectral characteristic (Fig.19a). Because the thickness is not known, the value of the coefficient *m* cannot be found; therefore, the thickness can be calculated only via numerical solution of the equation $h = \frac{C_l k(h)}{2f}$.

The bisection (half-division) method (Demidovich, B.P. & Maron, I.A., 1970), which involves the comparison of the expression

$$h - \frac{C_1 k(h)}{2f}$$

to zero at some values of *h* lying in the range h_{min}–h_{max}, can be used as one such method of solution. If the expression resulting from the substitution of *h* equal to the middle of the segment (h_{min}, h_{max}) (i.e., $h = (h_{max} - h_{min})/2 + h_{min}$) is positive (negative), the interval from h_{min} to *h* (from *h* to h_{max}) is chosen at the next iteration. Hence, the tested-object thickness can be calculated quite accurately after 10 – 15 iterations. The solution of the numerical problem by the bisection method yielded *h* = 0.3 m.

3.3 Multiplicative processing of spectral characteristics

The examples considered above show that the use of dispersion characteristics fundamentally allows determination of both the thickness and the velocity of acoustic vibrations in compact concrete BCs. However, these examples prove that this method is quite laborious and requires unwieldy preliminary calculations. In addition, multiple experiments intended for searching for optimal methods for testing concrete BCs have shown that the techniques proposed above also have other limitations.

First, it is not always possible to mount transducers in an optimal zone on the surface of an actual BC. This can be due to several reasons: a BC is not always a symmetric article of the parallelepiped type; it is not always possible to find the lines (point) of symmetry, or to set a transducer on the lines (point) of symmetry because of a rough, porous, or damp surface or the presence of structural reinforcing elements on the surface, etc. As a result, one has to mount transducers not at prescribed places but where it is merely possible. The negative consequences of this effect were shown above.

Second, an actual concrete BC is not so ideal in its internal structure as a model used in calculation of spectral characteristics. Apart from a coarse-grained filler and reinforcing bars, actual BCs often contain larger inhomogeneities—air cavities and flaws. The presence of the latter in BCs has a negative effect on the spectral characteristic. Hence, during testing of actual compact concrete articles, it is not always possible to unambiguously determine the value of a resonance peak, as was successfully done for a block of 80×50×30 cm in Figs. 18a and 18b (when the frequency $f_w = 8025$ Hz was determined) and in Figs. 18c and 18d ($f_l = 6730$ Hz).

We succeeded in solving these problems using the developed multichannel multiplicative testing method, which is a modification of eigenfrequency methods (the IE and resonance methods). This resonance–multiplicative method (RMM) implies a concept of multichannel testing of BCs. This means that testing must be performed either simultaneously or sequentially (with storing of the results in the instrument's memory) at several points of the BC surface. Then, the spectral characteristics obtained as a result of multichannel testing were subjected to multiplicative processing. As an example of using the RMM, let us consider testing of a compact reinforced-concrete block with dimensions of 130×60×45 cm.

The concrete structure and the dimensions of the reinforced block are such that the maximum value of the longitudinal-wave mode cannot be determined from the block's spectral characteristics that correspond to the position of the receiver on the longitudinal axis of symmetry (Figs. 19a, 19b). This maximum can be determined only after the multiplicative processing of the characteristics shown in Figs. 19a and 19b; for the characteristic shown in Fig. 6c, the frequency $f_w = 6323$ Hz corresponding to the maximum is determined unambiguously. Similarly, the positions of resonances in the spectral characteristics shown in Figs. 6d and 6e, which correspond to positions of transducers on the transverse axis of symmetry, can be determined only after multiplicative processing of these characteristics (the frequency $f_l = 4462$ Hz corresponding to the maximum in Fig. 19d is determined unambiguously). Table 2 presents the experimental data for a concrete block with dimensions of 130×60×45 cm.

	h, m	*m*	*k*	*F*, Hz	C_l. m/s
Along length	0.45	2.89	1.003	4462	4003
Along width	0.45	1.33	1.432	6323	3973

Table 2. Calculated and experimental data for a block of 130 x 60 x 45 cm

Calculations yield the following results for velocities: C_{lw} = 3973 m/s and C_{ll} = 4003 m/s. The spread of the results and calculations is <1%. Calculations yield the following results for velocities: C_{lw} = 3973 m/s and C_{ll} = 4003 m/s. The spread of the results and calculations is <1%.

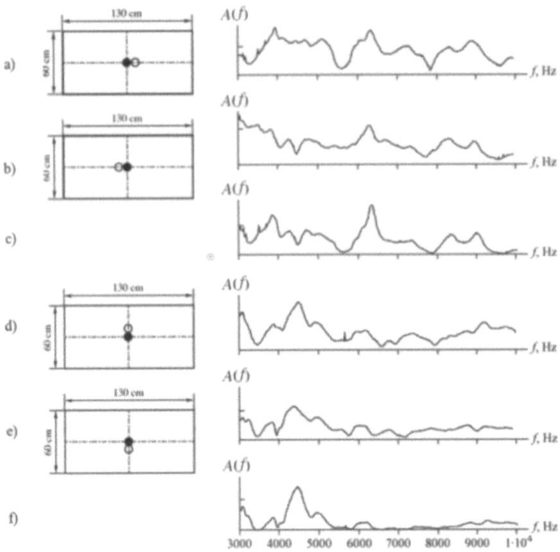

Fig. 19. Experimental spectral characteristics for a compact block with dimensions of 130×60×45 cm.

For comparison, Fig. 20 shows two spectral characteristics of the column obtained at a single point with the IE and resonance methods. In the first case, free vibrations were excited by an impact of a steel ball 10 mm in diameter against the longitudinal axis of symmetry of the column; in the second case, forced vibrations were excited by an emitting piezoelectric transducer (PET) at the same surface point. In both cases, a receiving PET was placed on the same axis at a point 15 cm lower than the point at which vibrations were excited. Measurements were per-formed at a height of 0.5 m from the floor level.

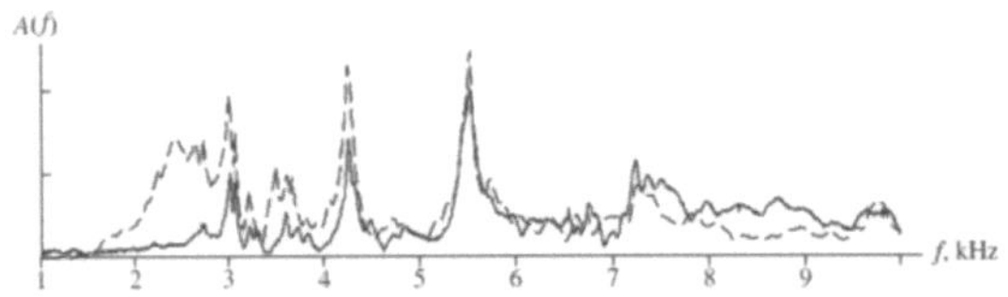

Fig. 20. Spectra of free and forced vibrations in a column.

As is seen from the plots, both spectral characteristics coincide quite well. On this basis, it can be concluded that any of the proposed methods can be used to obtain a spectral characteristic. When the column surface is dry and hard, the impact excitation is more expedient because it ensures an appreciably higher measurement rate.

Therefore, all the subsequent experimental data for this column were obtained using the IE method. Using the spectral characteristics shown in Fig. 20, it is difficult to draw an unambiguous conclusion about the position of the main resonance peak. To obtain more precise information, we obtained a series of spectral characteristics corresponding to different positions of the impactor–receiver pair on the longitudinal axis of symmetry. The distance between the impactor and receiver was always 15 cm.

Figures 21a and 21b show two spectral characteristics corresponding to positions of the impactor at heights of 0.5 and 0.8 m from the floor level (the corresponding positions of the PETs on the column are shown in Figs. 22a and 22b).

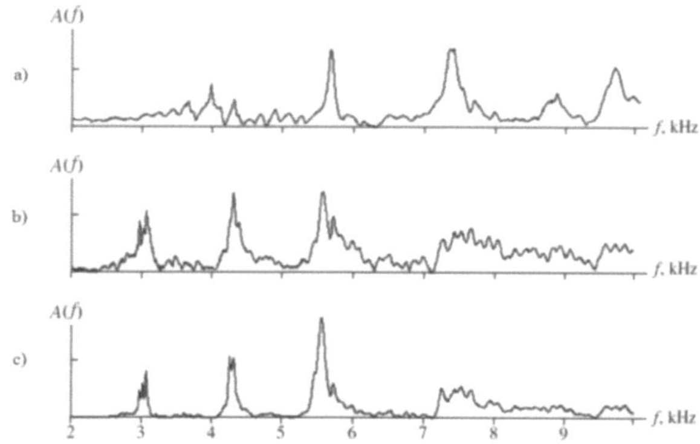

Fig. 21. Experimental spectral characteristics of a column with a square cross section of 60x60 cm measured with transducers positioned on the longitudinal axis of symmetry.

The fact that these characteristics, as well as all others, are identical indicates that the internal structure of the column is homogeneous. It is still difficult to unambiguously determine the position of the maximum at the resonance frequency even from these characteristics. However, after the resonance- multiplicative processing of the characteristics (a) and (b), the value of the maximum is determined unambiguously: Fig. 21c shows the result of the multiplication of characteristics, from which we obtain the resonance frequency f = 5524 Hz, the compactness m = 1, k = 1.733, and the calculated velocity C_l = 3825 m/s.

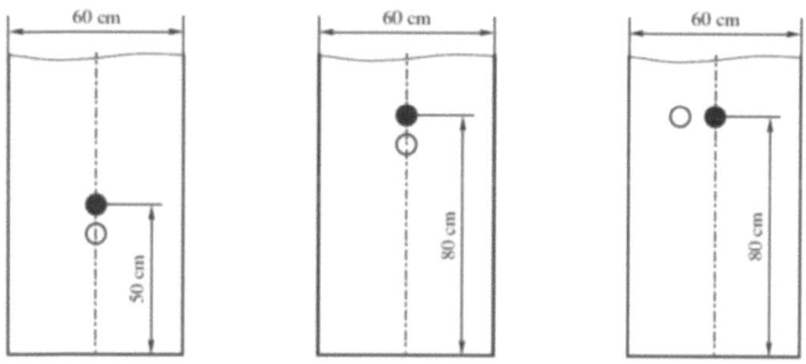

(a) (b) (c)

Fig. 22. Positions of the impactor and receiver during measurements on the column.

Figure 22c shows a situation where the impactor lies on the axis of the column, while the receiver does not. The corresponding spectral characteristic (Fig. 23) shows that the main resonance peak in this characteristic corresponding to the column thickness is not distinguished.

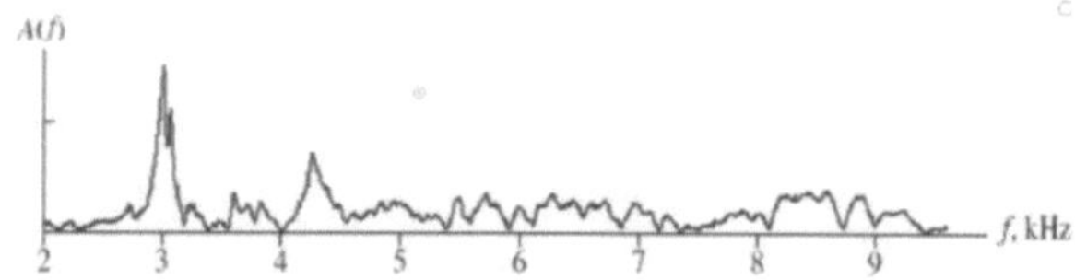

Fig. 23. Experimental spectral characteristic of a column with a square cross section of 60×60 cm.

This confirms the earlier conclusion that the axis of symmetry is the optimal position of the transmitter and receiver for effectively exciting the resonance of the first mode of a longitudinal wave. Thus, it can be concluded that the application of multiplicative processing to two spectral characteristics allows unambiguous determination of characteristics if the PETs are placed on axes of symmetry.

An increase in the number of spectral characteristics always improves the testing results. This is observed in the following example of measurement of the acoustic-wave propagation velocity in a reinforced-concrete block with dimensions of 50x50x25 cm with a computerized multifunctional flaw detector.

These measurements were performed by the resonance method. The spectral characteristics corresponding to four different positions of the transmitter-receiver pair on the block surface are shown in Figs. 24a–24d.

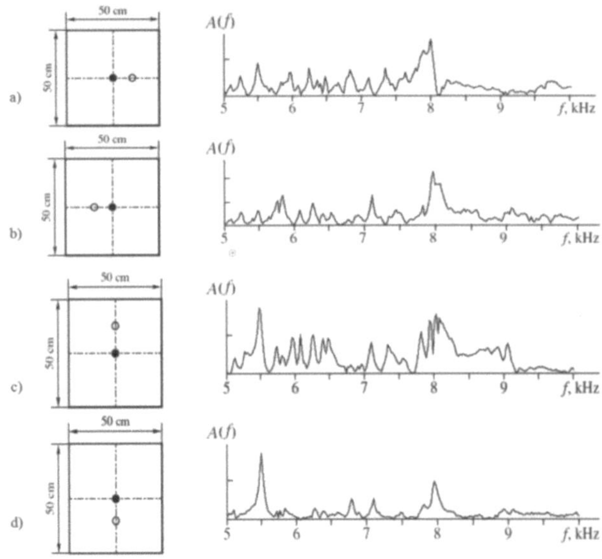

Fig. 24. Experimental spectral characteristics of a compact block with dimensions of 50x50x25 cm.

Because the block is symmetric (its length and width are identical), only one resonance of the first mode of a longitudinal wave is expected in the spectral characteristic, but, owing to a complex structure of the reinforced-concrete block, additional measurements are necessary. For this reason, the characteristics obtained on both the longitudinal and transverse axes of symmetry were included in the resonance–multiplicative processing.

The result of the resonance–multiplicative processing of all four characteristics is shown in Fig. 24e. As is seen, a resonance peak at a frequency f = 7963 Hz is unambiguously determined in this characteristic. The experimental data for the block with dimensions of 5x50x25 cm are listed in Table 3.

	h, m	m	k	F, Hz	C_l. m/s
Along length	0.25	2	1.095	7963	3636
Along width	0.35	1	1.095	7963	3636

Table 3. Calculated and experimental data for a block of 50 x 50 x 25 cm

When the zones of optimal arrangement of transducers are inaccessible for design reasons, an increase in the number of multiplied characteristics also allows detection of the resonance of a longitudinal wave's first mode.

Figure 25 shows four spectral characteristics of a block of 80x50x30 cm additionally measured at different points of the block surface near the longitudinal axis of symmetry. We see that a departure from the axis of symmetry complicates an unambiguous interpretation of the spectrum using individual characteristics. Figure 25e shows the result of multiplication of these characteristics, which is a clearly pronounced single resonance peak corresponding to the desired thickness of the measured compact article.

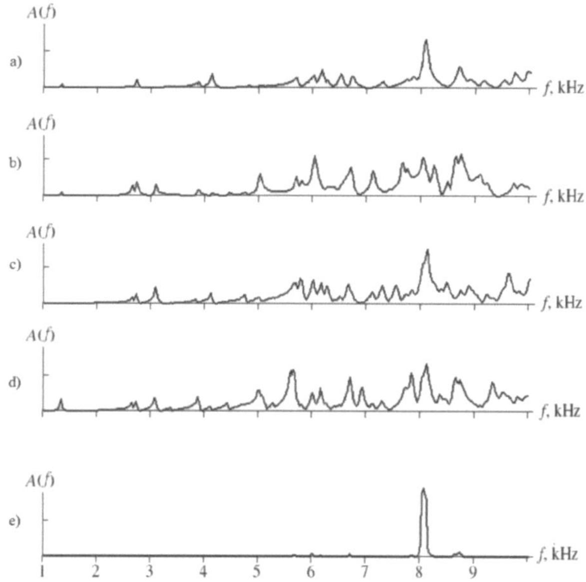

Fig. 25. (a–d) Spectral characteristics for a block with dimensions of 80x50x30 cm at four different positions of the transmitter and (e) the result of their multiplication

4. Conclusion

Note that, at this stage of investigation, the proposed RMM was used for large-size compact BCs of the most widespread standard shapes—parallelepipeds and disks (blocks, columns, supports, etc.). The developed algorithms for testing compact BCs, which involve construction of dispersion characteristics and determination of zones for optimal arrangement of transducers, allow adaptation of the proposed method to testing of arbitrarily shaped compact articles. However, in this case, it is necessary to construct an individual dispersion characteristic and identify particular optimal zones for positioning transducers for each new shape of an article.

Hence, as a result of studies performed at the Moscow Power Engineering Institute, a multichannel RMM for testing large-size compact concrete articles with unambiguous spectrum interpretation has been developed. This method allows determination of both the thickness of a tested article and the velocity of acoustic vibrations in it.

Work are executed with partial support of the Federal target program "Scientific and scientific-pedagogical personnel of innovative Russia"

5. References

Avramenko, S.L. & Kachanov, V.K. (2007). Computer Simulation of Dimensional and Structural Resonances in Articles of Complex Shapes in Ultrasonic Testing by the Resonance Method, *Proceedings of 13th Int. Scientific and Technical Conf. of Students and Postgraduates*, Moscow, MPEI, vol. 1

Avramenko, S.L. (2008). Acoustic Hardware–Software System for Thickness Measurements of Concrete Articles, *Bulletin MPEI*, no. 1, pp. 124–127

Basov, K.A. (2002). *ANSYS in Examples and Problems*, Komp'yuter Press, Moscow

Bolotin, V.V. (1999). *Vibrations in Technology: Handbook. Vibrations of Linear Systems*, vol. 1, Mashinostroenie, Moscow

Carino, N.J. (2001). The Impact-Echo Method, *An Overview, Proceedings of the 2001 Structures Congress & Exposition*, Washington, DC: American Society of Civil Engineers, Reston, Virginia, 2001 ,P.C. Chang, Ed

Carino, N.J.; Sansalone, M. and Hsu, N.N. (1986). *Flaw Detection in Concrete by Frequency Spectrum Analysis of Impact-Echo Waveforms, International Advances in Nondestructive Testing*, Gordon & Breach Science Publishers, New York.

Chigarev, A.V.; Kravchuk, A.S., & Smalyuk, A.F. (2004). *ANSYS for Engineers: Handbook*, Mashinostroenie-1, Moscow

Demidovich, B.P. & Maron, I.A. (1970) *Foundations of Calculus Mathematics*, Nauka, Moscow

Ermolov, I.N. & Lange, Yu.V. (2004). *Nondestructive Testing: Handbook. Ultrasonic Testing*, vol. 3, Mashinostroenie, Moscow

Ferraro, C.C. (2003) *Advanced Nondestructive Monotoring and Evaluation of Damage in Concrete Materials*, a Thesis Presented to the Graduate School of the University of Floridian Partial Fulfillment of the Requirements for the Degree of Master of Engineering University of Florida, University of Florida

Glikman A.G. (2009). Spectral Seismic Prospecting—Sources and consequences, available: http://www.newgeophys.spb.ru/ru/book2/index.shtml

Kachanov, V.K. & Sokolov, I.V. (2007). Problems of Ultrasonic Testing of Extended Complexly Structured Articles with Strong Attenuation of Signals. *Rus. J. Nondestruct. Test.* vol. 43, no. 8, pp. 555–563

Kachanov, V.K., Sokolov, I.V., & Avramenko, S.L. (2008). Acoustic Resonance–Multiplicative Method for Thickness Measurement of Extended Concrete Building Constructions. *Measuring Tech.*, no. 5, pp. 15–18

Kaplun, A.B.; Morozov, E.M., & Olfer'eva, M.A. (2003). *ANSYS as a Tool for Engineers: Practical Manual*, Editorial URSS, Moscow

Korobov, A.I. (2003). Resonance Acoustic Spectroscopy of Solids. *Methodological Elaboration of Special Training for a Chair of Acoustics,* Moscow: Mosk. Gos. Univ.

Sansalone, M. & Carino, N.J.(1986). Impact-Echo: A Method for Flaw Detection in Concrete Using Transient Stress Waves, NBSIR 86–3452, National Bureau of Standards

Sansalone, M. and Streett, W.B. (1997). *Impact-Echo: Nondestructive Testing of Concrete and Masonry*, Bulibrier Press, Jersey Shore, PA

Skuchik, E. (1971). *The Foundations of Acoustics*, Springer, NY:, vol. 1.

Analysis of Axisymmetric and Non-Axisymmetric Wave Propagation in a Homogeneous Piezoelectric Solid Circular Cylinder of Transversely Isotropic Material

Michael Yu. Shatalov
Sensor Science and Technology (SST) of the CSIR Material Science and Manufacturing, P.O. Box 395, Pretoria 0001, CSIR and
Department of Mathematics and Statistics of Tshwane University of Technology, P.B. X680, Pretoria 0001 FIN-40014, TUT
South Africa

1. Introduction

Ultrasonic non-destructive evaluation relies on thorough understanding of the propagating and evanescent waves in the material under investigation. This study could be useful in such applications as, for example, the identification of defects of finite size across the circumference of a piezoelectric rod, since reflections of waves from such defects are generally non-axisymmetric in nature (Rose, 1999). Also, results obtained in this paper can serve as a basis for design of piezoelectric transducers, for which standing wave modes resulting from reflections of travelling waves at cross-section boundaries of the cylinder play an important role. The broad development of the finite (FEM) and boundary element methods (BEM) also needs some reference results obtained from exact solutions. In this case the proposed investigation of the non-axisymmetric propagating and evanescent waves in a piezoelectric cylinder could help to create reliable test beds of FEM and BEM.

The study of wave propagation in systems with cylindrical geometry has been undertaken by a number of investigators. While the systems considered so far have been both isotropic and anisotropic in nature, a relatively larger literature exists for isotropic materials. The first investigator of waves propagating in a solid isotropic cylinder was Pochhammer (Pochhammer, 1876). Other investigators of the subject can be found in standard texts (Achenbach, 1984), (Graff, 1991), and (Rose, 1999). For cylinders composed of anisotropic materials Chree (Chree, 1890) investigated propagation of the axisymmetric waves. Much later Mirsky (Mirsky, 1964) investigated a problem of non-axisymmetric wave propagation in transversely isotropic circular solid and hollow cylinders. Other contributors to the subject were (Armenakas & Reitz, 1973), (Frazer, 1980), (Nayfeh & Nagy (1995), (Berliner & Solecki, 1996), (Niklasson & Datta, 1998), and (Honarvar, et al., 2007), just to name a few. Numerical results for the dispersion of axisymmetric guided waves in a composite cylinder with a transversely isotropic core were presented by (Xu & Datta, 1991).

For piezoelectric cylinders the analysis of vibrations of circular shells was performed by Paul (Paul, 1966). Investigation of axisymmetric waves in layered piezoelectric rods with open circuit electric field conditions and their composites was carried out in (Nayfeh et al., 2000). The axisymmetric problem of the wave propagation in a piezoelectric transversely isotropic rod was analysed for rigid sliding and elastic simply supported mechanical boundary conditions and different types of electric field conditions in (Wei & Su, 2005). Several papers were devoted to development of numerical and finite element methods of investigation of piezoelectric cylinders. In (Siao et al., 1994) the authors solved the problem of wave propagation in a laminated piezoelectric cylinder via the FEM. Their paper contains tables and graphs of dispersion curves for real and imaginary values of wavenumbers. Unfortunately the data contains a misprint in a scale factor for the dimensionless wave number, making it quite difficult to compare their results with the results of the present paper. This misprint was further corrected in (Bai et al., 2004), who in their paper studied the electromechanical response of a laminated piezoelectric hollow cylinder by means of a semi-analytical FEM formulation. In (Shatalov & Loveday, 2004), and (Bai et al., 2006) the authors also studied the phenomenon of end reflections of waves in a semi-infinite layered piezoelectric cylinder.

Our approach parallels that of (Mirsky, 1964) and (Berliner & Solecki, 1996). It is also similar to (Winkel et al., 1995) who analytically solved the problem of wave propagations in an infinite cylindrical piezoelectric core rod immersed into an infinite piezoelectric cladding material. In (Winkel et al., 1995) the authors focused on the pure guided waves with real solutions of the determining bi-cubic equation. It was found in the process of investigation that this approach is applicable to the general problem of propagating and evanescent waves with real and complex solutions. In our approach the three dimensional equations of elastodynamics together with the quasi-electrostatic Gauss law are solved in terms of seven displacements and three electric potentials, each satisfying the Helmholtz equations. In contrast to (Winkel et al., 1995) we propose a simple approach to solution of the problem which has an additional advantage that in the limiting case of small electric and electromechanical constants our results automatically coincide with the classical results of (Mirsky, 1964) and (Berliner & Solecki, 1996) for a transversely isotropic cylinder. This evidence gives one confidence in the correctness of the obtained results. Imposing the allowed mechanical and electric boundary conditions on the cylinder surface, the characteristic or dispersion equation is derived in the form of a determinant of the fourth order in the non-axisymmetric case and a determinant of the third order in the axisymmetric case.

On the basis of the obtained results a numerical algorithm for displaying the dispersion curves is developed. The algorithm is based on calculation of the logarithm of modulus of the left hand side of the dispersion equation on a discrete mesh in the "wavenumber - frequency" ($k-\omega$) plane. If the left hand side value of the dispersion equation tends to zero the logarithm tends to minus infinity. This method produces sharp negative spikes on the surface plot and automatically provides a picture of configuration of the dispersion curves. Similar approach of simultaneous qualitative displaying of the dispersion equation solutions was used in (Honarvar et al., 2006) that produced a 3-D cross-section of the real part of left hand side of the equation. The main advantage of our approach is that the local minima of the logarithm of modulus of the left hand side of the dispersion equation's matrix give proper approximations of the roots which can be further used as guess values of solutions of the dispersion equation. In the present chapter we present the dispersion curves of solid cylinders made from two piezoelectric materials – PZT-4 and PZT-7A for open- and short-circuit electric boundary conditions and for different circumferential wavenumbers m = 1, 2, 3 and for the axisymmetric case. The curves demonstrate substantial influence of the electric boundary conditions and electromechanical coupling on behaviour of the dispersion curves. These results obtained from exact solution of the problem could serve as the reference data and help researchers to create reliable FEM techniques for analysis of vibration of piezoelectric bodies.

2. Formulation of problem

In this section we establish the equations of motion and boundary conditions, and obtain solution of the problem within the framework of the following assumptions and approximations: linear elasticity, linear constitutional model of piezoelectricity, quasi-static approximation of the electric field, axial polarization of the piezoelectric material, neglecting of thermal effects, absence of free charges in the material, and boundary and body forces. Axis Oz coincides with the axis of the cylinder, r,θ,z - radius, polar angle and axial coordinate, and u,v,w - radial, tangential and axial displacements correspondingly.

Equations of motion and boundary conditions of the cylindrical piezoelectric body are derived from the Hamilton variational principle. In order to apply Hamilton's principle we define the Lagrangian of the system as follows:

$$L = K - U + W \tag{1}$$

Where K is the kinetic energy, U is the strain energy and W is the electric energy of the system:

$$K = \frac{\rho}{2}\int_{-\infty}^{\infty}\int_{0}^{2\pi}\int_{0}^{a}\left(\dot{u}^2 + \dot{v}^2 + \dot{w}^2\right) r\, dr\, d\theta\, dz\,,$$

$$U = \frac{1}{2}\int_{-\infty}^{\infty}\int_{0}^{2\pi}\int_{0}^{a}\left(T_1S_1 + T_2S_2 + T_3S_3 + T_4S_4 + T_5S_5 + T_6S_6\right) r\, dr\, d\theta\, dz\,, \tag{2}$$

$$W = \frac{1}{2}\int_{-\infty}^{\infty}\int_{0}^{2\pi}\int_{0}^{a}\left(D_1E_1 + D_2E_2 + D_3E_3\right) r\, dr\, d\theta\, dz$$

In these expressions u, v, w are the radial, tangential and axial displacements correspondingly, $T_{1,2,\ldots,6}$ are the stresses, $S_{1,2,\ldots,6}$ are the straiuns, $D_{1,2,3}$ are the electric displacements and $E_{1,2,3}$ are the electric fields. These components in the Voight notation are:

$$T_1 = c_{11}^E S_1 + c_{12}^E S_2 + c_{13}^E S_3 - e_{31}E_3\,,$$

$$T_2 = c_{12}^E S_1 + c_{11}^E S_2 + c_{13}^E S_3 - e_{31}E_3\,,$$

$$T_3 = c_{13}^E S_1 + c_{13}^E S_2 + c_{33}^E S_3 - e_{33}E_3\,, \tag{3}$$

$$T_4 = c_{44}^E S_4 - e_{15}E_2\,, \quad T_5 = c_{44}^E S_5 - e_{15}E_1\,, \quad T_6 = c_{66}^E S_{66}\,,$$

$$D_1 = \varepsilon_{11}^S E_1 + e_{15} S_5\,,\quad D_2 = \varepsilon_{11}^S E_2 + e_{15} S_4\,,\quad D_3 = \varepsilon_{33}^S E_3 + e_{31} S_1 + e_{31} S_2 + e_{32} S_2$$

where c_{11}^E, c_{12}^E, c_{13}^E, c_{33}^E, c_{44}^E, $c_{66}^E = (c_{11}^E - c_{12}^E)/2$ are the elastic stiffnesses at constant electric field, e_{15}, e_{31} are the piezoelectric constants or electromechanical coupling factors and ε_{11}^S, ε_{33}^S are the clamped dielectric constants at constant strain.

$$S_1 = \frac{\partial u}{\partial r},\quad S_2 = \frac{1}{r}\left(u + \frac{\partial v}{\partial \theta}\right),\quad S_3 = \frac{\partial w}{\partial z},\quad S_4 = \frac{\partial v}{\partial z} + \frac{1}{r}\frac{\partial w}{\partial \theta},\quad S_5 = \frac{\partial u}{\partial z} + \frac{\partial w}{\partial r},$$

$$S_6 = \frac{1}{r}\left(\frac{\partial u}{\partial \theta} - v\right) + \frac{\partial v}{\partial r},\quad E_1 = -\frac{\partial \phi}{\partial r},\quad E_2 = -\frac{1}{r}\frac{\partial \phi}{\partial \theta},\quad E_3 = -\frac{\partial \phi}{\partial z} \tag{4}$$

Substituting (2) – (3) into (1) the Lagrangian

$$L = \int_{-\infty}^{\infty}\int_{0}^{2\pi}\int_{0}^{a} \Lambda \, dr\, d\theta\, dz \tag{5}$$

is represented by the Lagrangian density Λ:

$$\Lambda = \Lambda\left(\dot{u},\dot{v},\dot{w};u'_r,v'_r,w'_r,\phi'_r;u'_\theta,v'_\theta,w'_\theta,\phi'_\theta;u'_z,v'_z,w'_z,\phi'_z;u,v\right) \tag{6}$$

The Euler-Lagrange equations for Lagrangian (6) are:

$$\frac{\partial}{\partial t}\left(\frac{\partial \Lambda}{\partial \dot{u}}\right) + \frac{\partial}{\partial r}\left(\frac{\partial \Lambda}{\partial u'_r}\right) + \frac{\partial}{\partial \theta}\left(\frac{\partial \Lambda}{\partial u'_\theta}\right) + \frac{\partial}{\partial z}\left(\frac{\partial \Lambda}{\partial u'_z}\right) - \frac{\partial \Lambda}{\partial u} = 0,$$

$$\frac{\partial}{\partial t}\left(\frac{\partial \Lambda}{\partial \dot{v}}\right) + \frac{\partial}{\partial r}\left(\frac{\partial \Lambda}{\partial v'_r}\right) + \frac{\partial}{\partial \theta}\left(\frac{\partial \Lambda}{\partial v'_\theta}\right) + \frac{\partial}{\partial z}\left(\frac{\partial \Lambda}{\partial v'_z}\right) - \frac{\partial \Lambda}{\partial v} = 0,$$

$$\frac{\partial}{\partial t}\left(\frac{\partial \Lambda}{\partial \dot{w}}\right) + \frac{\partial}{\partial r}\left(\frac{\partial \Lambda}{\partial w'_r}\right) + \frac{\partial}{\partial \theta}\left(\frac{\partial \Lambda}{\partial w'_\theta}\right) + \frac{\partial}{\partial z}\left(\frac{\partial \Lambda}{\partial w'_z}\right) = 0,$$

$$\frac{\partial}{\partial r}\left(\frac{\partial \Lambda}{\partial \phi'_r}\right) + \frac{\partial}{\partial \theta}\left(\frac{\partial \Lambda}{\partial \phi'_\theta}\right) + \frac{\partial}{\partial z}\left(\frac{\partial \Lambda}{\partial \phi'_z}\right) = 0 \tag{7}$$

Possible boundary conditions on the outer cylindrical surface obtained from the Hamilton principle are as follows:

$$\left.\frac{\partial \Lambda}{\partial u'_r}\right|_{r=a} = T_1\big|_{r=a} = 0 \quad \text{(cylindrical surface is free in the radial direction) or} \tag{8a}$$

$$u\big|_{r=a} = 0 \quad \text{(cylindrical surface is clamped in the radial direction)} \tag{8b}$$

$$\left.\frac{\partial \Lambda}{\partial v'_r}\right|_{r=a} = T_6\big|_{r=a} = 0 \quad \text{(cylindrical surface is free in the tangential direction) or} \tag{9a}$$

$$v\big|_{r=a} = 0 \quad \text{(cylindrical surface is clamped in the tangential direction)} \tag{9b}$$

$$\left.\frac{\partial \Lambda}{\partial w'_r}\right|_{r=a} = T_5\big|_{r=a} = 0 \quad \text{(cylindrical surface is free in the axial direction) or} \tag{10a}$$

$$w\big|_{r=a} = 0 \quad \text{(cylindrical surface is clamped in the axial direction)} \tag{10b}$$

$$\left.\frac{\partial \Lambda}{\partial \phi'_r}\right|_{r=a} = D_1\big|_{r=a} = 0 \ \text{(open-circuit conditions on the cylindrical surface) or} \tag{11a}$$

$$\phi\big|_{r=a} = 0 \quad \text{(close-circuit electric boundary conditions on the cylindrical surface)} \tag{11b}$$

The Euler-Lagrange equation could be also presented in the standard form of the Navier equations of the linear dynamical elasticity (Redwood, 1960) and the Gauss law in the cylindrical coordinates:

$$\frac{\partial T_1}{\partial r} + \frac{1}{r}\frac{\partial T_6}{\partial \theta} + \frac{\partial T_5}{\partial z} + \frac{1}{r}(T_1 - T_2) = \rho \ddot{u} \,,$$

$$\frac{\partial T_6}{\partial r} + \frac{1}{r}\frac{\partial T_2}{\partial \theta} + \frac{\partial T_4}{\partial z} + \frac{1}{r}T_6 = \rho \ddot{v} \,,$$

$$\frac{\partial T_5}{\partial r} + \frac{1}{r}\frac{\partial T_4}{\partial \theta} + \frac{\partial T_3}{\partial z} + \frac{1}{r}T_5 = \rho \ddot{w} \,, \tag{12}$$

$$\frac{\partial D_1}{\partial r} + \frac{1}{r}\left(D_1 + \frac{\partial D_2}{\partial \theta}\right) + \frac{\partial D_3}{\partial z} = 0$$

Further we will assume that the outer cylindrical surface is free, hence, using (8a), (9a) and (10a) the boundary conditions are

$$T_1\big|_{r=a} = T_5\big|_{r=a} = T_6\big|_{r=a} = 0 \tag{13}$$

Electric boundary conditions will be considered in both open- and close-circuit form (11a) or (11b) and one of these conditions will be added to boundary conditions (13).
In the explicit form the system (9) or (12) of equations is as follows:

$$c_{11}^E\left(\frac{\partial^2 u}{\partial r^2} + \frac{1}{r}\frac{\partial u}{\partial r} - \frac{u}{r^2} - \frac{1}{r^2}\frac{\partial v}{\partial \theta}\right) + \frac{c_{12}^E}{r}\frac{\partial^2 v}{\partial r \partial \theta} + \left(c_{13}^E + c_{44}^E\right)\frac{\partial^2 w}{\partial r \partial z} + c_{44}^E\frac{\partial^2 u}{\partial z^2}$$

$$+c_{66}^E\left(\frac{1}{r^2}\frac{\partial^2 u}{\partial \theta^2} + \frac{1}{r}\frac{\partial^2 v}{\partial r \partial \theta} - \frac{1}{r^2}\frac{\partial v}{\partial \theta}\right) - e_{15}\frac{\partial E_1}{\partial z} - e_{31}\frac{\partial E_3}{\partial r} = \rho \ddot{u} \,,$$

$$\frac{c_{11}^E}{r^2}\left(\frac{\partial u}{\partial \theta} + \frac{\partial^2 v}{\partial \theta^2}\right) + \frac{c_{12}^E}{r}\frac{\partial^2 u}{\partial r \partial \theta} + \frac{c_{13}^E + c_{44}^E}{r}\frac{\partial^2 w}{\partial \theta \partial z} + c_{44}^E\frac{\partial^2 v}{\partial z^2}$$

$$+c_{66}^E\left(\frac{1}{r^2}\frac{\partial u}{\partial \theta}+\frac{1}{r}\frac{\partial^2 u}{\partial r\partial \theta}+\frac{\partial^2 v}{\partial r^2}+\frac{1}{r}\frac{\partial v}{\partial r}-\frac{v}{r^2}\right)-e_{15}\frac{\partial E_2}{\partial z}-\frac{e_{31}}{r}\frac{\partial E_3}{\partial \theta}=\rho\ddot{v}\,, \tag{14}$$

$$\left(c_{13}^E+c_{44}^E\right)\left(\frac{1}{r}\frac{\partial u}{\partial z}+\frac{\partial^2 u}{\partial r\partial z}+\frac{1}{r}\frac{\partial^2 v}{\partial r\partial \theta}\right)+c_{44}^E\left(\frac{\partial^2 w}{\partial r^2}+\frac{1}{r}\frac{\partial w}{\partial r}+\frac{1}{r^2}\frac{\partial^2 w}{\partial \theta^2}\right)+c_{33}^E\frac{\partial^2 w}{\partial z^2}$$

$$-e_{15}\left(\frac{\partial E_1}{\partial r}+\frac{E_1}{r}-\frac{1}{r}\frac{\partial E_3}{\partial \theta}\right)-e_{33}\frac{\partial E_3}{\partial z}=\rho\ddot{w}\,,$$

$$e_{15}\left(\frac{\partial^2 u}{\partial r\partial z}+\frac{1}{r}\frac{\partial u}{\partial r}+\frac{1}{r}\frac{\partial^2 v}{\partial \theta\partial z}+\frac{\partial^2 w}{\partial r^2}+\frac{1}{r}\frac{\partial w}{\partial r}+\frac{1}{r^2}\frac{\partial w}{\partial \theta}\right)+e_{31}\left(\frac{1}{r}\frac{\partial u}{\partial z}+\frac{\partial^2 v}{\partial r\partial z}+\frac{1}{r}\frac{\partial^2 w}{\partial \theta\partial z}\right)$$

$$+\varepsilon_{11}^S\left(\frac{\partial E_1}{\partial r}+\frac{E_1}{r}+\frac{1}{r}\frac{\partial E_2}{\partial \theta}\right)+\varepsilon_{33}^S\frac{\partial E_3}{\partial z}=0$$

The explicit form of boundary conitions (13) is:

$$T_1\big|_{r=a}=\left[c_{11}^E\frac{\partial u}{\partial r}+c_{12}^E\left(\frac{1}{r}\left(u+\frac{\partial v}{\partial \theta}\right)\right)+c_{13}^E\left(\frac{\partial w}{\partial z}\right)+e_{31}\left(\frac{\partial \phi}{\partial z}\right)\right]_{r=a}=0\,,$$

$$T_5\big|_{r=a}=\left[c_{44}^E\left(\frac{\partial u}{\partial z}+\frac{\partial w}{\partial r}\right)+e_{15}\frac{\partial \phi}{\partial r}\right]_{r=a}=0\,,\quad T_6\big|_{r=a}=c_{66}^E\left[\frac{1}{r}\left(\frac{\partial u}{\partial \theta}-v\right)+\frac{\partial v}{\partial r}\right]_{r=a}=0 \tag{15}$$

$$D_1\big|_{r=a}=\left[-\varepsilon_{11}^S\frac{\partial \phi}{\partial r}+e_{15}\left(\frac{\partial u}{\partial z}+\frac{\partial w}{\partial r}\right)\right]_{r=a}=0\ \text{ or }\ \phi\big|_{r=a}=0$$

3. Analytic solution of the problem

We seek the solution of the problem (14) – (15) in terms of harmonic travelling waves along z - axis in terms of several displacement and electric potentials first introduced by (Mirsky, 1964) and further used by (Winkel et al 1995) and (Shatalov et al., 2009) as follows:

$$u=u(r,\theta,z,t)=\left[\frac{\partial \varphi(r,\theta)}{\partial r}+\frac{1}{r}\frac{\partial \psi(r,\theta)}{\partial \theta}\right]e^{i(\omega t+kr)}\,,$$

$$v=v(r,\theta,z,t)=\left[\frac{1}{r}\frac{\partial \varphi(r,\theta)}{\partial \theta}-\frac{\partial \psi(r,\theta)}{\partial r}\right]e^{i(\omega t+kr)}\,, \tag{16}$$

$$w=w(r,\theta,z,t)=\chi(r,\theta)e^{i(\omega t+kr)}\,,\qquad \phi=\phi(r,\theta,z,t)=\tau(r,\theta)e^{i(\omega t+kr)}$$

where $\varphi(r,\theta)$, $\psi(r,\theta)$, $\chi(r,\theta)$ and $\tau(r,\theta)$ are unknown potentials.
After substitution (16) into (14) we obtain the following system of equations:

$$\frac{\partial}{\partial r}\left\{\left[c_{11}^E\nabla^2\varphi+\left(\rho\omega^2-k^2c_{44}^E\right)\varphi\right]+\left[ik\left(c_{13}^E+c_{44}^E\right)\chi\right]+\left[ik\left(e_{15}+e_{31}\right)\tau\right]\right\}$$

$$+\frac{1}{r}\frac{\partial}{\partial\theta}\left\{\left[c_{66}^E\nabla^2\psi+\left(\rho\omega^2-k^2c_{44}^E\right)\psi\right]\right\}=0,$$

$$\frac{1}{r}\frac{\partial}{\partial\theta}\left\{\left[c_{11}^E\nabla^2\varphi+\left(\rho\omega^2-k^2c_{44}^E\right)\varphi\right]+\left[ik\left(c_{13}^E+c_{44}^E\right)\chi\right]+\left[ik\left(e_{15}+e_{31}\right)\tau\right]\right\}$$

$$-\frac{\partial}{\partial r}\left\{\left[c_{66}^E\nabla^2\psi+\left(\rho\omega^2-k^2c_{44}^E\right)\psi\right]\right\}=0, \tag{17}$$

$$\left[ik\left(c_{13}^E+c_{44}^E\right)\nabla^2\varphi\right]+\left[c_{44}^E\nabla^2\chi+\left(\rho\omega^2-k^2c_{33}^E\right)\chi\right]+\left[e_{15}\nabla^2\tau-k^2e_{33}\tau\right]=0,$$

$$\left[ik\left(e_{15}+e_{31}\right)\nabla^2\varphi\right]+\left[e_{15}\nabla^2\chi-k^2e_{33}\chi\right]-\left[\varepsilon_{11}\nabla^2\tau-k^2\varepsilon_{33}\tau\right]=0$$

where $\nabla^2=\frac{\partial^2}{\partial r^2}+\frac{1}{r}\frac{\partial}{\partial r}+\frac{1}{r^2}\frac{\partial^2}{\partial\theta^2}$ is the two dimensional Laplace operator in polar coordinates.

The first two equations of systems (17) are satisfied if

$$c_{66}^E\nabla^2\psi+\left(\rho\omega^2-k^2c_{44}^E\right)\psi=0, \tag{18}$$

$$\left[c_{11}^E\nabla^2\varphi+\left(\rho\omega^2-k^2c_{44}^E\right)\varphi\right]+\left[ik\left(c_{13}^E+c_{44}^E\right)\chi\right]+\left[ik\left(e_{15}+e_{31}\right)\tau\right]=0$$

First equation of system (18) is separated from the second equation as well as from the third and fourth equations of system (17). It could be represented as the Helmholtz equation in the polar coordinates:

$$\nabla^2\psi+\zeta_0^2\psi=0 \tag{19}$$

where $\zeta_0^2=\frac{\rho\omega^2-k^2c_{44}^E}{c_{66}^E}$.

Second equation of system (18) must be compatible with the third and fourth equations of system (17), i.e. we have the following system of equations:

$$\left[c_{11}^E\nabla^2\varphi+\left(\rho\omega^2-k^2c_{44}^E\right)\varphi\right]+\left[ik\left(c_{13}^E+c_{44}^E\right)\chi\right]+\left[ik\left(e_{15}+e_{31}\right)\tau\right]=0,$$

$$\left[ik\left(c_{13}^E+c_{44}^E\right)\nabla^2\varphi\right]+\left[c_{44}^E\nabla^2\chi+\left(\rho\omega^2-k^2c_{33}^E\right)\chi\right]+\left[e_{15}\nabla^2\tau-k^2e_{33}\tau\right]=0, \tag{20}$$

$$\left[ik\left(e_{15}+e_{31}\right)\nabla^2\varphi\right]+\left[e_{15}\nabla^2\chi-k^2e_{33}\chi\right]-\left[\varepsilon_{11}\nabla^2\tau-k^2\varepsilon_{33}\tau\right]=0$$

To solve this system of equations we try to represent them in the form of the Helmholtz equation as (19):

$$\nabla^2\varphi+\zeta^2\varphi=0,\ \nabla^2\chi+\zeta^2\chi=0,\ \nabla^2\tau+\zeta^2\tau=0 \tag{21}$$

where ζ^2 is unknown parameter which will be calculated after substitution (21) in (20). Substituting $\nabla^2\varphi=-\zeta^2\varphi$, $\nabla^2\chi=-\zeta^2\chi$ and $\nabla^2\tau=-\zeta^2\tau$ into (20) we obtain the following system of equation for determination of parameter ζ^2:

$$\begin{gathered}
\left[\left(\rho\omega^2 - k^2 c_{44}^E\right) - c_{11}^E \zeta^2\right]\varphi + \left[ik\left(c_{13}^E + c_{44}^E\right)\right]\chi + \left[ik\left(e_{15} + e_{31}\right)\right]\tau = 0,\\
\left[-ik\left(c_{13}^E + c_{44}^E\right)\zeta^2\right]\varphi + \left[\left(\rho\omega^2 - k^2 c_{33}^E\right) - c_{44}^E \zeta^2\right]\chi + \left[-\left(k^2 e_{33}\right) - e_{15}\zeta^2\right]\tau = 0,\\
\left[-ik\left(e_{15} + e_{31}\right)\zeta^2\right]\varphi + \left[-\left(k^2 e_{33}\right) - \zeta^2\right]\chi + \left[\left(k^2 \varepsilon_{33}\right) + \zeta^2 \varepsilon_{11}\right]\tau = 0
\end{gathered} \tag{22}$$

System (22) has nontrivial solutions if the determinant of the system is zero:

$$\begin{vmatrix}
\left(\rho\omega^2 - k^2 c_{44}^E\right) - c_{11}^E \zeta^2 & ik\left(c_{13}^E + c_{44}^E\right) & ik\left(e_{15} + e_{31}\right)\\
-ik\left(c_{13}^E + c_{44}^E\right)\zeta^2 & \left(\rho\omega^2 - k^2 c_{33}^E\right) - c_{44}^E \zeta^2 & -\left(k^2 e_{33}\right) - e_{15}\zeta^2\\
-ik\left(e_{15} + e_{31}\right)\zeta^2 & -\left(k^2 e_{33}\right) - \zeta^2 & \left(k^2 \varepsilon_{33}\right) + \zeta^2 \varepsilon_{11}
\end{vmatrix} = 0 \tag{23}$$

Hence the unknown values of ζ^2 are calculated from this determining equation which could be rewritten in the form of the bi-cubic determining equation:

$$\left(\zeta^2\right)^3 + b_1\left(\zeta^2\right)^2 + b_2\left(\zeta^2\right) + b_3 = 0 \tag{24}$$

where

$$b_1 = \frac{-\rho\omega^2 B_1 + k^2 B_2}{B_0}, \qquad b_2 = \frac{\varepsilon_{11}\left(\rho\omega^2\right)^2 - k^2\left(\rho\omega^2\right)B_3 + k^4 B_4}{B_0},$$

$$b_3 = \frac{k^2\left(k^2 c_{44}^E - \rho\omega^2\right)\left[k^2 B_5 - \left(\rho\omega^2\right)\varepsilon_{33}\right]}{B_0}, \qquad B_0 = c_{11}^E\left(e_{15}^2 + \varepsilon_{11} c_{44}^E\right),$$

$$B_1 = e_{15}^2 + \varepsilon_{11}\left(c_{11}^E + c_{44}^E\right), \qquad B_3 = \varepsilon_{11}\left(c_{33}^E + c_{44}^E\right) + \varepsilon_{33}\left(c_{11}^E + c_{44}^E\right) + \left(e_{15} + e_{31}\right)^2 + 2e_{15}e_{33},$$

$$B_2 = \varepsilon_{11}\left(c_{11}^E c_{33}^E - 2c_{13}^E c_{44}^E - \left(c_{13}^E\right)^2\right) + 2e_{15}\left(c_{11}^E e_{33} - c_{13}^E e_{31}\right) + c_{44}^E\left(e_{31}^2 + \varepsilon_{33} c_{11}^E\right) - 2c_{13}^E e_{15}^2,$$

$$B_4 = -\left(c_{13}^E\right)^2 \varepsilon_{33} - 2c_{13}^E\left[e_{33}\left(e_{15} + e_{31}\right) - c_{44}^E \varepsilon_{33}\right] + c_{33}^E\left[\left(e_{15} + e_{33}\right)^2 + c_{44}^E \varepsilon_{11} + c_{11}^E \varepsilon_{33}\right] + c_{11}^E \varepsilon_{33}^2 - 2c_{44}^E e_{31} e_{33},$$

$$B_5 = c_{33}^E \varepsilon_{11} + e_{33}^2 \tag{25}$$

Hence, in general case we have three different solutions of bi-cubic equation (24) ζ_1^2, ζ_2^2 and ζ_3^2:

$$\zeta_1^2 = -\frac{b_1}{3} + (A + B), \quad \zeta_{2,3}^2 = -\frac{b_1}{3} - \frac{A + B}{2} \pm i\frac{A - B}{2}\sqrt{3} \tag{26}$$

where

$$A = \sqrt[3]{-\frac{q}{2} + \sqrt{Q}}, \quad B = \sqrt[3]{-\frac{q}{2} - \sqrt{Q}}, \tag{27}$$

$$Q=\frac{p^3}{27}+\frac{q^2}{4}, \qquad p=-\frac{b_1^3}{3}+b_2, \qquad q=\frac{2b_1^3}{27}-\frac{b_1b_2}{3}+b_3$$

Hence, we have three independent Helmholtz equations:

$$\nabla^2\varphi_j+\zeta_j^2\varphi_j=0, \quad (j=1,2,3) \tag{28}$$

It follows from equations (20) - (22) that functionals $\chi_j=\chi_j(r,\theta)$ and $\tau_j=\tau_j(r,\theta)$ in expressions (17) – (18) and (20) – (22) could be calculated as follows:

$$\chi_j=\chi_j(r,\theta)=\eta_j\varphi_j(r,\theta), \quad \tau_j=\tau_j(r,\theta)=\mu_j\varphi_j(r,\theta) \tag{29}$$

where coefficients η_j, μ_j are obtained from system of equations (22) after substitution expressions (29):

$$\left[ik\left(c_{13}^E+c_{44}^E\right)\right]\eta_j+\left[ik\left(e_{15}+e_{31}\right)\right]\mu_j=-\left[\left(\rho\omega^2-k^2c_{44}^E\right)-c_{11}^E\zeta_j^2\right],$$

$$\left[\left(\rho\omega^2-k^2c_{33}^E\right)-c_{44}^E\zeta_j^2\right]\eta_j+\left[-\left(k^2e_{33}\right)-e_{15}\zeta_j^2\right]\mu_j=\left[ik\left(c_{13}^E+c_{44}^E\right)\zeta_j^2\right], \tag{30}$$

$$\left[-\left(k^2e_{33}\right)-\zeta_j^2\right]\eta_j+\left[\left(k^2\varepsilon_{33}\right)+\zeta_j^2\varepsilon_{11}\right]\mu_j=\left[ik\left(e_{15}+e_{31}\right)\zeta_j^2\right]$$

For example, we can find η_j from the first and second equations of (30) and μ_j from the third equation of this system:

$$\eta_j=\eta_j(k,\omega)=-\frac{i}{k}\frac{\zeta_j^4c_{11}^Ee_{15}+\zeta_j^2\left[k^2B_6-\left(\rho\omega^2\right)e_{15}\right]+k^2\left(k^2c_{44}^E-\rho\omega^2\right)e_{33}}{k^2B_7+\zeta_j^2B_8+\rho\omega^2B_9},$$

$$\mu_j=\mu_j(k,\omega)=\frac{\zeta_j^2\left[\eta_je_{15}-ikB_9\right]+k^2\eta_je_{33}}{\zeta_j^2\varepsilon_{11}+k^2\varepsilon_{33}}, \quad (j=1,2,3) \tag{31}$$

where

$$B_6=c_{11}^Ee_{33}-c_{13}^Ee_{15}+\left(c_{13}^E+c_{44}^E\right)e_{31}, \qquad B_7=\left(c_{13}^E+c_{44}^E\right)e_{33}-c_{13}^E\left(e_{15}+e_{31}\right),$$

$$B_8=c_{13}^Ee_{15}-c_{44}^Ee_{31}, \quad B_9=e_{15}+e_{31} \tag{32}$$

Finally the solution of system (14) using representation (16) is as follows:

$$u=u(r,\theta,z,t)=\left[\frac{1}{r}\frac{\partial\psi(r,\theta)}{\partial\theta}+\frac{\partial\varphi_1(r,\theta)}{\partial r}+\frac{\partial\varphi_2(r,\theta)}{\partial r}+\frac{\partial\varphi_3(r,\theta)}{\partial r}\right]e^{i(\omega t+kr)},$$

$$v=v(r,\theta,z,t)=\left[-\frac{\partial\psi(r,\theta)}{\partial r}+\frac{1}{r}\frac{\partial\varphi_1(r,\theta)}{\partial\theta}+\frac{1}{r}\frac{\partial\varphi_2(r,\theta)}{\partial\theta}+\frac{1}{r}\frac{\partial\varphi_3(r,\theta)}{\partial\theta}\right]e^{i(\omega t+kr)},$$

$$w=w(r,\theta,z,t)=\left[\eta_1\varphi_1(r,\theta)+\eta_2\varphi_2(r,\theta)+\eta_3\varphi_3(r,\theta)\right]e^{i(\omega t+kr)}, \tag{33}$$

$$\phi=\phi(r,\theta,z,t)=\left[\mu_1\varphi_1(r,\theta)+\mu_2\varphi_2(r,\theta)+\mu_3\varphi_3(r,\theta)\right]e^{i(\omega t+kr)}$$

Solution of the Helmholtz equations (19): $\nabla^2\psi+\zeta_0^2\psi=0$ and (28): $\nabla^2\varphi_j+\zeta_j^2\varphi_j=0$, $(j=1,2,3)$ for the solid cylinder is:

$$\psi=\psi(r,\theta)=A_0W_m(\zeta_0 r)\sin(m\theta+\theta_0)\,,\ \varphi_j=\varphi_j(r,\theta)=A_jW_m(\zeta_j r)\cos(m\theta+\theta_0) \tag{34}$$

where m is a positive integer in the non-axisymmetric case, A_j - arbitrary constants, representing the circumferential wave number, $W_m(\zeta r)=J_m(\zeta r)$ is the Bessel function of the first kind if ζ is real or complex and $W_m(\zeta r)=I_m(|\zeta|r)$ is the Bessel function of the second kind if ζ is pure imaginary number.
Mechanical boundary conditions correspond to the assumption of absence of external forces on the cylindrical boundary:

$$T_1\big|_{r=a}=T_5\big|_{r=a}=T_6\big|_{r=a}=0 \tag{35}$$

There are two types of electric boundary conditions:

$$D_1\big|_{r=a}=0 \ \text{ or } \ \phi\big|_{r=a}=0 \tag{36}$$

where the first expression corresponds to the open-circuit condition and the second expression – to the close-circuit condition.
After substitution (34) into (33), (4), and (3) we obtain two systems of four linear homogeneous algebraic equations in the unknown amplitudes $A_{0,1,2,3}$. Each of these systems of equations has a non-trivial solution if and only if its main determinant equals zero:

$$Det(k,\omega)=\begin{vmatrix} a_{11} & a_{12} & a_{13} & a_{14}\\ a_{21} & a_{22} & a_{23} & a_{24}\\ a_{31} & a_{32} & a_{33} & a_{34}\\ a_{41} & a_{42} & a_{43} & a_{44}\end{vmatrix}=0 \tag{37}$$

where $a_{j1,j2}=a_{j1,j2}(k,\omega)$ $(j1,j2=1,2,3,4)$ are given in the Appendix. Equation (37) forms the dispersion equation.
Remark on limiting case of small electro-mechanical coupling coefficients. If electro-mechanical coupling coefficients $e_{15}=e_{31}=e_{33}=0$ the determining equation (24) is converted into

$$\left[\left(\xi_j^2\right)-k^2\frac{\varepsilon_{33}}{\varepsilon_{11}}\right]\left[\left(\xi_j^2\right)^2+b_4\left(\xi_j^2\right)+b_5\right]=0 \tag{38}$$

where

$$b_4=\frac{-\left(\rho\omega^2\right)B_{11}+k^2B_{12}}{B_{10}},\qquad b_5=\frac{\left[\left(\rho\omega^2\right)-k^2c_{33}^E\right]\left[\left(\rho\omega^2\right)-k^2c_4^E\right]}{B_{10}},$$

$$B_{10}=c_{11}^Ec_{44}^E\,,\ B_{11}=c_{11}^E+c_{44}^E\,,\ B_{12}=c_{11}^Ec_{33}^E-\left(c_{13}^E\right)^2-2c_{13}^Ec_{44}^E \tag{39}$$

Hence in (17) the electric and mechanical parts are separated and the determining equation $(\xi_j^2)^2 + b_4(\xi_j^2) + b_5 = 0$ describes the wave dynamics in the passive transversely isotropic medium. This equation coincides with the corresponding equation (Berliner & Solecki, 1996) and hence, all the presented results are converted into the well known results (Mirsky, 1964) and (Berliner & Solecki, 1996) in the limiting case of small electro-mechanical coupling coefficients. Additional dispersion line, corresponding to $(\xi_j^2)\varepsilon_{11} - k^2\varepsilon_{33} = 0$ could be considered as an artefacts in this case.

Furthermore if we suppose that $e_{15} = e_{31} = 0$ in the first expression (31) we obtain that

$$\eta_j = \frac{i}{k}\frac{(\rho\omega^2) - (\zeta_j^2)c_{11}^E - k^2 c_{44}^E}{c_{13}^E + c_{44}^E} \tag{40}$$

which also coincides with the known result (Berliner & Solecki, 1996). Hence, the results obtained contain the classical results of investigation of a passive transversely isotropic material as a particular limiting case.

4. Axisymmetric case

Axisymmetric vibrations of the piezoelectric cylinder could be considered as a particular case of the general problem with $m = 0$, $v = 0$, $E_2 = 0$, $T_4 = T_6 = 0$ and all variables are θ-independent.

In this case system of equations (14) is rewritten as follows:

$$c_{11}^E\left(\frac{\partial^2 u}{\partial r^2} + \frac{1}{r}\frac{\partial u}{\partial r} - \frac{u}{r^2}\right) + \left(c_{13}^E + c_{44}^E\right)\frac{\partial^2 w}{\partial r\,\partial z} + c_{44}^E\frac{\partial^2 u}{\partial z^2} - e_{15}\frac{\partial E_1}{\partial z} - e_{31}\frac{\partial E_3}{\partial r} = \rho\ddot{u},$$

$$\left(c_{13}^E + c_{44}^E\right)\left(\frac{1}{r}\frac{\partial u}{\partial z} + \frac{\partial^2 u}{\partial r\,\partial z}\right) + c_{44}^E\left(\frac{\partial^2 w}{\partial r^2} + \frac{1}{r}\frac{\partial w}{\partial r}\right) + c_{33}^E\frac{\partial^2 w}{\partial z^2} - e_{15}\left(\frac{\partial E_1}{\partial r} + \frac{E_1}{r}\right) - e_{33}\frac{\partial E_3}{\partial z} = \rho\ddot{w},$$

$$e_{15}\left(\frac{\partial^2 u}{\partial r\,\partial z} + \frac{1}{r}\frac{\partial u}{\partial r} + \frac{\partial^2 w}{\partial r^2} + \frac{1}{r}\frac{\partial w}{\partial r}\right) + e_{31}\left(\frac{1}{r}\frac{\partial u}{\partial z} + \frac{\partial^2 v}{\partial r\,\partial z}\right) + \varepsilon_{11}^S\left(\frac{\partial E_1}{\partial r} + \frac{E_1}{r}\right) + \varepsilon_{33}^S\frac{\partial E_3}{\partial z} = 0 \tag{41}$$

Boundary conitions (15) are:

$$T_1\big|_{r=a} = \left[c_{11}^E\left(\frac{\partial u}{\partial r}\right) + c_{12}^E\left(\frac{u}{r}\right) + c_{13}^E\left(\frac{\partial w}{\partial z}\right) + e_{31}\left(\frac{\partial \phi}{\partial z}\right)\right]_{r=a} = 0,$$

$$T_5\big|_{r=a} = \left[c_{44}^E\left(\frac{\partial u}{\partial z} + \frac{\partial w}{\partial r}\right) + e_{15}\frac{\partial \phi}{\partial r}\right]_{r=a} = 0, \tag{42}$$

$$D_1\big|_{r=a} = \left[-\varepsilon_{11}^S\frac{\partial \phi}{\partial r} + e_{15}\left(\frac{\partial u}{\partial z} + \frac{\partial w}{\partial r}\right)\right]_{r=a} = 0 \text{ or } \phi\big|_{r=a} = 0$$

The variables are changed as follows:

$$u = u(r,z,t) = \left[\frac{\partial \varphi(r)}{\partial r}\right] e^{i(\omega t + kr)}, \qquad w = w(r,z,t) = \chi(r) e^{i(\omega t + kr)},$$

$$\phi = \phi(r,z,t) = \tau(r) e^{i(\omega t + kr)} \tag{43}$$

and system of equations (20) is obtained. All relationships (21) – (32) are true for the axisymmetric case and the final representation of solutions of system (41) is:

$$u = u(r,z,t) = \left[\frac{\partial \varphi_1(r)}{\partial r} + \frac{\partial \varphi_2(r)}{\partial r} + \frac{\partial \varphi_3(r)}{\partial r}\right] e^{i(\omega t + kr)},$$

$$w = w(r,z,t) = \left[\eta_1 \varphi_1(r) + \eta_2 \varphi_2(r) + \eta_3 \varphi_3(r)\right] e^{i(\omega t + kr)}, \tag{44}$$

$$\phi = \phi(r,z,t) = \left[\mu_1 \varphi_1(r) + \mu_2 \varphi_2(r) + \mu_3 \varphi_3(r)\right] e^{i(\omega t + kr)}$$

Solution of the Helmholtz equations (28): $\nabla^2 \varphi_j + \zeta_j^2 \varphi_j = 0$, $(j = 1, 2, 3)$ is:

$$\varphi_j = \varphi_j(r) = A_j W_m(\zeta_j r) \tag{45}$$

where A_j - arbitrary constants, representing the circumferential wave number, $W_m(\zeta_j r) = J_m(\zeta_j r)$ is the Bessel function of the first kind if ζ_j is real or complex and $W_m(\zeta_j r) = I_m(|\zeta_j| r)$ is the Bessel function of the second kind if ζ_j is pure imaginary number.

Boundary conditions are obtained from determinant (37) by means of elimination of third row and fourth colimn.

5. Numerical results and discussion

In this section we present the dispersion curves for non-axisymmetric waves with circumferential wavenumbers $m = 1,2,3$ and axisymmetric waves resulting from the characteristic equation (38). Two piezoelectric materials are chosen, PZT-4 and PZT-7A, to illustrate the influence of electro-mechanical coupling coefficients on the configuration of the dispersion curves. The relevant material parameters for PZT-4 and PZT-7A are given in the Table 1.

In order to obtain the dispersion curves we made use of a method similar to the novel method (Honarvar et al., 2008) where the dispersion curves are not produced as a result of solving of the dispersion equation by a traditional iterative find-root algorithm but are obtained by a zero-level cut in the velocity-frequency plane. In this chapter we modify this approach calculating the logarithm of modulus of determinant (37) on the mesh $(k_{j1}, \omega_{j2}), j1 = 1,\ldots,N_1; j2 = 1,\ldots,N_2$ (Shatalov et al., 2009). In those points where the real and imaginary parts of determinant (37) are close to zero substantial negative spikes occurwhich are displayed on a surface plot and give a picture of the configuration of the dispersion curves. The main advantage of this approach is that the local minima of the $[N_1 \times N_2]$ - matrix of the logarithms are the proper guess values of the dispersion equation's

Geometric and Material Constants	PZT-4	PZT-7A	SI Units
Radius (a)	1	1	m
(ρ)	$7.5 \cdot 10^3$	$7.6 \cdot 10^3$	kg / m^{-3}
c_{11}^E	$13.9 \cdot 10^{10}$	$14.8 \cdot 10^{10}$	N / m^{-2}
c_{12}^E	$7.78 \cdot 10^{10}$	$7.62 \cdot 10^{10}$	N / m^{-2}
c_{13}^E	$7.43 \cdot 10^{10}$	$7.42 \cdot 10^{10}$	N / m^{-2}
c_{33}^E	$11.5 \cdot 10^{10}$	$13.1 \cdot 10^{10}$	N / m^{-2}
c_{44}^E	$2.56 \cdot 10^{10}$	$2.53 \cdot 10^{10}$	N / m^{-2}
e_{15}	12.7	9.2	C / m^{-2}
e_{31}	-5.2	-2.1	C / m^{-2}
e_{33}	15.1	9.5	C / m^{-2}
ε_{11}^S	$730 \cdot \varepsilon_0$	$460 \cdot \varepsilon_0$	$C^2 m^{-2} N^{-1}$
ε_{33}^S	$635 \cdot \varepsilon_0$	$235 \cdot \varepsilon_0$	$C^2 m^{-2} N^{-1}$

Table. Material constants and geometric parameters for PZT-4 and PZT-7A
$\left(\varepsilon_0 = 8.85 \cdot 10^{-12}\ C^2 m^{-2} N^{-1}\right)$

roots. Hence all roots of equation (37) could be found for values of wavenumbers (k_{j1}) on the real axis and on or near the imaginary axis as a function of frequency (ω_{j2}). Another advantage of this method is that it is much faster than the traditional root finding methods and as fast as the Honarvar method (Honarvar et al., 2008).

The main disadvantage of this approach is that the roots of characteristic arguments $(\xi_j), (j = 0,1,2,3)$ are also displayed on the surface plots as obvious artefacts. An elaborate discussion of these artefacts is given by Yenwong-Fai, (Yenwong-Fai, 2008). These artefacts could be simply detected and eliminated from the dispersion plots by program tools.Our algorithm, as it has been implemented, does not search for branches of the dispersion relation well away from the real and imaginary axes for k. It would be relatively straightforward in principle to locate these additional branches.

Dispersion curves of bending waves $(m = 1)$ in the cylinder made from PZT-4 with the short-circuit lateral (cylindrical) surface are depicted in Fig. 1 in dimensionless coordinates $\left(k \cdot a \div \Omega = \omega / (a \cdot V_s)\right)$, where $V_s = \sqrt{c_{44}^E / \rho}$, and a is the outer radius of the cylinder. The picture of the dispersion curves is obtained by a method described above for real (propagating waves) and pure imaginary values of the wavenumber in the limits: $\mathrm{Re}(k \cdot a), \mathrm{Im}(k \cdot a) \in (0,8]$, $\Omega = \omega / (a \cdot V_s) \in (0,14]$ with resolution 500 (250 pixels for real and 250 pixels for imaginary $(k \cdot a)$) × 250 ($\Omega = \omega / (a \cdot V_s)$) pixels. The same resolution is used for Fig. 2 – 17.

The first dispersion curve of the propagating waves (real values of the wavenumber) tends to an asymptote of the surface wave propagation. It is joined to the second curve which tends to the asymptote of the shear waves through the domain of the evanescent waves.

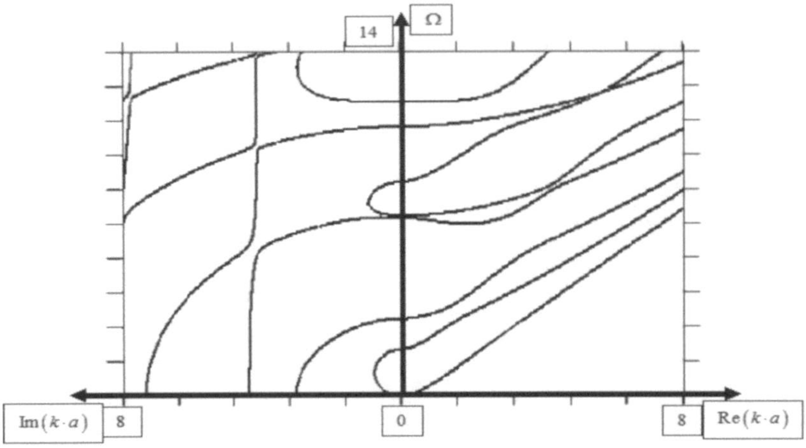

Fig. 1. PZT-4 cylinder with short-circuit lateral surface (m = 1)

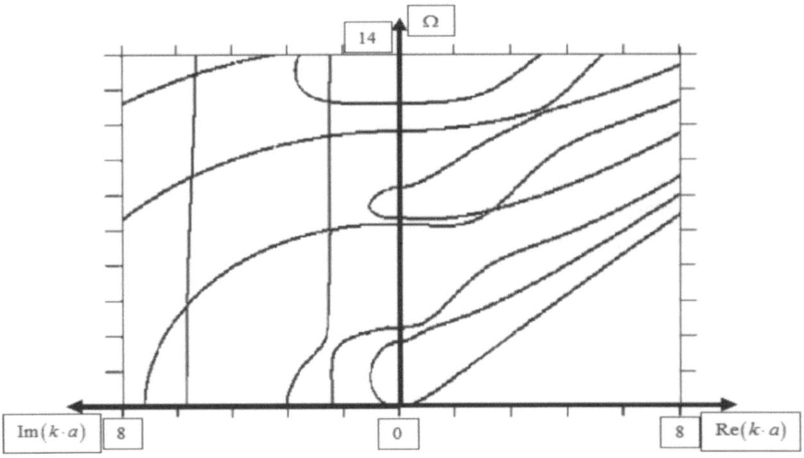

Fig. 2. PZT-4 cylinder with open-circuit lateral surface (m = 1)

Dispersion curves of bending waves $(m=1)$ in the cylinder made from PZT-4 with the open-circuit lateral surface are demonstrated in Fig. 2. It is obvious that the electric boundary conditions substantially influence both propagating and evanescent waves. For example, in the case of the open-circuit lateral surface the dispersion curves are even steeper than the corresponding curves in the case of the short-circuit lateral surface.

In Fig. 3 a conceptual case of reduced electro-mechanical coupling coefficients $\left(e_{15}=e_{31}=0,\ e_{33}=10^{-3}e_{33(PZT-4)}\right)$ is shown.

In Fig. 4 and 5 the dispersion curves of PZT-7A material are presented for the open- and short-circuit lateral surfaces respectively.

In comparison with PZT-4 this material has lower values of the electro-mechanical coupling coefficients but practically the same elastic coefficients and mass density.

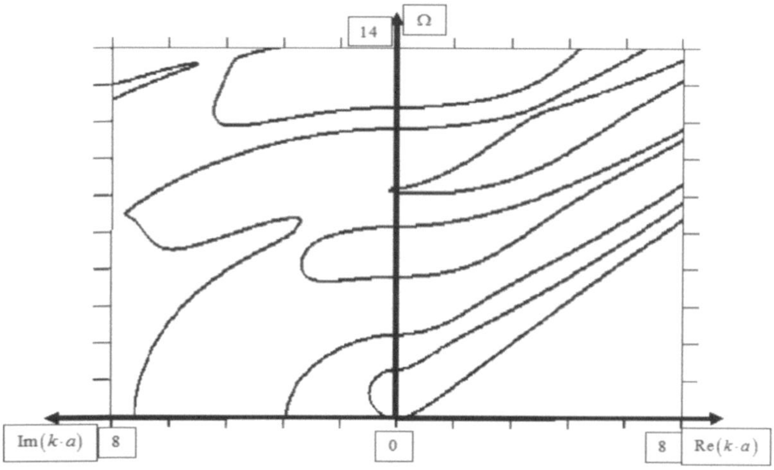

Fig. 3. PZT-4 cylinder with reduced electro-mechanical coupling
$\left(m=1,\ e_{15}=e_{31}=0,\ e_{33}=10^{-3}e_{33(PZT-4)}\right)$

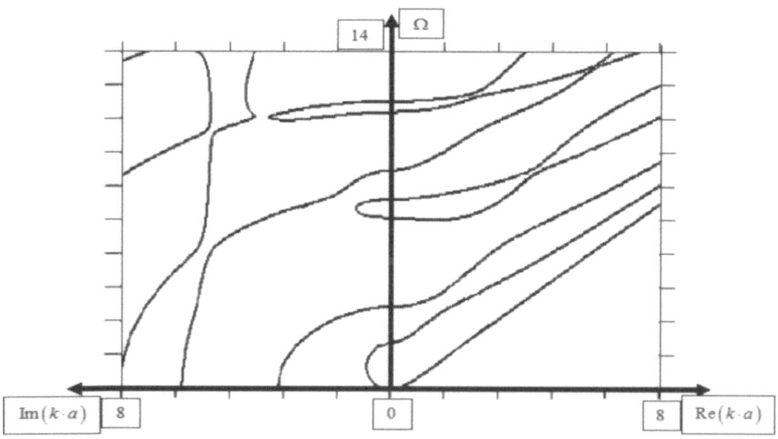

Fig. 4. PZT-7A cylinder with short-circuit lateral surface (m = 1)

It follows from Fig. 1 - 5 that the first fundamental mode of the bending waves is not sensitive to the nature of the electric boundary conditions on the lateral cylindrical surface. Furthermore it is practically not sensitive to the measure of electro-mechanical coupling of the material. The higher order modes are more sensitive to the nature of the electric boundary condition as well as to the measure of the electro-mechanical cross-coupling. Fig. 3 - 5 shows that dispersion curves differ quite substantially from the curves in Fig. 1 and 2. This difference is explained mainly by the factor that the electro-mechanical coupling coefficients of PZT-7A are less than the corresponding factors of PZT-4. It is reflected in undulating behaviour of the propagating higher modes as well as the values of the cut-off frequencies. The PZT-4 cylinder with short-circuit lateral surface demonstrates a negative slope of the fourth branch in a quite broad range of wavenumbers (Fig. 1). Substantial dependence of dispersion curves on electric boundary conditions is obvious from the behaviour of the curves for the evanescent waves.

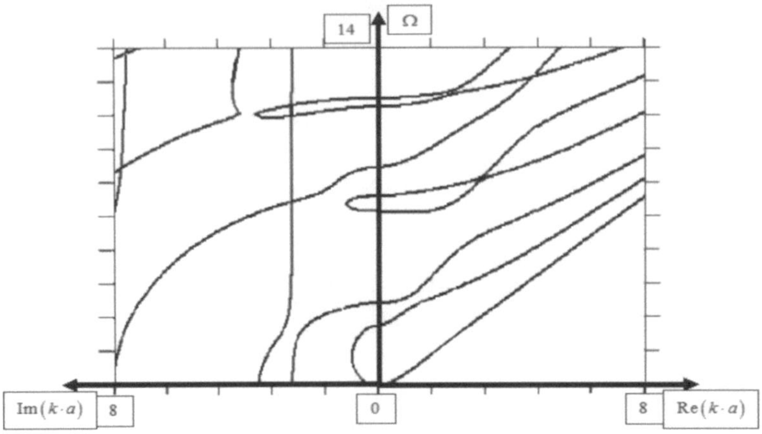

Fig. 5. PZT-7A cylinder with open-circuit lateral surface (m = 1)

Dispersion curves of non-axisymmetric waves with the circumferential wavenumber $m = 2$ in the cylinders made from PZT-4 and PZT-7A with the open- and close-circuit lateral surface are depicted in Fig. 6 - 9. Again as for the case $m = 1$ the substantial difference in the dispersion curves behaviour is explained by different types of the electric boundary conditions as well as by the difference in the electro-mechanic coupling coefficients.
Dispersion curves of non-axisymmetric waves with the circumferential wavenumber $m = 3$ in the cylinder made from PZT-4 and PZT-7A with the open- and close-circuit lateral surface are depicted in Fig. 10-13.
The dispersion curves of higher circumferential wavenumbers (m = 2, 3) are sensitive to the nature of the electric boundary condition as well as to the measure of the electro-mechanical cross-coupling for both propagating and evanescent waves. These dispersions curves obtained from the exact solution of the problem could be used as references data for developing of reliable finite elements for approximate solution of the problems of wave propagation in piezoelectric structures.

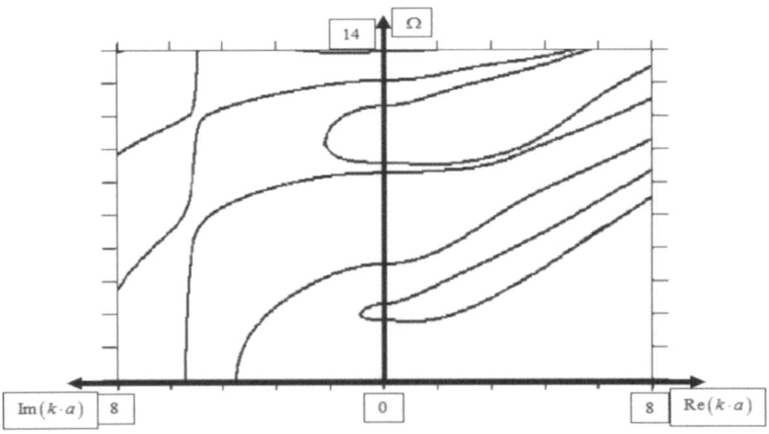

Fig. 6. PZT4 cylinder with short-circuit lateral surface (m = 2)

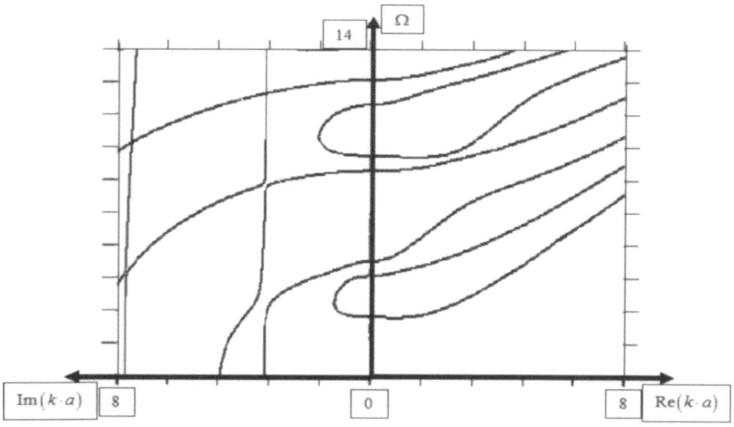

Fig. 7. PZT4 cylinder with open-circuit lateral surface (m = 2)

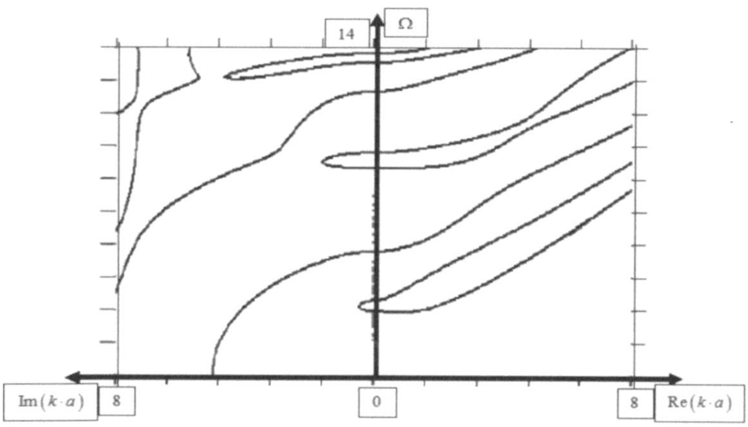

Fig. 8. PZT7A cylinder with short-circuit lateral surface (m = 2)

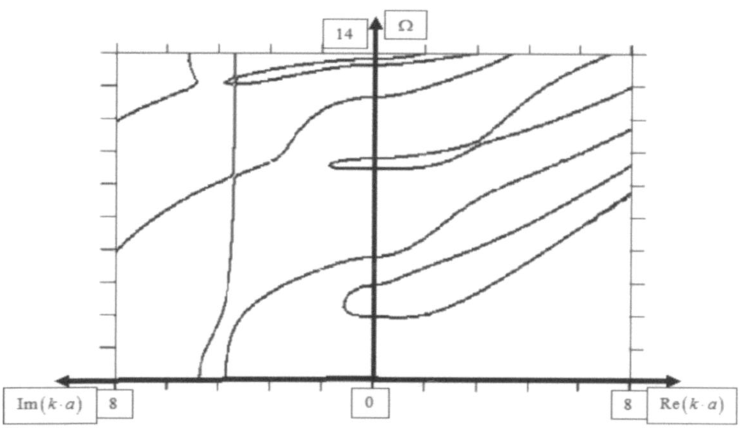

Fig. 9. PZT7A cylinder with open-circuit lateral surface (m = 2)

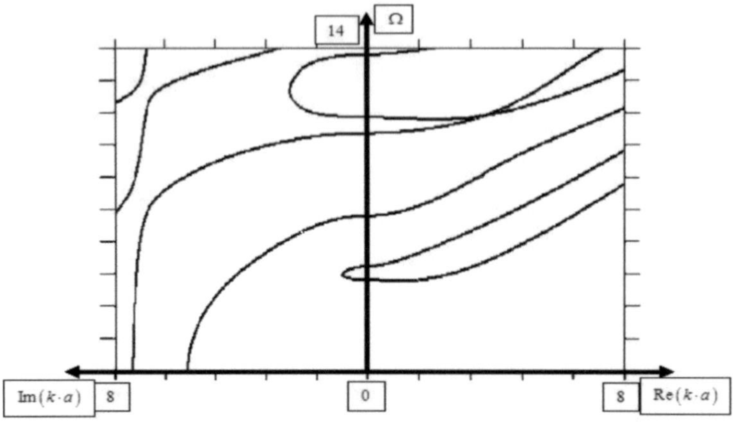

Fig. 10. PZT-4 cylinder with short-circuit lateral surface (m = 3)

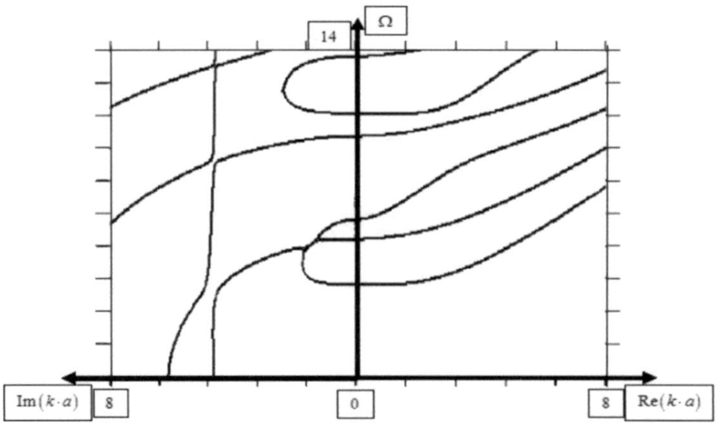

Fig. 11. PZT-4 cylinder with open-circuit lateral surface (m = 3)

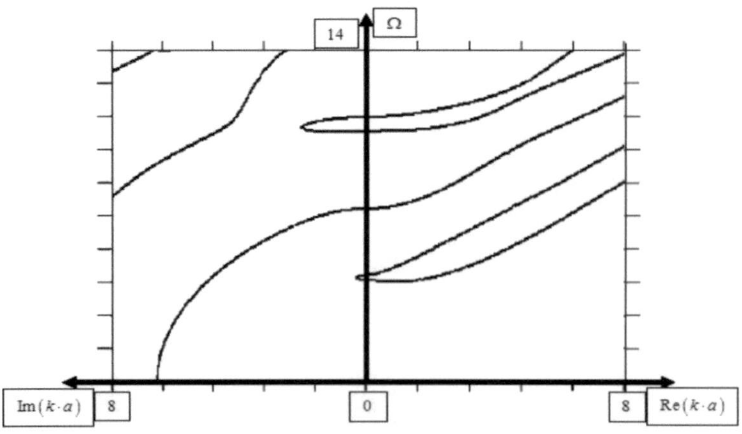

Fig. 12. PZT-7A cylinder with short-circuit lateral surface (m = 3)

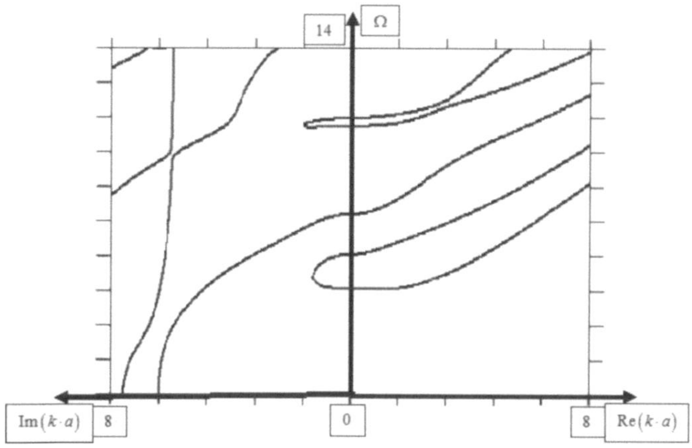

Fig. 13. PZT-7A cylinder with open-circuit lateral surface (m = 3)

Dispersion curves of axisymmetric waves in the cylinder made from PZT-4 and PZT-7A with the short-circuit and free lateral (cylindrical) surface are depicted in Fig. 14 - 17 in the same dimensionless coordinates $\left(k\cdot a \div \Omega = \omega/(a\cdot V_s)\right)$, where $V_s = \sqrt{c_{44}^E/\rho}$, and a is the outer radius of the cylinder. The picture of the dispersion curves is obtained by a method described above for real (propagating waves) and pure imaginary values of the wavenumber in the limits: $\mathrm{Re}(k\cdot a), \mathrm{Im}(k\cdot a) \in (0,8]$, $\Omega = \omega/(a\cdot V_s) \in (0,28]$ with resolution 500 (250 pixels for real and 250 pixels for imaginary $(k\cdot a)$) × 250 ($\Omega = \omega/(a\cdot V_s)$) pixels.

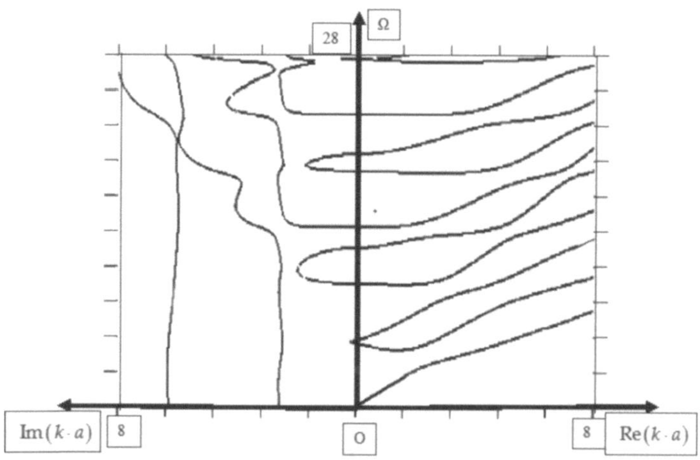

Fig. 14. Axisymmetric dispesion curves in PZT-4 cylinder with short-circuit lateral surface

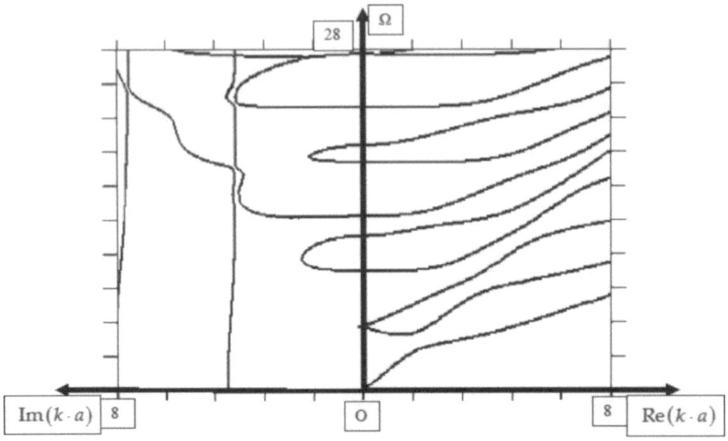

Fig. 15. Axisymmetric dispesion curves in PZT-4 cylinder with open-circuit lateral surface

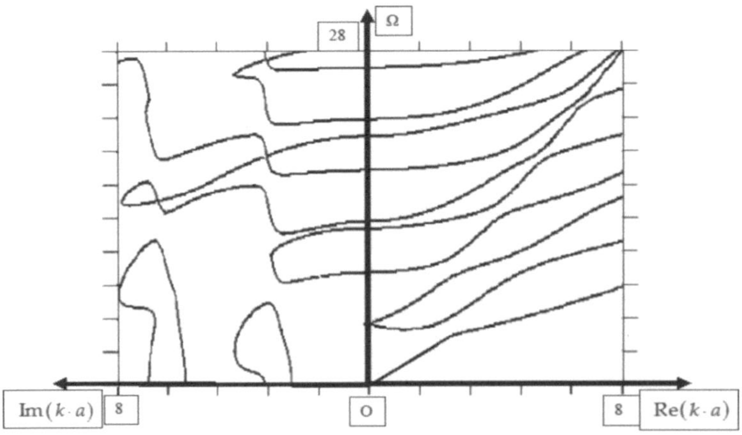

Fig. 16. Axisymmetric dispesion curves in PZT-7A cylinder with short-circuit lateral surface

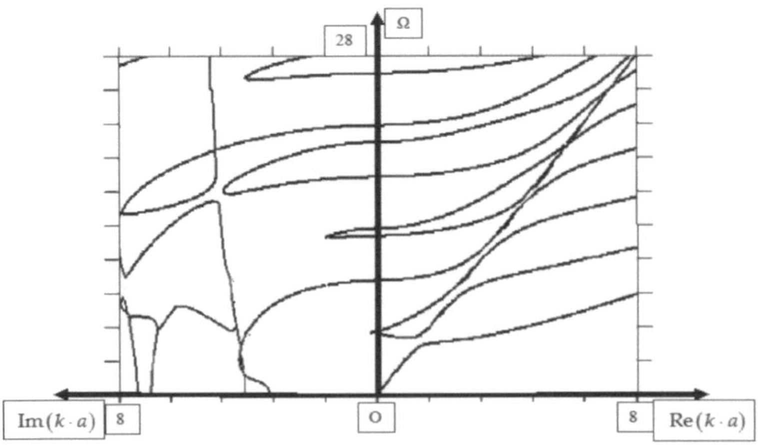

Fig. 17. Axisymmetric dispesion curves in PZT-7A cylinder with open-circuit lateral surface

Appendix

Coefficients of main determinant (38) for the case $\mathrm{Re}(\xi_j) \neq 0, \quad (j = 0,1,2,3)$:

$$a_{11} = \frac{2\xi_1 c_{66}^E}{a} J_{m+1}(\xi_1 a) + \left[ik(\eta_1 c_{13}^E + \mu_1 e_{13}) - \xi_1^2 c_{13}^E + \frac{2c_{66}^E}{a^2} m(m-1) \right] J_m(\xi_1 a),$$

$$a_{21} = -\left[(\eta_1 + ik) c_{44}^E + \mu_1 e_{15} \right] \left[\xi_1 J_{m+1}(\xi_1 a) - \frac{m}{a} J_m(\xi_1 a) \right],$$

$$a_{31} = \frac{2mc_{66}^E}{a} \left[\xi_1 J_{m+1}(\xi_1 a) - \frac{m-1}{a} J_m(\xi_1 a) \right],$$

$$a_{12} = \frac{2\xi_2 c_{66}^E}{a} J_{m+1}(\xi_2 a) + \left[ik(\eta_2 c_{13}^E + \mu_2 e_{13}) - \xi_2^2 c_{13}^E + \frac{2c_{66}^E}{a^2} m(m-1) \right] J_m(\xi_2 a),$$

$$a_{22} = -\left[(\eta_2 + ik) c_{44}^E + \mu_2 e_{15} \right] \left[\xi_2 J_{m+1}(\xi_2 a) - \frac{m}{a} J_m(\xi_2 a) \right],$$

$$a_{32} = \frac{2mc_{66}^E}{a} \left[\xi_2 J_{m+1}(\xi_2 a) - \frac{m-1}{a} J_m(\xi_2 a) \right],$$

$$a_{13} = \frac{2\xi_3 c_{66}^E}{a} J_{m+1}(\xi_3 a) + \left[ik(\eta_3 c_{13}^E + \mu_3 e_{13}) - \xi_3^2 c_{13}^E + \frac{2c_{66}^E}{a^2} m(m-1) \right] J_m(\xi_3 a),$$

$$a_{23} = -\left[(\eta_3 + ik) c_{44}^E + \mu_3 e_{15} \right] \left[\xi_3 J_{m+1}(\xi_3 a) - \frac{m}{a} J_m(\xi_3 a) \right],$$

$$a_{33} = \frac{2mc_{66}^E}{a} \left[\xi_3 J_{m+1}(\xi_3 a) - \frac{m-1}{a} J_m(\xi_3 a) \right],$$

$$a_{14} = -\frac{2mc_{66}^E}{a} \left[\xi_0 J_{m+1}(\xi_0 a) - \frac{m-1}{a} J_m(\xi_0 a) \right],$$

$$a_{24} = -\frac{imkc_{44}^E}{a} J_m(\xi_0 a),$$

$$a_{34} = -\frac{2c_{66}^E}{a} \left[\xi_0 J_{m+1}(\xi_0 a) - \left(\frac{\xi_0^2}{2} - \frac{m(m-1)}{a^2} \right) J_m(\xi_0 a) \right].$$

For electric boundary condition $D_1|_{r=a} = 0$, $\left(\mathrm{Re}(\xi_j) \neq 0\right)$:

$$a_{41} = -\left[(\eta_1 + ik) e_{15} - \mu_1 \varepsilon_{11}^S \right] \left[\xi_1 J_{m+1}(\xi_1 a) - \frac{m}{a} J_m(\xi_1 a) \right],$$

$$a_{42} = -\left[(\eta_2 + ik) e_{15} - \mu_2 \varepsilon_{11}^S \right] \left[\xi_2 J_{m+1}(\xi_2 a) - \frac{m}{a} J_m(\xi_2 a) \right],$$

$$a_{43} = -\left[(\eta_3 + ik)e_{15} - \mu_3\varepsilon_{11}^S\right]\left[\xi_3 J_{m+1}(\xi_3 a) - \frac{m}{a}J_m(\xi_3 a)\right],$$

$$a_{44} = -\frac{imke_{15}}{a}J_m(\xi_0 a).$$

For electric boundary condition $\phi|_{r=a} = 0$, $\left(\mathrm{Re}(\xi_j) \neq 0\right)$:

$$a_{41} = \mu_1 J_m(\xi_1 a),\quad a_{42} = \mu_2 J_m(\xi_2 a),\quad a_{43} = \mu_3 J_m(\xi_3 a),\quad a_{44} = 0.$$

Coefficients of main determinant (38) for the case $\mathrm{Re}(\xi_j) = 0,\ (j = 0,1,2,3)$:

$$a_{11} = -\frac{2|\xi_1|c_{66}^E}{a}I_{m+1}(|\xi_1|a) + \left[ik\left(\eta_1 c_{13}^E + \mu_1 e_{13}\right) + |\xi_1|^2 c_{13}^E + \frac{2c_{66}^E}{a^2}m(m-1)\right]I_m(|\xi_1|a),$$

$$a_{21} = \left[(\eta_1 + ik)c_{44}^E + \mu_1 e_{15}\right]\left[|\xi_1| I_{m+1}(|\xi_1|a) + \frac{m}{a}I_m(|\xi_1|a)\right],$$

$$a_{31} = -\frac{2mc_{66}^E}{a}\left[|\xi_1| I_{m+1}(|\xi_1|a) + \frac{m-1}{a}I_m(|\xi_1|a)\right],$$

$$a_{12} = -\frac{2|\xi_2|c_{66}^E}{a}I_{m+1}(|\xi_2|a) + \left[ik\left(\eta_2 c_{13}^E + \mu_2 e_{13}\right) + |\xi_2|^2 c_{13}^E + \frac{2c_{66}^E}{a^2}m(m-1)\right]I_m(|\xi_2|a),$$

$$a_{22} = \left[(\eta_2 + ik)c_{44}^E + \mu_2 e_{15}\right]\left[|\xi_2| I_{m+1}(|\xi_2|a) + \frac{m}{a}I_m(|\xi_2|a)\right],$$

$$a_{32} = -\frac{2mc_{66}^E}{a}\left[|\xi_2| I_{m+1}(|\xi_2|a) + \frac{m-1}{a}I_m(|\xi_2|a)\right],$$

$$a_{13} = -\frac{2|\xi_3|c_{66}^E}{a}I_{m+1}(|\xi_3|a) + \left[ik\left(\eta_3 c_{13}^E + \mu_3 e_{13}\right) + |\xi_3|^2 c_{13}^E + \frac{2c_{66}^E}{a^2}m(m-1)\right]I_m(|\xi_3|a),$$

$$a_{23} = \left[(\eta_3 + ik)c_{44}^E + \mu_3 e_{15}\right]\left[|\xi_3| I_{m+1}(|\xi_3|a) + \frac{m}{a}I_m(|\xi_3|a)\right],$$

$$a_{33} = -\frac{2mc_{66}^E}{a}\left[|\xi_3| I_{m+1}(|\xi_3|a) + \frac{m-1}{a}I_m(|\xi_3|a)\right],$$

$$a_{14} = \frac{2mc_{66}^E}{a}\left[|\xi_0| I_{m+1}(|\xi_0|a) + \frac{m-1}{a}I_m(|\xi_0|a)\right],$$

$$a_{24} = \frac{imkc_{44}^E}{a}I_m(|\xi_0|a),$$

$$a_{34} = \frac{2c_{66}^E}{a}\left[|\xi_0| I_{m+1}(|\xi_0|a) - \left(\frac{|\xi_0|^2}{2} + \frac{m(m-1)}{a^2}\right)I_m(|\xi_0|a)\right].$$

For electric boundary condition $D_1\big|_{r=a}=0$, $\left(\operatorname{Re}\left(\xi_j\right)=0\right)$:

$$a_{41}=\left[\left(\eta_1+ik\right)e_{15}-\mu_1\varepsilon_{11}^S\right]\left[\left|\xi_1\right|I_{m+1}\left(\left|\xi_1\right|a\right)+\frac{m}{a}I_m\left(\left|\xi_1\right|a\right)\right],$$

$$a_{42}=\left[\left(\eta_2+ik\right)e_{15}-\mu_2\varepsilon_{11}^S\right]\left[\left|\xi_2\right|I_{m+1}\left(\left|\xi_2\right|a\right)+\frac{m}{a}I_m\left(\left|\xi_2\right|a\right)\right],$$

$$a_{43}=\left[\left(\eta_3+ik\right)e_{15}-\mu_3\varepsilon_{11}^S\right]\left[\left|\xi_3\right|I_{m+1}\left(\left|\xi_3\right|a\right)+\frac{m}{a}I_m\left(\left|\xi_3\right|a\right)\right],$$

$$a_{44}=\frac{imke_{15}}{a}I_m\left(\left|\xi_0\right|a\right).$$

For electric boundary condition $\phi\big|_{r=a}=0$, $\left(\operatorname{Re}\left(\xi_j\right)=0\right)$:

$$a_{41}=\mu_1 I_m\left(\left|\xi_1\right|a\right),\quad a_{42}=\mu_2 I_m\left(\left|\xi_2\right|a\right),\quad a_{43}=\mu_3 I_m\left(\left|\xi_3\right|a\right),\quad a_{44}=0.$$

In the axisymmetric case we assume $m=0$ in the above mentioned expressions and remove third row and fourth column in determinant (37).

7. Conclusion

The characteristic equation of non-axisymmetric propagating and evanescent waves of a piezoelectric cylinder of transversely isotropic material was developed. The results were numerically illustrated for sample PZT-4 and PZT-7A cylinders for the first three circumferential wavenumbers (m = 1, 2, 3) and for the axisymmetric case. It was shown that the dispersion curves are sensitive both to the electric boundary conditions and to the measure of electro-mechanical coupling. This effect was revealed more strongly in the higher order modes.

8. Acknowledgements

I acknowledge the support of my co-authors Prof. Arthur G. Every and our student Alfred S. Yenwong-Fai participating in the investigation of the non-axisymmetric case of the piezoelectric cylinder vibrations (Shatalov, et al. 2009). I also want to thank Mr. Yuri M. Shatalov who investigated the axisymmetric case under my supervision.

9. References

Achenbach, J. (1984). *Wave Propagation in Elastic Solids*, New York, North-Holland

Armenakas, A. & Reitz, E. (1973). Propagation of harmonic waves in orthotropic circular cylindrical shells, *J. Appl. Mech.* 40, 168-174

Bai, H.; Shah, A.; Dong, S. & Taciroglu, E. (2006). End reflections in layered piezoelectric circular cylinder. *International Journal of Solids and Structures*, 43, 6309-6325.

Bai, H.; Taciroglu; E., Dong, S. & Shah, A., (2004). Elastodynamic Green's function for a laminated piezoelectric cylinder, *International Journal of Solids and Struc*tures, 41, 6335–6350

Berliner, M. & Solecki, R. (1996). Wave propagation in fluid-loaded transversely isotropic cylinders. Part 1. Analytical formulation, *J. Acoust. Soc. Am.*, 99, 1841-1847

Chree, C. (1890). *Q. J. Math.*, 24, 340-354.

Crandall, S.; Karnop, E.; Kurtz, J. & Pridemore-Brown, D. (1968). *Dynamics of Mechanical and Electromechanical Systems*, Krieger Publishing Co, Malabar, Florida

Every A., & Neiman V. (1992). Reflection of electroacoustic waves in piezoelectric solids: Mode conversion into four bulk waves, *J. Appl. Phys.* 71, 6018-6024

Frazer, W. (1980). Separable equations for a cylindrical anisotropic elastic waveguide, *J. Sound Vib.* 72, 151-157

Graff, K. (1991). Wave *Motion in Elastic Solids*, New York: Dover

Hagood, N.; Chung, W. & Flotow, A. (1990). Modelling of piezoelectric actuator dynamics for active structural control, *J. of Intell. Mater. Syst. And Struct.* 1 (3), 327-354

Honarvar, F.; Enjilela, E. & Sinclair, A. (2008). An alternative method for plotting dispersion curves, *Ultrasonics*, doi:10.1016/ j.ultras.2008.07.002

Honarvar, F.; Enjilela, E.; Sinclair, A. & Minerzami, S. (2007). Wave propagation in transversely isotropic cylinders, *International Journal of Solids and Structures*, 44, 5236-5246

Mirsky, I. (1964). Wave propagation in transversely isotropic circular cylinders. Part 1: Theory, *J. Acoust. Soc. Am.* 36, 2106-2122

Nayfeh, A.; Abdelrahman, W. & Nagy, P. (2000). Analysys of axisymmetric waves in layered piezoelectric rods and their composites, *J. Acoust. Soc. Am.* 108 (4), 1496-1504

Nayfeh. A. & Nagy, P. (1995). General study of axisymmetric waves in layered anisotropic fibres and their composites, *J. Acoust. Soc. Am.*, 99, 931-941

Niklasson, A. & Datta, S. (1998). Scattering by an infinite transversely isotropic cylinder in a transversely isotropic medium, *Wave motion*, 27 (2), 169-185

Paul, H. (1966). Vibrations of circular cylindrical shells of piezoelectric silver iodide crystals, *J. Acoust. Soc. Am.* 40, 1077-1080

Pochhammer, L. (1876). *J. reine angew*, Math. 81, 324

Rose, J. (1999). *Ultrasonic Waves in Solid Media*, Cambridge University Press

Shatalov, M. & Loveday, P. (2004). Electroacoustic wave propagation in transversely isotropic piezoelectric cylinders, *Proceedings of the South African Conference on Applied Mechanics (SACAM04)*

Shatalov, et al. (2009). Analysis of non-axisymmetric wave propagation in a homogeneous piezoelectric solid circular cylinder of transversely isotropic material, *International Journal of Solids and Structures* 46, 837-850

Siao, J.; Dong, S. & Song, J. (1994). Frequency spectra of laminated piezoelectric cylinders, *ASME Journal of Vibrations and Acoustics*, 116, 364-370

Wei, J., Su, X., 2005. Wave propagation in a piezoelectric rod of 6mm symmetry. International Journal of Solids and Structures, 42, 3644-3654

Winkel, V.; Oliveira, J.; Dai, J. & Jen, C. (1995). Acoustic wave propagation in piezoelectric fibres of hexagonal crystall symmetry, *IEEE Trans. Ultrason. Ferroelectr. Freq. Control*, 42, 949-955

Xu, P. & Datta, S. (1991). Characterization of fibre-matrix interface by guided waves: Axisymmetric case, *J. Acoust. Soc. Am*, 89 (6), 2573-2583

Yenwong-Fai, A. (2008). Wave propagation in a piezoelectric solid cylinder of transversely isotropic material, *Master's thesis*, University of Witwatersrand, Johannesburg, South Africa

18

Propagation of Thickness-Twist Waves in an Infinite Piezoelectric Plate

Zheng-Hua Qian[1], Feng Jin[2], Jiashi Yang[3] and Sohichi Hirose[1]
[1]*Tokyo Institute of Technology,*
[2]*Xi'an Jiaotong University,*
[3]*University of Nebraska-Lincoln,*
[1]*Japan*
[2]*China*
[3]*USA*

1. Introduction

Thickness-twist waves in piezoelectric plates are widely used as the operating modes for acoustic wave resonators, filters, sensors, and other devices (Mindlin, 1965; Mindlin, 1967; Pearman, 1968; Bleustein, 1966; Bleustein, 1969; Yang et al., 2008a; Yang & Guo, 2006). Acoustic waves in these devices are driven by an alternating voltage applied to the electrodes deposited on the plates. Therefore there has been continuing research effort on the effect of electrodes (Ballato & Lukaszek, 1974; Kosinski, 2003; Wang & Shen, 2005; Yang et al., 2005). Recently, due to the need of device miniaturization, there has been growing research interest on electrode configurations. Electrodes of varying thickness have been used to adjust the vibration distribution in plates (Pao et al., 2007; Yang et al., 2007a; Wang et al., 2008). Electrode shape has also been analyzed for design optimization. In Yang et al. (2007b) it was shown that electrodes with corners cause field concentration and should be avoided in general. Optimal electrode size and shape were determined in Mindlin (1968) and Yang et al. (2008b). On the other hand, a vibrating elastic (or piezoelectric) body when put in contact with a viscous fluid changes its resonant frequencies due to the inertia and viscosity of the fluid. Equivalently, the speed of a propagating wave in a body in contact with a fluid is also affected by the fluid. This effect has been used to make various fluid sensors for measuring fluid viscosity or density (Kanazawa & Gordon, 1985; Josse et al., 1990; Reed et al., 2001; Kim et al., 1991; Vogt et al., 2004; Guo & Sun, 2008; Peng et al., 2006). More references can be found in a review article (Benes et al., 1995). For fluid sensor applications, vibration modes of a body without a normal displacement at its surface are of general interest. Thickness-shear and thickness-twist modes in a plate (Kanazawa & Gordon, 1985; Josse et al., 1990; Reed et al., 2001), torsional modes of a circular shaft (Kim et al., 1991; Vogt et al., 2004), and anti-plane surface waves (Guo & Sun, 2008; Peng et al., 2006) are modes with tangential surface displacements only and have been used for fluid sensor applications. To establish the relation between wave frequency and fluid density or viscosity, a coupled problem of fluid-structure interaction needs to be solved. This usually presents complicated mathematical problems. So far only limited theoretical results have been obtained.

In the devices studied in the above references, the electrodes are directly deposited on or attached to the surfaces of the plates. In device applications, unattached electrodes at some distance from the plate surfaces have been occasionally used (Detaint et al., 1989; Huang et al., 2001; Umeki et al., 2004). Unattached or separated electrodes offer new design parameters and certain advantages over electrodes deposited on the plate surfaces. Various undesirable effects of deposited (or attached) electrodes disappear when the electrodes are unattached. This includes, e.g., the residual stress in the electrodes and the plates, the frequency effect of electrode irregularity from manufacturing, and the field concentration at electrode edges. There are limited theoretical results on waves in piezoelectric plates with unattached electrodes. Plates of general material anisotropy with unattached electrodes were considered in Syngellakis and Lee (1993) and the results are rather complicated due to strong anisotropy. Numerical techniques had to be used to obtain quantitative results which were only given for attached electrodes. An approximate theoretical analysis on vibrations of a finite plate with unattached electrodes was given in Tiersten (1995). Pure thickness-shear modes (waves propagating in the plate thickness direction bouncing back and forth between the two surfaces) were studied in Yang et al. (2009).

In this Chapter, we study a simple and yet very useful case of thickness-twist waves propagating in an infinite piezoelectric plate with unattached electrodes. A theoretical analysis is performed using the exact equations of piezoelectricity. In order to make readers understand the problems under question easily, we first introduce the propagation of thickness-twist waves in an infinite piezoelectric plate with attached very thin electrodes. This part is mainly a reproduction of the work by Bleustein (1969), which can be used as a basic starting for those who want to conduct the research work on thickness-twist waves in an infinite piezoelectric plate. Some basic knowledge concerning thickness-twist waves and linear piezoelectricity will be introduced in this part. Then we analyze the propagation of thickness-twist waves in an infinite piezoelectric plate with air gaps between the plate surfaces and two electrodes. Dispersion relations of the waves are obtained and plotted. Results show that the wave frequency or speed is sensitive to the air gap thickness. This effect can be used to manipulate the behavior of the waves and has implications in acoustic wave devices. Last we study theoretically the propagation of thickness-twist waves in an infinite piezoelectric plate with unattached electrodes and viscous fluids between the plate surfaces and the electrodes. Based on the theories of linear piezoelectricity and viscous fluids, an equation that determines the dispersion relations of the waves is obtained, showing the dependence of the phase velocity on material and geometric parameters. Due to the viscosity of the fluid, the dispersion relations are complex in general, representing damped waves with attenuation. The dispersion relations obtained can reduce to the results of a few special cases with known results. The results are useful for developing and designing fluid sensors for measuring fluid viscosity or density.

2. Governing equations

2.1 Piezoelectric plate with attached electrodes

Consider the infinite piezoelectric plate shown in Fig. 1. The ceramic plate is poled in the x_3 direction determined by the right-hand rule from the x_1 and x_2 axes. The electrodes are shorted. The structure allows the following anti-plane motion (Bleustein, 1969)

$$u_1 = u_2 = 0\,,\; u_3 = u(x_1, x_2, t)\,, \tag{1}$$

$$\varphi = \varphi(x_1, x_2, t)\,, \tag{2}$$

where u is the displacement vector and φ is the electric potential.

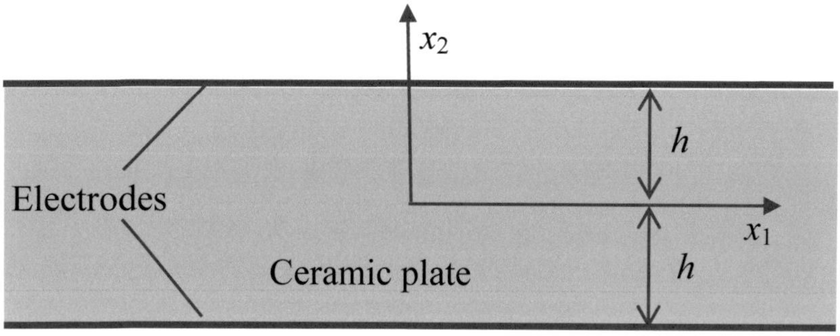

Fig. 1. A piezoelectric plate with attached electrodes

The nontrivial strain and electric field components are

$$\begin{Bmatrix} 2S_{13} \\ 2S_{23} \end{Bmatrix} = \nabla u, \quad \begin{Bmatrix} E_1 \\ E_2 \end{Bmatrix} = -\nabla \varphi\,, \tag{3}$$

where $\nabla = \mathbf{i}_1\partial_1 + \mathbf{i}_2\partial_2$ is the 2-D gradient operator. E is the electric field. The nontrivial components of the stress tensor T_{ij} and the electric displacement vector D_i are

$$\begin{Bmatrix} T_{13} \\ T_{23} \end{Bmatrix} = c_{44}\nabla u + e_{15}\nabla \phi, \quad \begin{Bmatrix} D_1 \\ D_2 \end{Bmatrix} = e_{15}\nabla u - \varepsilon_{11}\nabla \varphi, \tag{4}$$

where c_{44}, e_{15} and ε_{11} are the relevant elastic, piezoelectric and dielectric constants.

A ψ can be introduced[5] through $\varphi = \psi + eu/\varepsilon$, where $e = e_{15}$ and $\varepsilon = \varepsilon_{11}$ are the relevant piezoelectric and dielectric constants. The governing equations for u and ψ in the piezoelectric plate are

$$\left.\begin{aligned} \bar{c}\nabla^2 u &= \rho \ddot{u} \\ \nabla^2 \psi &= 0 \end{aligned}\right\}, \tag{5}$$

where $\bar{c} = c + e^2/\varepsilon$ and $c = c_{44}$ is the relevant elastic constants. ∇^2 is the 2-D Laplacian.

2.2 Piezoelectric plate with unattached electrodes

Consider the infinite piezoelectric plate with unattached electrodes and asymmetric air gaps, as shown in Fig. 2. The piezoelectric plate is the same as that in Fig. 1. The governing equations for thickness-twist wave in such configuration are

$$\left.\begin{aligned} \bar{c}\nabla^2 u &= \rho \ddot{u} \\ \nabla^2 \psi &= 0 \end{aligned}\right\}, \; -h \le x_2 \le h \tag{6}$$

$$\nabla^2 \varphi^{\mathrm{T}} = 0\,,\; h \le x_2 \le h + g^{\mathrm{T}}\,, \tag{7}$$

$$\nabla^2 \varphi^B = 0\,, \qquad -h - g^B \le x_2 \le -h\,, \tag{8}$$

where φ^T and φ^B are the electric potentials in the top and bottom gaps, respectively.

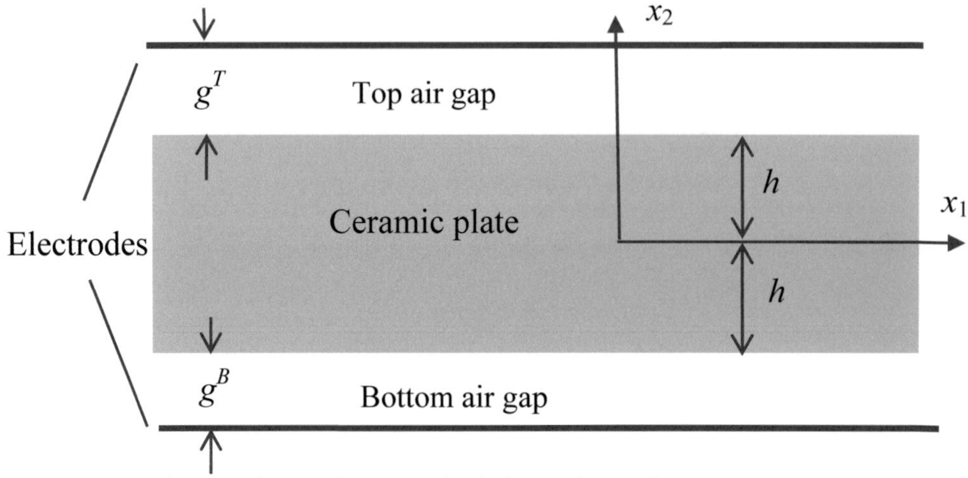

Fig. 2. A piezoelectric plate with unattached electrodes and asymmetric air gaps

2.3 Piezoelectric plate in contact with viscous fluids

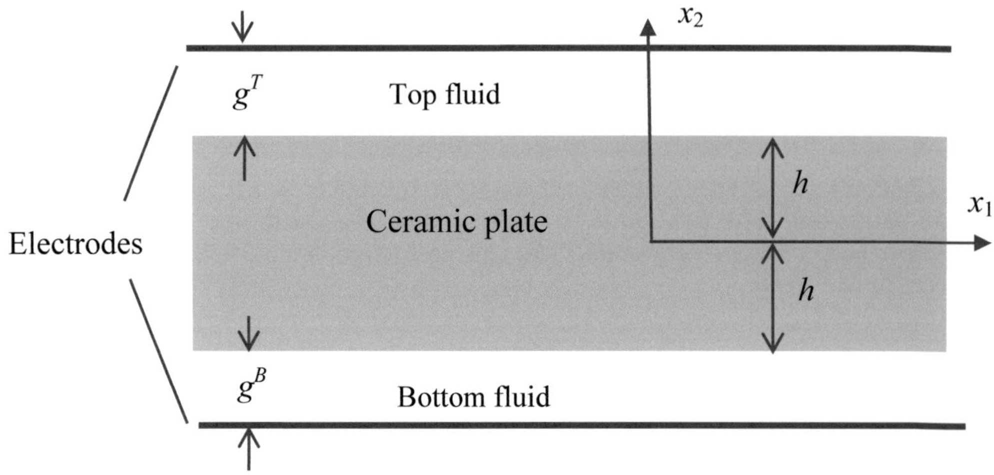

Fig. 3. A piezoelectric plate with unattached electrodes and gaps filled with fluids

Consider the infinite piezoelectric plate with unattached electrodes, as shown in Fig. 3. There are two gaps between the electrodes and the plate. In our problem the gaps are filled with two different viscous fluids. This structure includes the cases when the electrodes are on the plate surfaces or at infinity as special cases. The piezoelectric plate is the same as that in Figs. 1 and 2. The governing equations for thickness-twist wave in such configuration are

$$\begin{matrix} \bar{c}\nabla^2 u = \rho \ddot{u} \\ \nabla^2 \psi = 0 \end{matrix} \quad -h < x_2 < h\,, \tag{9}$$

$$\begin{aligned} &\frac{\partial v^{\mathrm{T}}}{\partial t} - \frac{\mu^{\mathrm{T}}}{\rho^{\mathrm{T}}}\nabla^2 v^{\mathrm{T}} = 0 \\ &\nabla^2 \varphi^{\mathrm{T}} = 0 \end{aligned} \qquad h \le x_2 \le h + g^{\mathrm{T}}, \tag{10}$$

$$\begin{aligned} &\frac{\partial v^{\mathrm{B}}}{\partial t} - \frac{\mu^{\mathrm{B}}}{\rho^{\mathrm{B}}}\nabla^2 v^{\mathrm{B}} = 0 \\ &\nabla^2 \varphi^{\mathrm{B}} = 0 \end{aligned} \qquad -h - g^{\mathrm{B}} \le x_2 \le -h, \tag{11}$$

where v^T and v^B are the fluid velocity fields, ρ^T and ρ^B are the fluid mass densities, and μ^T and μ^B are the fluid viscosities. The superscripts "T" and "B" are for the top and bottom fluids, respectively.

3. Propagating wave solution and discussion

3.1 Piezoelectric plate with attached electrodes

The wave propagation specified by Eq. (5) must satify the following boundary conditions

$$\varphi = 0, \quad T_{23} = 0, \quad x_2 = \pm h. \tag{12}$$

Consider the following propagating waves in the plate

$$\begin{aligned} u &= \left(A_1 \sin \xi_2 x_2 + A_2 \cos \xi_2 x_2\right) \cos\left(\xi_1 x_1 - \omega t\right), \\ \psi &= \left(A_3 \sinh \xi_1 x_2 + A_4 \cosh \xi_1 x_2\right) \cos\left(\xi_1 x_1 - \omega t\right), \end{aligned} \tag{13}$$

where A_1 through A_4 are undetermined constants, ω is the wave frequency, and ξ_1 and ξ_2 are waves numbers in the x_1 and x_2 directions. Eq. (13) satisfies Eq. (5) when

$$\xi_2^2 = \xi_1^2\left(\frac{v^2}{v_{\mathrm{T}}^2} - 1\right), \tag{14}$$

where the wave speed v is given by $\xi_1^2 = \omega^2 / v^2$ and $v_{\mathrm{T}}^2 = \bar{c} / \rho$ is the speed of plane shear waves propagating in the x_1 direction.

Substitution of Eq. (13) into the boundary conditions at $x_2 = \pm h$ in Eq. (12) yields four linear, homogeneous algebraic equations for A_1 through A_4. For nontrivial solutions of the undetermined constants, the determinant of the coefficient matrix of the linear equations has to vanish, which leads to the following dispersion relation of the waves

$$\frac{\tan \xi_2 h}{\tanh \xi_1 h} = \bar{k}^{-2}\frac{\xi_2}{\xi_1}, \text{ (AS)} \tag{15}$$

$$\frac{\tan \xi_2 h}{\tanh \xi_1 h} = -\bar{k}^{2}\frac{\xi_1}{\xi_2}, \text{ (S)} \tag{16}$$

where AS denotes the antisymmetric mode and S the symmetric mode.

3.2 Piezoelectric plate with unattached electrodes

The wave propagation specified by Eqs. (6-8) must satisfy the following boundary and continuity conditions

$$\begin{aligned} &\varphi^{T}=0, \quad x_2=h+g^{T}, \\ &\varphi=\varphi^{T}, \quad D_2=D_2^{T}, \quad T_{23}=0, \quad x_2=h, \\ &\varphi=\varphi^{B}, \quad D_2=D_2^{B}, \quad T_{23}=0, \quad x_2=-h, \\ &\varphi^{B}=0, \quad x_2=-h-g^{B}. \end{aligned} \tag{17}$$

Consider the following propagating waves in the plate

$$\begin{aligned} u&=(A_1\sin\xi_2 x_2+A_2\cos\xi_2 x_2)\cos(\xi_1 x_1-\omega t), \\ \psi&=(A_3\sinh\xi_1 x_2+A_4\cosh\xi_1 x_2)\cos(\xi_1 x_1-\omega t), \end{aligned} \tag{18}$$

and the waves in the gaps

$$\varphi^{T}=A_5\sinh\xi_1(x_2-h-g^{T})\cos(\xi_1 x_1-\omega t), \tag{19}$$

$$\varphi^{B}=A_6\sinh\xi_1(x_2+h+g^{B})\cos(\xi_1 x_1-\omega t), \tag{20}$$

where A_1 through A_4 are undetermined constants, ω is the wave frequency, ξ_1 and ξ_2 are defined in Eq. (14). A_5 and A_6 are also undetermined constants. Eqs. (19) and (20) satisfy Eqs. (7) and (8), respectively, as well as the boundary conditions at the top and bottom electrodes, i.e., the first and last equations in Eq. (17).

Substitution of Eqs. (18), (19) and (20) into the remaining boundary and continuity conditions at $x_2 = \pm h$ in Eq. (17) yields six linear, homogeneous algebraic equations for A_1 through A_6. For nontrivial solutions of the undetermined constants, the determinant of the coefficient matrix of the linear equations has to vanish, which leads to the following dispersion relation of the waves:

$$\begin{aligned} &\frac{n^2\tanh\xi_1 g^{T}+\tanh\xi_1 h-\bar{k}^2\frac{\xi_1}{\xi_2}\tan\xi_2 h}{n^2\tanh\xi_1 g^{B}+\tanh\xi_1 h-\bar{k}^2\frac{\xi_1}{\xi_2}\tan\xi_2 h} \\ &\qquad=-\frac{n^2\tanh\xi_1 g^{T}+\mathrm{ctanh}\,\xi_1 h+\bar{k}^2\frac{\xi_1}{\xi_2}\mathrm{ctan}\,\xi_2 h}{n^2\tanh\xi_1 g^{B}+\mathrm{ctanh}\,\xi_1 h+\bar{k}^2\frac{\xi_1}{\xi_2}\mathrm{ctan}\,\xi_2 h}, \end{aligned} \tag{21}$$

where $n^2 = \varepsilon/\varepsilon_0$ is the refractive index in the x_1 direction, ε_0 is the permittivity of free space, and $\bar{k}^2=e^2/(\varepsilon\bar{c})$ is a dimensionless number (electromechanical coupling factor). With Eq. (14), Eq. (21) determines v versus ξ_1 or ω versus ξ_1.

We examine some special cases below. When the gaps have the same thickness, i.e., $g^T=g^B=g$, Eq. (11) factors into two equations that determine two groups of waves. One may be called anti-symmetric waves and the other symmetric. The displacement u of the anti-symmetric and symmetric waves are odd and even functions of x_2, respectively. The corresponding dispersion relations are

$$n^2\tanh\xi_1 g+\tanh\xi_1 h-\bar{k}^2\frac{\xi_1}{\xi_2}\tan\xi_2 h=0, \quad \text{(AS)} \tag{22}$$

$$n^2 \tanh \xi_1 g + \operatorname{ctanh} \xi_1 h + \bar{k}^2 \frac{\xi_1}{\xi_2} \operatorname{ctan} \xi_2 h = 0 \,. \text{ (S)} \tag{23}$$

In particular, if the gaps are not present, i.e., when $g^T=g^B= 0$, Eqs. (22) and (23) reduce exactly to Eqs. (15) and (16).

3.3 Piezoelectric plate in contact with viscous fluids

The wave propagation specified by Eqs. (9-11) must satify the following boundary and continuity conditions

$$\begin{aligned}
&\varphi^{\mathrm{T}} = 0, \quad v^{\mathrm{T}} = 0 \quad \text{at} \quad x_2 = h + g^{\mathrm{T}}, \\
&\varphi = \varphi^{\mathrm{T}}, \quad D_2 = D_2^{\mathrm{T}}, \quad T_{23} = T_{23}^{\mathrm{T}}, \quad v = v^{\mathrm{T}} \quad \text{at} \quad x_2 = h, \\
&\varphi = \varphi^{\mathrm{B}}, \quad D_2 = D_2^{\mathrm{B}}, \quad T_{23} = T_{23}^{\mathrm{B}}, \quad v = v^{\mathrm{B}} \quad \text{at} \quad x_2 = -h, \\
&\varphi^{\mathrm{B}} = 0, \quad v^{\mathrm{B}} = 0 \quad \text{at} \quad x_2 = -h - g^{\mathrm{B}}.
\end{aligned} \tag{24}$$

We look for solutions representing waves propagating in the x_1 direction in the following form

$$\begin{aligned}
u_3 &= \left(A_1 \sin \xi_2 x_2 + A_2 \cos \xi_2 x_2\right) \exp \mathrm{i}\left(\xi_1 x_1 - \omega t\right) \\
\psi &= \left(A_3 \sinh \xi_1 x_2 + A_4 \cosh \xi_1 x_2\right) \exp \mathrm{i}\left(\xi_1 x_1 - \omega t\right)
\end{aligned} \quad -h < x_2 < h, \tag{25}$$

$$\begin{aligned}
v^{\mathrm{T}} &= A_7 \sinh \xi_2^{\mathrm{T}}\left(x_2 - h - g^{\mathrm{T}}\right) \exp \mathrm{i}\left(\xi_1 x_1 - \omega t\right) \\
\varphi^{\mathrm{T}} &= A_5 \sinh \xi_1\left(x_2 - h - g^{\mathrm{T}}\right) \exp \mathrm{i}\left(\xi_1 x_1 - \omega t\right)
\end{aligned} \quad h \le x_2 \le h + g^{\mathrm{T}}, \tag{26}$$

$$\begin{aligned}
v^{\mathrm{B}} &= A_8 \sinh \xi_2^{\mathrm{B}}\left(x_2 + h + g^{\mathrm{B}}\right) \exp \mathrm{i}\left(\xi_1 x_1 - \omega t\right) \\
\varphi^{\mathrm{B}} &= A_6 \sinh \xi_1\left(x_2 + h + g^{\mathrm{B}}\right) \exp \mathrm{i}\left(\xi_1 x_1 - \omega t\right)
\end{aligned} \quad -h - g^{\mathrm{B}} \le x_2 \le -h, \tag{27}$$

where ω is the wave frequency, and ξ_1 and ξ_2 are the wave numbers in the x_1 and x_2 directions, as defined in Eq. (14). A_1 through A_8 are undetermined constants. (25)-(27) already satisfy the boundary conditions in (24) at $x_2 = h+g^T$ and $x_2 = -h-g^B$. (25)-(27) also satisfy (9)-(11) provided that

$$\begin{aligned}
\left(\xi_2^{\mathrm{T}}\right)^2 &= \xi_1^2 - \mathrm{i}\omega \frac{\rho^{\mathrm{T}}}{\mu^{\mathrm{T}}} = \xi_1^2 \left(1 - \mathrm{i} \frac{v}{\xi_1} \frac{\rho^{\mathrm{T}}}{\mu^{\mathrm{T}}}\right), \\
\left(\xi_2^{\mathrm{B}}\right)^2 &= \xi_1^2 - \mathrm{i}\omega \frac{\rho^{\mathrm{B}}}{\mu^{\mathrm{B}}} = \xi_1^2 \left(1 - \mathrm{i} \frac{v}{\xi_1} \frac{\rho^{\mathrm{B}}}{\mu^{\mathrm{B}}}\right),
\end{aligned} \tag{28}$$

Substitution of (25)-(27) and their stress and electronic displacement components into the eight continuity conditions in (24) at $x_2 = \pm h$ results in the following eight linear, homogeneous equations for A_1 through A_8:

$$\begin{aligned}
&A_5 \sinh\left(-\xi_1 g^{\mathrm{T}}\right) = A_3 \sinh \xi_1 h + A_4 \cosh \xi_1 h + \tfrac{e}{\varepsilon}\left(A_1 \sin \xi_2 h + A_2 \cos \xi_2 h\right), \\
&-\varepsilon^{\mathrm{T}} A_5 \xi_1 \cosh \xi_1 g^{\mathrm{T}} = -\varepsilon\left(A_3 \xi_1 \cosh \xi_1 h + A_4 \xi_1 \sinh \xi_1 h\right),
\end{aligned} \tag{29a}$$

$$\bar{c}\left(A_1\xi_2\cos\xi_2 h - A_2\xi_2\sin\xi_2 h\right) + e\left(A_3\xi_1\cosh\xi_1 h + A_4\xi_1\sinh\xi_1 h\right) = \mu^{\mathrm{T}}\xi_2^{\mathrm{T}}A_7\cosh\xi_2^{\mathrm{T}}g^{\mathrm{T}}, \tag{29b}$$

$$-\left(A_1\sin\xi_2 h + A_2\cos\xi_2 h\right)\mathrm{i}\omega = A_6\sinh\xi_1 g^{\mathrm{B}} = \tag{29c}$$

$$\left(A_1\sin\xi_2 h - A_2\cos\xi_2 h\right)\mathrm{i}\omega = A_8\sinh\xi_2^{\mathrm{B}}g^{\mathrm{B}}, \tag{29d}$$

$$\bar{c}\left(A_1\xi_2\cos\xi_2 h + A_2\xi_2\sin\xi_2 h\right) + e\left(A_3\xi_1\cosh\xi_1 h - A_4\xi_1\sinh\xi_1 h\right) = \mu^{\mathrm{B}}\xi_2^{\mathrm{B}}A_8\cosh\xi_2^{\mathrm{B}}g^{\mathrm{B}}, \tag{29e}$$

$$-\varepsilon^{\mathrm{B}}A_6\xi_1\cosh\xi_1 g^{\mathrm{B}} = -\varepsilon\left(A_3\xi_1\cosh\xi_1 h - A_4\xi_1\sinh\xi_1 h\right), \tag{29f}$$

$$-A_3\sinh\xi_1 h + A_4\cosh\xi_1 h + \tfrac{e}{\varepsilon}\left(-A_1\sin\xi_2 h + A_2\cos\xi_2 h\right). \tag{29g}$$

Eq. (29) can be written as

$$\begin{bmatrix} M_{11} & M_{12} \\ M_{21} & M_{22} \end{bmatrix}\begin{bmatrix} A_3 \\ A_4 \end{bmatrix} = 0, \tag{30}$$

where

$$M_{11} = -\frac{\frac{\varepsilon}{\varepsilon_{\mathrm{T}}}\tanh\xi_1 g^{\mathrm{T}} + \frac{\varepsilon}{\varepsilon_{\mathrm{B}}}\tanh\xi_1 g^{\mathrm{B}}}{2\tan\xi_2 h} - \frac{\tanh\xi_1 h}{\tan\xi_2 h} - \frac{\tan\xi_2 h}{2}\left(\tfrac{\varepsilon}{\varepsilon_{\mathrm{B}}}\tanh\xi_1 g^{\mathrm{B}} - \tfrac{\varepsilon}{\varepsilon_{\mathrm{T}}}\tanh\xi_1 g^{\mathrm{T}}\right) + \frac{e^2}{\varepsilon\bar{c}}\frac{\xi_1}{\xi_2} + \frac{\mathrm{i}\omega\mu^{\mathrm{T}}\xi_2^{\mathrm{T}}}{\bar{c}\xi_2\tanh\xi_2^{\mathrm{T}}g^{\mathrm{T}}}\left(\tfrac{\varepsilon}{\varepsilon_{\mathrm{T}}}\tanh\xi_1 g^{\mathrm{T}} + \tanh\xi_1 h\right), \tag{31}$$

$$M_{12} = \frac{\tanh\xi_1 h}{2\tan\xi_2 h}\left(\tfrac{\varepsilon}{\varepsilon_{\mathrm{B}}}\tanh\xi_1 g^{\mathrm{B}} - \tfrac{\varepsilon}{\varepsilon_{\mathrm{T}}}\tanh\xi_1 g^{\mathrm{T}}\right) + \tan\xi_2 h + \frac{\tanh\xi_1 h}{2\operatorname{ctan}\xi_2 h}\left(\tfrac{\varepsilon}{\varepsilon_{\mathrm{T}}}\tanh\xi_1 g^{\mathrm{T}} + \tfrac{\varepsilon}{\varepsilon_{\mathrm{B}}}\tanh\xi_1 g^{\mathrm{B}}\right) + \frac{e^2}{\varepsilon\bar{c}}\frac{\xi_1}{\xi_2}\tanh\xi_1 h + \frac{\mathrm{i}\omega\mu^{\mathrm{T}}\xi_2^{\mathrm{T}}}{\bar{c}\xi_2\tanh\xi_2^{\mathrm{T}}g^{\mathrm{T}}}\left(\tfrac{\varepsilon}{\varepsilon_{\mathrm{T}}}\tanh\xi_1 g^{\mathrm{T}}\tanh\xi_1 h + 1\right), \tag{32}$$

$$M_{21} = -\frac{\frac{\varepsilon}{\varepsilon_{\mathrm{T}}}\tanh\xi_1 g^{\mathrm{T}} + \frac{\varepsilon}{\varepsilon_{\mathrm{B}}}\tanh\xi_1 g^{\mathrm{B}}}{2\tan\xi_2 h} - \frac{\tanh\xi_1 h}{\tan\xi_2 h} + \frac{\tan\xi_2 h}{2}\left(\tfrac{\varepsilon}{\varepsilon_{\mathrm{B}}}\tanh\xi_1 g^{\mathrm{B}} - \tfrac{\varepsilon}{\varepsilon_{\mathrm{T}}}\tanh\xi_1 g^{\mathrm{T}}\right) + \frac{e^2}{\varepsilon\bar{c}}\frac{\xi_1}{\xi_2} + \frac{\mathrm{i}\omega\mu^{\mathrm{B}}\xi_2^{\mathrm{B}}}{\bar{c}\xi_2\tanh\xi_2^{\mathrm{B}}g^{\mathrm{B}}}\left(\tfrac{\varepsilon}{\varepsilon_{\mathrm{B}}}\tanh\xi_1 g^{\mathrm{B}} + \tanh\xi_1 h\right), \tag{33}$$

$$M_{22} = \frac{\tanh\xi_1 h}{2\tan\xi_2 h}\left(\tfrac{\varepsilon}{\varepsilon_{\mathrm{B}}}\tanh\xi_1 g^{\mathrm{B}} - \tfrac{\varepsilon}{\varepsilon_{\mathrm{T}}}\tanh\xi_1 g^{\mathrm{T}}\right) - \tan\xi_2 h - \frac{\tanh\xi_1 h}{2\operatorname{ctan}\xi_2 h}\left(\tfrac{\varepsilon}{\varepsilon_{\mathrm{T}}}\tanh\xi_1 g^{\mathrm{T}} + \tfrac{\varepsilon}{\varepsilon_{\mathrm{B}}}\tanh\xi_1 g^{\mathrm{B}}\right) - \frac{e^2}{\varepsilon\bar{c}}\frac{\xi_1}{\xi_2}\tanh\xi_1 h - \frac{\mathrm{i}\omega\mu^{\mathrm{B}}\xi_2^{\mathrm{B}}}{\bar{c}\xi_2\tanh\xi_2^{\mathrm{B}}g^{\mathrm{B}}}\left(\tfrac{\varepsilon}{\varepsilon_{\mathrm{T}}}\tanh\xi_1 g^{\mathrm{T}}\tanh\xi_1 h + 1\right). \tag{34}$$

The dispersion relation of the waves, i.e., ω versus ξ_1, is determined by the following equation:

$$M_{11}M_{22} - M_{12}M_{21} = 0\,, \tag{35}$$

which is equivalent to that the determinant of the coefficient matrix of (29) vanishes. Expression (35) is a long equation when expanded. Below we consider the special case when the structure is symmetric about $x_2 = 0$, i.e.,

$$g^{T} = g^{B} = g\,,\ \mu^{T} = \mu^{B} = \mu\,,\ \rho^{T} = \rho^{B} = \rho^{L}\,,\ \varepsilon^{T} = \varepsilon^{B} = \varepsilon\,. \tag{36}$$

In this case, $\xi_2^{T} = \xi_2^{B} = \xi$. The waves separate into antisymmetric and symmetric ones about $x_2 = 0$. They are odd or even functions of x_2, respectively. (35) factors into two equations determining the dispersion relations of the antisymmetric (AS) and symmetric (S) waves:

$$\frac{\tanh \xi_1 h}{\tan \xi_2 h} + \frac{\varepsilon_{11}}{\varepsilon}\frac{\tanh \xi_1 g}{\tan \xi_2 h} - \frac{e_{15}^2}{\varepsilon_{11}\bar{c}_{44}}\frac{\xi_1}{\xi_2} - \frac{\mathrm{i}\omega\mu\xi}{\bar{c}_{44}\xi_2 \tanh \xi g}\left(\frac{\varepsilon_{11}}{\varepsilon}\tanh \xi_1 g + \tanh \xi_1 h\right) = 0\,, \text{(AS)} \tag{37}$$

$$\frac{\tan \xi_2 h}{\tanh \xi_1 h} + \frac{\varepsilon_{11}}{\varepsilon}\frac{\tanh \xi_1 g}{\operatorname{ctan} \xi_2 h} + \frac{e_{15}^2}{\varepsilon_{11}\bar{c}_{44}}\frac{\xi_1}{\xi_2} + \frac{\mathrm{i}\omega\mu\xi}{\bar{c}_{44}\xi_2 \tanh \xi g}\left(\frac{\varepsilon_{11}}{\varepsilon}\tanh \xi_1 g + \operatorname{ctanh} \xi_1 h\right) = 0\,. \text{(S)} \tag{38}$$

(37) or (38) can be viewed as an equation for v (or ω) with the use of (28). We make the following observations from (37) or (38):

i. The waves are dispersive due to the presence of ξ_1 in (37) or (38), in addition to v (or ω). Waves in plates without contacting a fluid are already dispersive in general (Bleustein, 1969). Fluids cause additional dispersion.
ii. Due to the presence of the fluid, ξ_1 is complex as suggested by (28). Then the dispersion relations determined by (37) or (38) are complex in general. These are damped waves or waves with attenuation as expected. This is fundamentally different from the case without fluids (Bleustein, 1969) where many real branches of dispersion relations can be obtained. If we keep ξ_1 real, (37) or (38) determines a complex v (or ω). If we keep v (or ω) real, (37) or (38) determines a complex ξ_1. In this case the phase velocity is calculated by $v=\omega/\mathrm{Re}(\xi_1)$, and the imaginary part of ξ_1 describes attenuation in the propagation direction.
iii. If the fluids are not present, we have air gaps. we can reduce (37) and (38) exactly into Eqs. (22) and (23).
iv. If we remove the gaps, i.e., $g^{T} = g^{B} = g = 0$, then the reduced results in (iii) further reduce to Eqs. (15) and (16) exactly.

4. Numerical results

4.1 Dispersion curves for thickness-twist waves in an infinite piezoelectric plate

Following Bleustein (1969), we introduce the following dimensionless frequency Ω and the dimensionless wave number Z by:

$$\Omega^2 = \omega^2 \Big/ \left(\frac{\pi^2 \bar{c}}{4\rho h^2}\right), \quad Z = \xi_1 \Big/ \left(\frac{\pi}{2h}\right). \tag{39}$$

Then the dispersion relations (15) and (16) can be rewritten as

$$\frac{\tan(\pi/2)(\Omega^2 - Z^2)^{1/2}}{\tanh(\pi/2)Z} = \frac{(\Omega^2 - Z^2)^{1/2}}{\bar{k}^2 Z} \text{(AS)} \tag{40}$$

$$\frac{\tan(\pi/2)(\Omega^2 - Z^2)^{1/2}}{\tanh(\pi/2)Z} = \frac{-\bar{k}^2 Z}{(\Omega^2 - Z^2)^{1/2}} \text{(S)} \tag{41}$$

The roots of Eqs. (40) and (41) determine the dispersion relation for the thickness-twist waves propagating in the x_1 direction. We note that the nondimensional frequency, Ω, must be real and positive, however the nondimensional wave number, Z, may be real, imaginary, or complex. It is clear from Eqs. (40) and (41) that if Z is a root (for a given Ω), so also is -Z. Segments of the first few positive real and imaginary branches of Eqs. (40) and (41) (computed for PZT-5H) are plotted as dashed lines and solid lines in Fig. 4, respectively.

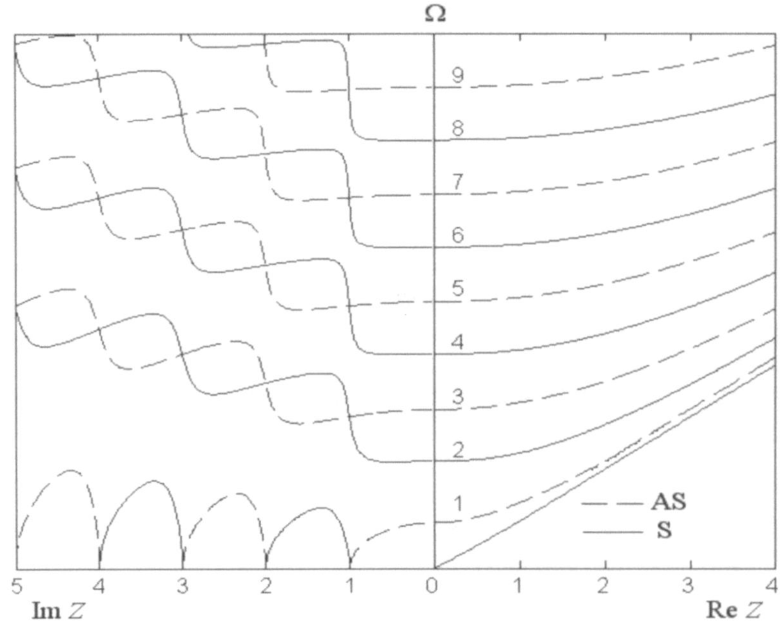

Fig. 4. Antisymmetric and symmetric real and pure imaginary branches

4.2 Effect of air gaps on dispersion curves

We here introduce a thickness ratio m=g/h. Numerical results for Ω versus Z are obtained by solving Eqs. (22) and (23) using PZT-5H, and are shown in Figs. 5 and 6, respectively, for m=0 and m=0.0001. The case of m=0 is determined by Eqs. (15) and (16) and was first given in Bleustein (1969). The figures show that the frequencies of short waves with a large wave number Z are more sensitive to m than long waves with a small Z. For a very small m=0.0001, the effect on short waves is already significant. The gaps raise the frequencies of short waves with a large Z.

To examine the effect of the gaps on long waves with a small Z, we plot the case of m=0.1 in Fig. 7 for anti-symmetric and symmetric waves together. The anti-symmetric waves are more sensitive to the air gaps, especially the lower-order modes. The symmetric waves are almost unaffected. The seemingly vertical line when Z is imaginary is in fact due to several branches of the dispersion relation. Those almost straight lines parallel to the vertical axis correspond to the non-horizontal solid lines closest to the vertical axis in the imaginary part

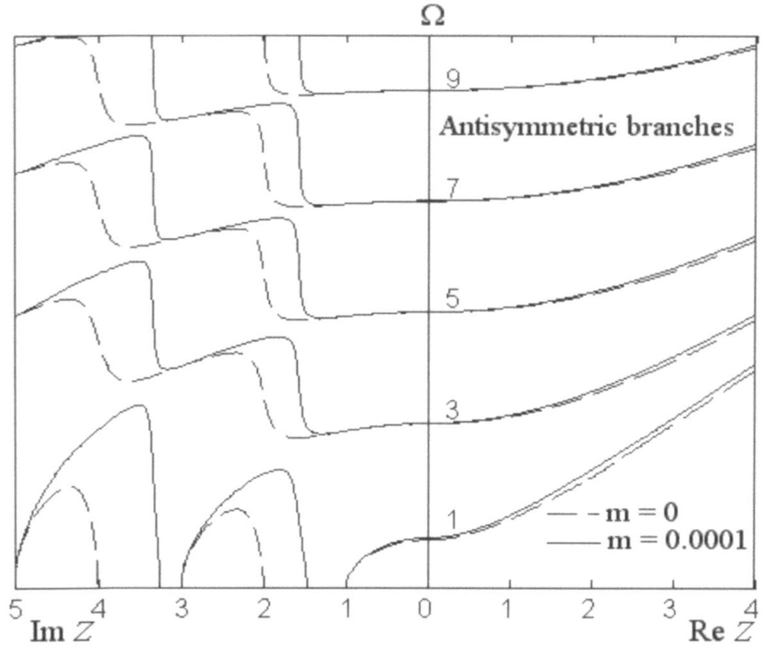

Fig. 5. Effect of air gaps on short anti-symmetric waves

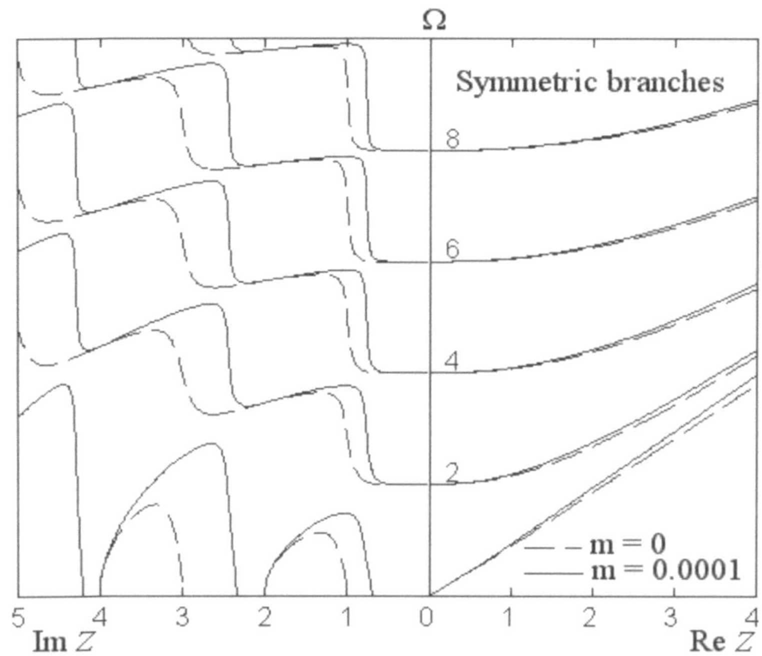

Fig. 6. Effect of air gaps on short symmetric waves

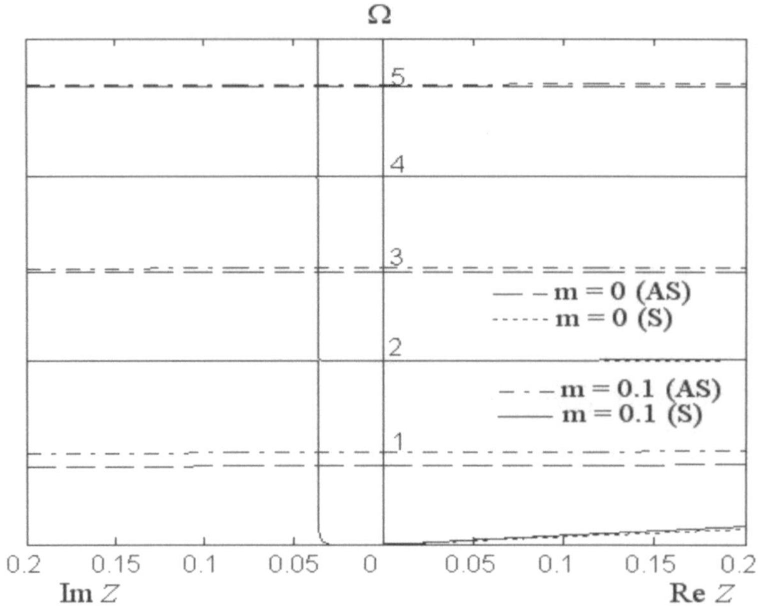

Fig. 7. Effect of air gaps on long waves

of Fig. 6. Again the gaps raise the frequencies. In Yang et al. (2009) it was found that for pure thickness-shear waves with Z=0 in quartz plates the gaps also raise the frequencies. This seems to be a common effect and can be explained as follows. When the electrodes are father apart, the shortening effect of the electrodes becomes less. Therefore stronger electric fields and the related piezoelectric stiffening effect may develop in the plate and hence the frequencies become higher.

The dependence on and sensitivity to m can be used in the design of acoustic wave devices to adjust the behavior of the waves. For example, gaps of varying thickness can be used to realize energy trapping of thickness waves in a plate (Yang et al., 2009), a phenomenon in which the waves are confined in a portion of the plate. Energy trapping is crucial to device mounting in applications.

4.3 Viscosity-induced frequency shift

To see the effect of the fluid viscosity on the wave propagation characteristics in further detail, we examine a specific case. Consider the antisymmetric waves described by Eq. (37) when g = ∞. When the fluid has a low viscosity, i.e., μ is small, we use an iteration (or perturbation) procedure. As the lowest order of approximation, we neglect the effect of the fluid viscosity by setting μ = 0 and in Eq. (37) and obtain

$$\frac{\tanh \xi_1 h}{\tan \xi_2^{(0)} h} + \frac{\varepsilon}{\varepsilon} \frac{1}{\tan \xi_2^{(0)} h} - \frac{e^2}{\varepsilon \bar{c}} \frac{\xi_1}{\xi_2^{(0)}} = 0, \tag{42}$$

where, for a given ξ_1, we have denoted the wave frequency when the viscosity is neglected by $\omega^{(0)}$ and

$$\left(\xi_2^{(0)}\right)^2 = \left(\omega^{(0)}\right)^2 / v_T^2 - \xi_1^2. \tag{43}$$

Given a wave number ξ_1, Eq. (42) determines a series of frequencies $\omega^{(0)}(\xi_1)$ for guided waves in the plate. When the fluid viscosity is considered, these wave frequencies are perturbed and are determined by Eq. (37). Let the corresponding frequencies from Eq. (37) be denoted by

$$\omega = \omega^{(0)} + \Delta\omega \,. \tag{44}$$

Substituting Eq. (44) into Eq. (37), we obtain the following first-order modification of the wave frequencies due to the fluid viscosity:

$$\frac{\Delta\omega}{\omega^{(0)}} \simeq \frac{v_{\mathrm{T}}^2 \xi_2^{(0)}}{\omega^{(0)}} \frac{\mathrm{i}\mu\xi^{(0)}}{\bar{c}} \frac{\tan(\xi_2^{(0)}h)\left(\frac{\varepsilon}{\varepsilon} + \tanh(\xi_1 h)\right)}{\frac{\varepsilon}{\varepsilon} + \tanh(\xi_1 h) - \frac{e^2}{\varepsilon\bar{c}}\frac{\xi_1 h}{\cos^2(\xi_2^{(0)}h)}}\,, \tag{45}$$

where

$$(\xi^{(0)})^2 = \xi_1^2 - i\omega^{(0)}\frac{\rho^{\mathrm{L}}}{\mu}\,. \tag{46}$$

Equation (45) determines the frequency perturbation due to the fluid viscosity. It is a complex number in general. In addition to its real part representing frequency shifts, its imaginary part describes the damping effect of the viscosity. The viscosity-induced frequency shift can be used for viscosity measurement.

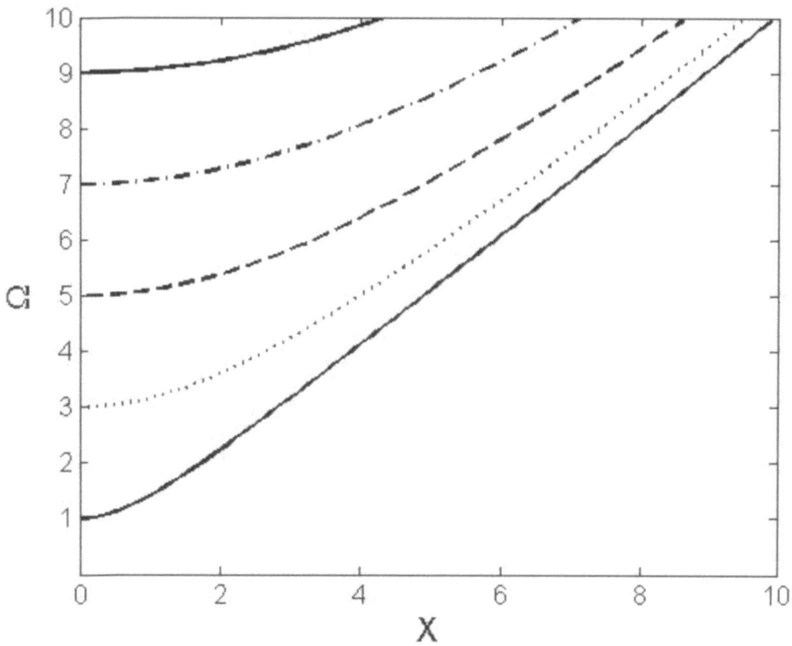

Fig. 8. Dispersion curves of Eq. (37) when fluid viscosity is neglected

For numerical results we consider a ceramic plate of PZT-5H in water, a low viscosity fluid. Using the dimensionless wave number X and the dimensionless frequency Ω defined in Eq.

(39), we plot the dispersion curves determined by Eq. (42) in Fig. 8. These are odd thickness-twist waves. Only the first five branches are shown. In applications long waves (X<<1) of the lowest branch (the fundamental mode) are used most often. Therefore in Fig. 9 we present the frequency shift calculated from the real part of Eq. (45) for long, fundamental thickness-twist waves. In acoustic wave sensors, thermal noise is usually of the order of 10^{-6} in terms of the relative frequency shift. Therefore the frequency shift in Fig. 9 is considered a clear signal.

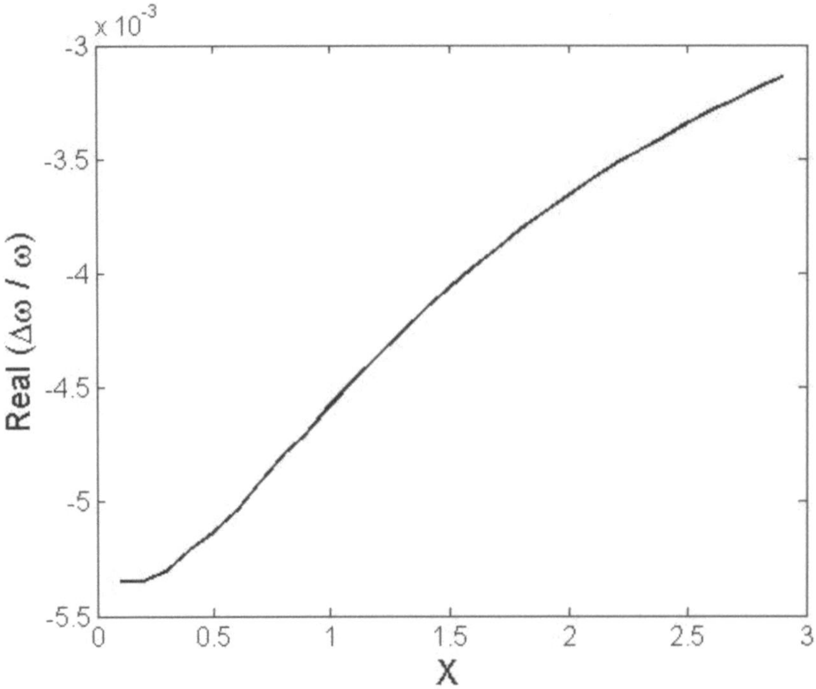

Fig. 9. Frequency shift of the lowest thickness-twist wave due to fluid viscosity

5. Conclusion

In this Chapter, we study the propagation of thickness-twist waves in an infinite piezoelectric ceramic plate. Three cases are taken into account: 1) Propagation of thickness-twist waves in a piezoelectric ceramic plate with attached electrodes; 2) Propagation of thickness-twist waves in a piezoelectric ceramic plate with unattached electrodes; 3) Propagation of thickness-twist waves in a piezoelectric ceramic plate in contact with viscous fluids. The first part is used as a basic introduction for those who want to conduct the research work on thickness-twist waves in an infinite piezoelectric ceramic plate. We mainly focus on the latter two parts. A simple solution for thickness-twist waves propagating in ceramics plates with unattached electrodes and air gaps is obtained. In the case of symmetric gaps the waves separate into symmetric and antisymmetric ones. The wave frequency is sensitive to the gap thickness, especially for short waves. In the examples calculated the gaps raise the wave frequencies. The dependence of frequency on air gaps provides a new design factor for acoustic wave devices. Equations determining the dispersion relations for thickness-twist waves in a piezoelectric ceramic plate with unattached electrodes and viscous fluids under the electrodes are obtained. The dispersion relations are complex in general, representing damped waves. The equations reduce to the case of a plate with unattached electrodes and empty air gaps (Qian et al., 2009) and the case of a plate with attached electrodes (Bleustein, 1969) as special cases. The viscosity-induced frequency shift is studied in detail by the the lowest order of perturbation approximation . The results are useful for developing plate fluid sensors.

6. References

Ballato, A. & Lukaszek, T. J. (1974). Mass-Loading of Thickness-Excited Crystal Resonators Having Arbitrary Piezo-Coupling. *IEEE Transactions on Sonics and Ultrasonics*, Vol. SW-21, pp. 269-274.

Benes, E.; Gröschl, M.; Burger, W. & Schmid, M. (1995). Sensors based on piezoelectric resonators. *Sensors and Actuators A*, Vol. 48, pp. 1-21.

Bleustein, J. (1966). Thickness-twist and face-shear vibrations of a contoured crystal plate. *Int. J. Solids Struct.*, Vol. 2, pp. 351-360.

Bleustein, J. L. (1969). Some simple modes of wave propagation in an infinite piezoelectric plate. *J. Acoust. Soc. Am.*, Vol. 45, pp. 614-620.

Detaint, J.; Carru, H.; Schwartzel, J.; Joly, C.; Capelle, B. & Zarka, Z. (1989). Air-gap determination of the cut-off frequency of unelectroded plates and effects of local thickness modifications in plate resonators, Proc. of the 43rd Annual Symposium on Frequency Control, pp. 563-565.

Guo, F. L. & Sun, R. (2008). Propagation of Bleustein-Gulyaev wave in 6mm piezoelectric materials loaded with viscous liquid. *Int. J. Solids Struct.*, Vol. 45, pp. 3699-3710.

Huang, M. H.; Shen, D. Z. & Yang, M. S. (2001). Effects of longitudinal wave on the resonance behavior of an electrode-separated piezoelectric sensor in liquids. *Analytica Chimica Acta*, Vol. 440, pp. 109-118.

Josse, F.; Shana, Z. A.; Radtke, D. E. & Haworth, D. T. (1990). Analysis of piezoelectric bulk-acoustic- wave resonators as detectors in viscous conductive liquids. *IEEE Trans. U. F. F. C.*, Vol. 37, pp. 359-368.

Kanazawa, K. K. & Gordon II, J. G. (1985). The oscillation frequency of a quartz resonator in contact with a liquid. *Analytica Chimica Acta*, Vol. 175, pp. 99-105.

Kim, J. O.; Wang, Y. Z. & Bau, H. H. (1991). The effect of an adjacent viscous fluid on the transmission of torsional stress waves in a submerged waveguide. *J. Acoust. Soc. Am.*, Vol. 89, pp. 1414-1422.

Kosinski, J. A. (2003). Thickness vibrations of flat piezoelectric plates with massy electrodes of unequal thickness. Proceedings of IEEE Ultrasonics Symposium, pp. 70-73, 2003.

Mindlin, R. D. (1965). Thickness-twist vibrations of an infinite, monoclinic, crystal plate. *Int. J. Solids Struct.*, Vol. 1, pp. 141-145.

Mindlin, R. D. (1967). Bechmann's number for harmonic overtones of thickness/twist vibrations of rotated Y-cut quartz plates. *J. Acoust. Soc. Am.*, Vol. 41, pp. 969-973.

Mindlin, R. D. (1968). Optimal sizes and shapes of electrodes for quartz resonators. *J. Acoust. Soc. Am.*, Vol. 43, pp. 1329-1331.

Pao, S. Y.; Huang, R. & Lam, C. S. (2007). Energy trapping effect of stepped electrode in miniature AT-cut quartz resonator, Proceedings of Piezoelectricity, Acoustics Waves and Device Applications, pp. 152-154.

Pearman, G. T. (1968). Thickness-twist vibrations in beveled AT-cut quartz plates. *J. Acoust. Soc. Am.*, Vol. 45, pp. 928-934.

Peng, F.; Liu, H. & Hu, S. Y. (2006). Love wave propagation in a layered piezoelectric structure immersed in a viscous fluid. *Key Eng. Mat.*, Vol. 306-308, pp. 1211-1216.

Qian, Z. H.; Kishimoto, K. & Yang, J. S. (2009). Propagation of thickness-twist waves in a piezoelectric ceramic plate with unattached electrodes. *Ultrasonics*, Vol. 49, pp. 501-504.

Reed, C. E.; Kanazawa, K. K. & Haufman, J. H. (1990). Physical description of a viscoelastically loaded AT-cut quartz resonator. *J. Apl. Phys.*, Vol. 68, pp. 1993-2001.

Syngellakis, S. & Lee, P. C. Y. (1993). Piezoelectric wave dispersion curves for infinite anisotropic plates. *J. Appl. Phys.*, Vol. 73, pp. 7152-7161.

Tiersten, H. F. (1995). The influence of an asymmetric air gap on the vibration characteristics of quartz trapped energy resonators, Proc. of IEEE International Frequency Control Symposium, pp. 740-745.

Umeki, M.; Sato, T.; Uehara, H. & Okazaki, M. (2004). 622MHz high frequency fundamental composite crystal resonator with an air-gapped electrode, Proc. of IEEE International Frequency Control Symposium and Exposition, pp. 365-368.

Vogt, T. K.; Lowe, M. J. S. & Cawley, P. (2004). Measurement of the material properties of viscous liquids using ultrasonic guided waves. *IEEE Trans. U. F. F. C.*, Vol. 51, pp. 737-747.

Wang, J. & Shen, L. J. (2005). Exact thickness-shear resonance frequency of electroded piezoelectric crystal plates. *Journal of Zhejiang University Science*, Vol. 6A, pp. 980-985.

Wang, J.; Shen, L. J. & Yang, J. S. (2008). Effects of electrodes with continuously varying thickness on energy trapping in thickness-shear mode quartz resonators. *Ultrasonics*, Vol. 48, pp. 150-154.

Yang, J. S.; Zhou, H. G. & Zhang, W. P. (2005). Thickness-shear vibration of rotated Y-cut quartz plates with relatively thick electrodes of unequal thickness. *IEEE Trans. U. F. F. C.*, Vol. 52, pp. 918-922.

Yang, J. S. & Guo, S. H. (2006). Effects of piezoelectric coupling on Bechmann's number for thickness-twist waves in a plate of hexagonal crystals. *IEEE Trans. U. F. F. C.*, Vol. 53, pp. 1960-1962.

Yang, J. S.; Xue, H.; Fang, H. Y.; Hu, Y. T.; Wang J. & Shen, L. J. (2007a). Effects of electrodes with varying thickness on energy trapping in thickness-shear quartz resonators. *IEEE Trans. U. F. F. C.*, Vol. 54, pp. 892-895.

Yang, J. S.; Xue, H. & Hu, Y. T. (2007b). Finite element analysis of stress field concentration near the edge of an electrode. *Ferroelectrics Letters Section*, Vol. 34, pp. 108-112.

Yang, J. S.; Chen, Z. G. & Hu, Y. T. (2008a). Vibration of a thickness-twist mode piezoelectric resonator with asymmetric, non-uniform electrodes. *IEEE Trans. U. F. F. C.*, Vol. 55, pp. 841-848.

Yang, Z. T.; Yang, J. S. & Hu, Y. T. (2008b). Optimal electrode shape and size of doubly rotated quartz plate thickness mode piezoelectric resonators. *Appl. Phys. Lett.*, Vol. 92, Art. No. 103516.

Yang, Z. T.; Guo, S. H.; Hu, Y. T. & Yang, J. S. (2009). Thickness-shear vibration of rotated Y-cut quartz plates with unattached electrodes and asymmetric air gaps. *Philosophical Magazine Letters*, Vol. 89, pp. 313-321.

IV. Wave Propagation in Plasma.

19

Interplay of Kinetic Plasma Instabilities

M. Lazar[1,2], S. Poedts[2] and R. Schlickeiser[3]
[1]*Research Department - Plasmas with Complex Interactions, Ruhr-Universität Bochum, D-44780 Bochum*
[2]*Centre for Plasma Astrophysics, Celestijnenlaan 200B, 3001 Leuven*
[3]*Institut für Theoretische Physik, Lehrstuhl IV: Weltraum- und Astrophysik, Ruhr-Universität Bochum, D-44780 Bochum*
[1,3]*Germany*
[2]*Belgium*

1. Introduction

Undulatory phenomena are probably among the most fascinating aspects of our existence. It is well known that plasma is the most dominant state of ionized matter in the Universe. Moreover, it can excite and sustain any kind of oscillatory motion, acoustic or electromagnetic (light) waves. A realistic perspective upon the dynamics of space or laboratory plasmas reveals a constant presence of various kinetic anisotropies of plasma particles, like beams or temperature anisotropies. Such anisotropic plasma structures give rise to growing fluctuations and waves. The present chapter reviews these kinetic instabilities providing a comprehensive analysis of their interplay for different circumstances relevant in astrophysical or laboratory applications.

Kinetic plasma instabilities are driven by the velocity anisotropy of plasma particles residing in a temperature anisotropy, or in a bulk relative motion of a counter streaming plasma or a beam-plasma system. The excitations can be electromagnetic or electrostatic in nature and can release different forms of free energy stored in anisotropic plasmas. These instabilities are widely invoked in various fields of astrophysics and laboratory plasmas. Thus, the so-called magnetic instabilities of the Weibel-type (Weibel; 1959; Fried; 1959) can explain the generation of magnetic field seeds and the acceleration of plasma particles in different astrophysical sources (e.g., active galactic nuclei, gamma-ray bursts, Galactic micro quasar systems, and Crab-like supernova remnants) where the nonthermal radiation originates (Medvedev & Loeb; 1999; Schlickeiser & Shukla; 2003; Nishikawa et al.; 2003; Lazar et al.; 2009c), as well as the origin of the interplanetary magnetic field fluctuations, which are enhanced along the thresholds of plasma instabilities in the solar wind (Kasper at al.; 2002; Hellinger et al.; 2006; Stverak et al.; 2008). Furthermore, plasma beams built in accelerators (e.g., in fusion plasma experiments) are subject to a variety of plasma waves and instabilities, which are presently widely investigated to prevent their development in order to stabilize the plasma system (Davidson et al.; 2004; Cottril et al.; 2008).

2. Nonrelativistic dispersion formalism

Plasma particles (electrons and ions) are assumed to be collision-less, with a non-negligible thermal spread, and far from any uniform fields influence, $\mathbf{E}_0 = 0$ and $\mathbf{B}_0 = 0$. This assumption allows us to develop the most simple theory for the kinetic plasma instabilities, but the results presented here can also be extended to the so-called high-beta plasmas (where beta corresponds to the ratio of the kinetic plasma energy to the magnetic energy) since recent analysis has proven that these instabilities are only slightly altered in the presence of a weak ambient magnetic field (Lazar et al.; 2008, 2009b, 2010).
We here investigate small amplitude plasma excitations using a linear kinetic dispersion formalism, based on the coupled system of the Vlasov equation and the Maxwell equations. The standard procedure starts with the linearized Vlasov equation (Kalman et al.; 1968)

$$\frac{\partial F_a}{\partial t} + \mathbf{v} \cdot \frac{\partial F_a}{\partial \mathbf{r}} = -q_a [\mathbf{E} + \frac{\mathbf{v} \times \mathbf{B}}{c}] \cdot \frac{\partial F_{a,0}}{\partial \mathbf{v}}, \tag{1}$$

where $F_a(\mathbf{r},\mathbf{v}, t)$ denotes the first order perturbation of the equilibrium distribution function $F_{a,0}(\mathbf{v})$ for particles of kind a. The unperturbed distribution function is normalized as

$$\int_{-\infty}^{\infty} d\mathbf{v}\, F_{a,0}(\mathbf{v}) = 1, \tag{2}$$

and is considered to be anisotropic (the free energy source), implying that

$$\frac{\partial F_{a,0}(\mathbf{r},\mathbf{v},t)}{\partial \mathbf{v}} \nparallel \mathbf{v} \tag{3}$$

and the non-vanishing term

$$(\mathbf{v} \times \mathbf{B}) \cdot \frac{\partial F_{a,0}}{\partial \mathbf{v}} \neq 0, \tag{4}$$

becomes responsible for the unstable solutions (Davidson et al.; 1972). Ohm's law defines the current density, **J**, and the conductivity tensor, $\tilde{\sigma}$, by

$$\mathbf{J} \equiv \tilde{\sigma} \cdot \mathbf{E} = \sum_a q_a \int_{-\infty}^{\infty} d\mathbf{v} \mathbf{v} F_a(\mathbf{r},\mathbf{v},t), \tag{5}$$

and using Maxwell's equations

$$\nabla \times \mathbf{B} = \frac{1}{c}\frac{\partial \mathbf{E}}{\partial t} + \frac{4\pi}{c}\mathbf{J}, \tag{6}$$

$$\nabla \cdot \mathbf{B} = 0, \tag{7}$$

$$\nabla \times \mathbf{E} = -\frac{1}{c}\frac{\partial \mathbf{B}}{\partial t}, \tag{8}$$

$$\nabla \cdot \mathbf{E} = 4\pi\rho = 4\pi \sum_a q_a \int_{-\infty}^{\infty} d\mathbf{v} F_a(\mathbf{r},\mathbf{v},t), \tag{9}$$

we find the solution of Vlasov equation (1) to be

$$F_a = -\frac{iq_a}{\omega}\left(\mathbf{E} + \frac{\mathbf{v}\cdot\mathbf{E}}{\omega - \mathbf{k}\cdot\mathbf{v}}\mathbf{k}\right)\cdot\frac{\partial F_{a,0}}{\partial \mathbf{v}}. \tag{10}$$

Here, we examine large-scale spatial and temporal variations in the sense ofWentzel-Kramer-Brillouin (WKB) approximation and treat plasma wave perturbations as a superposition of plane waves in space (Fourier components) and harmonic waves in time (Laplace transforms). Thus, the analysis is reduced to small amplitude excitations with a sine variation of the form ~ $\exp(-i\omega t + \mathbf{k}\cdot\mathbf{r})$. Since we consider an infinitely large, homogeneous and stationary plasma, we choose the wave-number k to be real, but the Laplace transform in time gives rise to complex frequencies $\omega = \omega_r + \imath\omega_i$, implying also a complex index of refraction, $N = kc/\omega$.
Now, substituting F_a from Eq. (10) into the Eq. (9) provides the wave equation for the linearized electric field, which admits nontrivial solutions only for

$$\det\left|\frac{\omega^2}{c^2}\epsilon_{ij} + k_i k_j - k^2\delta_{ij}\right| = 0 \tag{11}$$

where the dielectric tensor has the components, $\epsilon_{ij} \equiv \delta_{ij} + (4\pi\imath/\omega)\sigma_{ij}$, explicitly given by

$$\epsilon_{ij} = \delta_{ij} + \sum_a \frac{\omega_{p,a}^2}{\omega^2}\left[\int_{-\infty}^{\infty} d\mathbf{v} v_i \frac{\partial F_{a,0}}{\partial v_j} + \int_{-\infty}^{\infty} d\mathbf{v} v_i v_j \frac{\mathbf{k}\cdot\partial F_{a,0}/\partial\mathbf{v}}{\omega - \mathbf{k}\cdot\mathbf{v}}\right]. \tag{12}$$

3. Counterstreaming plasmas with intrinsic temperature anisotropies

In order to analyze the unstable plasma modes and their interplay we need a complex anisotropic plasma model including various forms of particle velocity anisotropy. Thus, we consider two counter streaming plasmas (see Fig. 1) with internal temperature anisotropies described by the distribution function (Maxwellian counterstreams)

$$f_0(v_x, v_y, v_z) = v_{th}^{-2} v_{th\perp}^{-1} 2\pi^{3/2} \exp\left[-v_x^2 + v_z^2 v_{th}^2\right]\left\{\exp\left[-(v_y + v_0)^2 v_{th,y}^2\right] + \exp\left[-(v_y - v_0)^2 v_{th,y}^2\right]\right\}. \tag{13}$$

Recent investigations have proved that such a model is not only appropriate for a multitude of plasma applications but, in addition, it can be approached analytically very well.
For the sake of simplicity, in what follows we neglect the contribution of ions, which form the neutralizing background, and the electron plasma streams are assumed homogeneous and symmetric (charge and current neutral) with the same densities, $\omega_{p,e,1} = \omega_{p,e,2} = \omega_{p,e}$, equal but opposite streaming velocities, $v_1 = v_2 = v_0$, and the same temperature parameters, i.e., thermal velocities, $v_{th,x,1} = v_{th,z,1} = v_{th,x,2} = v_{th,z,2} = v_{th}$, $v_{th,y,1} = v_{th,y,2} = v_{th,y}$. Furthermore, for each stream, the intrinsic thermal distribution is considered bi-Maxwellian, and the temperature anisotropy is defined by $A_1 = A_2 = A = T_y/T_x = (v_{th,y}/v_{th})^2$. Taking the counterstreaming plasmas symmetric, a condition frequently satisfied with respect to their mass center at rest, provides simple forms for the dispersion relations, and solutions are purely growing exhibiting only a reactive part, $\mathrm{Re}(\omega) = \omega_r \to 0$ and $\mathrm{Im}(\omega) = \Gamma > 0$, and, therefore, a negligible resonant Landau dissipation of wave energy on plasma particles. The anisotropic

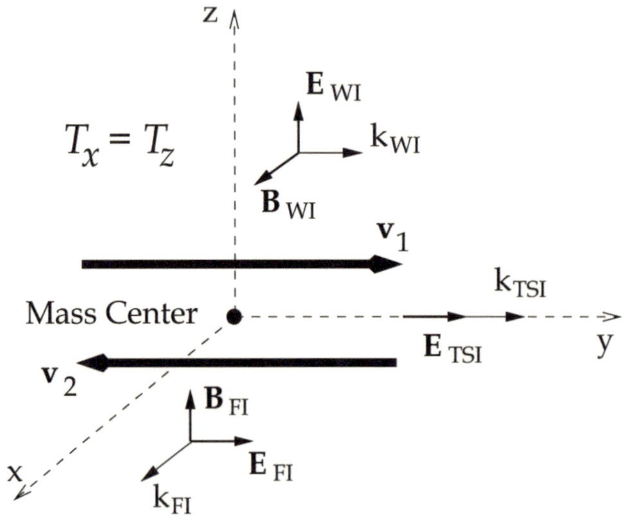

Fig. 1. Sketch of two plasma counter streams moving along y-axis and the instabilities developing in the system: the electromagnetic Weibel instability (WI) driven by an excess of transverse kinetic energy, and the electrostatic two-stream instability (TSI) both propagating along the streams, and the filamentation instability (FI) propagating perpendicular to the streams.

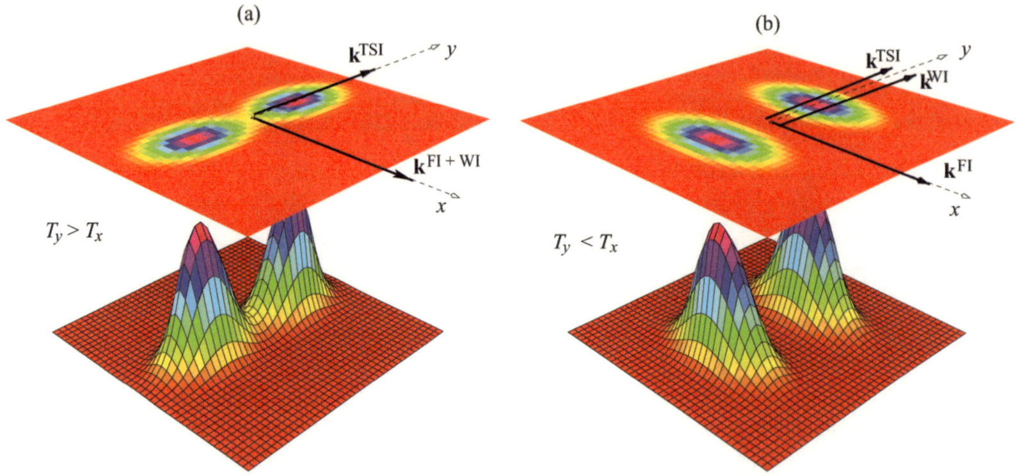

Fig. 2. Sketch of the distribution functions for two symmetric counterstreaming plasmas, and the wave-vectors for the unstable modes expected to develop when (a) $T_x = T_z < T_y$ and (b) $T_x = T_z > T_y$.

counterstreaming distribution functions are illustrated in Fig. 2, for two representative situations: (a) $T_x = T_z < T_y$ and (b) $T_x = T_z > T_y$.

Such a plasma system is unstable against the excitation of the electrostatic two-stream instability as well as the electromagnetic instabilities of the Weibel-type. We limit our analysis to the unstable waves propagating either parallel or perpendicular to the direction of streams. The orientation of these instabilities is given in the Figures 1 and 2.

4. Unstable modes with $\mathbf{k} \parallel \hat{\mathbf{y}}$

First we look for the unstable modes propagating along the streaming direction, $k = k_y$, and due to the symmetry of our distribution function (13), the dispersion relation (11) simplifies to

$$\left(\frac{\omega^2}{c^2}\epsilon_{xx} - k_y^2\right)\left(\frac{\omega^2}{c^2}\epsilon_{zz} - k_y^2\right)\epsilon_{yy} = 0. \tag{14}$$

This equation admits three solutions, viz. two electromagnetic modes

$$\frac{k_y^2 c^2}{\omega^2} = \epsilon_{xx} \quad \frac{k_y^2 c^2}{\omega^2} = \epsilon_{zz} = 0, \tag{15}$$

and one electrostatic mode

$$\epsilon_{yy} = 0, \tag{16}$$

where the dielectric tensor components are provided by Eq. (12), with our initial unperturbed distribution function given in Eq. (13).
In a finite temperature plasma there is an important departure from the cold plasma model, where no transverse modes could interact with the electrons for wave vectors parallel to the streaming direction, $\mathbf{k} \parallel \hat{\mathbf{y}}$, as no electrons move perpendicularly to the streams. These electrons are introduced here by a non-vanishing transverse temperature of the plasma counter-streams. Furthermore, the electromagnetic modes of Weibel-type and propagating along the streaming direction can be excited only by an excess of transverse kinetic energy, $T_x = T_z > T_y$ (Bret et al.; 2004). These modes are characterized in the next.

4.1 The Weibel instability ($\mathbf{k} \cdot \mathbf{E} = 0$, $v_{th} > v_{th,y}$)

Thus, let we consider symmetric counterstreams with an excess of transverse kinetic energy, $T_x = T_z > T_y$, and described by a bi-Maxwellian distribution as given in Eq. (13) and schematically shown Fig. 2 (a). Due to the symmetry of the system (see in Fig. 2 a) the two branches of the transverse modes (Eq. 15) are also symmetric and will be described by the same dispersion relation (Okada et al.; 1977; Bret et al.; 2004; Lazar et al.; 2009c)

$$\frac{k_y^2 c^2}{\omega^2} = \epsilon_{xx}^{WI} = \epsilon_{zz}^{WI} = 1 - \frac{\omega_{p,e}^2}{\omega^2}\left\{1 - \frac{1}{A}\left[1 + \frac{1}{2}\left(f_1 Z(f_1) + f_2 Z(f_2)\right)\right]\right\}, \tag{17}$$

which is written in terms of the well-known plasma dispersion function (Fried & Conte; 1961)

$$Z(f) = \pi^{-1/2}\int_{-\infty}^{\infty} dx \frac{\exp(-x^2)}{x - f}, \quad with \quad f_{1,2} = \frac{\omega \mp k_y v_0}{k_y v_{th,y}}. \tag{18}$$

Numerical solutions of Eq. (17) are displayed in Fig. 3: the growth rates of the Weibel instability are visibly reduced in a counterstreaming plasma and the wave number cutoff is also diminished according to

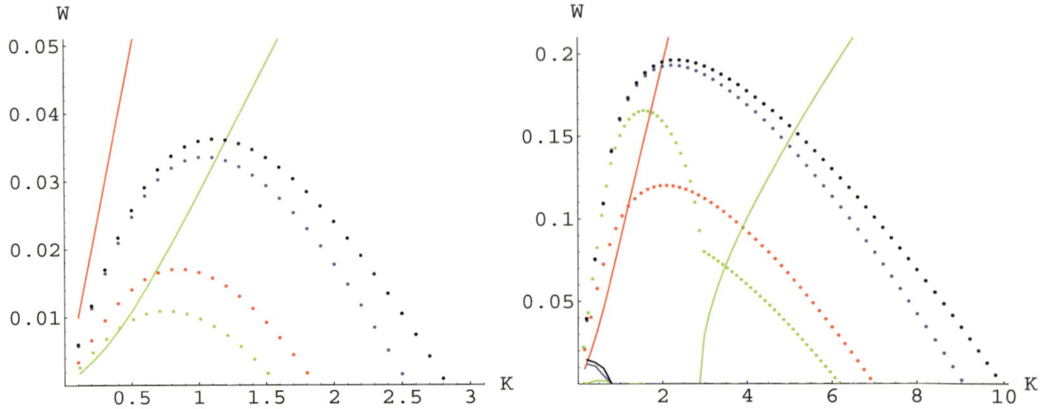

Fig. 3. Numerical solutions of equation (17): with dotted lines are plotted the growth rates, W= ω_i/ω_{pe}, and with solid lines the real frequency,W = ω_r/ω_{pe}, for $v_{th,y} = c/30 = 10^7$ m/s, three different streaming velocities, $v_0 = c/10$ (red), $c/30$ (green), $c/100$ (blue), 0 (black), and two anisotropies (a) $v_{th}/v_{th,y} = 3$ and (b) $v_{th}/v_{th,y} = 10$ (K = kc/ω_{pe}, and $c = 3 \times 10^8$ m/s is the speed of light in vacuum).

$$k_{y,c}^{WI} = \frac{\omega_{p,e}}{c}\left\{\frac{1}{A}\left[1+\frac{v_0}{v_{th,y}}\mathrm{Re}Z\left(\frac{v_0}{v_{th,y}}\right)\right]-1\right\}^{1/2} < k_{y,c}^{WI}(v_0=0). \quad (19)$$

Here, we have taken into account that, for a real argument, the real part of plasma dispersion function is negative: $\mathrm{Re}\,Z(x) = -2\exp(-x^2)\int_0^x dt \exp(t^2) < 0$, and $xZ(x) - xZ(-x) = 2x\mathrm{Re}Z(x)$.

This wave number cutoff must be a real (not complex) solution of Eq. (17) in the limit of $\Gamma(k) = \Im\omega(k) = 0$. For $v_0 = 0$ we simply recover the cutoff wave number of the Weibel instability driven by a temperature anisotropy without streams. According to Eq. (19), in the presence of streams ($v_0 \neq 0$) the threshold of the Weibel instability ($v_{\mathrm{th}}^2 / v_{\mathrm{th},y}^2 = T/T_y > 1$) grows to

$$\frac{v_{\mathrm{th}}^2}{v_{\mathrm{th},y}^2} > 1 + \frac{v_{\mathrm{th}}^2 v_0}{v_{\mathrm{th},y}^3}\left|\mathrm{Re}Z\left(\frac{v_0}{v_{th,y}}\right)\right|. \quad (20)$$

We remark in Fig. 3 that the Weibel instability is purely growing ($\omega_r = 0$) not only in a non-streaming plasma ($v_0 = 0$), but in the presence of streams as well. This is, however, valid only for small streaming speeds. Otherwise, for energetic streams with a sufficiently large bulk velocity, larger than the thermal speed along their direction, $v_0 > v_{th,y}$, the instability becomes oscillatory with a finite frequency $\omega_r \neq 0$. As the temperature anisotropy is also large, both these regimes can be identified, the purely growing regime for small wave numbers, and the oscillatory growing regime for large wave numbers (see Fig. 3 b, and Lazar et al. (2009a) for a supplementary analysis).

4.2 Two-stream instability (k × E = 0)

The two-stream instability is an electrostatic unstable mode propagating along the streaming direction and described by the dispersion relation (16), where the dielectric function reads

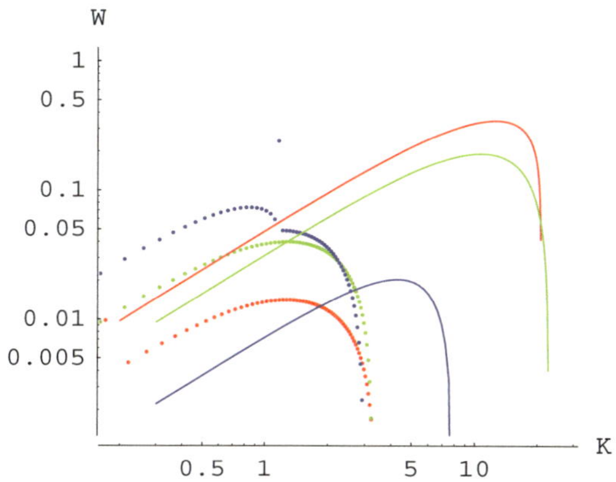

Fig. 4. For a given anisotropy $v_{th}/v_{th,y}$=5 and streaming speed $v_0 = c/20$, the growth rates (W = ω_i/ω_{pe} versus K = kc/ω_{pe}) of the Weibel instability (dotted lines) increase and those of the two-stream instability (solid lines) decrease with parallel thermal spread of plasma particles: $v_{th,y} = c/100$ (red lines), $c/30$ (green), $c/20$ (blue).

$$\epsilon_{yy}^{TSI} = 1 + \frac{\omega_{p,e}^2}{k_y^2 v_{th,y}^2}\left[2 + f_1 Z(f_1) + f_2 Z(f_2)\right] = 0. \tag{21}$$

The two-stream instability is inhibited by the thermal spread of plasma particles along the streaming direction. The growth rates can be markedly reduced by increasing $v_{th,y}$. Thus, the two-stream instability has a maximal efficiency in the process of relaxation only for a

$$k_{y,c}^{TSI} = \frac{\omega_{p,e}}{v_0}\left(1 + \frac{3}{2}\frac{v_{th,y}^2}{v_0^2}\right)^{1/2} \simeq \frac{\omega_{p,e}}{v_0}\left(1 + \frac{3}{4}\frac{v_{th,y}^2}{v_0^2}\right). \tag{22}$$

For a negligible thermal spread, $v_0 \gg v_{th,y}$ (i.e. cold plasmas), the cutoff wave number will depend only on the streaming velocity $k_{y,c}^{TSI} \to \omega_{p,e} / v_0$.

The instability is purely growing because the streams are symmetric, otherwise it is oscillatory. The growth rates, solutions of Eq. (21), are displayed in Fig. 4 in comparison to the Weibel instability growth rates ($v_{th} > v_{th,y}$), for conditions typically encountered in intergalactic plasma and cosmological structures formation (Lazar et al.; 2009c).

We can extract the first remarks on the interplay of these two instabilities from Fig. 1:

1. When the thermal speed along the streams is small enough, i.e. smaller than the streaming speed, the two-stream instability grows much faster than theWeibel instability (the growth rates of the two-stream instability are much larger than those of the Weibel instability).
2. While the two-stream instability is not affected by the temperature anisotropy, the Weibel instability is strictly dependent on that.
3. While the thermal spread along the streams inhibits the two-stream instability, in the presence of a temperature anisotropy, the same parallel thermal spread enhances the Weibel instability growth rates. In this case, the Weibel instability has chances to arise before the two-stream instability can develop.

Otherwise, the two-stream instability develops first and relaxes the counterstreams to a plateau anisotropic distribution with two characteristic temperatures (bi-Maxwellian). If this thermal anisotropy is large enough, it is susceptible again to relax through aWeibel excitation. How large this thermal anisotropy could be depends not only on the initial bulk velocity of the streams but on their internal temperature anisotropy as well.
Whether it develops as a primary or secondary mechanism of relaxation, the Weibel instability seems therefore to be an important mechanism of relaxation for such counterstreaming plasmas. This has important consequences for experiments and many astrophysical scenarios, providing for example, a plausible explanation for the origin of cosmological magnetic field seeds (Schlickeiser & Shukla; 2003; Lazar et al.; 2009c).

5. Unstable modes with $\mathbf{k} \perp \hat{\mathbf{y}}$

There is also another important competitor in this puzzle of kinetic instabilities arising in a counterstreaming plasma, and this is the filamentation instability which is driven by the bulk relative motion of plasma streams and propagates perpendicular to the streams, $\mathbf{k} \perp \hat{\mathbf{y}}$. In this case, we can choose without any restriction of generality, the propagation direction along x-axis, $k = k_x$, and in this case the dispersion relation (11) becomes

$$\epsilon_{xx}\left(\frac{\omega^2}{c^2}\epsilon_{yy} - k_x^2\right)\left(\frac{\omega^2}{c^2}\epsilon_{zz} - k_x^2\right) = 0. \tag{23}$$

This equation admits three branches of solutions, one electrostatic and two symmetric electromagnetic modes, but only the electromagnetic mode is unstable and this is the filamentation instability.

5.1 Filamentation instability ($E = E_y$, $k = k_x$)

The filamentation instability does not exist in a nonstreaming plasma and has originally been described by Fried (1959). The mechanism of generation is similar to that of the Weibel in stability: any small magnetic perturbation is amplified by the relative motion of two counter-streaming plasmas without any contribution of their intrinsic temperature anisotropy. This instability is also purely growing and has the electric field oriented along the streaming direction. Therefore, for a simple characterization of the filamentation instability, first we assume the streams thermally isotropic, $A \equiv T_y/T_x = 1$, with isotropic velocity distributions of Maxwellian type. The dispersion relation (23) provides then for the electromagnetic modes

$$\frac{k_x^2 c^2}{\omega^2} = \epsilon_{yy}^{\mathrm{FI}} = 1 + \frac{\omega_{p,e}^2}{\omega^2}\left[2\frac{v_0^2}{v_{th}^2} + \left(1 + 2\frac{v_0^2}{v_{th}^2}\right)\frac{\omega}{k_x v_{th}} Z\left(\frac{\omega}{k_x v_{th}}\right)\right]. \tag{24}$$

The unstable purely growing solutions describe the filamentation instability, and the growth rates are numerically derived and displayed with solid lines in Fig. 5. We should observe that they are restricted to wave-numbers less than a cutoff given by

$$k_{x,c}^{\mathrm{FI}} = \sqrt{2}\frac{\omega_{pe}}{c}\frac{v_0}{v_{th}}. \tag{25}$$

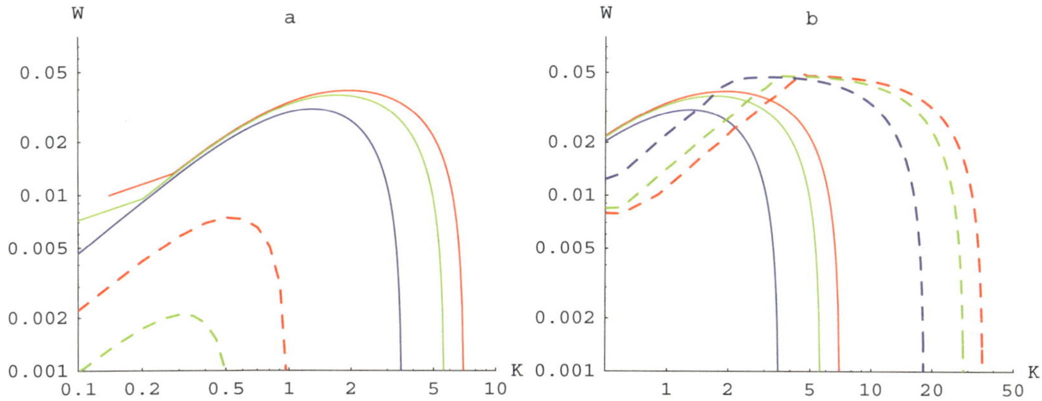

Fig. 5. The growth rates of the filamentation instability (solid lines) as given by Eq. (24) for a streaming speed $v_0 = c/20$, and different parallel thermal spread of plasma particles: $v_{th,y} = c/100$ (red lines), $c/80$ (green), $c/50$ (blue). The growth rates of the cumulative filamentation-Weibel instability given by (26) are shown with dashed lines for (a) $v_{\rm th}/v_{{\rm th},y} = 5$ and (b) $v_{\rm th}/v_{{\rm th},y} = 1/5$. The coordinates are scaled as $W = \omega_i/\omega_{pe}$ versus $K = kc/\omega_{pe}$.

5.2 Cumulative filamentation-Weibel instability ($E = E_y$, $k = k_x$, $A \neq 0$).

Since plasma streams exhibit an internal temperature anisotropy (see Fig. 2, a and b) the filamentation instability can be either enhanced by the cumulative effect of theWeibel instability when $T_y > T_x$ (see Fig. 5, b), or in the opposite case of $T_y < T_x$, the effective velocity anisotropy of plasma particle decreases and the instability is suppressed (see Fig. 5, a).

For streams with a finite intrinsic temperature anisotropy, $A \neq 0$, the dispersion relation (23) provides for the electromagnetic modes

$$\frac{k_x^2 c^2}{\omega^2} = \epsilon_{yy}^{\rm FWI} = 1 - \frac{\omega_{p,e}^2}{\omega^2}\left\{1 - \left(A + 2\frac{v_0^2}{v_{th}^2}\right)\left[1 + \frac{\omega}{k_x v_{th}} Z\left(\frac{\omega}{k_x v_{th}}\right)\right]\right\}. \tag{26}$$

In this case the unstable purely growing solutions describe the cumulative filamentation-Weibel instability, and the growth rates are displayed with dashed lines in Fig. 5. Again, we remark that the unstable solutions are restricted to wave-numbers less than a cutoff value which is given by

$$k_{x,c}^{FWI} = \frac{\omega_{pe}}{c}\left(A - 1 + 2\frac{v_0^2}{v_{th}^2}\right)^{1/2}. \tag{27}$$

The condition of existence for Eq. (27) provides the threshold of the cumulative filamentation-Weibel instability:

$$v_{th} < \left(v_{th,y}^2 + 2v_0^2\right)^{1/2} \equiv v_{th,c}. \tag{28}$$

For interested readers, supplementary analysis of this instability can be found in the recent papers of Bret et al. (2004, 2005a,b); Bret & Deutsch (2006); Lazar et al. (2006); Stockem & Lazar (2008); Lazar (2008); Lazar et al. (2008, 2009d, 2010). Here we continue to consider symmetric counterstreams making a simple description of this instability and compare to the other unstable modes discussed above.

5.2.1 $A = T_y/T_x < 1$

As plasma streams are transversally hotter, the effective anisotropy of the particle velocity distribution with respect to their mass center at rest decreases, and the growth rates of the cumulative filamentation-Weibel instability become also smaller (see Fig. 5 a). This instability is inhibited by a surplus of transverse kinetic energy (Lazar et al.; 2006; Stockem & Lazar; 2008). Furthermore, it has two competitors in the process of relaxation: the two-stream instability and theWeibel instability, both propagating parallel to the streams and described in the sections above.

For a complete characterization of their interplay, the growth rates of these three instabilities are displayed in Fig. 6 for the same conditions used in Fig. 4 but, for clarity, only two cases are plotted: $v_{th,y} = c/100$ (red lines), $c/30$ (green). Thus, the filamentation (cumulative filamentation-Weibel) growth rates (plotted with dashed lines) are smaller than the Weibel instability growth rates (dotted line), which are, in turn, smaller than those of the filamentation instability (solid lines). Moreover, when thermal spread of plasma particles is large enough, the surplus of kinetic energy transverse to the streams compensates the opposite particle velocity anisotropy due to bulk (counterstreaming) motion along the streams, and the effective anisotropy of plasma particles vanishes. In this case, the filamentation instability is completely suppressed: no growth rates are found in Fig. 6 for $v_{th,y} = c/30$ (no green dashed line). That is confirmed by the threshold condition (28): the instability exists only for $v_{th} < v_{th,c} \equiv \left(v_{th,y}^2 + 2v_0^2\right)^{1/2} \simeq c/14$, and in Fig. 6 this condition is satisfied only in the case of $v_{th} = A^{-1/2} v_{th,y} = \sqrt{5}c/100 < c/14$ (red dashed line), but not for $v_{th} = A^{-1/2} v_{th,y} = \sqrt{5}c/30 > c/14$ (no green dashed line).

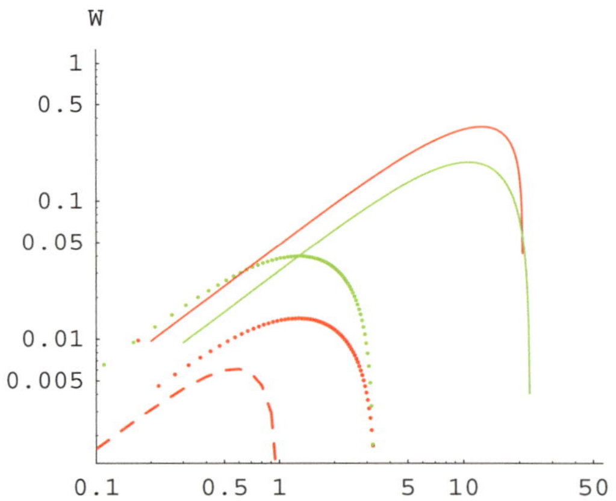

Fig. 6. The growth rates ($W = \omega_i/\omega_{pe}$ versus $K = kc/\omega_{pe}$) of the Weibel instability (dotted lines), the two-stream instability (solid lines), and the filamentation (cumulative filamentation-Weibel) instability (dashed line) for the same plasma parameters considered in Fig. 4: $v_0 = c/20$, $v_{th,y} = c/100$ (red lines), $c/30$ (green). The excess of transverse kinetic energy , $v_{th}/v_{th,y} = 5$, diminishes the growth rates of the filamentation instability (red dashed line), or even suppresses the instability (no green dashed line).

In section **4.2.** we have shown that the thermal spread of plasma particles along the streams prevents a fast developing of the two-stream instability, which, in general, is the fastest mechanism of relaxation. Furthermore, here it is proved that kinetic effects arising from the perpendicular temperature of the streams could stabilize the non-resonant filamentation mode. These results have a particular importance for the beam-plasma experiments, specifically, in the fast ignition scenario for inertial confinement fusion, where these instabilities must be avoided.

5.2.2 $A = T_y/T_x > 1$

In the opposite case, when plasma streams exhibit an excess of parallel kinetic energy, $A = T_y/T_x > 1$, theWeibel effect due to the temperature anisotropy cumulates to the filamentation instability given by the relative motion of counterstreaming plasmas, and yields an enhancing of the growth rate (see Fig. 5 b).

In this case, there is only one competitor for the cumulative filamentation-Weibel instability, and this is the two-stream electrostatic instability. The growth rates of these two instabilities are plotted in Figures 7 and 8 for several representative situations.

In Fig. 7 we consider a situation similar to that from Fig. 4 but this time with an excess of parallel kinetic energy. Thus, for a given anisotropy, $v_{th,y}/v_{th} = 5$, the growth rates of the two-stream instability (solid lines), are inhibited by the parallel thermal spread of plasma particles and decrease. The growth rates of the filamentation instability (dashed lines) are relatively constant, but the instability is constrained to smaller wave-numbers according to Eq. (27).

On the other hand, in Fig. 8 we change and follow the variation of the growth rates with the anisotropy: the streaming velocity is higher but still not relativistic, $v_0 = c/10$, $v_{th} = c/100$, and the anisotropy takes three values, $v_{th,y}/v_{th} = 1$ (red lines), 4 (green), and 10 (blue). In this case the cumulative filamentation-Weibel instability becomes markedly competitive, either extending to larger wave-numbers according to Eq. (27), or reaching at saturation, maximums growth rates comparable or even much larger than those of the two-stream

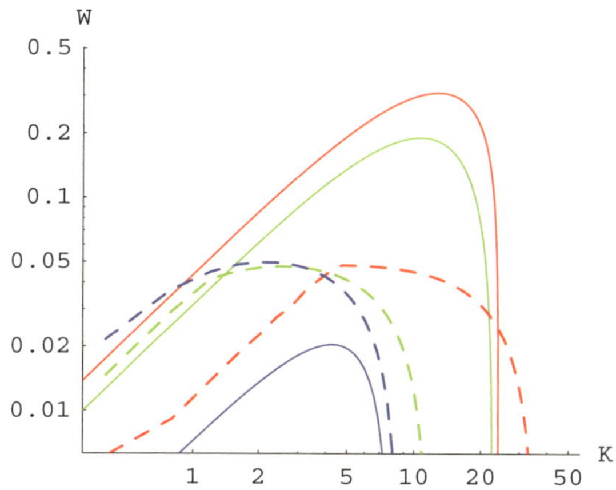

Fig. 7. The growth rates ($W = \omega_i/\omega_{pe}$ versus $K = kc/\omega_{pe}$) of the two-stream instability (solid lines), and the filamentation instability (dashed lines) for the same plasma parameters considered in Fig. 4: $v_0 = c/20$, $v_{th,y} = c/100$ (red lines), $c/30$ (green), $c/20$ (blue), but an opposite temperature anisotropy, $v_{th}/v_{th,y} = 1/5$.

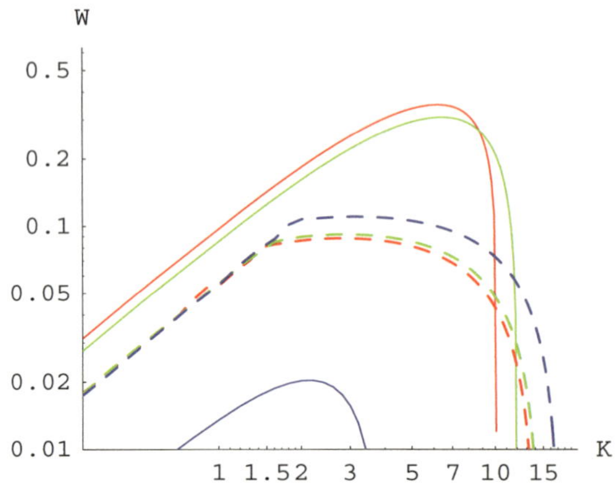

Fig. 8. The same as in Fig. 7 but for: $v_0 = c/10$, $v_{th} = c/100$ and the anisotropy $v_{th,y}/v_{th} = 1$ (red lines), 4 (green), 10 (blue).

instability. The main reason for that is clear, the two-stream instability is inhibited by increasing T_y, and in this case, the cumulative filamentation-Weibel instability can provide the fastest mechanism of relaxation for such counterstreaming plasmas.

This instability can explain the origin of the magnetic field fluctuations frequently observed in the solar wind, and which are expected to enhance along the temperature anisotropy thresholds.

6. Discussion and summary

In this chapter, we have described the interplay of kinetic plasma instabilities in a counterstreaming plasma including a finite and anisotropic thermal spread of charge carriers. Such a complex and anisotropic plasma model is maybe complicated but it allows for a realistic investigation of a wide spectra of plasma waves and instabilities. Small plasma perturbations, whether they are electrostatic or electromagnetic, can develop and release the free energy residing in the bulk relative motion of streams or in thermal anisotropy. Two types of growing modes have been identified as possible mechanisms of relaxation: an electrostatic growing mode, which is the two-stream instability, and two electromagnetic growing modes, which are the Weibel instability and the filamentation instability, respectively. The last two can cumulate leading either to enhancing or quenching the electromagnetic instability.

The most efficient wave mode capable to release the excess of free energy and relax the counterstreaming distribution, will be the fastest growing wave mode, and this is the mode with the largest maximum growth rate. Thus, first we have presented the dispersion approach and the dispersion relations of the unstable modes, and then we have calculated numerically their growth rates for various plasma parameters. Possible applications in plasma astrophysics and fusion experiments have also been reviewed for each case in part.

When the intrinsic temperature anisotropy is small, the two stream electrostatic instability develops first and relaxes the counterstreams to an anisotropic bi-Maxwellian plasma, which is unstable against the excitation of Weibel instability.

If the intrinsic temperature anisotropy increases, the electromagnetic instabilities can be faster than the two-stream instability. This could be the case of a plasma hotter along the

streaming direction, when the two-stream instability is inhibited, but the contributions of the filamentation and Weibel instabilities cumulate enhancing the magnetic instability. Otherwise, if the plasma kinetic energy transverse to the streams exceeds the parallel kinetic energy, the anisotropy in velocity space decreases and becomes less effective, and the filamentation instability is reduced or even suppressed. However, in this case a Weibel-like instability arises along the streaming direction, and if the temperature anisotropy is large enough, this instability becomes the fastest mechanism of relaxation with growth rates larger than those of the two-stream and filamentation instabilities.
We have neglected any influence of the ambient stationary fields, but the results presented here are also appropriate for the weakly magnetized (high-beta) plasmas widely present in astrophysical scenarios.

7. References

Bret, A. & Deutsch, C. (2006). Stabilization of the filamentation instability and the anisotropy of the background plasma, *Phys. Plasmas*, 13, 022110.

Bret, A.; Firpo, M.-C. & Deutsch, C. (2004). Collective electromagnetic modes for beam-plasma interaction in the whole *k* space, *Phys. Rev. E*, 70, 046401.

Bret, A.; Firpo, M.-C. & Deutsch, C. (2005). Electromagnetic instabilities for relativistic beam-plasma interaction in whole k space: Nonrelativistic beam and plasma temperature effects, *Phys. Rev. E*, 72, 016403.

Bret, A.; Firpo, M.-C. & Deutsch, C. (2005). Characterization of the initial filamentation of a relativistic electron beam passing through a plasma, *Phys. Rev. Lett.*, 94, 115002.

Cottrill, L.A.; Langdon, A.B.; Lasinski, B.F. et al. (2008). Kinetic and collisional effects on the linear evolution of fast ignition relevant beam instabilities, *Phys. Plasmas*, 15, 082108.

Davidson, R.C.; Hammer, D.A.; Haber, I. & Wagner, C.E. (1972). Nonlinear development of electromagnetic instabilities in anisotropic plasmas, *Phys. Fluids*, 15, 317.

Davidson, R.C.; Kaganovich, I.; Qin, H. & Startsvev, E.A. (2004), Collective instabilities and beam-plasma interactions in intense heavy ion beams, *Phys. Rev. ST Accel. Beams*, 7, 114801.

Fried, B. D. (1959), Mechanism for instability of transverse plasma waves, *Phys. Fluids*, 2, 337

Fried, B.D. & Conte, S.D. (1961). *The Plasma Dispersion Function*, Academic Press, New York.

Hellinger, P.; Travnicek, P.; Kasper, J. C. & A. J. Lazarus (2006). Solar wind proton temperature anisotropy: Linear theory and WIND/SWE observations, *Geophys. Res. Lett.*, 33, L09101 .

Kalman, G.; Montes, C. & Quemada, D. (1968). Anisotropic plasma temperature instabilities, *Phys. Fluids*, 11, 1797.

Kasper, J. C.; Lazarus, A. J. & Gary S. P. (2002).Wind/SWE observations of firehose constraint on solar wind proton temperature anisotropy, *Geophys. Res. Lett.*, 29, 1839.

Lazar, M. (2008). Fast magnetization in counterstreaming plasmas with temperature anisotropies, *Phys. Lett. A*, 372, 2446.

Lazar, M.; Dieckmann, M.E. & Poedts, S. (2009). Resonant Weibel instability in counterstreaming plasmas with temperature anisotropies, *J. Plasma Phys.*, published online, doi:10.1017/S0022377809008101

Lazar. M., Schlickeiser, R. & Poedts, S. (2009). On the existence of Weibel instability in a magnetized plasma. Parallel wave propagation, *Phys. Plasmas*, 16, 012106.

Lazar. M., Schlickeiser, R., Poedts, S. & Tautz, R. C. (2008). Counterstreaming magnetized plasmas with kappa distributions. I. Parallel wave propation, *Mon. Not. R. Astron. Soc.*, 390, 168.

Lazar. M., Schlickeiser, R., & Shukla, P. K. (2006). Cumulative effect of the filamentation and Weibel instabilities in counterstreaming thermal plasmas, *Phys. Plasmas*, 13, 102107.

Lazar, M.; Schlickeiser, R.; Wielebinski, R. & Poedts, S. (2009). Cosmological effects of Weibeltype instabilities, *Astrophys. J.*, 693, 1133-1141.

Lazar, M.; Smolyakov, A.; Schlickeiser, R. & Shukla, P.K. (2009). A comparative study of the filamentation andWeibel instabilities and their cumulative effect. Nonrelativistic theory, *J. Plasma Phys.*, 75, 19.

Lazar. M., Tautz, R. C., Poedts, S. & Schlickeiser, R. (2008). Counterstreaming magnetized plasmas with kappa distributions. I. Perpendicular wave propation, *Mon. Not. R. Astron. Soc.*, DOI: 10.1111/j.1365-2966.2009.15647.x (in press) .

Medvedev, M. & Loeb, A. (1999). Generation of magnetic fields in the relativistic shock of gamma-ray burst sources, *Astrophys. J.*, 526, 697-706.

Nishikawa, K.-I.; Hardee, P; Richardson, G. *et al.* (2003). Particle acceleration in relativistic jets due to Weibel instability, *Astrophys. J.*, 595, 555-563.

Okada, T., Yabe, T. & Niu, K. (1977). , *J. Phys. Soc. Jpn.*, 43, 1042

Schlickeiser, R. & Shukla, P.K. (2003). Cosmological magnetic field generation by the Weibel instability, *Astrophys. J.*, 599, L57-L60.

Stockem, A., & Lazar, M. (2008). Revision of "Cumulative effect of the filamentation and Weibel instabilities in counterstreaming thermal plasmas" , *Phys. Plasmas*, 15, 014501.

Stverak, S.; Travnicek, P.; Maksimovic, M. et al. (2008). Electron temperature anisotropy constraints in the solar wind, *J. Geophys. Res.*, 113, A03103.

Weibel, E.S. (1959). Spontaneously growing transverse waves in a plasma due to an anisotropic velocity distribution *Phys. Rev. Lett.*, 2, 83.

20

Plasma Wave Propagation with Light Emission in a Long Positive Column Discharge

Guangsup Cho[1] and John V. Verboncoeur[2]
[1]*Department of Electrophysics, Kwangwoon University 447-1 Nowon Wallgye, Seoul 139-701*
[2]*Department of Nuclear Engineering, University of California, Berkeley CA 94720,*
[1]*Korea*
[2]*USA*

1. Introduction

The fine-tube cold cathode fluorescent lamp (CCFL) [1-2] and external electrode fluorescent lamp (EEFL) [3-9], have been widely used as a backlight source in liquid crystal display (LCD) modules. In a notebook PC, the glass tube diameter of the lamp is a small as 1 *mm*. As the display area becomes larger in LCD TVs, the tube diameter is 3~4 *mm* and the lamp length is up to 1~2 *m*. However, the mechanism for radiation of light along such a long positive column is not well understood.
The characteristics of light emission in the EEFLs have been reported, specifically the propagation of a light-emitting along a tube driven by high voltage AC [8-9]. In the long positive column lamp, the light emits first from the side of high voltage propagates to the ground side with the propagation time of ~μs and the propagation velocity of 10^5~10^6 *m/s*. The previous observations have not been reported elsewhere to our knowledge since the fluorescent lamps have been introduced commercially in the 1940's.
However, those observations of light propagation would not be recognized even if they have been found first in the history of fluorescent lamps. Even more, those observations are misapprehended as the kinds of unstable waves during the lamp operation of lamp, such as striation, snaking, flickering, and streaming, etc. Generally, a moving light source in the long positive column plasma has been associated with “striation”. A striation is a quasi-standing wave which can be observed visually [10- 12]. However, striations as well as the other unstable waves are not observed in commercial lamps, and would be considered a major defect for hot cathode fluorescent tubes as well as CCFL. In the manufacture of those lamps, inadequate vacuum during the evacuation process results in impurity gases such as H_2O, CO_2, O_2, and N_2, etc. These impurities cause the perturbations of unstable discharge, and those lamps are discarded in the production line. Striations can sometimes appear in aging lamps or in the operation with an unstable inverter system. However, we have never observed striations in the normally manufactured lamps of our test samples. Our observation of optical signal is quite different from striations, as it is not visible to the naked eye and moves on the micro-second timescale. Rather than the instability, they are related to the stable process of distribution of plasma and fields that generate light in these discharges.

In the previous studies, the propagation velocity of light-emitting wave-front along the positive column has been observed u_p~$(10^{+5}$~$10^{+6})$ *m/s*. In Ref. [8], the propagation speed u_p was explained as the electron drift velocity, u_p~u_d. However, typical electron drift velocities driven by an electric field are $u_d \leq 10^{+4}$ *m/s* in a typical glow discharge plasma. Therefore, the propagation velocity of light emission cannot be related to the electron drift velocity. In Ref. [9], the propagation velocity has been explained as the electron plasma wave propagation having an electron thermal speed u_p~u_e. However, the electron plasma wave has not been described completely whether the wave can propagate a long distance of about 100~1,000 *mm* and a long time of a few *μs* since the damping time of an electron plasma wave is very short, about 10^{-9} s. In those respects, the propagation of optical emission could be neither the electron drift nor the electron plasma wave of electron thermal speed.

In this study, the propagation of light emission along a fine tube CCFL will be observed through the experimental measurement of the optical signals. CCFL can be operated with DC-voltage as well as AC-voltage while EEFL is driven only with AC-voltage. Through the operation of CCFL with DC and DC-ripple voltage, the propagation of light emission will be verified as new phenomena in a positive column discharge. The mechanism of these phenomena is the other scope in this study. However, the phenomena of light-propagation will provide a clue for the understanding of positive column plasma and a new motivation for the study of lamp physics. In this respect it is necessary to observe this phenomena more precisely again.

In Sec. 2, optical signal measurements are conducted according to the voltage of DC, rippled-DC, and AC in the operation of CCFLs. In the sub-section of Sec. 2-1, the optical propagation with DC-voltage operation is represented. In Sec. 2-2 with AC-voltage operation in the CCFL without coating the phosphor, the optical signal measurements are conducted according to the voltage variations. In this experiment we have obtained two different types of light propagation along the positive column. The propagation of light emission is analyzed with the plasma wave perturbation theory in Sec. 3. Discussions and conclusions are in Sec. 4 and Sec. 5, respectively.

2. Observation of light emission along the positive column

For the observation of light propagation along the tube, two kinds of operation as DC and AC voltage are involved. The DC-voltage operation with rippled voltage is the first attempt to observe the optical signals in the experiment of Sec. 2-1. In the AC-voltage operation of Sec. 2-2, we try to measure the light signals radiating directly from the discharge plasma through the naked glass without coating a phosphor inside glass wall. Additionally, the variation of optical signal according to the lamp current is represented in the AC-operation.

2-1. DC-Operation

In this experiment, the CCFL discharge tube is borosilicate glass, 0.5 *mm* thick with 4.0 *mm* outer diameter, and 699 *mm* length for 30" backlight units. The gas composition is typical for CCFLs, filled with 95 % neon and 5 % argon, at a total pressure of 60 *Torr*, and also containing 2 *mg* of mercury in liquid phase. The inner surface of the glass tube is coated with RGB tri-phosphor. Cylindrical Ni-electrodes are inserted at the ends sealed with the connection of the lead line of electrode.

The circuit system in this study has been designed to measure the optical signal with the conventional operation method as shown in Figure 1. For the operation of CCFL, two kinds of power supply systems are used as DC power and modulate-DC power as shown in the schematic drawing of the experimental circuit system. For DC and modulate-DC operation, a high voltage is applied to one end electrode connected with a 200 $k\Omega$ ballast resistor in series. The other electrode of lamp is grounded. The modulate-DC voltage is rippled $\pm\Delta V$ on the DC voltage V_o as $V(t)=V_o+\Delta V\sin\omega t$. The lamp voltage is measured at the electrode side of high voltage. The lamp current is measured at the grounded electrode with ammeter. The optical signals are measured at three equally spaced points, ①~③, along the tube surface with a photo-sensor, which is a light-to-voltage conversion amplifier using an optical fiber. The spacing between the measuring points is ΔL=305 *mm* and the length between ① and ③ is L=610 *mm* along the entire tube of 699 *mm* length.

Figure 2 for the operation with DC-voltage, shows the oscilloscope signals of voltage, current, and optical intensity measured at point ① where the luminance is 10,000 cd/m^2. The bottom line is the base value of zero for the DC-voltage, the DCcurrent, and the optical intensity. The lamp voltage noted with solid line is DC 1.24 kV after the voltage drops by 8.40 kV at the 200 $k\Omega$ ballast resistor from the DC 9.64 kV supplied at the DC-power source. The lamp current measured at the grounded electrode is 4.2 mA. The lamp power is 5.2 W. The optical signal as well as the current signal is constant on time due to the DC-voltage operation mode as it is expected. In the DC-voltage mode, the propagation of the optical signal cannot be observed since all the signals of light emission along the tube are the same. Since the light originates from atomic excitation due to the impact of high-energy electrons, the uniform light emission signals along the tube do not show any difference in the DC-voltage operation.

In order to observe the propagation of light-emission, a modulate-DC voltage is applied. The signals of voltage, current, and optical pulse, are represented with solid lines in Fig. 3(a). All the signals of straight line in the DC operation of Fig. 2 are also noted with dashed lines. The optical signal is measured at point ① where the luminance is 10,000 cd/m^2. Using the switching diode in the circuit of power supply, the high AC-voltage of 64.5 kHz is rectified to be a rippled high DC-voltage, $V_o+\Delta V\sin\omega t$ with the ripple frequency $f=\omega/2\pi$=129 kHz. The lamp voltage of rippled DC is averaged V_o=1.21 kV, while the output voltage from power source is 9.61 kV, so the voltage drops by 8.40 kV at the 200 $k\Omega$ ballast resistor. The ripple voltage is ΔV=110 V with $\Delta V/V_o$~10 %. Lamp current is also rippling as $I_o+\Delta I\sin\omega t$ with I_o~4.2 mA and ΔI~1.1 mA. The average lamp power is 5.1 W. The optical signal observed at position ① shows the wave rippled with the optical intensity of $O(t)\sim O_o+\Delta O\sin\omega t$ with $\Delta O/O_o$~10 % and the same frequency of 129 kHz as the rippling voltage and current.

In order to obtain the information of light-emission propagation for the operation of rippled DC voltage, the optical signals are observed at different points along the tube. Figure 3(b) shows the optical signals at points ①~③ with the time delay for each signal. The optical signal at ① is measured when the luminance is 10,000 cd/m^2 at ①. When the luminance at ① is 10,000 cd/m^2, the luminance at ② and ③ is slightly low as 9,600 cd/m^2 at ②, and 9,200 cd/m^2 at ③. The optical intensity at each measuring point is proportional to the luminance. When the optical intensity at ① is averaged to be O_o(①)~1, the intensity at ② and ③ is to be decayed to be O_o(②)~0.96 and O_o(③)~0.92, respectively. The time interval of optical signals between ①~③ is the propagation time t_p of light emission from ① to ③. The propagation time is t_p=1.82 μs. If the optical intensity from ① to ③ is decayed as $O_o(t)=O_o\exp(\ t/t_D)$, the characteristic decay time t_D of optical intensity from ① to ③ is t_D~22 μs from the calculation of O_o(③)/O_o(①)~0.92~exp(-t_p/t_D) with t_p=1.82 μs. The decay time t_D~22 μs is longer than the optical propagation time t_p=1.82 μs as well as the period of ripple voltage, $t_f=1/f=7.75$ μs with f=129 kHz. The propagation speed of light emission is $u_p=3.35\times10^5$ m/s which is calculated with $u_p=L/t_p$ as the ratio of the length between ①~③ of L=610 mm to the signal propagation time of t_p=1.82 μs. If the operation voltage increases, the luminance as well as the light emission propagation speed increases.

2.2 AC-operation of CCFL without phosphor coating

In order to directly observe a light emission from the plasma discharge with CCFLs, the inner wall of the glass tube is not coated with phosphor. The CCFL discharge tube is borosilicate glass, 0.5 mm thick with 3.4 mm outer diameter, and 900 mm length. The gas composition is typical for CCFLs, filled with 95 % neon and 5 % argon, at a total pressure of 60 $Torr$, and also containing 2 mg of mercury in liquid phase.

The circuit system in this study has been designed to measure the optical signal with the conventional operation method as shown in Figure 4. The high voltage of $V(t)=V_o\sin(\omega_o t)$ with $\omega_o=2\pi f_o$ is applied to one electrode through the ballast capacitor C=22 pF, and the other electrode is grounded. AC power is supplied through a DC-AC inverter with driving frequency f_o=65 kHz with the period 15 μs. The optical signals are measured at equally spaced points, ①~⑤, along the tube surface with a photosensor, which is a light-to-voltage conversion amplifier using an optical fiber. The spacing between the measuring points is ΔL=220 mm and the length between ① and ⑤ is L=880 mm along the entire tube of 900 mm length. The lamp current is measured at the ground, and the circuit voltage V_a is measured at the point a and the lamp voltage V_b is at the point b.

Figure 5 shows the I-V curves represented with the applied voltage V_a and the lamp voltage V_b. It shows the typical glow discharge process, the primary discharge before Townsend breakdown, and the normal glow discharge after Townsend breakdown. As the circuit voltage V_a is increased, the current of the primary discharge increases to a few mA. It increases to 2.6 mA just before the breakdown voltage V_a=2 kV, and it increases abruptly to be several tens of mA after breakdown. However, the lamp voltage V_b is increased to be 1.8 kV at Townsend breakdown and the voltage falls to below 0.98 kV after Townsend breakdown. The curve of lamp current versus lamp voltage V_b shows the property of negative resistance after Townsend breakdown. In this experiment, the optical signals are measured at the low current range noted with Figs. 6(a) and 6(b) before Townsend breakdown and at the high current of Figs. 6(c) and 6(d) after Townsend breakdown.

Figure 6 shows the oscilloscope signals measured at points ①~⑤ when the CCFL driving voltage is varied. The applied voltage and current corresponding to Figs. 6(a)~6(d) are noted in the I-V curve of Fig. 5. Fig. 6(a) is the applied voltage V_a=1.8 kV, the lamp voltage V_b=1.6 kV, and the lamp current 2.5 mA. Fig. 6(b) is just before Townsend breakdown with V_a=2.0 kV, V_b=1.8 kV, and the lamp current 2.6 mA. Fig. 6(c) is just after Townsend breakdown with V_a=2.3 kV, V_b=0.98 kV, and I=31.0 mA. Fig. 6(d) is V_a=2.45 kV, V_b=0.97 kV, and I=32.6 mA. All the voltage and current data are in root mean square values. Since the peaks of the optical signals correspond to the voltage peaks of $\pm V_o$ in two polarities, the frequency of the optical signal on the oscilloscope is $2f_o$=130 kHz, twice that of the driving voltage.

Figs. 6 show the optical signals of the discharges before and after Townsend breakdown. In Figs. 6(a) and 6(b) before Townsend breakdown, the optical signals have a time shift and a decay of intensity as the measuring point goes from ① to ⑤. The optical intensity corresponding to the negative polarity of voltage is a little higher than that of the positive polarity at the same measuring point. In Figs. 6(c) and 6(d) after Townsend breakdown, the time shifts of each optical signal measured at the points ①~⑤ are smaller and the deviations of optical intensity at each point and for both polarities are also smaller than those of Figs. 6(a) and 6(b). However, we have two types of optical signals. For the first type, the propagation of the optical signal is slow and the optical intensity decays as the measuring point from ① to ⑤ as shown in Figs. 6(a) and 6(b). For the second type, the propagation time is short and the intensity is nearly the same value between measuring points as shown in Figs. 6(c) and 6(d).

In the optical signals of Fig. 6(a), the total propagation time from ① to ⑤ of L=880 mm length is t_p~5.31 μs with the mean propagation velocity $u_p \sim L/t_p \sim 1.7\times10^{+5}$ m/s. The propagation speed slows slightly from ① to ⑤ but is nearly constant. In the optical signals of Fig. 6(b), the propagation time from ① to ⑤ is t_p~4.00 μs with the mean propagation velocity $u_p \sim 2.2\times10^{+5}$ m/s. As the current and voltage increase before breakdown, the propagation speed u_p increases. After the breakdown discharge of Figs. 6(c) and 6(d), the propagation speed is $u_p \sim 2.6\times10^{+6}$ m/s with the propagation time t_p~0.34 μs in Fig. 6(c) and the propagation speed is $u_p \sim 4.6\times10^{+6}$ m/s with t_p~0.19 μs in Fig. 6(d). The propagation speed after breakdown is faster by a factor of 10.

For the explanation of the optical propagation according to the two types of discharge, before and after the Townsend breakdown, the plasma diffusion and the electron plasma wave theory is adopted in the positive column plasma. In plasma diffusion, we calculate the diffusion speed along the discharge tube from the diffusion equation. For the electron plasma wave, the driving frequency f_o~65 kHz is treated as a source of the perturbation into the positive column plasma of the normal glow discharge after the Townsend breakdown. The key issues of this study will be whether a plasma wave of this driving frequency can be propagated, and thc physical effects that cause this wave to be propagated through the positive column filled with high density plasma.

3. Analysis of plasma propagation along the positive column

For the purpose of understanding the propagation of light emission, the experimental observations in the previous sections are summarized as:

i. Light always propagates from the high voltage side to the ground. The direction of propagation is independent of the electron drift velocity u_d whose directions are varied according to the electric field with the polarity of applied voltage.
ii. The speed of light propagation is $u_p \sim (10^5\text{-}10^6)$ m/s according to the lamp current. At the low current ~1 mA, the velocity is $u_p \sim 10^5$ m/s. As the current increases at the normal glow, the propagation velocity also increases over 10^6 m/s.
iii. The optical signals represent the type of ripple wave as $O(t)=O_o+dO\sin\omega t$. The ripple frequency is the frequency of voltage pulses. In DC-operation, the ripple frequency is the frequency of DC-voltage pulse. When the operation frequency is ω_o in AC-voltage, the frequency of optical ripple wave is $\omega=2\omega_o$, 2-times of AC-operation frequency. With respect to the ripple ratio dO/O_o, we have two types of optical signals according to the

current. At the low current before breakdown, the ripple ratio can be very large over 50 % and the signals decay along the z-direction of positive column as $O(t, z)=O_o+dO\exp(-z/z_D)\sin\omega t$. The scale of decay is about the lamp length L as z_D~L. At the normal glow of high current, the optical signals would not represent the decay with the large scale of decay as $z_D>L$ and the ripple ratio is about dO/O_o~20-30 %.

For the interpretation of experimental results summarized above, our models for the analysis are listed below as:

i. Plasma is generated at high voltage side of $z=0$ and diffused toward the ground electrode of $z=L$. At the steady state, the gradient of plasma density is formulated by the continuous generation of plasma at the high voltage side. The density is high at $z=0$ and is decreased toward $z=L$.

ii. Perturbation is generated at the high voltage side by the high voltage $V(t)=V_o\sin\omega_o t$ with the frequency $f_o=\omega_o/2\pi$. The wave is generated initially at the high voltage side with the frequency $\omega=2\pi(2f_o)$~10^5 s^{-1} for the operation frequency $f_o=65$ kHz. In this analysis we will show the wave always propagates to the z-direction with the propagation velocity and the decay. We will find what kinds of effect are for the propagation.

iii. The wave cannot be observed with the naked eye. It means the wave length is sub-millimeter. Then the phase velocity is determined with the wave frequency ω~10^5 s^{-1}. The group velocity of propagation should be ~(10^5-10^6) m/s according to the experimental data. For the presentation of wave decay along the tube, we have the characteristic decay length, which should be above the scale of lamp length, about 1 m or 10 m according to the lamp current.

In the previous reports [8-9], the fatal results of their analysis are that the waves are dependent of the electron drift velocity u_d whose direction is changed with the voltage polarity. Therefore, they do not explain the direction of propagation observed in the experiments where the waves always propagate from high voltage side to the ground without decay or with a small decay.

3.1 Electron plasma fluid equations and plasma parameters

The electron plasma waves have been described in Ref. [13-22] where all the waves cannot be propagated due to the high damping due to the elastic collision of electron. For the difference from Ref. [8-19], this report includes plasma density gradient along the positive column, which will be an important factor for the propagation of electron plasma waves without damping. The electron equation of motion and continuity equation are with $J\approx -neu_d$:

$$n\left[\frac{\partial \vec{u}}{\partial t}+\left(\vec{u}\cdot\vec{\nabla}\right)\vec{u}\right]=-\nu_e\left(\vec{J}/e+D_a\vec{\nabla}n+n\vec{u}\right) \tag{1}$$

$$\frac{\partial n}{\partial t}+\vec{\nabla}\cdot\left(n\vec{u}\right)=\nu_{iz}n \tag{2}$$

In the above electron fluid equation, the ambi-polar diffusion is considered [14-19]. Before manipulating the equations, the plasma parameters in this study are estimated follows.

In the general fluorescent lamps for LCD-TVs, the typical gas composition is filled with Ne (95 %)+Ar (5 %), at a total pressure 60 $Torr$, and also containing 2 mg of mercury in liquid phase. The neutral atom density is n_n~10^{+24} m^{-3} from the calculation of n_n~$3.25\times10^{+22}$P(*Torr*) m^{-3} [13-14]. The plasma is mostly Hg-ions with the plasma density n_o~$(10^{+16}$~$10^{+17})$ m^{-3} and the electron plasma temperature T_e~1 *eV* in a typical positive column of Hg discharge. The mean electron energy is $\varepsilon=3T_e/2$~1.5 *eV* and the electron thermal velocity is $u_e=(2\varepsilon/m)^{1/2}$~$7\times10^{+5}$ *m/s*. An electron collision frequency is v_e~$n_n\sigma u_e$~$2.3\times10^{+10}$ s^{-1} with σ(Ne)~2×10^{-20} m^2 at ε=1.5 *eV* [23]. The electron mean free path is λ_e~$(1/n_n\sigma)$~43 μm. The ionization collision frequency is v_{iz}~$n_n\sigma^{iz}u_e$~10^{+6} s^{-1} where the ionization collision cross section is very sensitive to the value of electron thermal energy T_e.

Generally, the diffusion coefficient of electron, ambi-polar, and ion, are estimated to be D_e~10^{+1}>D_a~$(10^{-2}$~$10^{-3})$>D_i~10^{-5} (m^2/s). In the calculation of D_e, we obtain D_e~35 m^2/s with T_e=1 *eV* and v_e~10^{+10} s^{-1}. When the mercury ion mass (Hg)~200 m_p (proton mass) and the ion thermal velocity u_i~$2\times10^{+2}$ *m/s* for T_i~0.03 *eV* and ion-neutral atom cross section σ^{in}~10^{-18} m^2 [15], we have the ion-neutral atom collision frequency $v_i=n_n\sigma^{in}u_i$~$4\times10^{+8}$ s^{-1}. Then, we obtain $D_a=T_e/Mv_i$~2.6×10^{-3} m^2/s.

The plasma density n_o is verified and the electron drift velocity u_d is calculated with the experiment data. When the operation frequency is f_o=60 *kHz*, the lamp current is about 1 *mA* or less before breakdown and the current is about 10 *mA* for the normal glow after breakdown. From the current $i=dQ/dt$~Qf_o, we have the total charge $Q=i/f_o$~$\{(1\text{-}10)/(60\times10^3)\}\times10^{-7}$ *C* corresponding to the current i=(1-10) *mA*. Then, the total plasma particles are $N=Q/e$. The plasma density is $n_o=N/(\pi r_o^2L)$ ~$(1.4\times10^{16}$~$1.4\times10^{17})$ m^{-3} with the tube inner radius r_o~1.5×10^{-3} *m* and the tube length L~1 *m*. The typical glow discharge plasma density is verified to be n_o~10^{16} m^{-3} before breakdown for the low current i~1 *mA* and to be n_o~10^{17} m^{-3} after breakdown for the current ~10 *mA* for the normal glow plasma. With the plasma density n_o, the electron drift velocity is calculated from $J=n_oeu_d=i/A$ and the tube cross section $A=\pi r_o^2$~7×10^{-6} m^2. Before breakdown for the low current i~1 *mA* and n_o~1.4×10^{16} m^{-3}, we have u_d~5×10^4 *m/s*. Even the plasma after breakdown for i~10 *mA* and n_o~1.4×10^{17} m^{-3}, we also have the electron drift velocity u_d~5×10^4 *m/s*. The typical value of drift velocity is known to be u_d~10^4 *m/s*.

3.2 Plasma diffusion along the tube

For the analysis of the experimental results, plasma diffusion is considered in a fine tube positive column. When a high voltage is applied to the electrode, the plasma of high density is generated at the high voltage electrode, and the electrons and ions diffuse along the positive column. Even if electron and ion fluxes are made up of identical ambi-polar diffusion fluxes and drift fluxes due to the electric current driven by the external field E, we have the diffusion equation for each plasma species of electrons and ions including the plasma generation in the continuity equation with the ionization collision v_{iz} [13-14]:

$$\left(\frac{\partial n}{\partial t}\right) - D_a\nabla^2 n = v_{iz}n \tag{3}$$

The motion of a charged particle is considered with radial and longitudinal components. Considering diffusion with time dependence in the z-direction only for a bounded plasma within the tube of inner radius r_o, the solution can be written $n(t, z, r)=n_r(r)n_z(t, z)$. The radial component of Bessel's equation of order zero becomes $\nabla_r^2 n_r+(c_o/D_a)n_r=0$ with a constant co

coupling to the z-component. The solution to the radial component is the Bessel function of order zero, $n_r(r) \sim J_0(r/[D_a/c_o]^{1/2})$. To satisfy the boundary condition $n_r(r_o)=0$ at the tube radius r_o, we have $\alpha = r_o/[D_a/c_o]^{1/2}$ which is the zero of the Bessel function $J_o(\alpha)=0$. The coupling constant $c_o=D_a(\alpha/r_o)^2$ depends on the value of α. According to the diffusion mode in a cylindrical cross section [13-19], the first zero of J_o is at $\alpha=2.4$.
Considering the diffusion only in the z-direction, we obtain the continuity equation

$$\frac{\partial n_z}{\partial t} - D_a \nabla_z^2 n_z = -\left(\frac{\alpha}{r_o}\right)^2 D_a n_z + v_{iz} n_z \tag{4}$$

One of the solutions for Eq. (2) is the time dependant Einstein's diffusion equation [16-18]:

$$n_z(t,z) = ct^{-1/2} e^{-(z^2/4D_a t)} e^{-(\alpha/r_o)^2 D_a t} e^{v_{iz} t} \tag{5}$$

where c is a normalization constant.
If the radial diffusion effect and ionization collisions are not included as r_o becomes infinite and $v_{iz}=0$, the solution becomes the well known solution of Einstein's diffusion equation [16-18], which is one dimensional diffusion away from a point source in an infinite medium. The solution represents the spread of the Gaussian formulation with the z-function. As time increases, the density of $n_z(t, z)$ will decrease and the distribution tends to become increasingly uniform.
The diffusion velocity is calculated as $u_n=-D_a(\nabla_z n)/n=z/2t$. Using the approximation of $(\alpha/r_o)2D_a>>v_{iz}$ and $z>>r_o$ since the tube radius is $r_o=1.2$ *mm* and the positive column length is $0<z<1$ *m* in this experiment, the time t at the peak value of density is obtained directly from Eq. (5) with $dn/dt=0$:

$$t \sim \frac{r_0}{2\alpha D_a} z \tag{6}$$

The above relation of Eq. (5) can be simply checked again from the derivative of exponential terms as $(z^2/4D_a t)+(\alpha/r_o)2D_a t$ in Eq. (4) with $(\alpha/r_o)2D_a>>v_{iz}$ and $z>>r_o$. Then we have $u_n=-D_a(\nabla_z n)/n=z/2t\sim(\alpha/r_o)D_a\sim40$ *m/s* with $\alpha=2.4$, $r_o=1.2\times10^{-3}$ *m*, and $D_a\sim10^{-2}$ m^2/s. In this section we verify the diffusion velocity u_n due to the plasma density gradient along the long positive column. When a high voltage applied to the electrode with the frequency $\omega=4\pi f_o\sim10^5$ s^{-1}, the plasma is generated at the high voltage side and it diffuses very slowly and travels only about 0.1 *mm* in the period about 10 μs of voltage pulse. Therefore, the plasma density gradient is formulated along the positive column in the steady state, a high density at the high voltage side and a low density at the ground side.

3.3 Electron plasma waves

The electron plasma wave [13-22] is represented with the perturbations of the wave varying as $f_1=f_1\exp[i(kz-\omega t)]$ where k is the propagation vector in the z-direction. The electron density n, the electron plasma fluid velocity $\vec{u}$, and the electric field $\vec{E}$, are separated into two parts: an equilibrium part indicated by the subscript o, and a perturbation part indicated by a subscript 1: $n=n_0+n_1$, $\vec{u}=\vec{u}_0+\vec{u}_1$, $\vec{E}=\vec{E}_0+\vec{E}_1$.

The equilibrium values of plasma are obtained at the steady state of electron fluid. When $d\vec{u}_0/dt\sim 0$ in the force balance of Eq. (1), the electron plasma fluid velocity $\vec{u}_0$ is represented

$$\vec{u}_0 = -\frac{\vec{J}}{n_0 e} - D_a \frac{\vec{\nabla} n_0}{n_0} = \vec{u}_d + \vec{u}_n \tag{7}$$

The electron drift velocity is $\vec{u}_d = -\mu_e \vec{E}_0 = -J/(n_0)e$ in the electric field $\vec{E}_0$. The electron fluid diffusion velocity of density gradient is $\vec{u}_n = -D_a \nabla n_o / n_o$.

Considering the plasma density in the continuity equation with steady state of $\partial n_o/\partial t\sim 0$, we have $\nabla^2 n_o + (v_{iz}/D_a) n_o = 0$. The plasma is generated at $z=0$ by the high voltage applied to the electrode and diffused into the positive column to z=L of lamp length. One of the solution $n_o(z) = n_o \cos\{(v_{iz}/D_a)^{1/2} z\}$. The corresponding diffusion velocity is $u_n = (v_{iz} D_a)^{1/2} \tan\{(v_{iz}/D_a)^{1/2} z\}$. The averaged magnitude of diffusion velocity is estimated to be $u_n \sim (v_{iz} D_a)^{1/2} \sim 10^2$ m/s with $D_a \sim 10^{-2}$ m^2/s and $v_{iz} \sim 10^6$ s^{-1}. The diffusion velocity ~10^2 m/s is too slow to diffuse into long positive column within micro-second of voltage pulse duration time so that the plasma diffusion signal is not fitted to the velocity of optical signal propagation having $10^5\sim 10^6$ m/s in this experiments. It takes about 10^{-2} s diffusing along the long positive column of 1 m length with this diffusion velocity. The distance of diffusion is about 0.6 mm during the period ~7 μs of a high voltage pulse.

The time var iation of equilibrium quantities are $d\vec{u}_0/dt\sim 0$, $\partial n_o/\partial t = 0$. The spatial variation of equilibrium plasma density is considered as $\nabla n_o = -(n_o/D_a)\,\vec{u}_n$ which provides the diffusion velocity $\vec{u}_n$ in this analyses. If we assume the velocity u_o as a constant value along the positive column for the simplicity, we have $\nabla u_o = 0$. When we consider $\nabla u_o = 0$, this effect are not affected on our results which will be discussed later.

Using the equilibrium values as $d\vec{u}_0/dt\sim 0$, $\partial n_o/\partial t = 0$, $\nabla u_o = 0$, and $\nabla n_o = -(n_o/D_a)\,\vec{u}_n$, the perturbed equations of u_1 and n_1 are

$$n_0\left(\frac{\partial u_1}{\partial t} + u_0 \nabla u_1\right) = -\nu_e\left[\frac{J_1}{e} + D\vec{\nabla} n_1 + n_1 \vec{u}_0 + n_0 u_1\right] \tag{8}$$

$$\frac{\partial n_1}{\partial t} + \vec{u}_0 \cdot \vec{\nabla} n_1 + n_0 \vec{\nabla}\cdot\vec{u}_1 + \vec{u}_1 \cdot \vec{\nabla} n_0 = \nu_{iz} n_1 \tag{9}$$

Using $J_1/(n_0)e = (-\vec{u}_d)_1 \approx \mu_e \vec{E}_1 = -\mu_e \vec{\nabla} V_1 = -ik\mu_e V_1 = i\mu_e e n_1/(\varepsilon_0 k)$ and $\nabla^2 V_1 \approx e n_1/\varepsilon_0$, we have $-k^2 V_1 = e n_1/\varepsilon_0$ by the assumption of stationary ions. Using $\partial/\partial t = -i\omega$ and $\nabla = ik$, the perturbed equations are

$$n_0 u_1(-i\omega + iku_0 + \nu_e) = -n_1 \nu_e\left[i\mu_e \frac{n_0 e}{\varepsilon k} + ikD + u_0\right] \tag{10}$$

$$n_1(-i\omega + iku_0 - \nu_{iz}) = -n_0 u_1\left(ik - \frac{u_n}{D_a}\right) \tag{11}$$

Combining above two equations, the dispersion equation of the electron plasma waves is obtained by the factoring out the first-order quantities:

$$\omega^2 - \omega_{pe}^2 - \kappa u_e^2 k^2 - 2\omega k u_o + i\omega \nu_e - iku_n \nu_e - i\frac{u_n \nu_e}{\kappa \lambda_D{}^2 k} - \frac{u_n u_o \nu_e{}^2}{\kappa u_e{}^2} = 0 \tag{12}$$

In the above equation, we use $\omega_{pe}{}^2 \gg v_e v_{iz}$ and $v_e D_a \gg u_o{}^2$ since $v_e D_a = \kappa u_e{}^2$ with $\kappa = mv_e/Mv_i$ and the relation $\lambda_D \omega_{pe} = u_e$ with the Debye shielding length λ_D.
If the electron flow $u_o = u_d + u_n$ and the electron collision frequencies v_e and v_{iz} are ignored, we have the typical electron plasma wave as the oscillation electrons at ω_{pe} are streaming into adjacent layers of plasma with thermal velocities u_e [14-19].
With an explicit expression for $\omega = \omega_r + i\omega_i$, the imaginary part of ω is expressed as

$$\omega_i = -\frac{\nu_e}{2}\frac{\omega_r - ku_n(1 + 1/\kappa \lambda_D{}^2 k^2)}{\omega_r - ku_o} \tag{13}$$

For the propagation wave without damping as $\omega_i = 0$, we have the wave,

$$\omega_r = ku_n(1 + 1/\kappa \lambda_D{}^2 k^2) \tag{14}$$

For the general electron plasma waves without density gradient as $\nabla n_o = 0$ and $u_n = 0$ in the above equation, we can verify all the waves are fast damping with the damping frequency $\omega_i \sim -v_e/2 \sim -10^{-10}\ s^{-1}$. The wave frequency ω_r is independent of the drift velocity u_d whose direction is changed by the polarities of applied voltage.
For the description of wave propagation along the longitudinal direction z of positive column, corresponding to the optical signals observed in this experiment, we would rather take the explicit expression with $k = k_r + ik_i$ than $\omega = \omega_r + i\omega_i$. The imaginary $\omega_i < 0$ shows the time decay as $\exp(\omega_i t)$ while the imaginary of $k_i > 0$ represents the scale of decay in the z-direction as $\exp(ikz) = \exp(ik_r z)\exp(-k_i z)$. When the scale of decay in the z-direction is z_D, we have $k_i = 1/z_D$ for $\exp(-z/z_D)$. In this experiment the decay scale of z_D is about 1~10 m. In a low current of ~1 mA we have $z_D \sim 1\ m$ which is $k_i \sim 1$. In the normal glow plasma of current ~10 mA we have a large decay scale as $z_D \sim 10\ m$ which is the case of $k_i \sim 10^{-1}$.
With an explicit expression for $k = k_r + ik_i$, the imaginary part of k is expressed as

$$k_i = \frac{\nu_e}{2}\frac{\omega - k_r u_n(1 + 1/\kappa \lambda_D{}^2 k_r{}^2)}{\omega u_o + \kappa k_r u_e{}^2} \tag{15}$$

For the propagation wave without damping longitudinally with $k_i = 0$ in the above equation, we have the same result obtained from $\omega_i = 0$ in Eq. (13) as

$$\omega = k_r u_n(1 + 1/\kappa \lambda_D{}^2 k_r{}^2) \tag{16}$$

The wave is approximated to be $\omega \sim u_n \lambda / 2\pi\kappa\lambda_D{}^2 \sim 10^5\ s^{-1}$ with $k_r = 2\pi/\lambda$, $u_n \sim 10^2$, $\kappa \sim 10^{-2}$. Then, we have the wave length $\lambda \sim 10(2\pi\lambda_D{}^2) \sim 0.1\ mm$.
The real part of k is given by

$$\omega^2 - \omega_{pe}^2 - \kappa u_e^2(k_r^{\ 2} - k_i^{\ 2}) - 2\omega k_r u_o + k_r u_n v_e - \frac{k_i}{k_r^{\ 2}}\frac{u_n v_e}{\kappa \lambda_D^{\ 2}} - \frac{u_n u_o v_e^{\ 2}}{\kappa u_e^{\ 2}} = 0 \quad (17)$$

From the derivative of $d\omega$ and dk_r in the above equation of dispersion relation, the propagation velocity of wave is calculated in the group velocity $u_p=d\omega/dk_r$ with ω in Eq. (16) and $k_i=0$;

$$u_g = \frac{d\omega}{dk_r} = \frac{u_e^{\ 2} + u_o u_n / \kappa(\lambda_D k_r)^2 - u_n v_e / k_r}{u_n / \kappa(\lambda_D k_r)^2 - u_d} \approx \frac{u_e^{\ 2}}{u_n}\kappa(\lambda_D k_r)^2 \sim (10^5 - 10^6)\,(m/s) \quad (18)$$

In the approximation of above equation, we have $u_e > u_n/\kappa(\lambda_D k_r)^2 > u_d \sim 10^4$ with $u_n \sim 10^2$, $\kappa \sim 10^{-2}$, $k_r = 2\pi/\lambda$, which is $(u_e/u_n) > (\lambda/2\pi\lambda_D)^2/\kappa > (u_d/u_n) \sim 10^2$. Using $(\lambda/2\pi\lambda_D) > 10\kappa^{1/2} \sim 1$ with $\lambda > 2\pi\lambda_D$, we have the wavelength of $\lambda \sim 0.1$ *mm*. Finally, the group velocity of wave propagation is $u_g \sim u_e^{\ 2}\kappa(2\pi\lambda_D/\lambda)^2/u_n \sim 10^6$ *m/s* or less with $u_e \leq 10^6$ *m/s*.

The wave corresponds to the optical signals as $O(t, z) = O_o + dO\sin\omega t$. The perturbed wave $f(t, z)$ describes dO. We have the wave as $O(t, z) = O_o + dO\sin\{2\pi(zu_g t)/\lambda_g\}\sin\{2\pi(z - u_p t)/\lambda\}\exp(-k_i z)$. The frequency is $\omega = 4\pi f_o \sim 130$ *kHz*$\sim 10^5$ s^{-1}. The wave length is $\lambda \sim 2\pi\lambda_D \sim 0.1$ *mm*. The propagation velocity is the group velocity $u_g = d\omega/dk \sim 10^6$ *m/s* of modulated wave whose wavelength is $\lambda_g \sim 2\pi$ *m*, the scale of lamp length. Since the value of k_i can be zero or small enough for the wave to propagate along the whole length of lamp, the scale of decay along the longitudinal direction is $z_D = k_i^{-1}$ which can be larger than the lamp length.

4. Discussions

The experimental observations summarized in the beginning of previous section, are explained with the analysis of wave propagation as follows:

i. The wave always propagates along the tube from the high voltage side to the ground. The direction of propagation is independent of the electron drift velocity. The propagation of wave is in the direction of diffusion velocity u_n since the group velocity $d\omega/dk \sim u_e^2/u_n$ as well as the phase velocity $\omega/k \sim u_n$ is related to the diffusion velocity u_n. In the previous reports [8-9], the direction of wave has been represented in the direction of electron drift velocity u_d as $\omega/k \sim u_d$ and $d\omega/dk \sim u_e^2/u_d$.

ii. In the observation, the light propagation speed is $u_p \sim 10^5$-10^6 *m/s* according to the lamp current. At the low current of ~1 *mA*, the velocity is $u_p \sim 10^5$ *m/s*. As the current increases at the normal glow, the propagation velocity also increases over 10^6 *m/s*. The propagation velocity in the analysis is the group velocity of $d\omega/dk \sim u_e^2/u_n$ which is proportional to the square of electron thermal velocity u_e. Generally in Townsend breakdown, the most important parameter in Townsend discharge is the electron energy $T_e = mu_e^2/3$. For instance, in Hg-discharge with rare gases, the electron temperature is $T_e < 1$ *eV* before the breakdown and it can be $T_e > 1$ *eV* at the normal glow and it can be increased with the discharge current.

iii. In the analysis, the wave describes the optical signals as $O(t, z) = O_o + dO\exp(-z/z_D)\sin\omega t$. Since the propagation velocity corresponds to the group velocity, and it becomes fast to be ~10^6 *m/s* as the current is increased, the wave would not decay for the short time of ~μs. If the propagation velocity is ~10^5 *m/s*, the propagation time of ~5 μs is long enough for the wave to be in decay. However, the scale of decay is the length of lamp even for the velocity ~10^5 *m/s*.

The mechanism of light propagation is explained with the results of analysis in the previous Sec. 3 as follows:

i. Plasma is generated at high voltage side of z=0 and diffused toward the ground electrode of z=L. At the steady state, the gradient of plasma density is formulated by the continuous generation of plasma at the high voltage side. The plasma density is high at z=0 and is decreased toward z=L. There are two kinds of electron flow in the fluid velocity u_o. One is the electron drift u_d~10^4 *m/s* driven by the electric field. Another is the electron diffusion velocity u_n driven by the gradient of plasma density ∇n_0. The electron flow of diffusion with the velocity u_n~10^2 *m/s*, due to the plasma density gradient in the steady state, is the origin of plasma wave propagation along the positive column.
ii. The wave is generated by the pulse of high voltage which is the role of the perturbation on the positive column plasma having a density gradient. Initially, the wave is generated at the high voltage side with the frequency of voltage pulse. The perturbed wave propagates along the slope of plasma density to the z-direction.
iii. The wavelength of wave is about ten times of Debye length as λ~$10\lambda_D$ ~0.1 *mm* so that it cannot be observed with the naked eye. The scale of modulated group wave is the scale of lamp length so that the group wave has no nodal point along the positive column. The wave of decay along the tube has the decay length above the scale of lamp length, about 1 *m* or 10 *m* according to the lamp current. Therefore, the wave seems to be a flowing tide, rising from ocean tiding toward beach.

5. Conclusion

The observation of light propagation along a long positive column is the first report in 70-years history of long discharge lamps such as neon-sign lamps, fluorescent lamps for general lighting, and CCFLs and EEFLs for LCD-TVs. The mechanism of light propagation has not been understood even in the previous reports [8-9].

While the wave is not generated in the DC-operation of constant voltage, the wave is observed only in the operation with DC-rippled voltage. Through the operation of DC-rippled voltage, the optical signals observed in this experiment are verified not to be a kind of unstable standing waves like a striation but it is clearly a new kind of wave propagating along the positive column.

In the AC-operation, the propagation velocity of light emission depends on the lamp current. At the low current before breakdown, the propagation velocity is about 10^5 *m/s* and the wave decays with the scale of lamp length. As the current increases after Townsend breakdown of normal glow, the propagation velocity is increased with the order of 10^6 *m/s*. The wave of high current will not show the decay during propagation.

The origin of propagating wave in this observation is the electron diffusion wave. When the plasma is generated by the high field at the side of electrode applied a high voltage, the plasma diffuses very slowly toward the ground electrode. In the steady state, the gradient of plasma density is formulated along the positive column, a high density at the high voltage electrode and a low density at the ground electrode. When the perturbed wave is generated due to the high voltage pulses which roles a perturbation at the side of high voltage electrode, the wave slides on the slope of plasma density along the positive column and propagates toward the ground electrode.

The substance of wave is the electron diffusion flow enforced by the plasma density gradient formulated along the whole positive column. The phase velocity as well as the group velocity of wave is proportional to the electron diffusion velocity. The wavelength of phase is a sub-millimeter so that the wave cannot be seen with naked eyes. Since the wavelength of modulated group wave is the scale of lamp length, the wave is propagating like a flowing tide rising from the high voltage electrode toward the grounded electrode.

6. Acknowledgments

This work was supported by Brain Korea 21 under a research grant in 2009.

7. References

1. J. Waymouth, Electrical Discharge Lamps, MIT Press, 1971.
2. R. Y. Pai, SID'97 Digest, 1997, p. 447.
3. T. S. Cho, Y. M. Kim, N. O. Kwon, S. J. Kang, E. H. Choi, and G. S. Cho, "Effect of Electrode Length on Capacitively Coupled External Electrode Fluorescent Lamp," Jpn. J. Appl., Phys., Vol. 41, No. 3B, 2002, pp. L355-L357.
4. T. S. Cho, H. S. Kim, Y. G. Kim, J. J. Ko, J. G. Kang, E. H. Choi, G. S. Cho, H. S. Uhm, "Characteristic Properties of Fluorescent Lamps Operated Using Capacitively Coupled Electrodes," Jpn, J. Appl. Phys., Vol. 41, No. 12, 2002, pp. 7518-7521.
5. T. S. Cho, N. O. Kwon, Y. M. Kim, H. S. Kim, S. J. Kim, J. G. Kang, E. H. Choi, and G. S. Cho, " Capacitive coupled electrodeless discharge backlight driven by square pulses", IEEE Trans. Plasma Sci., Vol. 30, No. 5, 2002, pp. 2005-2009.
6. G. S. Cho, N. O. Kim, Y. M. Kim. S. J. Kim, T. S. Cho, B. S. Kim, J. G. Kang, E. H. Choi, U. W. Lee, S. C. Yang, and H. S. Uhm, "Self-discharge Synchronizing Operations in the External Electrode Fluorescent Multi-lamps Backlight," J. Phys. D, Appl. Phys., Vol. 36, 2003, pp. 2526-2530.
7. G. S. Cho, M. J. Shin, J. M. Jeong, "Glass tube of high dielectric constant and low dielectric loss for external electrode fluorescent lamps.", Journal of Applied physics, Vol. 102, No. 11, 2007, p. 113307.
8. G. S. Cho, S. H. Ahn, J. M. Jeong, "Plasma propagation in external electrode fluorescent lamps." J. phys. D: Appl. Phys, Vol. 40, No. 13, 2007, pp. 3945 3950.
9. G. S. Cho, J. H. Kim, J. M. Jeong, "Electron plasma wave propagation in external electrode fluorescent lamps.", Applied Physics Letters, Vol. 92, No. 2, 2008, p. 021502.
10. A. Garscadden, D. A. Lee, "Forward and backward-wave moving striations in constricted discharge", Int. J. Electronic Control, Vol. 20, No. 6, 1966, p. 567.
11. D. A. Lee, P. Bletzinger, A. Garscadden, "Wave nature of moving striations", J. Appl. Phys., Vol. 37, No. 1, 1966, p. 377.
12. M. Hirsh and H. J. Oskam, "Gaseous electronics", Vol. 1, Electric Discharges, (Academic Press,1978), Part 2.2, p. 65.
13. Francis F. Chen, Introduction to Plasma Physics and Controlled Fusion, Vol. 1: Plasma Physics, 2nd ed. (Plenum, New York, 1984), Chap. 1, p. 13.
14. Y. Raizer, "Gas Discharge Physics", (Springer-Verlag Berlin Heidelberg 1991), Chap. 2, p. 32.
15. M. Lieberman and A. Lichtenberg, "Principles of Plasma Discharges and Materials Processing", (Wiley & Sons, N.Y. 2005), Chap. 10, p. 334.

16. E. Nasser, "Fundamentals of Gaseous Ionization and Plasma Electronics", (Wiley-Interscience, 1970), Chap. 6, p. 183.
17. E. Nasser, "Fundamentals of Gaseous Ionization and Plasma Electronics", (Wiley-Interscience, 1970), Chap. 4, p. 112, and Appendix D, p. 448.
18. L. Loeb, "Fundamental processes of Electrical Discharge in Gases", (John Wiley & Sons, Inc., 1939), Chap. 4, p. 177.
19. A. von Engel, "Ionized Gases", 2nd ed. (Oxford University Press, 1965), Chap. 4, p. 122.
20. Blake E. Cherrington, "Gaseous Electronics and Gas Lasers", (Pergamon Press Inc. U.S.A., 1979), Chap. 8, p. 149.
21. N. A. Krall and A. W. Trivelpiece, "Principles of Plasma Physics", (McGraw-Hill, N. Y. 1973), Chap. 6.
22. B. Cherrington, "Gaseous Electronics and Gas Lasers", (Pergamon Press, 1979), Chap. 7, p. 116.
23. A. Salop and H. H. Nakano, *Phys. Rev.*, A 2, p. 127, 1970.

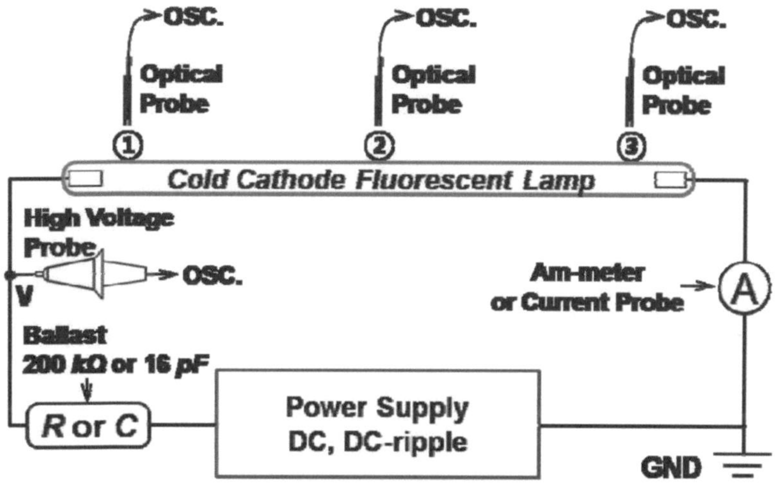

Fig. 1. Experimental circuit for measuring the optical signals in CCFL operated with DC or AC electric power supply. The high voltage is supplied to the left side of electrode and the other side is grounded. Measuring points of the optical signal are labeled with ①~③.

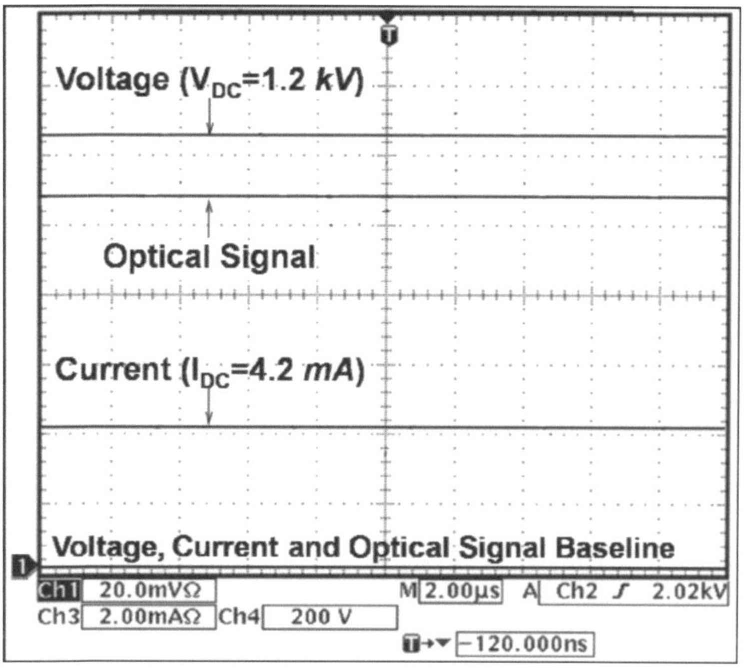

Fig. 2. Oscilloscope signals of lamp voltage, lamp current, and optical probe intensity for the operation with the DC voltage. The solid bottom line is the zero base line. All the signals are constant on time.

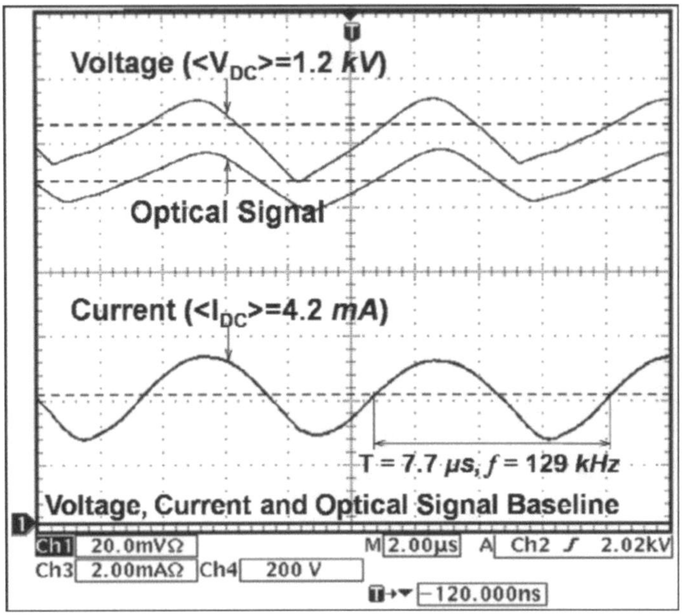

Fig. 3(a). Oscilloscope signals of lamp voltage 1.2 *kV*, lamp current 4.2 *mA*, and optical probe intensity for the operation with the rippled-DC voltage as $V(t)=V_o+\Delta V\sin\omega t$ with $\Delta V/V_o$~10 %. The solid bottom line is the zero base line of voltage, current, and optical intensity. The straight dashed-lines are obtained from the DC operation in Fig. 2. All the signals show the rippled sinusoidal wave with the operation DC frequency 129 *kHz* and the period of 7.7 *μs*. The optical intensity is $O(t)$~$O_o+\Delta O\sin\omega t$ with $\Delta O/O_o$~10 %.

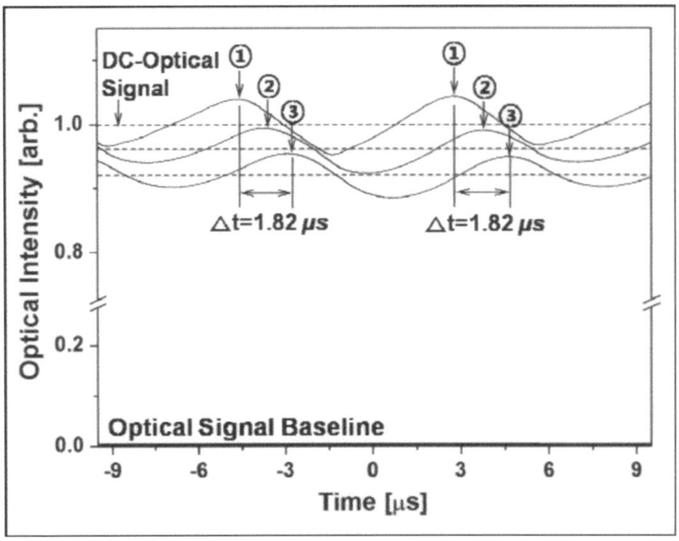

Fig. 3(b). The optical signals measured at points ①~③ when the luminance at ① is 10,000 *cd/m*^{2} for the operation with the rippled-DC voltage as shown in Fig. 3(a). The propagation time of optical emission from ① to ③ is t_p=1.82 *μs*. The propagation velocity is u_p=3.35×10^5 *m/s* traveling the distance 610 mm from ① to ③. During the propagation, the optical intensity decays slightly with the characteristic decay time t_D~22 *μs*.

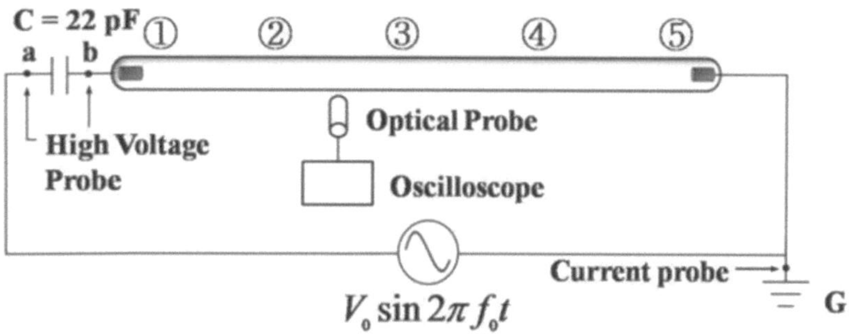

Fig. 4. Experimental circuit for measuring the optical signals at points ①~⑤ along the CCFL. Lamp length is 900 *mm*, with 220 *mm* between each point, and 880 *mm* between points of ① and ⑤.

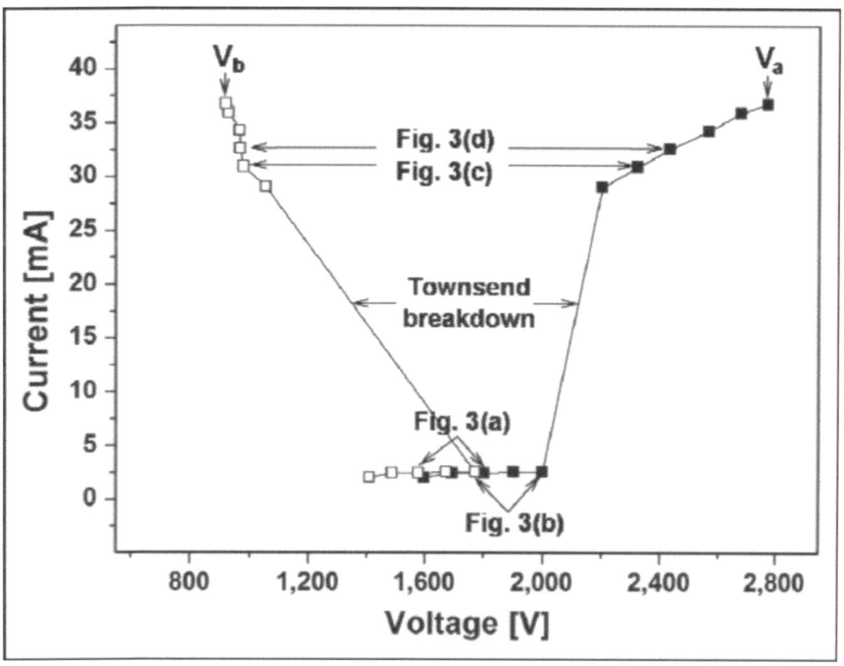

Fig. 5. Current and voltage curves with the applied voltage V_a and the lamp voltage V_b measured at the point a and b in the circuit of Fig. 1, respectively. The optical signals of Figs. 3 are measured at the following I-V: Fig. 3(a) is with I=2.5 *mA*, V_a=1.8 *kV*, V_b=1.6 *kV*, Fig. 3(b) is with I=2.6 *mA*, V_a=2.0 *kV*, V_b=1.8 *kV*, Fig. 3(c) is with I=31.0 *mA*, V_a=2.32 *kV*, V_b=0.98 *kV*, and Fig. 3(d) is with I=32.6 *mA*, V_a=2.45 *kV*, V_b=0.97 *kV*, respectively.

a) 2.5 *mA*, t_p=5.31 *μs*, u_p=1.66×10⁵ *m/s*

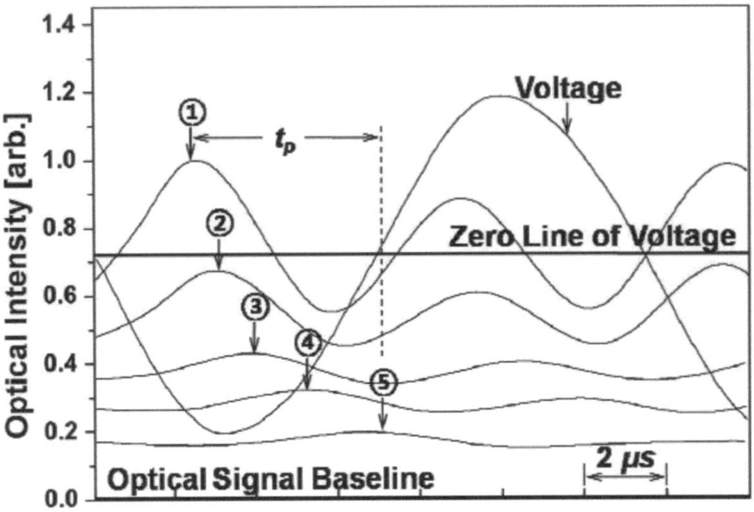

b) 2.6 *mA*, t_p=4.00 *μs*, u_p=2.20×10⁵ *m/s*

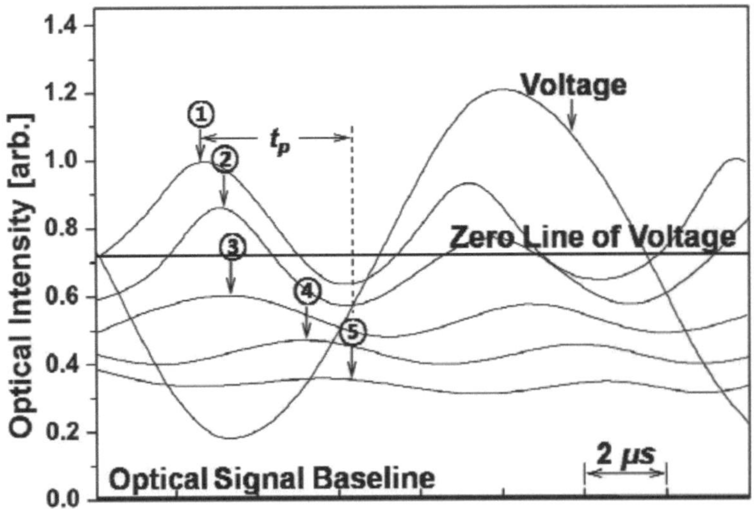

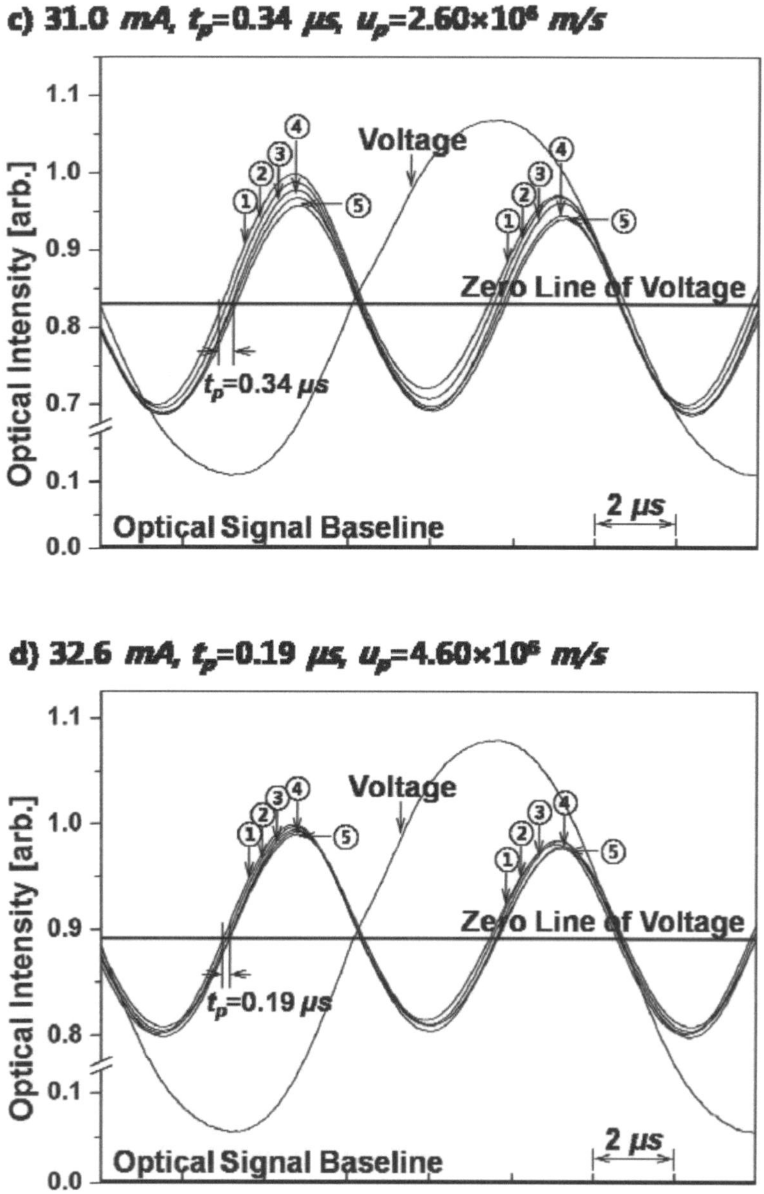

Fig. 6. Oscilloscope signals of the lamp voltage V_b and the optical probe pulses measured at each point ①~⑤. The signals at (a) and (b) are before Townsend breakdown and those at (c) and (d) are after Townsend breakdown as they are noted in Fig. 2. If the voltage signal is $V(t)=V_o\sin\omega_o t$ with $\omega_o=2\pi f_o$ for the applied frequency f_o=65 *kHz*, then the corresponding optical signal is represented by $O(t)=O_o+O_1\sin\{\omega_{op}(t-\Delta t)\}$ with the frequency of $\omega_{op}=2\omega_o$ for the two polarities of voltage peaks $\pm V_o$. In the optical signals at (a) and (b), the optical intensity O_o decays remarkably with the measuring points from ① to ⑤. The decay rate of

optical signal from ① to ⑤ is quite slow at (c) and (d). In the figures from (a) to (d), the propagation time Δt of the optical signal from the peak of ① to ⑤ becomes short so that the propagation velocity u_p becomes fast. Before Townsend breakdown of low current I=2.5 mA and I=2.6 mA in figures (a) and (b), the propagation time from ① to ⑤ is 5.31 μs and 4.00 μs, and the mean propagation velocity is u_p=880 $mm/$5.31 μs=1.66×10^5 m/s at (a) and u_p=2.20×10^5 m/s at (b). After Townsend breakdown of high current I=31.0 mA and I=32.6 mA in figures (c) and (d), the propagation time from ① to ⑤ is 0.34 μs and 0.19 μs, and the mean propagation velocity is u_p=2.60×10^6 m/s at (c) and u_p=4.60×10^6 m/s at (d).

V. Wave Propagation in Media with Chemical Reactions.

21

Influence of Heavy Water on Waves and Oscillations in the Belousov-Zhabotinsky Reaction

Oliver Klink[1], Wolfgang Hanke[1],
Edeltraud Gerbershagen[1] and Vera Maura Fernandes de Lima[2]
[1]University of Hohenheim, Institute of Physiology, Stuttgart,
[2]FUNREI, Campus Divinopolis, MG,
[1]Germany
[2]Brazil

1. Introduction

Excitable media allow excitation waves to travel through them, which play an important role in a variety of natural processes, in physics, chemistry and live-sciences. (Sagues & Epstein, 2003; Tabony, 2006). By far the best understood model system for all these processes is the Belousov-Zhabotinsky (BZ) reaction (Belousov, 1959; Zaitkin & Zhabotinsky, 1970). This oscillating chemical reaction, among others, exhibits propagation of two-dimensional waves in thin layers of reaction solution (Luengviriya et al., 2006) or in gels (Yamaguchi et al, 2006), typically at velocities of some millimeter per minute. Also, in a stirred bulk solution, regulary oscillations in time can be observed (Dolzmann et al., 2007).
From the beginning a variety of detailed kinetic models have been constructed to explain the oscillatory behavior of the bulk BZ reaction. The first and simplest is known as the FKN-model (Field et al., 1972), which involves 18 elementary steps and 21 chemical species. It can be simplified, however, assuming three key species: $HBrO_2$ as exchange intermediate, Br- as control intermediate and Mox, i.e. the oxidized form of the catalyst, as regeneration intermediate. Later other, sometimes much more complex model have been composed.
According to the propagation of waves in thin layers of excitale medium, similar theories have been build up which, however, have to include diffusion of the involved species.
The oscillating BZ-reaction is excepted to be a useful model for biological fluctuations (i.e. Winfree, 1987). Here it is dogma that the substitution of normal (light) water by heavy-water slows down every biologcal rhythem and process. The influence of deuterated compounds on the global kinetics of BZ reaction in stirred bulk systems was discussed by Karavaev et al. for Ru-catalyzed systems (1986). It was found that the substitution of H_2O by D_2O as solvent caused a complete suppression of the oscillating regime. In other studis (Hsu & Jwo, 1999; Rossi et al., 2007, 2009 an increase induction period of the oscillations was found. In any case a deuteration reaction and the bromation of the organic compound of the system in the theoretical framework can be made responsible for the effects described. This means, putting the effects into the rate constants.
Not very much is known about the effects of heavy water on propagating waves of the BZ-reaction. Theoretically also a slowing down of the wave fronts might be predicted, but in

addition to the situation of the oscillating system, here diffusion of the involved species has to be taken into account, and this is lower in heavy water due to the increased viscosity (see materials section).
First we have done some more simple experiments with an ocillating BZ-reation using normal-water and heavy-water, and then have investigated the behaviour of propagating waves in BZ-systems in more detail.

2. Materials and methods

2.1 General

All salts and solvents including heavy water were obtained from Sigma at a quality of p.a. or better. Normal water was from a double laboratory distillery.
Wave propagation experiments in fluids and gels were done in Petri-dishes (35mm and 100mm diameter), bulk oscillation experiments were done in laboratory beakers with a volume of 500ml.
In the wave-propagation experiments, the Petri-dishes were fixed on a light table (white light), a colour video-camera was mounted over them and the video traces were recorded on a computer using the proper hardware In some cases a blue filter between light panel an Petri-dish was used to enhance the contrast.
In the bulk-oscillation experiments the beaker with the solution was fixed on a laboratoy stirrer/heater. On one side of the beaker a blue LED (480nm) was mounted, on the other side a photomultiplier. The data from the photomultiplier (intesity of transmitted light) were stored on a computer using the proper A/D-converter and software. More details about the set-ups used have been published previously (Hanke, 1999; Dolzmann et al., 2007).
In all procedures described below, experiments were done identically for D_2O and H_2O. D_2O was used in all steps of the D_2O experiments, thus the remaining concentration of H_2O in the D_2O experiments was neglectable (far below 1% according to our calculations). To have the neccessary physico-chemical information available, the data for normal- and heavy water are summarized in tabel 1.

	D_2O	H_2O
Freezing point (°C)	3.8	0.0
Boiling point (°C)	101.4	100.0
Density (at 20°C, g/ml)	1.106	0.998
Temp. of max. density (°C)	11.6	4.0
Viscosity (at 20°C,mPas)	**1.25**	**1.005**
Surface tension (at 25°C, µJ)	7.193	7.197
Heat of fusion (cal/mol)	1,515	1.436
Heat of vaporization (cal/mol)	10.864	10.515
pH (at 25°C)	7.41	7.00

Table 1. Physico-chemical properties of heavy water and normal (light) water. Of special interest concerning this text is the about 20% higher viscosit of heavy water.

2.2 Oscillating bulk reaction

The solution of the BZ-reaction was composed from the following ingredients. 1 M sulphuric acid (H_2SO_4), 275 mM malonic acid ($C_3H_4O_4$), 8.5 mM ammoniumcernitrate

($CeH_8N_8O_{18}$), 62 mM sodium bromate ($BrNaO_3$), and 0.4 mM ferroin. The BZ-mixture was given to a beaker and stirred slowly on a magnetic-stirring/heating device at constant rate and controlled temperature (22 ^oC+- 0.5 ^oC). Usually 500ml of the given solution was prepared. Oscillations in theses experiments start spontaneously.

3. Waves in thin layers of fluids

The same solution as used in bulk experiments was used in the thin fluid layer experiments. A layer of 2mm hight was given into a Petri-dish. Then this dish was treated as described for the gels below. In some experiments higher ferroin concentrations were used, as stated in the experimental section.

3.1 Preparation of gels for the BZ-reaction

The silica gels for the BZ-reaction were made in Peri-dishes (35mm and 100mm diameter) slightly modified according to the procedure described by Yamaguchi et al. (Yamaguchi et al., 1991). A gel-solution containing 9% $Na_2Si_2O_3$, 130 mM H_2SO_4 and 3.3 mM Ferroin was made. In some control experiments, different concentrations of silicat were used (see experimental part), to obtain different viscosities of the gels. A 5mm thick layer of this solution was given into the Petri-dishes. After polymerisation of the gel, the surface of the gels was washed with 330 mM H_2SO_4 two times, and then the gels were covered (5mm high) with a reaction solution containing 330 mM malonic acid, 330 mM $NaBrO_3$, and 330 mM H_2SO_4. The dishes were then immediately mounted in the set-up, it was waited until waves showed up spontaneously and then the video recording was started. The gels were run at room temperature, 20°C +-1°C.

3.2 Data evaluation

From all experiments (waves in fluids and gels, and in bulk experiments) video traces were stored on a computer in the avi-format using a standard colour video-camera and a proper frame grabber. The avi-files were changed to stacks of jpg-files by the software VirtualDub^R and imported to the software ImageJ (NIH), alternativly the software LabView ^{R} was used. Using this software, the propagation velocity of the wave-fronts at was calculated. Also, from the stacks photos of series of frames were constructed as shown in the figures and single snap-shots were extracted.

4. Experimental results

In a first set of experiments we have investigated the oscillating BZ-reaction (stirred bulk-system without diffusion, convection or bouyancy) in plain light-water and in plain heavy-water. Traces of transmitted blue light through both systems are compared in figure 1. As known from previous studies (Dolzmann et al., 2007) and as can be seen in the figure, upper trace, in normal water there is a short induction period (not shown) followed by a quite stabel oscillation (up to hours). At the beginnig the peaks of the blue phase are slightly higher, decreasing to a stable value, but then the amplitude as well as period are stable. In the case of heavy water, shown in the lower trace, there is a longer induction period (not shown), followed by some peaks of increasing amplitude. The amplitude stabilizes then for a smaller period, but the period becomes quite big on a short time scale.

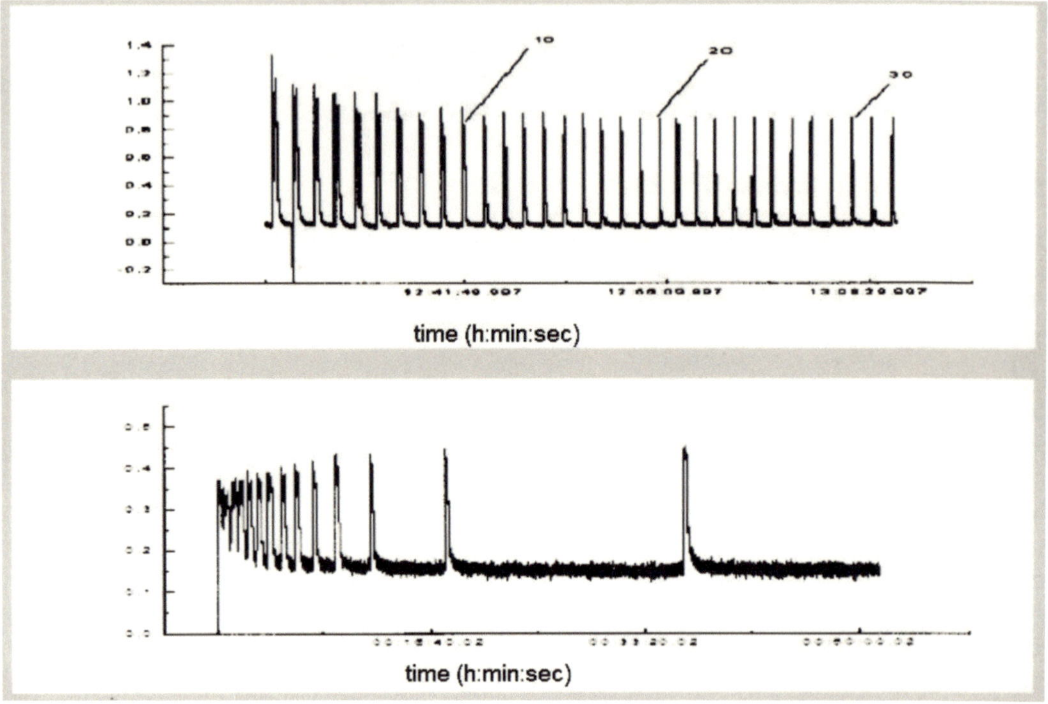

Fig. 1. Traces of oscillations of BZ bulk reactions in normal water, upper trace, and in heavy water, lower trace. Significant differences in amplitude and temporal behavior can be seen. The y-axis gives the transmitted light on an arbitrary scaling.

We have shown in a previous study (Dolzmann et al., 2007) that the time course of the single colour changes in the two catalyst system we used has a typical doulbe peak shape at the beginning. In figure 2 it is verified that this behaviour under heavy water is not changed in principal.

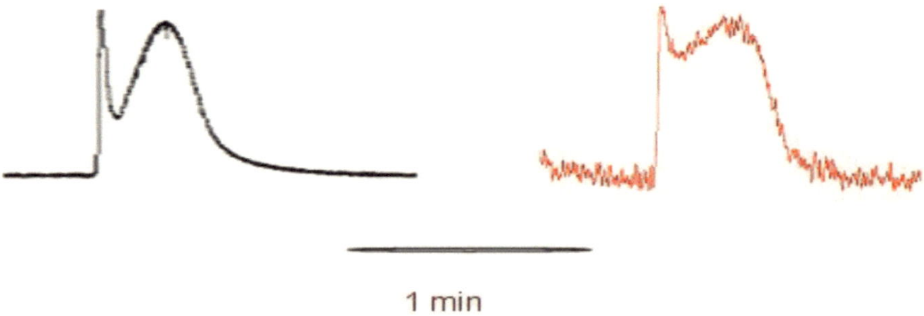

Fig. 2. Colour changes of the BZ-bulk reaction on an expanded time scale (compared to figure 1). Left shows the situation in normal water, the right side depicts it in heavy water. The global shape of both recordings is about the same.

Fig. 3. Sequences of frames from waves in fluid BZ-systems. The upper part is from normal water, the lower one from heavy water. Temporal spacing between consecutive frames is 4 seconds; the dishes have a diameter of 100mm. As can be seen, the activity in heavy water is lower over all.

Next we have investigated the behaviour of propagating waves in thin layers of fluids. In principle, in heavy water stable propagating wave fronts were observed, too. The velocity of these fronts was about 3% less compared to the situation in normal water. However, first, the number of ecitation centers in heavy water always was lower than in normal water. Second, opposite to normal water, very often the reaction stopped in heavy water after waves had covered the whole dish, but could be reinitiated by shaking the dish and thus starting the reaction again.

To depict the general behaviour, in figure 3 sequences of video frames are shown for both situations.

In addition to the slightly lower propagation velocity in heavy water and the reduced number of excitation centers, very often in the dishes global background colour changes could be observed, see figure 3 background of the frames under heavy water and figure 4, moving much faster than normal waves or possibly being bulk fluctuations.

Assuming the big waves to be propagating bulk-fluctuatuions, in one and the same system in the presence of heavy water, propagating waves and temporal oscillation would coexsist. This was never observed in norml water under comparable conditions.

Finally we investigated propagating waves in gels of BZ medium in normal and in heavy water. Comparable to figure 3 the situation is depicted in figure 5. This time drastic changes in the behaviour of wave propgation could be observed.

Different to the situation in fluid systems, in gels under heavy water the propagation velocity was drastically reduced, 20%, compared to normal water. The waves did not invade the complete gels but died away, creating open fronts and spiral fragments. A more detailed example is shown in figure 6.

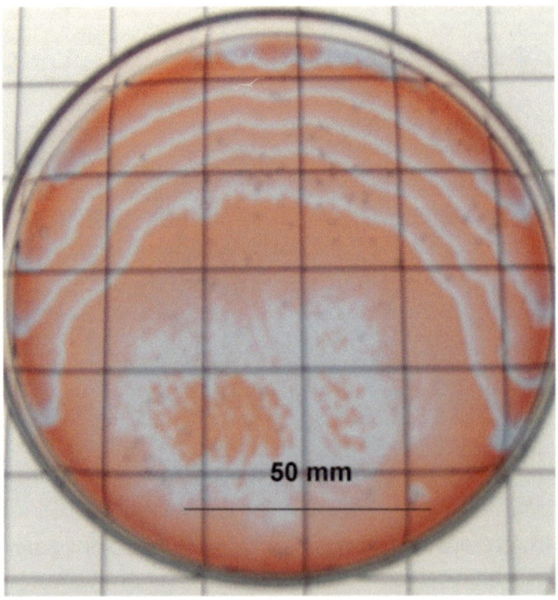

Fig. 4. BZ waves in heavy water. As well some normal wave fronts can be seen as also the global colour change in the lower part of the dish.

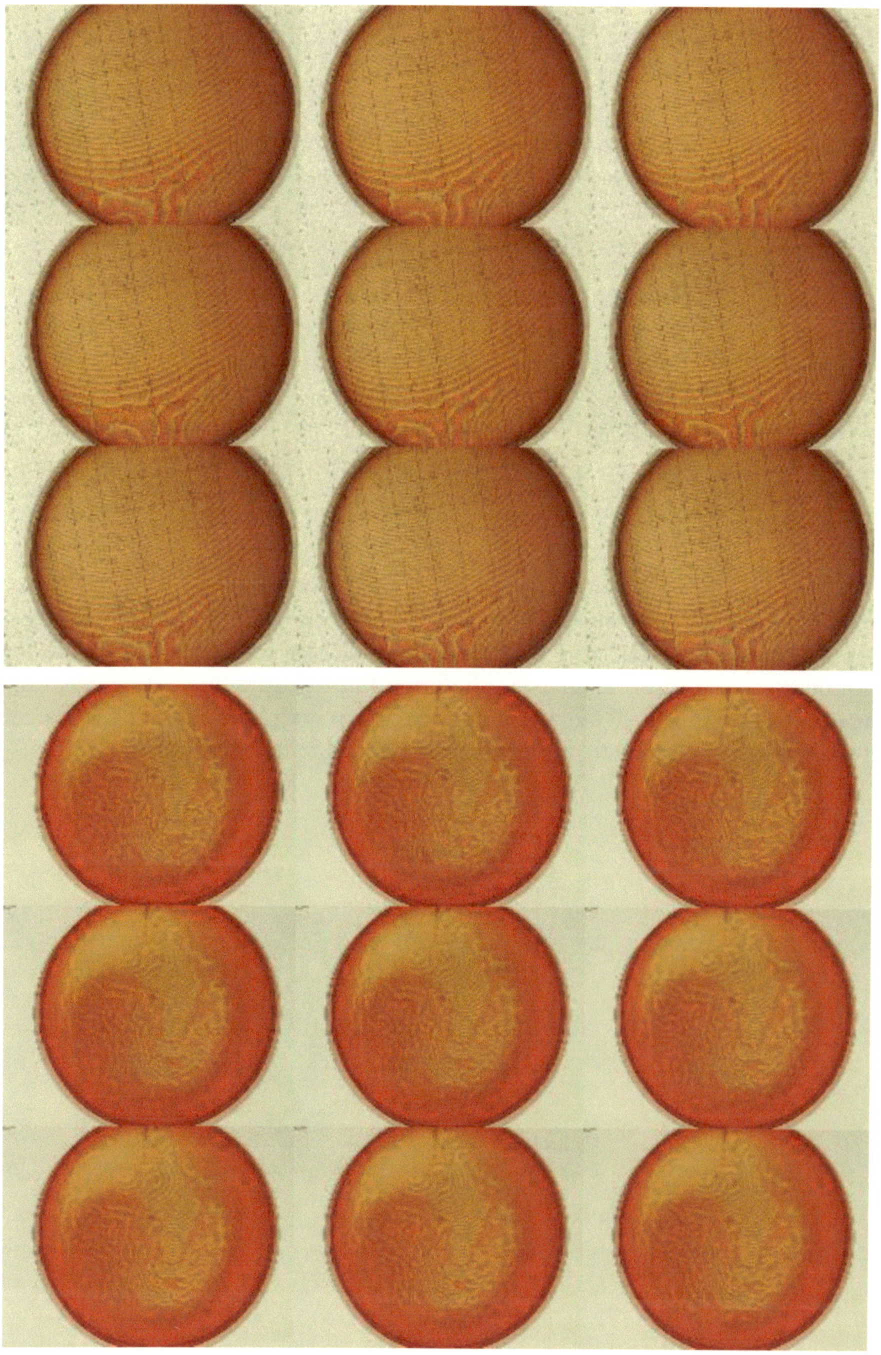

Fig. 5. Sequences of propagating BZ waves in gels, the upper part shows the situation in normal water, the lower one in heavy water. The temporal spacing between consecutive frames is 4 seconds, the diameter of the dishes is 1100mm. Waves in heavy water are not invading the whole gel and are creating open fronts and spiral fragments obviously.

Fig. 6. Open fronts and spiral fragments in propagating BZ waves in gels made with heavy water.

To learn about the influence of the viscosity of the medium we finally did a series of experiments in gels with increasing concentrationof silicate making the gels more stiff and inducing a higher viscosity. The result is shown in figure 7 in a block diagram, the velocity of propagating waves in gels of BZ medium decreases with increasing silicate concentration, meaning with increasing viscosity and thus decrising diffusion coefficient.

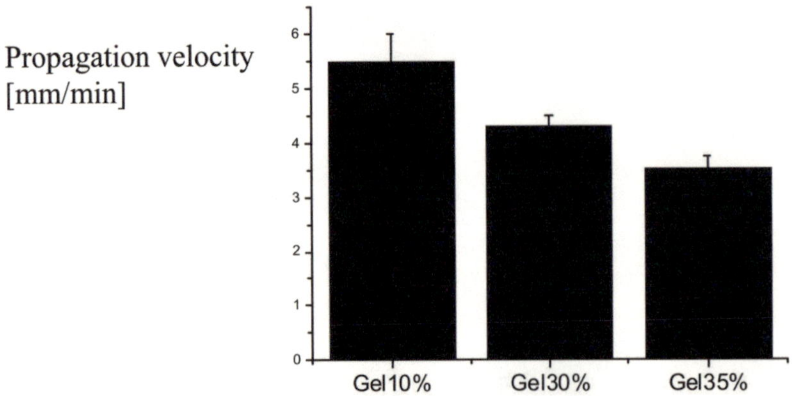

Fig. 7. Velocity of propagating BZ waves in gels made in normal water with increasing amounts of silicate as given in the figure. The propagation velocity decreases with incresing silicate concentration.

5. Discussion

Our experiments in BZ systems oscillating in time demonstrate a reduced excitability, due to the delayed onset of the oscillations. The temporal behaviour results confirm data from previous papers, arguing that the main difference in heavy water in this case is due to reduced rate constants in the bromation of the organic species (Rossi et al., 2007, 2009). The rate constants in the two catalyst system used describing the interactions between the catalysts are obviously not effected. All this also can be easily verified in doing simulations of the classical reaction schemes using for example the environment MathLab-Simulink, which will be published later.

All the data about the effect of heavy water on oscillating BZ reaction also confirm the statement of it slowing down all biological oscillations and rhythems, although there are by sure totally different basic mechanisms are involved.

Different to oscillating systems in which due to the stirring diffusion, convection and bouyancy do not exsist, in case of wave propagation all these phenomena must be taken into account. At least the diffusion of involved species is neccessarily included in mathematical models and simulations of wave propagation. On mechanistic reasons one can assume that in case of heavy water, compared to normal water, the diffuison coefficients of all species are reduced, see table in materials and methods. This must be true in addition to the reduced rate constants, see above, which also will be present. Amazingly, in fluid systems the wave propagation velocity was not that much reduced in heavy water, but some other effects pointing out a reduced excitability. In addition, there was the fundamental problem of the coexsistance of waves and oscillations in the same system, which has not been described previously.

In gel systems under heavy water conditions, the situation is obviously close to the region in the phase diagram were propagation is no longer possible. First, the velocity was drastically reduced and second, and in the presentation much more obvious, waves are not invading the complete accessible area but die away upon travelling and thereby produce open ended fronts and spiral fragments. Consequently we must assume that the interplay of reduced rate constants and reduced diffusion is inducing the described situation. Our finding that more stiff gels significantly reduce wave propagation velocity but does not induce a brake down of wave fronts seems to verify this statement.

In future investigations, especially in simulations of wave propagation these theories will be tested in detail.

6. References

Belousov, B.P. Eine periodische Reaktion und ihr Mechanismus (translated Russian to German). in: Sbornik referatov po radiacionnojj medicine za 1958 g Moskau, 147: 145 (1959)

Dolzmann, K., Kaltenbach, S., Hanke, W. and Fernandes de Lima, V.M.: Biphasic optical signals of an oscillating non-stationary Belousov-Zhabotinsky bulk reaction and its similarities to some neuophysiological events. Int. J. Bifurcation and Chaos, 17, No 4, 1329-1335 (2007)

Field, R.J., Korös, E., Noyes, R.M., Oscillations in chemical systems. II. Thorough analysis of temporal oscillation in the bromate-cerium-malonic acid system. J. Am. Chem. Soc. 94,8649-8664 (1972)

Hanke, W. Inverse dispersion relation of BZ waves in non-stationary gels. Int. J. Bifurcation and Chaos, Vol 9, N 10, 2009-2014 (1999)

Hsu, M.-C. and Jwo, J.-J., Kinetic Study of the Isotope Exchange reactions of malonic acid and its derivatives in various acidic D_2O media using 1H NMR spectroscopy.

Implication in the Belousov-Zhabotinsky reaction. Int. J. Chem. Kin., 31, 455-461 (1999)

Karavaev, A.D., Kazakov, V.P. and Tolstikov, G.A., Deutrium effects in auto-oscillation chemoluminescence of the Belousov-Zhabothinsky reaction. React. Kinet. Catal. Lett., Vol. 32, No. I, 21-26 (1986)

Luengviriya, C., Storb, U., Hauser, M.J.B. and Mueller, S.C. An elegant method to study an isolated spiral wave in a thin layer of a batch Belousov.Zhabotinsky reaction under oxygen free conditions. Phys. Chem. Chem. Phys., 8, 1425-1429 (2006)

Rossi, F. Eugenio Simoncini, E., Marchettini, N. and Tiezzi, E., Deuterium isotope effect on the induction period of the cerium catalyzed Belousov–Zhabotinsky reaction. Chemical Physics Letters, V 470, I1-3, 147-150 (2009)

Rossi, F., Rustici, M., Rossi, C., and Tiezzi, E., Isotopic effect on the kinetics of the Belousov-Zhabotinsky reaction Int. J. Mol. Sci., 8, 943-949 (2007)

Sagues, F. and Epstein, I.R. Nonlinear chemical dynamics. Dalton Trans. 1201-1217 (2003)

Tabony, J. Historical and conceptual background of self-organisation by reactive processes. Biol. Cell, 98, 589-602 (2006)

Winfree, A.T. (1987) The timing of biological clocks. Scientific American Books, New York

Yamaguchi, T. et.al. Gel systems for the Belousov-Zhabotinsky reaction. in: Journal of Physical Chemistry, Vol. 95: 5831 – 5837 (1991)

Zaikin A.N. and Zhabotinsky, A.M.: Concentration wave propagation in two-dimensional liquid-phase self-oscillating system. Nature, 225, 535-537 (1970)

VI. Wave Propagation in Nonlinear Media.

22

Wave Propagation and Dynamic Fracture in Laser Shock-Loaded Solid Materials

Thibaut De Rességuier[1], Jean-Paul Cuq-Lelandais[1], Michel Boustie[1], Emilien Lescoute[1] and Laurent Berthe[2]
[1]*Laboratoire de Combustion et de Détonique (CNRS, ENSMA)*
[2]*Laboratoire Procédés et Ingénierie en Mécanique et Matériaux (Arts et Métiers, ParisTech)*
France

1. Introduction

Shock wave loading of solid materials results in specific damage processes at high strain rates. The most widely studied of these processes is probably spall fracture (e.g. Antoun et al., 2002, and references therein), which arises from tensile stresses generated by the interaction of release waves within the material upon reflection of a compressive pulse from a free surface or from an interface with a layer of lower acoustic impedance. If such tensile stresses exceed the dynamic strength of the material, they cause the nucleation and growth of micro-voids or micro-cracks which may eventually coalesce to form a macroscopic fracture and lead to the ejection of one or several fragments (spalled layers) from the sample. Spall damage and wave propagation are tightly coupled. On one hand, the creation of new free surfaces accompanying damage development induces stress relaxation which gives rise to recompression waves. Such waves can be detected in time-resolved velocity (Antoun et al., 2002; Tollier et al., 1998) or piezoelectric (De Rességuier et al., 1997) measurements, and their analysis provides very rich data on the fracture process (location, time and tensile stress at damage initiation, rate of the damage growth, thickness of the spalled layer…). On the other hand, spall fracture results from wave interaction, so that post-test observations of the residual damage in recovered samples (location, sizes and shapes of the damages zones, fracture surface morphology…) may provide key information on the propagation of compression and release waves prior to failure. In this chapter, we illustrate this second, more original statement with experimental results obtained under laser driven shock loading. Intense irradiation of an absorbing target by a high power pulsed laser produces the vaporization of a thin layer of material, transformed into a plasma cloud, whose expansion toward the laser source induces by reaction a compressive pulse into the solid target. The main specificity of this technique is the very short time of pressure application (typically a few ns) compared to other shock generators such as plate impacts or explosive loading, where the duration of the pressure load is usually of µs-order. This difference makes laser shocks less destructive than those more conventional techniques, and favours sample recovery for post-shock analyses of residual damage.
In a first example, spall fracture observed in laser shock-loaded single crystal quartz provides very clear evidence of the strong effect of crystal anisotropy on stress wave

propagation, which is confirmed and explained by three-dimensional (3D) numerical simulations. In a second application, laser shock-induced damage in aluminium samples of different aspect ratios reveals edge effects associated to highly 2D propagation of the stress waves. In a third case, the fracture surface morphology observed in iron samples spalled under laser shock loading is used to investigate a shock-induced phase transformation occurring in this metal and its influence on wave propagation.

2. Anisotropic wave propagation in laser shock-loaded single crystal quartz

The crystalline structure of α-quartz is schematically shown in Fig. 1. It is unchanged under a 180° rotation around the electrical axis *x*, or under a 120° rotation around the optical axis *z*. The *y* axis, usually called mechanical axis, completes the (*x*,*y*,*z*) Cartesian coordinates. To explore the dependence of the dynamic behaviour on the initial direction of shock application relatively to those crystallographic orientations, we have determined experimentally the directions of the mechanical (*y*) and optical (*z*) axes in our samples (originally cut normally to the *x* axis), using uniaxial birefringent properties of quartz. For normal incidence of light on *x*-cut quartz, extraordinary transmitted rays are polarised parallel to the optical axis *z*, whereas ordinary transmitted rays are polarised parallel to the *y* axis. Thus, when an *x*-cut crystal set between two crossed polarisers is rotated around the *x* axis, the incident light stops being transmitted as soon as either the *y* or the *z* axis of the crystal becomes parallel to the transmission axis of one of the polarisers. That way we determine two particular axes *y* or *z*, normal to each other. Finally, to distinguish between those two axes, we use the fact that for normal incidence on *z*-cut quartz, the intensity of the transmitted light does not depend on the crystal orientation between the crossed polarisers.

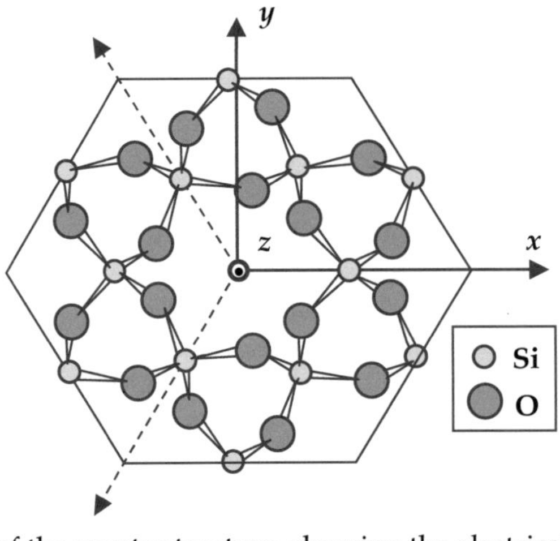

Fig. 1. Schematic view of the quartz structure, showing the electrical (*x*), mechanical (*y*) and optical (*z*) axes in the atomic arrangement of SiO_4 tetraedra.

The laser shock experiments have been performed in the *Laboratoire pour l'Utilisation des Lasers Intenses* (LULI, UMR #7605, Ecole Polytechnique, France). The samples are about 5 mm-thick, covered with a ~5 µm-thick gold coating. Shock loading is achieved by focussing a laser pulse of 1.06 µm-wavelength, ~2.2 ns-duration and ~90 J-energy onto a spot of diameter ranging from 1.2 to 4 mm, which leads to intensities of about 0.3 to 3.6 TW/cm² (1

TW = 10^{12} W). The gold layer is ablated into a plasma cloud. Its expansion induces a short compressive pulse in the quartz sample, due to recoil action. Laser-matter interaction computations indicate that the corresponding loading pressure, which increases with laser intensity, ranges from about 10 to 80 GPa. After each shot, the recovered samples are examined, some are embedded in a resin, sectioned through a plane containing the laser axis, polished and observed in optical microscopy.

Fig. 2 shows a partial view of the free surface of a 5 mm-thick quartz sample recovered after a laser shock of 42 GPa (1.56 TW/cm^2) applied onto a spot of diameter 1.6 mm in the opposite surface, along the x axis of the crystal (i.e. the irradiated surface was normal to the x axis). The gold coating has been removed from an elliptical region (dark area) much wider than the irradiated spot, and two cracks can be seen near both extremities of that zone. The dimensions of the ellipsoid are ~ 7.8 × 2.5 mm, the angle between its major axis and the z axis of the crystal is about 36°. Given the lateral dimensions of the samples (15 mm), this elliptical damage cannot be due to edge effects. Instead, it is likely to result from specifically orientated wave propagation due to the anisotropic structure of the crystal.

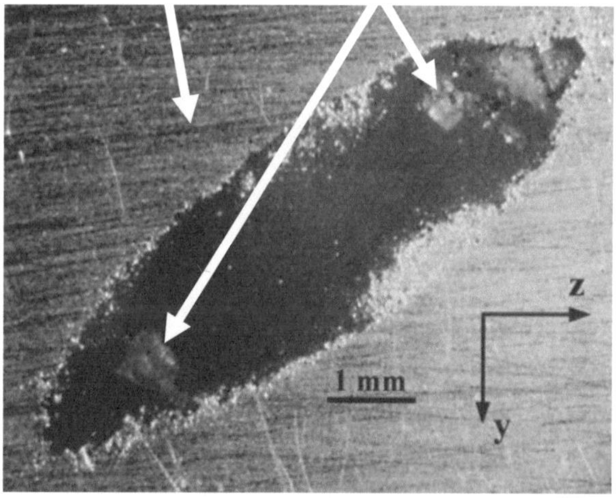

Fig. 2. Free surface of a 5 mm-thick quartz sample recovered after a laser shock of ~42 GPa applied onto a 1.6 mm-diameter spot in the opposite surface, along the x axis.

When the shock is applied along the y axis, the spalled zone near the free surface remains roughly circular, although slightly 'D'-shaped (Fig. 3a), but it is shifted in the z direction, about 1.6 mm away from the irradiated spot (while no damage was observed in the $-z$ direction), as can be confirmed by a cross section along a y-z plane (Fig. 3b). That lateral shift, which has been observed in all similar shots, indicates an oblique propagation of the stress pulse when applied along the y axis. Finally, in all samples shocked along the z axis, a single, roughly circular spalled zone opposite to the irradiated spot can be observed (Fig. 4), like would be evidenced in a brittle, isotropic target such as a glass plate (De Rességuier & Cottet, 1995; Li et al., 1998).

To comfort our interpretation of the above observations as effects of quartz anisotropy, we have performed 3D Lagrangian computations with the explicit code *Radioss*. Due to the fast

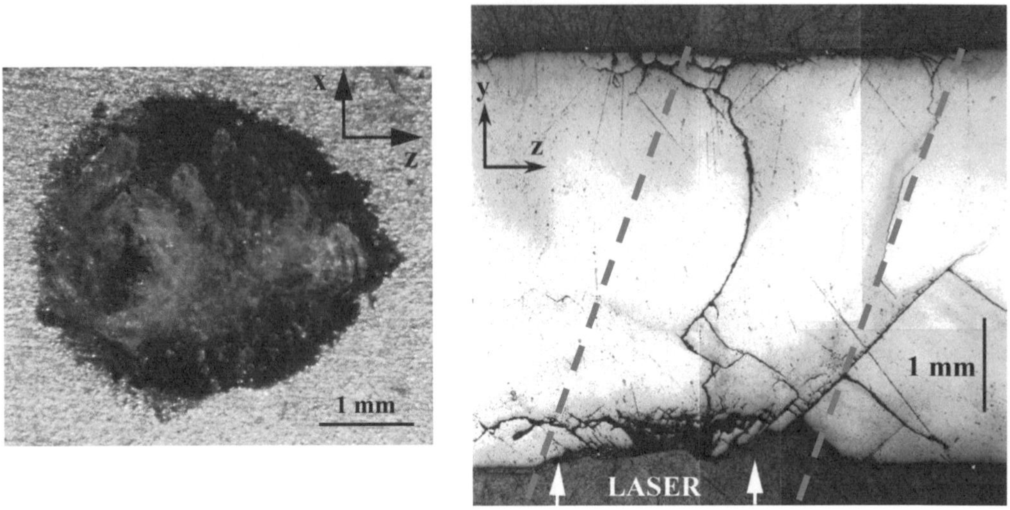

(a) (b)

Fig. 3. Optical micrograph of the free surface **(a)** and cross section in a *y-z* plane **(b)** of a 4.5 mm-thick quartz sample recovered after a laser shock of ~30 GPa applied onto a 2.1 mm-diameter circular spot along the *y* axis.

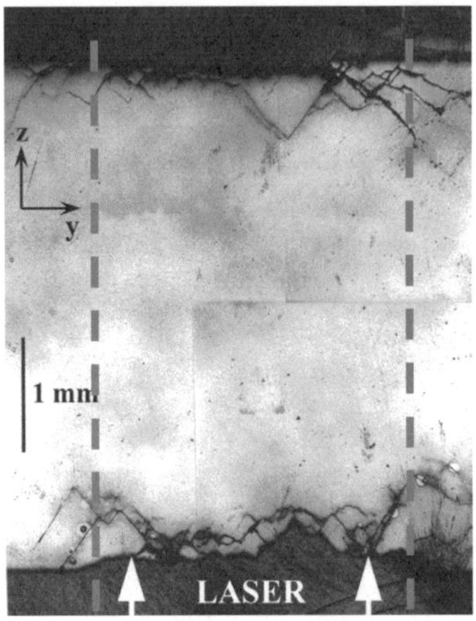

Fig. 4. Cross section in a *y-z* plane of a 4.5 mm-thick quartz sample recovered after a laser shock of ~25 GPa applied onto a 2.1 mm-diameter circular spot along the *z* axis.

decay of short laser shocks with propagation distance, the peak compressive stress rapidly attenuates down to the Hugoniot elastic limit (~6 GPa in *x*-cut quartz), so that the phenomena observed near the free surface depend essentially on the elastic response. To account for this elastic behaviour, we have implemented into *Radioss* a constitutive law based on the stiffness matrix of trigonal crystals like quartz (Royer & Dieulesaint, 1996):

$$C_{\alpha\beta} = \begin{vmatrix} C_{11} & C_{12} & C_{13} & C_{14} & 0 & 0 \\ C_{12} & C_{11} & C_{13} & -C_{14} & 0 & 0 \\ C_{13} & C_{13} & C_{33} & 0 & 0 & 0 \\ C_{14} & -C_{14} & 0 & C_{44} & 0 & 0 \\ 0 & 0 & 0 & 0 & C_{44} & C_{14} \\ 0 & 0 & 0 & 0 & C_{14} & (C_{11}-C_{12})/2 \end{vmatrix} \quad (1)$$

where subscripts α and β refer to the crystallographic directions x,y,z (Fig. 1) : 1 stands for xx, 2 for yy, 3 for zz, 4 for yz, 5 for xz, and 6 for xy. The x,y,z axes, attached to each finite element, are allowed to rotate with shock-induced motion. The stresses T_α are computed from the strains S_β according to Hooke's law, neglecting the piezoelectric contribution:

$$T_\alpha = C_{\alpha\beta} \; S_\beta \quad (2)$$

A simple criterion assuming instantaneous stress relaxation in the elements where the tensile stress reaches a constant threshold value has been added to handle spall fracture. As discussed elsewhere (De Rességuier et al., 2005), realistic simulations of our shock experiments, where very steep pressure pulses (with rise times of a few ns imposing a very fine mesh) propagate in large volumes, would require a great number of cells beyond our computational capabilities. Instead, we use a relatively coarse grid (~150 000 elements) which leads to an artificial spreading of the stress pulses and therefore limits the quantitative consistency of the predictions.

Fig. 5 illustrates the computed response of a 5 mm-thick quartz sample to the same laser shock as in Fig. 2, applied onto a 1.6 mm-diameter circular spot in the centre of the bottom surface, along the x axis of the crystal. Lateral dimensions of 10 mm have been chosen, lower than the dimensions of the real target, but enough to minimize edge effects. The distribution of longitudinal stress σ is shown at two different times. The upper part of the sample has been set 'transparent' (drawn as points) to visualize the emergence of the pulse at different depths. The threshold tensile stress for rupture has been set to 2.5 GPa in order to obtain reasonable residual damage. Actually, a lower, more realistic spall strength of 1.5 GPa (De Rességuier et al., 2003) would lead to overestimate the damage, because the duration of the stress pulse is artificially increased by the mesh coarseness. After 0.51 µs (Fig. 5a), a compressive pulse ($\sigma < 0$) has propagated from the irradiated surface through 3 mm. Its amplitude has decayed down to about 2 GPa, and the initially circular shape of the compressed zone has transformed into an elliptical shape, with a major axis at a 35° angle from the z axis. When that pulse reflects from the free (upper) surface, after about 0.91 µs, tensile stresses are induced in an elliptical zone (Fig. 5b). Their amplitude is too low to produce fracture inside the sample, but it is probably high enough to eject the thin gold layer (not accounted for in the simulation). Later, tension is increased by lateral waves issued from the edges of the ellipsoid, so that spallation takes place near the extremities, in the form of two penny-shaped cracks, about 170 µm below the free surface (Fig. 5c). All these results are in good agreement with the experimental observations reported above.

Fig. 6 depicts the response of a 5 mm-thick quartz sample to the same laser shock applied in the centre of the bottom surface, along the y axis (Fig. 6a) or the z axis (Fig. 6b). The reflection of the pulse from the free surface induces tensile stresses and spall damage. Again, the upper part of the sample is set 'transparent' to visualize the stress and crack distributions 200 µm below the surface. In Fig. 6a, the pulse has propagated along an

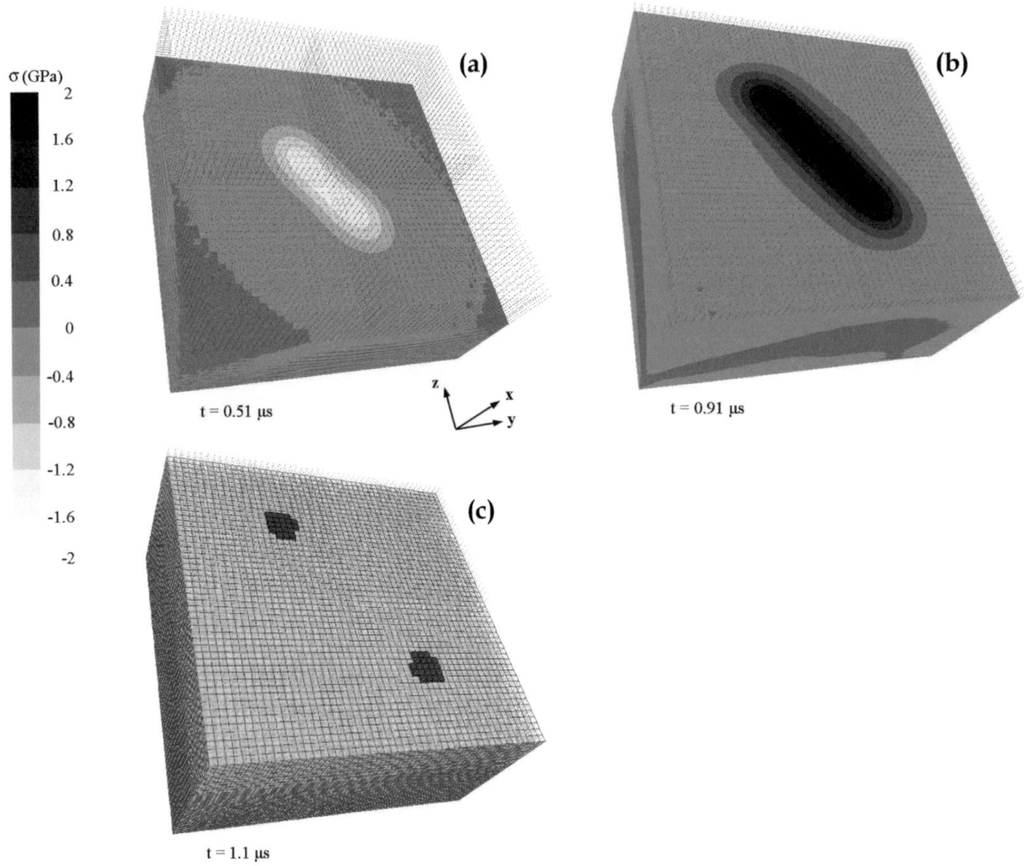

Fig. 5. Computed response of a 5 mm-thick quartz sample to a laser shock of ~42 GPa applied on a 1.6 mm-diameter spot in the bottom surface (not shown), along the x axis. The distribution of longitudinal stress is plotted 0.51 μs (a) and 0.91 μs (b) after the shot. The residual spall damage consists in two separate cracks, appearing as dark grey elements (c).

oblique direction, with an angle of about 23° from the y axis, so that damage occurs in a roughly circular (slightly 'D'-shaped) region shifted about 2 mm from the centre in the z direction. In Fig. 6b, spallation occurs in a circular zone, still in the centre of the target (i. e. above the loaded spot). Again, these predictions match the observed behaviour.

Thus, the analysis of the residual spall damage in laser shock-loaded quartz samples shows that the anisotropy of such trigonal crystals has a key influence on stress wave propagation, which strongly depends on the direction of load application versus crystal orientation. A simple elastic description involving the rigidity matrix of quartz is shown to account for those effects that were never evidenced previously.

3. Two-dimensional wave propagation in laser shock-loaded aluminium

In most laser shock experiments, the diameter of the irradiated spot is much larger than the target thickness (typically 5 times greater), which ensures conditions of uniaxial strain in the central zone near the laser axis, where edge effects do not affect the material response (e.g.

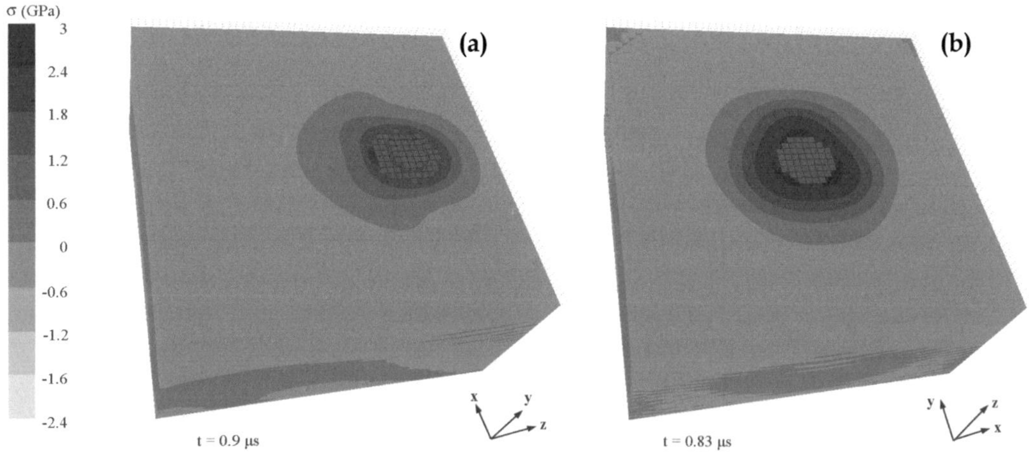

Fig. 6. Computed stress distribution in a 5 mm-thick quartz sample submitted to a laser shock of ~42 GPa applied on a 1.6 mm-diameter spot in the bottom surface (not shown), along the *y* axis **(a)** or the *z* axis **(b)**. The grey elements have been deleted after reaching the spall strength.

Boustie & Cottet, 1991; De Rességuier et al., 1997, Tollier et al., 1998). In thicker samples, such edge effects include lateral release waves issued from the periphery of the loaded area (Fig. 7), following the planar compression front and inducing a pressure decay in the transverse direction, in addition to the classical hydrodynamic attenuation of the pulse during its propagation along the symmetry axis. In such conditions, the space and time distribution of the tensile stresses induced by release waves interactions is strongly conditioned by the aspect ratio *L/D* between the target thickness *L* and the laser spot diameter *D*. Then, the observation of the resulting spall damage in the recovered samples coupled with 2D numerical simulations can be used to reconstitute the pressure wave propagation history (Cuq-Lelandais et al., 2009).

To illustrate this statement, we present laser shock experiments performed in the LULI, on aluminium samples of thickness 1 or 2 mm. Laser pulses of 3 ns duration and intensities ranging from 0.3 to about 30 TW/cm^2 are focused on spots of diameter ranging from 1 to 4 mm. A Velocity Interferometer System for Any Reflector (VISAR) (Barker & Hollenbach, 1972) is used to measure the velocity of the free surface, in a point opposite to centre of the loaded spot. Fig. 8 shows micrographs of this free surface in 1 mm-thick samples recovered from laser shots of increasing intensity applied on spots of two different diameters corresponding to an aspect ratio of 0.25 (top) or 1 (bottom). Gradual evolution from undamaged to bulged surface is seen at low intensity, then a crater appears for higher intensities, characteristic of the ejection of a spalled layer. Unlike in quartz (see Section 2) all damaged zones are roughly circular, like the irradiated spots. For a given aspect ratio, their diameter is shown to increase with laser intensity. Fig. 9 shows cross sections of two 1 mm-thick samples (top-d and bottom-b in Fig. 8) subjected to two laser shots of similar intensity (i.e. similar shock pressure) applied on spots of different diameters. While a spalled layer has been ejected from the first target, the second one has a bulged free surface. This difference clearly illustrates the dependency of spall damage on the aspect ratio *L/D*.

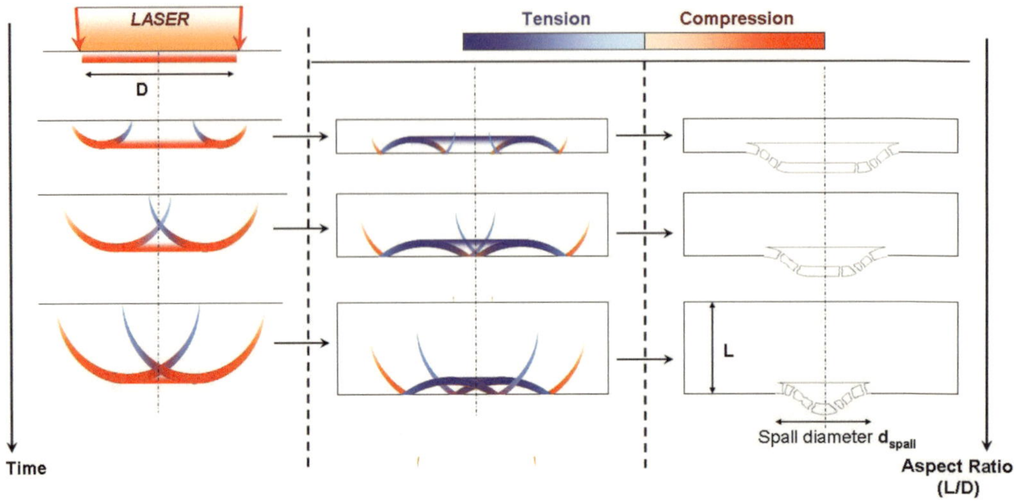

Fig. 7. Schematic description of the propagation of a laser shock of moderate intensity in samples of increasing aspect ratio, illustrating the role of edge effects in wave interactions and their consequences on the spallation process.

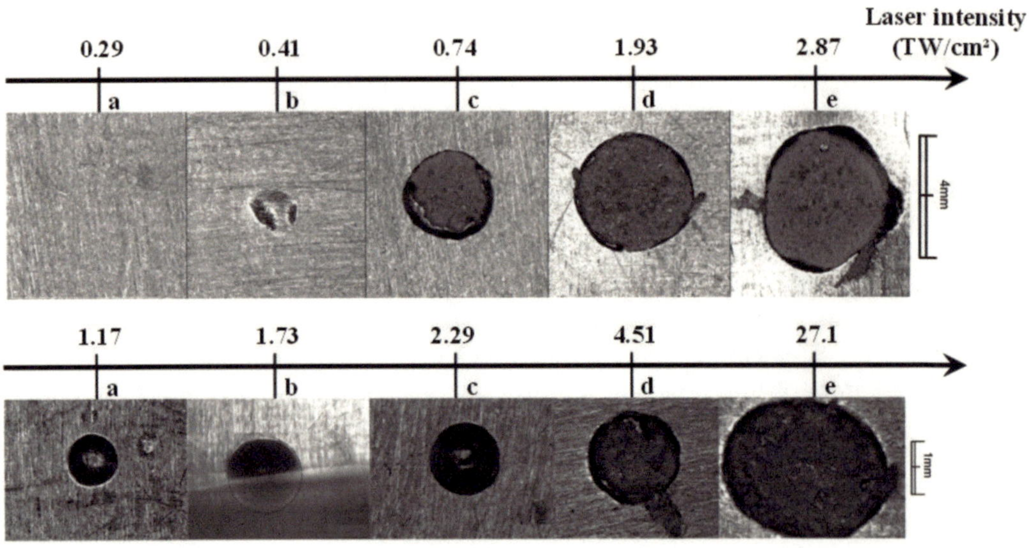

Fig. 8. Overview of the free surface in 1 mm-thick aluminium targets recovered from laser shocks of increasing intensities with two different laser spot diameters : D = 4 mm, L/D = 0.25 (top) ; D = 1 mm, L/D = 1 (bottom).

To correlate the above observations to stress wave propagation, 2D numerical simulations have been performed with the *Radioss* code. Aluminium is modelled with a Mie-Grüneisen equation of state, a Johnson-Cook constitutive law (Johnson & Cook, 1983) and a constant tensile strength to account for spall fracture. Laser shock loading is modelled with a pressure pulse applied as an input boundary condition over the irradiated spot. The temporal shape of this pressure load is inferred from 1D laser-matter interaction computations performed with the *FILM* code of the LULI (Fig. 10a). A scale factor is used to

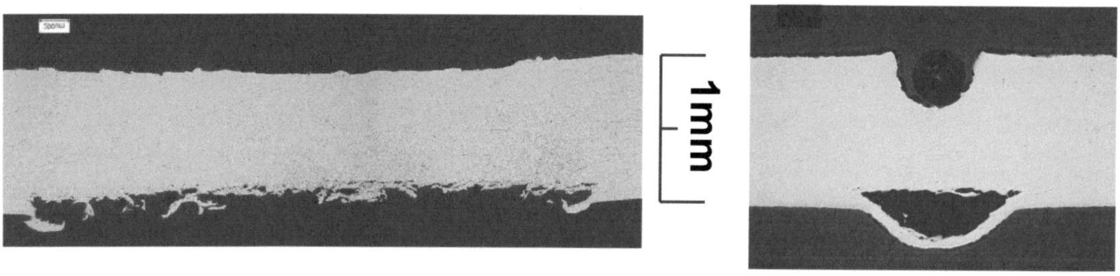

Fig. 9. Sections of 1 mm-thick aluminium samples recovered after laser shots (applied on the upper surface) of 1.9 TW/cm² on a 4 mm-spot (left) and 1.7 TW/cm² on a 1 mm-spot (right).

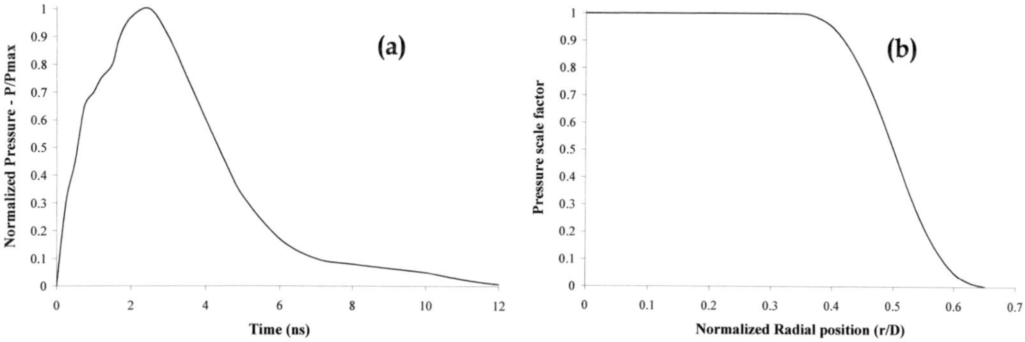

Fig. 10. Temporal **(a)** and spatial **(b)** shapes of the pressure distribution imposed as an input boundary condition over the loaded surface in the computations.

account for the radial decrease of loading pressure with laser energy in the peripheral region of the irradiated area (Fig. 10b). Fig. 11 shows the computed distribution of axial stress at different times for two different geometrical configurations, with a 50 GPa peak loading pressure representative of a ~2 TW/cm² laser shot. A planar shock front parallel to the irradiated surface propagates along the laser axis (Fig. 11-t1) while spherical waves are generated from the edges of the loaded spot (Fig. 11-t2), both outward (compression waves) and inward (release waves). The interaction of these lateral release waves with the planar unloading wave following the laser shock creates tension zones just behind the main compression front. For the higher aspect ratio, these lateral waves meet on the symmetry axis where their crossing generates a new tension pulse that propagates along this axis (Fig. 11b-t3 and t4).

Finally, the reflection of the incident pressure pulse from the free surface produces tensile loading in a planar region close to that surface, as classically occurring in 1D geometry. Thus, the spatial and temporal distribution of dynamic tension results from complex interactions between many waves. As soon as it reaches locally the threshold tensile strength of the material, damage occurs in the form of element deletion accompanied by instantaneous stress relaxation (Fig. 11-t5). Although the pressure load is the same for both cases, the resulting damage differs, either bulging or ejection of a spalled layer (Fig. 11-t7), depending on the aspect ratio. These predictions are consistent with the experimental results presented above (Fig. 9).

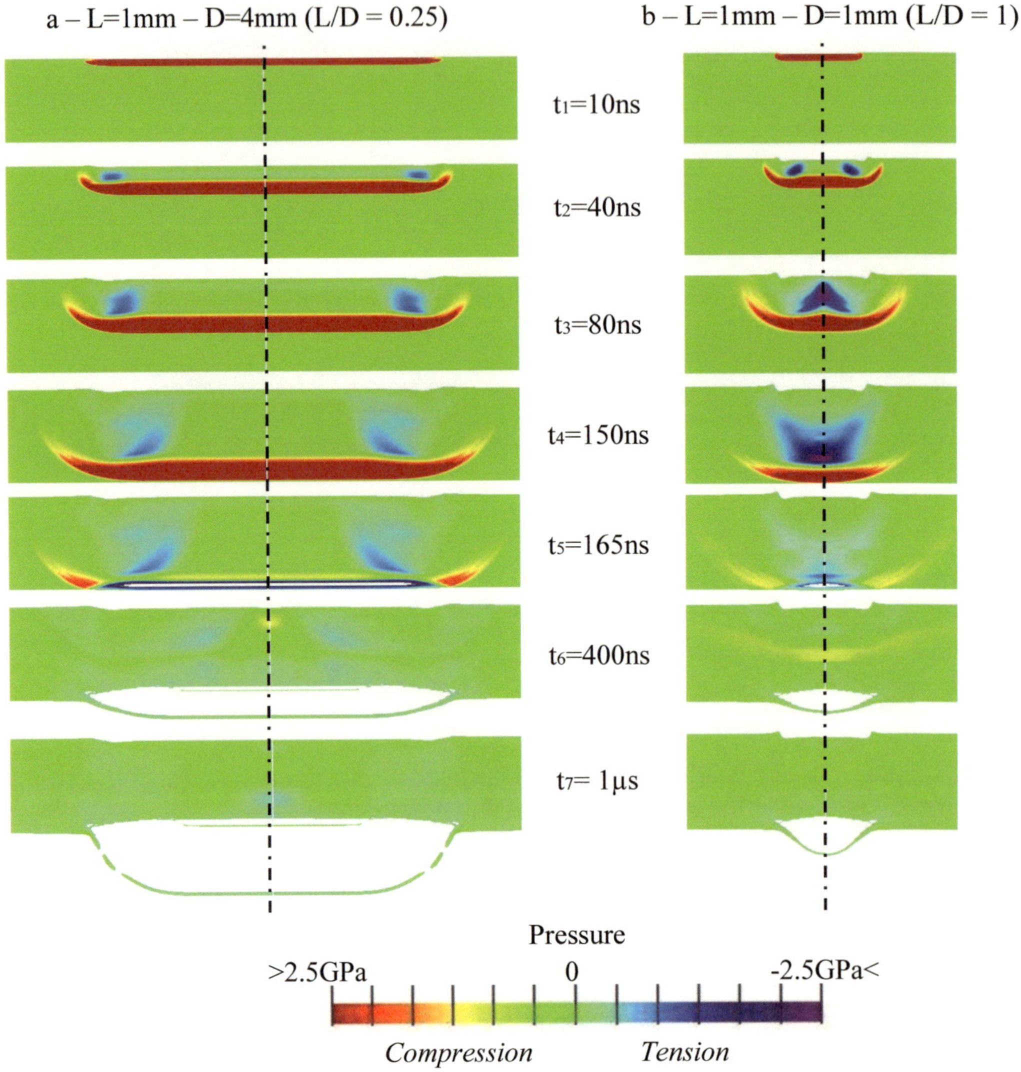

Fig. 11. Computed distribution of axial stress at successive times in 1 mm-thick aluminium targets subjected to a 50 GPa laser shock with aspect ratios L/D = 0.25 **(a)** and L/D = 1 **(b)**.

Contributions of 2D effects can also be evidenced in the free surface velocity profiles measured on the central axis. Fig. 12 shows a comparison between experimental VISAR signals recorded behind 1 mm-thick samples subjected to 50 GPa laser shocks applied on two spots of different diameters (solid lines) and the computed profiles resulting from the 2D simulations above (dotted lines). All curves start with a sharp acceleration upon shock breakout, followed by a progressive deceleration due to the reflection of the unloading wave, interrupted by oscillations characteristic of the spall process. Based on an acoustic approximation, the dynamic tensile strength of the material can be estimated from the so-called velocity pullback (Antoun et al., 2002). This pullback (Δu in Fig. 12) is 350 m/s in both records, which leads to a spall strength of 2.5 GPa. A good overall agreement is found between the measured and calculated velocity profiles, except for the final oscillations, corresponding to wave reverberation through the spalled layer, much too strong in the computations. This discrepancy points out the limitations of the very simple failure criterion, not accounting for the damage dynamics. Peak velocity is found to depend strongly on the aspect ratio, since it is almost twice higher for the 4 mm-spot than for the 1 mm-spot. This is due to the additional attenuation of the pressure pulse by the lateral release waves when they reach the central zone before the shock breaks up at the free surface (see Fig. 7). A 1D simulation has been performed with the same pressure load and target thickness. It is found to match exactly the 2D simulation for the smaller aspect ratio (L/D = 0.25), which shows that edge effects have not affected wave propagation along the central axis in that case.

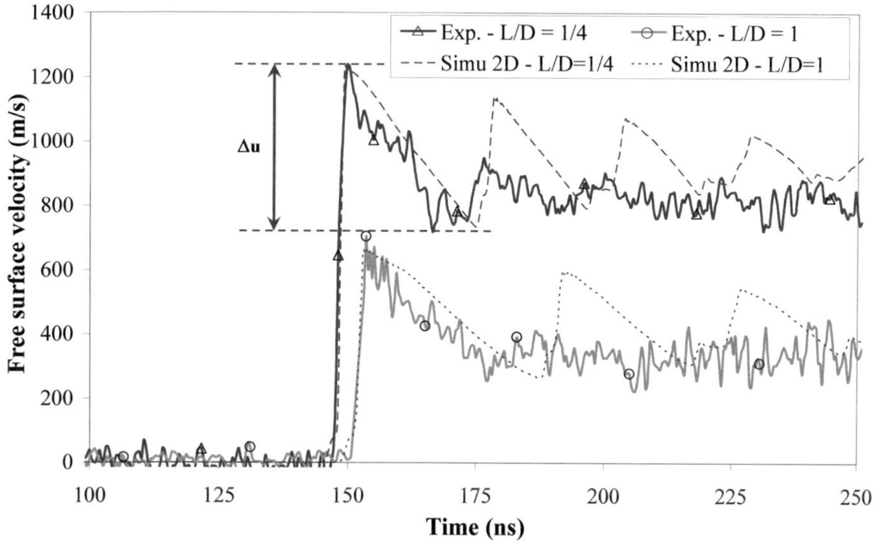

Fig. 12. Comparison of the velocity profiles measured (solid lines) and calculated (dotted lines) at the free surface of 1 mm-thick aluminium samples subjected to laser shocks of 50 GPa applied on spots of diameter 4 mm (L/D = 0.25) or 1 mm (L/D = 1).

The diameter of the spalled zones measured in the free surfaces shown in Fig. 8 is plotted in Fig. 13 for various loading pressures and both aspect ratios, compared to the results of 2D simulations. A good consistency is obtained between experiments and calculations. A zero diameter means that no spallation could be observed. In the computations, this absence of spall damage is represented by the vertical dashed line. In a 1D geometry, one would expect a spalled zone of same diameter as the laser spot. This is observed for loading pressures of 40 GPa with L/D = 1 (D = 1 mm), and about 60 GPa with L/D = 0.25 (D = 4 mm). Above those pressure values, the spalled zone is larger than the loaded spot (e.g. Fig. 9b), because of the effects of the spherical compressive waves propagating outward from the edges of this spot (see Fig. 7). Below those values, the diameter of the spalled zone decreases with loading pressure, because of the attenuation of the compressive pulse by the lateral release waves. Thus, simple examination of the free surface of spalled samples provides indications about the shock pressure applied onto the opposite surface and the subsequent stress wave propagation.

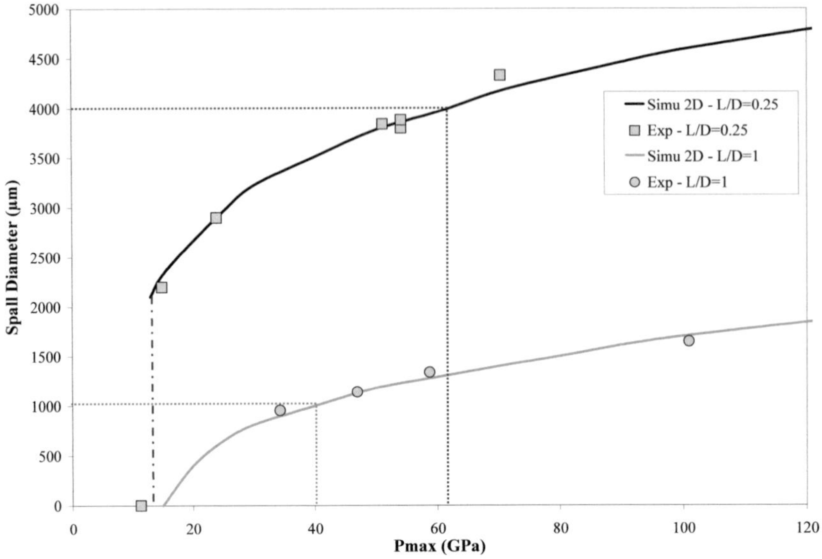

Fig. 13. Comparison of the measured and calculated diameters of the spalled zone in the free surface of 1 mm-thick samples after laser shocks of increasing intensity applied on spots of diameter 4 mm (L/D = 0.25) or 1 mm (L/D = 1).

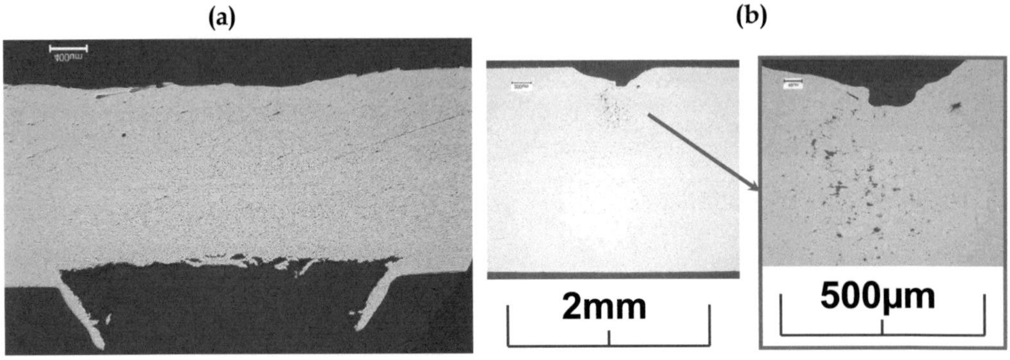

Fig. 14. Cross sections of 2 mm-thick aluminium samples recovered after laser shocks of about 2 TW/cm^2 applied on spots of diameter 4 mm **(a)** or 1 mm **(b)** in the upper surface.

Finally, 2D effects can lead to damage mechanisms different from spallation (Boustie et al., 2007). Fig. 14 shows cross sections of 2 mm-thick aluminium targets subjected to laser shocks of same intensity applied on spots of different diameters. The first sample (Fig. 14a), of smaller aspect ratio (L/D = 0.5), is typical of the ejection of a planar spalled layer. The second one (L/D = 2) presents no sign of spall fracture near the free surface, but damage appears in the form of pores down to about 500 µm beneath the loaded zone (Fig. 14b). Such damage results from the interaction of the lateral release waves issued from the edges of the laser spot (see Fig. 11b). When they meet in the central zone near the symmetry axis, their crossing produces tensile stresses higher than the dynamic strength of the metal, which causes void nucleation and growth. Meanwhile, the compressive pulse emerging at the free surface after both axial and lateral decay is too low to produce spallation.

Overall, like in the previous section, post-shock analysis of the residual damage in laser shock-loaded samples combined with numerical simulations can be used to investigate wave propagation, especially the complex 2D effects that become prominent (even in an isotropic solid like aluminium) as soon as the diameter of the loaded surface is of the same order as the sample thickness.

4. Phase transformation and wave propagation in laser shock-loaded iron

When shocked above 13 GPa, iron undergoes a polymorphic (solid-solid) transformation from body-centered-cubic (α) to hexagonal-close-packed (ε) structure. This transition was first inferred from wave profile measurements several decades ago (Minshall, 1955; Bancroft et al., 1956), then corroborated under quasi-static compression (Jamieson & Lawson, 1962). Although it has been reported that tiny quantities of ε phase could be quenched to ambient pressure upon release following ultra-short, fs-duration laser-driven shocks (Sano et al., 2003), the transformation is essentially reversible, so the ε structure can only be detected during the time of pressure application, which makes direct identifications challenging (Kalantar et al., 2005; Hawreliak et al., 2006). One indirect way to characterize this phase change under shock conditions is to study the subsequent formation of double wave structures from initially steep shock fronts (Barker & Hollenbach, 1974; Duvall & Graham, 1977; Boettger & Wallace, 1997). Another one is to analyze the residual distribution of specific micro-twins following the excursion into the high pressure phase (Smith, 1958; Johnson et al., 1962) and leading to a significant increase in the material residual hardness. Here, we use spall fracture as a third indirect approach to study this phase transition and its effects on wave propagation in laser shock-loaded iron.

Again, the experiments have been performed in the LULI. A pulsed laser beam of 1.06 µm-wavelength, 4.5 ns-duration, and ~1000 J-energy is focused onto a 2.8 mm-diameter spot on pure polycrystalline iron foils of thicknesses ranging from 150 µm to 520 µm, set in a vacuum chamber to avoid laser breakdown in air. The resulting irradiation at about 5×10^{12} W/cm^2 produces the vaporization of a thin layer of material whose expansion induces by reaction a compressive pulse into the solid target. The amplitude and shape of this pressure pulse are inferred from 1D computations of laser-matter interaction with the code *FILM* of the LULI. The resulting load consists of a sharp compression within about 1 ns to a peak pressure of ~130 GPa followed by slower unloading in about 8 ns. Like in Section 3, a VISAR interferometer is used to measure the velocity of the opposite free surface. The recovered samples are embedded in a resin, sectioned through a plane containing the laser axis, etched with a Nital solution to reveal grain boundaries and defects, and observed in optical microscopy.

The 150 µm-thick target was fully pierced by the laser shot, so that nothing could be recovered from the shocked zone. The thicker samples remained attached to the target holder. Fig. 15 shows cross sections of shock-loaded samples of initial thicknesses 520 µm (a), 400 µm (b) and 250 µm (c). The laser shock has been applied onto the lower surface, and its reflection from the upper free surface has lead to the ejection of a ~90 to 110 µm-thick spalled layer. In addition, high shear stresses induced in the periphery of the loaded zone have cut out the 250 µm-thick foil, so the cross section is split into three parts.

Higher magnification micrographs of two of those cross sections after etching are presented in Fig. 16, showing detailed views below the spall craters (upper surfaces). In the two thicker samples, spallation occurs in the form of straight cracks, both intra- and inter-

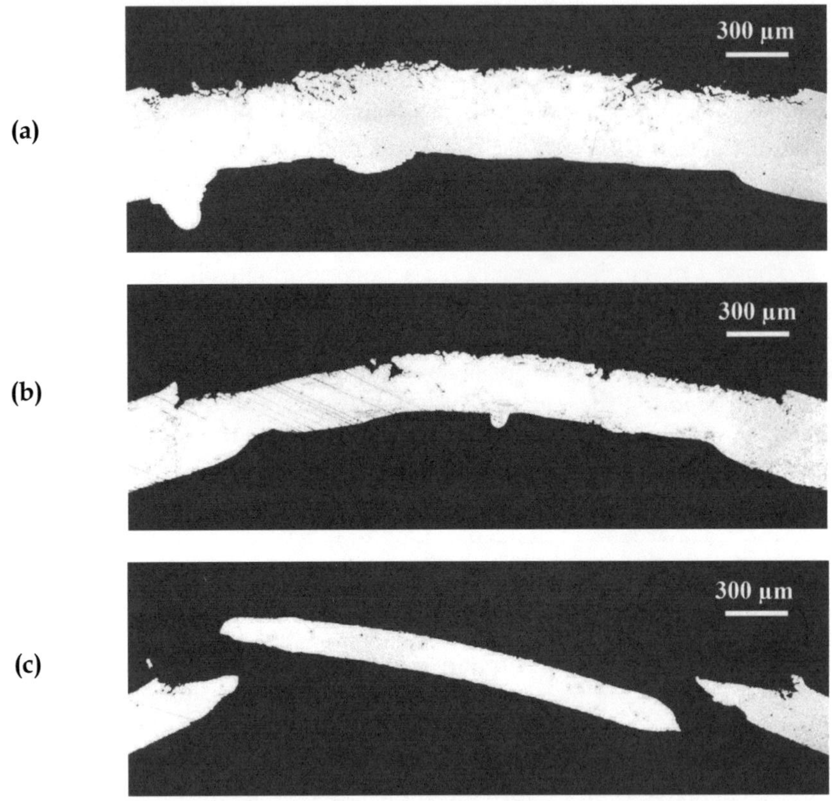

Fig. 15. Cross sections in iron samples of initial thickness 520 μm **(a)**, 400 μm **(b)**, and 250 μm **(c)**, recovered after laser shocks of about 130 GPa applied onto the bottom surface. The ejection of a spalled layer results in a ~90 to 110 μm-deep crater in the upper surface.

granular, consistent with the brittle fracture behaviour of iron at high strain rates (Zurek & Meyers, 1996), distributed over a depth of several tens of μm (Fig. 16a). In the 250 μm-thick sample, the spall plane is remarkably smooth (Fig. 15c and 16b). Such "smooth spall", already reported under explosive and impact loading, is interpreted as a sign that fracture took place after completion of the α–ε–α cycle (Ivanov & Novikov, 1986; Novikov et al., 1994; Voltz et al., 2006). It is usually attributed to the formation of a rarefaction shock resulting from the steepening of the release wave upon reverse transformation back to the α phase, which produces a drop in density and an increase in sound speed. As mentioned earlier, the change to the ε phase should also result in the formation of specific micro-twins. Actually, such micro-twins can be seen in the first grains at the bottom of Fig. 16a (400 μm-thick sample), suggesting that significant amount of ε-phase has been formed up to about 100 μm away from the spall crater, while extensive twinning is observed up to the spall plane in Fig. 16b (250 μm-thick sample). This confirms that the spalling region has remained in the α-phase in the thicker sample (rough fracture surface) while it has undergone the α–ε–α cycle before fracture in the thinner one (smooth fracture surface). Such difference provides information on the decay of the laser driven pressure pulse during its propagation from the irradiated surface, since it indicates that peak pressure has remained higher than the 13 GPa-threshold required for the phase transformation over a propagation distance of about 300 μm.

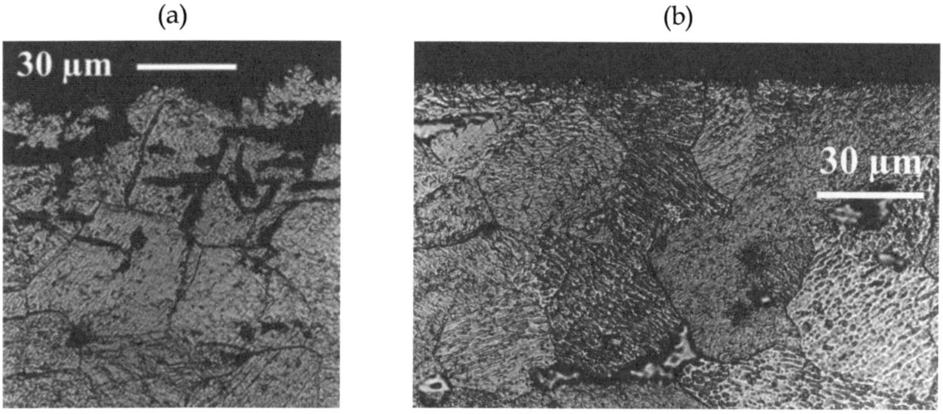

Fig. 16. Detailed views of the spall craters (upper surfaces) in cross sections of laser shock-loaded iron samples of initial thickness 400 μm **(a)** and 250 μm **(b)** after etching. The 250 μm-thick sample exhibits a remarkably smooth fracture surface, as well as a dense distribution of micro-twins up to that surface.

VISAR records of the free surface velocity are shown in Fig. 17. A vertical shift of -200 m/s has been added between the shots for clarity. They all show an elastic precursor followed by a slower plastic front which does not split into the two-wave structure usually observed over larger propagation distances. The peak free surface velocity drops rapidly while the target thickness increases from 150 μm to 250 μm, then it decreases slowly with propagation distance. As already discussed in Section 3, fracture-induced stress relaxation gives rise to compressive waves that propagate from the cracks and produce a re-acceleration of the free surface called the spall pulse (Antoun et al., 2002). Thereafter, wave reverberation within the spalled layers generates damping oscillations in the velocity profiles. These profiles are compared to the results of 1D simulations with the *SHYLAC* code, involving a two-phase equation of state with a description of the kinetics of the phase change (De Rességuier & Hallouin, 2001), a constitutive law accounting for time-dependent twinning (De Rességuier & Hallouin, 1998), and a simple spall criterion assuming instantaneous rupture and subsequent relaxation as soon as the tensile stress reaches a threshold value. This comparison, which has been discussed in details elsewhere (De Rességuier & Hallouin, 2008), shows a good overall consistency. According to the model predictions, the decay of the pressure pulse should yield to full transition from α to ε iron in the first ~200 μm below the loaded surface, then the fraction of ε phase should decrease from 1 to 0 through the next ~200 μm-thick layer. Hence the tensile stresses leading to spall fracture upon reflection of the compressive pulse from the free surface would be induced in a region of untransformed α iron in the 520 μm-thick sample, in a material having undergone very little (less than 8 %) α–ε phase transformation in the 400 μm-thick sample, and in a zone which was almost fully transformed in the two thinner targets. This is fully consistent with the difference in fracture surface morphology reported above.

Finally, a smooth fracture surface similar to that observed in the 250 μm-thick sample can be seen in the left part of the spall crater in Fig. 15b, in the 400 μm-thick sample. As stated above, the α–ε phase transition should have propagated only up to about 100 μm beneath the spall plane (Fig. 16a). However, some hotter spot in the laser beam may have lead to a somewhat inhomogeneous spatial distribution of the loading pressure applied onto the irradiated surface, so that the depth of the transformed region may have reached the spall plane in this left part of the cross section, which would account for the smooth spall.

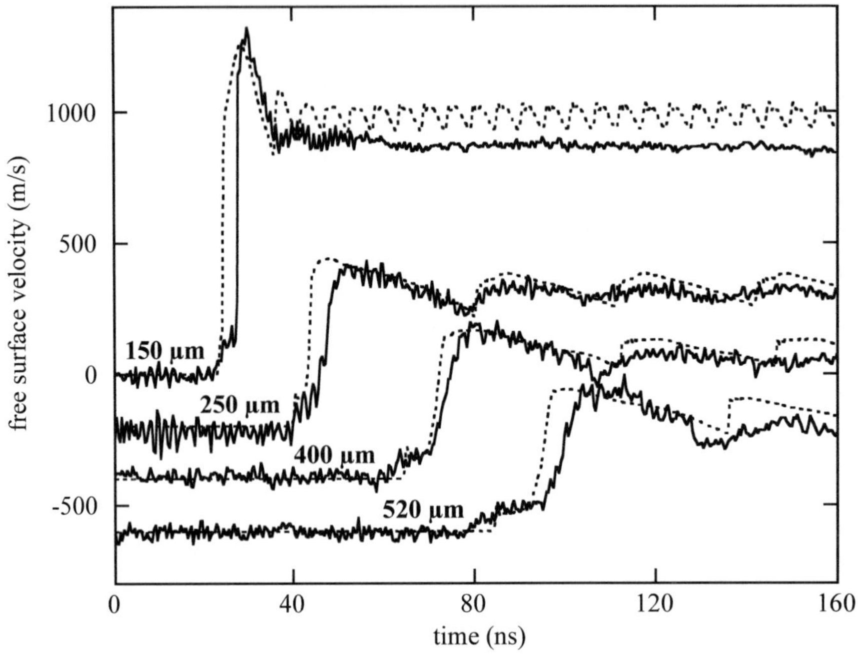

Fig. 17. Solid lines are free surface velocity profiles measured in iron samples of different thicknesses (bold numbers) subjected to laser shocks of about 130 GPa. For clarity, some records have been shifted vertically (offsets of -200 m/s, -400 m/s and -600 m/s have been subtracted from the velocity for thicknesses of 250 μm, 400 μm and 520 μm, respectively. Dotted lines result from computer simulations of the experiments.

Thus, the change in fracture surface morphology observed in iron samples of increasing thicknesses recovered after laser shock-induced spallation provides a signature of the $\alpha-\varepsilon$ phase transformation occurring in this metal, which can be used to infer the decay of shock pressure with propagation distance, as well as the spatial distribution of the pressure applied onto the loaded surface.

5. Conclusion

Dynamic fracture in shock-loaded materials is governed by the propagation, reflection and interactions of stress waves. Post-shock analyses of the residual damage observed in samples recovered from laser shock experiments, less destructive than more conventional techniques, can provide valuable insight into key aspects of wave propagation prior to fracture, such as the effects of structural anisotropy, the role of lateral waves associated to edge effects, or the influence of polymorphic phase transformations on the response to shock loading. Such observations can be based on very simple techniques like optical microscopy. Still, their interpretation should be comforted by time-resolved measurements and numerical simulations since the phenomenology leading to the final damage may include several complex, coupled processes.

6. References

Antoun, T.; Seaman, L.; Curran, D.R.; Kanel, G.I.; Razorenov, S.V. & Utkin, A.V. (2002). *Spall Fracture*, Ed. Springer, ISBN 0-387-95500-3

Bancroft, D.; Peterson, E.L. & Minshall, S. (1956). Polymorphism of iron at high pressure, *J. Appl. Phys.* 27(3), pp. 291-298

Barker, L.M. & Hollenbach, R.E. (1972). Laser interferometer for measuring high velocities of any reflecting surface, *J. Appl. Phys.* 43, pp. 4669-4675

Barker, L.M. & Hollenbach, R.E. (1974). Shock wave study of the α–ε phase transition in iron, *J. Appl. Phys.* 45, pp. 4872-4887

Boettger, J.C. & Wallace, D.C. (1997). Metastability and dynamics of the shock-induced phase transition in iron, *Physical Review B* 55(5), pp. 2840-2849

Boustie, M. & Cottet, F. (1991). Experimental and numerical study of laser induced spallation into aluminum and copper targets, *J. Appl. Phys.* 69, pp. 7533–7538

Boustie, M.; Cuq-Lelandais, J.P.; Bolis, C.; Berthe, L.; Barradas, S.; Arrigoni, M.; De Rességuier, T. & Jeandin, M. (2007). Study of damage phenomena induced by 2D effects into materials under laser driven shocks, *J. Phys. D* 40, pp. 7103-7108

Cuq-Lelandais, J.P.; Boustie, M.; Berthe, L.; Combis, P.; Sollier, A.; De Rességuier, T.; Lescoute, E.; Gay, E.; Soulard, L. & Bontaz, J. (2009). Two-dimensional investigation of laser shock induced spallation in sub-picosecond regime, Proceedings of the 9th DYMAT Conference, Brussels, Belgium, September 2009, EDP Sciences, pp. 625-631

De Rességuier, T. & Cottet, F. (1995). Experimental and numerical study of laser induced spallation in glass, *J. Appl. Phys.* 77(8), pp. 3756-3761

De Rességuier, T.; Couturier, S., David, J. & Niérat, G. (1997). Spallation of metals subjected to intense laser shocks, *J. Appl. Phys.* 82(5), pp. 2617-2623

De Rességuier, T. & Hallouin, M. (1998). Stress relaxation and precursor decay in laser shock-loaded iron, *J. Appl. Phys.* 84(4), pp. 1932-1938

De Rességuier, T. & Hallouin, M. (2001). Interaction of two laser shocks inside iron samples, *J. Appl. Phys.* 90(9), pp. 4377-4384

De Rességuier, T.; Berterretche, P. & Hallouin, M. (2005). Influence of quartz anisotropy on shock propagation and spall damage, *Int. J. of Impact Engineering* 31(5), pp. 545-557

De Rességuier, T. & Hallouin, M. (2008). Effects of the α–ε phase transition on wave propagation and spallation in laser shock-loaded iron, *Phys. Rev. B* 77(17), 174107

Duvall, G.E. & Graham, R.A. (1977). Phase transition under shock wave loading, *Rev. Mod. Phys.* 49, pp. 523-579

Hawreliak, J.; Colvin, J.D.; Eggert, J.H.; Kalantar, D.H.; Lorenzana, H.E.; Stölken, J.S.; Davies, H.M.; Germann, T.C.; Holian, B.L.; Kadau, K.; Lomdahl, P.S.; Higginbotham, A.; Rosolankova, K.; Sheppard, J. & Wark, J.S. (2006). An analysis of the X-ray diffraction signal for the α–ε transition in shock-compressed iron: simulation and experiment, *Phys. Rev. B* 74, 184107

Ivanov, A.G. & Novikov, S.A. (1986). Rarefaction shock waves in iron from explosive loading, *Comb. Expl. Shock Waves* 22(3), pp. 343-350

Jamieson, J.C. & Lawson, A.W. (1962). X-ray diffraction studies in the 100 kbar pressure range, *J. Appl. Phys.* 33(3), pp. 776-780

Johnson, G.R. and Cook, W.H. (1983). A constitutive model and data for metals subjected to large strains, high strain rates and high temperatures, Proceedings of the 7th Int. Symp. on Ballistics, La Hague, pp. 541-547

Johnson, P.C.; Stein, B.A. & Davis, R.S. (1962). Temperature dependence of shock-induced phase transformation in iron, *J. Appl. Phys.* 33, pp. 557-562

Kalantar, D.H.; Belak, J.F.; Collins, G.W.; Colvin, J.D.; Davies, H.M.; Eggert, J.H.; Germann, T.C.; Hawreliak, J.; Holian, B.L.; Kadau, K.; Lomdahl, P.S.; Lorenzana, H.E.; Meyers, M.A.; Rosolankova, K.; Schneider, M.S.; Sheppard, J.; Stölken, J.S. & Wark, J.S., Direct observation of the $\alpha-\varepsilon$ transition in shock-compressed iron via nanosecond X-ray diffraction (2005). *Phys. Rev. Lett.* 95(7), 075502

Li X.-Z., Nakano M., Yamauchi Y., Kishida K. and Tanaka K.A. (1998). Microcracks, spall and fracture in glass : a study using short pulsed laser shock waves, *J. Appl. Phys.* 83(7), pp. 3583-3594

Minshall, S. (1955). Investigation of polymorphic transition in iron at 130 kbar, *Phys. Rev.* 98(3), p. 271

Novikov, S.A.; Pogorelov, A.P. & Sinitsyna, L.M. (1994). The formation of smooth spalls in steel during the interaction of glancing detonation waves, *Comb. Expl. Shock Waves* 30(4), pp. 537-539

Royer, D. & Dieulesaint, F. (1996). *Ondes élastiques dans les solides (Tome 1, Propagation libre et guidée)*, ed. Masson, Paris

Sano, T.; Mori, H.; Ohmura, E. & Miyamoto, I. (2003). Femtosecond laser quenching of the ε phase of iron, *Appl. Phys. Lett.* 83(17), pp. 3498-3500

Smith, C.S. (1958). Metallographic studies of metals after explosive shock, Trans. Metall. Soc. AIME 212, pp. 574-589

Tollier, L.; Fabbro, R. & Bartnicki, E. (1998). Study of the laser-driven spallation process by the VISAR interferometry technique II: experiment and simulation of the spallation process, *J. Appl. Phys.* 83(3), pp. 1231-1237

Voltz, C.; Buy, F. & Roy, G. (2006). Iron damage and spalling behavior below and above shock-induced $\alpha-\varepsilon$ phase transition *Shock Compression of Condensed Matter*, CP845, Proceedings of APS Shock Conference, August 2005, Baltimore, USA, pp. 678-681

Zurek, A.K. & Meyers, M.A. (1996). Microstructural aspects of dynamic failure, in *High Pressure Shock Compression of Solids II*, eds. L. Davison, D.E. Grady and M. Shahinpoor, Springer, New York, pp. 25-70

23

Nonlinear Waves in Transmission Lines Periodically Loaded with Tunneling Diodes

Koichi Narahara
Yamagata University
Japan

1. Introduction

We review the properties of nonlinear waves on the electrical transmission lines periodically loaded with tunneling diodes (TDs) termed—TD lines.In general, the transmission equation of a linear dispersive line allows both sinusoidal and exponential solutions. However, exponential solutions are usually discarded because they diverge at infinity, and therefore, do not satisfy any physically meaningful boundary conditions. In a TD line, once the input pulse crosses the peak voltage of the loaded TDs, the exponential wave develops atsmaller voltagesthan the peak, and the ordinary sinusoidal wave is coupled to it atgreater voltages. The sinusoidal part can continuously unite with the leading exponential wave, so that the exponential wave can be formed along the dispersive line without violating physical boundary conditions. By developing exponential waves, a TD line exhibits technologically useful properties that result in the generation and management of short pulses.When an impulse is input to a TD line, the resulting exponential wave is much steeper than the input; therefore, the input pulse experiences significant shortening.

Moreover, when a rising step pulse, whose bottom and top voltage levels lie in the voltage ranges below and above the peak voltage is input, the pulse edge oscillates on the line.

This chapter discusses nonlinear wave propagation on TD lines in detail. We first define TD lines and describes their circuit configuration and principle of operation. We employ an idealized model of a TD line, which makes a comprehensive description of the operating principles possible. It also gives us design criteria for a short-pulse generator or oscillator using TD lines. We then examine the validity of the analytically obtained design criteria through numerical integration of the transmission equations of a TD line. We alsocharacterize the line using full-wave calculations for monolithically integrated lines. Next, severalexperimental results are described, which are obtained by the time-domain measurements using TD lines breadboarded with Esaki diodes. Finally, we describe the potential of TD lines for the management of short pulses.

2. Fundamentals of TD lines

Figure 1(a) shows a circuit diagram of a TD line, where L, R, C, and G represent the series inductor, series resistor, shunt capacitor, and shunt TD of the unit cell, respectively. The typical current-voltage relationshipof a TD is shown by the dotted curve in Fig. 1(b). For simplicity, we idealize it as the solid curve in Fig. 1(b) (Richer, 1966). The voltage levels where TDs exhibit a negative differential resistance (NDR) degenerate to a threshold V_{pk} and the finite conductance above V_{pk} is neglected. We can then write the current through TDs as$G = G_0\theta(V_{pk} - V)$, where $\theta(V)$ shows the Heaviside function. Hereafter, we consider the situations where an impulse or a step pulse is input to the line, such that they cross V_{pk}. Figure 2 shows the signal applications. For convenience, we call the voltage range greater (less) than V_{pk} regions I(II). The present simplified TD model shows that the pulse is influenced by finite shunt conductance in region II and is completely loss free in region I.

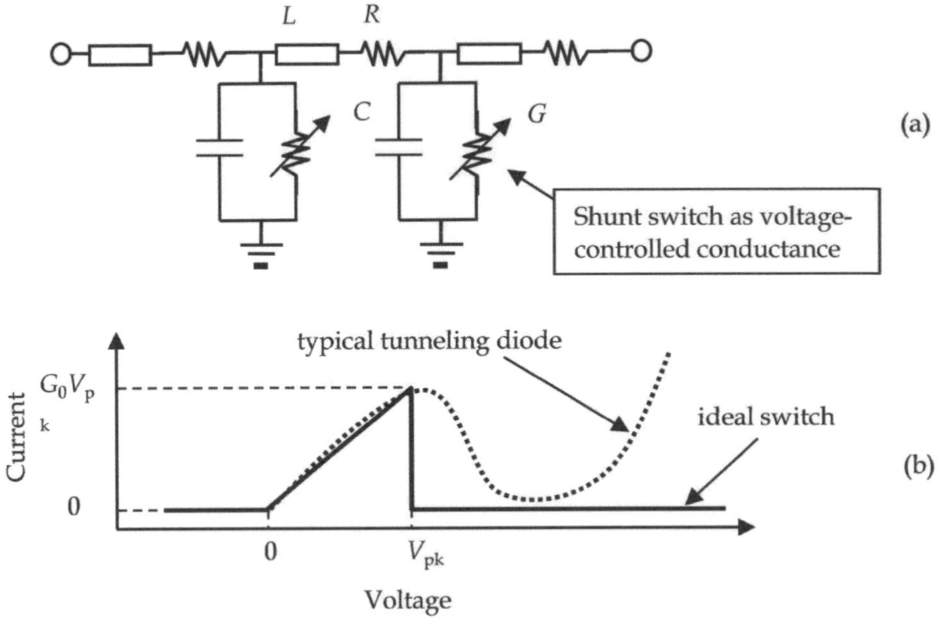

Fig. 1. Circuit configuration of TD lines. (a) The unit cell of TD lines and (b) current-voltage relationship of TDs.

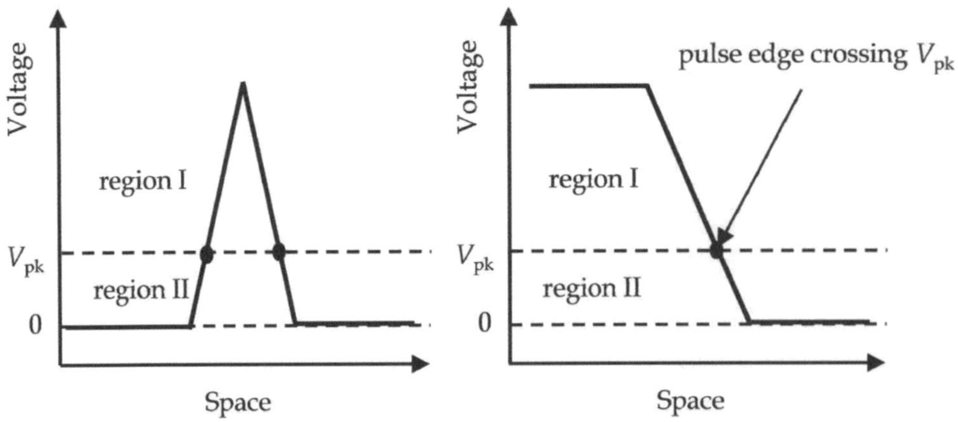

Fig. 2. Signal applications.

When denoting the voltage and current at the nth node as V_n and I_n, respectively, the transmission equation of the line is given by

$$l\frac{dI_n}{dt} = \frac{V_n - V_{n+1}}{d} - rI_n, \tag{1}$$

$$c\frac{dV_n}{dt} = \frac{I_{n-1} - I_n}{d} - gV_n, \tag{2}$$

where d is the length of the unit cell. Moreover, $l = L/d$, $c = C/d$, $r = R/d$, and $g = G_0/d$ are the series inductance per unit length, shunt capacitance per unit length, and shunt conductance per unit length, respectively.The line is linear and dispersive when considering only regions I and II. Thus, it is meaningful to analyze the dispersion curves in each region. The phase velocities normalized by $1/(lc)^{0.5}$ of the modes having wave number k are explicitly given by

$$v_{I,\sin} = \frac{1}{2kd}\sqrt{-\alpha^2 + 16\sin^2\frac{kd}{2}}, \tag{3}$$

$$v_{I,\exp} = \frac{1}{2kd}\left(-\alpha + \sqrt{\alpha^2 + 16\sinh^2\frac{kd}{2}}\right), \tag{4}$$

$$v_{II,\exp} = \frac{1}{2kd}\left(-\alpha - \beta + \sqrt{(\alpha - \beta)^2 + 16\sinh^2\frac{kd}{2}}\right), \tag{5}$$

where $v_{I,\sin}$, $v_{I,\exp}$, and $v_{II,\exp}$ show the normalized phase velocities of the sinusoidal mode in region I, the exponential mode in region I, and the exponential mode in region II, respectively. Moreover, α and β are normalized series resistance and shunt conductance given by

$$\alpha = rd\sqrt{\frac{c}{l}}, \tag{6}$$

$$\beta = gd\sqrt{\frac{l}{c}}. \tag{7}$$

We employ the sign convention: $\omega t - nkd$for positive propagation. Thus, in region I, $k$must be negative, while in region II, it must be positive for exponential modes.
Figure 3 shows the dispersion curves for α=0.18 and β=2.0. The horizontal axis shows the wave number with a unit of the inverse of d, and the vertical axis shows the normalized phase velocities. For steady or quasisteady pulse propagation, the velocity of the pulse edge in region I must be the same as that in region II. When the pulse propagates forward, the exponential mode in region II can only couple with the sinusoidal mode in region I because no exponential counterpart is present. At point P, the velocity and steepness are coincident, so that the forward pulse occupies the region around this point in Fig. 3. Note that the wavenumber at P is relatively large, so that for the forward-propagating pulse, short-wavelength oscillatory waves should be observed in region I, which is supported by an exponential wave developed in region II. According to this mechanism, when an impulse is applied to the TD line, it finally reaches the cross-point of $v_{I,\sin}$ and $v_{II,\exp}$. Initially, a wave

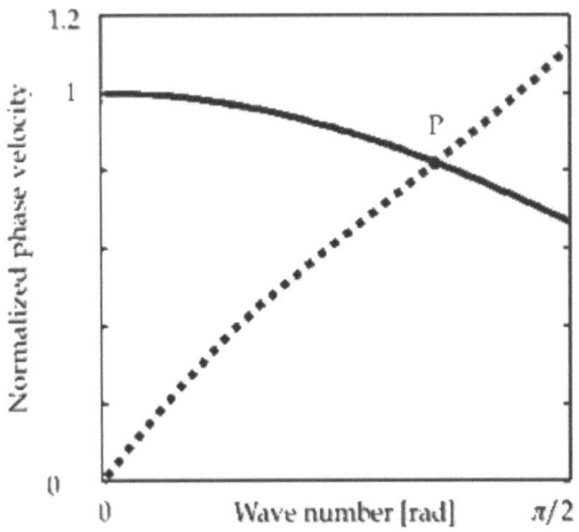

Fig. 3. Dispersion of TD line. The solid and dotted curves represent the normalized velocities in regions I and II, respectively.

component of an input pulse having a specific wave number must have different velocities between region I and region II. Thus, for coincidence of the velocities in both the regions I and II, the wave component experiences a wavelength shortening. As a result, the width of the incident pulse is greatly reduced (Narahara et al., 2005). Because the wave number at the cross point of the curves $v_{I,sin}$ and $v_{II,exp}$ becomes larger for greater β, the greater the conductance, the more the pulse is shortened.

Similar pulse shortening is possible in Schottky-contacted TD lines. It is well known that a Schottky-contacted line exhibits several useful properties for high-speed electronics (Scott, 1970). When a pulse is input such that the nonlinearity of the Schottky varactors compensates for dispersion, the line generates multiple solitonic pulses, whose widths are generally smaller than that of the input. By extracting the largest pulse, a Schottky-contacted line operates as a good short pulse generator (Kintis et al., 2008). Moreover, when a step pulse is input such that both the nonlinearity and dispersion sharpen the edge, the edge finally results in a shock, by which a sub-picosecond temporal transient is observed (Rodwell et al., 1994). It is found that the larger soliton travels more than the smaller one in Schottky-contacted TD lines; therefore, we obtain highly shortened pulse by detecting only the largest soliton (Narahara et al., 2006).

On the other hand, when a step pulse is applied to the line, the pulse edge oscillates, as shown in Fig. 4. The spatial position on the line is shown horizontally, and the voltage is shown vertically. Figure 4(a) shows the behavior of the forward pulse. Because the forward pulse, combining a sinusoidal mode in region I and an exponential mode in region II, does not have permanent profile, it is not stable—it becomes attenuated and finally disappears, leaving only the exponential pulse edge is in region II, as shown in Fig. 4(b). This edge develops exponential modes in region I to form a stable pulse, and then starts traveling backward, as shown in Fig. 4(c). When the backward stable pulse reaches the input, it is reflected as in Fig. 4(d), and again begins to travel forward as an unstable sinusoidal-exponential wave (Narahara, 2006). This process continues permanently with proper boundary condition.

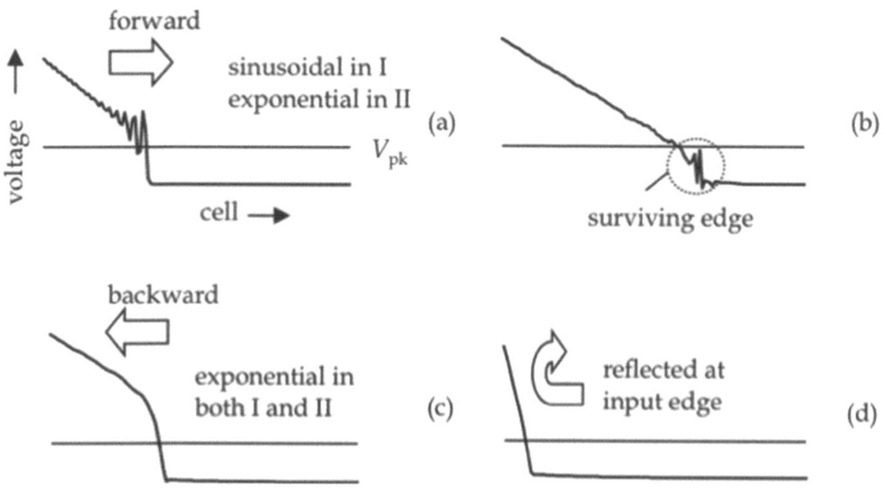

Fig. 4. Step pulse in TD line.

3. Numerical characterization of TD lines

3.1 Transmission line analysis

To validate the above given qualitative discussion and to observe the pulse shortening, we numerically solved Eqs.(1) and (2). Figure 5 shows the typical behavior of a pulse propagating along a TD line. We set c, l, V_{pk}, β, and d to 6.0 pF/mm, 3.0 nH/mm, 0.2 V, 0.75, and 50 μm, respectively. For the present parameters, the phase velocity at a long wavelength is 0.025 c_0 (c_0: the velocity of light in vacuum) and the cutoff frequency is 47.5 GHz.

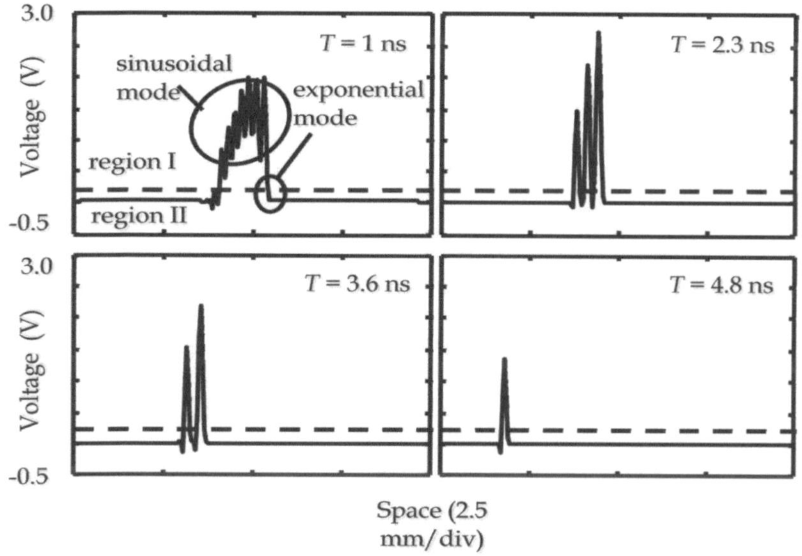

Fig. 5. Impulse in TD line.

Input, an impulse of a Gaussian form with an amplitude of 1.5 V and a full-width at half-maximum (FWHM) of 100.0 ps is supplied. The waveforms at 1.0, 2.3, 3.6 and 4.8 ns after the

Gaussian impulse starts to be input are shown in Fig. 5. Oscillation, which is caused by the development of the sinusoidal wave in region I, is observed at every temporal point. Moreover, the wave number of the oscillation decreases as time progresses. Finally, the sinusoidal wave in region I becomes monocycled at 4.8 ns, so that the waveform becomes a single short pulse, whose FWHM is estimated to be 12.5 ps.

Figure 6 shows the behavior of a propagating step pulse. We set c, l, V_{pk}, α, β, and d to 6.0 pF/mm, 3.0 nH/mm, 0.2 V, 0.02, 2.0, and 50 μm, respectively. The waveforms on a switch line after an elapse of every hundred picoseconds are shown sequentially from top to bottom in Fig. 6. For forward pulses, which moves toward the output as specified by S_F in Fig.6, the pulse exhibits eminent oscillatory behavior in region I, which proves the development of the sinusoidal wave in region I. Moreover, the pulse is gradually attenuated and disappears at around point P in Fig.6. Then, the pulse starts to propagate backward, as shown by S_B in Fig.6. For the pulses in S_B, no oscillatory behavior is observed, unlike the

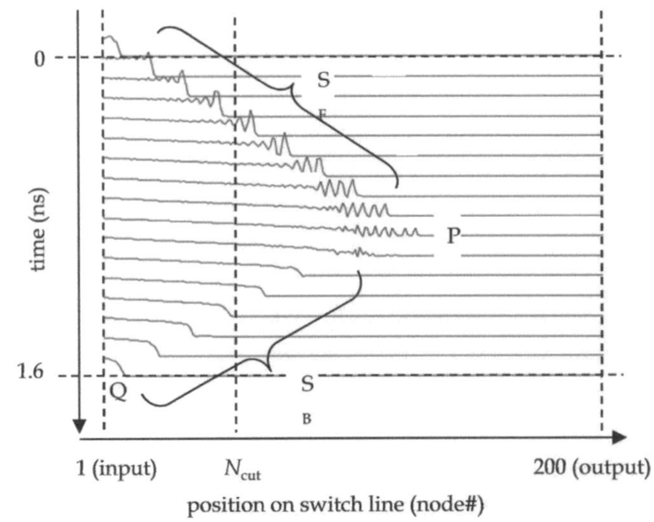

Fig. 6. Numerically obtained step pulse response of TD line.

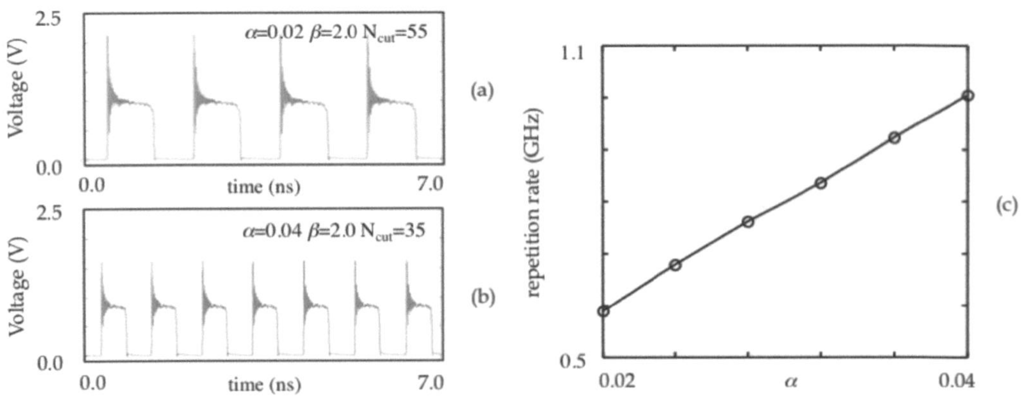

Fig. 7. Dependence of edge's oscillating period on normalized resistance α.

pulses in S_F. This observation results from the fact that exponential waves develop in both regions I and II and establish a stable propagation. The backward pulse reaches the input edge at Q in Fig.6 and then is reflected to start forward propagation again in quasisteady manner. The oscillating pulse edge is numerically confirmed. To see this more clearly, the seven-nanosecond slice at node N_{cut} in Fig.6 is shown in Fig. 7. Figures 7(a) and (b) show the temporal variations of the pulse voltage at N_{cut}=55 on a TD line with α=0.02 and at N_{cut} =35 with α=0.04, respectively. Both exhibit periodic pulse trains, indicating that the pulse edge oscillates. Present calculations simulate the case where the voltage source is directly coupled with a TD line, so that the input port of the line functions as a fixed-value boundary for the traveling pulses. Thus, in the calculations, the backward pulse is totally reflected at the input port, so that the back-and-forth edge motion continues permanently. Practically, the design of the interconnection between the pulse source and a switch line is very important to guarantee permanent oscillation. It is interesting to note that the repetition rate increases when α increases. Figure 7(c) shows this dependence. The repetition rate is vertically plotted for six different values of α. The repetition rate increases proportionally with increasing α. This is because the increase in α contributes to the decrease in the round-trip distance. Increases in β and V_{pk} have the same effect. Moreover, when the amplitude of the input pulse increases, the round-trip time becomes longer, reducing the oscillation frequency. By this simple arrangement, the device can act as a voltage-controlled oscillator. By proper design of loss elements such as electrode loss, switch conductance, and signal application, we can obtain an oscillator having the required oscillation frequency (Narahara, 2006).

3.2 Full-wave analysis

Above all, resonant tunneling diodes (RTDs) exhibit the most excellent performance in submillimeter and terahertz bands (Brown et al., 1991, Ohashi et al., 2005). To design monolithically integrated RTD lines, a full-wave solver of Maxwell's equations must be required.Afinite-difference time-domain (FDTD) method (Taflove, 1995) best suits to the present purpose. In an FDTD, a circuit element such as a capacitor, an inductor, and anonlinear device, are usually implemented as field-dependent conductance and capacitance in a single Yee cell (Yee, 1966). For example, when the conduction current density flowing in the device is denoted by $\vec{J}_L$, the temporal evolution of the electromagnetic fields is calculated on the basis of an extended Ampere's law as

$$\varepsilon \frac{\partial \vec{E}}{\partial t} = \vec{\nabla} \times \vec{H} - \vec{J}_L, \tag{8}$$

where $\vec{E}$, $\vec{H}$, and ε are the electric field, magnetic field, and dielectric constant, respectively. By the single-cell implementation of the lumped device, Eq. (8) is converted as follows:

$$E_z^n = E_z^{n-1} + \frac{\Delta t}{\varepsilon}\left(\vec{\nabla} \times \vec{H}^{n-1/2}\right)_z - \frac{\Delta t}{\Delta y\, \Delta x} I_{L,z}{}^{n-1/2}, \tag{9}$$

where Δx, Δy, and Δt show the cell size in the x, y, and t directions, respectively. The superscripts show the temporal positions, by which we represent the alternative evaluations of electrical and magnetic fields in FDTD, and $\vec{I}_L$ shows the device current, which is assumed to flow in the z direction, and is equal to $J_L \Delta x \,\Delta y$. Moreover, when $\vec{I}_L$ depends on the terminal voltage, $\vec{I}_L$ becomes a function of $E_z \Delta z$ at the corresponding cell. Note that the argument of $\vec{I}_L$ is evaluated at time n for numerical stability (Luebbers et al., 1993); therefore, we need root-finding routines such as the Newton–Raphson method, to obtain the temporal advanced electrical fields. The situations are more complicated when the device size is significant compared with that of the circuit. When a device occupies N adjacent cells, the difference equations to be solved become

$$E_z^n[i] = E_z^{n-1}[i] + \frac{\Delta t}{\varepsilon}\left(\vec{\nabla} \times \vec{H}^{n-1/2}[i]\right)_z - \frac{\Delta t}{\Delta y\,\Delta x} I_{L,z}^{\;n-1/2}\left(E_z^n[1], E_z^n[2], \cdots, E_z^n[N]\right), \tag{10}$$

where $X[i]$ (X=E_z, H) represents the field at the ith cell occupied by the device. For the present case, we have to find roots of a multiple-variable function. This root-finding procedure is very time consuming, whereas the analysis of a TD line needs many distributed TD models; therefore, some simplification is unavoidable. To eliminate the performance-limiting root-finding procedures, we approximate the voltage dependence of the device current by a piecewise-linear (PWL) function (Narahara et al., 2008a). The key is the fact that $E_z^n[i]$ is solved by hand in Eq. (10), when $\vec{I}_L^{\;n-1/2}$ is a linear function of the arguments.When we put I_j=$I_L(V_j)$ for Mdifferent voltages V_j, as shown in Fig.8, I_L is approximated by the following PWL function:

$$I_L(V) = \frac{I_{j+1} - I_j}{V_{j+1} - V_j}(V - V_j) + I_j, \tag{11}$$

for $V_j<V<V_{j+1}$. Substituting I_L from Eq. (10), we obtain

$$\mathbf{E}_z^{\,n} = \mathbf{A}[j]^{-1}\mathbf{S}[j], \tag{12}$$

where $\mathbf{E}_z^{\,n}$ shows the column vector $(E_z^{\,n}[1], E_z^{\,n}[2], \cdots, E_z^{\,n}[N])^t$ and

$$\mathbf{A}[j]_{kl} = \delta_{kl} + \frac{\Delta t\,\Delta z}{\varepsilon\,\Delta x\,\Delta y}\frac{I_{j+1} - I_j}{V_{j+1} - V_j}, \tag{13}$$

$$\mathbf{S}[j]_k = E_z^{\,n-1}[k] + \frac{\Delta t}{\varepsilon}(\vec{\nabla} \times \vec{H}\,[k])_z - \frac{\Delta t}{\varepsilon \Delta x \Delta y}\frac{I_j V_{j+1} - I_{j+1} V_j}{V_{j+1} - V_j},. \tag{14}$$

$\mathbf{A}[j]_{kl}$ (k,l=1,2,...,N) and $\mathbf{S}[j]_k$ (k=1,2,...,N) show the (k,l)th entry of $\mathbf{A}[j]$, and the kth component of $\mathbf{S}[j]$, respectively. After obtaining $E_z^{\,n}[i]$ using Eq.(12), we have to check if the terminal voltage V is really in the range $V_j<V<V_{j+1}$ with $V = \Delta z \sum_{i=1}^{N} E_z^{\,n}[i]$. If not, the procedure is repeated with other$j$ values, until $V$is within (V_j, V_{j+1}).

In the following, we show the results of a sample three-dimensional FDTD calculations that demonstrateoscillating pulse edge along aTD line. The calculation setup is illustrated in Fig.9. Based on a coplanar waveguide, TDs are placed every 30 μm beneath the signal line. The total cell size is 100×100×100 and the spatial increments $\Delta x, \Delta y, \Delta z$ are all set to 10 μm. Each TD is represented by a singleE_z node. We simplify the TD current as a PWL function with M=27 to model an InP-based RTD (Sugiyama, 1995). V_1 (V_{27}) is taken to be sufficiently small (large) for the inclusion of any voltages needed for the calculations. The peak and valley voltages are set to 0.4 and 0.45 V, respectively. The input is a step pulse with amplitude 1.5 V and a 5 ps rise time. The pulse edge achieves steady oscillation with a period of about 3.8 ps. Figure 10 shows this steady oscillation of the edge. The spatial variations of line voltage are plotted and recorded at 0.38 ps intervals. The FDTD calculation successfully demonstrates the potential of monolithically integrated TD lines for generating submillimeter waves.

4. Experimental characterization of TD lines

4.1 Impulse response

The measured circuit is built on a standard breadboard (Narahara et al., 2008b). The shunt electronic switches are NEC 1S1763 Esaki diodes. The peak current and voltage, which corresponds to V_{pk} in the ideal switch model, are typically 6.0 mA and 60 mV, respectively, and the typical parasitic capacitance is 30.0 pF. Series inductances and shunt capacitances are implemented using 1.0 μH inductors (TDK SP0508) and capacitors (TDK FK24C0G1). To

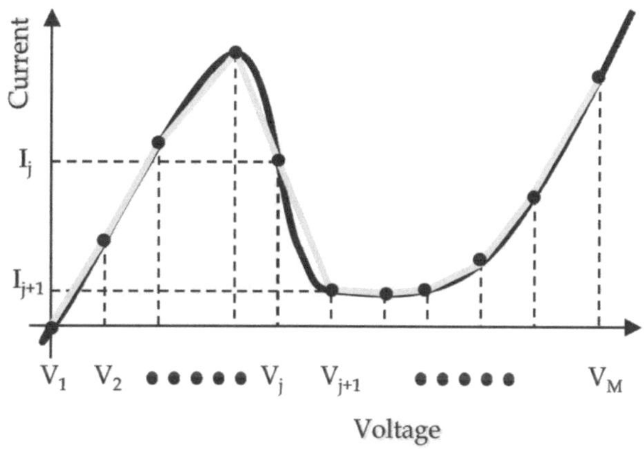

Fig. 8. Piece-wise-linear model of current-voltage relationship of TDs.

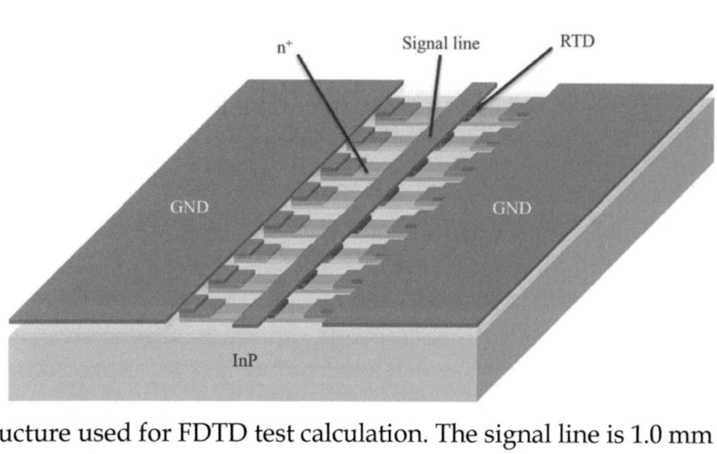

Fig. 9. Line structure used for FDTD test calculation. The signal line is 1.0 mm long, and 30.0 μm wide. The spacing between the signal line and adjacent ground (GND) is set to 30.0 μm. A step pulse is input at the end of the signal line.

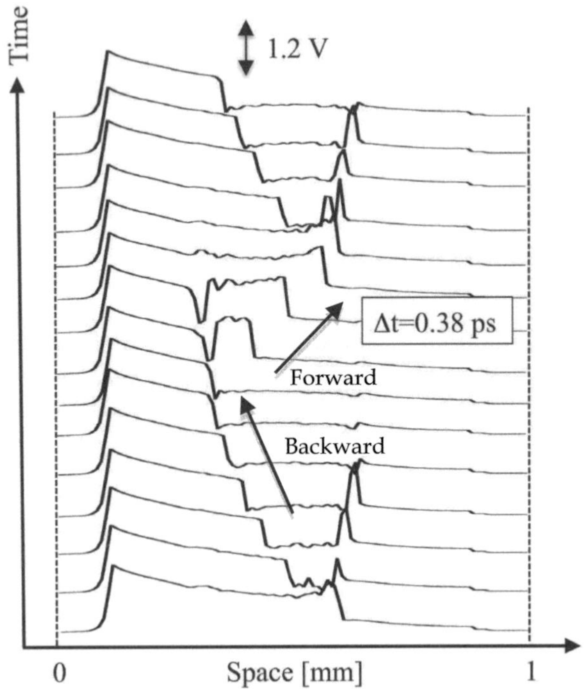

Fig. 10. Results of FDTD calculation.

see the β dependence of the degree of the pulse compression, we prepared two different capacitors, one with a capacitance of 10 nF, and another with 4.7 nF. The test switch line is fed by a pulse signal generated by an Agilent 81150A function generator. The input pulse is a Gaussian pulse with a full-width-at-half-maximum (FWHM) of 1.0 μs. The generator output impedance is set to 5 Ω, and the other end of the test line is short-circuited. The signals along the test switch line are detected and monitored in the time domain using an Agilent DSO90254A oscilloscope.

Figure 11(a) shows the voltage waveforms monitored at cells n=1, 10, 20, 30, 40, and 44. The shunt capacitance is set to 10 nF, such that β=1.0. We can see that the pulse is split in two at n>20. The symbols P_i (i=1, 2, 3, and 4) in Fig.11(a) show the positions of the second pulses developed at n>20. The lifetime of the first pulse is greater than that of the second; therefore, a single short pulse remains at n=44. To determine the degree of compression, we compared the waveform at n=1 (thin solid curve) with that measured at n=44 (thick solid curve) in Fig.11(b). The left vertical axis measures the voltage at n=1, while the right-vertical axis measures the voltage at n=44. Although the amplitude is attenuated, the 1.0-μs wide pulse is compressed to give a width of 0.2 μs after propagation. The attenuated amplitude can result from the parasitic resistances of the inductors and capacitors. It is established that the finite resistive elements do not prevent the switch line from achieving pulse compression.

The results shown in Fig. 12(a) were obtained by the same measurements to obtain those shown in Fig. 11, but with a shunt capacitance of 4.7 nF (β= 1.45). Figure 12(a) shows more compression than seen in Fig. 11(a). This result is consistent with the prediction that a pulse is compressed more for greater β. To quantify the degree of compression, thefundamental frequency of the resulting single pulse is evaluated. For β=1.0, the fundamental frequency is calculated to be 2.9 MHz, while for β= 1.45, it is 4.2 MHz. This means that the pulse width becomes 70 % smaller for 4.7 nF capacitors than for 10.0 nF over time.Figure 12(b) compares the compressed pulses for both cases. The amplitudes are normalized to make them identical for clear comparison.

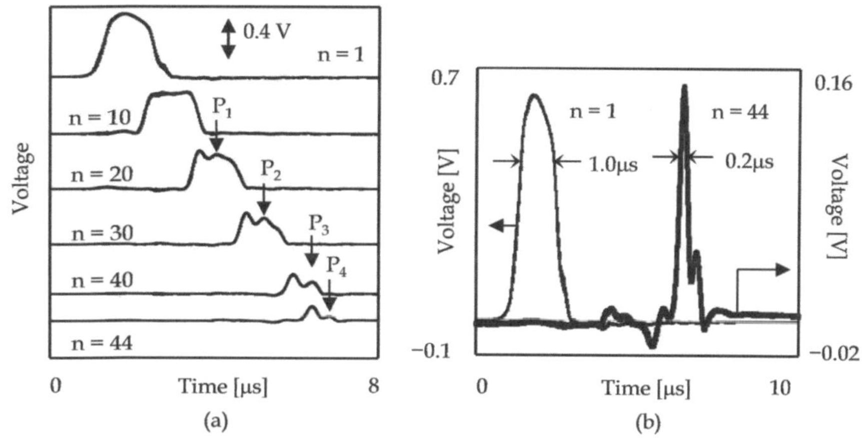

Fig. 11. Measured impulse response of TD line.

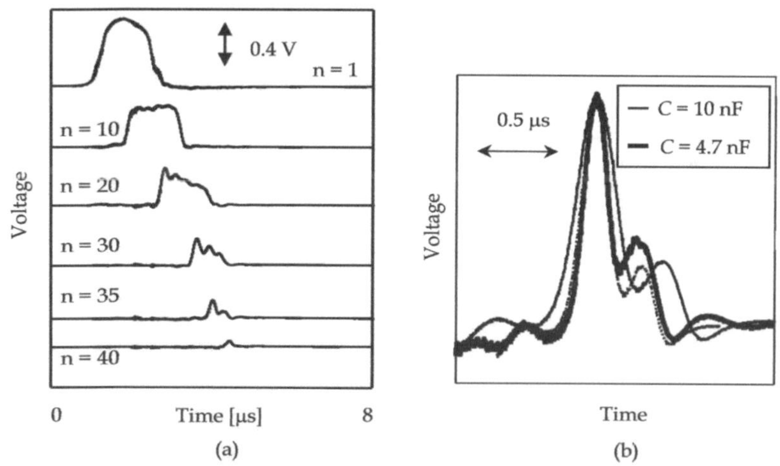

Fig. 12. Measured dependence of shortened pulse width on line capacitance.

The thin and thick pulses represent the compressed pulses for 10.0 and 4.7 nF capacitors, respectively, and the dotted pulse represents the one obtained by artificially compressing the solid pulse by 70 % along the horizontal axis.For the first peak, the dotted pulse fits the thick-solid line well. These observations strongly suggest that the compressed pulses are generated in the test TD line.

4.2 Step pulse response

For step pulse applications, the series inductance, resistance, and shunt capacitance were implemented using 1.0 μH inductors (TDK SP0508), 1.0 Ω resistors (Tyco Electronics

CFR25J), and 470 pF capacitors (TDK FK18C0G1), respectively. To monitor the waveforms using the oscilloscope, we input a pulse having finite duration instead of a step pulse. The input pulse had rise, and fall times, and duration of 1.0 μs, 1.0 μs, and 30.0 μs, respectively. The output impedance of the function generator was 50 Ω.

Figure 13 shows the voltage waveforms monitored at the first cell. Figure 13(a) shows a measured waveform resulting from a single sweep with the oscilloscope. Although the input signal was a simple pulse, we can observe a cycle of short-period pulses. Because of the weakness of the coherence between the input step pulse and the short-period pulses, the temporal positions of the short-period pulses varied for different single sweeps. A typical measured waveform of the short-period pulse is shown in the inset of Fig. 13(a). It exhibits trapezoidal shape with oscillation.

When the pulse edge passes a point, the voltage at that point remains constant until the edge returns. Moreover, the forward edge is carried by the sinusoidal-exponential hybrid mode. These observations are consistent with the measured pulse shape.

Figure 14(a) shows the measured waveform monitored at the first cell, which was averaged 1024 times with the oscilloscope. As mentioned above, the short-period pulses have weak coherence with the step input. Through the waveform averaging, we can extract the coherent part. It is interesting to note that the averaged amplitude of the short-period pulse becomes smaller, i.e., the degree of decoherence increases, with time after the rising edge of

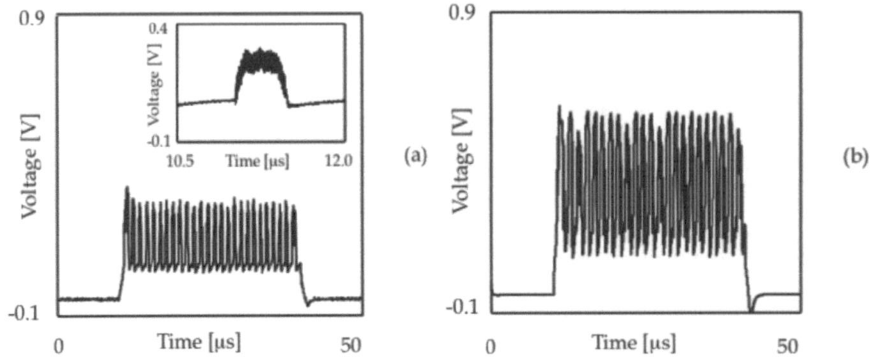

Fig. 13. Experimentally obtained step pulse response of TD line.

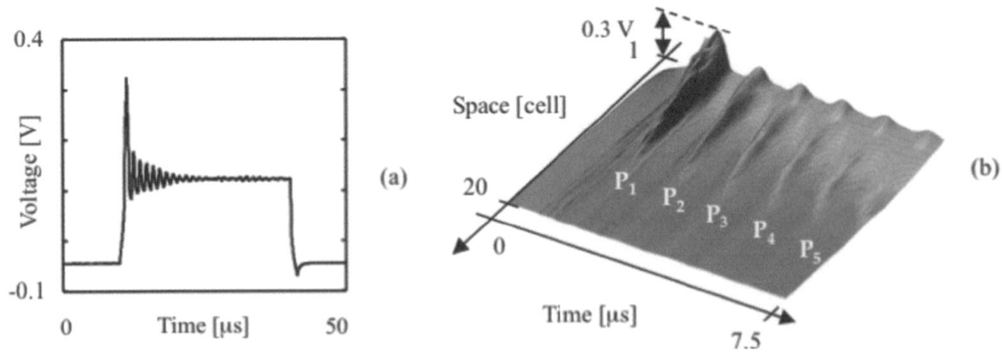

Fig. 14. Observation of oscillating pulse edge.

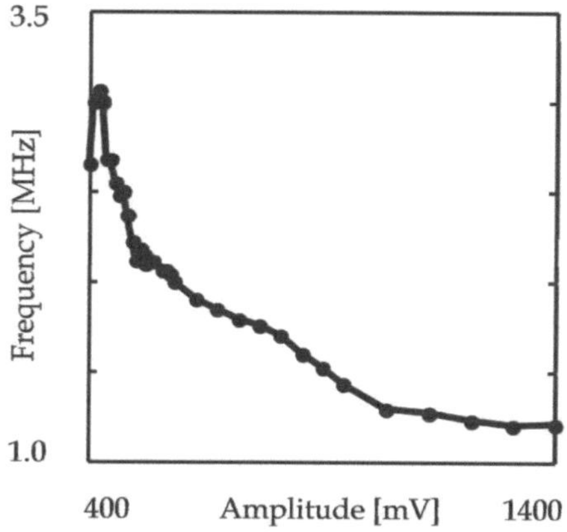

Fig. 15. Measured dependence of edge's oscillating frequency on pulse amplitude.

the input step pulse. By using the coherent parts, it is possible to compare the phase of the voltage waveform at a cell with that at other cells. The spatio-temporal distribution of the averaged voltage wave is shown in Fig. 14(b). The waveforms at the first 20 cells with a 7.5 μs duration are plotted. The edge first travels forward. Then, at the turning point P_i (i=1, 2, 3, 4) in Fig. 14(b), the wave edge starts to travel backward. The backward wave front is reflected at the input port, starts to travel forward again, and reaches the turning point P_{i+1}. This spatio-temporal voltage distribution clearly shows that the oscillating motion of the edge is established in the test switch line.

To examine the dependence of the oscillation frequency on the input amplitude, we carried out frequency-domain measurements, the results of which are shown in Fig.15. We monitored the voltage in the first section using the spectrum analyzer. The frequency of the resulting lowest spectral peak, which corresponds to the inverse of the turn-around time of the oscillating pulse edge, is shown as a function of the input amplitude. Consistently, it is observed that the oscillating frequency increases as the amplitude decreases (Yokota et al., 2009).

5. Applications

This section discusses the potential of TD lines for pulse management. We show two such applications. One is the detection of pulse amplitudes on a one-dimensional TD line. The other utilizes a TD line in two dimensions—the control of pulse propagation orientation on a two-dimensional TD line.

As mentioned above, we consider only pulses that are greater than V_{pk}; therefore, the pulse is attenuated partially at the bottom. It is then expected that a pulse of large amplitude can travel on the line longer than the one with small amplitude. This phenomenon is illustrated in Fig. 16.

Pulses with amplitudes A_1, A_2, and A_3 ($A_1>A_2>A_3$) travel on the TD line by X_1, X_2, and X_3 ($X_1>X_2>X_3$), respectively. When we monitor the waveform at M_1, we detect the pulse regardless of the amplitude. However, because the smallest pulse disappears before it reaches M_2, we can detect only those pulses of amplitudes of A_1 and A_2. Similarly, at M_3, we can detect only the largest pulse. By specifying the monitoring cell on the switch line, we can set thepeak voltage level to detect pulses as required.

Figure 17 shows the measured results. The waveforms were monitored at a fixed cell for two pulses having different amplitudes. By arranging the input amplitudes, only the larger pulse can be detected. Figures 17(a) and (b) show the voltage waveforms monitored at cells n=1 and 5, respectively. The dashed waveforms in Figs. 17(a) and (b) are for the pulse of amplitude 720 mV; the solid waveforms are for the pulse of amplitude 490 mV. Only the dashed pulse is detected at the output. It is thus established that only the larger pulse is detected at the output.

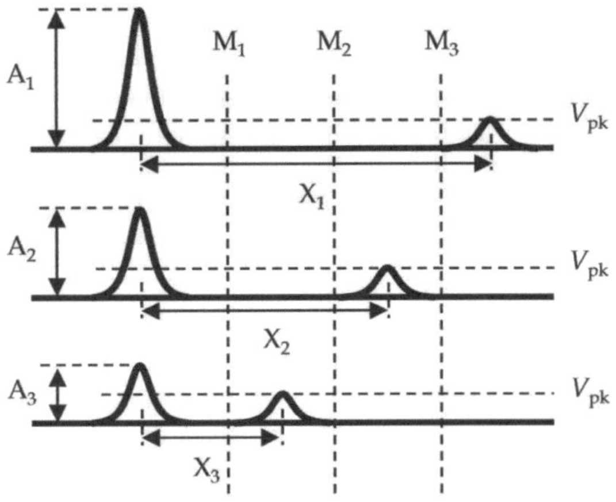

Fig. 16. Pulse control using one-dimensional TD line.

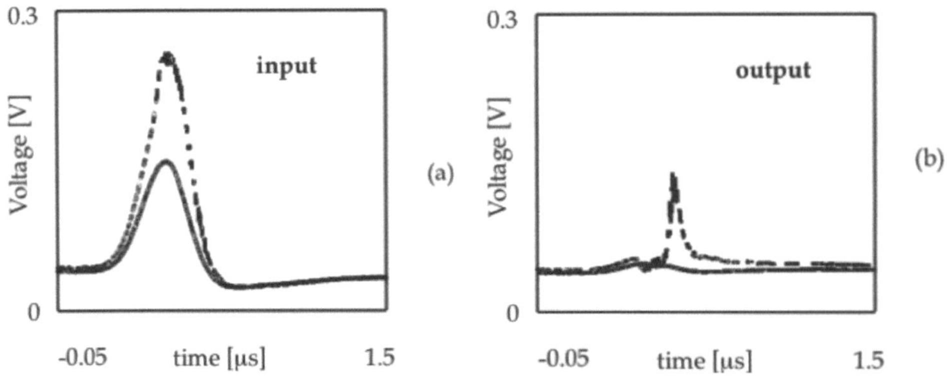

Fig. 17. Experimental demonstration of voltage-level detection.

Another example of the TD line applications requires a two-dimensional TD line. Figure 18 showsa unit cell of two-dimensional TD lines. When several points are excited by the voltage pulses in a two-dimensional line, the propagating pulses overlap and gain amplitude. As mentioned above, the larger pulse survives longer than the smaller pulses; therefore, as time progresses, the overlapped pulses have finite amplitude. The simplest application of this phenomenon is to control pulse propagation orientation by the arrangement of cells where the pulses are applied. To examine the validity of this application, we numerically solved the transmission line equations of a two-dimensional switch line. We set C, L, V_{pk}, and G_0 as 1.0 pF, 1.0 nH, 0.1 V, and 0.04 S, respectively. The total cell size was 100×100. Each edge cell was terminated using a 50 Ω resistor. An impulse having a Gaussian form with amplitude 1.0 V and FWHM 100.0 ps was input at cells located at (45, 45) and (55, 55).

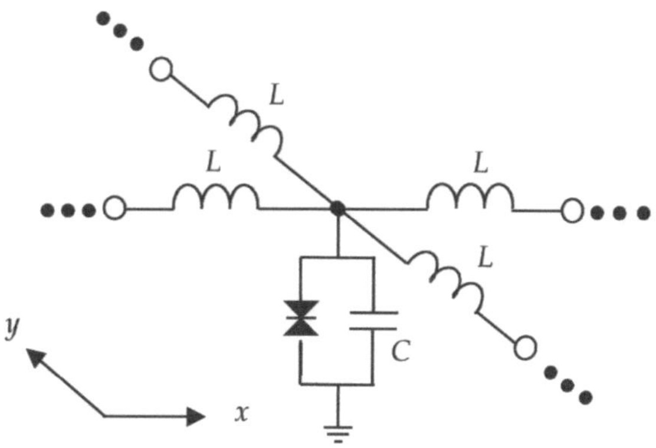

Fig. 18. Two-dimensional TD line.

Figure 19 shows the numerical results of pulse propagation. Waveforms at times 0.4, 0.8, 1.2, and 1.6 ns passed after the pulse incidence are shown in Figs.19(a), (b), (c), and (d), respectively. As observed in Fig. 19(b), each pulse starts isotropic propagation and overlaps the other pulses. As shown in Fig.19(c), the parts of the overlapped pulses with large amplitude survive. The remainders of the pulses on the bisection of the line connecting two excitation cells are shown in Fig.19(d). By changing the positions of the two excitation cells, we can control the propagation orientation of the surviving pulses as required.

For a pulse with a wave vector (k_x, k_y), the dispersions of the two-dimensional switch line are given by

$$v_{I,\sin} = \frac{2}{kd}\sqrt{\sin^2\frac{k_x d}{2} + \sin^2\frac{k_y d}{2}}, \tag{15}$$

$$v_{II,\exp} = -\frac{\beta}{2kd} + \frac{1}{2kd}\sqrt{\beta^2 + 16\sinh^2\frac{k_x d}{2} + 16\sinh^2\frac{k_y d}{2}}, \tag{16}$$

where $v_{\mathrm{I,sin}}$, and $v_{\mathrm{II,exp}}$ show the normalized phase velocities of the sinusoidal mode in region I, and the exponential mode in region II, respectively. Because the cutoff frequency increases monotonically with $\theta = \tan^{-1}(k_x / k_y)$ up to $\pi/4$ and is $\sqrt{2}$ times as large as at $\theta = 0$, it becomes possible for the intersection frequency, where $v_{\mathrm{I,sin}}$is coincident with $v_{\mathrm{II,exp}}$, in a two-dimensional switch line to surpass the upper limit achieved in a one-dimensional switch line; therefore, we can expect that the pulse shortening is more effectively established in two dimensions.

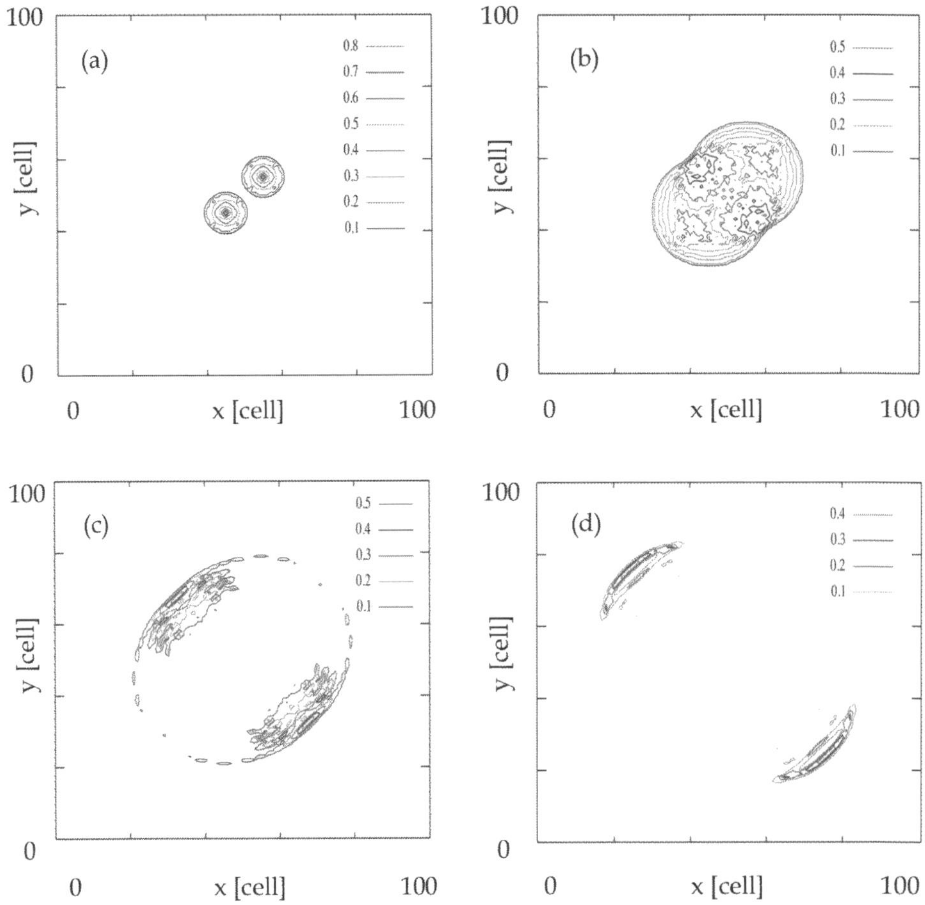

Fig. 19. Fundamental property of two-dimensional TD line.

6. Conclusions

This chapter reviews the pulse propagation characteristics of TD lines.We found that a pulse wave propagates on the line either by theunstable exponential-sinusoidal hybrid mode or stable exponential-exponential mode. Through thesepeculiar wave propagation properties, an input impulse experiences width shortening and an edge of the step pulse oscillates. These provide efficient methods for generating short pulses and continuous waves. Moreover, TD lines can be used to manage pulse amplitude and propagation orientation. Test TD lines using Esaki diodes successfully demonstrated the pulse generation and management.Our approach could be scaled from its current MHz form into microwave, millimeter-wave, and terahertz forms, for implementation with state-of-the-art RTDs.

7. References

Brown, E.R.; Soderstrom, J.R.;Parker, C.D.; Mahoney, L.J.; Molvar, K.M. &McGill, T.C. (1991). Oscillations up to 712 GHz in InAs/AlSb resonant-tunneling diodes, Applied Physics Letters, Vol.58, 2291-2293, ISSN:0003-6951.

Kintis, M.; Xing, L.; Flavia, F.; Sawdai, D.; Kwok, L.K. &Gutierres, A. (2008). An MMIC pulse generator using dual nonlinear transmission lines, Microwave and Wireless Components Letterts, Vol.17, No.6, 454-457, ISSN: 1531-1309.

Luebbers, R.; Beggs, J. & Chamberlin, L. (1993). Finite difference time domain calculation of transients in antennas with nonlinear loads, IEEE Transacions on Antennas and Propagation, Vol.41, No.5, 566-573, ISSN:0018-926X.

Narahara, K.; Otsuji, T. &Sano, E. (2006). Pulse compression by quasi-steady propagation along switch lines, Japanese Journal of Applied Physics, Vol.45, No.7, 5692-5695, ISSN: 0021-4922.

Narahara, K.; Otsuji, T. &Sano, E.(2006). Generation of electrical-short-pulse using Schottky line periodically loaded with electronic switches, Journal of Applied Physics, Vol.100, No.2, 024511.1-5, ISSN: 0021-8979.

Narahara, K. (2006). Electromagnetic continuous-wave generation using switch lines, Journal of Applied Physics, Vol.100, No.6, 064908.1-5, ISSN: 0021-8979.

Narahara, K. & Yokota, A. (2008a). Full-wave analysis of quasi-steady propagation along transmission lines periodically loaded with resonant tunneling diodes, Japanese Journal of Applied Physics, Vol.47, No.2, 1126-1129, ISSN: 0021-4922.

Narahara, K. & Yokota, A. (2008b). Experimental characterization of short-pulse generation using switch lines, IEICE Electronics Express, Vol.5, No.22, 973-977, ISSN: 1349-2543.

Ohashi, N.; Hattori, S.; Suzuki, S. &Asada, M. (2005). Experimental and theoretical characteristics of sub-terahertz and terahertz oscillations of resonant tunneling diodes integrated with slot antennas, Japanese Journal of Applied Physics, Vol.44, No.11, 7809-7815, ISSN: 0021-4922.

Richer, I. (1966). The switch-line: A simple lumped transmission line that can support unattenuated propagation. IEEE Transactions on Circuit Theory, Vol.13, No.4, 388-392, ISSN: 0018-9324.

Rodwell, M.J.W.; Allen, S.T.; Yu, R.Y.; Case, M.G.; Bhattacharya, U.; Reddy, M.; Carman, E.; Kamegawa, M.; Konishi, Y.; Pusl,J. &Pullela, R. (1994). Active and nonlinear wave propagation devices in ultrafastelectronics and optoelectronics, Proceeding of the IEEE, Vol.82, No.7, 1037-1059, ISSN: 0018-9219.

Scott, A. (1970). *Active and nonlinear wave propagation in electronics*, Wiley-Interscience, ISBN: 471767905 New York.

Sugiyama, H.; Matsuzaki, H.; Oda, Y. & Yokoyama, H. (2005). Metal-organic vapor-phase epitaxy growth of InP-based resonant tunneling diodes with a strained $In_{0.8}Ga_{0.2}As$ well and AlAs barriers, Japanese Journal of Applied Physics, Vol.44, No.10, 7314-7318, ISSN: 0021-4922.

Taflove, A. (1995). *Computational electrodynamics: the finite-difference time-domain method*, Artech House, ISBN: 1580538320,London.

Yee, K.S. (1966). Numerical solution of initial boundary value problems involving Maxwell's equations in isotropic media, IEEE Transactions on Antennas and Propagation, Vol.14, No.4, 302-307, ISSN:0018-926X.

Yokota, A. &Narahara, K. (2009). Characterization of oscillating pulse edges in switch lines for development of widely tunable voltage-controlled oscillators, Japanese Journal of Applied Physics, Vol.48, No.8, 084502-084506, ISSN: 0021-4922.

VII. Wave Propagation in Disperse Media.

24

Wave Velocity Dispersion and Attenuation in Media Exhibiting Internal Oscillations

Marcel Frehner[1], Holger Steeb[2] and Stefan M. Schmalholz[3]
[1]*Department for Geodynamics and Sedimentology, University of Vienna,*
[2]*Mechanics - Continuum Mechanics, Ruhr-University Bochum,*
[3]*Geological Institute, ETH Zurich,*
[1]*Austria*
[2]*Germany*
[3]*Switzerland*

1. Introduction

Modeling the propagation of acoustic or fully elastic (i.e. seismic) waves is an important tool for interpreting, understanding and predicting real-world measurements in various industrial disciplines. In this context, modeling can be anything from three-dimensional full waveform modeling using numerical techniques (e.g. finite difference method, finite element method, spectral method; Kelly & Marfurt, 1990; Martin & Komatitsch, 2009) to analytical treatment of wave phenomena (e.g. Liu et al., 2000; Aki & Richards, 2002; Korneev, 2008) and can be in both, time and frequency domain. In its simplest form, the propagation of waves through a heterogeneous medium is described with a linear elastic material behavior (Aki & Richards, 2002). Such models, including and resolving heterogeneities on various scales with different elastic material parameters, are already able to describe a number of wave propagation phenomena, e.g. reflection and transmission coefficients at interfaces, scattering at heterogeneities (Frehner et al., 2008) or propagation of surface waves. As the number of investigated heterogeneities increases, the complexity of the numerical model and the computational cost also increases. For example, Saenger et al. (2007) use an advanced parallel finite difference scheme to model wave propagation on the pore-scale resolving the entire pore structure of a rock or Lee et al. (2009) use a spectral element method comprising an unstructured numerical mesh to investigate the effect of topography on seismic wave amplitudes. These are only two examples of a numerical model with a basic linear elastic constitutive equation containing a large number of heterogeneities with very complex geometries.

When the number of heterogeneities and therefore the geometrical complexity of a model becomes too large and/or the characteristic size of the heterogeneities is much smaller than the wavelength under consideration, effective medium theories are applied. Such models upscale physical processes on the microscale (i.e. much smaller than the wavelength) by introducing coarse-grained/homogenized macroscopic effective material parameters. For example, Korneev et al. (2004) added an additional term describing viscous damping to the elastic wave equation to take into account attenuation. However, the physical process acting on the microscale causing the attenuation is not described. The widely accepted effective medium model for fully saturated porous media is the Biot model (Biot, 1962). More advanced models for partially saturated porous media can be separated into two groups: (1) models based on Biot's theory for fully saturated media, but applying spatially varying pore fluid properties representing a partial saturation on the mesoscale (i.e. larger than the pore size and smaller than the wavelength under consideration; White, 1975; Dutta & Ode, 1979; Quintal et al., 2009) and (2) macroscale effective medium models for three-phase media (i.e. solid rock, wetting and non-wetting fluid) considering a homogenized partial saturation on the pore-scale, including in particular capillary pressure effects (Santos et al., 1990; Smeulders & van Dongen, 1997; Wei & Muraleetharan, 2002). Common to all of these effective medium models for fully or partially saturated porous rocks is that in their derivation the individual phases are usually mixed and the material properties are averaged over a so-called representative elementary volume. By doing so, processes taking place at the interfaces between the individual phases (e.g. surface tension effects or scattering at the pore structure) are ignored. More advanced mixture theory-based models, which take specific surface areas between wetting and non-wetting phases into account, are still limited to quasi-static processes. However, such effects can be re-included by introducing additional effective material parameters.

Within a medium containing any kind of heterogeneity or a number of heterogeneities, oscillations with a characteristic resonance frequency, depending on the mass and internal length of the heterogeneity, can occur. When excited, heterogeneities can self-oscillate with their natural frequency (Carstensen & Foldy, 1947; MacPherson, 1957; van Wijngaarden, 1972; Anderson & Hampton, 1980). If the external excitation force is an acoustic or seismic wave, this process is called resonant scattering (Werby & Gaunaurd, 1990; Hassan & Nagy, 1997) and has application in non-destructive testing of materials (e.g. Schultz et al., 2006). Korneev (2009) even demonstrated that the resonance frequency of an object in the subsurface with a large impedance contrast to its surrounding can be measured with an active seismic experiment at the Earth surface. The oscillations are more easily detectable in late arrivals when they are not masked by high-energy body waves. Other mechanisms can cause oscillations within a heterogeneous medium. For example, the dynamical behavior of non-wetting fluid blobs or fluid patches in idealized residually saturated pore spaces were studied (Dvorkin et al., 1990; Graham & Higdon, 2000a; Graham & Higdon, 2000b). Thereby, one of the main results is the oscillatory movement of the fluids when an external driving force is applied (Hilpert et al., 2000). The restoring force driving the oscillations is the surface tension force or capillary force. The fact that isolated oil blobs in residually saturated pore spaces can exhibit a resonance frequency, motivated the suggestion of a new enhanced oil recovery method (EOR) termed "wave stimulation of oil" or "vibratory mobilization" (Beresnev & Johnson, 1994; Iassonov & Beresnev, 2003; Beresnev et al., 2005; Li et al., 2005; Hilpert, 2007; Pride et al., 2008). Another example of oscillatory behavior within a medium is given by Urquizu and Correig (2004). They showed that under certain circumstances a seismic wave pulse propagating through a layered medium can be described mathematically with a differential equation for an oscillator.

All of these oscillatory phenomena within a medium should have an effect on acoustic or seismic waves propagating through such a heterogeneous medium. Presumably, the wave will show a strongly frequency-dependent propagation behavior (i.e. velocity dispersion and frequency-dependent attenuation). However, such oscillatory phenomena are not included in current effective medium theories, such as the Biot theory (Biot, 1962). Oscillations are usually described on the microscale (i.e. the scale of one single oscillator; Dvorkin et al., 1990; Graham & Higdon, 2000; Hilpert et al., 2000; Beresnev et al., 2005), not taking into account the macroscale, i.e. continuous scale. Frehner et al. (2009) presented a model to combine the dynamics of wave propagation and rock-internal oscillations caused by capillarity effects using an effective medium theory. However, the model was presented for internal undamped oscillations with only one particular resonance frequency.

In this chapter, we extend the effective medium model of Frehner et al. (2009) by introducing a viscous damping term and generalize the model for an arbitrary number of resonance frequencies (described e.g. by a probability density function of possible resonance frequencies in the limit case). We describe the simplified case of an acoustic medium exhibiting internal oscillations, i.e. a linear elastic medium with an elastic shear modulus equal to zero. In such a material only one type of wave (i.e. a compressional wave or P-wave) can propagate. The internal oscillations will be described by the equation of motion for a damped harmonic oscillator. Different to Frehner et al. (2009), we use as an example for such a medium water containing gas bubbles, where the dispersed gas bubbles can oscillate with their natural frequency. For that specific example, the elastic and viscous interaction forces between the phases can be derived analytically and a comparison with literature and experimentally obtained results can be performed.

2. Mathematical model of acoustic medium exhibiting internal oscillations

For simplicity reasons and because we want to understand first-order effects, we choose a basic model. By choosing an acoustic medium, we concentrate on only one type of wave (i.e. a compressional wave or P-wave, also called acoustic wave or sound wave). Because only one type of wave can propagate in an acoustic medium, we choose a one-dimensional model description (x is the spatial coordinate). Thus, we neglect amplitude decrease of the acoustic wave due to geometrical spreading and energy redirection in the different spatial directions.

The chosen model leads to a straightforward analogy, i.e. water (the acoustic medium) containing gas bubbles (internal oscillations). This particular case was studied by a number of researchers (e.g. Minnaert, 1933; Carstensen & Foldy, 1947; Fox et al., 1955; MacPherson, 1957; Silberman, 1957; van Wijngaarden, 1972; Commander & Prosperetti, 1989) and a large number of experimental data is available, as well as detailed theory. However, we choose a different approach for deriving our equations. While in the cited literature the dynamics of the bubble is considered in detail, we take the result of this work and simply assume that the bubble movement can be described by a damped harmonic oscillator. Fig. 1 shows the conceptual one-dimensional model of water (acoustic medium) containing gas bubbles (oscillators).

The governing set of equations for the gas-water-mixture can be derived by averaging the properties of the gas bubbles in a certain representative elementary volume and applying concepts of the mixture theory (Truesdell, 1957; Bowen, 1980; Bowen, 1982). In contrast to classical mixture theory, we would like to denote that only the fluid (i.e. water) phase is continuous, while the gas phase consists of individual disconnected bubbles. Therefore, no acoustic wave can propagate through the gas phase and we do not introduce a Darcy-like term to describe the flow of the gas phase.

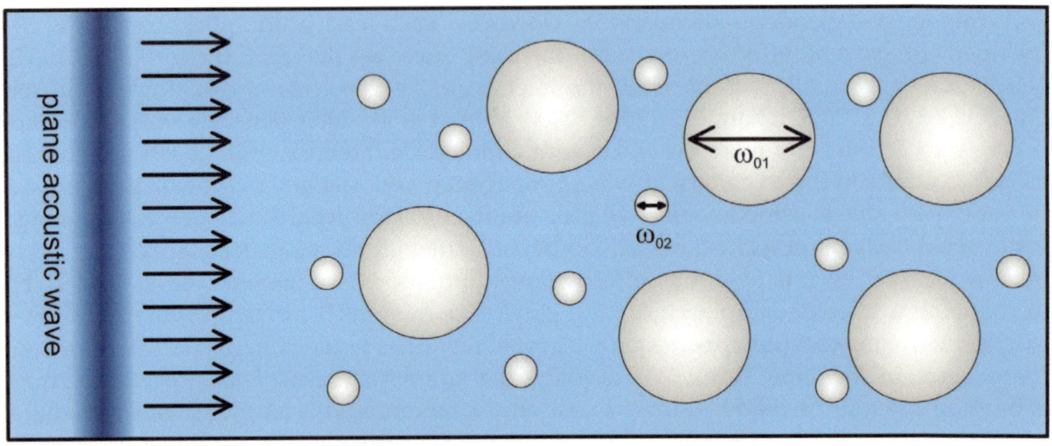

Fig. 1. Conceptual one-dimensional model of an acoustic medium (i.e. water) exhibiting internal oscillations (i.e. individual disconnected and compressible gas bubbles). Exemplarily, a model with two different internal resonance frequencies (ω_1 and ω_2) is displayed here. Only displacements and waves propagating in the horizontal (i.e. x-) direction are considered.

2.1 Oscillations of gas bubbles

First, we study the oscillatory behavior of one individual disconnected gas bubble (superscript g) in water (superscript f). Minnaert (1933) determined the angular eigenfrequency of a spherical gas bubble in water under isothermal conditions as

$$\Omega_{0j} = \frac{1}{r_j}\sqrt{\frac{3\gamma p_0}{\rho^{fR}}} , \tag{1}$$

depending on the radius r_j of the bubble, the heat capacity ratio γ of the gas, the initial gas pressure in the bubble p_0 and the density of bubble-free water ρ^{fR}. The index j denotes the particular bubble size under consideration. Superscript R denotes material parameters of a single-phase medium (i.e. bubble free water or pure gas). The initial gas pressure in the bubble p_0 is assumed to be the same for all bubble sizes and can be calculated from the water depth H as

$$p_0 = 10^5 + 9.81\rho^{fR} H . \tag{2}$$

The value 10^5 [Pa] is the atmospheric pressure at sea level, i.e. the pressure at the water surface. The value 9.81 [m/s^2] is the gravitational acceleration. Because we do not follow the derivations of Minnaert (1933) and we couple the bubble oscillations to a wave propagation equation, we can not expect the resonance frequency of our coupled wave propagation-oscillator model to be equal to Equation (1). However, we use the frequency of Minnaert (1933) (Equation (1)) as a reference frequency. Accordingly, the resonance frequency of our model is

$$\omega_{0j} = A\Omega_{0j} . \tag{3}$$

The scaling factor A will be determined, once the coupled system of equations is derived further down. According to the harmonic oscillator model, the restoring force for one oscillating gas bubble is

$$\hat{f}_j^g = -\hat{m}_j^g \omega_{0j}^2 w \,, \tag{4}$$

where $\hat{m}_j^g$ is the mass of one individual gas bubble with radius r_j and $w = u_j^g - u^f$ is the relative displacement of the center of mass of the gas bubble to the fluid (with u being displacement). The symbol ^ above parameters refers to the fact that Equation (4) is valid for one particular isolated gas bubble, i.e. at one particular position in the model, and is not a continuous formulation for the whole model yet. According to Batchelor (2000), the viscous interaction (i.e. damping) force of a single gas bubble surrounded by a viscous fluid is given by

$$\hat{b}_j^g = -4\pi r_j \eta^{fR} \dot{w} \,, \tag{5}$$

where η^{fR} is the dynamic viscosity of bubble-free water and $\dot{w}$ is the material time derivative of the relative displacement w, i.e. the relative velocity of the center of mass of the gas bubble to the fluid[1]. Note that the interacting forces (Equations (4) and (5)) are written in terms of relative displacement and velocity. We would like to remark, that, in contrast to Biot-type models or other mixture theory-based models, two physically different momentum interaction terms are introduced. A purely elastic term of oscillatory nature that scales with the volume of the bubble and a viscous term that scales with the specific surface of the bubble. Thus we are able to take into account damping with respect to the specific surface area of the gas bubbles.

The volume of one single gas bubble of radius r_j is given by $\hat{V}_j^g$ while the number n_j is introduced as the number of bubbles with the same radius r_j per representative elementary volume (REV), i.e. n_j carries the dimension [1/m³]. Finally, in order to describe the macroscopic behaviour of the gas-fluid-mixture in the REV, we introduce the volume fraction of the sum of gas bubbles with radius r_j as ϕ_j^g. Summing up all bubbles with different radii in the REV is done by summing up the volume fractions ϕ_j^g of bubbles with radius r_j. The result is the overall volume fraction of the gas phase ϕ^g. The following relations apply:

---

[1]The non-equilibrium interaction between the fluid and the gas bubble was generalized in Batchelor (2000) for bubbles consisting of fluids with an intrinsic viscosity leading to

$$\hat{f}_j^g = -4\pi r_j \eta^{fR} \frac{\eta^{fR} + \frac{3}{2}\eta^{gR}}{\eta^{fR} + \eta^{gR}} \dot{\hat{w}}$$

with the limits for

a) gas bubbles: $\lim \hat{f}_j^g \Big|_{\eta^{gR} \to 0} = -4\pi r_j \eta^{fR} \dot{\hat{w}}$

b) rigid spheres $\lim \hat{f}_j^g \Big|_{\eta^{gR} \to \infty} = -6\pi r_j \eta^{fR} \dot{\hat{w}}$

$$\phi_j^g = n_j \hat{V}_j^g \quad \text{and} \quad \phi^g = \sum_{j=1}^{m} \phi_j^g = \sum_{j=1}^{m} \left(n_j \hat{V}_j^g \right). \tag{6}$$

The value m is the number of different gas bubble sizes in the system. For example, in a model containing a large number of equally sized gas bubbles, m is equal to 1, in contrast to n_1, which is the absolute number of gas bubbles with that particular size. Also, partial densities ρ_j^g and ρ^g are introduced as

$$\rho_j^g = n_j \hat{m}_j^g = \phi_j^g \rho^{gR} \quad \text{and} \quad \rho^g = \sum_{j=1}^{m} \left(\phi_j^g \right) \rho^{gR} = \phi^g \rho^{gR}. \tag{7}$$

The partial density ρ_j^g is defined as the mass of gas inside gas bubbles of one particular radius r_j per REV and is therefore different to the true or so-called effective density of gas ρ^{gR}. The partial density ρ^g is defined as the total mass of gas per REV.

2.2 Acoustic medium containing oscillating gas bubbles

The summation of all momentum interaction terms $\hat{f}_j^g$ and $\hat{b}_j^g$ of the gas phase (Equations (4) and (5)) must be equal to the momentum interaction term of the water phase. Therefore, the following constraints must be fulfilled:

$$\sum_{j=1}^{m} \left(n_j \hat{f}_j^g \right) + \sum_{j=1}^{m} \left(n_j \hat{b}_j^g \right) + f^f + b^f = 0. \tag{8}$$

The water phase is assumed to behave like a purely acoustic medium. Thus, we neglect any viscous shear stresses. The balance of momentum in one dimension is given by:

$$\rho^f \ddot{u}^f = \frac{\partial \sigma}{\partial x} + f^f + b^f. \tag{9}$$

In Equation (9), ρ^f is the partial density of water, i.e. $\rho^f = \phi^f \rho^{fR}$ (with ϕ^f and ρ^{fR} being the volume fraction and the true density of water, respectively) and σ is the normal stress in the water phase in the only spatial direction x and is defined as positive for extensional stress. The constitutive equation for the stress in the (acoustic) water phase has the following form:

$$\sigma = \phi^f K^f \frac{\partial u^f}{\partial x}. \tag{10}$$

In Equation (10), K^f is the bulk modulus of water. The balance of momentum for the water phase (Equation (9)) becomes (including Equations (4), (5), (8) and (10))

$$\rho^f \ddot{u}^f = \phi^f K^f \frac{\partial^2 u^f}{\partial x^2} + \sum_{j=1}^{m} \left(n_j \hat{m}_j^g \omega_{0j}^2 \left(u_j^g - u^f \right) \right) + \sum_{j=1}^{m} \left(4\pi n_j r_j \eta^{fR} \left(\dot{u}_j^g - \dot{u}^f \right) \right) \tag{11}$$

for a spatially constant bulk modulus of water K^f and volume fraction of water ϕ^f. The viscous friction coefficient is introduced as

$$\bar{\eta}_j = 4\pi n_j r_j \eta^{fR} = \frac{3\phi_j^g \eta^{fR}}{r_j^2}, \tag{12}$$

Equation (12) shows the distinct size effect of the gas-water mixture. Keeping the volume fraction of the gas phase ϕ_j^g constant and decreasing the radius of the gas bubbles the viscous behavior, i.e. the attenuation mechanism in the system, becomes more and more important (scales with $1/r_j^2$). This is obvious, as for smaller bubbles the ratio of surface to volume becomes bigger and bigger. Using Equations (7) and (12), Equation (11) can be further simplified:

$$\rho^f \ddot{u}^f = \phi^f K^f \frac{\partial^2 u^f}{\partial x^2} + \sum_{j=1}^{m}\left(\rho_j^g \omega_{0j}^2 \left(u_j^g - u^f\right)\right) + \sum_{j=1}^{m}\left(\overline{\eta}_j \left(\dot{u}_j^g - \dot{u}^f\right)\right). \tag{13}$$

2.3 Gas-water-mixture

For the mixture of water and gas, the momentum interaction terms must vanish (Equation (8)). Also, we assume that the bulk modulus of the gas-water-mixture is equal to the Reuss average (Mavko et al., 2003) of the bulk moduli of the two constituents, i.e. K^g for gas and K^f for water:

$$\frac{1}{K^{\text{Reuss}}} = \frac{\phi^g}{K^g} + \frac{\phi^f}{K^f}. \tag{14}$$

The Reuss average of the bulk moduli is plotted in Fig. 2 for the material parameters shown in Table 1 and for a range of gas volume fractions. Using Equation (14), the pressure of the gas-water-mixture can be expressed in terms of the deformation of the water phase while the deformation of the gas phase is taken into account through the Reuss average of the bulk moduli. Taking these considerations into account, the total momentum balance equation for the gas-water-mixture can be written as

$$\rho^f \ddot{u}^f + \sum_{j=1}^{m}\left(\rho_j^g \ddot{u}_j^g\right) = K^{\text{Reuss}} \frac{\partial^2 u^f}{\partial x^2}. \tag{15}$$

Subtracting Equation (13) from Equation (15) leads to the partial balance of momentum for the sum of the gas bubbles:

$$\sum_{j=1}^{m}\left(\rho_j^g \ddot{u}_j^g\right) = \left(K^{\text{Reuss}} - \phi^f K^f\right)\frac{\partial^2 u^f}{\partial x^2} - \sum_{j=1}^{m}\left(\rho_j^g \omega_{0j}^2 \left(u_j^g - u^f\right)\right) - \sum_{j=1}^{m}\left(\overline{\eta}_j \left(\dot{u}_j^g - \dot{u}^f\right)\right). \tag{16}$$

Equation (16) is split into m different equations, one for each bubble size in the system. The first term on the right-hand side is distributed among these m equations according to the relative volume fraction of gas inside a particular bubble size. The m equations are:

$$\rho_j^g \ddot{u}_j^g = \frac{\phi_j^g}{\phi^g}\left(K^{\text{Reuss}} - \phi^f K^f\right)\frac{\partial^2 u^f}{\partial x^2} - \rho_j^g \omega_{0j}^2 \left(u_j^g - u^f\right) - \overline{\eta}_j \left(\dot{u}_j^g - \dot{u}^f\right) \quad \text{for } j = \{1,\ldots,m\}. \tag{17}$$

2.2 Monochromatic acoustic wave

Next, we analyze monochromatic acoustic waves propagating through a medium that is described by the partial balance Equations (17) and (13). We apply a standard harmonic ansatz of the form

$$u^{\beta} = U^{\beta}\exp\left(i\left(kx - \omega t\right)\right) \tag{18}$$

with β={g_1,...,g_m,f} representing the individual phases. In Equation (18) U^{β} is the amplitude of the wave, i is the imaginary unit (i.e. i^2=-1), ω is angular frequency, t is time and k is the complex wave number. For the model containing bubbles with m different radii r_j with i={1,...,m} with the volume fractions ϕ_j^g we obtain

$$\begin{gathered}
-\omega^2\rho_1^g U_1^g + \rho_1^g\omega_{01}^2\left(U_1^g - U^f\right) - i\omega\bar{\eta}_1\left(U_1^g - U^f\right) - \frac{\phi_1^g}{\phi^g}\left(\phi^f K^f - K^{\text{Reuss}}\right)k^2 U^f = 0, \\
\vdots \\
-\omega^2\rho_m^g U_m^g + \rho_m^g\omega_{0m}^2\left(U_m^g - U^f\right) - i\omega\bar{\eta}_m\left(U_m^g - U^f\right) - \frac{\phi_m^g}{\phi^g}\left(\phi^f K^f - K^{\text{Reuss}}\right)k^2 U^f = 0, \\
\left(-\omega^2\rho^f + \phi^f K^f k^2 + \sum_{j=1}^{m}\left(\rho_j^g\omega_{0j}^2 - i\omega\bar{\eta}_j\right)\right)U^f - \sum_{j=1}^{m}\left(\rho_j^g\omega_{0j}^2 - i\omega\bar{\eta}_j\right)U_j^g = 0.
\end{gathered} \tag{19}$$

Equation (19) can be written in matrix notation as (**A**-k^2**B**)**U**=0, resulting in the generalized eigenvalue problem

$$\det\left(\mathbf{A} - k^2\mathbf{B}\right) = 0\,. \tag{20}$$

The vector **U** is defined as $\mathbf{U} = \begin{bmatrix} U_1^g & \dots & U_m^g & U^f \end{bmatrix}$ and the matrices **A** and **B** are given as

$$\mathbf{A} = \begin{bmatrix}
\left(\omega_{01}^2 - \omega^2\right)\rho_1^g - i\omega\bar{\eta}_1 & \dots & 0 & -\rho_1^g\omega_{01}^2 + i\omega\bar{\eta}_1 \\
\vdots & \ddots & \vdots & \vdots \\
0 & \dots & \left(\omega_{0m}^2 - \omega^2\right)\rho_m^g - i\omega\bar{\eta}_m & -\rho_m^g\omega_{0m}^2 + i\omega\bar{\eta}_m \\
-\rho_1^g\omega_{01}^2 + i\omega\bar{\eta}_1 & \dots & -\rho_m^g\omega_{0m}^2 + i\omega\bar{\eta}_m & -\omega^2\rho^f + \sum_{j=1}^{m}\left(\rho_j^g\omega_{0j}^2 - i\omega\bar{\eta}_j\right)
\end{bmatrix}, \tag{21}$$

$$\mathbf{B} = \begin{bmatrix}
0 & \dots & 0 & \frac{\phi_1^g}{\phi^g}\left(\phi^f K^f - K^{\text{Reuss}}\right) \\
\vdots & \ddots & \vdots & \vdots \\
0 & \dots & 0 & \frac{\phi_m^g}{\phi^g}\left(\phi^f K^f - K^{\text{Reuss}}\right) \\
0 & \dots & 0 & -\phi^f K^f
\end{bmatrix}. \tag{22}$$

The solution of the generalized eigenvalue problem (Equation (20)) is the dispersion relation expressed by the frequency-dependent and complex wave number k, from which the phase velocity c and attenuation factor α of the acoustic wave can be calculated:

$$c = \frac{\omega}{\text{Re}\left(k\right)}\,, \quad \alpha = \text{Im}\left(k\right). \tag{23}$$

In Equation (23), Re(k) and Im(k) denote the real and imaginary part of the complex wave number k, respectively.

3. Dispersion and frequency-dependent attenuation

The solution of the generalized eigenvalue problem (Equation (20)) is frequency-dependent and complex. Therefore, also the phase velocity c and the attenuation α is frequency-dependent. However, low and high frequency limits for the phase velocity can be determined:

$$\lim_{\omega\to 0}(c) = c^{\text{Wood}} = \sqrt{\frac{K^{\text{Reuss}}}{\rho^{\text{eff}}}} \quad , \quad \lim_{\omega\to\infty}(c) = c^{fR} = \sqrt{\frac{K^{\text{f}}}{\rho^{fR}}} \,. \tag{24}$$

The low-frequency-limit for the phase velocity is the so-called Wood-limit (Mavko et al., 2003). At frequencies much smaller than the resonance frequency of the gas bubbles, water and gas move in phase. Therefore, in this regime, effective material parameters have to be used, i.e. the Reuss average of the bulk moduli of the two constituents and the effective density of the gas-water-mixture, which is given as

$$\rho^{\text{eff}} = \rho^f + \rho^g = \rho^f + \sum_{j=1}^{m} \rho_j^g = \phi^f \rho^{fR} + \sum_{j=1}^{m} \left(\phi_j^g\right)\rho^{gR} = \phi^f \rho^{fR} + \phi^g \rho^{gR} \,. \tag{25}$$

The effective density and the Wood-limit is plotted in Fig. 2 together with the Reuss average of the bulk moduli (Equation (14)) for the material parameters shown in Table 1 and for a range of gas volume fractions. The high-frequency-limit for the phase velocity (Equation (24)) is equal to the phase velocity of bubble-free water c^{fR}. At frequencies much larger than the resonance frequency, inertia prohibits a movement of the gas bubbles. Therefore, in this regime, the true material parameters of the fluid have to be used.

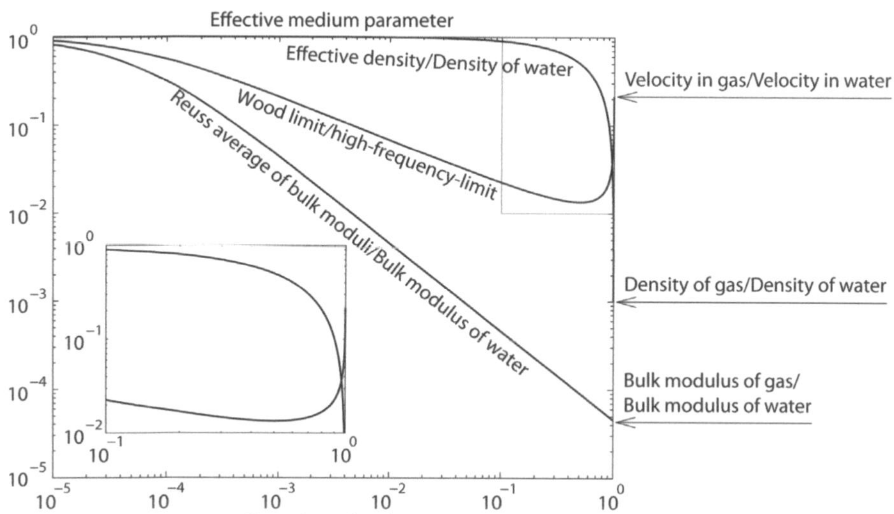

Fig. 2. Effective density ρ^{eff}, low-frequency-limit (i.e. Wood-limit) c^{Wood} and Reuss average of the bulk moduli K^{Reuss} for the material parameters shown in Table 1 and for a range of gas volume fractions. All values are dimensionless and normalization was done with the respective values for pure water, i.e. the high-frequency-limits. The gray box in the upper right corner is blown up and shown as the inlet in the lower left of the figure.

	Parameter	Symbol	Value [Unit]
Water	Density	ρ^{fR}	1000 [kg/m³]
	Bulk modulus	K^f	2.2x10⁹ [Pa]
	Viscosity	η^{fR}	1x10⁻³ [Pa s]
	Sound speed	c^{fR}	1483 [m/s]
Gas	Density	ρ^{gR}	1 [kg/m³]
	Bulk modulus	K^g	101x10³ [Pa]
	Heat capacity ratio	γ	1.4 [-]
	Sound speed	c^{gR}	362 [m/s]
Water depth		H	0 [m]

Table 1. Physical properties of the pure media used in this study, i.e. bubble-free water and pure gas, respectively, and the water depth that is common to all models in this study.

We define the resonance frequency ω_{0j} of our coupled wave propagation-oscillator model to be the frequency where the peak attenuation occurs. The resonance frequency is

$$\omega_{0j} = A\Omega_{0j} = \frac{c^{fR}}{c^{Wood}}\Omega_{0j} \,. \tag{26}$$

Equation (26) shows that the scaling factor A is not an arbitrary value, but is a combination of material parameters defined earlier, i.e. the ratio between high- and low-frequency-limit of the phase velocity.

3.1 Gas bubbles of only one size

First, we study a model containing gas bubbles of only one size. Consequently, only the first and the last equations of the set of Equations (19) are used and matrices **A** and **B** (Equations (21) and (22)) reduce to 2x2-matrices. Fig. 3 and Fig. 4 both show the phase velocity dispersion (a) and the frequency-dependent attenuation (b) of acoustic waves in such a model. The material parameters used for producing Fig. 3 and Fig. 4 are given in Table 1. In Fig. 3 the gas volume fraction is constant (ϕ^g=0.001) for all lines while the different lines represent different gas bubble sizes (r={7x10⁻⁵m, 1x10⁻⁴m, 2x10⁻⁴m, 5x10⁻⁴m, 1x10⁻³m}). These gas bubble sizes result in angular resonance frequencies Ω_0={2.93x10⁵Hz, 2.05x10⁵Hz, 1.02x10⁵Hz, 4.10x10⁴Hz, 2.05x10⁴Hz}. In Fig. 4 the gas bubble size is constant (r=2x10⁻⁴m, ω_0=1.02x10⁵Hz) for all lines while the different lines represent different gas volume fractions (ϕ^g={1x10⁻⁴, 5x10⁻⁴, 1x10⁻³, 5x10⁻³, 1x10⁻²}).
For a constant gas volume fraction the low- and high-frequency-limit for the phase velocity (Fig. 3a) is also constant. This is obvious because these limits (Equations (24)) only depend on the effective material properties of the gas-water-mixture (i.e. Reuss average of bulk moduli and effective density) and on the material properties of water, respectively, and not on the size of the gas bubbles. However, the size of the gas bubbles influences the phase velocity dispersion curves in the vicinity of the resonance frequency of the gas bubbles. For small gas bubbles, the attenuation becomes more important (Equation (12)) and the oscillations are damped more. Therefore, not such a strong excitation of the oscillations can take place and the transition from the low- to the high-frequency-limit of the phase velocity dispersion curve is smoother for small gas bubble sizes, i.e. not such strong minimum and maximum values. On the other hand, for larger bubble sizes, the damping is small and the

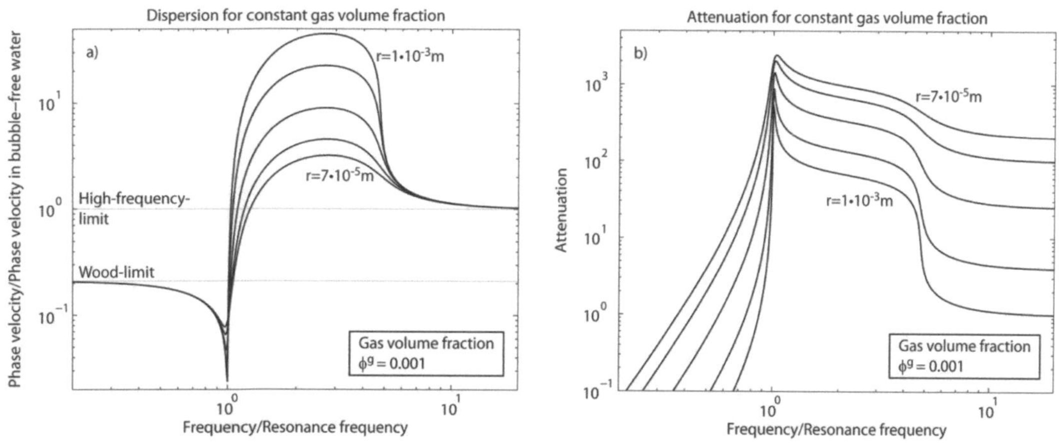

Fig. 3. Dispersion (a) and frequency-dependent attenuation (b) for water containing gas with a fixed volume fraction ϕ^g=0.001. Material parameters are given in Table 1. While the gas volume fraction is kept constant, for each curve the gas volume is contained in bubbles with equal radius (in meter) r={7x10^{-5}, 1x10^{-4}, 2x10^{-4}, 5x10^{-4}, 1x10^{-3}}. The frequency is normalized with the resonance frequency and the phase velocity is normalized with the phase velocity in bubble-free water, i.e. with the high-frequency-limit.

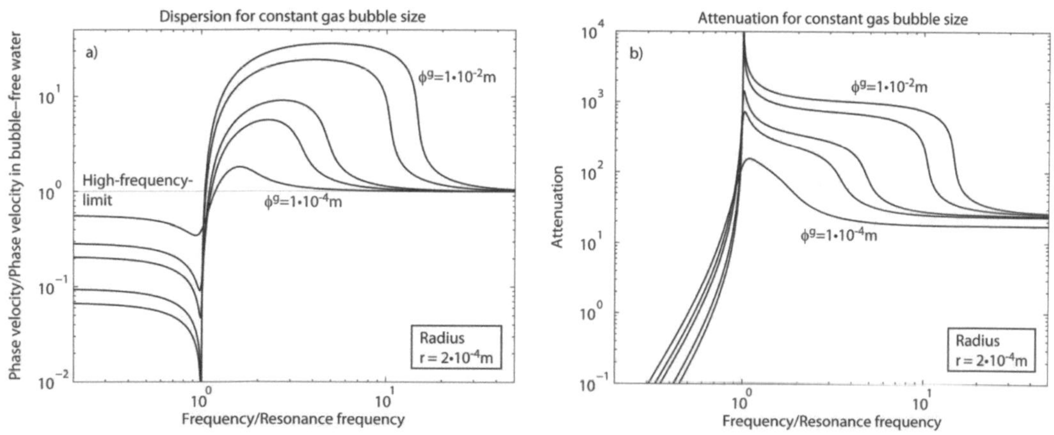

Fig. 4. Dispersion (a) and frequency-dependent attenuation (b) for water containing gas bubbles with a fixed radius r=2x10^{-4}m. Material parameters are given in Table 1. While the gas bubble size is kept constant, the gas volume fraction is different for the different lines (ϕ^g={1x10^{-4}, 5x10^{-4}, 1x10^{-3}, 5x10^{-3}, 1x10^{-2}}). The frequency is normalized with the resonance frequency and the phase velocity is normalized with the phase velocity in bubble-free water, i.e. with the high-frequency-limit.

oscillations are strongly excited. For all bubble sizes, at frequencies slightly lower than the resonance frequency, the phase velocity strongly decreases before it rapidly increases to values much larger than the high-frequency-limit. For increasing frequencies, the phase velocity decreases to eventually reach the high-frequency-limit. Such a behavior of the phase velocity around the resonance frequency was also described by van Wijngaarden (1972), Anderson and Hampton (1980) or Commander and Prosperetti (1989).

The effect of the gas bubble size on attenuation is shown in Fig. 3b. Around the resonance frequency, the attenuation of an acoustic wave is largest because energy is transferred from the water to the gas bubbles to excite the oscillations. At frequencies much larger than the resonance frequency, inertia prohibits a movement of the gas bubbles. Obviously, the non-moving gas bubbles damp a propagating acoustic wave. As described above, the attenuation depends on the gas bubble size (Equation (12)). At frequencies much smaller than the resonance frequency of the gas bubbles, water and gas move in phase. Therefore, there is no relative movement between gas and water and consequently no attenuation.
The phase velocity dispersion curves for different gas volume fractions but for constant gas bubble size (Fig. 4a) are qualitatively comparable. The major difference between these curves is the low-frequency-limit. A smaller gas volume fraction leads to a larger value of the Reuss average of the bulk moduli (Equation (14)) and therefore to a higher value of the low-frequency-limit of the phase velocity (Equation (24)). At the same time, the influence on the Wood-limit of the change of the effective density (Equation (25)) with changing gas volume fraction is only minimal in the range considered in Fig. 4. This can also be observed in Fig. 2. The attenuation of an acoustic wave (Fig. 4b) is stronger for larger gas volume fractions. This is intuitive, because the larger the gas volume fraction is, the more gas bubbles are present in the water and the stronger water and gas interact. However, the high-frequency-limit for the attenuation is only weakly influenced by the gas volume fraction with a slightly higher attenuation for larger gas volume fractions. From the equation for viscous friction term $\overline{\eta}$ (Equation (12)) it is clear that the bubble radius (Fig. 3b) has a much stronger influence on the attenuation than the gas volume fraction (Fig. 4b), i.e. quadratic vs. linear relation, respectively.

3.2 Gas bubbles with two different sizes

The model equation (Equations (13) and (17)) are written for an arbitrary number of gas bubble sizes. Exemplarily, we consider two different bubble sizes in the water, leading to two different resonance frequencies. The material parameters for gas and water are given in Table 1 and the two bubble sizes and volume fractions are given in Table 2. The volume fraction of the total number of bubbles of each bubble size is equal, i.e. half of the gas is contained in one bubble size, the other half is contained in the other bubble size. Fig. 5 shows the phase velocity dispersion (a) and frequency-dependent attenuation (b) for this model. The low- and high-frequency-limits for the phase velocity are the same as in Fig. 3, because the gas volume fraction is the same. However, the dispersion curve in the vicinity of the two resonance frequencies is quite different. The typical dispersion behavior for only one bubble size shown in Fig. 3 and Fig. 4 (i.e. decreasing velocity below resonance frequency, rapid increase to values larger than high-frequency-limit, decrease to high-frequency-limit) takes place twice, once at each of the two resonance frequencies. Therefore, two, rather than one, sets of minimum and maximum occur in the phase velocity dispersion curve (Fig. 5a). For comparison, the dispersion and the frequency-dependent attenuation of two models containing only either of the two bubble sizes with their respective volume fraction are also plotted in Fig. 5. Note that the scaling of the resonance frequencies is different for these models because the total gas volume fraction is different. As for the phase velocity dispersion, the attenuation (Fig. 5b) shows two maxima, one at each of the two resonance frequencies.

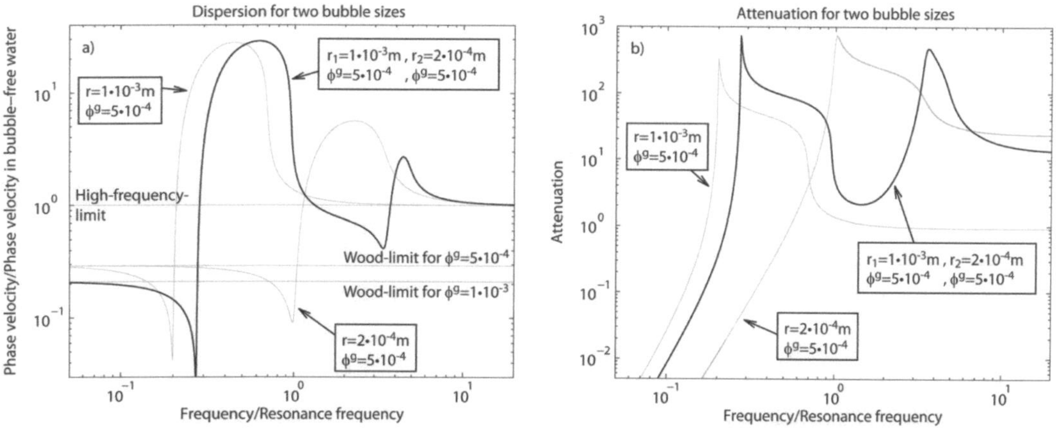

Fig. 5. Dispersion (a) and frequency-dependent attenuation (b) for water containing gas bubbles with two different radii r_1=1x10^{-3}m and r_2=2x10^{-4}m. Material parameters are given in Table 1 and Table 2. Plotted in light gray for comparison are the dispersion and the frequency-dependent attenuation of two models containing only either of the two bubble sizes with their respective volume fraction. The frequency is normalized with the resonance frequency of the larger bubble and the phase velocity is normalized with the phase velocity in bubble-free water, i.e. with the high-frequency-limit.

Parameter	**Value [Unit]**	
Gas volume fraction	ϕ^g=1x10^{-3} [-]	
	Gas bubble size 1	**Gas bubble size 2**
Bubble radius	r_1=1x10^{-3} [m]	r_2=2x10^{-4} [m]
Volume fraction of gas bubble sizes	ϕ_1^g =5x10^{-4} [-]	ϕ_2^g =5x10^{-4} [-]
Angular resonance frequency	ω_{01}=2.05x10^4 [Hz]	ω_{02}=1.02x10^5 [Hz]

Table 2. Model parameters for the model containing gas bubbles of two different sizes. Material parameters for gas and water are given in Table 1.

3.3 Probability function for gas bubble sizes

The model equation (Equations (13) and (17)) are written for an arbitrary number of gas bubble sizes. In principle, every gas bubble size distribution can be approximated by a combination of discrete size (i.e. radius) and volume fraction distributions, which can then be used in Equations (13) and (17). Exemplarily, we show results for bubble radii that lie in the range [0.1r_0...10r_0] and whose corresponding volume fractions are log-normal distributed around a central value r_0. The following equation describes the distribution of the volume fraction for the different bubble sizes:

$$\phi_j^g = \Phi \exp\left(\frac{-\left(\ln\left(r_j\right) - \ln\left(r_0\right)\right)^2}{s^2} \right) \quad \text{for } r_j \in \left[0.1r_0 \ldots 10r_0\right] \text{and } j = \{1,...,501\}. \qquad (27)$$

We set r_0=2x10^{-4}m and we pick 501 values for the bubble radius r_j in the range [0.1r_0...10r_0] in a way that the logarithm of r_j is equally spaced. The factor Φ is chosen in such a way that

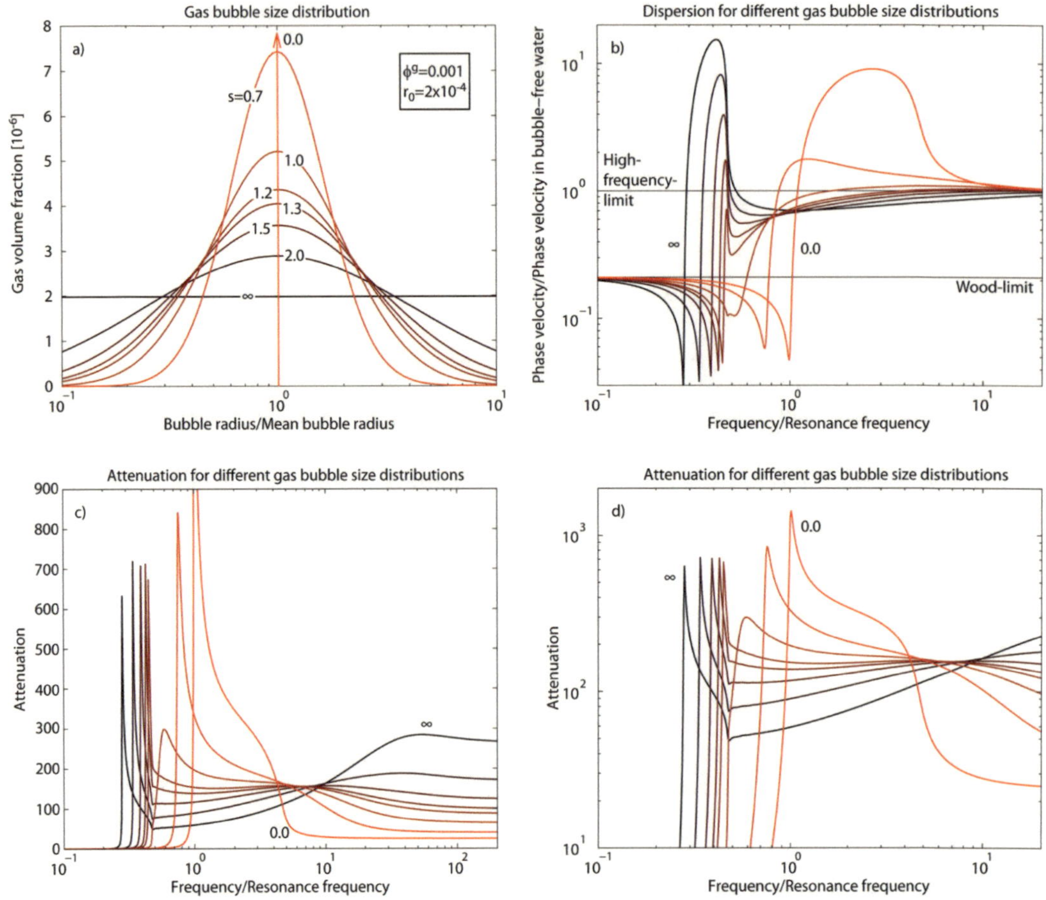

Fig. 6. a) Gas volume fraction versus bubble radius. The bubble radius is normalized with r_0=2x10^{-4}m. Along these curves, 501 discrete values are picked at equally spaced values for the logarithm of the bubble radius, i.e. the curves are discrete probability functions, rather than continuous ones. The sum of all 501 values is equal to ϕ^g=0.001. Labels correspond to the values of parameter s in Equation (27). The curve with label "0.0" corresponds to a model with only one bubble size. b), c) and d) Dispersion and two different representations of the frequency-dependent attenuation for water containing gas with a fixed volume fraction ϕ^g=0.001 and with bubble size distributions shown in a). Labels are only given for the end member curves. All other curves correspond to the curves labeled in a). Material parameters are given in Table 1. The frequency is normalized with the resonance frequency and the phase velocity is normalized with the phase velocity in bubble-free water, i.e. with the high-frequency-limit.

Equations (6) are fulfilled for a gas volume fraction ϕ^g=0.001. Equation (27) describes the discrete approximation of a log-normal probability function. The log-normal function is cut off at the values $0.1r_0$ and $10r_0$ because bubble radii only lie between these two values. In Fig. 6a, Equation (27) is plotted for different values of the parameter s (i.e. s={0.0, 0.7, 1.0, 1.2, 1.3, 1.5, 2.0, ∞}). Choosing s=∞ (infinity) results in a flat distribution where the volume fraction for each bubble size is equal. Choosing s=0 results in a Dirac-distribution where

only one bubble size is present, i.e. bubbles with radius r_0. In Fig. 6 the phase velocity dispersion (b) and the frequency-dependent attenuation (c and d) are shown for the different distribution functions shown in Fig. 6a. Note that curves labeled "0.0" correspond to a model with only one bubble size and are the same as the third curves in Fig. 3 and Fig. 4. The high- and low-frequency-limits for the phase velocity is equal for all bubble size distributions because the gas volume fraction ϕ_g is equal. Between the two limits, the different bubble size distributions lead to significantly different dispersion curves. Only the first and second curve (i.e. s={0.0, 0.7}), for which only one gas bubble size occurs or the gas bubble size is very narrowly distributed around r_0, respectively, show the characteristic features described above for Fig. 3a and Fig. 4a (i.e. maximum peak velocity and asymptotic approach of the high-frequency-limit). For wider gas bubble size distributions, the clear maximum peak disappears and the asymptotic approach of the high-frequency-limit is very flat or even from values smaller than the high-frequency-limit. For larger values of s (i.e. s={1.3, 1.5, 2.0, ∞}), the phase velocity exhibits a maximum value that is higher than the high-frequency-limit and drops below the high-frequency-limit for increasing frequencies.
The attenuation shows different characteristics in the high-frequency-range (Fig. 6c) and at frequencies around the central resonance frequency (Fig. 6d). The high-frequency-limit strongly resembles the behavior described for Fig. 3b, where models containing equally sized gas bubbles were analyzed for different bubble sizes. A wider gas bubble size distribution (larger value for parameter s) introduces more small gas bubbles compared to large ones. The small gas bubbles dominate the attenuation characteristics in the high-frequency-limit. In the vicinity of the central resonance frequency the attenuation curves show a distinct peak. This peak changes its maximum value with changing bubble size distribution. It is roughly constant for large values of s (i.e. s={1.2, 1.3, 1.5, 2.0, ∞}), has a minimum for s=1.0 and becomes larger for small values of s (i.e. s={0.0, 0.7}).

3.4 Comparison with existing models

We compare our phase velocity dispersion and attenuation curves with existing and published curves. First, we consider two curves of Commander and Prosperetti (1989). Figures 1 and 8 of Commander and Prosperetti (1989) show the dispersion and phase velocity, respectively, of water containing gas bubbles with only one size (r=9.94x10^{-4}m) and a gas volume fraction ϕ_g=3.77x10^{-4}. Also, it is shown that the theory of Commander and Prosperetti (1989) matches the experimental data of Silberman (1957) for the same model conditions. Fig. 7 compares the two figures of Commander and Prosperetti (1989) with our model for the same model parameters. The phase velocity dispersion matches the model of Commander and Prosperetti (1989) very well. All, the low-frequency-limit, the high-frequency-limits and the complicated dispersion behavior in the vicinity of the resonance frequency is almost identical. Also, the data of Silberman (1957) is equally well matched by our model in comparison with the model of Commander and Prosperetti (1989). Attenuation in Commander and Prosperetti (1989) is given in dB/cm, that is, a logarithmic unit with an arbitrary reference value. Therefore, we also plot the attenuation in logarithmic units and move the curve of Commander and Prosperetti (1989) vertically to adjust the two curves to the same reference value. Doing this does not change the vertical scale of either of the curves. It only changes the absolute values on the vertical axis. Our model matches the

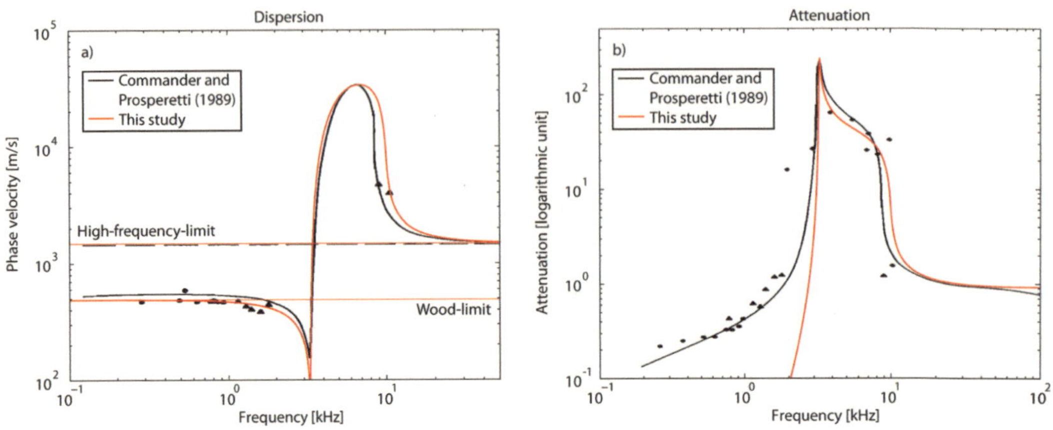

Fig. 7. Phase velocity dispersion (a) and frequency-dependent attenuation (b) for a model with one gas bubble size r=9.94x10^{-4}m and a gas volume fraction ϕ_g=3.77x10^{-4}. Plotted in black are Figures 8 and 1 of Commander and Prosperetti (1989). Data points are also taken from Commander and Prosperetti (1989), who took the data from Silberman (1957). All values are plotted as absolute values. In b), attenuation is plotted as a logarithmic unit, because in Commander and Prosperetti (1989) the attenuation is given in dB/cm.

model of Commander and Prosperetti (1989) very well in the frequency range at and above the resonance frequency. Here, also the attenuation data of Silberman (1957) is matched equally well in comparison with the model of Commander and Prosperetti (1989). However, in the frequency range below the resonance frequency the two models diverge from each other. The model of Commander and Prosperetti (1989) matches the data of Silberman (1957) significantly better compared to our model.

The second model we chose is the model by MacPherson (1957). Because no phase velocity dispersion curve is plotted in MacPherson (1957), we take Figure 10 of Anderson and Hampton (1980) who plot the phase velocity dispersion of MacPherson (1957). The model describes acoustic waves in water containing gas bubbles of one size. Fig. 8 compares Figure 10 of Anderson and Hampton (1980) with our model. In Anderson and Hampton (1980) it is unclear which bubble size is used for Figure 10. Therefore, we plot the two extreme values described by MacPherson (1957) (i.e. r=8x10^{-5}m and r=2.5x10^{-4}m) to have the possible range of bubble sizes. Also, in Figure 10 of Anderson and Hampton (1980) a gas volume fraction ϕ_g=1x10^{-3} is indicated. However, the Wood-limit (i.e. low-frequency-limit) is matched for a gas volume fraction ϕ_g=7x10^{-4}. We assume that the value given by Anderson and Hampton (1980) is only an approximation and we chose ϕ_g=7x10^{-4} for Fig. 8. The general trend of the phase velocity dispersion curve of Anderson and Hampton (1980) is well matched by our model, especially in the frequency range at and below the resonance frequency, where our model almost perfectly matches the curve of Anderson and Hampton (1980). However, in the frequency range above the resonance frequency, the two models diverge from each other.

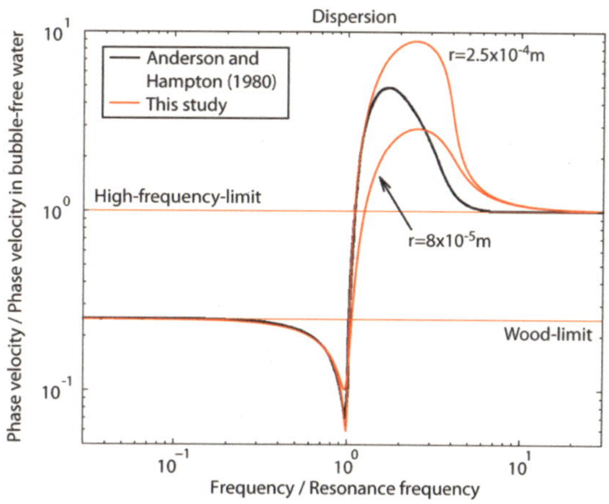

Fig. 8. Phase velocity dispersion for a model with one bubble size and a gas volume fraction ϕ_g=7x10^{-4}. Plotted in red are two dispersion curves for two different bubble radii r=8x10^{-5}m and r=2.5x10^{-4}m. Plotted in black is Figure 10 of Anderson and Hampton (1980). The frequency is normalized with the resonance frequency and the phase velocity is normalized with the phase velocity in bubble-free water, i.e. with the high-frequency-limit.

4. Discussion

There are many features that can generate oscillations and resonances in a medium through which acoustic and seismic waves propagate. Examples are air bubbles in water, holes in elastic plates, oil blobs in porous rocks or fluid-filled fractures in rocks. All oscillation processes exhibit a resonance frequency and the propagation of waves having frequencies in the range of about one order of magnitude around the resonance frequency will be considerably affected by these oscillations. The internal oscillations cause a dispersion and a frequency-dependent attenuation of the waves and consequently cause a frequency-dependent reflection and transmission behaviour of the medium exhibiting oscillations. Equations describing the coupling of wave propagation and oscillations are important, first, to better understand the general impact of oscillations on the wave propagation and, second, to apply these equations in inversion problems which aim to determine the size and distribution of heterogeneities causing oscillations inside a medium by analysing the waves that propagated through this medium. However, an exact mathematical treatment of the coupling between wave propagation and resonant oscillations is not trivial. For example, the propagation of acoustic waves in water with air bubbles may appear at first as a relatively simple process, but the process is actually quite complex and the exact mathematical treatment is complicated (e.g. Caflisch et al., 1985). An exact mathematical treatment of even more complicated processes, such as for example the wave propagation in partially saturated poroelastic solids with isolated fluid blobs, may be too complicated to be of practical use. Therefore, there is a need for a simple, more basic, mathematical description of wave propagation in media exhibiting oscillations.

We present here one basic mathematical approach to couple macroscopic wave propagation with microscopic oscillations. We assume that the oscillations (whatever the exact physical mechanism) can be effectively described with the equation for a damped linear oscillator.

We then couple the oscillator equation to the wave equation using the mixture theory. The coupling of the microscopic oscillations with the macroscopic wave propagation is done through moment interaction terms. We consider two interaction terms where one term scales with the volume and represents inertial coupling and the other term scales with the surface and represents the viscous coupling. With this approach it is easy to "switch" the coupling terms on and off and investigate their impact on the dispersion and attenuation. For example, Frehner et al. (2009) studied the coupling through only inertial terms and observed in a narrow frequency range around the resonance frequency a significant dispersion with a discontinuity of the phase velocity at the resonance frequency.

For acoustic waves propagating through water with air bubbles the dispersion and attenuation curves resulting from our approach agree well with the corresponding dispersion and attenuation curves resulting from more elaborated mathematical theories and also from laboratory experiments. This indicates that our approach captures the essential first-order effects of wave propagation in a medium with oscillations. Our approach also allows an easy extension to an arbitrary number of oscillators providing a particular distribution of resonance frequencies. Clearly, an exact mathematical treatment is always helpful and important to, for example, provide the range of applicability for the simplified equations.

Our approach can be applied to more complicated processes and we expect it to also capture the first-order effects on dispersion and attenuation. The equations resulting from our approach are more transparent and easier to use for inversion problems. Inversions are of great industrial and scientific interest because they allow to assess the distribution and amount of heterogeneities inside a medium by analyzing waves that propagated through the medium. An example is the determination of the bubble distribution in sea water near the ocean surface (e.g. Commander and McDonald, 1991).

5. Conclusions

Oscillations with a certain resonance frequency within a medium is a common physical phenomenon that takes place in a large number of media, e.g. porous or fractured rocks, water containing gas bubbles or heterogeneous media in general. Internal oscillations lead to a strongly frequency-dependent propagation behavior of waves that propagate through such media, i.e. velocity dispersion and frequency-dependent attenuation. However, such oscillations are usually not included in effective medium theories for heterogeneous media. We presented a one-dimensional continuum model for an acoustic medium exhibiting internal damped oscillations. It is a two- (or more) phase model with one connected elastic (i.e. acoustic) phase and one (or more) disconnected oscillating phase. The obvious application of this model is water containing oscillating gas bubbles. This application provided the material and model parameters used in this study. However, the material and model parameters could be adapted to other applications of a medium exhibiting internal oscillations. The presented model of water containing gas bubbles includes two physically based momentum interaction terms between the two phases, a purely elastic term of oscillatory nature that scales with the volume of the bubbles and a viscous term that scales with the specific surface of the bubble. Thus we are able to take into account damping with respect to the specific surface area of the gas bubbles. The model is capable of taking into account a large number of oscillators with different resonance frequencies. In the limit case, a continuous probability function for the resonance frequencies can be approximated with a discrete number of oscillators. Exemplarily, we showed volume fractions that are log-normally distributed with respect to bubble size. The results show that a certain distribution of resonance frequencies around a central value changes the phase velocity dispersion and frequency-dependent attenuation significantly compared to the case where only one bubble size is present. The dispersion and attenuation resulting from our approach agree well with the dispersion and attenuation (1) derived with a more exact mathematical treatment and (2) measured in laboratory experiments. Hence, our basic approach captures the first-order effects of oscillations on the propagation of acoustic waves.

6. References

Aki, K. & Richards, P. G. (2002). *Quantitative seismology*, University Science Books, ISBN 0-9357-0296-2, Sausalito

Anderson, A. L. & Hampton, L. D. (1980). Acoustics of gas-bearing sediments, 1. Background. *Journal of the Acoustical Society of America*, 67, 6, 1865-1889, ISSN 0001-4966

Batchelor, G. K. (2000). *An introduction to fluid dynamics*. Cambridge University Press, ISBN 0-2516-6396-2, Cambridge

Beresnev, I. A. & Johnson, P. A. (1994). Elastic-wave stimulation of oil production - A review of methods and results. *Geophysics*, 59, 6, 1000-1017, ISSN 0016-8033

Beresnev, I. A., Vigil, R. D., Li, W. Q., Pennington, W. D., Turpening, R. M., Iassonov, P. P. & Ewing, R. P. (2005). Elastic waves push organic fluids from reservoir rock. *Geophysical Research Letters*, 32, 13, L13303, ISSN 0094-8276

Biot, M. A. (1962). Mechanics of deformation and acoustic propagation in porous media. *Journal of Applied Physics*, 33, 4, 1482-1498, ISSN 0021-8979

Bowen, R. M. (1980). Incompressible porous media models by use of the theory of mixtures. *International Journal of Engineering Science*, 18, 9, 1129–1148, ISSN 0020-7225

Bowen, R. M. (1982). Compressible porous media models by use of the theory of mixtures. *International Journal of Engineering Science*, 20, 6, 697–735, ISSN 0020-7225

Caflisch, R. E., Miksis, M. J., Papanicolaou, G. C. & Ting, L. (1985). Effective equations for wave propagation in bubbly liquids. *Journal of Fluid Mechanics*, 153, April, 259-273, ISSN 0022-1120

Carstensen, E. L. & Foldy, L. L. (1947). Propagation of sound through a liquid containing bubbles. *Journal of the Acoustical Society of America*, 19, 3, 481-501, ISSN 0001-4966

Commander, K. W. & McDonald, R. J. (1991). Finite-element solution of the inverse problem in bubble swarm acoustics. *Journal of the Acoustical Society of America*, 89, 2, 592-597, ISSN 0001-4966

Commander, K. W. & Prosperetti, A. (1989). Linear pressure waves in bubbly liquids: Comparison between theory and experiments. *Journal of the Acoustical Society of America*, 85, 2, 732-746, ISSN 0001-4966

Dutta, N. C. & Ode, H. (1979). Attenuation and dispersion of compressional waves in fluid-filled porous rocks with partial gas saturation (White model). 1. Biot theory. *Geophysics*, 44, 11, 1777-1788, ISSN 0016-8033

Dvorkin, J., Mavko, G. & Nur, A. (1990). The oscillations of a viscous compressible fluid in an arbitrarily-shaped pore. *Mechanics of Materials*, 9, 2, 165-179, ISSN 0167-6636

Fox, F. E., Curley, S. R. & Larson, G. S. (1955). Phase velocity and absorption measurements in water containing air bubbles. *Journal of the Acoustical Society of America*, 27, 3, 534-539, ISSN 0001-4966

Frehner, M., Schmalholz, S. M. & Podladchikov, Y. (2009). Spectral modification of seismic waves propagating through solids exhibiting a resonance frequency: A 1-D coupled wave propagation–oscillation model. *Geophysical Journal International*, 176, 2, 589-600, ISSN 0956-540X

Frehner, M., Schmalholz, S. M., Saenger, E. H. & Steeb, H. (2008). Comparison of finite difference and finite element methods for simulating two-dimensional scattering of elastic waves. *Physics of the Earth and Planetary Interiors*, 171, 1-4, 112-121, ISSN 0031-9201

Graham, D. R. & Higdon, J. J. L. (2000). Oscillatory flow of droplets in capillary tubes. Part 1 and 2. *Journal of Fluid Mechanics*, 425, 31-53 and 55-77, ISSN 0022-1120

Hassan, W. & Nagy, P. B. (1997). Circumferential creeping waves around a fluid-filled cylindrical cavity in an elastic medium. *Journal of the Acoustical Society of America*, 101, 5, 2496-2503, ISSN 0001-4966

Hilpert, M. (2007). Capillarity-induced resonance of blobs in porous media: Analytical solutions, Lattice-Boltzmann modeling, and blob mobilization. *Journal of Colloid and Interface Science*, 309, 2, 493-504, ISSN 0021-9797

Hilpert, M., Jirka, G. H. & Plate, E. J. (2000). Capillarity-induced resonance of oil blobs in capillary tubes and porous media. *Geophysics*, 65, 3, 874-883, ISSN 0016-8033

Iassonov, P. P. & Beresnev, I. A. (2003). A model for enhanced fluid percolation in porous media by application of low-frequency elastic waves. *Journal of Geophysical Research B: Solid Earth*, 108, 3, ESE 2-1-ESE 2-9, ISSN 0148-0227

Kelly, K. R. & Marfurt, K. J. (1990). *Numerical Modeling of Seismic Wave Propagation*. Society of Exploration Geophysicists, ISBN 1-56080-011-9, Tulsa, Oklahoma

Korneev, V. (2008). Slow waves in fractures filled with viscous fluid. *Geophysics*, 73, 1, N1-N7, ISSN 0016-8033

Korneev, V. (2009). Resonant seismic emission of subsurface objects. *Geophysics*, 74, 2, T47-T53, ISSN 0016-8033

Korneev, V. A., Goloshubin, G. M., Daley, T. M. & Silin, D. B. (2004). Seismic low-frequency effects in monitoring fluid-saturated reservoirs. *Geophysics*, 69, 2, 522-532, ISSN 0016-8033

Lee, S.-L., Komatitsch, D., Huang, B.-S. & Tromp, J. (2009). Effects of topography on seismic-wave propagation: An example from northern Taiwan. *Bulletin of the Seismological Society of America*, 99, 1, 314-325, ISSN 0037-1106

Li, W. Q., Vigil, R. D., Beresnev, I. A., Iassonov, P. & Ewing, R. (2005). Vibration-induced mobilization of trapped oil ganglia in porous media: Experimental validation of a capillary-physics mechanism. *Journal of Colloid and Interface Science*, 289, 1, 193-199, ISSN 0021-9797

Liu, Y. B., Wu, R. S. & Ying, C. F. (2000). Scattering of elastic waves by an elastic or viscoelastic cylinder. *Geophysical Journal International*, 142, 2, 439-460, ISSN 0956-540X

MacPherson, J. D. (1957). The effect of gas bubbles on sound propagation in water. *Proceedings of the Physical Society. Section B*, 70, 1, 85-92, ISSN 0370-1301

Martin, R. & Komatitsch, D. (2009). An unsplit convolutional perfectly matched layer technique improved at grazing incidence for the viscoelastic wave equation. *Geophysical Journal International*, 179, 1, 333-344, ISSN 0956-540X

Mavko, G., Mukerji, T. & Dvorkin, J. (2003). *The rock physics handbook: Tools for seismic analysis in porous media*. Cambridge University Press, ISBN 0-5215-4344-4, Cambridge

Minnaert, M. (1933). On musical air bubbles and the sound of running water. *Philosophical Magazine*, 16, 104, 235-248, ISSN 1478-6435

Pride, S. R., Flekkoy, E. G. & Aursjo, O. (2008). Seismic stimulation for enhanced oil recovery. *Geophysics*, 73, 5, O23-O35, ISSN 0016-8033

Quintal, B., Schmalholz, S. M. & Podladchikov, Y. Y. (2009). Low-frequency reflections from a thin layer with high attenuation caused by interlayer flow. *Geophysics*, 74, 1, N15-N23, ISSN 0016-8033

Saenger, E. H., Ciz, R., Krüger, O. S., Schmalholz, S. M., Gurevich, B. & Shapiro, S. A. (2007). Finite-difference modeling of wave propagation on microscale: A snapshot of the work in progress. *Geophysics*, 72, 5, SM293-SM300, ISSN 0016-8033

Santos, J. E., Douglas, J., Corbero, J. & Lovera, O. M. (1990). A model for wave-propagation in a porous-medium saturated by a 2-phase fluid. *Journal of the Acoustical Society of America*, 87, 4, 1439-1448, ISSN 0001-4966

Schultz, T., Cattafesta, L. N. & Sheplak, M. (2006). Modal decomposition method for acoustic impedance testing in square ducts. *Journal of the Acoustical Society of America*, 120, 6, 3750-3758, ISSN 0001-4966

Silberman, E. (1957). Sound velocity and attenuation in bubbly mixtures measured in standing wave tubes. *Journal of the Acoustical Society of America*, 29, 8, 925-933, ISSN 0001-4966

Smeulders, D. M. J. & van Dongen, M. E. H. (1997). Wave propagation in porous media containing a dilute gas-liquid mixture: Theory and experiments. *Journal of Fluid Mechanics*, 343, 351-373, ISSN 0022-1120

Truesdell, C. A. (1957). Sulle basi della termomeccanica I & II, *Accademia Nazionale dei Lincei, Rendiconti della Classe di Scienze Fisiche, Matematiche e Naturali*, 8, 22, 33-38 and 158-166

Urquizu, M. & Correig, A. M. (2004). On the equivalence between stratified media and oscillators. *Geophysical Journal International*, 157, 1, 245-250, ISSN 0956-540X

van Wijngaarden, L. (1972). One-dimensional flow of liquids containing small gas bubbles. *Annual Review of Fluid Mechanics*, 4, 369-396, ISSN 0066-4189

Wei, C. F. & Muraleetharan, K. K. (2002). A continuum theory of porous media saturated by multiple immiscible fluids: I. Linear poroelasticity. *International Journal of Engineering Science*, 40, 16, 1807-1833, ISSN 0020-7225

Werby, M. F. & Gaunaurd, G. C. (1990). Resonance scattering from submerged elastic spheroids of high aspect ratios and its 3-dimensional interpretation. *Journal of the Acoustical Society of America*, 88, 2, 951-960, ISSN 0001-4966

White, J. E. (1975). Computed seismic speeds and attenuation in rocks with partial gas saturation. *Geophysics*, 40, 2, 224-232, ISSN 0016-8033

25

Pulse Wave Propagation in Bistable Oscillator Array

Kuniyasu Shimizu[1], Motomasa Komuro[2] and Tetsuro Endo[3]
[1]*Chiba Institute of Technology*
[2]*Teikyo University of Science and Technology*
[3]*Meiji University*
Japan

1. Introduction

A study of propagating wave phenomenon in coupled systems is one of the familiar topics of research. There exist propagating wave phenomena in various systems such as reaction-diffusion systems (Nishiura et al.; 2005; Comte et al.; 2001), coupled map lattice (Kaneko; 1993) and coupled oscillator systems (Hikihara et al.; 2001; Yamauchi et al.; 1999), etc.. They are important not only in a pure nonlinear science viewpoint but also from the viewpoint of various applications such as engineering purpose (Sato et al.; 2003) and biological information processing (Keener; 1987). A basic question concerning these systems is the condition under which propagating wave can emerge. It is known as propagation failure phenomenon that propagating wave fails to propagate below a certain critical coupling strength (Comte et al.; 2001).

In this chapter, a simple model of one-dimensinally coupled bistable oscillators is shown to exhibit wave propagation phenomena. The propagating wave consists of several adjacent oscillators oscillating with large amplitude, and the part of large amplitude oscillation in the array propagates with a constant speed. In particular, we pay attention to the formation mechanism of the propagating wave related to disappearance of a certain kind of standing wave. When a coupling strength is weak, a standing pulse wave exists. This solution disappears when the coupling strength exceeds a certain critical value, and around this point the propagating wave begins to exist. On the basis of the observation, one of the onset mechanisms of the propagating waves is investigated paying attention to global phase-space structure around the bifurcation point.

At first, we will introduce the ring of coupled bistable oscillator system, and derive the fundamental equation. Then, a typical propagating wave phenomenon and its characteristic features are stated briefly. Next, we make a study on the onset mechanism of the propagating wave for 6 coupled oscillator case. As a result, we have found that a global bifurcation of maps based on the heteroclinic tangle converts the fixed point (= standing wave) into the invariant circle (= propagating pulse wave).

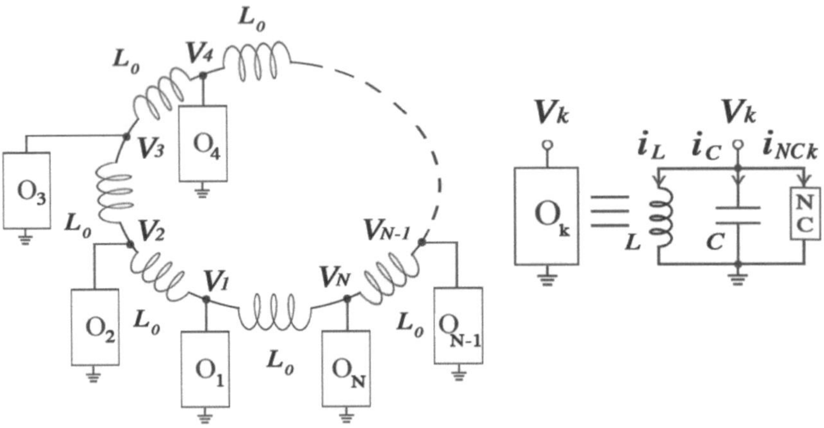

(a) A ring of *N*-coupled bistable oscillators

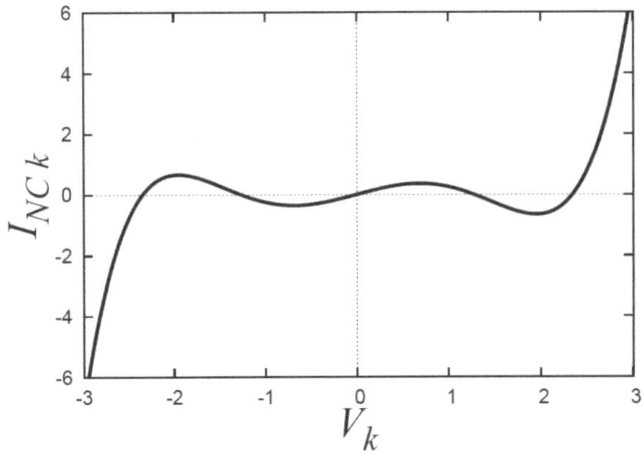

(b) The fifth-power nonlinearity on the *I-V* characteristics of NC

Fig. 1. A circuit model

2. Fundamental equation and its dynamics

Figure 1(a) shows a circuit model of the ring of *N*-coupled bistable oscillators. The current-voltage characteristics of negative conductance (= NC) in Fig.1(a) has the fifth-power nonlinearity as shown in Fig.1(b): $i_{NCk} = g_1 V_k - g_3 V_k^3 + g_5 V_k^5$, $g1, g3, g5 > 0$, $k = 1, 2, \ldots, N$. Then, Kirchhoff's current law gives the following equation:

$$\frac{1}{L}\int V_k dt + C\frac{dV_k}{dt} + g_1 V_k - g_3 V_k^3 + g_5 V_k^5 = \frac{1}{L_0}\int (V_{k-1} - 2V_k + V_{k+1})dt, \quad k = 1,2,\cdots,N. \tag{1}$$

Taking time derivative of (1), we obtain the following equation (V_0 [resp. V_{N+1}] is regarded as V_N [resp. V_1]).

$$\frac{d^2V_k}{dt^2} + \frac{g_1}{C}(1 - \frac{3g_3}{g_1}V_k^2 + \frac{5g_5}{g_1}V_k^4)\frac{dV_k}{dt} +$$
$$\left(\frac{1}{LC} + \frac{2}{L_0C}\right)V_k - \frac{1}{L_0C}(V_{k-1} + V_{k+1}) = 0.$$

By normalizing the time and voltage variables:

$$t = \tau/\sqrt{(1/LC) + (1/L_0C)},\ \ V_k = \sqrt[4]{g_1/5g_5}x_k,$$

and introducing new parameters:

$$\varepsilon \equiv g_1/\sqrt{(C/L) + (C/L_0)},\ \ \alpha \equiv L/(L + L_0),$$
$$\beta \equiv 3g_3/\sqrt{5g_1g_5}.$$

the equation of this system can be written in the following normalized equation ($\cdot = d/d\tau$, $\cdot\cdot = d^2/d\tau^2$).

$$\ddot{x}_k + \varepsilon(1 - \beta x_k^2 + x_k^4)\dot{x}_k + (1 - \alpha)x_k - \alpha(x_{k-1} - 2x_k + x_{k+1}) = 0 \tag{2}$$

If we introduce new variables as $y_k \equiv \dot{x}_k$, we obtain the following autonomous ordinary differential equation.

$$\begin{aligned} \dot{x}_k &= y_k \\ \dot{y}_k &= -\varepsilon(1 - \beta x_k^2 + x_k^4)y_k \\ &\quad -(1 - \alpha)x_k + \alpha(x_{k-1} - 2x_k + x_{k+1}) \end{aligned} \tag{3}$$

$$, k = 1, 2, \cdots, N,\ \ x_0 = x_N, x_1 = x_{N+1},$$

, where N is the number of oscillators. The x_k denotes the normalized output voltage of the kth oscillator, y_k denotes its derivative. The parameter ε (> 0) shows the degree of nonlinearity. The parameter α ($0 \leqq \alpha \leqq 1$) is a coupling factor; namely $\alpha = 1$ means maximum coupling, and $\alpha = 0$ means no coupling. The parameter β controls amplitude of oscillation. Each isolated oscillator has two steady-states, namely, no oscillation and periodic oscillation depending on the initial condition.

The analysis of modes based on the averaging method or perturbation method for weakly nonlinear cases was extensively performed in the past (Endo & Ohta; 1980; Yoshinaga & Kawakami; 1993). However, the solution for non-weak nonlinear cases seems not to be analyzed; in fact, it is complicated including the propagating pulse wave solution. In this study, we analyze the onset mechanism of the propagating pulse wave solution for non-weak nonlinear case via bifurcation theory. The propagating pulse wave solution consists of several adjacent oscillators oscillating with large amplitude, and the part of large amplitude oscillation in the ring array propagates with a constant speed as shown in Fig.2[1].

This propagating pulse wave solution has the following characteristic features from the results in (Shimizu et al.; 2008).

[1] Numerical integrations throughout this paper are carried out by 4th order Runge-Kutta method with a step size of 0.01 if not otherwise specified.

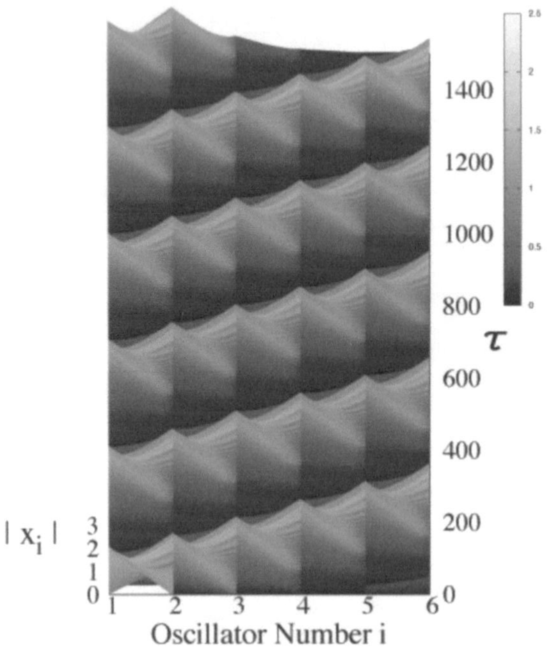

Fig. 2. A typical propagating pulse wave observed for the ring of 6 coupled oscillators (the absolute value of x-components are plotted for clarity). The initial condition is given as $x_1 = 2.0$, $y_2 = 1.3$ and all other variables are zero. ($\alpha = 0.1$, $\beta = 3.2$ and $\varepsilon = 0.36$)

- The direction of the propagating pulse wave depends on the initial condition.
- Propagating speed becomes faster for larger α.
- Robust against fluctuation and noise especially for comparatively large α.

Such propagating pulse wave seems to be observed for an arbitrary number of coupled oscillators[2] in comparatively large parameter regime. For instance, the existence region of several solutions including the propagating wave is shown in (Shimizu et al.; 2008) for the 100 coupled oscillator case by computer simulation. Hereafter, we will show one of the onset mechanisms of the propagating pulse wave for the ring of 6-coupled oscillators.

3. One of the onset mechanisms of propagating pulse wave

In this system, there exists a certain kind of standing pulse wave solution for small coupling strength α. The standing pulse wave solution is a periodic oscillation, one of which is shown in Fig.3. At first, we investigate this type of standing pulse wave, and then the transition from the standing pulse wave to the propagating pulse wave. Moreover, we will explain that the results obtained for the 6 coupled oscillator case may be extended to larger number of coupled oscillator cases.

[2] We confirmed the existence of the propagating pulse wave for the $N = 5$ to $N = 100$ cases via computer simulation.

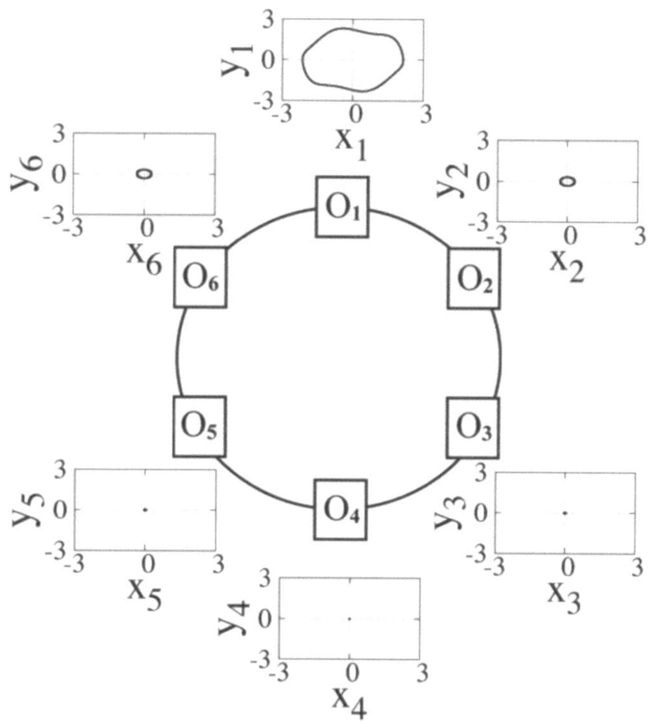

Fig. 3. A standing pulse wave (= a periodic solution) occurring in the ring of 6-coupled oscillators. Initial condition: $x_1(0) = 2.0$, $y_1(0) = 0.0$ and $x_k(0) = y_k(0) = 0.0$, $k = 2, 3, \dots, 6$. Parameters: $\alpha = 0.05$, $\beta = 3.2$ and $\varepsilon = 0.36$.

3.1 The standing pulse wave

Figure 3 shows one of the periodic solutions (the standing pulse wave) obtained from the initial condition: $x_1(0) = 2.0$, $y_1(0) = 0.0$ and $x_k(0) = y_k(0) = 0$, $k = 2, 3, \dots ,6$ for $\alpha = 0.05$, $\beta = 3.2$ and $\varepsilon = 0.36$. We choose $\alpha = 0.05$ in order to realize weak coupling so that there exists the standing pulse wave. Since this system has rotational symmetric property (due to ring coupling structure), other 5 periodic solutions obviously coexist for the same parameters. We define Poincare section as $y_1 = 0$[3], and trace these periodic solutions with respect to α (because the characteristic features of the propagating pulse wave mainly depends on coupling strength). Since the periodic solution becomes a fixed point on the Poincare section, this point becomes a curve when α is varied. In this manner, we can trace 6 periodic solutions as depicted in Fig.4 (Kawakami; 1984). The solid curves are mapped points corresponding to the above mentioned stable periodic solutions, namely the nodes (N_i, i = 1, 2, . . . , 6). The dotted curves represent their corresponding saddles ($\bar{S}_i$, i = 1, 2, . . . , 6)[4].
The node and the corresponding saddle coalesce at a certain value of α. This is called the Saddle-Node (SN) bifurcation point α_{SN}. In this system, there are 6 pairs of (N_i, S_i) curves

[3] We take mapped points when the flow penetrates the hyper-plane from + to –.

[4] This saddle is index 2, at least, for $\alpha \geqq 0.05$.

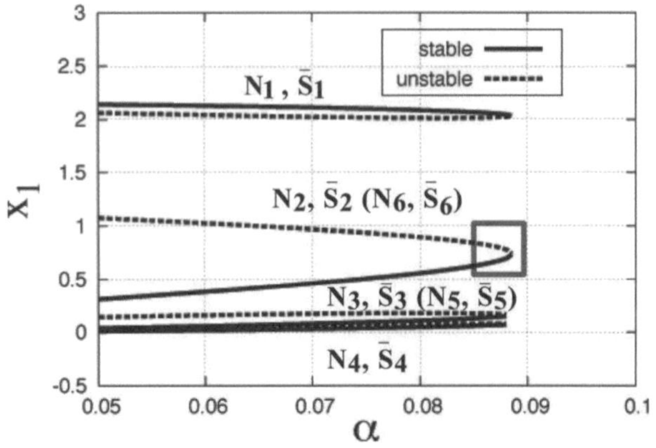

Fig. 4. The node Ni with its corresponding saddle $\bar{S}_i$ for i = 1, 2, ... , 6 showing the SN bifurcation in terms of α for β = 3.2 and ε = 0.36. The curve (N_2, $\bar{S}_2$) overlaps with the curve (N_6, $\bar{S}_6$), and (N_3, $\bar{S}_3$) with (N_5, $\bar{S}_5$), because they are symmetrically-placed. Subtle structure around the tip region surrounded by a square box is shown in Fig.6(a).

and they disappear simultaneously at the same SN bifurcation point. Namely, the aligned structure of the SN bifurcation points is formed at $\alpha_{SN} \doteqdot 0.088$ in this case. This is rough explanation of bifurcation diagram. In fact, the tip region of each curve presents more sophisticated bifurcations. This is explained later.

3.2 Propagating pulse wave based on the global bifurcation of maps

The SN bifurcation set of the periodic solution shown in Fig.3 and the existence regions of several solutions are depicted in Fig.5 in the α - β plane for ε = 0.36. Each solution is confirmed by direct computer simulation of (3). In the left hand side region of the SN bifurcation curve, there exist the 6 periodic solutions (abbr."PS" in Fig.5) and they disappear simultaneously at this SN bifurcation point approximately[5]. After the SN bifurcation, several types of solutions can be observed depending on the value of β. For instance, in the "W" region, the whole oscillation such as all oscillators oscillate with large amplitude can be observed. In the "Z" region, no oscillation exists. In the meantime, two different kinds of propagating pulse waves emerge. One of them ("PW1") is observed in the region filled with gray color. It should be noted that PW1 emerges right after the SN bifurcation for $3.14 \leqq \beta \leqq 3.25$. The other abbreviated as the "PW2" is shown with mesh pattern. The PW2 is different kind of propagating pulse wave from PW1 (A). In the following discussion, we focus on the dynamics of PW1 observed for $3.14 \leqq \beta \leqq 3.25$ and explain the onset mechanism of PW1 solution.

3.2.1 Global bifurcation related to the onset of PW1

For simplicity, we fix ε = 0.36 and β = 3.2. Figure 6 (a) presents the magnified bifurcation diagram in the square region of Fig.4. From this figure, it is noted that before the SN bifurcation a pitchfork bifurcation (PF) occurs. After the PF bifurcation, a stable node

[5] More accurately, the periodic solution disappears at α_{PF} in Fig.6(a). But, α_{PF} and α_{SN} are very close. So, we say the periodic solution disappears at α_{SN} approximately.

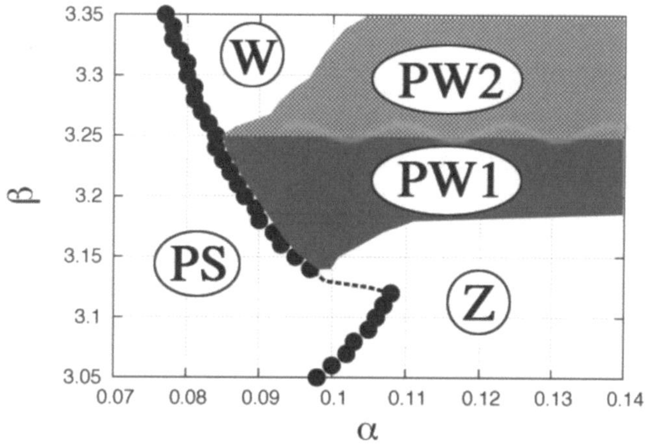

Fig. 5. The SN bifurcation set of the periodic solution in Fig.3 (black dots) and the existence regions of several solutions in the α - β plane for $\varepsilon = 0.36$. The "PS", "W" and "Z" are existence regions of the standing pulse wave, whole oscillation and no oscillation, respectively. In addition, there exist two kinds of propagating pulse waves. They are labeled as "PW1" and" PW2". For $3.12 < \beta < 3.14$ (dotted curve), more complicated bifurcation phenomena including Neimark-Sacker bifurcation is detected, but this region is of no concern with our discussion from now on. In the boundary region, PW1 and PW2 coexist.

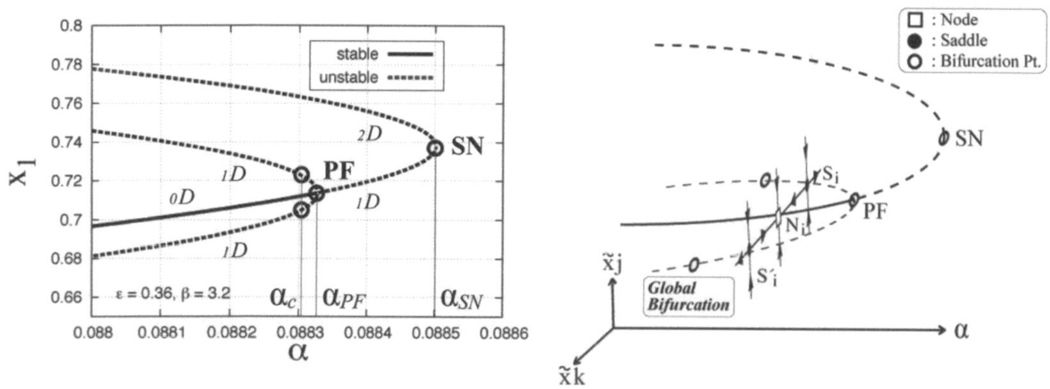

(a) Actual bifurcation diagram. (b) The 3D schematic diagram of (a)

Fig. 6. Subtle bifurcation diagram around the tip region surrounded by a square in Fig.4 for $\varepsilon = 0.36$ and $\beta = 3.2$. The bifurcation points are as follows: $\alpha_{SN} = 0.088501$, $\alpha_{PF} = 0.088328$ and $\alpha_c = 0.088302$. The solid curve denotes stable and the dotted curve denotes unstable fixed point. (a) Actual bifurcation diagram. The notation ${}_mD$ indicates that number of unstable direction of the fixed point is "m". (b) 3D schematic diagram around the tip region. The axis $\tilde{x}_j$ denotes the stable direction of saddles and the axis $\tilde{x}_k$ unstable direction of them. The axes $\tilde{x}_j$ and $\tilde{x}_k$ may not correspond to the actual state variables x_j and x_k directly.

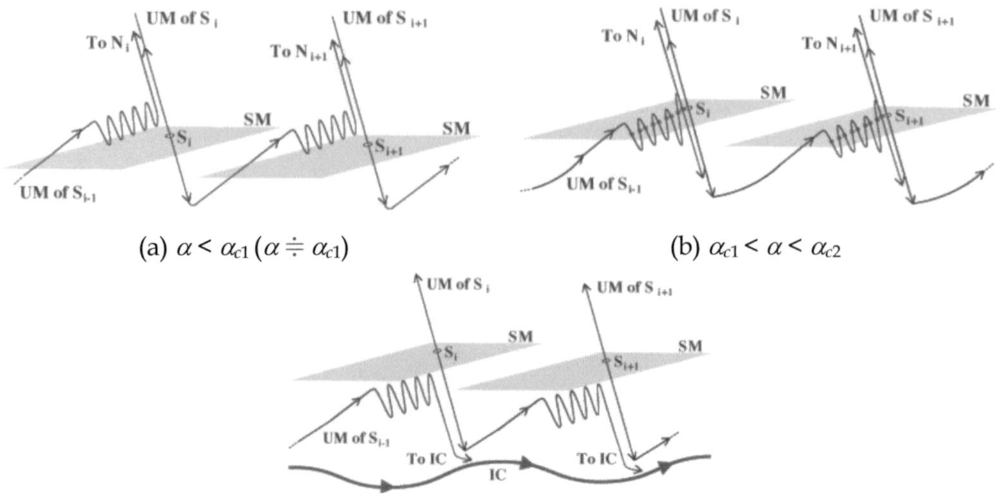

(a) $\alpha < \alpha_{c1}\,(\alpha \doteqdot \alpha_{c1})$ (b) $\alpha_{c1} < \alpha < \alpha_{c2}$

(c) $\alpha > \alpha_{c2}(\alpha \doteqdot \alpha_{c2})$

Fig. 7. A schematic diagram of the behavior of UM and SM around the bifurcation point. The stable direction of S_i (SM) denotes 10 dimensional subspace and the unstable direction of S_i denotes 1 dimensional curve.

becomes a saddle of index 1 and the corresponding saddle is an index 2 saddle. On the other hand, at the PF point two saddles appear in the backward direction as shown in Fig.6(a); namely, this is the subcritical PF bifurcation. Figure 6 (b) presents the 3D schematic diagram for better understanding the behavior of stable and unstable manifold of saddles in Fig.7. Here the stable direction denotes 10 dimensional subspace and the unstable direction denotes 1 dimensional curve[6].

Due to Kuznetsov (Kuznetsov; 1995), there exist many heteroclinic points formed by UM of S_i, S_i' (one dimensional in this case) and stable manifold (SM) of S_{i+1}, S_{i+1}' (ten dimensional in this case) around the bifurcation point for $\alpha_{c1} < \alpha < \alpha_{c2}$. Figures 7 (a), (b) and (c) present the schematic diagrams representing the connection of S_i,N_i, i = 1, 2, … ,6 for $\alpha < \alpha_{c1}$, $\alpha_{c1} < \alpha < \alpha_{c2}$, and $\alpha_{c2} < \alpha$. The UM of S_i waves severely as it approaches the SM of S_{i+1} for all cases. For $\alpha < \alpha_{c1}$, as seen in Fig.7 (a), UM of S_i finally converges to N_{i+1}. The UM and SM do not intersect. Therefore, a stable node corresponding to the periodic solution appears. For α much smaller than α_{c1}, the UM emanating from S_i goes to N_{i+1} directly without approaching near S_{i+1}. For $\alpha_{c2} < \alpha$, as seen in Fig.7 (c), UM of S_i eventually converges to invariant curve (= IC). The IC is the *propagating pulse wave just after birth* going left or right as shown in Fig.2. The UM also has no intersection with the SM. For α much larger than α_{c2}, the UM emanating from S_i goes to the IC without undulating near S_{i+1}. For $\alpha_{c1} < \alpha < \alpha_{c2}$, as seen in Fig.7 (b), UM of S_i goes two directions. The UM of S_i and SM of S_{i+1} intersect transversally. At $\alpha = \alpha_{c1}, \alpha_{c2}$, the heteroclinic tangency occurs; namely, UM is tangent to SM.

6 For example, for $\varepsilon = 0.36$, $\beta = 3.2$, $\alpha = 0.0883$, the eigenvalues of the saddles S_i and S_i' are as follows: 1.010, 0.925, 0.547, 0.402 ± 0.073i, 0.398, 0.345 ± 0.107i, 0.319 ± 0.164i and 0.142.

To clarify the above, we perform computer simulation in Fig.8 (a), (b) and (c)[7]. The UM can be presented in two dimensional phase space, because it is one dimensional. However, the SM cannot be presented in two dimensional phase space, because it is ten dimensional. Fortunately, the dominant eigenvalue of SM is 0.925 only and other nine eigenvalues are much smaller than unity as previously mentioned in the footnote. Therefore, we draw the stable eigenvector associated with this eigenvalue by the thick line for reference. Figure 8 (a), (b) and (c) demonstrate 3 cases. The waving thin curve presents the UM of S5. Since we cannot show whole ten dimensional stable manifold, we cannot obtain the critical values α_{c1}, α_{c2} from the figures. Therefore, we estimate whether or not the given α is within $\alpha_{c1} < \alpha < \alpha_{c2}$, by knowing the destination of UM. Namely, if it converges to N6, the given α is smaller than α_{c1}, and if it converges to IC, the given α is greater than α_{c2}, and if it is separated in two directions, the given α is in between α_{c1} and α_{c2}. In Fig.8 (a) the UM goes to N6. Therefore, in this case $\alpha < \alpha_{c1}$. In Fig.8 (b) UM of S_5 is separated in two directions along the UM of S_6, therefore, α is set between $\alpha_{c1} < \alpha < \alpha_{c2}$. In Fig.8 (c) UM goes to IC, therefore, α is set for $\alpha_{c2} < \alpha$.

In the same manner, the connection between S_i' and N_i, $i = 1, 2, \ldots ,6$ can be explained in the following. For $\alpha < \alpha_{c1}$, the UM of S_i' goes to N_{i-1}. For $\alpha_{c1} < \alpha < \alpha_{c2}$, the UM of S_i' and SM of S_{i-1}' intersect transversally. For $\alpha_{c2} < \alpha$, the UM of S_i' eventually converges to the IC corresponding to the propagating pulse wave just after birth. Therefore, the propagating direction depends on the initial condition.

Since α_{c1} and α_{c2} are very close, we recognize them as the same number: $\alpha_c = \alpha_{c1} = \alpha_{c2} = 0.088302$ for convenience. In addition, α_c (=0.088302) and α_{PF} (=0.088328) are close to α_{SN} (= 0.088501), therefore we say that the propagating pulse wave occurs at the SN bifurcation point approximately. For other values of β in $3.14 \leqq \beta \leqq 3.25$, we confirmed the same bifurcation structure ensuring generation of IC. For $\beta \geqq 3.26$ and $\alpha > \alpha_{PF}$ ($\fallingdotseq \alpha_{SN}$), one of the UM of each saddle goes to the stable node representing the whole oscillation (W in Fig.5). For $\beta \leqq 3.14$ and $\alpha > \alpha_{PF}$ ($\fallingdotseq \alpha_{SN}$), it goes to the stable node representing the zero solution (Z in Fig.5).

Practically, for $3.14 \leqq \beta \leqq 3.25$, if the initial condition is set on the periodic solution (standing wave solution) and increase α, the periodic solution persists up to $\alpha = \alpha_{PF}$ and for $\alpha > \alpha_{PF}$, the periodic solution jumps to the propagating wave solution (IC). On the contrary, if the initial condition is set on IC for $\alpha > \alpha_{PF}$ and then decrease α, the IC disappears at $\alpha = \alpha_c$ to become the periodic solution (standing wave solution). Namely, a hysteresis phenomenon between the standing wave solution and the propagating wave solution can be seen in $\alpha_c \leqq \alpha \leqq \alpha_{PF}$.

4. Conclusions

In this chapter, we revealed existence of the propagating wave in the ring of coupled bistable oscillator system. Then, we give one of the onset mechanisms of the propagating

---

[7] The shape of UM is obtained, together with the compensation algorithm, by repeating the mapping of which initial value is chosen on the unstable eigenvector(Parker & Chua; 1989). The method to obtain initial point on the unstable eigenvector is referred to (57) in (Katsuta & Kawakami; 1993). Numerical integrations for the compensation algorithm are carried out with 4th order Runge-Kutta method with step size = 0.001.

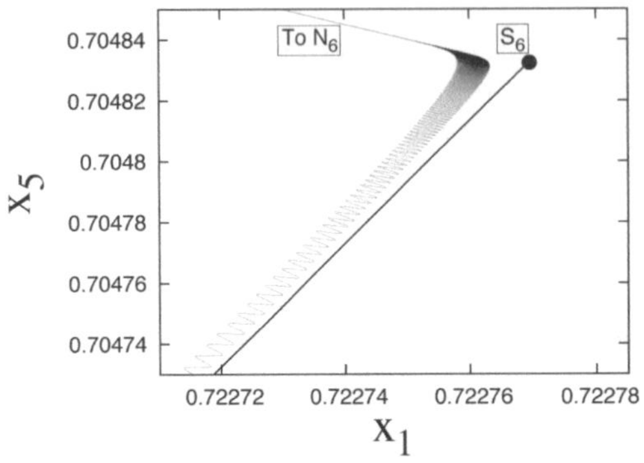

(a) $\alpha = 0.088302850$

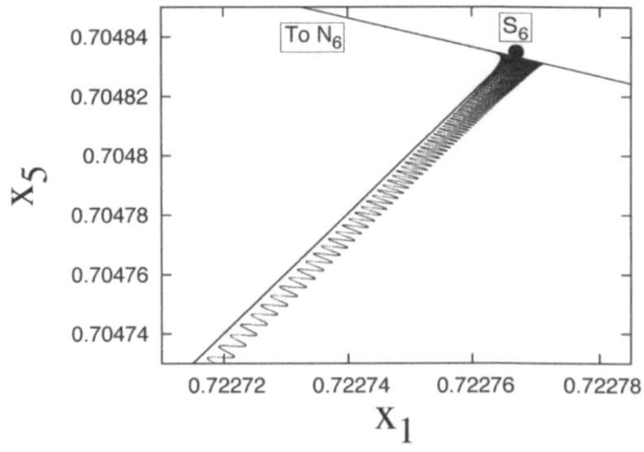

(b) $\alpha = 0.088302864$

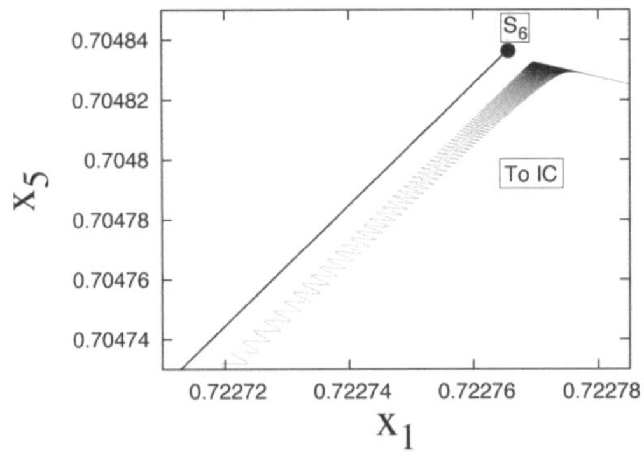

(c) $\alpha = 0.088302872$

Fig. 8. Three situation of UM of S_5 (thin curve) and a stable eigenvector of S_6 (thick line) corresponding to the largest stable eigenvalue.

pulse wave for the ring of 6 coupled bistable oscillators by exploring global phase-space structure around the bifurcation point. The transition from the standing pulse wave to the propagating pulse wave is due to the heteroclinic tangle shown in Fig.7. In our forthcoming paper, we will investigate these subjects more in detail.

5. References

Y. Nishiura, D. Ueyama and T. Yanagita, "Chaotic pulses for discrete reaction diffusion systems", SIAM J. Applied Dynamical Systems, vol.4, no.3, pp.723-754, 2005.

J.C. Comte, S. Morfu and P.Marqui´e, "Propagation failure in discrete bistable reactiondiffusion systems: Theory and experiments", Phys. Rev. E, vol.64, 027102, 2001.

K. Kaneko, "Chaotic traveling waves in a coupled map lattice", Phyisica D, vol.68, pp.299-317, 1993.

T.Hikihara, K.Torii and Y.Ueda, "Wave and basin structure in spatially coupled magnetoelastic beam system -Transitions between coexisting wave solutions", Int. Journal of Bifurcation and Chaos, vol.11, no.4, pp.999-1018, 2001.

M. Yamauchi, M. Wada, Y. Nishio and A. Ushida, "Wave propagation phenomena of phase states in oscillators coupled by inductors as a ladder", IEICE Trans.Fundamentals, vol.E82-A, no.11, pp.2592-2598, 1999.
M.Sato, B.E.Hubbard, A.J.Sievers, B.Ilic, D.A.Czaplewski and H.G.Craighead, "Observation of locked intrinsic localized vibrational modes in a micromechanical oscillator array", Phys.Rev.Lett., vol.90, no.4 (2003) 044102.
J.P. Keener, "Propagation and its failure in coupled systems of discrete excitable cells", SIAM J. Appl. Biol., vol.47, no.3, pp.556-572, 1987.
K.Shimizu, T. Endo and D.Ueyama, "Pulse Wave Propagation in a Large Number of Coupled Bistable Oscillators", IEICE Trans. Fundamentals, vol.E91-A, no.9, pp.2540-2545, 2008.
T. Endo and T. Ohta, "Multimode oscillations in a coupled oscillator system with fifthpower nonlinear characteristics", IEEE Trans. on circuit and systems, vol.cas-27, no.4, pp.277-283, 1980.
T. Yoshinaga and H. Kawakami, "Synchronized quasi-periodic oscillations in a ring of coupled oscillators with hard characteristics", Electronics and Communications in Japan, Part III, vol.76, no.5, pp.110-120, 1993.
H. Kawakami, "Bifurcation of periodic responses in forced dynamic nonlinear circuits: computation of bifurcation values of the system parameters", IEEE Trans. Circuits Syst., vol.CAS-31, no.3, pp.248-260, 1984.
T.S.Parker and L.O.Chua, Practical numerical algorithms for chaotic systems, Springer-Verlag, New York, 1989.
Y. Katsuta, H. Kawakami, "Bifurcations of equilibriums and periodic solutions in nonlinear autonomous system with symmetry", Electronics and Communications in Japan, Part III, vol.76, no.7, pp.1-14, 1993.
Y.A. Kuznetsov, "Elements of applied bifurcation theory", Springer-Verlag, New York, p.466, 1995.

A. The propagating pulse wave PW2

The PW2 is a certain kind of propagating pulse wave. The mapped points of PW1 and PW2 projected onto the (x_1, x_3, x_5) phase space are shown in Fig.9(a) and (b), respectively. Comparing both cases, each flow on the phase space moves along a different orbit. In addition, for PW1 the mapped points stay for a long time on several points (which correspond to the locus of the nodes N_i, $i = 1, 2, \ldots, 6$.). This is one of characteristic feature of PW1 originating in the heteroclinic tangle. On the other hand, for PW2 the mapped points no longer stay the locus for a long time. Therefore, we distinguish PW2 from PW1. The existence region of such solution is shown in Fig.5. It should be noted that the starting point of PW2 is no longer close to the existence region of PS. That is, between them the existence region of W is sandwiched. For example, for β = 3.26 and ε = 0.36, PS disappears via PF bifurcation at $\alpha_{PF} \fallingdotseq 0.083$. In contrast, PW2 begins to exist for $\alpha \geqq 0.087$. Namely, there exists a gap between them. Probably, it originates in the standing wave where two adjacent oscillators are oscillating and where other oscillators are not. This is confirmed by continuously changing the parameter β of Fig.9. Further research will be necessary to clarify the generation mechanism of PW2.

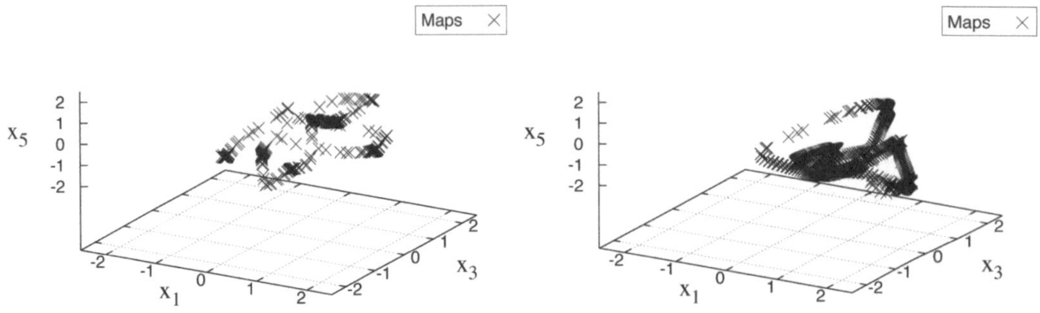

(a) Mapped points of PW1 (b) Mapped points of PW2

Fig. 9. Mapped points of PW1 and PW2 projected onto the (x_1, x_3, x_5) phase space. (a) PW1 (α = 0.089, β = 3.25 and ε = 0.36). The initial condition is given as x_1 = 2.0, y_2 = 1.3 and all other variables are zero. (b) PW2 (α = 0.089, β = 3.26 and ε = 0.36). The initial condition is given as x_1 = 1.7, x_2 = -2.2, x_3 = 0.9, x_4 = 0.2, x_5 = 0.1, x_6 = 0.5, y_1 = 1.8, y_2 = 0.4, y_3 = -2.3, y_4 = y_5 = 0.3 and y_6 = -0.3.

26

Circuit Analogs for Wave Propagation in Stratified Structures

Daniel Sjöberg
Lund University
Sweden

1. Introduction

To simulate or design a reasonably large system, fast and simple models are necessary. To verify the design versus the specifications, more detailed (and costly) calculations can be performed and final adjustments made. In wave propagation problems, circuit analogs provide a powerful, yet simple, means of computing the desired response of the system, such as reflection or transmission coefficients. The reason circuit analog models are good for wave propagation problems, is that they are exact for one-dimensional wave propagation, regardless of whether we consider acoustic or electromagnetic waves.

Typically, wave propagation through homogeneous media is modeled as a transmission line with propagation constant β and characteristic impedance Z, whereas obstacles such as thin sheets are modeled as lumped elements. If the sheets are lossless, the circuit models contain only reactive elements such as capacitors and inductors.

Modeling complex wave propagation problems with circuit analogs was to a large extent developed in conjunction with the development of radar technology during the Second World War. Many of the results from this very productive era are collected in the Radiation Laboratory Series and related literature, in particular (Collin, 1991; 1992; Marcuvitz, 1951; Schwinger & Saxon, 1968). Further development has been provided by research on frequency selective structures (Munk, 2000; 2003). In recent years, the circuit analogs have even been used in an inverse fashion: by observing that wave propagation through a material with negative refractive index could be modeled as a transmission line with distributed series capacitance and shunt inductance, *i.e.*, the dual of the standard transmission line, the most successful realization of negative refractive index material is actually made by synthesizing this kind of transmission line using lumped elements (Caloz & Itoh, 2004; Eleftheriades et al., 2002).

This chapter is organized as follows. In Section 2 we show that propagation of electromagnetic waves in any material, regardless how complicated, boils down to an eigenvalue problem which can be solved analytically for isotropic media, and numerically for arbitrary media. From this eigenvalue problem, the propagation constant and characteristic impedance can be derived, which generates a transmission line model. In Section 3, we show how sheets with or without periodic patterns can be modeled as lumped elements connected by transmission lines representing propagation in the surrounding medium. The lumped elements can be given a firm definition and physical interpretation in the low frequency limit, and in Section 4 we show how these low frequency properties provide some useful physical limitations on scattering characteristics. The calculation of circuit analogs in the general case using an optimization approach is treated in Section 5, and examples of the use of circuit analogs in design problems are given in Section 6. Finally, conclusions are given in Section 7.

2. Wave propagation in stratified structures

In this section, we show that the description of plane waves propagating through any homogeneous material at any angle of incidence, reduces to a simple eigenvalue problem from which we can compute the propagation constant and transverse wave impedance.

We consider a geometry where the material parameters are constants as functions of x and y, but may depend on z, which is considered as the main propagation direction. This corresponds to a laminated structure, z being the lamination direction. Our strategy is to eliminate the x and y dependence through a spatial Fourier transform, and then eliminate the field components along the z direction. This is motivated by the fact that the remaining field components, $\boldsymbol{E}_{\mathrm{t}} = E_x \hat{\boldsymbol{x}} + E_y \hat{\boldsymbol{y}}$ and $\boldsymbol{H}_{\mathrm{t}} = H_x \hat{\boldsymbol{x}} + H_y \hat{\boldsymbol{y}}$, are continuous across interfaces, and are thus easily matched at boundaries. The resulting equation (24) (or (25) for isotropic media) can be formulated as an algebraic eigenvalue problem by looking for solutions where the only z dependence is through a propagation factor $e^{-j\beta z}$. The wave number β corresponds to the eigenvalue, and the wave impedance is given by the eigenvectors.

2.1 Notation

We consider time harmonic waves using time convention $e^{j\omega t}$. The material is described through the mapping from the fields [$\boldsymbol{E}$,$\boldsymbol{H}$] to the fields [$\boldsymbol{D}$,$\boldsymbol{B}$]:

$$\begin{cases} \boldsymbol{D} &= \boldsymbol{\epsilon} \cdot \boldsymbol{E} + \boldsymbol{\xi} \cdot \boldsymbol{H} \\ \boldsymbol{B} &= \boldsymbol{\zeta} \cdot \boldsymbol{E} + \boldsymbol{\mu} \cdot \boldsymbol{H} \end{cases} \tag{1}$$

where the dyadics $\boldsymbol{\epsilon}$, $\boldsymbol{\xi}$, $\boldsymbol{\zeta}$, and $\boldsymbol{\mu}$ can be represented by 3×3 matrices. Other mappings for the material are possible, for instance from the fields [$\boldsymbol{E}$,$\boldsymbol{B}$] to [$\boldsymbol{D}$,$\boldsymbol{H}$]. In vacuum the relations are $\boldsymbol{D} = \epsilon_0 \boldsymbol{E}$ and $\boldsymbol{B} = \mu_0 \boldsymbol{H}$, where the permittivity and permeability of vacuum are denoted by $\epsilon_0 = 8.854 \cdot 10^{-12}$ F/m and $\mu_0 = 4\pi \cdot 10^{-7}$H/m, respectively. Materials are often classified according to the various symmetries of the material dyadics as in Table 1.

When choosing a particular direction z, it is natural to introduce a decomposition as (where the index t represents the x and y components)

$$\boldsymbol{E} = \boldsymbol{E}_{\mathrm{t}} + \hat{\boldsymbol{z}} E_z, \quad \boldsymbol{\epsilon} = \boldsymbol{\epsilon}_{\mathrm{tt}} + \hat{\boldsymbol{z}} \boldsymbol{\epsilon}_z + \boldsymbol{\epsilon}_{\mathrm{t}} \hat{\boldsymbol{z}} + \epsilon_{zz} \hat{\boldsymbol{z}} \hat{\boldsymbol{z}} \tag{2}$$

so that the transverse components of the $\boldsymbol{D}$ and $\boldsymbol{B}$ fields are (vector equations)

Type	$\boldsymbol{\epsilon}$, $\boldsymbol{\mu}$,	$\boldsymbol{\xi}$, $\boldsymbol{\zeta}$
Isotropic	All ~**1**	Both 0
An-isotropic	Some not ~**1**	Both 0
Bi-isotropic	All ~**1**	Both ~**1**
Bi-an-isotropic	All other cases	

Table 1. Classification of electromagnetic materials (**1** denotes the unit dyadic).

$$\boldsymbol{D}_{\mathrm{t}} = \boldsymbol{\epsilon}_{\mathrm{tt}} \cdot \boldsymbol{E}_{\mathrm{t}} + \boldsymbol{\epsilon}_{\mathrm{t}} E_z + \boldsymbol{\xi}_{\mathrm{tt}} \cdot \boldsymbol{H}_{\mathrm{t}} + \boldsymbol{\xi}_{\mathrm{t}} H_z \tag{3}$$

$$\boldsymbol{B}_{\mathrm{t}} = \boldsymbol{\zeta}_{\mathrm{tt}} \cdot \boldsymbol{E}_{\mathrm{t}} + \boldsymbol{\zeta}_{\mathrm{t}} E_z + \boldsymbol{\mu}_{\mathrm{tt}} \cdot \boldsymbol{H}_{\mathrm{t}} + \boldsymbol{\mu}_{\mathrm{t}} H_z \tag{4}$$

and the z components are (scalar equations)

$$D_z = \boldsymbol{\epsilon}_z \cdot \boldsymbol{E}_{\mathrm{t}} + \epsilon_{zz} E_z + \boldsymbol{\xi}_z \cdot \boldsymbol{H}_{\mathrm{t}} + \xi_{zz} H_z \tag{5}$$

$$B_z = \boldsymbol{\zeta}_z \cdot \boldsymbol{E}_{\mathrm{t}} + \zeta_{zz} E_z + \boldsymbol{\mu}_z \cdot \boldsymbol{H}_{\mathrm{t}} + \mu_{zz} H_z \tag{6}$$

Since the material parameters are assumed independent of x and y, it makes sense to represent the fields through a Fourier transform in the transverse variables x and y as

$$\boldsymbol{E}(\boldsymbol{r}) = \frac{1}{(2\pi)^2} \int\!\!\int_{-\infty}^{\infty} \boldsymbol{E}(z,\boldsymbol{k}_{\mathrm{t}}) e^{-\mathrm{j}(k_x x + k_y y)} dk_x dk_y \tag{7}$$

$$\boldsymbol{E}(z,\boldsymbol{k}_{\mathrm{t}}) = \int\!\!\int_{-\infty}^{\infty} \boldsymbol{E}(\boldsymbol{r}) e^{\mathrm{j}(k_x x + k_y y)} dx dy \tag{8}$$

where the transverse wave vector is $\boldsymbol{k}_{\mathrm{t}} = k_x \hat{\boldsymbol{x}} + k_y \hat{\boldsymbol{y}}$. The action of the curl operator on the Fourier amplitude is shown by

$$\begin{aligned}\nabla \times \boldsymbol{E}(\boldsymbol{r}) &= \frac{1}{(2\pi)^2} \int\!\!\int_{-\infty}^{\infty} \nabla \times [\boldsymbol{E}(z,\boldsymbol{k}_{\mathrm{t}}) e^{-\mathrm{j}(k_x x + k_y y)}] dk_x dk_y \\ &= \frac{1}{(2\pi)^2} \int\!\!\int_{-\infty}^{\infty} e^{-\mathrm{j}(k_x x + k_y y)} \left(-\mathrm{j}\boldsymbol{k}_{\mathrm{t}} + \hat{\boldsymbol{z}} \frac{\partial}{\partial z} \right) \times \boldsymbol{E}(z,\boldsymbol{k}_{\mathrm{t}}) dk_x dk_y \end{aligned} \tag{9}$$

We then obtain the decomposition

$$\begin{aligned}\left(-\mathrm{j}\boldsymbol{k}_{\mathrm{t}} + \hat{\boldsymbol{z}} \frac{\partial}{\partial z} \right) \times \boldsymbol{E}(z,\boldsymbol{k}_{\mathrm{t}}) &= -\mathrm{j}\boldsymbol{k}_{\mathrm{t}} \times \boldsymbol{E}(z,\boldsymbol{k}_{\mathrm{t}}) + \frac{\partial}{\partial z} \hat{\boldsymbol{z}} \times \boldsymbol{E}(z,\boldsymbol{k}_{\mathrm{t}}) \\ &= \underbrace{-\mathrm{j}\boldsymbol{k}_{\mathrm{t}} \times \boldsymbol{E}_{\mathrm{t}}(z,\boldsymbol{k}_{\mathrm{t}})}_{\text{parallel to } \hat{z}} \underbrace{- \mathrm{j}\boldsymbol{k}_{\mathrm{t}} \times \hat{\boldsymbol{z}} E_z(z,\boldsymbol{k}_{\mathrm{t}})}_{\text{orthogonal to } \hat{z}} + \underbrace{\frac{\partial}{\partial z} \hat{\boldsymbol{z}} \times \boldsymbol{E}_{\mathrm{t}}(z,\boldsymbol{k}_{\mathrm{t}})}_{\text{orthogonal to } \hat{z}} \end{aligned} \tag{10}$$

The result for the curl of the magnetic field is exactly the same.

2.2 Application to Maxwell's equations

We now apply the above decompositions with respect to z to Maxwell's equations. These are

$$\nabla \times \boldsymbol{H}(\boldsymbol{r}) = \mathrm{j}\omega \boldsymbol{D}(\boldsymbol{r}) \tag{11}$$

$$\nabla \times \boldsymbol{E}(\boldsymbol{r}) = -\mathrm{j}\omega \boldsymbol{B}(\boldsymbol{r}) \tag{12}$$

When considering the Fourier amplitudes of the electromagnetic fields and using the constitutive relations this turns into (in the following we suppress the arguments z and $\boldsymbol{k}_{\mathrm{t}}$ of the fields for brevity)

$$\left(-\mathrm{j}\boldsymbol{k}_{\mathrm{t}}+\hat{z}\frac{\partial}{\partial z}\right)\times\boldsymbol{H}=\mathrm{j}\omega\left(\epsilon\cdot\boldsymbol{E}+\xi\cdot\boldsymbol{H}\right) \tag{13}$$

$$\left(-\mathrm{j}\boldsymbol{k}_{\mathrm{t}}+\hat{z}\frac{\partial}{\partial z}\right)\times\boldsymbol{E}=-\mathrm{j}\omega\left(\zeta\cdot\boldsymbol{E}+\boldsymbol{\mu}\cdot\boldsymbol{H}\right) \tag{14}$$

Another way to write this is by using dyadics (identifying (13) as the first row and (14) as the second row, and writing **1** for the unit dyadic)

$$\frac{\partial}{\partial z}\begin{pmatrix}0 & -\hat{z}\times\mathbf{1}\\ \hat{z}\times\mathbf{1} & 0\end{pmatrix}\cdot\begin{pmatrix}\boldsymbol{E}\\ \boldsymbol{H}\end{pmatrix}=\begin{pmatrix}0 & -\mathrm{j}\boldsymbol{k}_{\mathrm{t}}\times\mathbf{1}\\ \mathrm{j}\boldsymbol{k}_{\mathrm{t}}\times\mathbf{1} & 0\end{pmatrix}\cdot\begin{pmatrix}\boldsymbol{E}\\ \boldsymbol{H}\end{pmatrix}-\mathrm{j}\omega\begin{pmatrix}\epsilon & \xi\\ \zeta & \mu\end{pmatrix}\cdot\begin{pmatrix}\boldsymbol{E}\\ \boldsymbol{H}\end{pmatrix} \tag{15}$$

The left hand side is orthogonal to $\hat{z}$, and the equations for the z components are then (using that the cross product $\boldsymbol{k}_t\times\boldsymbol{E}_t$ is necessarily in the z direction since both vectors are in the xy-plane, with the scalar value $\hat{z}\cdot(\boldsymbol{k}_t\times\boldsymbol{E}_t)=(\hat{z}\times\boldsymbol{k}_t)\cdot\boldsymbol{E}_t$)

$$\begin{pmatrix}0\\0\end{pmatrix}=\begin{pmatrix}0 & -\mathrm{j}\hat{z}\times\boldsymbol{k}_{\mathrm{t}}\\ \mathrm{j}\hat{z}\times\boldsymbol{k}_{\mathrm{t}} & 0\end{pmatrix}\cdot\begin{pmatrix}\boldsymbol{E}_{\mathrm{t}}\\ \boldsymbol{H}_{\mathrm{t}}\end{pmatrix}-\mathrm{j}\omega\begin{pmatrix}\epsilon_{\mathrm{z}} & \xi_{\mathrm{z}}\\ \zeta_{\mathrm{z}} & \mu_{\mathrm{z}}\end{pmatrix}\cdot\begin{pmatrix}\boldsymbol{E}_{\mathrm{t}}\\ \boldsymbol{H}_{\mathrm{t}}\end{pmatrix}-\mathrm{j}\omega\begin{pmatrix}\epsilon_{zz} & \xi_{zz}\\ \zeta_{zz} & \mu_{zz}\end{pmatrix}\begin{pmatrix}E_{z}\\ H_{z}\end{pmatrix} \tag{16}$$

from which we solve for the z components of the fields:

$$\begin{pmatrix}E_{z}\\ H_{z}\end{pmatrix}=\begin{pmatrix}\epsilon_{zz} & \xi_{zz}\\ \zeta_{zz} & \mu_{zz}\end{pmatrix}^{-1}\left[\begin{pmatrix}0 & -\omega^{-1}\hat{z}\times\boldsymbol{k}_{\mathrm{t}}\\ \omega^{-1}\hat{z}\times\boldsymbol{k}_{\mathrm{t}} & 0\end{pmatrix}-\begin{pmatrix}\epsilon_{\mathrm{z}} & \xi_{\mathrm{z}}\\ \zeta_{\mathrm{z}} & \mu_{\mathrm{z}}\end{pmatrix}\right]\cdot\begin{pmatrix}\boldsymbol{E}_{\mathrm{t}}\\ \boldsymbol{H}_{\mathrm{t}}\end{pmatrix} \tag{17}$$

The transverse part of (15) is

$$\begin{aligned}\frac{\partial}{\partial z}\begin{pmatrix}0 & -\hat{z}\times\mathbf{1}\\ \hat{z}\times\mathbf{1} & 0\end{pmatrix}\cdot\begin{pmatrix}\boldsymbol{E}_{\mathrm{t}}\\ \boldsymbol{H}_{\mathrm{t}}\end{pmatrix}&=\begin{pmatrix}0 & -\mathrm{j}\boldsymbol{k}_{\mathrm{t}}\times\hat{z}\\ \mathrm{j}\boldsymbol{k}_{\mathrm{t}}\times\hat{z} & 0\end{pmatrix}\begin{pmatrix}E_{z}\\ H_{z}\end{pmatrix}\\ &\quad-\mathrm{j}\omega\begin{pmatrix}\epsilon_{\mathrm{tt}} & \xi_{\mathrm{tt}}\\ \zeta_{\mathrm{tt}} & \mu_{\mathrm{tt}}\end{pmatrix}\cdot\begin{pmatrix}\boldsymbol{E}_{\mathrm{t}}\\ \boldsymbol{H}_{\mathrm{t}}\end{pmatrix}-\mathrm{j}\omega\begin{pmatrix}\epsilon_{\mathrm{t}} & \xi_{\mathrm{t}}\\ \zeta_{\mathrm{t}} & \mu_{\mathrm{t}}\end{pmatrix}\begin{pmatrix}E_{z}\\ H_{z}\end{pmatrix}\\ &=-\mathrm{j}\omega\begin{pmatrix}\epsilon_{\mathrm{tt}} & \xi_{\mathrm{tt}}\\ \zeta_{\mathrm{tt}} & \mu_{\mathrm{tt}}\end{pmatrix}\cdot\begin{pmatrix}\boldsymbol{E}_{\mathrm{t}}\\ \boldsymbol{H}_{\mathrm{t}}\end{pmatrix}+\mathrm{j}\omega\left[\begin{pmatrix}0 & -\omega^{-1}\boldsymbol{k}_{\mathrm{t}}\times\hat{z}\\ \omega^{-1}\boldsymbol{k}_{\mathrm{t}}\times\hat{z} & 0\end{pmatrix}-\begin{pmatrix}\epsilon_{\mathrm{t}} & \xi_{\mathrm{t}}\\ \zeta_{\mathrm{t}} & \mu_{\mathrm{tt}}\end{pmatrix}\right]\begin{pmatrix}E_{z}\\ H_{z}\end{pmatrix}\end{aligned} \tag{18}$$

Inserting the expressions for the z components of the fields implies

$$\frac{\partial}{\partial z}\begin{pmatrix}0 & -\hat{z}\times\mathbf{1}\\ \hat{z}\times\mathbf{1} & 0\end{pmatrix}\cdot\begin{pmatrix}\boldsymbol{E}_{\mathrm{t}}\\ \boldsymbol{H}_{\mathrm{t}}\end{pmatrix}=-\mathrm{j}\omega\begin{pmatrix}\epsilon_{\mathrm{tt}} & \xi_{\mathrm{tt}}\\ \zeta_{\mathrm{tt}} & \mu_{\mathrm{tt}}\end{pmatrix}\cdot\begin{pmatrix}\boldsymbol{E}_{\mathrm{t}}\\ \boldsymbol{H}_{\mathrm{t}}\end{pmatrix}+\mathrm{j}\omega\mathbf{A}\cdot\begin{pmatrix}\boldsymbol{E}_{\mathrm{t}}\\ \boldsymbol{H}_{\mathrm{t}}\end{pmatrix} \tag{19}$$

where **A** is the dyadic product[1]

[1] A dyadic product between two vectors $\boldsymbol{ab}$ is defined by its action on an arbitrary vector $\boldsymbol{c}$ as $(\boldsymbol{ab})\cdot\boldsymbol{c}=\boldsymbol{a}(\boldsymbol{b}\cdot\boldsymbol{c})$, *i.e.*, a vector parallel to $\boldsymbol{a}$ with amplitude $|\boldsymbol{a}|\,|\boldsymbol{b}\cdot\boldsymbol{c}|$. Thus, dyadic multiplication does not commute unless $\boldsymbol{a}$ is parallel to $\boldsymbol{b}$.

$$\mathbf{A}=\left[\begin{pmatrix}\mathbf{0} & -\omega^{-1}\boldsymbol{k}_{\mathrm{t}}\times\hat{z}\\ \omega^{-1}\boldsymbol{k}_{\mathrm{t}}\times\hat{z} & \mathbf{0}\end{pmatrix}-\begin{pmatrix}\boldsymbol{\epsilon}_{\mathrm{t}} & \boldsymbol{\xi}_{\mathrm{t}}\\ \boldsymbol{\zeta}_{\mathrm{t}} & \boldsymbol{\mu}_{\mathrm{t}}\end{pmatrix}\right]\begin{pmatrix}\epsilon_{zz} & \xi_{zz}\\ \zeta_{zz} & \mu_{zz}\end{pmatrix}^{-1}\left[\begin{pmatrix}\mathbf{0} & -\omega^{-1}\hat{z}\times\boldsymbol{k}_{\mathrm{t}}\\ \omega^{-1}\hat{z}\times\boldsymbol{k}_{\mathrm{t}} & \mathbf{0}\end{pmatrix}-\begin{pmatrix}\boldsymbol{\epsilon}_{z} & \boldsymbol{\xi}_{z}\\ \boldsymbol{\zeta}_{z} & \boldsymbol{\mu}_{z}\end{pmatrix}\right] \tag{20}$$

For an isotropic material, where $\boldsymbol{\epsilon}=\epsilon\mathbf{1}$, $\boldsymbol{\mu}=\mu\mathbf{1}$, $\boldsymbol{\xi}=\boldsymbol{\zeta}=\mathbf{0}$, this is (writing $\boldsymbol{a}=k_0^{-1}\,\boldsymbol{k}_{\mathrm{t}}\times\hat{z}=-k_0^{-1}\,\hat{z}\times\boldsymbol{k}_{\mathrm{t}}$ where $k_0=\omega/c_0$ is the wave number in vacuum)

$$\mathbf{A}=\begin{pmatrix}0 & -c_0^{-1}\boldsymbol{a}\\ c_0^{-1}\boldsymbol{a} & 0\end{pmatrix}\begin{pmatrix}\epsilon^{-1} & 0\\ 0 & \mu^{-1}\end{pmatrix}\begin{pmatrix}0 & c_0^{-1}\boldsymbol{a}\\ -c_0^{-1}\boldsymbol{a} & 0\end{pmatrix}=\frac{1}{c_0^{2}}\begin{pmatrix}\mu^{-1}\boldsymbol{aa} & 0\\ 0 & \epsilon^{-1}\boldsymbol{aa}\end{pmatrix} \tag{21}$$

Since $-\hat{z}\times(\hat{z}\times\boldsymbol{E}_{\mathrm{t}})=\boldsymbol{E}_{\mathrm{t}}$ for all transverse fields $\boldsymbol{E}_{\mathrm{t}}$, we can write (19) as

$$\frac{\partial}{\partial z}\begin{pmatrix}\boldsymbol{E}_{\mathrm{t}}\\ -\hat{z}\times\boldsymbol{H}_{\mathrm{t}}\end{pmatrix}=\begin{pmatrix}\mathbf{0} & -\hat{z}\times\mathbf{1}\\ \mathbf{1} & \mathbf{0}\end{pmatrix}\cdot\left[-\mathrm{j}\omega\begin{pmatrix}\boldsymbol{\epsilon}_{\mathrm{tt}} & \boldsymbol{\xi}_{\mathrm{tt}}\\ \boldsymbol{\zeta}_{\mathrm{tt}} & \boldsymbol{\mu}_{\mathrm{tt}}\end{pmatrix}+\mathrm{j}\omega\mathbf{A}\right]\cdot\begin{pmatrix}\mathbf{1} & \mathbf{0}\\ \mathbf{0} & \hat{z}\times\mathbf{1}\end{pmatrix}\cdot\begin{pmatrix}\boldsymbol{E}_{\mathrm{t}}\\ -\hat{z}\times\boldsymbol{H}_{\mathrm{t}}\end{pmatrix} \tag{22}$$

By keeping the vector product with $\hat{z}$ in the magnetic field, the vectors $\boldsymbol{E}_{\mathrm{t}}$ and $-\hat{z}\times\boldsymbol{H}_{\mathrm{t}}$ will be parallel to each other in isotropic media. Identifying the transverse electric and magnetic fields as vector voltage and vector current, *i.e.*,

$$\boldsymbol{E}_{\mathrm{t}}=\boldsymbol{V}\quad\text{and}\quad-\hat{z}\times\boldsymbol{H}_{\mathrm{t}}=\boldsymbol{I} \tag{23}$$

we find

$$\boxed{\frac{\partial}{\partial z}\begin{pmatrix}\boldsymbol{V}\\ \boldsymbol{I}\end{pmatrix}=\begin{pmatrix}\mathbf{0} & -\hat{z}\times\mathbf{1}\\ \mathbf{1} & \mathbf{0}\end{pmatrix}\cdot\left[-\mathrm{j}\omega\begin{pmatrix}\boldsymbol{\epsilon}_{\mathrm{tt}} & \boldsymbol{\xi}_{\mathrm{tt}}\\ \boldsymbol{\zeta}_{\mathrm{tt}} & \boldsymbol{\mu}_{\mathrm{tt}}\end{pmatrix}+\mathrm{j}\omega\mathbf{A}\right]\cdot\begin{pmatrix}\mathbf{1} & \mathbf{0}\\ \mathbf{0} & \hat{z}\times\mathbf{1}\end{pmatrix}\cdot\begin{pmatrix}\boldsymbol{V}\\ \boldsymbol{I}\end{pmatrix}} \tag{24}$$

In particular, for an isotropic material this simplifies to

$$\boxed{\frac{\partial}{\partial z}\begin{pmatrix}\boldsymbol{V}\\ \boldsymbol{I}\end{pmatrix}=\left[-\mathrm{j}\omega\begin{pmatrix}\mathbf{0} & \mu\mathbf{1}\\ \epsilon\mathbf{1} & \mathbf{0}\end{pmatrix}+\frac{\mathrm{j}\omega}{c_0^2}\begin{pmatrix}0 & \epsilon^{-1}\boldsymbol{bb}\\ \mu^{-1}\boldsymbol{aa} & 0\end{pmatrix}\right]\cdot\begin{pmatrix}\boldsymbol{V}\\ \boldsymbol{I}\end{pmatrix}} \tag{25}$$

where the unitless vector $\boldsymbol{a}=k_0^{-1}\,\boldsymbol{k}_{\mathrm{t}}\times\hat{z}$ defines the direction of the TE polarized transverse electric field (electric field transverse to the plane of incidence), and the unitless vector $\boldsymbol{b}=\hat{z}\times\boldsymbol{a}=k_0^{-1}\,\boldsymbol{k}_{\mathrm{t}}$ defines the direction of the TM polarized transverse electric field (electric field in the plane of incidence). The amplitude of both vectors is $|\boldsymbol{a}|=|\boldsymbol{b}|=|\boldsymbol{k}_{\mathrm{t}}|/k_0=\sin\theta$, where θ is the angle of incidence in vacuum.

Equation (24) is recognized as a linear dynamical system for the transverse field components. If the material parameters are constant with respect to z, the solution of (24) can be written using the exponential matrix as (where $\boldsymbol{V}_1=\boldsymbol{V}(z_1)$ and $\boldsymbol{V}_2=\boldsymbol{V}(z_2)$ etc)

$$\begin{pmatrix}\boldsymbol{V}_1\\ \boldsymbol{I}_1\end{pmatrix}=\exp\left\{(z_1-z_2)\begin{pmatrix}\mathbf{0} & -\hat{z}\times\mathbf{1}\\ \mathbf{1} & \mathbf{0}\end{pmatrix}\cdot\left[-\mathrm{j}\omega\begin{pmatrix}\boldsymbol{\epsilon}_{\mathrm{tt}} & \boldsymbol{\xi}_{\mathrm{tt}}\\ \boldsymbol{\zeta}_{\mathrm{tt}} & \boldsymbol{\mu}_{\mathrm{tt}}\end{pmatrix}+\mathrm{j}\omega\mathbf{A}\right]\cdot\begin{pmatrix}\mathbf{1} & \mathbf{0}\\ \mathbf{0} & \hat{z}\times\mathbf{1}\end{pmatrix}\right\}\cdot\begin{pmatrix}\boldsymbol{V}_2\\ \boldsymbol{I}_2\end{pmatrix}=\mathbf{P}(z_1,z_2)\cdot\begin{pmatrix}\boldsymbol{V}_2\\ \boldsymbol{I}_2\end{pmatrix} \tag{26}$$

This formal solution reveals an important structure, which generalizes to inhomogeneous media where the material parameters may depend on z: the transverse fields at $z=z_1$ can be written as a dyadic $\mathbf{P}$ operating on the fields at $z=z_2$, where z_1 and z_2 are arbitrary (although the dyadic of course depends on z_1 and z_2). This dyadic is called a *propagator*, and its existence is guaranteed by the linearity of the problem. We write the explicit form of this dyadic for isotropic media in (34), but first we must define a few properties.

2.3 Eigenvalue problem in infinite media

If the wave is propagating in a medium which is infinite in the z direction, it is natural to search for solutions on the form

$$\begin{pmatrix} \boldsymbol{V}(z) \\ \boldsymbol{I}(z) \end{pmatrix} = \begin{pmatrix} \boldsymbol{V}_0 \\ \boldsymbol{I}_0 \end{pmatrix} \mathrm{e}^{-\mathrm{j}\beta z} \tag{27}$$

This implies $\frac{\partial}{\partial z}[\boldsymbol{V}(z), \boldsymbol{I}(z)] = -\mathrm{j}\beta[\boldsymbol{V}_0, \boldsymbol{I}_0]\mathrm{e}^{-\mathrm{j}\beta z}$ which makes (24) turn into an algebraic eigenvalue problem (after dividing by $-\mathrm{j}\omega$ and the exponential factor $\mathrm{e}^{-\mathrm{j}\beta z}$)

$$\frac{\beta}{\omega}\begin{pmatrix} \boldsymbol{V}_0 \\ \boldsymbol{I}_0 \end{pmatrix} = \begin{pmatrix} \mathbf{0} & -\hat{z}\times\mathbf{1} \\ \mathbf{1} & \mathbf{0} \end{pmatrix} \cdot \left[\begin{pmatrix} \boldsymbol{\epsilon}_{\mathrm{tt}} & \boldsymbol{\xi}_{\mathrm{tt}} \\ \boldsymbol{\zeta}_{\mathrm{tt}} & \boldsymbol{\mu}_{\mathrm{tt}} \end{pmatrix} - \mathbf{A} \right] \cdot \begin{pmatrix} \mathbf{1} & \mathbf{0} \\ \mathbf{0} & \hat{z}\times\mathbf{1} \end{pmatrix} \cdot \begin{pmatrix} \boldsymbol{V}_0 \\ \boldsymbol{I}_0 \end{pmatrix} \tag{28}$$

Thus, the propagation constant β can be found from the eigenvalue problem (28), which can easily be solved numerically once the material model is specified along with the transverse wave vector $\boldsymbol{k}_{\mathrm{t}}$ (which occurs only in $\mathbf{A}$).

In addition, the field amplitudes $[\boldsymbol{V}_0, \boldsymbol{I}_0]$ are the eigenvectors of the same dyadic and can be determined up to a multiplicative constant. Independent of the normalization, the eigenvectors always provide a mapping between the transverse components of the electric and magnetic fields, *i.e.*,

$$\boldsymbol{V}_0 = \mathbf{Z} \cdot \boldsymbol{I}_0 \tag{29}$$

where the dyadic $\mathbf{Z}$ is the *transverse wave impedance* of the wave. For isotropic media corresponding to (25), we have

$$\frac{\beta}{\omega}\begin{pmatrix} \boldsymbol{V}_0 \\ \boldsymbol{I}_0 \end{pmatrix} = \begin{pmatrix} 0 & \mu - \mathrm{c}_0^{-2}\epsilon^{-1}\boldsymbol{bb} \\ \epsilon - \mathrm{c}_0^{-2}\mu^{-1}\boldsymbol{aa} & 0 \end{pmatrix} \cdot \begin{pmatrix} \boldsymbol{V}_0 \\ \boldsymbol{I}_0 \end{pmatrix} \tag{30}$$

This implies

$$\left(\frac{\beta}{\omega}\right)^2 \boldsymbol{V}_0 = \epsilon\mu\left(\mathbf{1} - \frac{\mathrm{c}_0^{-2}}{\epsilon\mu}\boldsymbol{aa}\right) \cdot \left(\mathbf{1} - \frac{\mathrm{c}_0^{-2}}{\epsilon\mu}\boldsymbol{bb}\right) \cdot \boldsymbol{V}_0 = \epsilon\mu\left(\mathbf{1} - \frac{\mathrm{c}_0^{-2}}{\epsilon\mu}(\boldsymbol{aa} + \boldsymbol{bb})\right) \cdot \boldsymbol{V}_0 = \left(\epsilon\mu - \mathrm{c}_0^{-2}\sin^2\theta\right)\boldsymbol{V}_0 \tag{31}$$

where we used $\boldsymbol{a} \cdot \boldsymbol{b} = 0$ and $|\boldsymbol{a}| = |\boldsymbol{b}| = \sin\theta$. The final result for the wave number is then

$$\boxed{\beta = \omega\sqrt{\varepsilon\mu - \mathrm{c}_0^{-2}\sin^2\theta}} \tag{32}$$

and the wave impedance, defined by the relation $\boldsymbol{V}_0 = \mathbf{Z} \cdot \boldsymbol{I}_0$, is

$$\boxed{\mathbf{Z} = \frac{\omega}{\beta}\mu\left(\mathbf{1} - \frac{\mathrm{c}_0^{-2}}{\varepsilon\mu}\boldsymbol{bb}\right) = \frac{\omega\mu}{\beta}\frac{\boldsymbol{aa}}{|\boldsymbol{a}|^2} + \frac{\beta}{\omega\varepsilon}\frac{\boldsymbol{bb}}{|\boldsymbol{b}|^2}} \tag{33}$$

In vacuum, we have $\omega\mu/\beta = \eta_0/\cos\theta$ and $\beta/(\omega\epsilon) = \eta_0 \cos\theta$, where $\eta_0 = \sqrt{\mu_0/\epsilon_0}$ is the intrinsic wave impedance of vacuum. Finally, the propagator dyadic for a slab of length ℓ of isotropic media is

$$\boxed{\mathbf{P} = \begin{pmatrix} \cos(\beta\ell)\mathbf{1} & \mathrm{j}\sin(\beta\ell)\mathbf{Z} \\ \mathrm{j}\sin(\beta\ell)\mathbf{Z}^{-1} & \cos(\beta\ell)\mathbf{1} \end{pmatrix}} \tag{34}$$

In microwave theory, this is recognized as the ABCD-matrix of a transmission line with propagation constant β and characteristic impedance $\mathbf{Z}$ (Pozar, 2005, p. 185). Note however that we have generalized it to include both TE and TM polarization, through the dyadic character of $\mathbf{Z}$. The important thing about the propagator dyadic is that since tangential electric and magnetic fields are continuous, we can find the total propagator dyadic for a layered structure by cascading:

$$\mathbf{P}_{\text{tot}} = \mathbf{P}_1 \cdot \mathbf{P}_2 \cdots \mathbf{P}_N = \begin{pmatrix} \mathbf{A} & \mathbf{B} \\ \mathbf{C} & \mathbf{D} \end{pmatrix}, \quad \text{where} \quad \begin{pmatrix} \boldsymbol{V}_1 \\ \boldsymbol{I}_1 \end{pmatrix} = \mathbf{P}_{\text{tot}} \cdot \begin{pmatrix} \boldsymbol{V}_2 \\ \boldsymbol{I}_2 \end{pmatrix} \tag{35}$$

The dyadic $\mathbf{P}_{\text{tot}}$ maps the total fields from one side of the layered structure to the other. Outside the structure, the total fields can be expressed in terms of the incident field amplitude $\boldsymbol{V}_{\text{inc}}$ using reflection and transmission dyadics $\mathbf{r}$ and $\mathbf{t}$ as (where we assume the same medium on both sides, with the characteristic impedance $\mathbf{Z}_0$, and use the fact that waves propagating along the positive z direction satisfy $\boldsymbol{V}_+ = \mathbf{Z}_0 \cdot \boldsymbol{I}_+$, whereas waves propagating in the negative z direction satisfy $\boldsymbol{V}_- = -\mathbf{Z}_0 \cdot \boldsymbol{I}_-$)

$$\begin{pmatrix} \boldsymbol{V}_1 \\ \boldsymbol{I}_1 \end{pmatrix} = \begin{pmatrix} (\mathbf{1}+\mathbf{r}) \cdot \boldsymbol{V}_{\text{inc}} \\ \mathbf{Z}_0^{-1} \cdot (\mathbf{1}-\mathbf{r}) \cdot \boldsymbol{V}_{\text{inc}} \end{pmatrix} \quad \text{and} \quad \begin{pmatrix} \boldsymbol{V}_2 \\ \boldsymbol{I}_2 \end{pmatrix} = \begin{pmatrix} \mathbf{t} \cdot \boldsymbol{V}_{\text{inc}} \\ \mathbf{Z}_0^{-1} \cdot \mathbf{t} \cdot \boldsymbol{V}_{\text{inc}} \end{pmatrix} \tag{36}$$

Solving for the reflection and transmission dyadics imply

$$\mathbf{r} = (\mathbf{A} + \mathbf{B} \cdot \mathbf{Z}_0^{-1} - \mathbf{Z}_0 \cdot \mathbf{C} - \mathbf{Z}_0 \cdot \mathbf{D} \cdot \mathbf{Z}_0^{-1}) \cdot (\mathbf{A} + \mathbf{B} \cdot \mathbf{Z}_0^{-1} + \mathbf{Z}_0 \cdot \mathbf{C} + \mathbf{Z}_0 \cdot \mathbf{D} \cdot \mathbf{Z}_0^{-1})^{-1} \tag{37}$$

$$\mathbf{t} = 2(\mathbf{A} + \mathbf{B} \cdot \mathbf{Z}_0^{-1} + \mathbf{Z}_0 \cdot \mathbf{C} + \mathbf{Z}_0 \cdot \mathbf{D} \cdot \mathbf{Z}_0^{-1})^{-1} \tag{38}$$

Thus, the concept of propagator dyadics enables a straight-forward analysis of layered structures, although the final results in terms of reflection and transmission coefficients may be complicated. In addition, thin sheets which are inhomogeneous in the xy-plane can also be modeled with corresponding propagator dyadics. This is explored in the next sections.

3. Lumped element models of scatterers

In real applications, relatively thick homogeneous slabs are often interlaced with thinner sheets, which may also be inhomogeneous in the transverse plane. Such scatterers can be modeled as lumped elements, the simplest of which corresponds to homogeneous, thin sheets. We are thus led to study the limit of the ABCD-matrix for a slab when its thickness ℓ becomes small. Denote the thickness of the sheet by t. Considering the factors in the propagator dyadic (34) and keeping factors to first order in βt, we find

$$\cos(\beta t) \to 1 \quad \text{and} \quad \sin(\beta t) \to \beta t \tag{39}$$

Thus, to first order in βt the ABCD-matrix is (using $\mathbf{Z}=\frac{\omega\mu}{\beta}\frac{\boldsymbol{aa}}{|\boldsymbol{a}|^2}+\frac{\beta}{\omega\epsilon}\frac{\boldsymbol{bb}}{|\boldsymbol{b}|^2}$)

$$\begin{pmatrix} \cos(\beta t)\mathbf{1} & \mathrm{j}\sin(\beta t)\mathbf{Z} \\ \mathrm{j}\sin(\beta t)\mathbf{Z}^{-1} & \cos(\beta t)\mathbf{1} \end{pmatrix} \to \begin{pmatrix} \mathbf{1} & \mathrm{j}\omega\mu t\left(\frac{\boldsymbol{aa}}{|\boldsymbol{a}|^2}+\frac{\beta^2}{\omega^2\epsilon\mu}\frac{\boldsymbol{bb}}{|\boldsymbol{b}|^2}\right) \\ \mathrm{j}\omega\varepsilon t\left(\frac{\beta^2}{\omega^2\epsilon\mu}\frac{\boldsymbol{aa}}{|\boldsymbol{a}|^2}+\frac{\boldsymbol{bb}}{|\boldsymbol{b}|^2}\right) & \mathbf{1} \end{pmatrix} \tag{40}$$

In order to treat the sheet as a lumped element, the reference planes T and T' in Figure 1 should coincide. This corresponds to back propagating the fields at T' by multiplying the dyadic above by the inverse of the corresponding dyadic for the background medium (denoted by index 0), or to first order in βt, subtracting the corresponding phase change in the off-diagonal elements. For instance, the upper right element should be replaced by (using $\beta^2 = \epsilon\mu - c_0^{-2}\sin^2\theta$ and $c_0^{-2} = \epsilon_0\mu_0$)

$$\mathrm{j}\omega\mu t\left[\frac{\boldsymbol{aa}}{|\boldsymbol{a}|^2}+(1-\frac{c_0^{-2}}{\epsilon\mu}\sin^2\theta)\frac{\boldsymbol{bb}}{|\boldsymbol{b}|^2}\right]-\mathrm{j}\omega\mu_0 t\left[\frac{\boldsymbol{aa}}{|\boldsymbol{a}|^2}+(1-\frac{c_0^{-2}}{\epsilon_0\mu_0}\sin^2\theta)\frac{\boldsymbol{bb}}{|\boldsymbol{b}|^2}\right]$$

$$=\mathrm{j}\omega(\mu-\mu_0)t\left(\frac{\boldsymbol{aa}}{|\boldsymbol{a}|^2}+\frac{\boldsymbol{bb}}{|\boldsymbol{b}|^2}\right)+\left(-\mathrm{j}\omega\mu t\frac{c_0^{-2}}{\epsilon\mu}\sin^2\theta+\mathrm{j}\omega\mu_0 t\sin^2\theta\right)\frac{\boldsymbol{bb}}{|\boldsymbol{b}|^2}$$

$$=\mathrm{j}\omega(\mu-\mu_0)t\mathbf{1}+\mathrm{j}\omega\mu_0 t\sin^2\theta\left(1-\frac{\epsilon_0}{\epsilon}\right)\frac{\boldsymbol{bb}}{|\boldsymbol{b}|^2} \tag{41}$$

In the final step we used that $\boldsymbol{a}$ and $\boldsymbol{b}$ are orthogonal and span the xy-plane, *i.e.*, $\frac{\boldsymbol{aa}}{|\boldsymbol{a}|^2}+\frac{\boldsymbol{bb}}{|\boldsymbol{b}|^2}=\mathbf{1}$. To first order, the result is then (using $\sin^2\theta = |\boldsymbol{a}|^2 = |\boldsymbol{b}|^2$)

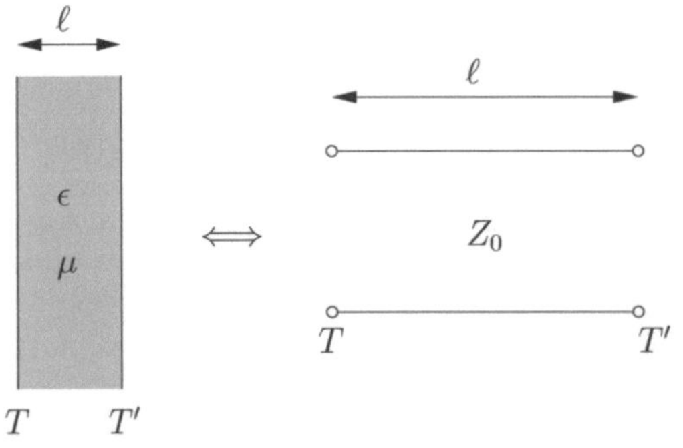

Fig. 1. Transmission line model of an isotropic slab.

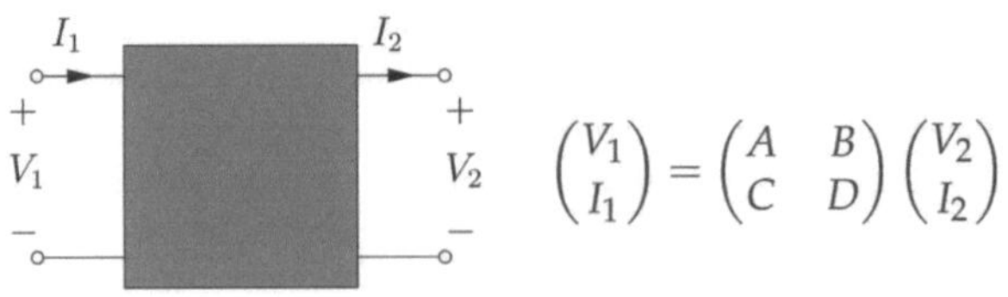

Fig. 2. Definition of ABCD matrix parameters for a general twoport network.

$$\begin{pmatrix} \mathbf{1} & \mathrm{j}\omega(\mu-\mu_0)t\mathbf{1}+\mathrm{j}\omega\mu_0 t(1-\frac{\epsilon_0}{\epsilon})\boldsymbol{bb} \\ \mathrm{j}\omega(\epsilon-\epsilon_0)t\mathbf{1}+\mathrm{j}\omega\epsilon_0 t(1-\frac{\mu_0}{\mu})\boldsymbol{aa} & \mathbf{1} \end{pmatrix} \tag{42}$$

Thus, a thin sheet of homogeneous material with permittivity ϵ and permeability μ can be modelled to first order as a series impedance Z and shunt admittance Y with the values

$$Z = \begin{cases} \mathrm{j}\omega(\mu-\mu_0)t & \text{TEpolarization} \\ \mathrm{j}\omega(\mu-\mu_0)t+\mathrm{j}\omega\mu_0 t(1-\epsilon_0/\epsilon)\sin^2\theta & \text{TMpolarization} \end{cases} \tag{43}$$

$$Y = \begin{cases} \mathrm{j}\omega(\epsilon-\epsilon_0)t+\mathrm{j}\omega\epsilon_0 t(1-\mu_0/\mu)\sin^2\theta & \text{TEpolarization} \\ \mathrm{j}\omega(\epsilon-\epsilon_0)t & \text{TMpolarization} \end{cases} \tag{44}$$

where t is the thickness of the sheet. An important special case is the *resistive sheet*, where $\epsilon = \epsilon' - \mathrm{j}\sigma/\omega$ and $\mu = \mu_0$. In the limit $\omega \to 0$, we then have

$$\boxed{Y \to \sigma t \quad \text{and} \quad Z \to 0} \tag{45}$$

regardless of polarization. The quantity $1/(\sigma t) = R_s$ is called the *sheet resistance*.

To see how sheets with a periodic pattern can be handled, we introduce the electric and magnetic polarizability per unit area γ_e/A and γ_m/A, such that $\epsilon_0\gamma_e \cdot \boldsymbol{E}_0$ is the static polarization induced in the sheet when subjected to a homogeneous field $\boldsymbol{E}_0$. The physical unit of γ_e/A and γ_m/A is length. The polarizability is in general a dyadic that can be represented as a 3×3 matrix, with the decomposition

$$\boldsymbol{\gamma}_{\mathrm{e}} = \boldsymbol{\gamma}_{\mathrm{ett}} + \hat{z}\boldsymbol{\gamma}_{\mathrm{ez}} + \boldsymbol{\gamma}_{\mathrm{et}}\hat{z} + \gamma_{\mathrm{ezz}}\hat{z}\hat{z} \tag{46}$$

with the corresponding decomposition for the magnetic polarizability. As shown in (Sjöberg, 2009a), the polarizability dyadics can be calculated from the solutions of the following static problems, where $\boldsymbol{E}_0$ and $\boldsymbol{H}_0$ are given constant vectors,

$$\nabla\cdot[\boldsymbol{\epsilon}\cdot(\boldsymbol{E}_0-\nabla\varphi_{\mathrm{e}})]=0 \tag{47}$$

$$\nabla\cdot[\boldsymbol{\mu}\cdot(\boldsymbol{H}_0-\nabla\varphi_{\mathrm{m}})]=0 \tag{48}$$

with periodic boundary conditions in the xy-plane and $\nabla\varphi_{e,m} \to 0$ as $z \to \pm\infty$. In these equations, ϵ and μ are the static permittivity and permeability dyadics, which may be anisotropic but are always symmetric and real-valued. The polarizability dyadics are then defined by (where U denotes the unit cell in the xy plane)

$$\int_{-\infty}^{\infty}\int_U(\boldsymbol{\epsilon}/\epsilon_0-\mathbf{1})\cdot(\boldsymbol{E}_0-\nabla\varphi_{\mathrm{e}})dSdz=\boldsymbol{\gamma}_{\mathrm{e}}\cdot\boldsymbol{E}_0 \tag{49}$$

$$\int_{-\infty}^{\infty}\int_U(\boldsymbol{\mu}/\mu_0-\mathbf{1})\cdot(\boldsymbol{H}_0-\nabla\varphi_{\mathrm{m}})dSdz=\boldsymbol{\gamma}_{\mathrm{m}}\cdot\boldsymbol{H}_0 \tag{50}$$

Generalizations of these equations to encompass the possibility of metal inclusions are given in (Sjöberg, 2009a). Using these quantities, the low frequency scattering against a low-pass sheet with periodic structure is (Sjöberg, 2009a)

$$\mathbf{t}=\mathbf{1}-\frac{\mathrm{j}k_0}{2}\left\{\eta_0^{-1}\mathbf{Z}_0\cdot\left[\frac{\gamma_{\mathrm{ett}}}{A}+\frac{\gamma_{\mathrm{mzz}}}{A}\boldsymbol{aa}\right]+\left[-\hat{z}\times\frac{\gamma_{\mathrm{mtt}}}{A}\cdot\hat{z}\times+\frac{\gamma_{\mathrm{ezz}}}{A}\boldsymbol{bb}\right]\cdot\mathbf{Z}_0^{-1}\eta_0\right.$$
$$\left.+\mathbf{Z}_0\cdot\left[\boldsymbol{a}\frac{\gamma_{\mathrm{mz}}}{A}-\frac{\gamma_{\mathrm{et}}}{A}\boldsymbol{a}\right]\cdot\hat{z}\times\mathbf{Z}_0^{-1}+\hat{z}\times\left[\boldsymbol{a}\frac{\gamma_{\mathrm{ez}}}{A}-\frac{\gamma_{\mathrm{mt}}}{A}\boldsymbol{a}\right]\right\} \tag{51}$$

$$\mathbf{r}=-\frac{\mathrm{j}k_0}{2}\left\{\eta_0^{-1}\mathbf{Z}_0\cdot\left[\frac{\gamma_{\mathrm{ett}}}{A}+\frac{\gamma_{\mathrm{mzz}}}{A}\boldsymbol{aa}\right]-\left[-\hat{z}\times\frac{\gamma_{\mathrm{mtt}}}{A}\cdot\hat{z}\times+\frac{\gamma_{\mathrm{ezz}}}{A}\boldsymbol{bb}\right]\cdot\mathbf{Z}_0^{-1}\eta_0\right.$$
$$\left.+\mathbf{Z}_0\cdot\left[\boldsymbol{a}\frac{\gamma_{\mathrm{mz}}}{A}-\frac{\gamma_{\mathrm{et}}}{A}\boldsymbol{a}\right]\cdot\hat{z}\times\mathbf{Z}_0^{-1}-\hat{z}\times\left[\boldsymbol{a}\frac{\gamma_{\mathrm{ez}}}{A}-\frac{\gamma_{\mathrm{mt}}}{A}\boldsymbol{a}\right]\right\} \tag{52}$$

The cross product with the z direction, $\hat{z}\,\times$, can be represented as a skew-symmetric matrix which is its own negative inverse. Thus, the expression $-\hat{z}\times\frac{\gamma_{\mathrm{mtt}}}{A}\cdot\hat{z}\times=(\hat{z}\times)^{-1}\cdot\frac{\gamma_{\mathrm{mtt}}}{A}\cdot\hat{z}\times$ is a similarity transform of γ_{mtt}/A.

In order to identify the circuit analog of these expressions, we compare with the simple networks in Figure 3 and compute their reflection and transmission coefficients. Assuming $Z_1=\mathrm{j}\omega L$ and $Y_2=\mathrm{j}\omega C$, all networks in Figure 3 have the same ABCD-matrix to first order in ω,

$$\begin{pmatrix} A & B \\ C & D\end{pmatrix}=\begin{pmatrix} 1 & Z_1 \\ Y_2 & 1\end{pmatrix}\Rightarrow\begin{cases} r=\frac{1}{2}(Z_1/Z_0-Y_2Z_0) \\ t=1-\frac{1}{2}(Z_1/Z_0+Y_2Z_0)\end{cases} \tag{53}$$

where Z_0 is the characteristic impedance of the surrounding medium. Comparison between the two expressions implies

$$\mathbf{Z}_1\cdot\mathbf{Z}_0^{-1}=\mathrm{j}k_0\left\{\left[-\hat{z}\times\frac{\gamma_{\mathrm{mtt}}}{A}\cdot\hat{z}\times+\boldsymbol{bb}\frac{\gamma_{\mathrm{ezz}}}{A}\right]\cdot\mathbf{Z}_0^{-1}\eta_0+\hat{z}\times\left[\boldsymbol{a}\frac{\gamma_{\mathrm{ezt}}}{A}-\frac{\gamma_{\mathrm{mtz}}}{A}\boldsymbol{a}\right]\right\} \tag{54}$$

$$\mathbf{Z}_0\cdot\mathbf{Y}_2=\mathrm{j}k_0\left\{\eta_0^{-1}\mathbf{Z}_0\cdot\left[\frac{\gamma_{\mathrm{ett}}}{A}+\boldsymbol{aa}\frac{\gamma_{\mathrm{mzz}}}{A}\right]+\mathbf{Z}_0\cdot\left[\boldsymbol{a}\frac{\gamma_{\mathrm{mzt}}}{A}-\frac{\gamma_{\mathrm{etz}}}{A}\boldsymbol{a}\right]\cdot\hat{z}\times\mathbf{Z}_0^{-1}\right\} \tag{55}$$

Using $\mathbf{Z}_1=\mathrm{j}\omega\mathbf{L}$ and $\mathbf{Y}_2=\mathrm{j}\omega\mathbf{C}$, this implies the sheet series inductance dyadic $\mathbf{L}$ and sheet shunt capacitance dyadic $\mathbf{C}$ is (which generalizes (43) and (44) to anisotropic materials)

$$\mathbf{L}=\frac{1}{c_0}\left\{\left[-\hat{z}\times\frac{\gamma_{\mathrm{mtt}}}{A}\cdot\hat{z}\times+\frac{\gamma_{\mathrm{ezz}}}{A}\boldsymbol{bb}\right]\eta_0+\hat{z}\times\left[\boldsymbol{a}\frac{\gamma_{\mathrm{ezt}}}{A}-\frac{\gamma_{\mathrm{mtz}}}{A}\boldsymbol{a}\right]\cdot\mathbf{Z}_0\right\} \tag{56}$$

$$\mathbf{C}=\frac{1}{c_0}\left\{\eta_0^{-1}\left[\frac{\gamma_{\mathrm{ett}}}{A}+\frac{\gamma_{\mathrm{mzz}}}{A}\boldsymbol{aa}\right]+\left[\boldsymbol{a}\frac{\gamma_{\mathrm{mzt}}}{A}-\frac{\gamma_{\mathrm{etz}}}{A}\boldsymbol{a}\right]\cdot\hat{z}\times\mathbf{Z}_0^{-1}\right\} \tag{57}$$

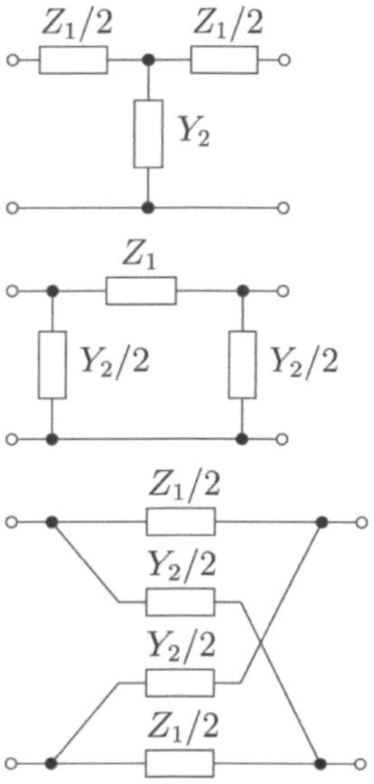

$$\begin{pmatrix} 1+Z_1Y_2/2 & Z_1(1+Z_1Y_2/4) \\ Y_2 & 1+Z_1Y_2/2 \end{pmatrix}$$

$$\begin{pmatrix} 1+Z_1Y_2/2 & Z_1 \\ Y_2(1+Z_1Y_2/4) & 1+Z_1Y_2/2 \end{pmatrix}$$

$$\frac{1}{1-Z_1Y_2/4}\begin{pmatrix} 1+Z_1Y_2/4 & Z_1 \\ Y_2 & 1+Z_1Y_2/4 \end{pmatrix}$$

Fig. 3. ABCD-matrices for symmetric T, Π, and trellis net.

These dyadics are represented by diagonal matrices if there is no coupling between TE and TM modes. For normal incidence on an isotropic slab with thickness t, the parameters take on the simple scalar values

$$L = (\mu - \mu_0)t \quad \text{and} \quad C = (\epsilon - \epsilon_0)t \tag{58}$$

Note that the circuit parameters defined in this section correspond to a low frequency expansion, where the sheet is considered thin in terms of wavelength. For higher frequencies, the method presented in Section 5 can be used.

4. Physical limitations

Circuit analogs appear in a very natural way when considering physical limitations of scattering against stratified structures. The methodology dates back to classical work on optimum matching (Fano, 1950), using clever integration paths in the complex plane for functions representing linear, causal, passive systems. In physics, the corresponding relations are known as sum rules, connecting an integral over all frequencies of some quantity to the static value of another (Nussenzveig, 1972). Often, the sum rules are derived from relations similar to the Kramers-Kronig's relations (de L. Kronig, 1926; Kramers, 1927). In this section, we only give the final results of other authors' work, and refer to the original papers for more in depth discussions.

The first paper to discuss physical limitations on scattering from planar structures was by (Rozanov, 2000). He derived the following limitation on the reflection coefficient R from any metal-backed planar structure (where $\lambda = c_0/f$ is the wavelength in vacuum):

$$\int_0^\infty \ln\left(\frac{1}{|r(\lambda)|}\right) d\lambda \leq 2\pi^2 \sum_i \mu_{s,i} d_i = 2\pi^2 \frac{L}{\mu_0} \tag{59}$$

Here, we identify the inductance as $L = \mu d$ instead of $(\mu - \mu_0)d$, since the reference plane of the reflection is at the top of the structure and not at the ground plane. The expression (59) demonstrates that the bandwidth over which the amplitude of the reflection coefficient is less than unity, is bounded above by the static permeability of the structure, which can be interpreted as the low frequency series inductance. The interesting part of this physical limitation is that it is valid for any realization of the structure, and provides a useful upper bound for absorbers. This is seen from the fact that the integral is bounded below by $(\lambda_2 - \lambda_1) \ln(1/r_0)$, where r_0 is the largest reflection level in the band $[\lambda_1,\lambda_2]$. Using the relative bandwidth $B = (\lambda_2 - \lambda_1)/\lambda_0$, where the center wavelength is $\lambda_0 = (\lambda_1 + \lambda_2)/2$, we find

$$B \ln\left(\frac{1}{r_0}\right) \leq 2\pi^2 \frac{\sum \mu_{s,i} d_i}{\lambda_0} = 2\pi^2 \frac{L}{\lambda_0 \mu_0} \tag{60}$$

Thus, the product of bandwidth and reflection level in logarithmic scale is bounded above by a factor proportional to the low frequency series inductance of the structure.
A similar bound was found by (Brewitt-Taylor, 2007) for the realization of artificial magnetic conductors, by studying the factor $P = (r - 1)/2$. Magnetic conductors are attractive in antenna design problems, and are characterized by a reflection coefficient $r \approx +1$, meaning P becomes small in the band of interest. The bound is

$$\int_0^\infty \ln\left(\frac{1}{|P(\lambda)|}\right) d\lambda \leq \pi^2 \sum_i \mu_{s,i} d_i = \pi^2 \frac{L}{\mu_0} \tag{61}$$

with similar interpretation as Rozanov's result and corresponding bandwidth bound. Our final example is of the transmission through a periodic low-pass screen (Gustafsson et al., 2009), where the following bound for a non-magnetic structure was derived (where t is the transmission coefficient)

$$\int_0^\infty \ln \frac{1}{|t(\lambda)|} d\lambda \leq \pi^2 \frac{\boldsymbol{E}_0 \cdot \boldsymbol{\gamma}_{\rm ett} \cdot \boldsymbol{E}_0}{2A |\boldsymbol{E}_0|^2} = \frac{\pi^2}{2} \frac{C}{\epsilon_0} \tag{62}$$

The factor $\boldsymbol{\gamma}_{\rm ett}/A$ is the capacitance dyadic in (57) for normal incidence, and similar physical bounds can be derived for antennas, materials and general scatterers (Sohl et al., 2007a; Gustafsson et al., 2007; Sohl et al., 2007a;b; 2008; Sohl & Gustafsson, 2008). When considering the physical limitations, it is noteworthy that the circuit parameters (or rather, the polarizability dyadics) can be bounded using variational principles as discussed in (Sjöberg, 2009b). These typically state that the polarizability of a given structure cannot decrease if we add more material; in particular, the electric polarizability of any body is always less than (or at most equal to) the polarizability of a circumscribed metallic body. If the metallic body has a simple shape (such as a sphere), its polarizability can be computed analytically, and hence a useful approximation of the polarizability of the original body is provided.

5. Computation of circuit analogs in the general case

So far, we have only demonstrated how to compute circuit analogs in the low frequency limit. Indeed, this is the primary region where we can give firm definitions and physical

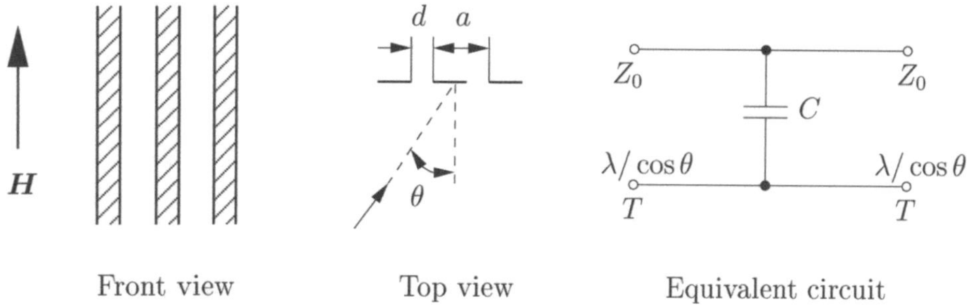

Fig. 4. Geometry and equivalent circuit for capacitive strips (TM polarization).

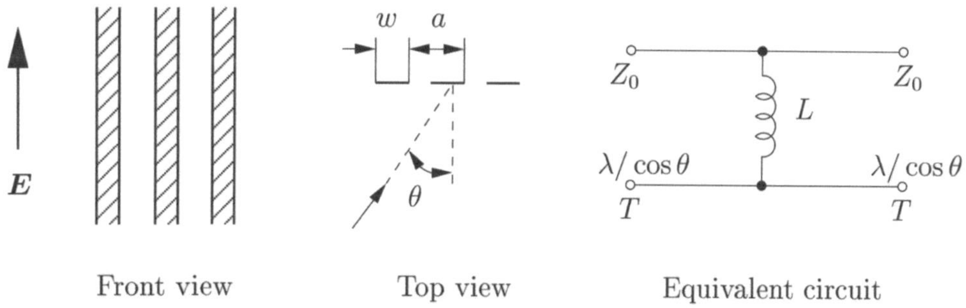

Fig. 5. Geometry and equivalent circuit for inductive strips (TE polarization).

interpretations of the circuit analogs, but analogs are still valuable as a modeling tool even for higher frequencies, in particular for structures of subwavelength size. The general procedure is the same as previously employed: the circuit analogs are computed to provide the same scattering characteristics as the full structure.

Many analytical expressions have been derived throughout the years in the microwave literature, in particular associated with the development of radar technology during the Second World War. Many of these results are collected in references such as (Collin, 1991; Marcuvitz, 1951; Schwinger & Saxon, 1968). The most explored geometry is that of metallic strips, see Figures 4 and 5. Depending on the polarization of the incident wave, the strips behave dominantly capacitive or inductive. Provided that the width of the strips and the distance between them can be considered small in terms of wavelengths, the circuit parameters in Figures 4 and 5 can be estimated as follows (Marcuvitz, 1951, pp. 280 and 284)

$$C \approx \frac{4a\epsilon_0}{\pi}\ln\frac{2a}{\pi d} \quad \text{and} \quad L \approx \frac{a\mu_0}{2\pi}\ln\frac{2a}{\pi w} \tag{63}$$

Note that the inductance L is now a shunt inductance, in contrast to the series inductance obtained by transmission through a thin slab. We can immediately interpret these results in order to gain some design intuition:

- To make the capacitance C large, the ratio d/a should be small, *i.e.*, the gap between the strips should be small.
- To make the inductance L large, the ratio w/a should be small, *i.e.*, the width of the strips should be small.

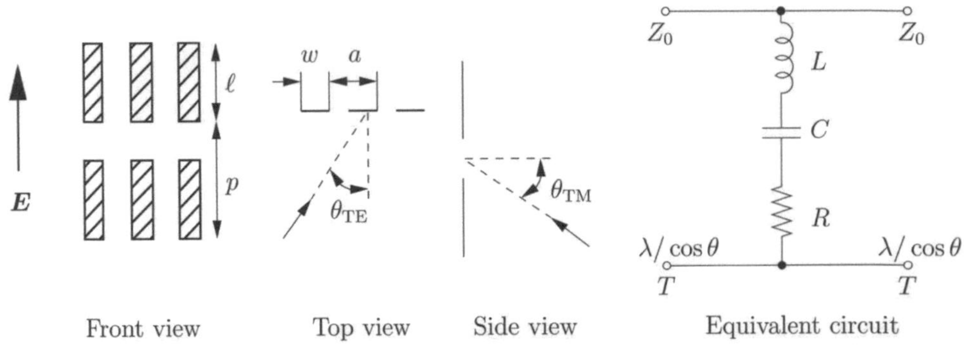

Fig. 6. Finite dipole. Observe that the incidence plane changes depending on whether we study TE or TM polarization, since the dipoles would be practically invisible for other cases.

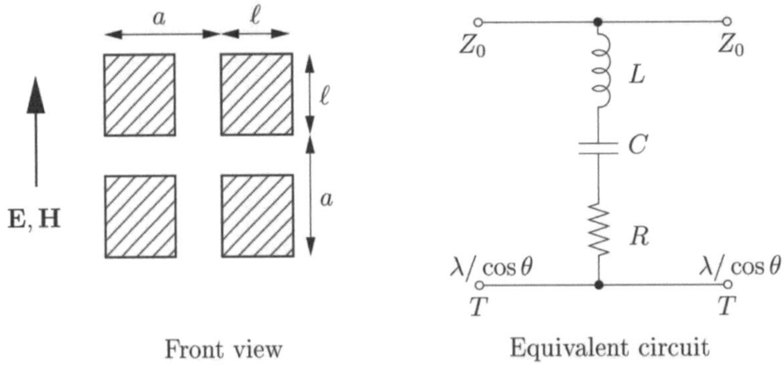

Fig. 7. A pattern of square patches. Compared to the finite dipoles in Figure 6, this structure is relatively independent of polarization, due to its higher symmetry.

When the strips are not infinite, a more sophisticated modeling must be made. For an array of finite strips as in Figure 6, we need to incorporate more elements in order to reflect the possibility of resonance, when the dipoles become approximately half a wavelength. If the metal strips are lossy, we also need to include a resistance in the circuit analog. It is not trivial to compute the exact circuit parameters in this case, and in practice a numerical approach is necessary. Also, the thin finite dipoles are polarization sensitive, being essentially invisible to electric fields orthogonal to their longest extension. By using square patches as in Figure 7, a higher symmetry is achieved and thereby less polarization sensitivity. On the other hand, since the patch is wide, its inductance tends to be smaller than the corresponding dipole.

Many practical structures can be considered to be thin and non-magnetic. In this case, the simple model in Figure 8 can be used, where the sheet is considered as a single shunt lumped element, possibly with a complicated frequency dependence. To compute the relevant circuit analog for an arbitrary such sheet, we can make a full wave simulation to compute scattering parameters such as the reflection coefficient $r(\omega)$. From this given data we can turn to the circuit model in Figure 8 to find the reflection coefficient

$$r = \frac{Y_0 - (Y + Y_0)}{Y + (Y + Y_0)} = \frac{1 - (Y + Y_0)/Y_0}{1 + (Y + Y_0)/Y_0} \Rightarrow \frac{Y + Y_0}{Y_0} = \frac{1 - r}{1 + r} \tag{64}$$

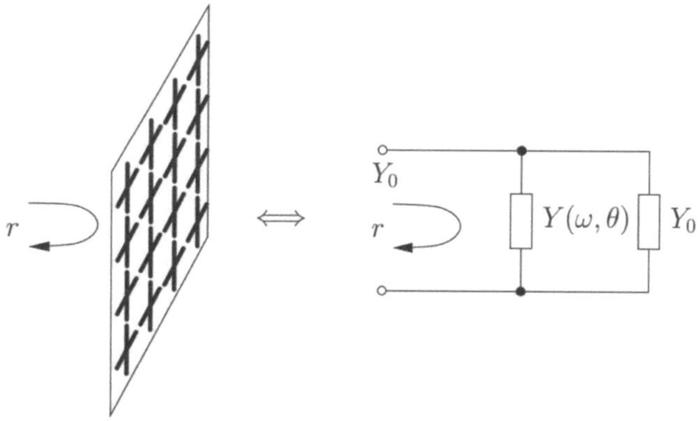

Fig. 8. Equivalent circuit model for reflection against a nonmagnetic sheet backed by free space. The shunt admittance Y is usually a function of frequency and angle of incidence, as well as polarization.

Thus, we can compute a numerical value for the normalized surface admittance $(Y + Y_0)/Y_0$ for each frequency ω. The next step is to match this data with a rational approximant $(Y + Y_0)/Y_0 = p(s)/q(s)$, where p and q are polynomials in $s = j\omega$. In order for the identified model $p(s)/q(s)$ to correspond to a realizable circuit, it is necessary that all zeros of $q(s)$ have negative real part. This can be achieved by using matlab's routine `invfreqs`.

To compute the numerical reflection coefficients $r(\omega)$, it is necessary to reduce the problem to a unit cell with periodic boundary conditions in the xy-plane, and plane wave ports in the z direction. These requisites are commonly available in most advanced simulation programs today.

6. Applications

The usefulness of circuit analogs resides in the fact that they are a compact representation of the scattering properties of a given structure. In particular, in design problems where many potential designs need to be evaluated, circuit analogs provide a convenient means of producing concept designs, which must later be refined and verified using full-wave simulations.

One design methodology, is to choose a general geometry for the sheets such as the finite dipoles in Figure 6, and then build a database of corresponding circuit parameters for different geometry parameters. The desired electromagnetic function, for instance absorption over a certain bandwidth, is then first treated as a circuit design problem. Once a circuit design is found, typically consisting of segments of transmission lines loaded by lumped elements, the realization is found by lookup in the previously calculated tables. This methodology works well when the individual sheets are about a quarter wavelength from each other.

The method breaks down when the sheets become closely spaced, say within a tenth of a wavelength. There is then substantial coupling between the sheets through the otherwise evanescent fields, so that the previous circuit models do not apply anymore. However, new circuit models can be applied to clusters of sheets, although the number of geometry parameters may become so large that it becomes difficult to build a reasonable database.

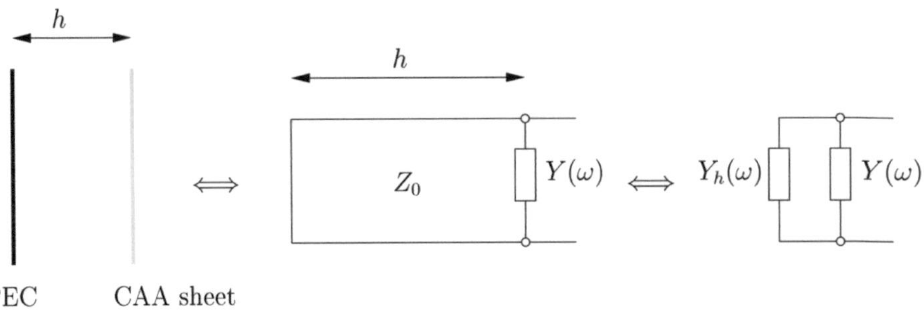

Fig. 9. Typical situation for a Circuit Analog Absorber (CAA) sheet. A CAA sheet is situated at height h above a metallic ground plane, and is modeled as a lumped shunt element in a corresponding transmission line. The shorted transmission line can be transformed to a lumped load $Y_h(\omega)$, shunting the sheet admittance $Y(\omega)$.

In (Sjöberg, 2008), this design methodology is demonstrated for electromagnetic absorbers as in Figure 9. A typical absorber structure is the Salisbury screen, where a resistive sheet of sheet resistance $R_s = 1/(\sigma t) = Z_0 = \sqrt{\mu_0/\epsilon_0} = 377\Omega$ is placed at height h above a ground plane. The short circuit of the transmission line provided by the ground plane is transformed to the location of the sheet as (Pozar, 2005, p. 60)

$$Y_h(\omega) = -jY_0 \cot(k_0 h \cos\theta) \tag{65}$$

where $k_0 = \omega/c_0$ is the free space wave number, θ is the angle of incidence, and Y_0 is the characteristic admittance of the transmission line (the inverse of Z in (33)). For normal incidence, it is seen that $Y_h(\omega_0) = 0$ for $k_0 h = \pi/2$, implying the load impedance as seen by the incident wave is $Z_L = 1/(1/R_s + Y_h) = R_s = Z_0$ at this frequency, providing zero reflection since the transmission line is then matched. The condition $k_0 h = \pi/2$ can be interpreted as h being equal to a quarter wavelength at the design frequency ω_0. The relative bandwidth is about 25% for −20dB reflectivity level (Knott et al., 2004, p. 316) corresponding to the bandwidth of the requirement $\cot(k_0 h) = 0$. However, the bound (60) given by (Rozanov, 2000) suggests that the upper bound on the bandwidth at −20dB well exceeds 100%, suggesting the design can be improved.

One improvement is to include a pattern in the resistive sheet, using the additional reactive elements to increase the bandwidth. Results of this method are depicted in Figures 10 and 11. It is seen that the design in Figure 10, based only on the pure circuit design and circuit analog representation of the sheet (details can be found in (Sjöberg, 2008)), produces a structure whose full wave characteristics are not optimal. A slight tuning of the geometry results in Figure 11, which has a better full wave result. The full wave simulations were made with the program PB-FDTD, which uses finite differences in the time-domain with periodic boundary conditions (Holter & Steyskal, 1999). We note that neither of these designs is close to the optimal limit by Rozanov. More advanced designs based on capacitive squares close to the ground plane can come close to the optimal limit (Kazem Zadeh & Karlsson, 2009). A design procedure taking oblique incidence into account can be found in (Munk et al., 2007).

Another design case which is made easier using circuit analogs is the construction of a linear-to-circular polarizer using meander lines, see (Young et al., 1973; Terret et al., 1984) and (Munk, 2003, pp. 306–326). In order to create circular polarization from linear

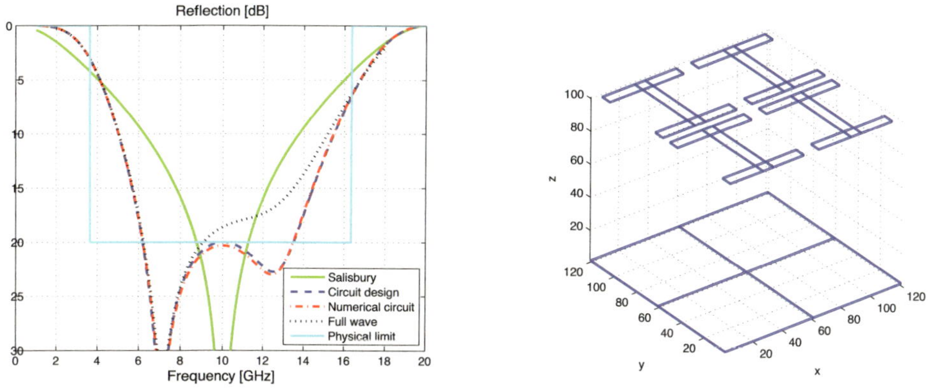

Fig. 10. Comparison between a Salisbury screen and a circuit analog absorber (total height is 7.5mm, normal incidence). The "ideal" circuit parameters computed from a circuit design are $R = 308\Omega$, $C = 80.4$fF, and $L = 3.15$nH. These parameters are approximately achieved by the geometry depicted on the right, where the optimization procedure described after equation (64) predicts $R = 311\Omega$, $C = 80.4$fF, and $L = 3.16$nH. These parameters are used in a circuit model, and the full wave results are also displayed. The physical limit (60) is represented by the square box, demonstrating that there is room for improvement on this design.

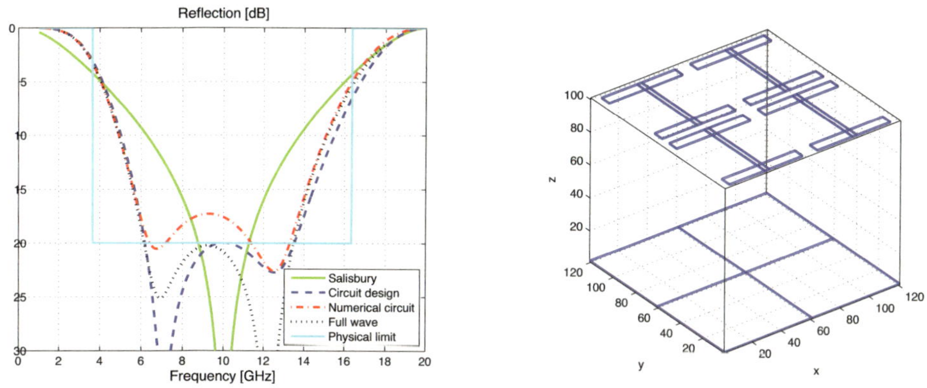

Fig. 11. A version of the CAA pattern in Figure 10 with more slender center part, and lower resistance per square. This provides higher inductance and therefore lower resonance frequency than the structure in Figure 10. The circuit parameters are $R = 319\Omega$, $C = 79.1$fF, and $L = 4.35$nH.

polarization, it is common to use a structure which delays the x-component 90°relative the y-component; by sending in the linear polarization with equal strength in x and y, the result is a circularly polarized wave. One idea to create this delay is to use the behavior of the metal strips in Figures 4 and 5, which are dominantly inductive for electric fields polarized along them, and dominantly capacitive for electric fields polarized orthogonally to them. In order to increase the reactive properties the lines are meandered, and several sheets are layered in order to achieve the total phase shift necessary. As is explained in detail in (Munk, 2003, pp. 306–326), it is necessary to adjust the outmost layers in order to reduce the reflection from the structure, leading to the final design geometry in Figure 12. It is seen in Figure 13 that the results are quite broad band. In the real application, further dielectric sheets need to be added to the design for mechanical rigidity.

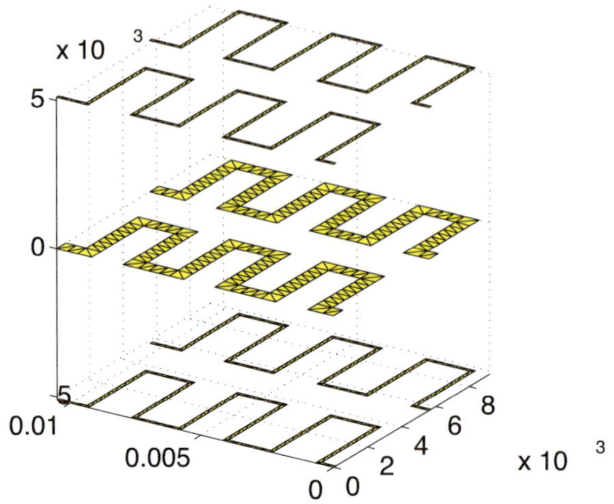

Fig. 12. Geometry of the layered meander structure, making up a linear-to-circular polarizer.

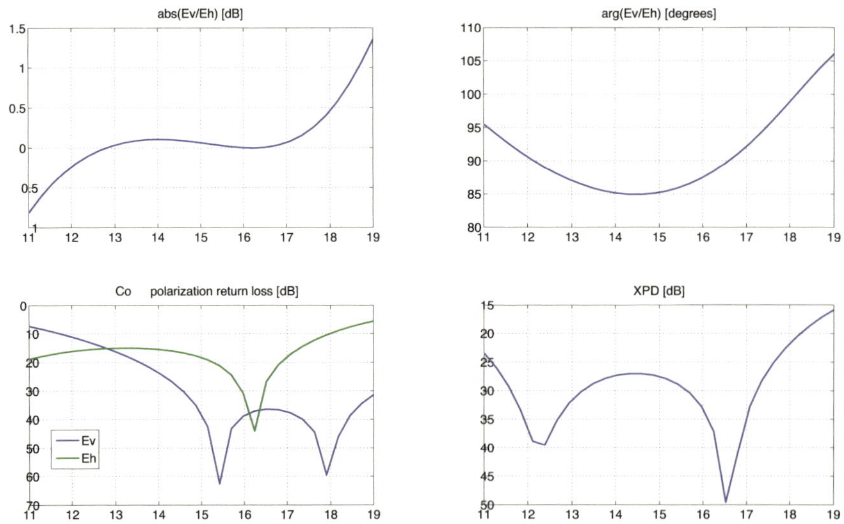

Fig. 13. Results for the layered meander structure (the x-axes in all subfigures are frequency in GHz). Upper left: the ratio between the vertical and horizontal polarization of the transmitted field. Upper right: the phase difference between the polarizations. Lower left: co-polarization reflection. Lower right: cross-polarization discrimination.

7. Conclusions

We have shown how relatively complex wave propagation problems can be efficiently modeled and designed using circuit analogs. Propagation of plane waves in any bianisotropic material can be modeled as propagation of voltage and current in two transmission lines, one for each polarization. The wave number and characteristic impedance for these transmission lines are determined from the algebraic eigenvalue problem (28). For isotropic media, there is no coupling between the TE and TM polarization, but in general we must allow for this coupling by using a dyadic transverse impedance.
Scatterers such as thin sheets, with or without a periodic pattern, may be modeled as lumped elements. If the sheets are non-magnetic, they can be reduced to a single shunt element. The circuit parameters can be determined in the static limit using the electric and magnetic polarizability per unit area. For higher frequencies, more advanced calculations taking the finite wavelength into account must be applied, for instance numerically.
The strength of the simple circuit analogs lies in their use in concept design, where a large number of possible realizations of a specified function must be evaluated. Using the simple models, reliable concept designs can be provided which are subsequently subjected to a more detailed (and costly) analysis and refined, to provide the final design.

8. References

Brewitt-Taylor, C. R. (2007). Limitation on the bandwidth of artificial perfect magnetic conductor surfaces, *Microwaves, Antennas & Propagation, IET* 1(1): 255–260.

Caloz, C. & Itoh, T. (2004). Transmission line approach of left-handed (LH) materials and microstrip implementation of an artificial LH transmission line, *IEEE Trans. Antennas Propagat.* 52(5): 1159–1166.

Collin, R. E. (1991). *Field Theory of Guided Waves*, second edn, IEEE Press, New York.

Collin, R. E. (1992). *Foundations for Microwave Engineering*, second edn, McGraw-Hill, New York.

de L. Kronig, R. (1926). On the theory of dispersion of X-rays, *J. Opt. Soc. Am.* 12(6): 547–557.

Eleftheriades, G. V., Iyer, A. K. & Kremer, P. C. (2002). Planar negative refractive index media using periodically loaded *L-C* loaded transmission lines, *IEEE Trans. Microwave Theory Tech.* 50(12): 2702–2712.

Fano, R. M. (1950). Theoretical limitations on the broadband matching of arbitrary impedances, *Journal of the Franklin Institute* 249(1,2): 57–83 and 139–154.

Gustafsson, M., Sohl, C. & Kristensson, G. (2007). Physical limitations on antennas of arbitrary shape, *Proc. R. Soc. A* 463: 2589–2607.

Gustafsson, M., Sohl, C., Larsson, C. & Sjöberg, D. (2009). Physical bounds on the all-spectrum transmission through periodic arrays, *EPL Europhysics Letters* 87(3): 34002 (6pp). URL: *http://stacks.iop.org/0295-5075/87/34002*

Holter, H. & Steyskal, H. (1999). Infinite phased-array analysis using FDTD periodic boundary conditions—pulse scanning in oblique directions, *IEEE Trans. Antennas Propagat.* 47(10): 1508–1514.

Kazem Zadeh, A. & Karlsson, A. (2009). Capacitive circuit method for fast and efficient design of wideband radar absorbers, *IEEE Trans. Antennas Propagat.* 57(8): 2307–2314.

Knott, E. F., Shaeffer, J. F. & Tuley, M. T. (2004). *Radar Cross Section*, SciTech Publishing Inc., 5601 N. Hawthorne Way, Raleigh, NC 27613.

Kramers, M. H. A. (1927). La diffusion de la lumière par les atomes, *Atti. Congr. Int. Fis. Como* 2: 545–557.

Marcuvitz, N. (1951). *Waveguide Handbook*, McGraw-Hill, New York.

Munk, B. (2000). *Frequency Selective Surfaces: Theory and Design*, John Wiley & Sons, New York.

Munk, B. (2003). *Finite Antenna Arrays and FSS*, John Wiley & Sons, New York.

Munk, B. A., Munk, P. & Pryor, J. (2007). On designing Jaumann and circuit analog absorbers (CA absorbers) for oblique angle of incidence, *IEEE Trans. Antennas Propagat.* 55(1): 186–193.

Nussenzveig, H. M. (1972). *Causality and dispersion relations*, Academic Press, London.

Poulsen, S. (2006). *Stealth radomes*, PhD thesis, Lund University, Department of Electroscience, Lund University, P.O. Box 118, S-221 00 Lund, Sweden.

Pozar, D. M. (2005). *Microwave Engineering*, third edn, John Wiley & Sons, New York.

Rozanov, K. N. (2000). Ultimate thickness to bandwidth ratio of radar absorbers, *IEEE Trans. Antennas Propagat.* 48(8): 1230–1234.

Saleh, B. E. A. & Teich, M. C. (1991). *Fundamentals of Photonics*, John Wiley & Sons, New York.

Schwinger, J. & Saxon, D. S. (1968). *Discontinuities in Waveguides*, Gordon and Breach Science Publishers, London.

Sjöberg, D. (2008). Analysis of wave propagation in stratified structures using circuit analogs, with application to electromagnetic absorbers, *Eur. J. Phys.* 29: 721–734.

Sjöberg, D. (2009a). Low frequency scattering by passive periodic structures for oblique incidence: low pass case, *J. Phys. A: Math. Theor.* 42: 385402.

Sjöberg, D. (2009b). Variational principles for the static electric and magnetic polarizabilities of anisotropic media with perfect electric conductor inclusions, *J. Phys. A: Math. Theor.* 42: 335403.

Sohl, C. & Gustafsson, M. (2008). A priori estimates on the partial realized gain of Ultra-Wideband (UWB) antennas, *Quart. J. Mech. Appl. Math.* 61(3): 415–430.

Sohl, C., Gustafsson, M. & Kristensson, G. (2007a). Physical limitations on broadband scattering by heterogeneous obstacles, *J. Phys. A: Math. Theor.* 40: 11165–11182.

Sohl, C., Gustafsson, M. & Kristensson, G. (2007b). Physical limitations on metamaterials: Restrictions on scattering and absorption over a frequency interval, *J. Phys. D: Applied Phys.* 40: 7146–7151.

Sohl, C., Larsson, C., Gustafsson, M. & Kristensson, G. (2008). A scattering and absorption identity for metamaterials: experimental results and comparison with theory, *J. Appl. Phys.* 103(5): 054906.

Terret, C., Levrel, J. R. & Mahdjoubi, K. (1984). Susceptance computation of a meander-line polarizer layer, *IEEE Trans. Antennas Propagat.* 32(9): 1007–1011.

Young, L., Robinson, L. A. & Hacking, C. A. (1973). Meander-line polarizer, *IEEE Trans. Antennas Propagat.* 21(3): 376–378.

VIII. Some Experimental Methods in Wave Propagation.

Field Experiments on Wave Propagation and Vibration Isolation by Using Wave Barriers

Seyhan Fırat[1], Erkan Çelebi[2], Günay Beyhan[3], İlyas Çankaya[4], Osman Kırtel[2] and İsa Vural[1]
[1]*Sakarya University, Technical Education Faculty, Construction Department*
[2]*Sakarya University, Engineering Faculty, Civil Engineering Department*
[3]*Sakarya University, Engineering Faculty, Geophysical Engineering Department*
[4]*Sakarya University, Technical Education Faculty, Electronics and Computer Department*
Turkey

1. Introduction

In most cases the major part of the vibration energy induced by dynamic sources transferred by the Rayleigh waves propagating in the region nearby soil surface may cause strong ground motions and stress levels that transmit the vibrations through the subsoil to the structures. Therefore, the permanent adversely effects of these excessive vibrations on the foundations, particularly supported on the soft soil deposits, cause structural damage to the adjacent structures. These vibrations give even disturbances to the nearby housing, sensitive electronic equipment, measuring installations and undesired actions on human comfort in the buildings. This type of vibrations in the frequency range from about 4-50 Hz may cause some structures to resonance with their vertical modes [1-2].

For an effective protection of the buildings from structural damage due to dynamic loads generated by man-made activities, such as rock drilling and blasting in road construction, working engine foundations in industrial areas, heavy and dense transport traffics due to increasing interconnections of residential regions etc., there are many possibilities to be considered as vibration screening systems. Especially, the development in passenger transport with high speed and the increased weight of high speed trains will cause strong ground and structural vibrations at the load path and in its neighborhood, particularly in intensively populated urban areas [3-4].

The special attention to this subject from the field of civil and railway engineering in association with the design of the railway track structures originates an increasing interest in methods, which can be classified as numerical, analytical or semi-empirical approaches for isolating vibrations. Published literature reveals several analytical studies [5-8] and some numerical models taking advances of both finite and boundary element approaches combined with analytical solutions for analysis of wave propagation problems in elastic medium with emphasis on soil-structure interaction due to moving loads [9-16].

The reduction of the structural response may be accomplished as: a) by adjusting the frequency contents of the excitation, b) by changing the location and direction of the vibratory source, c) by modifying the wave dissipation characteristics of the soil deposit, and d) by partially interrupting the spreading of waves into the structure or by providing the structure more damping by means of installation certain devices such as additional dampers or other base isolation systems. It is also possible to modify the dynamic transmitting behavior of local sub-soil through a complex mechanism of wave reflection and mode alteration around the vibration source by constructing a suitable wave barrier in the path of the propagating waves between the dynamic load and the affected structures to be protected. When the wave barrier is located nearby the vibratory source, such application is known as active (near field) isolation. If the barrier is situated away from the source but around the structure to be protected from incoming waves, such far field isolation is known as passive isolation.

Both open trench and solid barrier, such as an in-filled trench with suitable materials, can be useful as vibration measures. There is a wide range of construction types of wave barriers, varying from very stiff concrete walls or row of piles to very flexible gas mattresses or wave impeding barriers, where the latter is based on the cut-off frequency of a soil layer over rigid bedrock. Because of screening efficiency, without great difficulty to realize and low cost, both open and in-filled trenches are the most common in practical engineering applications as isolation measures. Many researches have primarily dealt with the development of different numerical techniques as a tool for analyzing the influences of different parameters on the vibration screening by means of trench barriers [16-20] to compare with the few experimental studies which are carried out full scale tests on site and laboratory model investigations only for particular cases [21-24]. The effect of soil heterogeneity and layering on the wave screening efficiency of vibration isolation systems under plane strain conditions are also investigated by using frequency domain formulations for numerical analysis [25, 26].

In this chapter, as an experimental study, electrodynamic shaker is used to produce vertical harmonic vibrations in the certain frequency range and accelerometers are used to obtain generated values that are stored on the computer by using signal calculator program. Two footings are constructed with clear distance where Rayleigh wave becomes dominant on site close to Sakarya city (Turkey). The first footing is used to produce the harmonic load and the other for accelerometers record and vice versa. A number of experiments are carried out on site in order to examine the screening efficiency of open and in-filled trench barriers, such as backfilled with water, bentonite (softer material than soil) and concrete (stiffer material than soil). The screening effectiveness of those barriers is determined from field measurements by comparing site data without barriers. Two different approaches are considered for vibration isolation, namely active and passive isolations.

2. Test site investigation

Site investigation is the procedure by which geophysical, geotechnical and other pertinent knowledge which might influence the construction or performance of a civil engineering or building project is gained.

Subsoil conditions can be explored by drilling and sampling, seismic surveying, excavation of test pits, and by the study of existing data. Extensive soil investigation will be necessary even for minor structures if the area is suspected of having deep fill, a high water table or swelling soil problems [27].

Properties of the local soil conditions should be determined to investigate isolation effect of the wave barriers accurately. Boring logs are conducted on the site for ground exploration and soil strata definition. Borings were located where site refraction tests indicated possible anomalies, e.g., water or air filled voids, fractures etc. Two additional borings were drilled for correlation purposes. Test borings in soil material were conducted using hollow stem auger. A hollow-stem auger consists of a continuous flight auger surrounding a hollow drill stem. The hollow-stem auger is advanced similar to other augers; however, removal of the hollow stem auger is not necessary for sampling. SPT and undisturbed samples are obtained through the hollow drill stem, which acts like a casing to hold the hole open. This increases usage of hollow-stem augers in soft and loose soils [28].

The case study site is 2.5 km^2 flat area. The test site consists of thick alluvial deposits that are transported by the river. Area covered by water was filled by floods of the Sakarya River that occurs nearly every two years. At present, almost all the area is developed to be a flat area and marsh is seldom seen. As indicated in this geological history, surface soil of the area is very young Holocene soil developed for recent 200 years.

2.1 Standard penetration test

The standard penetration test (SPT) is an in-situ dynamic penetration test designed to provide information on the geotechnical engineering properties of soil. The test uses a thick-walled sample tube, with an outside diameter of 50 mm and an inside diameter of 35 mm, and a length of around 650 mm. This is driven into the ground at the bottom of a borehole by blows from a slide hammer with a weight of 63.5 kg falling through a distance of 760 mm. The sample tube is driven 150 mm into the ground and then the number of blows needed for the tube to penetrate each 150 mm up to a depth of 450 mm is recorded. In cases where 50 blows are insufficient to advance it through a 150 mm interval the penetration after 50 blows is recorded. The blow count provides an indication of the density of the ground, and it is used in many empirical geotechnical engineering formulae.

The key reason of the test is to supply an indication of the relative density of granular deposits, for example sands and gravels from which it is virtually impossible to obtain undisturbed samples. The soil strength parameters which can be inferred are approximate, but may give a useful guide in ground conditions where it may not be possible to obtain borehole samples of adequate quality like gravels, sands, silts, clay containing sand or gravel and weak rock. In conditions where the quality of the undisturbed sample is suspect, e.g. very silty or very sandy clays, or hard clays, it is often advantageous to alternate the sampling with standard penetration tests to check the strength [29].

2.2 Soil classification

Soil is created by many processes out of a wide variety of materials. Because deposition is irregular, soils are notoriously variable, and often have properties which are undesirable from the point of view of a proposed structure. Soil classification systems are set up to allow the expected properties of the soil in a given condition to be conveyed in a shorthand form. The stability and performance of a structure founded on soil depend on the subsoil conditions, ground surface features, type of construction, and sometimes the meteorological changes. Soil, in the engineering sense, is the relatively soft and uncemented material which overlies the rock of the outer part of the Earth's crust [30].

The water table is generally high to be about 1 to 3 meters and it may come closer the ground surface in rainy season. The ground dominantly consists of gravely and silty sand having different densities and contains low plasticity silty and clay bandage at some places.

Site soils are characterized as clay, silty clays, silty gravel and gravel. Material properties of the test site are given in Table 1.

SUMMARY OF TESTING RESULTS											
Report No:	06/0327		Sieve Analysis		Atterberg Limits			Soil Classification	Water content	Hydrometer	
Project Name	Soil Structure Profile		+ # 4	- # 200	LL	PL	PI			-0,076	-0,005
Hole No	Specimen No	Depth (m)	%	%	%	%	%	TS1500	%	%	%
SK1	SPT 1	1.00-1.45	0	98	42	22	20	CI	32	98	46
	SPT2	2.00-2.45	0	87	27	20	7	CL	32	87	18
	UD1	2.50-3.00	0	85	26	20	6	CL	32	85	16
	SPT 3	3.00-3.45	50	21	NP	NP	NP	GM	9		
	SPT 4	4.50-4.95	59	5	NP	NP	NP	GW-GM	7		
	SPT 5	6.00-6.45	0	84	28	22	6	CL	26	84	14
	SPT 6	7.50-7.95	46	6	NP	NP	NP	GW-GM	12		
	SPT 7	9.00-9.45	62	5	NP	NP	NP	GW-GM	7		
	SPT 8	10.00-10.45	91	1	NP	NP	NP	GW	4		
SK2	SPT1	1.50-1.95	0	98	39	21	18	CI	28		
	UD1	2.50-3.00	0	95	40	21	19	CI	35		
	SPT2	3.00-3.45	0	93	38	22	16	CI	33		
	SPT3	4.50-4.95	54	5	NP	NP	NP	GW-GM	9		
	SPT4	6.00-6.45	0	88	28	21	7	CL	37	88	18
	SPT5	7.50-7.95	47	21	NP	NP	NP	GM	14		
	SPT6	9.00-9.45	81	6	NP	NP	NP	GW-GM	6		

Table 1. Material properties of the test site.

2.3 Seismic refraction and reflection tests

In reflection and refraction prospecting, body waves are the source of information used to image the Earth's interior. In reflection experiments, analysis is concentrated on energy arriving after the initial ground motion. Specifically, the analysis concentrates on ground movement that has been reflected off of subsurface interfaces. Subsurface structures can be complex in shape but like the refraction methods are interpreted in terms of boundaries separating material with differing elastic parameters. Refraction experiments are based on the times of arrival of the initial ground movement generated by a source recorded at a variety of distances. Later arriving complications in the recorded ground motion are discarded. These are then interpreted in term of the depths to subsurface interfaces and the speeds at which motion travels through the subsurface within each layer. These speeds are controlled by a set of physical constants, called elastic parameters that describe the material [31]. In this study, thumper is used for both seismic refraction and reflection tests as an energy source. Soil dynamic parameters that are obtained by the tests are given in Table 2.

Soil Dynamic Parameters						
Parameters	**Symbol**	**Unit**	**1. Layer**	**2. Layer**	**3. Layer**	**4. Layer**
P (Compression) Wave Velocity	C_P	m/s	370	580	1012	1739
S (Shear) Wave Velocity	C_S	m/s	133	220	341	570
Layer Thickness	h	m	0.7	1.2	1.8	2.3
Density	ρ	kN/m^3	13.6	15.2	17.5	18.9
Maximum Shear Module	G_{max}	kN/m^2	23545	72098	199075	625615
Elasticity Module	E	kN/m^2	67155	204174	571727	1319592
Poisson Ratio	ν	-	0.42	0.44	0.35	0.30
Soil Vibration Period	T_0	s	0.6			

Table 2. Soil dynamic parameters in the test site.

3. Measurements for vibration screening performance

Evaluation of the screening effectiveness precisely depends on the barrier material stiffness. Hence, a series of experiments are necessary to understand the propagating characteristics of the waves. The test layout for both active and passive isolation cases is shown in Fig. 1.
The layout consists of exciter, two concrete footing, wave barrier, and 2 measurement points. Electrodynamic shaker, which induces a sinusoidal motion, shown in Fig. 2 is used as a stationary vibration source to produce vertical harmonic force of maximum amplitude of 250 N in a frequency range of practical importance of 10 Hz to 95 Hz. Besides accelerometers are employed to obtain generated values that are stored on computer by using signal calculator program. The excitation frequency is increased progressively in Δf = 5 Hz steps. The noise in the signals recorded during the test was eliminated during signal processing by digital filtering with a band-pass filter.
The shaker is mounted on thin metal plate and placed centrically above the rigid square footing in order to excite only vertical vibrations. Two concrete surface footings with dimensions 1.0 m x 1.0 m x 0.5 m which are built on the site with clear distance of L_F = 25 m are used. For research purposes, a rectangular, D_t = 3 m long open trench is constructed symmetrical about the center line between these footings. The first footing is used to produce the harmonic load and the other for accelerometers record and vice versa. The installed source on footing is placed at a distance of 4 m from the trench for the measurement of active isolation case and a distance of L_t = 20 m for measurement of passive isolation. The vertical components of harmonic vibrations are recorded with those accelerometers located on these foundations, which corresponds to a time interval of Δt = 0.0005 sec. The displacement amplitudes are computed from the acceleration data.
The screening performance of the material stiffness of the trench compared to soil and the excitation frequency range are investigated by conducting a series of field tests of source and receiver isolation barrier, namely active and passive vibration screens. The considered parameters are summarized in Table 3.
The length of Rayleigh (λ_R) wave of the generated vibration is one of the most critical factors to determine the screening effectiveness of wave barriers. Provided that the minimum depth of open trench should be at least $0.6\lambda_R$ at a point $10\lambda_R$ away from such trench for active isolation and $1.33\lambda_R$ for the passive when the measurement point is located at a distance

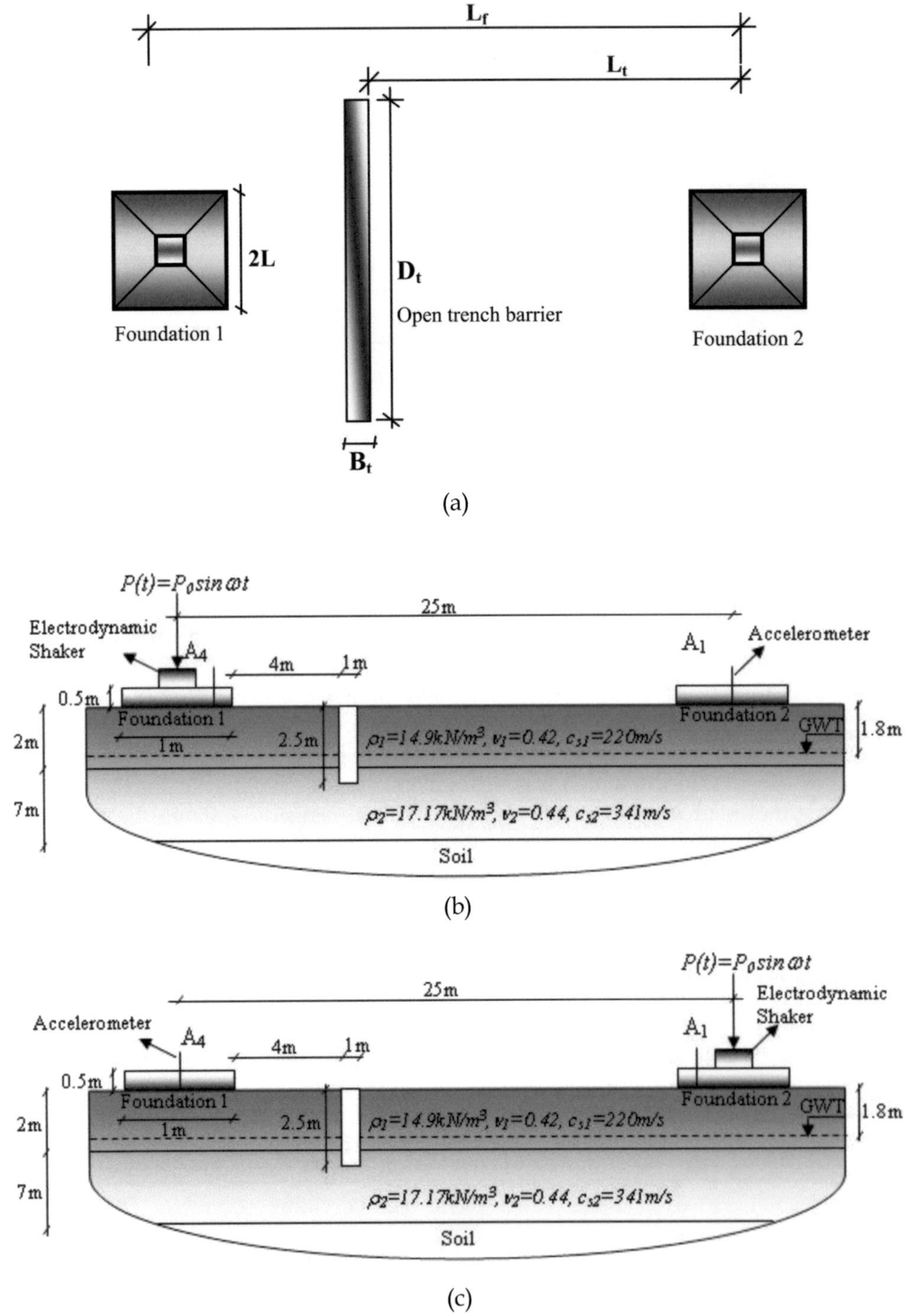

Fig. 1. Field test model for active and passive isolation cases: a) Plan view, b) Active isolation, c) Passive isolation.

(a) (b) (c) (d)

Fig. 2. Electrodynamic shaker and accelerometers placed on the foundations: a) Electrodynamic shaker placed on the foundation, b) Electrodynamic shaker and accelerometer, c) Measurements recorded foundation, d) Accelerometers placed on the foundation.

Material	Mass density, ρ (t/m^3)	Pressure wave velocity, C_p (m/s)	Shear wave velocity, C_s (m/s)	Poisson's ratio, ν	Geometry	Depth of trench, H_t (m)	Width of trench, B_t (m)
Bentonite (softer) trench	1.65	400	100	0.35	Rectangular	2.5	1.0
Concrete (stiffer) barrier	2.40	5000	2400	0.20	Rectangular	2.5	1.0
Water filled trench	0.98	-	-	0	Rectangular	2.5	1.0

Table 3. Material properties and geometric parameters of the in-filled trench barrier.

between $2\lambda_R$ and $7\lambda_R$ from the wave barrier. The trench width has to be built between $0.1\lambda_R$ and $0.5\lambda_R$ to accomplish such remarkable reduction in vertical soil vibrations [21-23]. The predominant values of the applied excitation frequencies in these experimental studies and the related Rayleigh wavelengths are given in Table 4 in order to determine the optimum geometrical parameters of the rectangular trench barrier an average for an effective protection and to avoid the difficulties in their practical applications such as instability of soil, high water table levels, and high costs.

Frequency of exciter (f) (Hz)	Wave length of Rayleigh waves (λ_R) (m)	Trench width, B_t (min.0.1λ_R) (m)	Active isolation		Passive isolation	
			Trench depth, H_t (min.0.6λ_R) (m)	Measurement point from trench, L_t (min.10λ_R) (m)	Trench depth, H_t (min.1.33λ_R) (m)	Measurement point from trench (min.2λ_R) (m)
25	7.92	0.79	4.75	79.2	10.53	15.84
50	3.96	0.40	2.37	39.6	5.26	7.92
75	2.64	0.26	1.58	26.4	3.51	5.28
100	1.98	0.20	1.19	19.8	2.63	3.96
All	All	1.00	2.50	20.0	2.50	5.00

Table 4. Rayleigh wavelength and minimum conditions for the screening effectiveness of an open trench barrier.

Four types of trench barriers are used to obtain better result of vibration control. For the case of in-filled trenches as shown in Fig. 3, the backfill material compared to soil is respectively considered as water, bentonite as softer material and concrete as stiffer material in place of the open trenches. For the sake of slope stability the trench walls are sealed by reinforced concrete in a width of 0.15 m.

In the experimental program, A_1 is denoted as observation point where foundation to be protected and A_4 as excited foundation for active isolation case. A_1 is donated as excited foundation and A_4 to be protected foundation for passive isolation case.

2.4 Data processing

The data is obtained experimentally on the site, which is unrefined, for the case of active and passive isolations then refined by using SeismoSignal 3.02 programme which is defined as band-pass filtration (See Fig. 4 and 5). Then, the filtrated data is reproduced in Matlab environment to obtain the graphs. These obtained data for A_1 and A_4 recorded stations of displacement-time history graphs are figured out for all harmonic loadings and consequently for both active and passive cases (Fig. 6).

3.1 A_1 measurement for active isolation

The resulting time histories of the vertical response at point A_1 for active isolation measures in the case of no trench, rectangular open trench and an in-filled trench are compared in Fig. 6. The wave propagation pattern of the transmitted vibrations in the case of both open and in-filled trench barriers is similar to the case of no trench. This general trend of the observed behavior changes only for an excitation frequency of 50 Hz. However, any time delay does not exist between the amplitudes of the spreading waves.

The amplitude reduction factor R_f is defined as the vertical displacement amplitude after the presence of the trench barriers relative to the amplitude on the undisturbed site (without trench barriers). An effective screening exists when the calculated reduction factor from the experimental data is less than 0.6 for the applied excitation frequencies.

(a) (b)

(c) (d)

Fig. 3. Trench barriers: a) Open trench, b) Water filled trench, c) Bentonite filled trench and d) Concrete filled trench.

The amplitude reduction factor of vertical displacement due to harmonic sinusoidal load with different frequencies applied for A_1 measurements are shown in Fig. 7. At almost all considered source frequencies the trench causes significantly amplification of the soil vibration (R_f is greater than 1.0). The influence of the distance (L_t) between the measurement point and the barrier location is significant for wave propagation. It should be over 10 times the wavelength of Rayleigh wave ($L_t = \min 10\lambda_R$) for a considerable reduction in the vibration level. In this study, the predominant values of applied excitation frequencies give Rayleigh wavelengths λ_R to vary between 1.98 m and 7.92 m, which result in inadequate screening (see Table 4). For insufficient distances (here, $L_t = 20$ m), strong wave interactions with wave interference effects occur between the vibratory source and affected foundation to be protected.

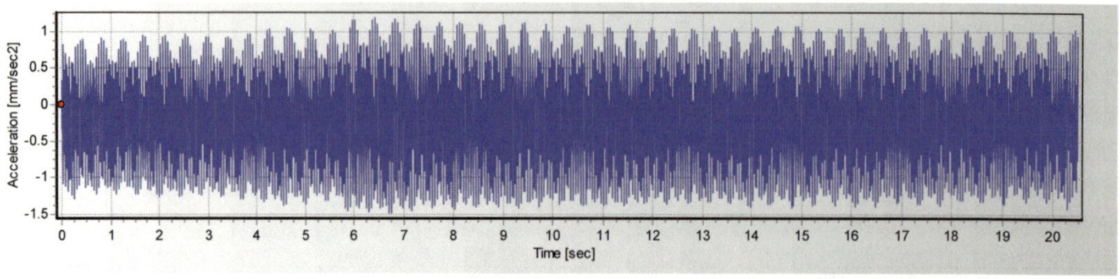

Fig. 4. A_1 active isolation for unrefined recorded data from accelerometer (25 Hz).

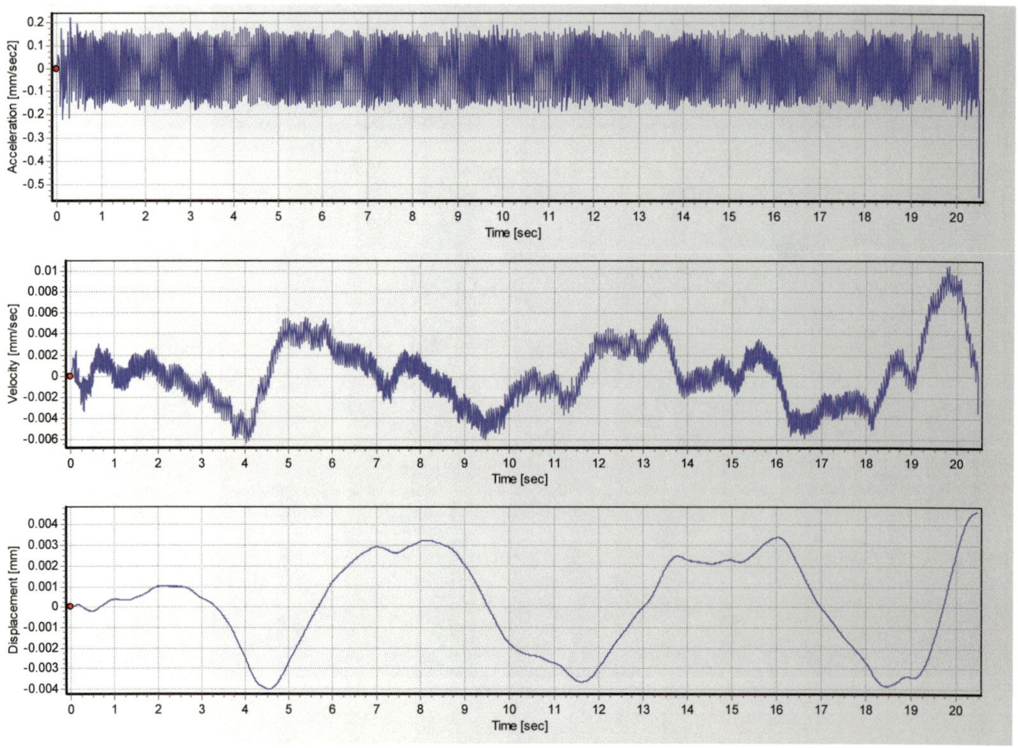

Fig. 5. A1 active isolation for refined recorded data for acceleration, velocity and displacement (25 Hz).

3.2 A_4 measurement for active isolation

The reduction factor as a function of excited frequencies for the different backfill material properties of the trench barriers to isolate vibrations at measurement point A_4, where it is located an accelerometer near the vibratory source on the foundation is obtained as shown in Fig. 8. Screening effects of installing rectangular open trench, water filled trench, bentonite and concrete trench barriers are compared at the same experimental site. Nevertheless, the measured data of the undisturbed site (without trench) is included in the comparison. From in-situ measurements of amplitude in case of soil medium with and

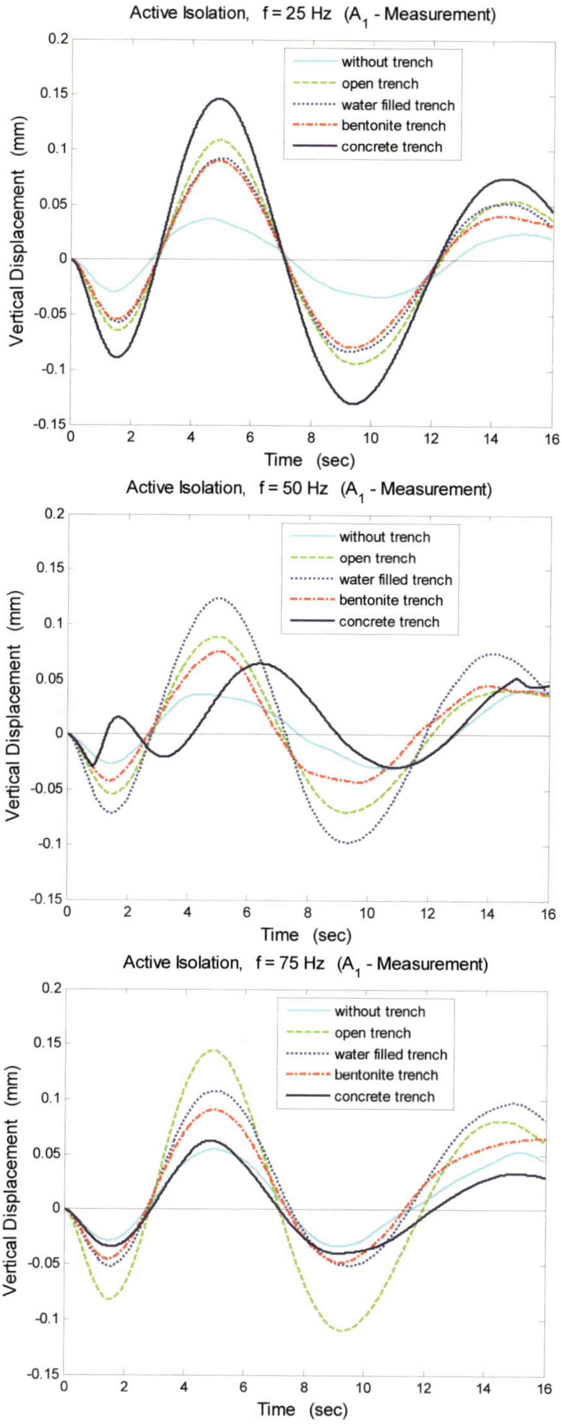

Fig. 6. Comparison of the vertical displacement time histories at point A_1 for active isolation measures due to three different frequencies of the exciter

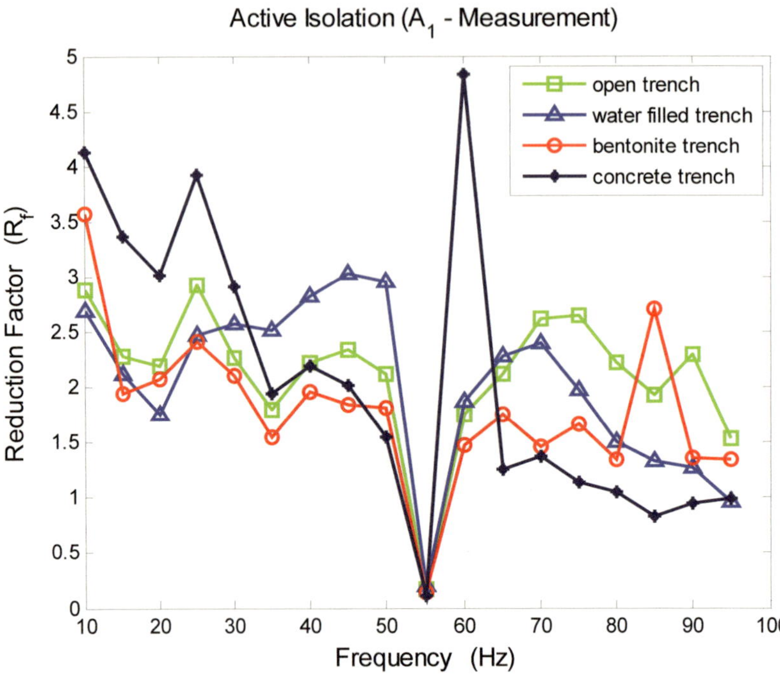

Fig. 7. Vertical amplitude reduction factor as a function of excited frequencies for active isolation at measurement point A_1

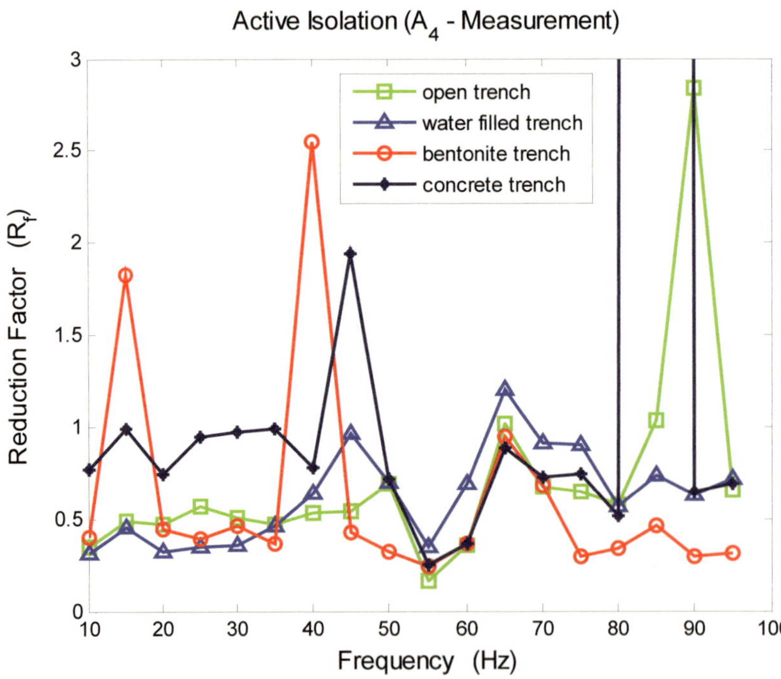

Fig. 8. Vertical amplitude reduction factor as a function of excited frequencies for active isolation at measurement point A_4

without any reduction measure, very effective vibration screening is observed for applied frequencies. For both 10 and 25 Hz frequencies of exciter, water filled trench gives a good isolation (R_f = 0.27) that achieve a reduction level up to 200% of the maximum vertical displacements at observation time about t = 5 sec compared in case of subsoil without trench barrier (R_f = 1.0). For these frequencies the isolation effect of bentonite and concrete trenches follows that of the water filled trench, respectively. Because of the traveling a longer propagation path surrounding the trench barrier, there is a certain time delay in the incoming waves to the source. Bentonite trench barrier gives the best isolation measures in high frequency values of 50, 75 and 95 Hz in Fig. 8. It reduces the maximum response respect to the original site from 0.16 mm to 0.05 mm (R_f = 0.31) at about t = 5 sec for excitation frequency of 50 Hz. Comparing the screening effects of bentonite trench with that of the concrete barrier at 75 Hz, vibration isolation by bentonite trench is reduced the maximum values about 2.5 times more than that of concrete barrier. The differences of the screening efficiency depend on propagating wave characteristics which occur after hitting an obstacle such as reflection, refraction and diffraction varied with the in-filled material properties of the trench barriers.

3.3 A_1 measurement for passive isolation

The Fig. 9 illustrates a significant isolation effect in the vertical displacement amplitudes in the case of both open and in-filled trench barrier. The maximum displacement (u_{zmax} = 0.2

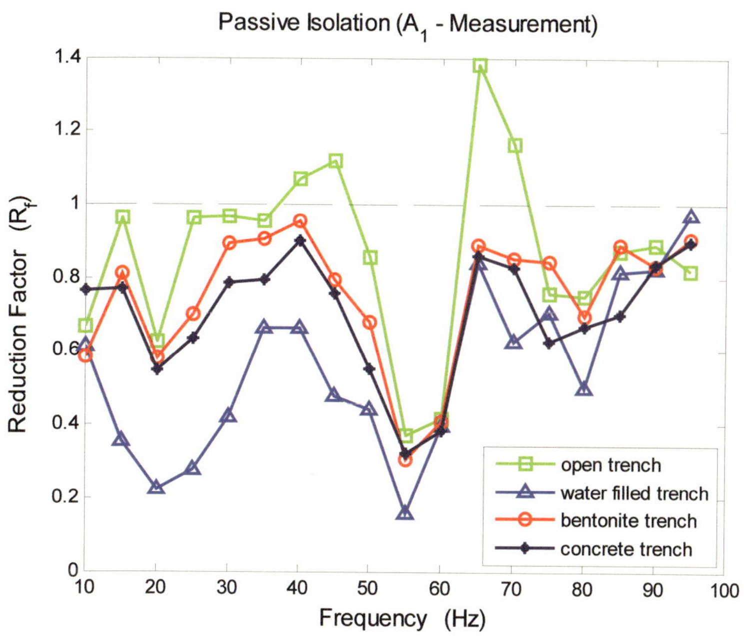

Fig. 9. Vertical amplitude reduction factor as a function of excited frequencies for passive isolation at measurement point A_1

mm) is obtained in the excitation frequency of 10 Hz with no trench case as expected (R_f = 1.0). It is observed that certain time delay occurs between the amplitudes of the spreading waves. At all considered source frequencies, the trench barriers cause significantly reduction of the soil vibrations ($0 < t < 10$ sec). Water filled trench gives the best screening effect ($0.2< R_f <0.6$) in the range of the excitation frequencies from 10 Hz to 60 Hz. It reduces the maximum vertical response from 0. 15 mm to 0.025 mm at t = 4 sec for applied frequency of 55 Hz. Concrete barrier, bentonite filled trench, open trench, and no trench follows in that case. The displacement values are scattering in low frequency but the values are identical in high frequency cases. Waves are traveling near the surface in high frequency. This causes to be identical for all isolation measures.

3.4 A_4 measurement for passive isolation

In Fig. 10 the resulting time history on vertical displacements at measurement point A_4 for passive isolation is shown for the cases of subsoil without any reduction measure as well as a trench barrier with various in-filled materials as reduction measures. The wave propagation form of the transmitted vibrations in the case of both open and in-filled trench barriers is almost similar to the case of no trench for low frequency values. When increasing the frequency values of the stationary exciter the wave pattern becomes irregular due to soil formations and complex mechanism of wave reflection varied with the in-filled material properties of the trench barriers. Soil layers are more inhomogeneous near to the ground surface. It is well known that waves penetrate to lower soil layers in low frequency values. Bentonite filled trench barrier gives the best isolation effect in the frequency of 10 Hz. The reduction efficiency of this trench barrier can reach around 40%. As shown in field test results, the isolation effect of water filled trench is more effective for excitation frequency of 25 Hz. It reduces the maximum response respect to the undisturbed field from 0.038 mm to 0.0175 mm at about t = 5 sec. In high frequency values water and bentonite filled trenches are effective. But in those cases waves are traveling near to surface and are named as noise type of waves. Comparing the screening effects of bentonite barrier with that of the water filled trench at 75 Hz of vibratory source, vibration isolation by water filled trench is reduced the maximum values about 20% more than that of bentonite barrier. It is anticipated that a softer material compared to soil is also performed as backfill material for an in-filled trench barrier.

Table 5 compares the presented data with the empirical formula [23], numerical solutions [16] and laboratory test results of Haupt [22]. For possible comparisons some values are normalized as H_t/λ_R (Trench Depth/Rayleigh Wave), B_t/λ_R (Trench Width/Rayleigh Wave) and L_t/λ_R (Distance from the Vibration Source/Rayleigh Wave) in terms of amplitude reduction ratio A_r which is the ratio of the vertical displacement amplitudes at the point in the presence and in the absence of the trench.

Wave characteristics such as reflection and diffraction at layer interfaces and the heterogeneous nature of the soil play significant role on the results with the material properties of the barrier especially for the experimental measurements. Also, it is not easy to make available close results with published data due to the nature of the soil (water table level, soil structure, layering effect, heterogeneity etc.).

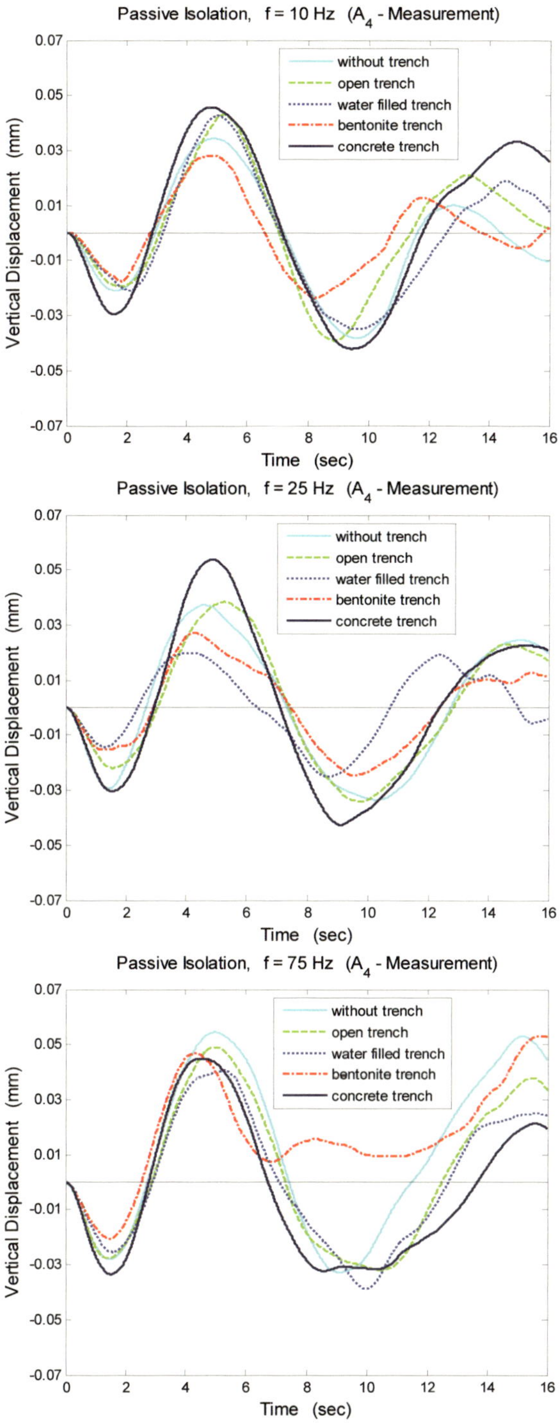

Fig. 10. Comparison of the vertical displacement time histories at point A_4 for passive isolation measures due to three different frequencies of the exciter

Vibration screening case	Normalized parameters H_t/λ_R	B_t/λ_R	L_t/λ_R	Reduction ratio (A_r) Present data	Ref. [16]	Ref. [22]	Ref. [23]
Open trench	0.64	0.26	5	0.21	0.28	0.35	0.27
Concrete barrier	0.96	0.40	7.5	0.65	0.54	0.47	0.50

Table 5. Comparison with presented and published experimental data, empirical formula and Boundary Element Method results on passive isolation.

4. Conclusions

A detailed investigation on the reduction of foundation vibrations due to a harmonic load which is produced by electrodynamic shaker using a trench barrier has been presented. The effectiveness of using open or in-filled trench as a reduction measure has been demonstrated through a site measurement study depending on the obtained results. Time dependent displacement values are reduced for both cases of active and passive isolations. In this case wave absorption plays very important role. Maximum displacements are obtained at 2-10 seconds.

Using open or in-filled trench barriers can reduce the vibrations of a structure and the resulting internal forces significantly. The use of an open trench is more effective than using an in-filled trench but its practical application is limited to relatively shallow depths. On the other hand, using softer backfill material increases the effectiveness of in-filled trench and allows for larger trench depth with no supporting measures of the vertical walls of the trench. The barriers have been found to be generally more effective in passive isolation compared to active isolation for both measurement points.

The current study aimed to provide a few general guidelines for the design of vibration isolation measures by means of trenches. It should be noted, however, that in many practical cases it seems to be appropriate to perform a more detailed investigation of the structure/soil/trench system under consideration similar as it has been done in this contribution. Designing the optimum trench with respect to its depth and width study should be performed for each particular case.

5. References

[1] C.J.C. Jones, J.R. Block. Prediction of ground vibration from freight trains. J. Sound Vibration 1996; 193 (1):205–213.

[2] A.T. Peplow, C.J.C. Jones, M. Petyt. Surface vibration propagation over a layered elastic half-space with an inclusion. Appl. Acoust 1999;56:283-296.

[3] V.V. Krylov. Vibration impact of high-speed trains effects of track dynamics. J. Acous. Soc. Am 1996; 100 (5):3121–3133.

[4] K.R. Massarsch. Settlements and damage caused by construction-induced vibration. In: Proceedings of International Workshop Wave 2000. Chouw and Schmid (eds), Bochum, Roterdam: Balkema, 13-15 December 2000, 299-315.

[5] R.G. Payton. Transient motion of an elastic half-space due to a moving surface line load. International Journal of Engineering Sciences 1967; 5:49-79.
[6] E. Kausel. Thin-layer method. Int. J. Numer. Meth. Eng. 1994;37:927-941.
[7] J.R. Barber. Surface displacements due to a steadily moving point force. J. Appl. Mech. 1996;63:245–251.
[8] X. Sheng, C.J.C. Jones, M. Petyt. Ground vibration generated by a harmonic load acting on a railway track. Journal of Sound and Vibration 1999;225 (1):3-28.
[9] G. Schmid, B. Verbic. Modellierung der Erschütterung aus dem Schienenverkehr mit der Randelelementmethode. In: H. Bachmann (Ed.), Erdbebensicherung bestehender Bauwerke und aktuelle Fragen der Baudynamik 1997; Tagungsband D-A-CH'97, SIA, Dokumentation DO145.
[10] M. Adam, G. Pflanz, G. Schmid. Two- and three-dimensional modeling of half-space and train-track embankment under dynamic loading. Soil Dyn. Earthquake Eng. 2000; 19 (8);559–573.
[11] B.Y. Yang, H.H. Hung. A 2.5D finite/infinite element approach for modeling visco-elastic bodies subjected to moving loads. Int. J. Numer. Meth. Eng. 2001; 240:1317-1336.
[12] C. Bode, R. Hirschauer, S.A. Savadis. Soil–structure interaction in the time domain using half-space Green's functions. Soil Dyn. Earthquake Eng. 2002;22 (4):283–295.
[13] E. Celebi, G. Schmid. Investigation of ground vibrations induced by moving loads. Engineering Structures 2005;27:1981-1998.
[14] J.O'Brien, D.C. Rizos. A 3D BEM-FEM methodology for simulation of high speed train induced vibrations. Soil Dyn. Earthquake Eng. 2005;25:289-301.
[15] E. Celebi. Three-dimensional modeling of train-track and sub-soil analysis for surface vibrations due to moving loads. Applied Mathematics and Computation 2006; 79:209–230.
[16] D.E. Beskos, G. Dasgupta, I.G. Vardoulakis. Vibration isolation using open or filled trenches part 1: 2-D homogeneous soil. Comput. Mech. 1986;1 (1):43–63.
[17] R. Klein, H. Antes, D. Le Houedec. Efficient 3D modeling of vibration isolation by open trenches. Computers & Structures 1997;Vol. 64, No. 1-4:809-817.
[18] G. Pflanz, K. Hashimoto, N. Chouw. Reduction of structural vibrations induced by a moving load. J. Appl. Mech. 2002;5:555–563.
[19] M. Adam, O. von Estorff b. Reduction of train-induced building vibrations by using open and filled trenches. Computers and Structures 2005;83:11–24.
[20] E. Celebi, S. Fırat, I. Cankaya. The effectiveness of wave barriers on the dynamic stiffness coefficients of foundations using boundary element method. Applied Mathematics and Computation 2006;180:683–699.
[21] R.D. Woods. Screening of surface waves in soils. J. Soil Mech. Found. Eng. Div. ASCE 1968; 94 (4):951–979.
[22] W. A. Haupt. Model tests on screening of surface waves. In: Proceedings of the 10th International Conference on Soil Mech. Found. Eng., Stockholm, 1981, vol. 3, 215-222.
[23] S. Ahmad, T.M. Al-Hussaini. Simplified design for vibration screening by open and infilled trenches. Journal of Geotechnical Engineering 1991;117 (1):67-88.
[24] H. Antes, R. Klein, D. Le Houedec, J.P. Region. Validation in-situ de la theorie des barrieres de discontinuite dans le sol. In : Troisieme Colloque National, Genie

Parasismique et Aspects Vibratoires dans le Genie Civil, 1993, Paris, Association Francaise du Genie Parasismique.
[25] K. L. Leung, D.E. Beskos, I. G. Vardoulakis. Vibration isolation using open and infilled trenches. Part 3: 2-D non-homogenous soil. Computational Mechanics 1990;7:137-148.
[26] K. L. Leung, I. G. Vardoulakis, D.E. Beskos, J. L. Tassoulas. Vibration isolation by trenches in continuously non-homogenous soil by the BEM. Soil Dynamics and Earthquake Engineering 1991; Vol. 10(3):172-179.
[27] F. H. Chen. Soil Engineering and testing, Design and Remediation, CRC Press, 2000.
[28] Soils and Foundations Handbook, State Materials Office, Gainesville, Florida, April 2004.
[29] http://en.wikipedia.org/wiki/Standard_penetration_test
[30] C. R. I. Clayton, M. C. Matthews and N. E. Simons. Soil Classification (Site Investigation), Second Edition, *Department of Civil Engineering, University of Surrey.*
[31] S. H. Danbom. Reflection Seismic Data Processing, Through the Looking Glass of Applied Exploration Geophysics, P.G. for Rice University ESCI 444, 2009.